Lecture Notes in Electrical Engineering

Volume 655

The book series *Lecture Notes in Electrical Engineering* (LNEE) publishes the latest developments in Electrical Engineering—quickly, informally and in high quality. While original research reported in proceedings and monographs has traditionally formed the core of LNEE, we also encourage authors to submit books devoted to supporting student education and professional training in the various fields and applications areas of electrical engineering. The series cover classical and emerging topics concerning:

- Communication Engineering, Information Theory and Networks
- Electronics Engineering and Microelectronics
- Signal, Image and Speech Processing
- Wireless and Mobile Communication
- Circuits and Systems
- Energy Systems, Power Electronics and Electrical Machines
- Electro-optical Engineering
- Instrumentation Engineering
- Avionics Engineering
- Control Systems
- Internet-of-Things and Cybersecurity
- Biomedical Devices, MEMS and NEMS

For general information about this book series, comments or suggestions, please contact leontina.dicecco@springer.com.

To submit a proposal or request further information, please contact the Publishing Editor in your country:

China

Jasmine Dou, Associate Editor (jasmine.dou@springer.com)

India, Japan, Rest of Asia

Swati Meherishi, Executive Editor (Swati.Meherishi@springer.com)

Southeast Asia, Australia, New Zealand

Ramesh Nath Premnath, Editor (ramesh.premnath@springernature.com)

USA, Canada:

Michael Luby, Senior Editor (michael.luby@springer.com)

All other Countries:

Leontina Di Cecco, Senior Editor (leontina.dicecco@springer.com)

**** Indexing: The books of this series are submitted to ISI Proceedings, EI-Compendex, SCOPUS, MetaPress, Web of Science and Springerlink ****

More information about this series at http://www.springer.com/series/7818

P. Satish Rama Chowdary ·
V. V. S. S. S. Chakravarthy ·
Jaume Anguera · Suresh Chandra Satapathy ·
Vikrant Bhateja
Editors

Microelectronics, Electromagnetics and Telecommunications

Proceedings of the Fifth ICMEET 2019

Springer

Editors
P. Satish Rama Chowdary
Department of Electronics and Communication Engineering
Raghu Institute of Technology
Visakhapatnam, Andhra Pradesh, India

V. V. S. S. S. Chakravarthy
Department of Electronics and Communication Engineering
Raghu Institute of Technology
Visakhapatnam, Andhra Pradesh, India

Jaume Anguera
Department of Electronics and Telecommunication Engineering
Universitat Ramon Llull
Barcelona, Spain

Suresh Chandra Satapathy
School of Computer Engineering
KIIT University
Bhubaneswar, Odisha, India

Vikrant Bhateja
Department of Electronics and Communication Engineering
Shri Ramswaroop Memorial Group of Professional Colleges (SRMGPC)
Lucknow, Uttar Pradesh, India

ISSN 1876-1100 ISSN 1876-1119 (electronic)
Lecture Notes in Electrical Engineering
ISBN 978-981-15-3827-8 ISBN 978-981-15-3828-5 (eBook)
https://doi.org/10.1007/978-981-15-3828-5

This Springer imprint is published by the registered company Springer Nature Singapore Pte Ltd.
The registered company address is: 152 Beach Road, #21-01/04 Gateway East, Singapore 189721, Singapore

Organizing Committee

Chief Patrons

Sri. Raghu Kalidindi, Chairman, Raghu Engineering Institutions
Smt. Rama Devi Kalidindi, Secretary, Raghu Engineering Institutions

Patrons

Sri. Rahul Varma Kalidindi, Director, Raghu Engineering Institutions

Chairman

Prof. S. Satyanarayana, Principal, Raghu Institute of Technology

Programme Coordinator

Dr. P. Satish Rama Chowdary, Vice Principal, Raghu Institute of Technology

Publication Committee

Dr. Jaume Anguera, Department of Electronics and Telecommunication, Universitat Ramon Llull, Barcelona, Spain; Senior Member IEEE

Dr. P. Satish Rama Chowdary, Professor and Vice Principal, Department of ECE, Raghu Institute of Technology
Dr. V. V. S. S. S. Chakravarthy, Professor, Department of ECE, Raghu Institute of Technology
Dr. Suresh Chandra Satapathy, LMCSI, Senior Member IEEE and Professor, School of Computer Engineering, Kalinga Institute of Industrial Technology (KIIT), Deemed to be University, Bhubaneswar-24, Odisha
Dr. Vikranth Bhateja, Associate Professor, Department of ECE, SRMGPC, Lucknow, India

Technical Advisory Committee

Prof. Xin-She Yang, Middlesex University, London, UK
Prof. Jaume Anguera, Universitat Ramon Llull, Spain
Prof. Nihad I. Dib, Electrical Engineering Department, Jordan University of Science and Technology, Irbid, Jordan
Prof. Ganapati Panda, IIT Bhubaneswar
Dr. Seyedali Mirjalili, Institute for Integrated and Intelligent Systems—Griffith University, Queensland, Australia
Dr. Aurora Andujar, Fractus, Spain
Prof. R. K. Mishra, Department of Electronic Science, Berhampur University, Odisha, India
Col. Prof. Dr. G. S. N. Raju, Vice Chancellor, CUTMAP
Dr. K. Jayanthi, Professor, Department of ECE, Pondicherry Engineering College
Dr. T. Srinivas, IISC, Bengaluru
Prof. N. Bheema Rao, NIT, Warangal
Prof. G. Sasibhushana Rao, Department of ECE, AUCOE, Visakhapatnam
Dr. Ch. Srinivasa Rao, Professor, Department of ECE, JNTUK Vizianagaram
Dr. Babita Majhi, Department of CS, GG Central University, Bilaspur
Mr. V. Balaji, Head-Wireless Deployment, Reliance Corporate Park, Navi Mumbai
Dr. Lakshminarayana Sadasivuni, Andhra University (Retd.), Visakhapatnam
Dr. Tumati Venkateswara Rao, Professor and HOD, Department of ECE, Sir C. R. Reddy College of Engineering, Eluru
Prof. A. Mallikarjuna Prasad, JNTUK
Prof. P. Mallikarjuna Rao, Department of ECE, AUCOE, Visakhapatnam
Dr. Suresh Chandra Satapathy, Professor, School of Computer Engineering, Kalinga Institute of Industrial Technology (KIIT), Deemed to be University, Bhubaneswar
Dr. Sudheer Kumar Terlapu, Professor, Department of ECE, Shri Vishnu Engineering College for Women (A), Bhimavaram
Prof. P. Rajesh Kumar, Department of ECE, AUCOE, Visakhapatnam
Dr. Tummala Surya Narayana Murthy, Department of ECE, UCEV, JNTUK Vizianagaram

Dr. Vikranth Bhateja, Associate Professor, Department of ECE, SRMGPC, Lucknow, India
Dr. M. Vamsi Krishna, Associate Dean, CUTM

Proceedings Committee

Dr. V. V. S. S. S. Chakravarthy, Professor, Department of ECE, Raghu Institute of Technology
Dr. Sonali Dash, Associate Professor, Department of ECE, Raghu Institute of Technology
Mr. Anil Kumar Patnaik, Assistant Professor, Department of ECE, Raghu Institute of Technology
Ms. M. Divya, Assistant Professor, Department of ECE, Raghu Institute of Technology

Technical Programme Committee

Mr. B. S. S. V. Ramesh Babu, Associate Professor, Department of ECE, Raghu Institute of Technology
Mr. S. Suresh, Assistant Professor, Department of ECE, Raghu Institute of Technology
Mr. Anil Kumar Patnaik, Assistant Professor, Department of ECE, Raghu Institute of Technology
Mrs. P. Kusuma Vani, Assistant Professor, Department of ECE, Raghu Institute of Technology
Mr. Akash Kumar Gupta, Assistant Professor, Department of ECE, Raghu Institute of Technology
Ms. M. Divya, Assistant Professor, Department of ECE, Raghu Institute of Technology

Publicity Committee

Mr. P. V. K Durga Prasad, Associate Professor, Department of ECE, Raghu Institute of Technology
Mrs. K. Roopa, Assistant Professor, Department of ECE, Raghu Institute of Technology
Mr. M. Pavan Kumar, Assistant Professor, Department of ECE, Raghu Institute of Technology

Dr. Ch. Vijay, Assistant Professor, Department of ECE, Raghu Institute of Technology

Finance Committee

Mr. S. Suresh, Assistant Professor, Department of ECE, Raghu Institute of Technology
Mrs. A. V. S. Swathi, Assistant Professor, Department of ECE, Raghu Institute of Technology
Mr. K. Phanindra, Assistant Professor, Department of ECE, Raghu Institute of Technology
Mr. K. Chandra Sekhar, Assistant Professor, Department of ECE, Raghu Institute of Technology
Mr. V. K. Kiran, Assistant Professor, Department of ECE, Raghu Institute of Technology
Mr. P. Pydi Reddy, Assistant Professor, Department of ECE, Raghu Institute of Technology
Mrs. J. Lavanya, Assistant Professor, Department of ECE, Raghu Institute of Technology
Mr. T. Kiran Kumar, Assistant Professor, Department of ECE, Raghu Institute of Technology

Website Committee

Dr. V. V. S. S. S. Chakravarthy, Professor, Department of ECE, Raghu Institute of Technology
Mr. Anil Kumar Patnaik, Assistant Professor, Raghu Institute of Technology
Mr. D. V. S. A. Raju, Web Developer, REI
Mr V. Narsinga Rao, Technical Assistant, REI
Mr. Sravan Kumar, System Admin, RES
Mr. Bochha Chandramouli, Student
Mr. Dannina Harsha Sai Manoj Kumar, Student

Hospitality Committee

Mr. G. Santosh Kumar, Assistant Professor, Department of ECE, Raghu Institute of Technology
Mr. Ch. Tejesh Kumar, Assistant Professor, Department of ECE, Raghu Institute of Technology

Mr. N. Srinivasa Gupta, Assistant Professor, Department of ECE, Raghu Institute of Technology
Mrs. S. Aruna Kumari, Assistant Professor, Department of ECE, Raghu Institute of Technology
Ms. P. Bhagyalakshmi, Assistant Professor, Department of ECE, Raghu Institute of Technology
Mr. S. Ramu, Assistant Professor, Department of ECE, Raghu Institute of Technology
Mrs. T. L. Spandana, Assistant Professor, Department of ECE, Raghu Institute of Technology
Mr. D. Arun, Assistant Professor, Department of ECE, Raghu Institute of Technology
Mr. Ch. Yoshaya, Assistant Professor, Department of ECE, Raghu Institute of Technology
Mr. B. Suresh Chandra, Assistant Professor, Department of ECE, Raghu Institute of Technology
Mr. K. Srinivas, Assistant Professor, Department of ECE, Raghu Institute of Technology

Special Session Chairs

Dr. T. Sudheer Kumar Terlapu, Department of ECE, Shri Vishnu Engineering College for Women, Bhimavaram, India
Dr. Subba Rao M., Department of ECE, Shri Vishnu Engineering College for Women, Bhimavaram, India

Preface

This volume contains high-quality research papers that were presented at the **5th International Conference on Microelectronics, Electromagnetics and Telecommunications (ICMEET)** held at Department of Electronics and Communication Engineering, Raghu Institute of Technology, Visakhapatnam, during 6–7 December 2019. The ICMEET aims to bring together academic scientists, researchers and research scholars to discuss the recent developments and future trends in the fields of microelectronics, electromagnetics and telecommunications. Raghu Institute of Technology is hosting this fifth edition of ICMEET after successful completion of the second edition earlier during 2016–2017. The ICMEET received 352 submissions. Each paper was peer-reviewed by at least two well-qualified reviewers and the members of the Programme Committee. Finally, 81 papers were accepted for publication in this proceedings. Several special sessions were offered by eminent professors on many cutting-edge technologies. Several eminent researchers and academicians delivered talks, addressing the participants in their respective field of proficiency. A topic on deep learning, distributed learning and their application for engineering problems has been discussed by Prof. Ganapati Panda, Former Deputy Director, IIT BBSR. A thorough discussion on the opportunities for engineering aspirants in foreign countries in the field of electrical and electronics along with an insight on the research opportunities in the latest fields of electronics is presented by Prof. Naeem Hannoon from Malaysia. An introduction to the latest advancements in the artificial intelligence, machine learning applications and data modelling is presented in his lecture by Prof. Suresh Chandra Satapathy. Prof Ch. Srinivasa Rao from JNTUK Kakinada graced the event and shared his valuable research to the attendees in the field of communications. Also, Dr. T. S. N. Murthy from the Department of ECE, JNTUK Vizianagaram campus, chaired various technical sessions and provided valuable suggestions to the participants.

We would like to express our appreciation to the members of the Programme Committee for their support and cooperation in this publication. We are also thankful to Springer team for providing a meticulous service for the timely production of this volume. Our heartfelt thanks to our Chairman Sri. Kalidindi Raghu and other management members of Raghu Educational Society for extending wholehearted support to host this in their campus. Special thanks to all guests who have honoured us in their presence in the inaugural day of the conference. Our thanks are due to all special session chairs, track managers and reviewers for their excellent support. Last but certainly not least, our special thanks go to all the authors who submitted papers and all the attendees for their contributions and fruitful discussions that made this conference a great success.

Visakhapatnam, India — Prof. P. Satish Rama Chowdary
Visakhapatnam, India — Prof. V. V. S. S. S. Chakravarthy
Barcelona, Spain — Prof. Jaume Anguera
Vijayawada, India — Prof. Suresh Chandra Satapathy
Lucknow, India — Dr. Vikrant Bhateja

About This Book

This book is a collection of best papers presented in the 5th International Conference on Microelectronics, Electromagnetics and Telecommunications (ICMEET 2019), an international colloquium, which aims to bring together academic scientists, researchers and research scholars to discuss the recent developments and future trends in the fields of microelectronics, electromagnetics and telecommunications. Microelectronics research investigates semiconductor materials and device physics for developing electronic devices and integrated circuits with data-/energy-efficient performance in terms of speed, power consumption and functionality. This book discusses various topics like Internet of Things (IoT), smart systems, machine learning techniques, evolutionary computing tools, smart antennas, analogue, digital and mixed-signal circuits, bio-medical circuits and systems, RF circuit design, microwave and millimetre wave circuits, OFDM and massive OFDM, nanoelectronics, VLSI circuits and systems, SoC and NoC, MEMS and NEMS, VLSI digital signal processing, wireless communications, cognitive radio and data communication.

Contents

About the Editors

Dr. P. Satish Rama Chowdary is Professor and Vice Principal, Department of Electronics & Communication Engineering at Raghu Institute of Technology, Visakhapatnam. He has 19 years of teaching and 2 years of industrial experience along with 5 years of research experience. He is currently guiding three PhD scholars in the field of machine learning, IoT and artificial intelligence. His research interests are computational electromagnetics, Fractal Antennas and Image Processing. He has earlier co-edited proceedings of ICMEET-2016 published by LNEE. He has also Chaired several technical and special sessions in FICTA-2018, SCI, 3rd ICMEET, IC3T Conferences. He is a member of IEEE, ISOI, IACQER, and SCRS.

Dr. V. V. S. S. S. Chakravarthy is a Professor in the Department of Electronics & Communication Engineering at Raghu Institute of Technology, Visakhapatnam. His research interests include Computational Intelligence, Smart Antenna, Data Modelling, Machine Learning and Evolutionary Computing Tools. He has 17 years of teaching and industrial experience with 5 years of research experience. He is a member of IEEE and also serving as EC members in the Vizag Bay Sub-section. He has three research scholars working in the field of smart antennas, evolutionary computing tools and RF energy harvesting. He served as a Chair in 3rd ICMEET, IC3T, 2nd SCI Conferences. He is a member of professional bodies like Instrumentation Society of India, IISc Bangalore and Soft Computing Research Society.

Dr. Jaume Anguera, IEEE Fellow, is CTO and co-founder of the technology company Fractus Antennas (Barcelona). Professor at Ramon LLull University where he teaches subjects related to antennas and is a member of the GRITS research group. Inventor of more than 130 patents, most of them licensed to telecommunication companies and author of more than 230 scientific widely cited papers and international conferences. His h-index is 47 with more than 6500 citations. The most salient contribution is Virtual Antenna™ technology (antenna booster technology) where he is the lead inventor together with his peer colleague Dr. A. Andújar. Author of 6 books. He has participated in more than 21 competitive research projects financed by the Spanish Ministry, by CDTI, CIDEM (Generalitat de Catalunya) and the European Commission for an amount exceeding € 6M being principal researcher in most of them. He has taught more than 20 antenna courses around the world (USA, China, Korea, India, UK, France, Poland, Czech Republic, Tunisia, Spain). With over 20 years of R&D experience, he has developed part of his professional experience with Fractus in South Korea in the design of miniature antennas for large Korean companies such as Samsung and LG. He has received several national and international awards (ex. 2004 Best Ph.D Thesis COIT, 2004 IEEE New Faces of Engineering, 2011 Alé Vinarossenc - Vinaròs, 2014 Finalist European Patent Award). He has directed the master/doctorate thesis to more than 100 students, many of them have received awards for their thesis. Founder of the ciènciaprop® scientific dissemination program. His biography appears in Who'sWho in Science and Engineering. He is associate editor of the IEEE Open Journal on Antennas and Propagation and reviewer in several IEEE scientific journals.

Dr. Suresh Chandra Satapathy is currently working as a Professor, School of Computer Engineering, Kalinga Institute of Industrial Technology (KIIT), Bhubaneswar, Odisha, India. He obtained his Ph.D. in Computer Science and Engineering from JNTU Hyderabad and M.Tech. in CSE from NIT, Rourkela, Odisha, India. He has 29 years of teaching experience. His research interests include data mining, machine intelligence and swarm intelligence. He has acted as a program chair of many international conferences and edited 6 volumes of proceedings from Springer LNCS and AISC series. He is currently guiding 8 Ph.D. scholars. Dr. Satapathy is also a Senior Member of IEEE.

Prof. Vikrant Bhateja is an Associate Professor, Department of Electronics & Communication Engineering, Shri Ramswaroop Memorial Group of Professional Colleges (SRMGPC), Lucknow, and also the Head (Academics & Quality Control) in the same college. His areas of research include digital image and video processing, computer vision, medical imaging, machine learning, pattern analysis and recognition, neural networks, soft computing and bio-inspired computing techniques. He has more than 90 quality publications in various international journals and conference proceedings. Professor Bhateja has been on TPC and chaired various sessions from the above domain in international conferences of IEEE and Springer. He has been the track chair and served in the core-technical/editorial teams for international conferences: FICTA 2014, CSI 2014 and INDIA 2015 under Springer-AISC Series and INDIACom-2015, ICACCI-2015 under IEEE. He is an

Associate Editor in International Journal of Convergence Computing (IJConvC) and also an editorial board member of the International Journal of Image Mining (IJIM) under Inderscience Publishers. At present, he is the guest editor for two special issues floated in the International Journal of Rough Sets and Data Analysis (IJRSDA) and International Journal of System Dynamics Applications (IJSDA) under IGI Global publications.

A Robust Architecture Based on Adaptive Recursive Filter for Gigabit Communications

Kothapalli Roopa, R. V. Siva Krishna Addagattu, and Kuppili Sunil Kumar

Abstract This paper proposes a novel design which is capable of faster data recovery for high-speed communications such as serializer and deserializer (SerDes). The SerDes is an essential and widely used component in high-speed gigabit communications such as PCIe, Ethernet, passive optical networks (PON), MIPI and HD video streaming interfaces. The high-speed serializer/deserializer is the dominant implementation of I/O interfaces at speeds of 2.5 Gbps and higher. The SerDes works with a source-synchronous interface in which no synchronization clock will be present while transmitting/receiving the data. The SerDes receivers must have the clock and data recovery (CDR) circuit which dynamically extracts the clock and data from the receiving differential serial data. The design in this paper proposes a Nyquist sampling-based architecture that will simultaneously capture the serial data with high redundancy and without any bit loss. The architecture uses an algorithm which also features the adaptive sampling rate independent of the bit duration. The algorithm is capable of estimating the interleaving window between successive bits and significantly analyzes the samples of successive bits and dynamically filters the noisy samples and recovers the bit information and also has the ability to adjust the offset deviations occurred while sampling the serial data. The algorithm is implemented and verified on the SerDes serial receiver at 25 Gbps data rate at 14 nm technology node.

Keywords SerDes · LVDS · PON · CDR · Channel noise · SNR · BER

K. Roopa (✉)
Department of Electronics and Communications Engineering, Raghu Institute of Technology (RIT), Visakhapatnam, India
e-mail: roopa.kothapalli@gmail.com

R. V. Siva Krishna Addagattu
Qualcomm Technology India Pvt. Ltd., Bangalore, India

K. Sunil Kumar
Dr. B. R, Ambedkar University, Srikakulam, India

P. S. R. Chowdary et al. (eds.), *Microelectronics, Electromagnetics and Telecommunications*, Lecture Notes in Electrical Engineering 655,
https://doi.org/10.1007/978-981-15-3828-5_1

1 Introduction

1.1 High-Speed Digital Communication

The chip density is increasing day by day, and the demand for the high-speed circuits has skyrocketed. Previously, high-speed data communication was meant to be for limited applications and offered at higher prices. As the chip density increases day by day, the larger complex circuits can be accommodated in the integrated circuits which brought the high-speed circuits [1] into daily life for almost all each and every application. This has increased the demand for the circuits which can perform high-speed gigabit communications. Eventually, the design requirements and design challenges, pushing the limits of the current technology and reducing the time to market, have become the inevitable task for the engineers.

Low-Voltage Differential Signaling (LVDS)
At first glance, it might appear that one of the disadvantages of utilizing LVDS in an application as opposed to a customary single-finished information transmission technique is that it requires double the same number of wires to transmit a similar number of channels. In contrast, the inverted pair of wires used in the place of single wire will help us to increase the data speed [2]. The two wires for the same signal will carry bit information inverted to each other. This will help to recover the data at the receivers when being transmitted at gigabits per second. In reality, the LVDS interface can undoubtedly reduce the wires between the transmitter and receiver for higher data speeds. On the contrary, the throughput will be the same when compared with the multiple bits sent via the parallel interface at lower data rates and data sent serially using LVDS at higher data rates.

The LVDS can be implemented for multiple configurations as shown in Fig. 1. The point to point configuration for end-to-end interface and multibit configuration to interface multiple devices is significantly more efficient unlike bus interface. The multiple channels can be utilized for LVDS interface to increase the overall throughput. The LVDS is being utilized in major high-speed applications such as mobile industry processor interface (MIPI), high-definition media interface (HDMI), modem interface and PCIe.

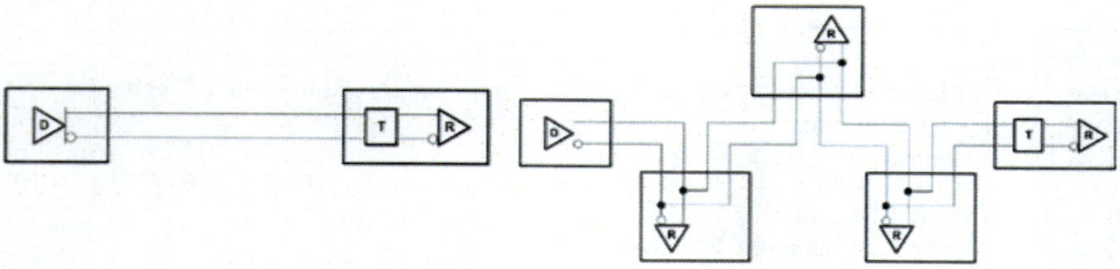

(a) (b)

Fig. 1 LVDS **a** point-to-point configuration, **b** multidrop configuration

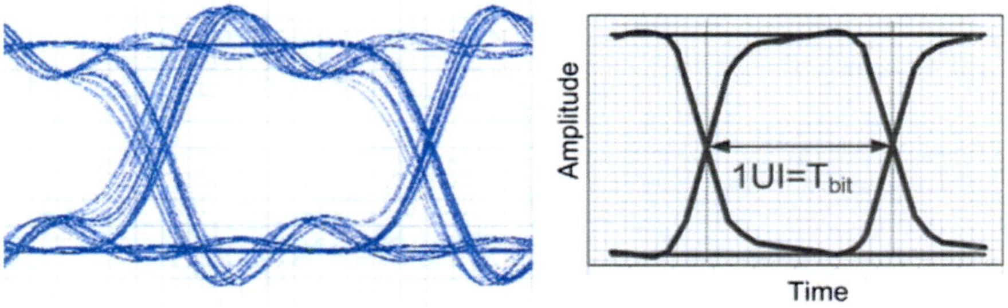

Fig. 2 Eye diagram of high-speed differential signaling

The LVDS is likely to be more redundant to environmental noise and demonstrated robustness toward common-mode noise and also less affected by the noise-related problems such as cross talk from the neighboring channels. This will also enable the use of LVDS in low-power applications with higher data rates. Eye diagrams are used to characterize a high-speed signal source or transmitter (receiver testing usually requires bit error rate testing). A typical eye diagram test setup is shown in Fig. 2. Eye amplitude is the difference between the one and zero levels. The data receiver logic circuits will determine whether a received data bit is a "0" or "1," based on the eye amplitude.

The vertical opening is the main characteristic while recovering the bit information from the differential signals which define the eye height of an eye diagram. The channel noise, cross talk, signal-to-noise (SNR) ratio, jitter and additional noise parameters will influence the eye amplitude. All these noise parameters will be involved and try to reduce the eye amplitude which will cause the eye to close in high-speed communications. If the eye amplitude reduced, it is very difficult to recover the clock and data information using CDR. This paper addresses this problem and proposes a possible solution to recover the data under these eye closing circumstances.

2 Theory

2.1 SerDes *Overview*

The basic architecture of the *SerDes* is shown in Fig. 3. The architecture is mainly consisting of parallel in serial out (PISO) and serial in parallel out (SIPO) shift register to convert the parallel data into serial and serial data into parallel, respectively [3]. The architecture of *SerDes* can be classified into four categories (1) parallel clock *SerDes*, (2) bit interleaved *SerDes*, (3) embedded clock *SerDes* and (4) 8b/10b SerDes which is exclusively used for passive optical network (PON) applications.

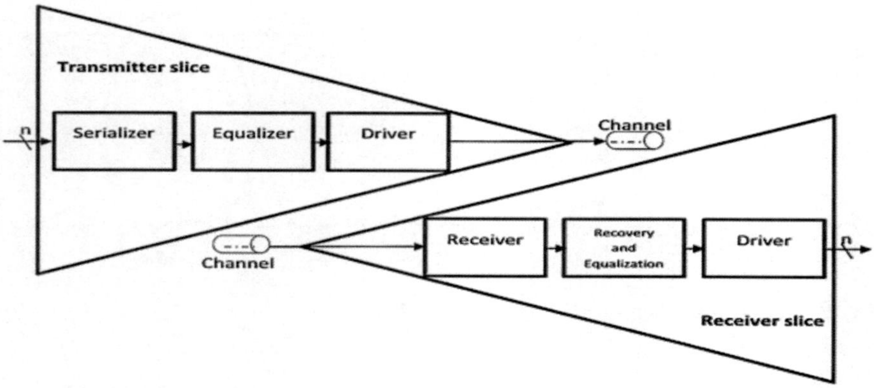

Fig. 3 Architecture of *SerDes*

The parallel in serial out (PISO) is typically consisting of parallel interface with the number of flip flops to hold the data, clock input for the flip flops and a serial output line where the serial data is available. The PISO module utilizes the internal or external phase-locked loop (PLL). The basic element in parallel in serial out (PISO) is a shift register which will convert the parallel data into serial out for one bit per clock.

However, the design will require various synchronizing schemes when the data is transferred from one clock domain to another clock domain to avoid metastability and also to prevent the data loss.

The serial in parallel out (SIPO) performs an opposite function to parallel in serial out (PISO). The serial in parallel out (SIPO) used in *SerDes* slightly differs from conventional SIPO. The SIPO used in *SerDes* will not be given any the same clock which was given at the PISO as there will be no reference clock provided along with the data. The *SerDes* internal clock management system will recover the clock information from the serial data after analyzing the serial bits, and then, the CMS will provide the corresponding reference clock to the phase lock loop (PLL). The PLL will lock the reference clock and generate the serial clock which will be given to serial in parallel out (SIPO) shift register. Further, the SIPO will shift the serial high-speed data and then converts it into parallel bit information. The clock management system (CMS) and phase-locked loop (PLL) performance should be good enough so that the low harmonic frequencies in the data stream and frequency offset can be nullified.

2.2 *Clock and Data Recovery (CDR)*

In the cutting-edge high-speed data circuits, specifically the gigabit transceivers, solid impedances and cutoff between intersymbol interference (ISI), is completely dependent on frequency. The intersymbol interference (ISI) is considered as one of the major causes for performance degradation of theses gigabit transceivers by effecting in eye opening and timing jitter which leads to data mismatch [4] and bit error rate (BER). The major challenge in the gigabit communications is that there is no dedicated clock is allotted to the signal in high-speed SerDes transceivers. The clock and data recovery (CDR) plays a vital role [5] in gigabit communications. As there will not be any reference clock sent along with the high-speed serial data, the CDR must analyze the serial bit pattern and extract the clock information from the incoming serial data. Clock and data recovery (CDR) units are observing the transitions and choosing the minimal sampling phase for the data at the midpoint between edges. It retrieves clock data from the actual receiving data stream and uses this extracted clock to recollect the data waveform and absorb the data. CDR is nonlinear which significantly limits jitter and noise inside the SerDes [6] circuit. There is numerous clock recovery circuit (CRC) design approaches existing like traditional CRC, over-sampled CRC, source-synchronous links, etc. where the data stream must guarantee transitions. Most of all clock and data recovery circuits employ phase-locked [7] loop circuits.

3 Proposed Design

This paper proposed a Nyquist sampling-based architecture that will simultaneously capture the serial data with high redundancy and without any bit loss. The architecture uses an algorithm which also features the adaptive sampling rate independent of the bit duration. The algorithm is capable of estimating the interleaving window between successive bits and significantly analyzes the samples of successive bits and dynamically filters the noisy samples and recovers the bit information and also has the ability to adjust the offset deviations occurred while sampling the serial data. The block diagram of the design is shown in Fig. 4. The serial data is sampled by the sampler and collects a fixed number of samples for each serial bit. The sampler in the design is independent of the bit duration and also scalable. This scalability feature does not limit the number of samples to be fixed.

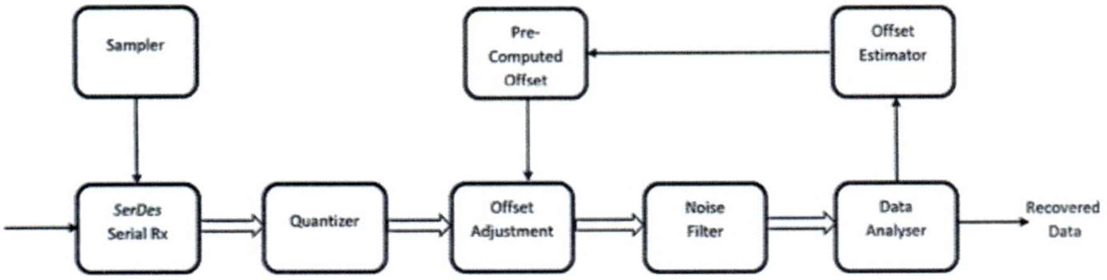

Fig. 4 Block diagram of proposed design

The number of samples required to sample the data can be varied from one scenario to another scenario considering several factors such as bandwidth, signal-to-noise ratio (SNR), bit error rate (BER) and noise present in the data. The sampling frequency should be significantly higher than the speed of the incoming data. The sampling frequency of the serial data can be chosen depending upon the number of samples that need to be collected for bit analysis in the noisy medium. The major feature of this architecture is that it will be suitable for high-speed source-synchronous interfaces where no reference clock will be distributed along with the serial data. The quantizer collects the sampled information of each bit and then identifies the samples related to transitional window and active window where the actual portion of the samples will be utilized for data recovery.

The transitional window samples are used to adjust offset deviations between the successive bits while sampled by the sampler. This technique effectively tunes the number of transitional samples required for each bit depending on the offset deviations caused by the sample. This will help the design to be more robust. The quantizer categorizes the samples into two groups such as (i) *transitional window* and (ii) *active window*.

Transitional Window

The transitional window is the key area that will compute the offset caused by the residual components while sampling the information. This transitional window further subdivided into pre- and post-computed offset windows. The pre-computed offset window will hold the previous bit offset deviation which will affect the sampling duration of the current bit. This bit-to-bit offset will be summarized and results in data mismatch due to the offset residual components [8]. The post-computed offset window which was computed successfully after the previous data will be buffered as a pre-computed offset to utilize during the next bit sampling window for tuning the transitional window. This will dynamically tune the complete sampling window for successive bits when they were retrieved at higher data rates.

The pre- and post-computed window size is scalable and can be chosen at which the rate of data being transmitted and offset deviation between the residual components generated while sampling (Fig. 5).

The post-computed offset will be taken into account when the next serial data is getting sampled, and the pre-computed window and number of samples collected in the active window will be adjusted dynamically such that offset deviation will be

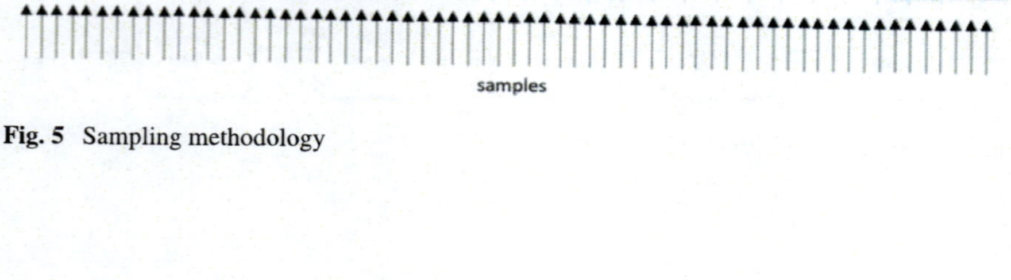

Fig. 5 Sampling methodology

nullified. In this way, the proposed design will address the offset deviation between successive bits during the high-speed reception.

$$m = M - (\Delta m_{\rm p} + \Delta m_{\rm q}) \quad (1)$$

$$\text{rxd\_int}\,(n+1) = \text{'1'when} \sum_{k=0}^{m-1} x(n-k) \geq \tfrac{M}{2} \quad \text{'0' otherwise} \quad (2)$$

$$\text{rxd\_int}\,(n) = \text{'1' when} \sum_{k=0}^{m-1} x(n-k) \geq \tfrac{M}{2} \quad \text{'0' otherwise} \quad (3)$$

where

m	Active window samples
M	Total number of samples collected for differential serial bit
$\Delta m_{\rm p}$	Pre-computed offset values
$\Delta m_{\rm q}$	Post-computed offset values.

Active window
The active window is the group of samples that will be used to extract the bit information. The high-speed interfaces will communicate in differential signaling (LVDS). Thus, both positive and negative samples will be combined together. Then, the samples will be processed further to recover the bit information. The noise filter will be filtering the non-differential samples which should be a compliment in nature. Then, the inverted samples will be processed, and the final bit information will be evaluated after analyzing both inverted samples collected from the active window. This will enhance the robustness of the design which enables the design for better performance and faster data recovery.

4 Results

The proposed design is successfully implemented on a *SerDes* receiver at the data speed 25 Gbps. The SerDes has 40-bit parallel input bus operating at 625 MHz which will match the throughput of the serial data 25 Gbps. Similarly, it has the 40-bit parallel output data bus where the received serial data is latched. The reference clock of 156.25 MHz will be supplied to the PLL for generating the internal transmitter and receiver clocks. The *SerDes* serial receiver while receiving the data at 25 Gpbs data speed, each bit has duration of 37.253 ps. Each bit was sampled at 1 ps sampling clock resulting in 37 samples per bit (Fig. 6).

The 0.253 ps residual cumulatively will create an offset of 25 ps after receiving every 100 bits. This 25 ps offset is significantly capable of data corrupt or mismatch.

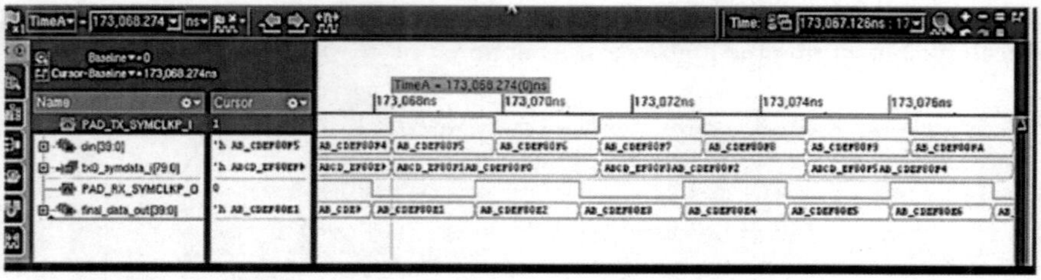

Fig. 6 SerDes recovered data with the proposed algorithm

To avoid this, the above algorithm was implemented on the receiver samples. The algorithm was dynamically self-tuned and adjusted the offset of each consecutive bit, and the offset of 0.253 ps per bit was successfully nullified without inserting delay.

5 Conclusion

This paper focuses on challenges that arise while receiving high-speed serial data only as it has more complex issues unlike transmission of the data. The SerDes works with a source-synchronous interface in which no synchronization clock will be present while transmitting/receiving the data. The SerDes receivers must have the clock and data recovery (CDR) circuit which dynamically extracts the clock and data from the receiving differential serial data. The algorithm is capable of estimating the interleaving window between successive bits and significantly analyzes the samples of successive bits and dynamically filters the noisy samples and recovers the bit information and also has the ability to adjust the offset deviations occurred while sampling the serial data. This will dynamically tune the complete sampling window for successive bits when they were retrieved at higher data rates.

References

1. Chang E, Narevsky N, Han J, Alon E (2018) An automated SerDes frontend generator verified with a 16NM instance achieving 15 GB/S at 1.96 PJ/Bit. In: Symposium on VLSI circuits, 2018. Honolulu, IEEE, pp 153–154
2. Gupta HS, Parmar R, Dave R (2009) High speed LVDS driver for SERDES. In: International conference on emerging trends in electronic and photonic devices and systems. IEEE, Varanasi, pp 92–95
3. Hou C, Wang Z, Huang K, Zhang C, Wang Z (2014) A 20 GHz PLL for 40 Gbps SerDes application with 4-bit switch-capacitor adaptive controller. In: International conference on electron devices and solid-state circuits. IEEE, Chengdu, pp 1–2
4. Stauffer DR, Mechler JT, Sorna MA, Dramstad K, Ogilvie CR, Mohammad A, Rockrohr JD (2008) High speed SerDes devices and applications. Springer, USA

5. Razavi B (2002) Challenges in the design of high-speed clock ad data recovery circuits, communications magazine. IEEE, pp 94–101
6. Zhihui Z, Yuan W, Junlei Z, Hailing Y, Song J (2010) A clock and data recovery circuit for 3.125 Gb/s RapidIO SerDes. In: International conference of electron devices and solid-state circuits (EDSSC). Hong Kong, IEEE, pp 1–4
7. Prinzie J et al (2019) A low noise fault tolerant radiation hardened 2.56 Gbps clock-data recovery circuit with high speed feed forward correction in 65 nm CMOS. In: 10th Latin American symposium on circuits & systems (LASCAS). m IEEE, Armenia, Colombia, pp 63–66
8. Roosevelt G, Roper W, Romanko T (2011) Optimizing high speed serial communication using Honeywell Rad Hard SerDes. In: NASA/ESA conference on adaptive hardware and systems (AHS). San Diego, CA, pp 215–219

DGS-Based T-Shaped Patch Antenna for 5G Communication Applications

Akash Kumar Gupta, Anil Kumar Patnaik, S. Suresh, P. Satish Rama Chowdary, and M. Vamshi Krishna

Abstract Technologies are advancing day by day after the successful implementation of 4G networks. Now, mobile technology is footed into 5G communication. To provide antenna solutions for 5G communications, a T-shaped multiband antenna has been proposed. The T-shaped microstrip patch antenna is intended to operate on 28/38 GHz frequency. T-shaped antenna has a compact size and planar geometry with high gain. To increase the bandwidth of the antenna, defected ground structures are used. These structures are formed by etching rectangular slots in ground.

Keywords T-shaped antenna · Defected ground structure · Multiband antenna · Dual-band frequencies · Gain · Radiation pattern

1 Introduction

The present smart world is living in 4G communications to provide user high-speed data services along with high-quality voice and video communications. As smartphone users are increasing day by day, it increases the need for optimum mobile network coverage, to avoid data congestion, provide high data rates. Conventional 4G communication operates on the band of frequencies 2–8 GHz providing a bandwidth in range of 5–20 MHz to fulfill consumer needs. The future communication

A. K. Gupta (✉) · A. K. Patnaik · S. Suresh · P. S. R. Chowdary
Raghu Institute of Technology (RIT), Visakhapatnam, Andhra Pradesh 531162, India
e-mail: akgupta452@gmail.com

A. K. Patnaik
e-mail: akp266@gmail.com

S. Suresh
e-mail: suresh.sangum@gmail.com

P. S. R. Chowdary
e-mail: satishchowdary@ieee.org

M. Vamshi Krishna
Centurion University of Technology and Management, Paralakhemundi, Odisha, India

P. S. R. Chowdary et al. (eds.), *Microelectronics, Electromagnetics and Telecommunications*, Lecture Notes in Electrical Engineering 655,
https://doi.org/10.1007/978-981-15-3828-5_2

technology is footing in 5G communications. The prime need for a communication system is to provide high gain antennas that are capable of delivering 5G frequencies and bandwidths.

The various 5G antennas are fabricated with geometrical shapes of circular [1], rectangular [2] patch antenna. The existing circular patch antenna covers space-borne radar applications that are operating over a frequency range of 9–10, 10.45 and 23.15–23.6 GHz; it does not cover mobile frequency bands like 28/38 GHz.

2 Design Requirements

The 5G communications are about to use 28/38 GHz frequency in the milli-meter wave spectrum. The 28/38 GHz has many advantages in terms of lower atmospheric absorptions and very lower attenuations [3, 4]. To provide antenna solutions at 28 GHz frequency, microstrip patch antennas are a convenient solution. The conventional microstrip patch antennas have many advantages like low cost, low profile, lightweight, small size, ease of designing in planar and non-planner structures, ease of fabrication, high gains. But, has a disadvantage like narrow bandwidth.

The bandwidth can be improved by defected ground structures, by using slots. Defected ground structures are formed by etching off a required geometry in the ground plane. These defects cause induced perturbations which creates an interruption in the uniform ground plane, and distribution, continuity of surface currents and causes controlled excitation of the feed line. DGS symmetrical or unsymmetrical structures function as resonant gaps and are etched directly underneath the patch. If defects are well aligned, then it allows effective coupling with the microstrip feed line. DGS structures are responsible for changes in the inductive and capacitive response of the transmission line. DGS works to enhance the effective reactance of transmission line by improving inductance and capacitance of the structure which reflects in multiple resonant frequencies to produce a multiple frequency band antenna [5], or it will improve the notch band capability of an antenna or a filter [6]. The designed antenna with DGS structures can be tunable to selected resonant frequency by choosing the suitable geometrical shape of the defects and placing them at appropriate positions on the ground plane.

3 Antenna Design

The basic structure of a patch antenna consists of the metallic ground plane, the dielectric substrate, and a radiating patch. To achieve efficiency and gain, dielectric substrate should be chosen carefully. In general, dielectric materials with low loss tangent values are used to reduce dielectric losses.

A low dielectric constant material reduces the total loss patch antenna without affecting the characteristic impedance. When a high dielectric constant material used, the patch losses are increased with a characteristic impedance. The thicker substrate increases radiation losses as they are severely affected by surface wave propagation. Thin substrates can reduce the surface waves and also are able to suppress the excitation of other undesired higher order modes. This reduces radiation power losses. Hence, thinner substrate provides a low loss, highly flexible structures [7]. The proposed T-shaped microwave patch antenna was designed on Rogers RT Duroid 5880 dielectric material. For designing thin substrate of Rogers RT Duroid 5880 having a dielectric constant (ε_r) of 2.2, loss tangent (tan δ) of 0.0009 used.

3.1 Basic T-shaped Patch Antenna Design

The basic structure of a patch antenna consists of layers of metal, dielectric, and a geometrical conducting patch. A rectangular patch 8 mm × 3.1 mm is designed on a substrate of Rogers RT Duroid 5880 with a height of 0.8 mm. The patch is excited with feed line having matched the impedance of 50 Ω. The antenna is designed and simulated in ANSYS Electromagnetics Suite. The design of the basic T-shaped antenna simulated for 28 GHz frequency [8] (Figs. 1 and 2).

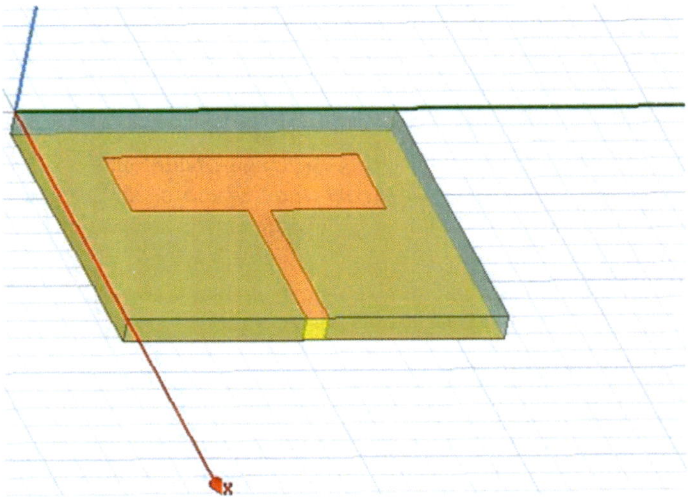

Fig. 1 T-shaped patch antenna top view

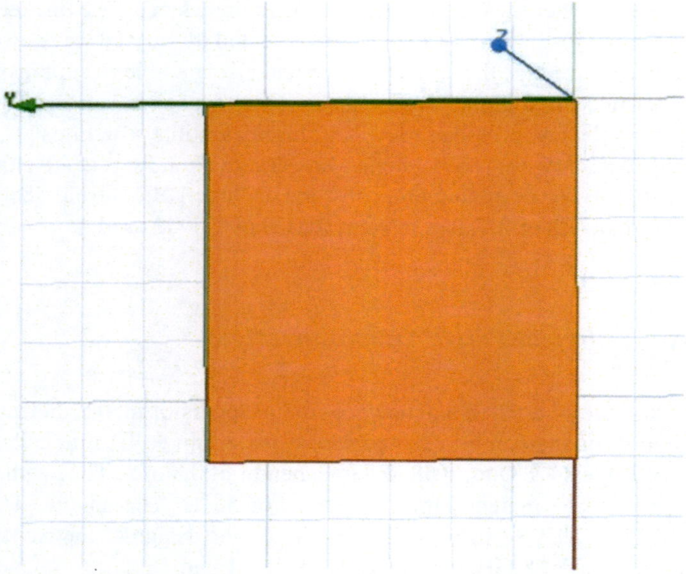

Fig. 2 T-patch bottom view-ground plane

3.2 *DGS Structure*

A defected ground structure is adapted to improve bandwidth and coupling. A partial ground structure was used with reduced ground length. Additionally, five rectangular slots are made on the ground plane for proper coupling and to reduce leakage currents. The rectangular slots with 0.62 mm × 0.2 mm are etched away in the ground plane (Figs. 3 and 4; Table 1).

4 Results and Discussions

The antenna designs are simulated in ANSYS Electromagnetics Suite, and the results are summarized below. The T-shaped patch antenna was simulated for 28 GHz frequency and inspected for return loss characteristics (S11), radiation pattern, normalized gain, efficiency.

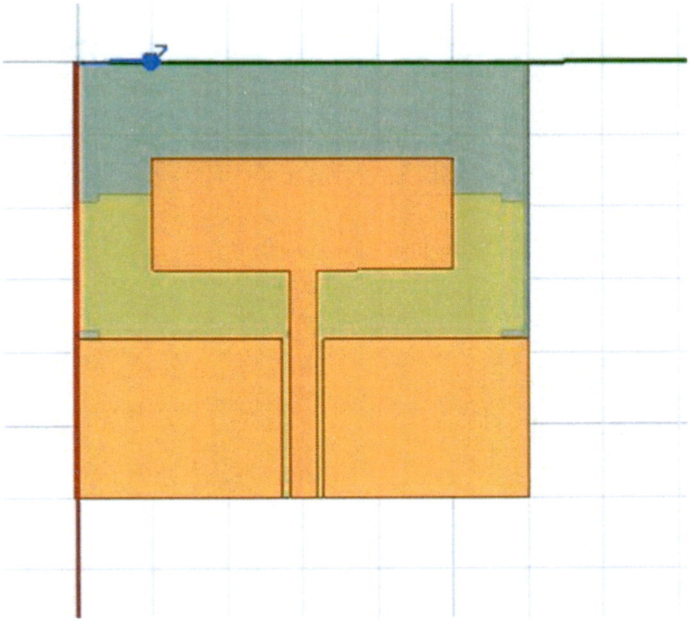

Fig. 3 DGS-based T-shaped patch antenna top view

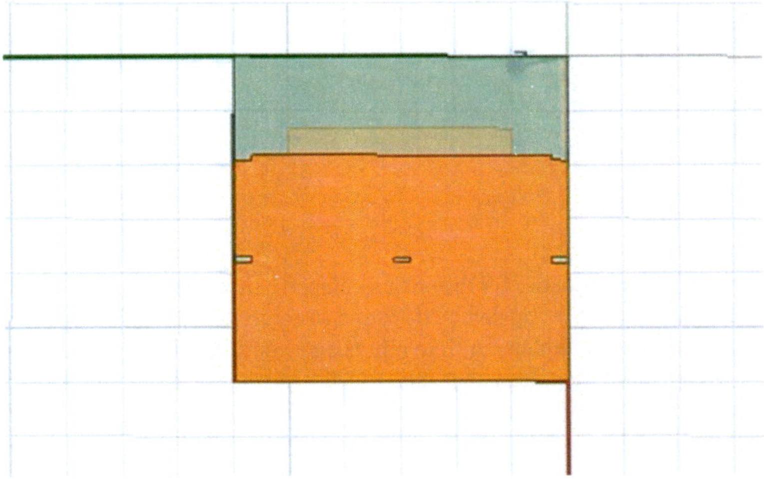

Fig. 4 DGS-based T-shaped patch antenna bottom view with slots

Table 1 Rectangular patch antenna design parameters

Parameters	Length in mm
Patch width	8
Patch length	3.1
Substrate height	0.8
The width of one side	3.3
Gap	1.9
Length of CPW ground	4.35
Length of partial ground	8.35
Feedline impedance	50 Ω

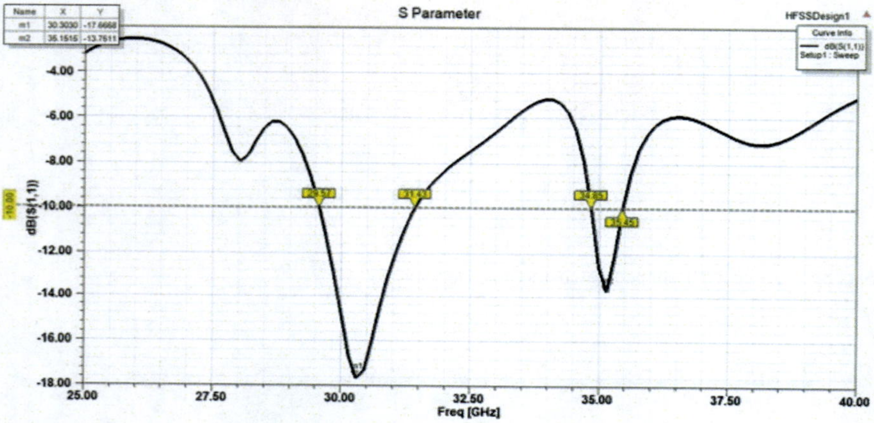

Fig. 5 Return loss characteristics of an antenna

4.1 Return Loss

The return loss characteristics for the basic T-shaped patch shows that the antenna resonating 25.757 GHz. By using defected ground structures, the resonant frequency can be improved, and it facilitates multiple inputs and multiple output applications. The return loss characteristics show that the DGS T-shaped patch antenna is resonating for multiple frequencies at 30.3 and 35.15 GHz (Fig. 5).

4.2 Radiation Pattern

The radiation pattern of the antenna is shown in Fig. 6. It shows mostly radiation in a broadside pattern.

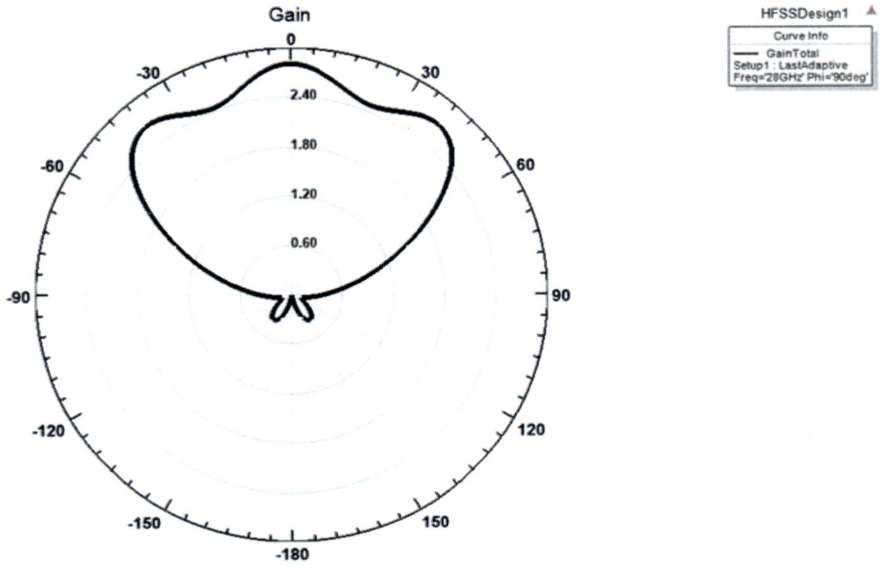

Fig. 6 Radiation pattern

4.3 *Gain and Efficiency*

The basic T-shaped patch antenna has a gain of −12 dB at a resonant frequency range of 25. The rectangular antenna with DGS has a maximum gain of 17.66 and 13.75 dB resonating at 30.30 and 35.1515 GHz (Fig. 7).

4.4 *Current Density of E-Field Distribution*

The current distribution over the T-shaped patch antenna is shown in Fig. 8. The surface current distribution was observed at a resonating frequency of 35.15 GHz. The current density is large at slot edges and in the ground plane. This also indicates the efficient coupling of slots with the feed line and leads to large bandwidth [9].

5 Conclusion

A high gain and large bandwidth T-shaped DGS-based antenna is presented which provides multiple resonant frequencies. Hence, it is suitable for MIMO applications. Using rectangular slots in the ground plane increases designs bandwidth and provides efficient coupling between the feed line patch and ground. The work was

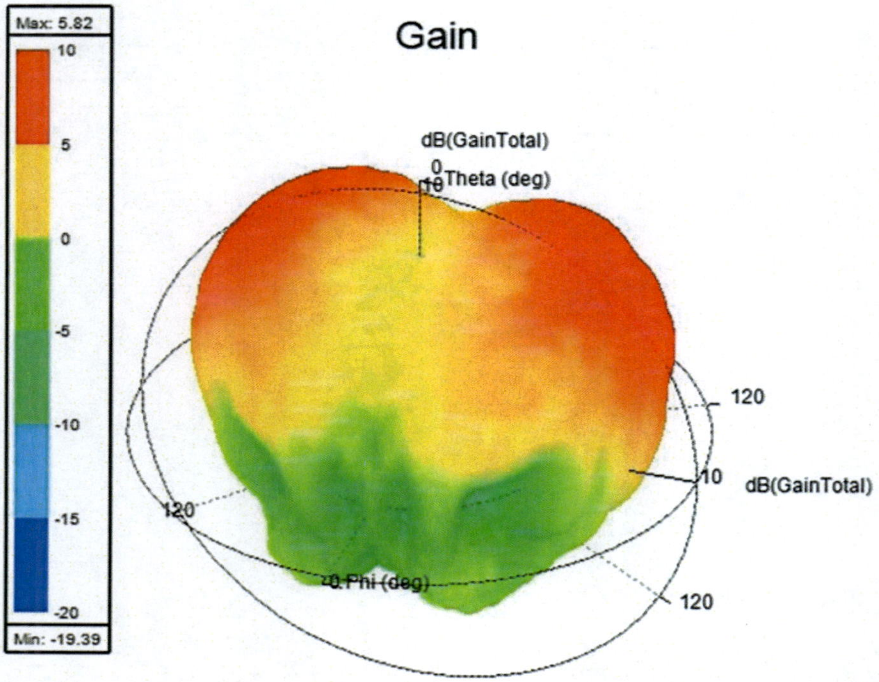

Fig. 7 Gain 3D plot

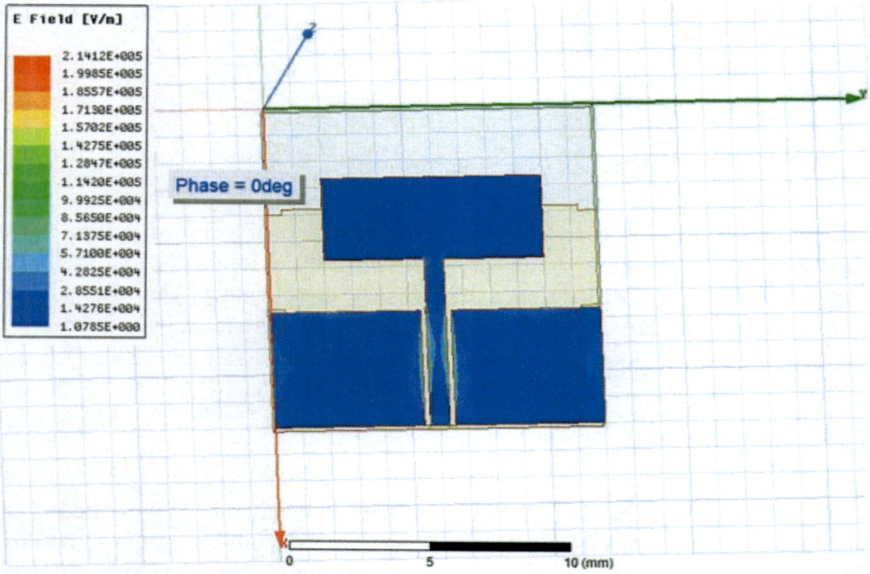

Fig. 8 Current density over the patch

initiated by designing a basic T-shaped patch antenna. When the design is simulated in ANSYS Electromagnetic Suite, it resonated at a single frequency of 25.757 GHz. By employing the DGS technique, five rectangular slots are formed in the ground plane with optimized separation in between them. This design modification improves coupling and bandwidth. The DGS-based T-shaped patch antenna provides multiple resonant frequencies at 30.30 and 35.1515 GHz with gains of 17.66 and 13.75 dB. This performance provides suitable solutions for 5G communications.

References

1. El Gholb Y, El Bakkali M et al (2018) Wide-band circular antenna for 5G applications. In: 4th international conference on optimization and applications (ICOA). IEEE Xplore
2. Rahman SU et al (2017) Design of rectangular patch antenna array for 5G wireless communication. In: Progress in electromagnetics research symposium—Spring (PIERS)
3. Rappaport TS et al (2013) Millimeter-wave mobile communications for 5G cellular: it will work! IEEE Access, pp 335–349
4. Zhao Q, Li J (2006) Rain attenuation in millimeter wave ranges. In: 7th international symposium antennas, propagation & EM theory, pp 1–4
5. Weng LH, Guo YC, Shi XW et al (2008) An overview on defected ground structure. Prog Electromagn Res B 7:173–189
6. Verma AK, Kumar A (2011) Synthesis of microstrip lowpass filter using defected ground structures. IET Microw Antennas Propag 5(12):1431–1439
7. Zhang J, Liu KC (1988) Microstrip antenna theory, and engineering. National Defense Industry Press, China
8. Jilani SF, Alomainy A (2018) Millimeter-wave T-shaped MIMO antenna with defected ground structures for 5G cellular networks. IET Microw Antennas Propag 12(5):672–677
9. Sim CYD, Chung WT, Lee CH (2010) Compact slot antenna for UWB applications. IEEE Antennas Wirel Propag Lett 9:63–66

Cascaded Operation of Hybrid Multilevel Inverter with Optimum Switching Angle Control for Power Quality Enhancement

Hejeebu Prasad, R. Kameswara Rao, and S. Kranthi Kumar

Abstract Hybrid multilevel inverters (MLI) are popular for high-voltage and high-power applications with a reduced number of devices, for the purpose of cost optimization, ease of design, control, and maintenance. The minimum total harmonic distortion (THD) levels indicate power quality improvement to the maximum level. THD can be minimized by optimum switching angle control. In this paper, two 19-level hybrid multilevel inverters are cascaded to produce output voltage of 37 levels with optimum switching angle control for enhancing the power quality by reducing the THD. The system is simulated in MATLAB/Simulink. The THD of simulated output of is compared with the calculated values.

Keywords THD · Multilevel inverters · Hybrid MLI · Optimum switching angles · Power quality

1 Introduction

The conventional multilevel inverters require a large number of switching devices and with high control complexity. Hence, the hybrid multilevel inverters can mitigate the disadvantages with lesser number of switching devices and ease of control. The research for multilevel inverters for design and control has been going on from the past few decades. In the year 1999, Leon M. Tolbert published his work on multilevel inverters for transformerless approach with carrier-based methods [1]. The THD minimization by switching angle control using selective harmonic elimination pulse width modulation is presented by Rai [2] and Espinosa [3]. A nineteen-level

H. Prasad (✉) · R. Kameswara Rao · S. Kranthi Kumar
Department of Electrical and Electronics Engineering, Raghu Engineering College, Visakhapatnam 531162, India
e-mail: hejeebuprasad@gmail.com

R. Kameswara Rao
e-mail: rkameswara@gmail.com

S. Kranthi Kumar
e-mail: skk.eee.rec@gmail.com

P. S. R. Chowdary et al. (eds.), *Microelectronics, Electromagnetics and Telecommunications*, Lecture Notes in Electrical Engineering 655,
https://doi.org/10.1007/978-981-15-3828-5_3

multilevel inverter is proposed by Venkataramaiah Jammala et al. using only 12 switching devices in 2018 [4]. The authors Hejeebu Prasad and R. Kameswara Rao presented an optimization algorithm for finding switching angles for reducing THD to possible minimum levels so as to improve the power quality levels [5].

1.1 Scope

This paper presents the cascading (series connection) output of hybrid multilevel inverters with coordinated control using optimum switching angles. Two 19-level hybrid multilevel inverters are cascaded with the coordinated operation to produce a 37-level output voltage. The system is simulated, and results are analyzed in MATLAB/Simulink and presented in this paper.

2 Cascaded Hybrid Multilevel Inverter

The cascading of the hybrid multilevel inverters holds the advantages of reducing the number of device required and easy maintenance and control. It adds the advantage of upgrading the number of levels as well as reducing the THD to further possible level.

2.1 19-Level Hybrid Multilevel Inverter

The schematic circuit of 19-level hybrid MLI is illustrated in Fig. 1. It basically consists of two H-bridges. Bridge 1 is fed from DC supply through a source selection circuit which can change DC voltage supplied to Bridge 1, i.e., 0, 1, 2 and 3 V. The Bridge 2 is directly fed from invariable 3V DC voltage source.

The transformers have been used to cascade the output voltage as the H-bridges are fed from same DC supply. Alternative circuits are given to reduce the number of transformers and transformerless approach subjecting to filtering requirement and increased requirement of sources, respectively.

The output of Bridge 2 is stepped up twice to produce 0, 6 or −6 V. The Bridge 1 output maintains its output level to produce 0, 1, 2, 3, −1, −2 or −3 V. These two outputs are cascaded to give output of either of levels 9 V through −9 V including 0 V. The switching states and corresponding outputs are given in Table 1. This switching table works for all the alternative circuits.

The switching devices should be of fully controllable solid-state devices such as IGBTs, MOSFETs or GTO-SCRs. It is recommended to use MOSFET or IGBT with reverse bias protection by internal diodes. The protective diodes create a freewheeling path for reactive currents during zero output state of H-bridges. This provision protects from transient over voltages (notches). The selection of components and other devices necessary for the system are based on the nature of load and load demands. Some other protective devices like fuse and isolators have to be used in practical systems.

Alternative Schemes of Connection

Figures 2 and 3 are the alternative circuits for Fig. 1.

Figure 2 is the alternative with a single source and a single transformer. This reduces the number of transformers. The disadvantage with this version is that the output contains switching over voltages called notches in the output voltage. These notches are produced because of non-ideal solid-state devices and reactance in the transformer. This version requires filtering to suppress the notches.

Figure 3 gives the transformerless approach with separate DC sources. Since each H-bridge is fed from isolated DC source, the outputs of the H-bridges can be cascaded directly.

Subject to the availability of voltage sources, if the number of voltage sources is only one, and filterless design is required, the schematic illustrated in Fig. 1 is preferred. If the high pass filters are used intentionally for other reasons, the alternative schematic shown in Fig. 2 is recommended. If the number of sources are sufficient (multiples of 3), the alternative scheme shown in Fig. 3 is recommended.

Features of the 19-Level Hybrid Multilevel Inverter

- Only 12 active switches are required for the design.
- It only requires six digital variables to control since the complementary switching for the devices in the same leg is adopted.

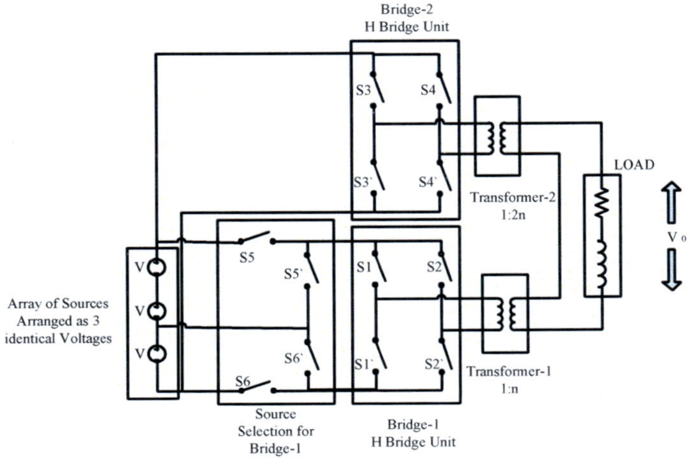

Fig. 1 19-level hybrid inverter with 12 switching devices and two transformers

Table 1 Switching states and output voltage of 19-level multilevel inverter

Level	*S*1	*S*2	*S*3	*S*4	*S*5	*S*6	Tr-1 primary	Tr-2 primary	Tr1-secondary	Tr2-secondary	Output V_0
0	0	0	0	0	0	0	0	0	0	0	0
1	1	0	0	0	0	1	V	0	nV	0	nV
2	1	0	0	0	1	0	2 V	0	2 nV	0	2 nV
3	1	0	0	0	1	1	3 V	0	3 nV	0	3 nV
4	0	1	1	0	1	0	−2 V	3 V	−2 nV	6 nV	4 nV
5	0	1	1	0	0	1	−V	3 V	−nV	6 nV	5 nV
6	0	0	1	0	0	0	0	3 V	0	6 nV	6 nV
7	1	0	1	0	0	1	V	3 V	nV	6 nV	7 nV
8	1	0	1	0	1	0	2 V	3 V	2 nV	6 nV	8 nV
9	1	0	1	0	1	1	3 V	3 V	3 nV	6 nV	9 nV
−1	0	1	1	1	0	1	−V	0	−nV	0	−nV
−2	0	1	1	1	1	0	−2 V	0	−2 nV	0	−2 nV
−3	0	1	1	1	1	1	−3 V	0	−3 nV	0	−3 nV
−4	1	0	0	1	1	0	2 V	−3 V	2 nV	−6 nV	−4 nV
−5	1	0	0	1	0	1	V	−3 V	nV	−6 nV	−5 nV
−6	1	1	0	1	0	0	0	−3 V	0	−6 nV	−6 nV
−7	0	1	0	1	0	1	−V	−3 V	−nV	−6 nV	−7 nV
−8	0	1	0	1	1	0	−2 V	−3 V	−2 nV	−6 nV	−8 nV
−9	0	1	0	1	1	1	−3 V	−3 V	−3 nV	−6 nV	−9 nV

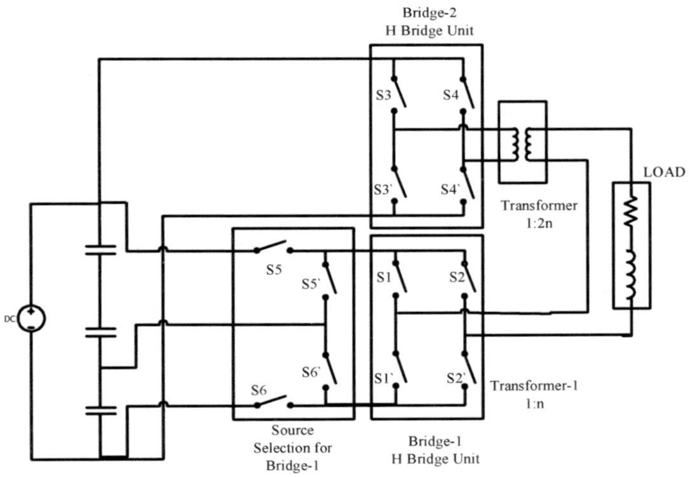

Fig. 2 Alternative schematic with a single source and single transformer

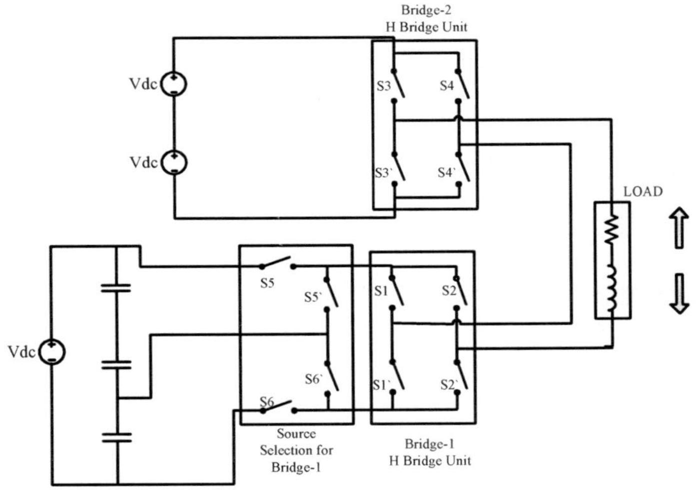

Fig. 3 Alternative schematic connection for transformerless approach by using separate DC sources

- Fault identification and maintenance is easy due to less number of devices.
- Gate circuit losses and the requirement of gating circuitry are less compared to conventional multilevel inverters.

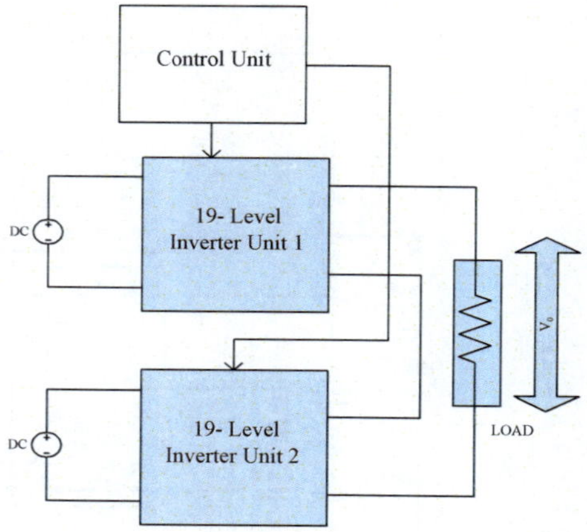

Fig. 4 Cascaded connection of two 19-level hybrid multilevel inverters with coordinated control

2.2 Cascading Hybrid Inverter Units

The outputs of two 19-level inverter units are cascaded (output terminals are connected in series) as shown in Fig. 4. Each inverter gives an output of 9 V through −9 V including zero output. The cascaded connection gives 18 V through −18 V including a zero output. Thus, the whole system produces a 37-level output.

The two inverters are operated with co-ordination. One of the inverters is operated at optimum switching angles given in Row 1 of Table 2. The other inverter unit uses Row 2 of Table 2.

3 Optimum Switching Angle Control

The THD of output voltage waveform can be minimized to possible levels with optimum switching angle control.

3.1 Methodology

The following assumptions are made for this strategy.

- The output voltage is symmetric about time axis.
- The output voltage is symmetric about the midpoint of half cycle.
- The rise of voltage in each level shift (transition in output) is uniform.

With the above considerations, the THD can be mathematically expressed as Eq. (1) (refer to [5])

Table 2 Optimum switching angles for 37-level multilevel inverter

Row 1	1.7189	4.5837	7.4485	10.886	13.751	17.188	20.053	23.491	27.502
Row 2	30.9397	34.3775	38.3882	42.3989	46.4096	51.5662	56.1499	63.5983	72.1927

$$\text{THD}(\%) = \frac{\sqrt{\sum_{n=3}^{\infty}\sum_{p=1}^{k}\frac{\cos^2(n\alpha_p)}{n^2}}}{\sum_{p=1}^{k}\cos\alpha_p} \times 100 \tag{1}$$

where

n Harmonic number odd integer (3, 5, 7….∞) (maximum of 999 considered for converging the optimum solution)
p output level number from 1 to k.
α_{p} Switching angle for level shift (transition instant)
k Number of switching transitions in the quarter cycle. The 'k' value is given by Eq. (2)

$$k = \frac{m-1}{2} \tag{2}$$

where

m number of levels in output voltage = 37.

Using Eq. (1), the optimum switching angles can be found using iterative algorithms by using digital computers. The switching angles are found by using MATLAB scripts for 37-level output and shown in Table 2. The corresponding theoretical THD is 2.259%.

Row 1 is implemented in inverter unit 1, and Row 2 is implemented in inverter unit 2.

Advantages with Optimum Switching Angle Control

- THD is minimized to the possible level.
- It does not require analog circuitry for gating signal generation.
- The gating signal generation can be done with small embedded development boards like Arduino or Raspberry Pi Boards hence reducing the cost.

4 Simulation

A simulation model is built in MATLAB/Simulink to verify the output and also to analyze the output waveforms using fast fourier transform tool (FFT tool).

The main system of 37-level cascaded hybrid multilevel inverter, the subsystem of 19-level inverter unit and output level of each subsystem and load voltage and load current are shown in Figs. 5, 6 and 7, respectively.

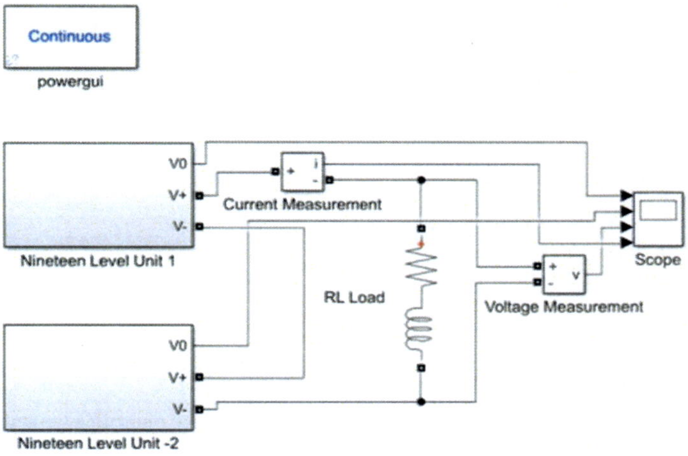

Fig. 5 Simulink model main system

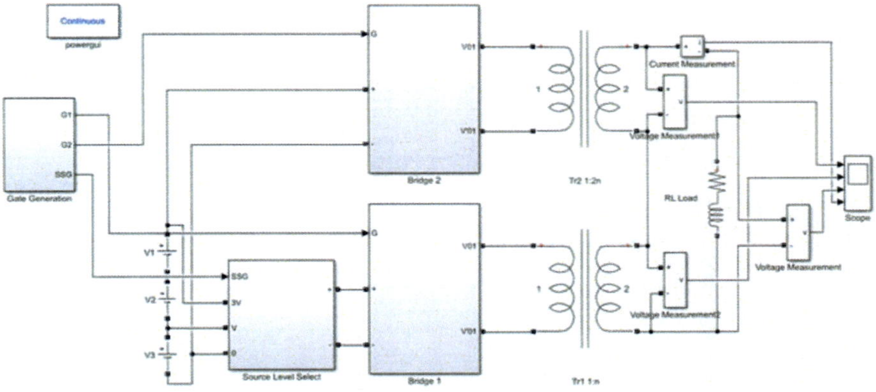

Fig. 6 Subsystem 19-level unit

5 Results and Conclusion

The load voltage output waveform is analyzed using FFT tool. The FFT window is shown in Fig. 8.

Conclusion

It is found that output voltage has a total harmonic distortion of 2.43%. The theoretical THD for the optimum switching angles for 37-level is found to be 2.259% (refer to Sect. 3.1 Methodology). The deviation of THD from theoretical value to simulated value can be due to various reasons. Some of the possible reasons are listed below.

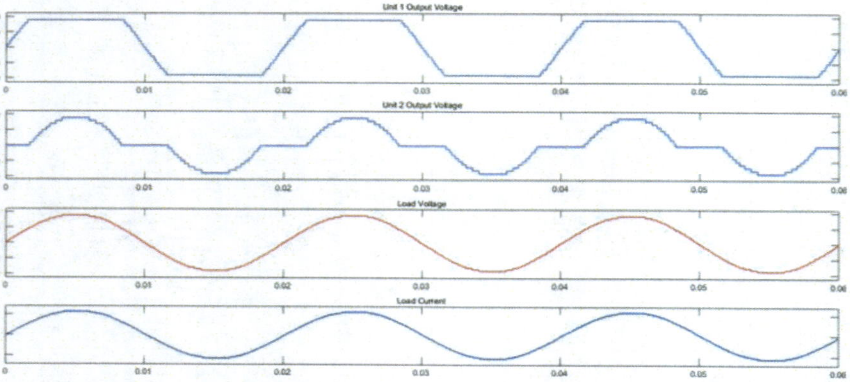

Fig. 7 Output waveforms of unit 1 output voltage, unit 2 output voltage load voltage and load current

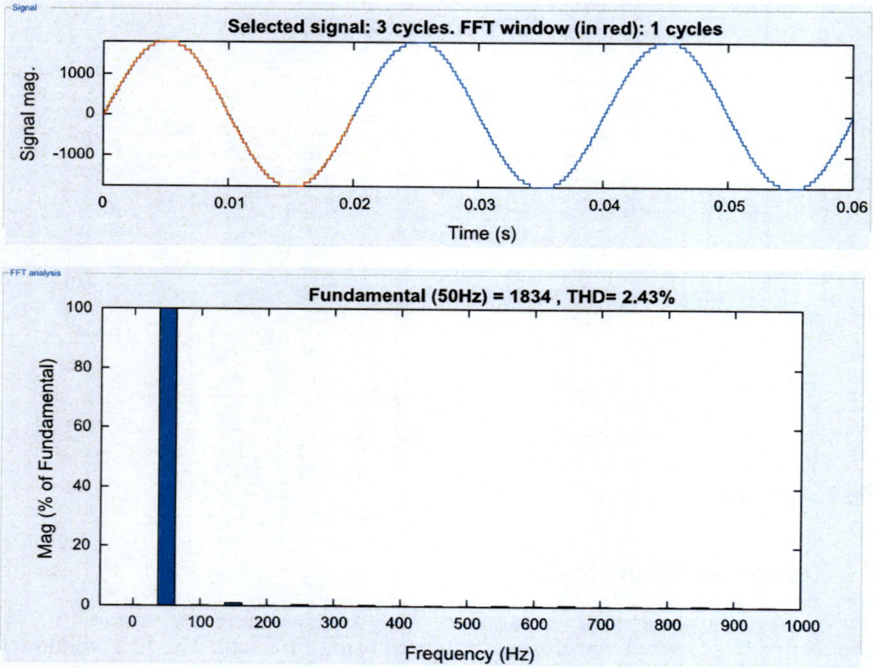

Fig. 8 FFT window for output voltage analysis

- Unequal voltage level transition
- The voltage drop across each switching devices due to internal resistance
- Resistance and reactance in the transformers used which causes voltage regulation.

IEEE Standards 519-2014 [6] recommends the THD value to be maintained for above 69–161 kV is 2.5%. Hence, the proposed scheme and control technique can be used for high-voltage applications[1] for the range cited, i.e., from 69 to 161 kV.

6 Future Scope

Every development will lead to create some new challenges to be accomplished. Some challenges are found in this system. The inverter unit 1 is delivering output for a longer duration, while unit 2 is operated for lesser time. This leads to the frequency of charging and discharging cycles of DC energy storage (batteries or super capacitors) is more for unit 1 than for unit 2. Hence, balancing the charge on the DC storage systems of both the units is an important issue. The research work will be continued to accomplish the challenge.

References

1. Leon M, Tolbert TG (1999) Novel multilevel inverter carrier-based PWM method. IEEE Trans Ind Appl 35(5):1097–1107
2. Rai SK, She-Pwm PC (2016) Based multilevel T-Type inverter topology for single-phase photovoltaic applications. In: IEEE international conference on power electronics, drives and energy, pp 1–4
3. Espinosa CAL, Portocarrero I, Izquierdo M (2012) Minimization of THD and angle calculation for multilevel inverters. Int J Eng Technol IJET-IJENS 12:83–86. 1211905-8282-IJET-IJENS 05
4. Jammala VR, Yellasiri S (2018) Development of new hybrid multilevel inverter using modified carrier SPWM switching strategy. IEEE Trans Power Electron 33(10):8192–8197
5. Prasad H, Rao RK, Optimum switching angles for multilevel inverters for minimization of THD. In: IEEE international conference on intelligent systems and green technology (ICISGT-2019), Visakhapatnam, India
6. IEEE Recommended practice and requirements for harmonic control in electric power systems. IEEE 519TM-(2014)

[1] Though the model is simulated for 1100 V peak output voltage, the THD depends on the shape and pattern of the output voltage. The voltage should be close to the sinusoidal waveform. Low THD indicates good power quality and closeness to sinusoidal wave shape.

Performance Comparison Analysis of GTTPC and AHTPC Technique for WBAN with Mobile Scenario

M. Raj Kumar Naik and P. Samundiswary

Abstract An efficient energy transmission is a key factor to be considered for long-term and low-power operations for WBAN. Reactive and proactive are two important available transmit power control techniques that are chosen based on the channel condition but are present with drawbacks in terms of delays, errors and overhead. As a result, an adaptive hybrid transmit power control (AHTPC) algorithm is used against proactive or reactive technique to overcome its limitations. However, the AHTPC which is used also possesses some additional delays and overhead. Thus, the game theory-based adaptive hybrid transmit power control (GTTPC) technique is developed for mobile condition in WBAN. In this algorithm, both received signal strength indicator (RSSI), signal-to-interference noise ratio (SINR) and packet reception rate (PRR) values are taken for choosing the appropriate channel for controlling power. It is noted through the simulation results that the proposed GTTPC technique achieves reduced overhead, delay and energy consumption with enhanced delivery ratio and residual energy compared to AHTPC techniques.

Keywords Power control · RSSI · SINR · PRR · Adaptive hybrid

1 Introduction

The major constraint in WBANs is their limited power supply [1]. This constraint has given a path for future research direction to propose an efficient power control algorithms to improve energy efficiency with better QoS, especially for health care application [2]. It is also observed that major energy waste in WBANs is caused by interference from inter-network WBANs due to which degradation in throughput occurs because of drop in the signal-to-interference and noise ratio (SINR) [3–5]. Most of the authors in the past did not follow any energy-saving technique. So, energy-efficient TPC techniques are developed for WBAN to maximize the energy

M. Raj Kumar Naik (✉) · P. Samundiswary
Department of Electronics Engineering, Pondicherry University, Kalapet, Puducherry 605014, India
e-mail: raj.meghavath@gmail.com

P. S. R. Chowdary et al. (eds.), *Microelectronics, Electromagnetics and Telecommunications*, Lecture Notes in Electrical Engineering 655,
https://doi.org/10.1007/978-981-15-3828-5_4

efficiency. Several researchers propose different reactive TPC approaches where the TPL is changed adaptively based on the measurement of RSSI values at the sink or AP [6–10]. The authors in [11 and 12] dealt with proactive approaches. In which, the channel situation is anticipated using RSSI values and channel gain measurements. Both the proactive and reactive approaches have certain drawbacks [13]. Due to the drawback in the above approach, AHTPC algorithm was developed and which also possess some overheads and delay. Therefore, energy-efficient TPC technique is developed in this paper for mobile condition in WBAN to maximize the energy efficiency of wearable device using game theory for WBAN network.

The remaining part of this paper is arranged as follows: adaptive hybrid technique is dealt in Sect. 2. In Sect. 3, adaptive hybrid using game theory is discussed in detail. In Sect. 4, the simulation results are presented. Finally, the conclusion along with future work is drawn in Sect. 5.

2 Adaptive Hybrid Technique for Power Control (AHTPC)

An adaptive hybrid TPC (AHTPC) algorithm for IEEE 802.15.6 WBAN is proposed. Initially, the MODE flag is set to *R* (reactive technique), the AP measures the RSSI and PDR values and save the values in a channel sample matrix in the format <TS_i, $RSSI_i$, PDR_i> where TS represents time slots of the corresponding period. Two threshold values are maintained: the upper bound (UB) and the lower bound (LB). If the difference of $RSSI_i$, $RSSI_{i+1}$ values are greater than the value of another threshold δ, then the proactive mode is triggered by changing the MODE flag to *P* (proactive technique) and if it is less RTPC is executed. So, AHTPC technique contains both proactive and reactive approaches also has some drawbacks. In order to overcome the above drawbacks, a game theory-based adaptive hybrid TPC (AHTPC) algorithm for IEEE 802.15.6 WBAN is proposed for mobile condition that is discussed in this paper.

3 Game Theory Adaptive Hybrid TPC Technique (GTTPC)

In this work, GTTPC technique for WBAN is developed by considering RSSI, SINR and PRR for estimation of channel condition for data transmission in the network. The GTTPC algorithm consists of selection and initialization of neighbour nodes and utility function estimation. So, the flowchart is given in Fig. 1. During this stage, the nodes which are used in the network will identify its neighbour nodes by its node id during that particular time slot. Then, the received node stores its information for the estimation of minimum TPL along with the residual energy used by it. So, the node which is using minimum TPL is chosen to transmit the data by using a game

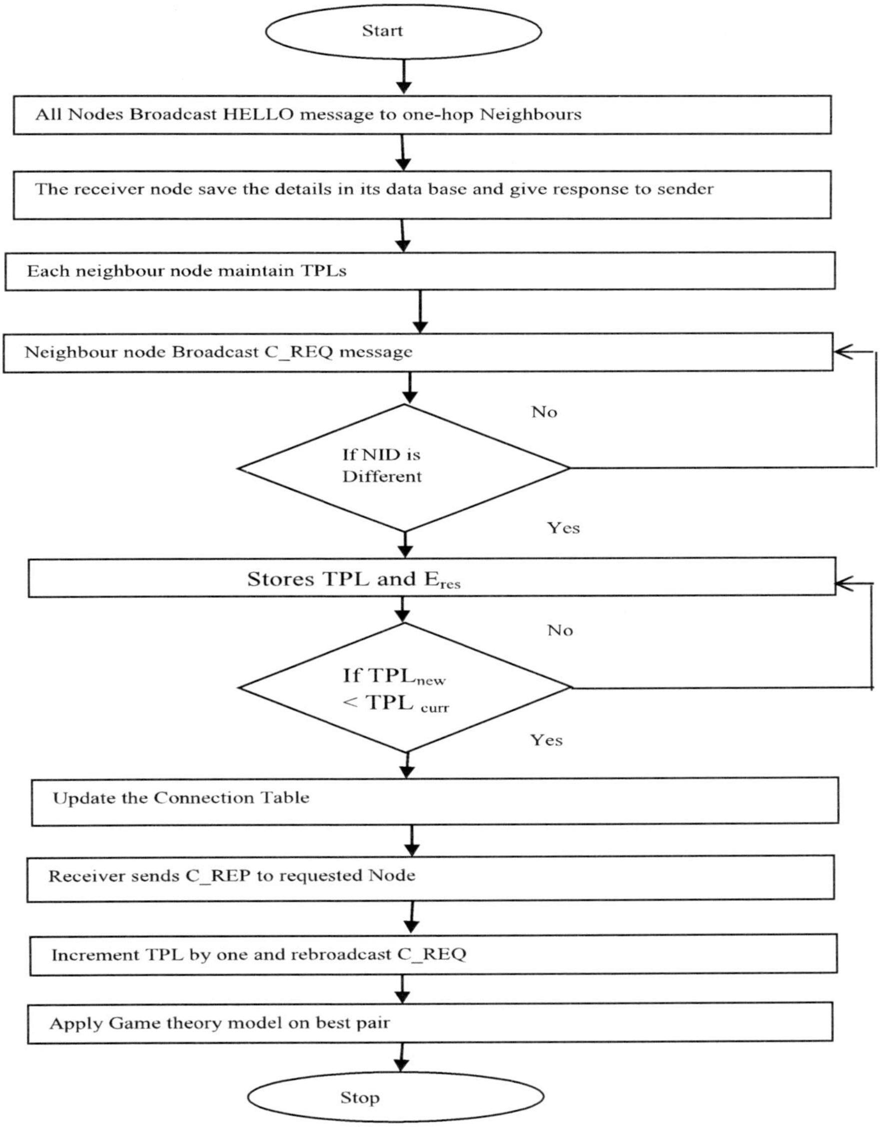

Fig. 1 Flowchart

theory model. In the estimation of the utility function, game theory is considered by parameters like throughput, SINR and PRR along with the residual energy. Thus, the game theory concept is utilized by choosing the node with minimum TPL which is suitable for mobile condition in WBANs.

4 Simulations Results

GTTPC algorithm is developed and simulated by using NS-2. In the simulation, mobile condition with two scenarios is considered. In scenario-1, the subject with hands raised and legs stretched, and in scenario-2, the subject with slow movement.

The simulation settings and their parameters are summarized in Table 1.

Scenario-1 The Subject with hands raised and Legs Stretched—The simulation topology for scenario-1 is shown in Fig. 2.

Table 1 Simulation parameters

Parameters	Values
Number of nodes Size of topology	12 50 m × 50 m
MAC protocol	IEEE 802.15.6
Simulation time	10–50 s
Traffic source	CBR
Channel model	CM3 and CM4
Propagation	Two ray ground
Initial energy	12 m Joules
Transmission power	0.5 m watts
Receiving power Speed	0.3 m watts 1 m/s

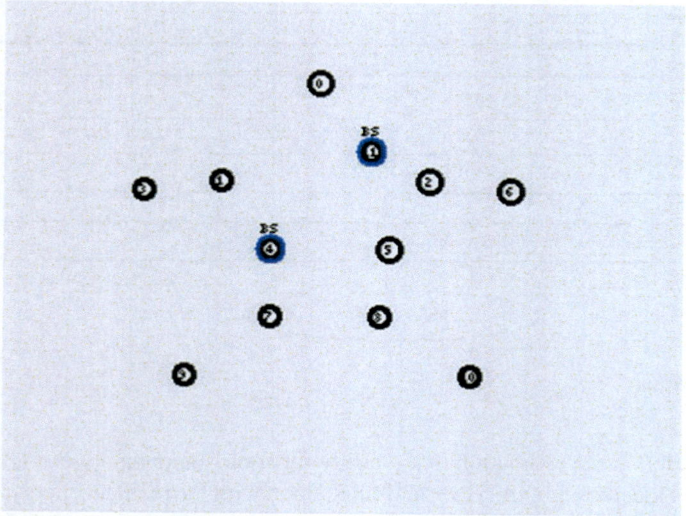

Fig. 2 Simulation topology of scenario-1

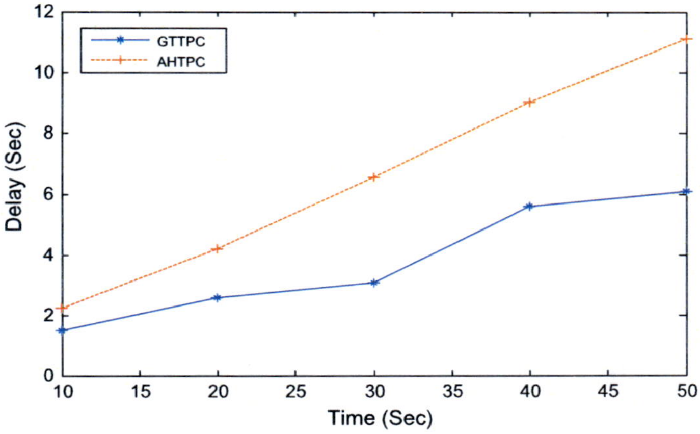

Fig. 3 Delay versus time

Figure 3 shows the delay response of GTTPC and AHTPC by varying the time. The game theory concept is used for path selection of source node to minimize delay. Hence, the delay of GTTPC is 43.11% lesser than that of AHTPC. As the packets transmission in the system will take longer time to reach the destination, so the delay is increased with increase in time.

It is portrayed through Fig. 4 that the delivery ratio response of GTTPC and AHTPC is decreased when varied with time. This is due to the reason that the path length required to transmit data is increased when the time increases resulting in more loss of packets. Hence, the delivery ratio of GTTPC is 43.01% improved compared to that of AHTPC.

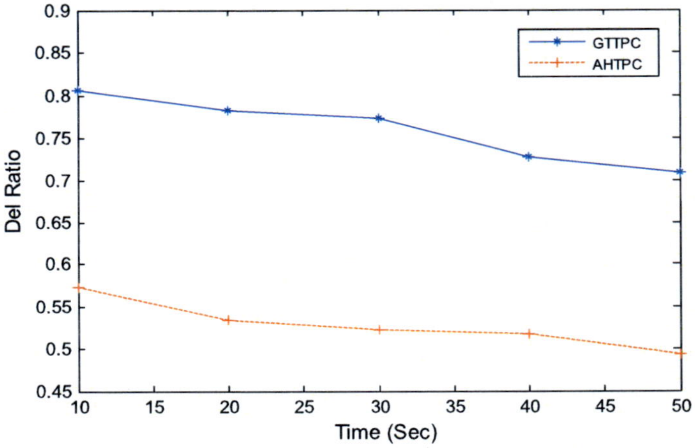

Fig. 4 Delivery ratio versus time

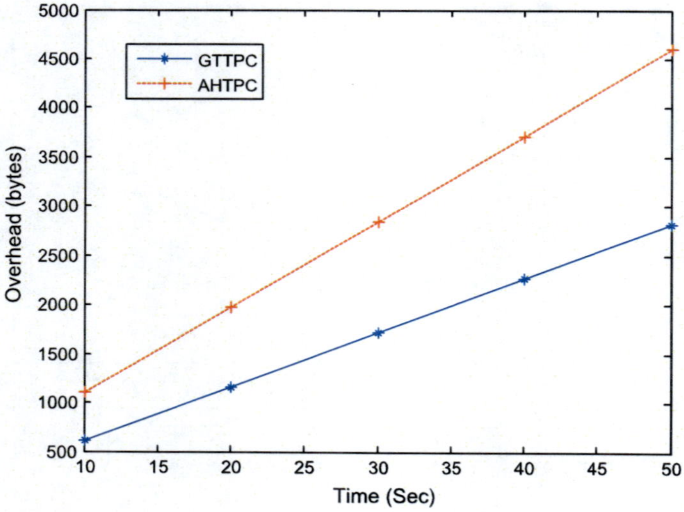

Fig. 5 Overhead versus time

The overhead response of GTTPC and AHTPC is inferred by Fig. 5 for varying the time. The overhead of GTTPC is 39.60% lesser compared to that of AHTPC. The reason for increase in overhead is due to the transmission of more routing control packets along with the data packets when the time is increased to achieve proper link.

It is observed from Fig. 6 that the energy consumption of GTTPC is 5.71% lesser compared to that of AHTPC. It is also noticed that as the time is increased, energy consumption of the system is increased because more power is consumed by the nodes due to which they dissipate at faster rate as a result more energy is consumed.

Figure 7 illustrates the residual energy response with respect to varying time. As the time increases, the nodes which are in close proximity to sink consume less energy results in more residual energy and vice versa. Hence, GTTPC is 0.69% higher than AHTPC.

Scenario-2 The Subject with Slow movement—The simulation topology for scenario-2 is shown in Fig. 8.

Figure 9 illustrates the delay response of GTTPC and AHTPC by varying the time. The GTTPC is 35.04% lesser compared to that of AHTPC because the packets take more time to reach the sink node due to increased number of hops which result in increased delay.

Figure 10 shows the delivery ratio response of GTTPC and AHTPC by varying time. This is because as the time increases, the more path length is required for data transmission resulting in more packet loss which in turn leads to decrease in delivery ratio. Hence, the delivery ratio of GTTPC is 29.41% enhanced compared to that of AHTPC.

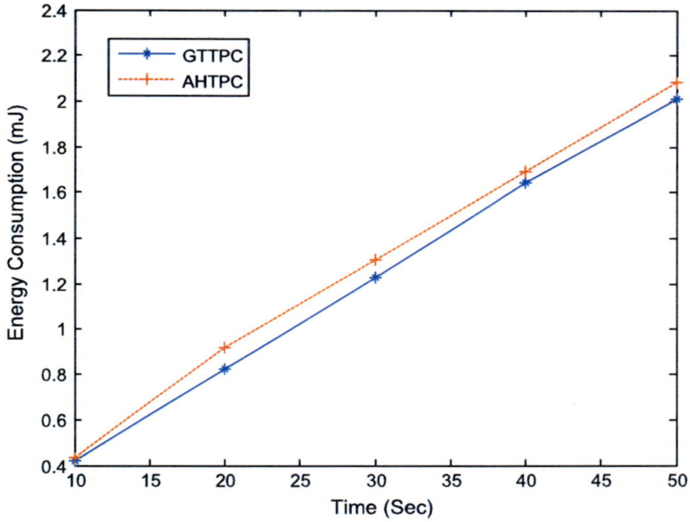

Fig. 6 Energy consumption versus time

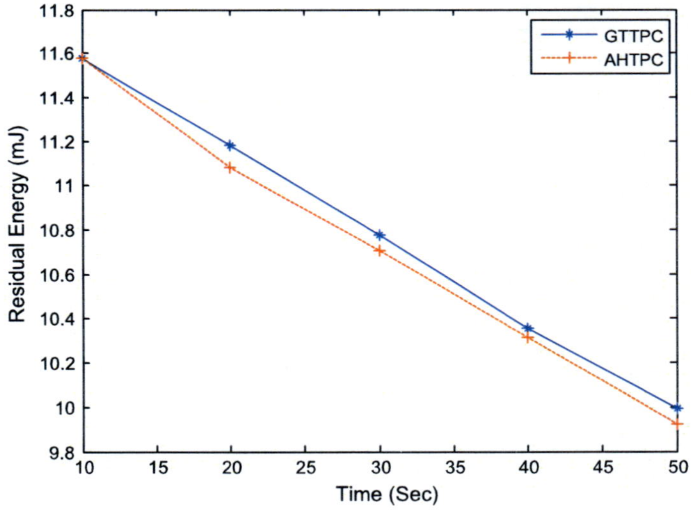

Fig. 7 Residual energy versus time

It is verified through Fig. 11 that the overhead response of GTTPC and AHTPC is examined by varying the time. This is due to increased number of control packets which in turn increase overhead for increased time. So, GTTPC is 34.06% lesser than AHTPC.

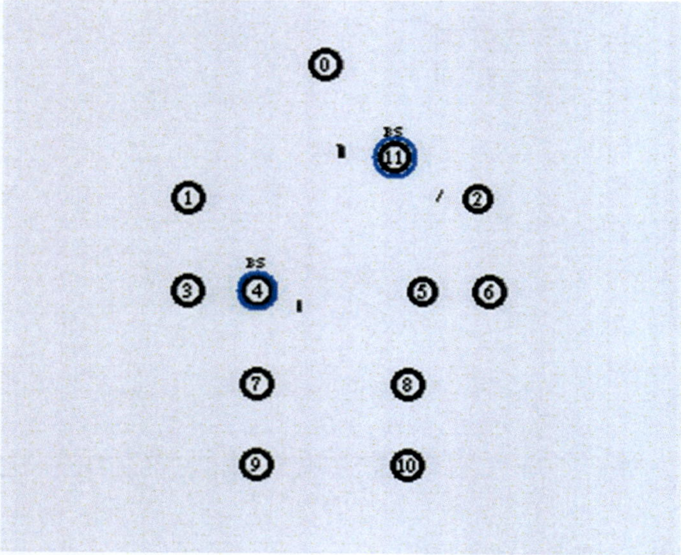

Fig. 8 Simulation topology for scenario-2

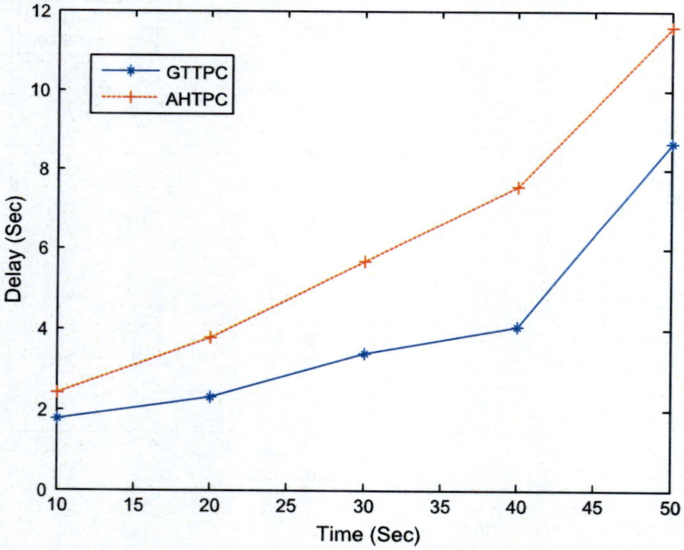

Fig. 9 Delay versus time

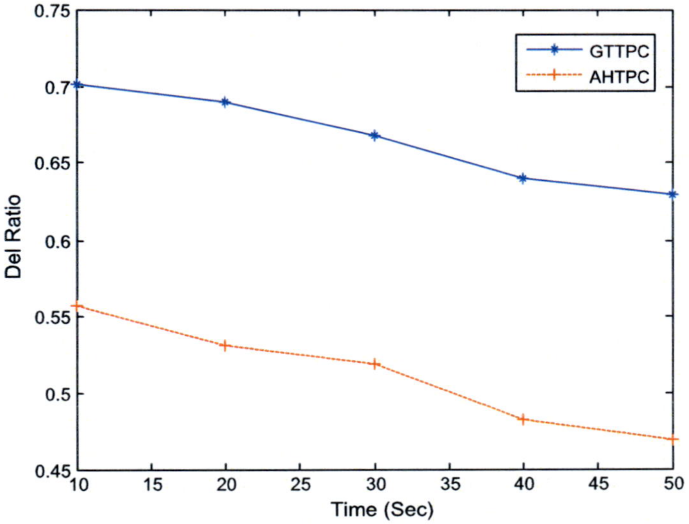

Fig. 10 Delivery ratio versus time

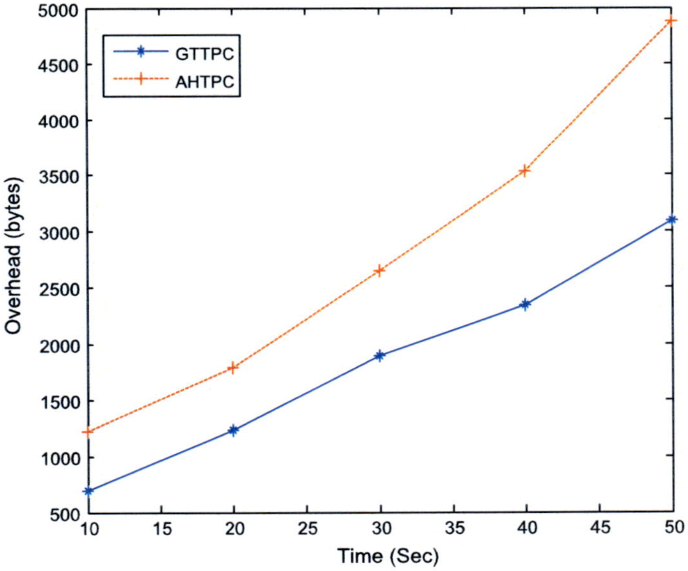

Fig. 11 Overhead versus time

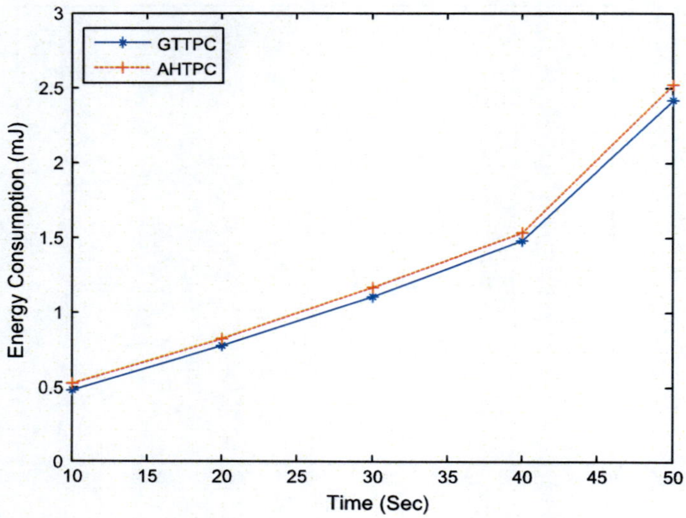

Fig. 12 Energy consumption versus time

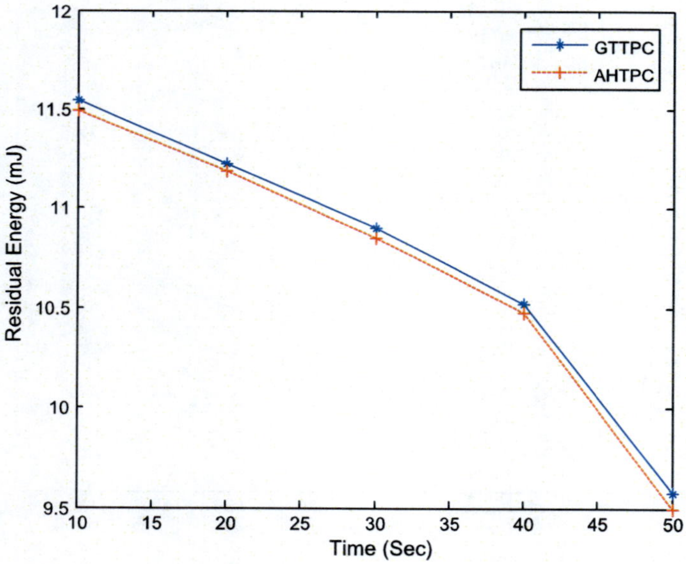

Fig. 13 Residual energy versus time

It is observed from Fig. 12 that energy consumption of GTTPC is 5.29% lesser compared to that of AHTPC. This is because the nodes used for longer time period will dissipate energy at higher rate with increase in time.

The residual energy response of GTTPC and AHTPC is illustrated in Fig. 13. As time increases the residual energy is decreased because the energy required to transmit data for the longest path will consume more energy. Thus, GTTPC is 0.76% higher than AHTPC.

5 Conclusion

In this paper, GTTPC with mobile condition is developed and analysed for two scenarios. In GTTPC algorithm, minimum TPL is selected along with channel index by forming a better pair to calculate utility function through estimation of certain parameters like SINR, throughput and residual energy to improve energy efficiency by controlling power in WBAN. The simulation results show that GTTPC algorithm achieves reduced energy consumption, delay and overhead with enhanced residual energy and delivery ratio compared to that of AHTPC.

References

1. Kwak KS, Ullah S, Ullah N (2010) An overview of IEEE 802.15. 6 standard. In: Proceedings of international symposium on applied sciences in biomedical and communication technologies, Italy, pp 1–6
2. Thotahewa KMS, Khan JY, Yuce MR (2014) Power efficient ultra wide band based wireless body area networks with narrowband feedback path. IEEE Trans Mobile Comput 13(8):1829–1842
3. Gu N, Jiang Y, Zhang J, Zheng H (2013) An implementation of WBAN module based on NS-2. In: Proceedings of international conference on computer sciences and applications, China, pp 114–118
4. Arfaoui A, Kribeche Vesilo A (2018) Game-based adaptive risk management in wireless body area networks. In: Proceedings of international wireless communications and mobile computing conference, Cyprus, pp 1087–1093
5. Gao W, Jiao B, Yang G, Hu W, Liu J (2014) Transmission power control for IEEE 802.15.6 body area networks. J Electron Telecommun Res Inst 36(2):313–316
6. Sodhro AH, Li Y, Shah MA (2016) Energy-efficient adaptive transmission power control for wireless body area networks. IET Commun 10(1):81–90
7. Moulton B, Hanlen L, Chen J, Croucher G, Mahendran L, Varis A (2010) Body-area-network transmission power control using variable adaptive feedback periodicity. In: Proceedings of communications theory workshop, Australia, pp 139–144
8. Vallejo M, Recas J, Ayala JL (2015) Proactive and reactive transmission power control for energy-efficient on-body communications. Sensors 15(3):5914–5934
9. Xiao S, Dhamdhere A, Sivaraman V, Burdett A (2009) Transmission power control in body area sensor networks for healthcare monitoring. IEEE J Sel Areas Commun 27(1):37–48
10. Newell G, Vejarano G (2016) Human-motion based transmission power control in wireless body area networks. In: Proceedings of IEEE third world forum on internet of things, USA, p 16
11. Di Franco F, Tachtatzis C, Atkinson RC, Tinnirello I, Glover IA (2015) Channel estimation and transmit power control in wireless body area networks. IET Wirel Sens Syst 5(1):11–19

12. Smith DB, Hanlen LW, Miniutti D (2012) Transmit power control for wireless body area networks using novel channel prediction. In: Proceedings of wireless communications and networking conference, China, pp 684–688
13. Lee W, Lee B-D, Kim N (2014) Hybrid transmission power control for wireless body sensor systems. Int J Distrib Sens Netw 20:1–9

A Comprehensive Study and Evaluation of Recommender Systems

A. Vineela, G. Lavanya Devi, Naresh Nelaturi, and G. Dasavatara Yadav

Abstract This paper presents a brief study within the field of recommender systems and describes the current generation of recommender system tools and evaluation metrics. Recommender system comprises of three methods, namely content-based filtering, collaborative filtering, and hybrid filtering algorithms. It addresses two common scenarios in collaborative filtering: rating prediction and item recommendation. There are many well-known accuracy metrics which replicate evaluation goals. This paper describes a few framework and libraries of recommender system that implements a state-of-the-art algorithmic rule furthermore as series of evaluation metric. We tend to find which recommender system tool performs quicker than different, whereas achieving competitive evaluating performance with steering for the comprehensive evaluation and choice of recommender algorithm.

Keywords Recommender system · Performance evaluation metrics · Machine learning · Framework · Libraries

1 Introduction

Recommender system is a part of information filtering system which predicts the 'ratings' or 'preference' for a user or an item. There is huge amount of data which is stored over the Internet and offered to the user; recommendations help users to get potential information to the user's interest. A recommendation engine is a marketing

A. Vineela (✉) · G. Lavanya Devi · N. Nelaturi · G. Dasavatara Yadav
Department of Computer Science and System Engineering, AU College of Engineering (A), Andhra University, Visakhapatnam, India
e-mail: alladavineela4@gmail.com

G. Lavanya Devi
e-mail: lavanyadevig@yahoo.co.in

N. Nelaturi
e-mail: naresh855@gmail.com

G. Dasavatara Yadav
e-mail: dasavatara@gmail.com

P. S. R. Chowdary et al. (eds.), *Microelectronics, Electromagnetics and Telecommunications*, Lecture Notes in Electrical Engineering 655,
https://doi.org/10.1007/978-981-15-3828-5_5

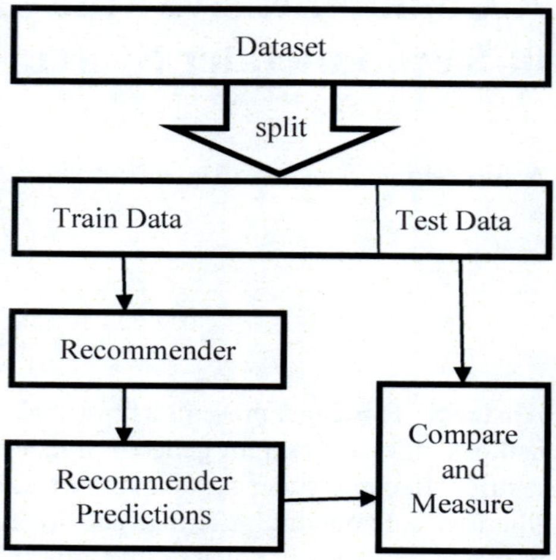

Fig. 1 Evaluation structure of recommender system

tool which increases profits in e-commerce industry. Machine learning algorithms for recommender system is to develop a function which will predict the utility of the user's and items to one other. In this article, we have performed a comparative study of some recommender system tools regarding performance and speed based on different evaluation metrics [1–4].

2 Evaluation Structure

In the process of evaluating recommender system any dataset is taken as input and later it is split into test and train data based on random seed value. The train data is given as input to the recommender which predicts recommendations and gives recommender predictions as output which will be compared with the test data and measured using the evaluation metrics (Fig. 1).

3 Overview of Evaluation Metrics

Evaluation metrics in recommender system can be classified into four types: hidden data evaluation, prediction accuracy metrics, decision support metrics, and rank aware metrics [5, 6].

3.1 Hidden Data Evaluation

The basic idea of hidden data evaluation is using existing data collected from users, use data to simulate behavior and test whether recommender can predict behavior. The prerequisite of hidden data evaluation will like to gather data of rating of movies, play counts of songs, number of clicks on news article, etc. We can use datasets to estimate recommender performance by hiding some data and ask recommender to predict.

What do we measure?

- How close is prediction to rating?
- Are the items being there in recommender list?
- We will cross-validate what if train–test split randomly selects mostly 'easy' or 'hard' users.
- Partition the data to K parts.
- Take each part to train and test data.

3.2 Prediction Accuracy Metrics

This metric is leave-one-out methodology which covers up a rating and tries to predict it.

Mean Absolute Error This is a measure of absolute deviance between the average of predictions and ratings which are given by the users in the system. Here, p indicates predicted ratings an R as actual ratings.

$$\text{Absolute Error} = |P - R| \tag{1}$$

$$\text{Mean Absolute Error} = \text{Average}(|P - R|) \tag{2}$$

$$\text{Prediction Rating} = \frac{\sum_{\text{ratings}} |P - R|}{\#\text{ratings}} \tag{3}$$

Mean Square Error This is used instead of mean absolute error (MAE). MSE gives more importance to cases with larger deviance from actual rating. If the MSE value is low, then recommender performance will be more and vice versa. The MSE for predicted ratings (P) and actual ratings (R) is given below.

$$\text{Mean Squared Error} = \frac{\sum_{\text{ratings}} (P - R)^2}{\#\text{ratings}} \tag{4}$$

Root Mean Squared Error Root mean squared error was an error metric which is used in Netflix. The result is the square root of mean squared error (MSE). The RMSE between predicted ratings (*P*) and actual ratings (*R*) is given below.

$$\text{Root mean Squared Error} = \sqrt{\frac{\sum_{\text{ratings}} (P - R)^2}{\#\text{ratings}}} \tag{5}$$

3.3 Decision Support Metrics

It measures how well a recommender helps users make good decisions. Good decisions are about choosing 'good' items and avoiding 'bad' ones.

Errors and Reversals Error is ad hoc measure of wrong predictions, e.g., determine that 3.5–5* = good, 1–2.5* = bad. Error is when a good movie (for a user) gets a bad prediction (or vice versa). It can also be used for top-*n* means; every time a bad movie appears in the top-*n*, it is an error, usually reported as total number (compared between algorithms), average error rate per user. Reversals are large mistakes, e.g., off by 3 points on a 5-point scale. Intuition is that these are likely really bad—lead to loss of confidence. It is reported as total or average rate.

Precision and Recall Precision is the percentage of items that are selected where 'relevant' means the items recommended. Recall is the percentage of relevant items that are selected. Precision is about returning mostly useful items, and recall is about not missing useful items; if these goals are in balance, then we call it as *F*-measure. Precision@*n* is the percentage of the top-*n* items that are 'good.' Recall@*n* is likewise significant to Precision@*n* [1].

$$\text{Precision} = \frac{|\text{relevant items recommended}|}{|\text{items in the recommendation list}|} \tag{6}$$

$$\text{Recall} = \frac{|\text{relevant items recommended}|}{|\text{relevant items}|} \tag{7}$$

$$F\text{-measure} = \frac{2\text{Precison.Recall}}{\text{Precision} + \text{Recall}} \tag{8}$$

$$\text{Precision}@n = \frac{N^{r@n}}{n} \tag{9}$$

$$\text{Recall}@n = \frac{N^{r@n}}{n} \tag{10}$$

3.4 Rank Aware Metrics

Rank aware top-n metrics require binary relevance and utility metrics. Binary relevance metrics need to know if an item is 'good' or not; this is almost similar to decision support metrics. Utility metrics need a measurement of absolute or relative 'goodness' of items (e.g., ratings).

Mean Reciprocal Rank It is a very simple metric which asks where is the first relevant item. It needs binary relevance judgement. The way we compute MRR is for each user u.

- List of recommendations was generated.
- Rank 1 is the first recommendation; we need to find rank k_u of its first relevant recommendation.
- Reciprocal rank $1/k_u$ should be computed

$$\text{MRR}(O, U) = \frac{1}{|U|} \sum_{u \in U} \frac{1}{k_u} \tag{11}$$

Mean Average Precision Average precision asks which fractions of recommendations are good. It requires fixed size of recommendations. It treats all errors equally. For each relevant item, we need to compute precision of list through that item and calculate the average sub-list of precisions.

$$\text{MAP}(O, U) = \frac{1}{|U|} \sum_{u \in U} \text{AP}(O(u)) \tag{12}$$

Discounted Cumulative Gain It measures the utility metric of item at each position in the list. If we have unary binary data, then we can use 1 for 'good' items and 0 for 'bad' items. Discount the utility of gain by position in the list and normalize by total achievable utility.

$$\text{DCG}(O, u) = \sum_{t} \frac{r_{ui}}{\text{disc}(i)} \tag{13}$$

$$\text{disc}(i) = \{^{1, i \leq 2}_{\log_2 i, x > 2} \tag{14}$$

where r_{ui} = ratings, O = order, disc(i) = discount(position), O_u = user order.

Normalized Discounted Cumulative Gain (nDCG) It is the normalization of DCG. Different users have different ratings and different possible gains. Normalize gain by best possible gain.

$$n\text{DCG}(O, u) = \frac{\text{DCG}(O, u)}{\text{DCG}(O_u, u)} \tag{15}$$

4 Overview of Recommender System Tools

4.1 MyMediaLite

MyMediaLite [7] is a quick and adaptable, multi-purpose library of recommender systems. It is aimed both at recommender system scientists and professionals. This library is for Common Language Runtime often called as .NET. In collaborative filtering MyMediaLite addresses two most common scenarios:

- rating prediction (e.g., on a scale of 1–5 stars), and
- item prediction from positive-only feedback (e.g., from clicks, likes, or purchase actions).

Evaluation MyMediaLite contains routines for processing evaluation metrics. For rating predictions, it evaluates mean average error (MAE) and root mean square error (RMSE); for item predictions, we get accuracy, mean reciprocal rank (MRR), precision-at-*N* (prec@*N*), mean average precision (MAP), and normalized discounted cumulative gain (NDCG). Internal *k*-fold cross-validation is supported for rating predictions and also preliminary support for hyperparameter selection using grid search and Nelder–Mead.

4.2 Case Recommender

Case recommender is a framework which was developed to give adaptability and extensibility in research environments. Instead of using in large-scale business operations, case recommender is used mainly for research and education. Case recommender is a framework which was implemented in python and being accepted for a number of recommendation algorithms. It gives both implicit and explicit feedback. A customized recommender system algorithm can be constructed using case recommender by providing a rich set of components. Case recommender has distinctive types of item recommendation and rating prediction methods and different metrics validation and evaluation. Case recommender can be installed through *$pip install caserecommender*. Requirements for case recommender: Python, SciPy, NumPy, pandas, scikit-learn [8].

Evaluation Item recommendation includes precision, recall, NDCG, MAP. Rating prediction includes MAE and RMSE.

4.3 LibRec

LibRec is an open source recommender system library dependent on Java 1.8 and higher version with GPL-3.0. LibRec contains a large set of recommended system algorithms which may essentially solve rating and ranking issues. Personalized recommendations can be obtained using LibRec which was a typical application of machine learning. Contrasted from version 2.0, the variant 3.0 remakes the entire structure of program and the interface more sensible and progressively versatile. LibRec has great encapsulation; configurations can be loaded directly through command line to run the code where the calculations are printed on the terminal. A user can also run their LibRec code in eclipse IDE where all classes are saved by instances of the configuration class. A config file can also be created to provide a collection of recommended interfaces. It can be done by creating file: librec.conf. The assuring Java library has three main components: recommendations algorithms, generic interfaces, and data structures. It can run much faster than any other library. LibRec can be executed in any platform with MS Windows, Linux, and mackintosh OS [9].

Features The features are flexible configurations, great scalability, excellent modularities, efficient execution and performance, rich algorithms, and simple usage.

Evaluation Measures Rating prediction includes MAE, RMSE, MPE, and item recommendation includes Precision@5, Recall@5, Precision@10, Recall@10, AUC, MAP, NDCG, MRR.

5 Evaluation

Considering all the metrics of LibRec, MyMediaLite libraries and case recommender framework, The metric values are less for LibRec which means it have better performance than case recommender and MyMediaLite (Table 1).

6 Conclusion and Future Work

This work focuses on implementation of recommender systems with available frameworks and libraries. Though there are several frameworks and libraries to implement in recommender systems, user is often doubtful to choose framework and library with better performance. Further, some libraries are outdated and also some have no timely updates in their versions. This work can direct one to choose the best library or framework to implement recommender systems. The study mainly focuses on *K*-nearest neighborhood algorithm which includes item recommendations and rating predictions. These are evaluated with evaluation metrics such as decision support metrics,

Table 1 Metric values for MyMediaLite, case recommender, and LibRec

Metrics	MyMediaLite	Case recommender	LibRec
RMSE	0.95994	1.11693	1.00260
MAE	0.76158	0.85759	0.75656
NDCG	0.63841	0.66989	0.35573
PREC@5	0.39172	0.52125	0.04748
RECALL@5	0.12882	0.07732	0.06451
PREC@10	0.06725	0.48176	0.23741
RECALL@10	0.36499	0.13971	0.32255
F-measure	0.19388	0.13466	0.05469
MSE	0.94946	1.06584	1.00522
MRR	0.59032	–	–
MPE	0.27555	–	0.84180
AP	–	–	0.25046
MAP	–	0.63189	–

prediction accuracy metrics, and rank aware metrics. The entire study is carried out with the MovieLens dataset. Considering all the metrics of LibRec, MyMediaLite libraries, and case recommender framework, the metric values are less for LibRec which means it have better performance than case recommender and MyMediaLite. This study is helpful for those who are searching for good recommender system [10, 11].

The future work can be in many ways as the main motivation of this work is to analyze the recommender system tools and libraries in regard to their performance and accuracy. So, we can speed up the process by using different libraries and frameworks available at that time which can give best results than existing tools. One can start working on remaining machine learning algorithm and analyze the results of all the algorithm and can conclude which algorithm gives better performance. We can even work on a single library and compare performances of all the machine learning algorithms.

References

1. Shah K, Salunke A, Dongare S, Antala K (2017) Recommender system: an overview of different approaches to recommendations. ICIIECS 978-1-5090-3294-5
2. Jannach D, Zanker M, Felfernig A, Friedrich G (2011 Recommender system an introduction, 1st edn. USA
3. Patel B, Desai P, Panchal U (2017) Methods of recommender system—an review. In: International conference on innovations in information embedded and communication systems ICIIECS. 978-5090-3294-5
4. Ricci F, Rokach L, Shapira B (2011) Recommender systems handbook, 2nd edn. Springer, US

5. Chen M, Liu P (2017) Performance evaluation of recommender system. Int J Performability Eng 13(8):1246–1256
6. Lerato M, Esan OA, Ebunoluwa A-D, Ngwira SM, Zuva T (2015) A survey of recommender system feedback techniques, comparison and evaluation metrics. IEEE, 978-1-4673-9354-6
7. Gantner Z, Rendle S, Freudenthanler C (2011) MyMediaLite: a free recommender system library. In: RecSys 2011 proceedings Chicago, Illinois USA
8. da Costa A, Fressato E, Neto F, Manzato M, Campello R (2018) Case recommender: a flexible and extensible python framework for recommender systems. In: RecSys 18 proceedings, Vancouver, BC, Canada
9. Guo G, Zhang J, Sun Z, Yorke-Smith N, Librec: a Java library for recommender system. http://www.librec.net. Version 3.0, last accessed 2019/05/01
10. Schroder G, Thiele M, Lehner W (2011) Setting goals and choosing metrics for recommender system evaluations. ResearchGate
11. Sahu SP, Nautiyal A, Prasad M (2017) Machine learning algorithm for recommender system—a comparative analysis. IJCTR 6(2):97–100

LBPH-Based Face Recognition System for Attendance Management

P. Lavanya, G. Lavanya Devi, and K. Srinivasa Rao

Abstract Facial recognition is the mostly used biometric technique for identification and acknowledgment of people automatically. This technology is widely used in fields, such as surveillance systems, criminal identification, law enforcement and banking. On the other hand, facial recognition systems do not need any direct contact and acceptance of the person/user. Surveillance cameras at various locations have helped to identify crimes/criminals. Also, this technology can be further used for many user-friendly applications like finding the participations of people at various programs, attendance monitoring at workplaces, universities/colleges, to restrict unauthorized entries, etc. Traditional manual attendance marking as well as biometric fingerprint methods is a time-taking procedure for instructors as well as students in academic fields. This paper discusses about the usage of facial recognition technique to label the participation of the student automatically. To reduce the faults of other systems, face recognition system using LBPH is implemented.

Keywords Face recognition · Haar-cascade classifier · Face detection · OpenCV · LBPH

1 Introduction

There are many biometric techniques available among which facial recognition is the most commonly used biometric technique for identifying individuals. In the academic field, the students are monitored and given attendance to quantize their regularity. If there are fewer students, marking the attendance manually is easy as it takes less

P. Lavanya (✉) · G. Lavanya Devi · K. Srinivasa Rao
Department of Computer Science and Systems Engineering, AU College of Engineering (A), Andhra University, Visakhapatnam, Andhra Pradesh 530003, India
e-mail: lavanya.prathipati524@gmail.com

G. Lavanya Devi
e-mail: lavanyadevig@yahoo.co.in

K. Srinivasa Rao
e-mail: sri.kurapati@gmail.com

P. S. R. Chowdary et al. (eds.), *Microelectronics, Electromagnetics and Telecommunications*, Lecture Notes in Electrical Engineering 655,
https://doi.org/10.1007/978-981-15-3828-5_6

Fig. 1 Basic steps of face recognition

time, but it is a tedious and time-consuming process if the number of students is more. One of the automated methods implemented for attendance monitoring is biometric fingerprinting. In this case, the student has to wait in the line until his/her turn comes to give attendance. The technology of facial recognition overcomes these defects. By using Haar-cascade, the facial characteristics such as the eyes, the nose, the mouth and the distance between them, are extracted. Finally, recognition is performed where the detected face is compared with the faces in the database [1] (Fig. 1).

2 OpenCV

Open source computer vision (OpenCV) is one of the open source computer vision and software libraries for machine learning. It was designed for providing a common platform for computer vision applications and to accelerate the use of machine perception in company products. The library has over 2500 optimized algorithms, which can be used to detect and recognize the faces, to identify objects, to classify human behavior in videos, to monitor camera movements, to monitor moving objects and to extract 3D object models. OpenCV comes up with three face recognizers [2]. They are as follows:

- Eigenface recognizer
- Fisherface recognizer
- LBPH recognizer

Eigenface Recognizer This algorithm takes into account that not all the features or sections of the face are equally essential. The persons are detected by considering the characteristics such as forehead, nose, cheeks, eyes and how they differ. No two different persons have same characteristics. It uses PCA to reduce the dimensionality. The eigenvectors and eigenfaces are generated. The generated eigenface will be compared with the already stored images in the database.

Fisherface Recognizer It is an improved eigenfaces face recognizer version. Eigenfaces look at all the pictures at once and find the primary components of them together. It does not concentrate on the different features of two people. Fisherfaces algorithm extracts useful features that distinguish one person from the other.

Local Binary Patterns Histogram Face Recognizer (LBPH) LBPH regards the texture descriptor as helpful for facial symbolization, because face information can be divided as micro-texture pattern compositions. For the processing of the training, this algorithm also needs grayscale images. The face images are represented in the form of a data vector when LBP is combined with histograms.

2.1 Haar-cascade Classifier

Haar-cascade is an efficient technique proposed by Paul Viola and Michael Jones to detect objects. It is a machine-based learning strategy that trains a cascade feature from many positive and negative pictures. It is then used to identify items in other pictures. To train the classifier, the algorithm initially requires a lot of favorable pictures and adverse pictures. Then, we have to extract characteristics from it. Each characteristic is a single value acquired by subtracting amount of pixels from the sum of pixels under the black rectangle under the white rectangle (Fig. 2).

The extracted combination of features from the training part will be used to identify faces in a picture. The mixture of characteristics will be investigated to acknowledge a face in an unknown image. For example, the scale can be a 24 × 24 pixel square. It will try to match every combination function one by one in the block. If one of the characteristics does not appear in the block, it will not be checked as the machine concludes that there is no face in the block. A weighted sum of these weak classifiers is the final classifier. It is called weak because it alone cannot classify the picture, but it is a powerful classifier together with others. All the detected faces that concern the same person are merged and considered to be one at the end of the entire process. The accumulation of these weak classifiers produces a face detector that can very rapidly detect faces with adequate accuracy. It is also possible to create an own Haar-cascade classifier or use a predefined classifier.

3 Methodology

The face recognition system is implemented using local binary patterns histogram (LBPH). The detailed information of how the algorithm works and what are the integral steps it follows in achieving the expected result is explained below.

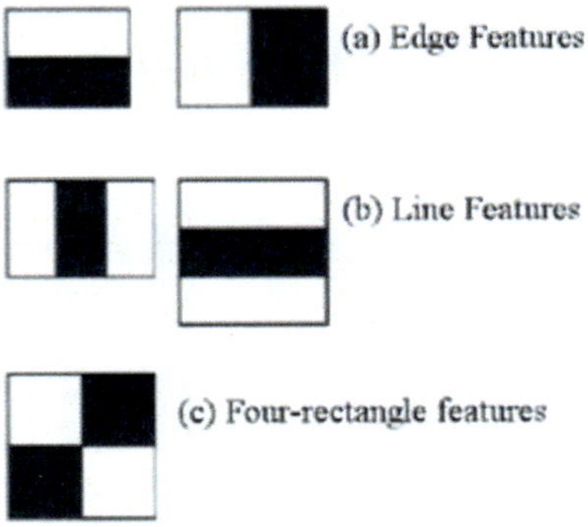

Fig. 2 Different Haar features

3.1 *Local Binary Patterns Histograms (LBPH)*

LBPH regards the texture descriptor as helpful for facial symbolization, because face information can be divided as micro-texture pattern compositions. Basically, three phases of LBPH are performed. The image is captured first and internally the image is converted into gray scale image, later with the help of Haar characteristics it will be checked whether it is a face or non-face image after that pixels are mapped, and the face is checked. LBPH is one of the most common algorithms for face recognition.

Local Binary Pattern (LBP) LBP is a straightforward and very effective texture operator that labels an image's pixels by thresholding each pixel's neighborhood and views the outcome as a binary number. It also says that the efficiency of detection is significantly increased when LBP is coupled with the targeted gradient (HOG) descriptor histograms. To represent the face pictures, a data vector is used (Fig. 3).

It scans the pixel values of each block here and places them in a 3×3 window. The main pixel has eight neighbors and the threshold value that will be taken. When comparing the limit value to the adjacent values, set the value to 1 if the adjacent value is bigger than the core pixel, otherwise to 0. By reading them in a clockwise direction, the binary values will be considered, and the corresponding decimal value will be updated in the central pixel [3, 4].

Step-by-Step Procedure The algorithm considers the following steps:

Parameters. LBPH considers the following four parameters.
Radius. Used to create a circular local binary pattern, it is the radius from the central pixel, normally it is taken as 1.
Neighbors. Number of sample points needed to create local circular binary pattern. So, it is usually set to 8.
Grid X. Represents the number of cells in the horizontal direction. It is usually set to 8.
Grid Y. Represents the number of cells in the vertical direction. The vectors dimensionality will be higher if the number of cells is more. Usually, it is set to 8.

Training the Algorithm First, it is necessary to train the algorithm. To do so, a dataset is needed that contains all the individual's facial pictures. The images are

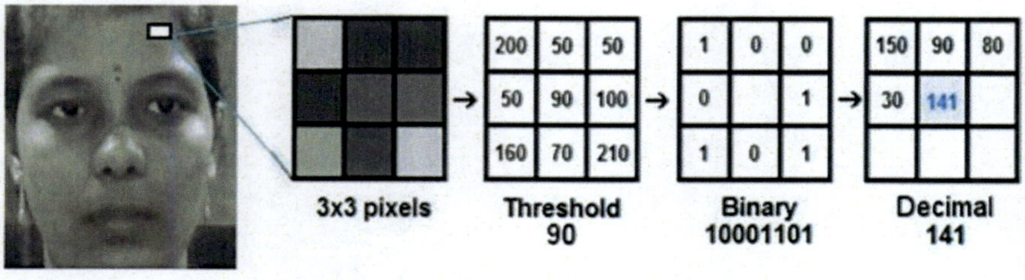

Fig. 3 Face detection using LBP operator

assigned either a name or an ID, so that this data is used by the algorithm to acknowledge an input picture and provide the precise output. After the training model is built, the identification method must be used further.

Applying the LBP Operation The first computational phase of the LBPH is to create an intermediate image that better portrays the original image by highlighting the facial characteristics. The algorithm utilizes a sliding window concept based on radius (R) and neighboring (p) parameters for this purpose.

Extracting Histograms The picture produced will be taken into account, and the Grid X and Grid Y parameters will be used to split the picture into various grids. Each histogram contains only 256 positions representing each pixel intensity event. The histograms will be concatenated, showing the characteristics of the original image in the final histogram. For removing contrast from the image, it is used.

3.2 Implementation

For implementing the algorithm, certain steps need to be performed. They are as follows:

Dataset Creation A face database is a set of people's faces used to train the system. The training set stores a pair of face labels that is a face image, along with the name of the person. The pictures are taken from multiple individual snapshots, consisting of all possible facial expressions, postures and illuminations (Fig. 4).

Grayscale Transformation Sometimes the captured image may not be clear due to the illumination, noise and clutter. In order to remove those noises, the captured image will be converted into grayscale before processing further.

Face Detection Face detection refers to the detection of faces in the image. It detects the faces from the picture and shows the number of faces detected. OpenCV provides a Haar-cascade classifier to identify faces. The cascade utilizes the AdaBoost algorithm to identify multiple facial characteristics, including the eyes, nose and mouth, used for facial identification. Viola P. and M. I. Jones have created a detection algorithm (Fig. 5).

Face Recognition and Attendance Marking The ultimate task is to recognize images of the face. It is a person's automatic picture or video identification or verification. It is one of the most commonly used methods because of its reliability and precision in the process of acknowledging and verifying the person's identity. Using LBPH technique, the recognized face features are acquired and compared to the database. If the face matches with the database, then the attendance is marked for the particular student. The attendance for the student is automatically updated into the database only after effective recognition has been achieved [5] (Fig. 6).

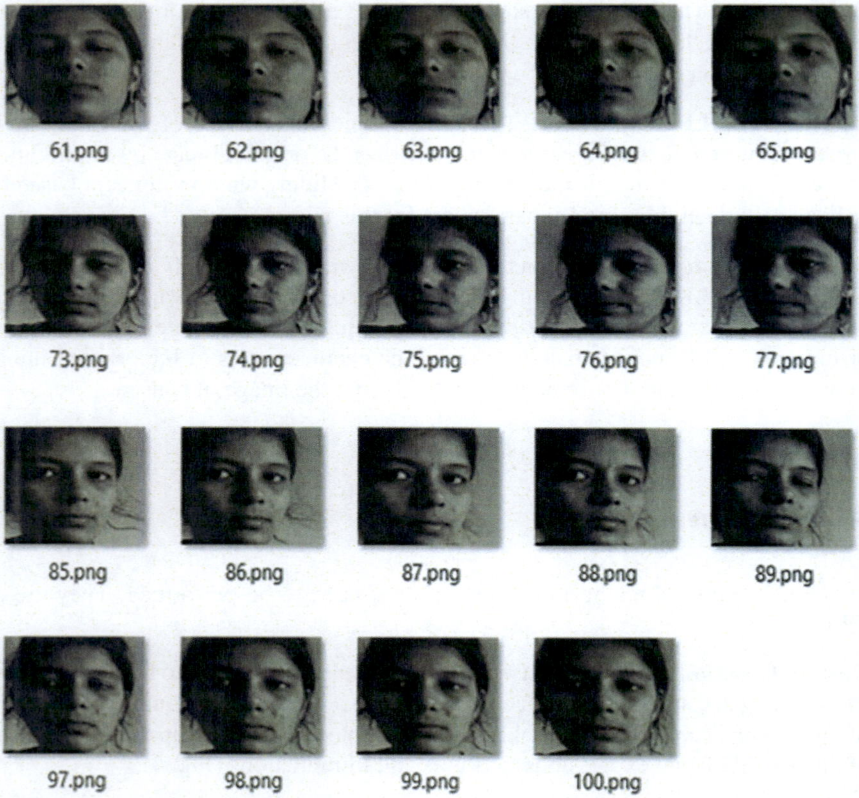

Fig. 4 Face dataset

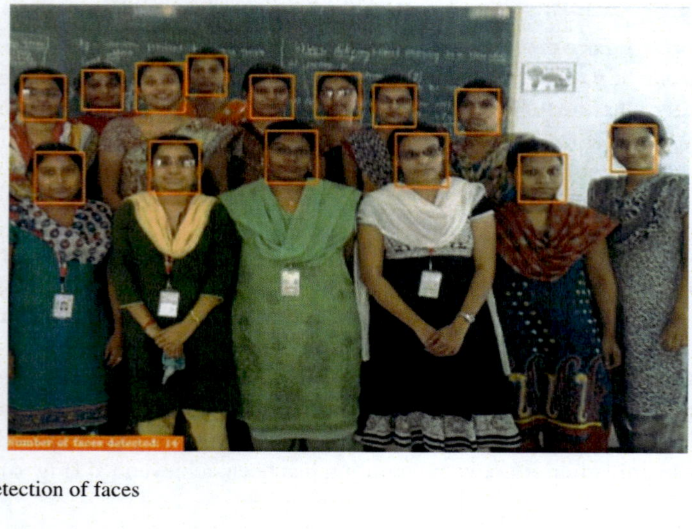

Fig. 5 Detection of faces

ID	01_03_19	05_03_19	11_03_19	19_03_19	20_03_19	10_04_19	05_06_19	06_06_19	07_06_19
17021D0520	0	1	1	1	0	1	1	1	0
17021D0511	0	0	0	0	0	0	0	0	0
17021D0515	0	0	1	0	0	1	0	1	0
317206417015	1	1	1	0	0	1	0	1	0
18025A0562	0	0	1	0	0	0	1	0	0
317206417011	0	0	0	0	1	1	0	0	0

Fig. 6 Updating the attendance in database

4 Conclusion

In this paper, attendance management system is implemented for student's attendance. It helps to reduce the drawbacks of the existing attendance systems. By using this, the time of both the lecturers and the students will not be wasted. The entire system is implemented in Python programming language. Face recognition techniques were used for the purpose of the student attendance. Compared to other face recognition algorithms LBPH gives better results. So it is used for implementing the attendance management system [6]. The current recognition system has been designed for frontal views of face images. In future, it should be designed in such a way that it can recognize non-frontal faces (side faces, covering with scarf, etc.). The enhancement of the system should be done in such a way that the accuracy rates must be high and the attendance reports of the student will be mailed to the staff and the messages to the parents will be sent automatically if he/she is absent.

References

1. Sabhanayagam T, Prasanna Venkatesan V, Senthamaraikannan K (2018) A comprehensive survey on various biometric systems. Int J Appl Eng Res 13(5):2276–2297. ISSN 0973-4562
2. Narang S, Jain K, Saxena M, Arora A (2018) Comparison of face recognition algorithms using Opencv for attendance system. Int J Sci Res Publ 8(2):268–273
3. Deeba F (2019) LBPH-based enhanced real-time face recognition. Int J Adv Comput Sci Appl 10(5):274–277
4. Ahmaed A (2018) LBPH based improved face recognition at low resolution. In: 2018 international conference on artificial intelligence and big data
5. Kutty NM, Mathai S (2017) Face recognition-a tool for automated attendance system 6:334–336
6. Tantak A, Sudrik A (2017) Face recognition for E-attendance for student and staff. IOSR J Comput Eng 19(02):89–94

Proposed Pipeline Clocking Scheme for Microarchitecture Data Propagation Delay Minimization

Wasim Ghder Soliman, P. V. S. Anusha, D. V. Rama Koti Reddy, N. Suresh Kumar, and B. Keerthi Priya

Abstract With the pipeline design, high data throughput is obtained. A pipeline works like an assembly line, before the prior data has finished, the new data can be processed. The core elements of the pipeline system are the flip-flops, and those flip-flops form the registers for the pipeline stages. In this paper, a proposed pipeline scheme is presented to avoid the halfway situation or unpredictable state due to the effect of flip-flops setup and hold times. A comparison with other pipeline schemes with respect to data propagation delay is also present. Conventional pipeline, wave pipeline and mesochronous pipeline systems are compared with the proposed pipeline system. The comparison process is considered with input pulses in the frequency range of 5 Hz–999 MHz and for three and four pipeline stages. The proposed pipeline system gives the best data propagation delay among the systems when the logic is introduced.

Keywords Pipeline stages · Data propagation delay · Flip-flops · Input pulses frequency

1 Introduction

Pipeline technique is the key element used in advance computing systems. It is used for different applications including parallel computing, advance computer architecture, high-speed ALU, fast fetching of I/O data and many more. Clock management

W. G. Soliman (✉)
Industrial Automation Engineering Department, Technical Engineering College, University of Tartous, Tartous, Syria
e-mail: wasimsoliman1987@gmail.com

W. G. Soliman · P. V. S. Anusha · D. V. Rama Koti Reddy
Instrument Technology Department, Andhra University, Visakhapatnam 530003, India

N. Suresh Kumar
GITAM Institute of Technology, GITAM (Deemed to be University), Visakhapatnam, India

B. Keerthi Priya
Gayatri Vidya Parishad College of Engineering (A), Visakhapatnam, India

P. S. R. Chowdary et al. (eds.), *Microelectronics, Electromagnetics and Telecommunications*, Lecture Notes in Electrical Engineering 655,
https://doi.org/10.1007/978-981-15-3828-5_8

affects the rate of data transfer. The data wave propagation is controlled on the basis of clock pulses [1, 2]. Many of flip-flops/latches are included in the pipelining. As the depth of pipeline reduced, the clock skew effects become worse, so to achieve high performance, an efficient clocking method is critical. The processor (in fast computers) computes the data quickly and performs the operations, simultaneously, with the help of the pipeline, but the conventional pipeline suffers clock skew problem. Poor data organization is caused by clock skew, and this leads to data loss at end element. Two main issues are faced in high-speed devices; they are jitter and clock skew [3]. Due to component mismatch, the clock skew occurs. There are two types of jitter, the first one is random which occurs due to power fluctuation, and the second one is deterministic due to environmental conditions. Hold issues between neighboring flip-flops are created because of the clock skew. The signal-to-noise ratio (SNR) is degraded by raising the noise floor due to jitter [4, 5]. In any digital circuitry, an essential task is to achieve good data propagation, and the clock is the key element to do that. Both data throughput and data propagation determine the speed of the digital device [6].

The benefit behind minimizing the data propagation delay is to obtain better pipeline system performance which can be used in the designing of reconfigurable microarchitectures. Those microarchitectures can replace the fixed microprocessor-based architecture [7, 8] to enhance the microarchitecture performance especially with new technologies like the Internet of Things (IoT) in the industry [9] and controllers design [10].

The paper is organized as follows. In Sect. 2, conventional pipeline, wave pipeline and mesochronous pipeline systems are explained. The proposed pipeline system is introduced in Sect. 3. A comparison between the pipeline schemes is presented in Sect. 4, and finally, Sect. 5 concludes the paper.

2 Pipeline Schemes

There are several clocking schemes which are proposed to avoid the tolerance caused by clock skew and to eliminate latching overheads [11, 12]. In the following subsections, a theoretical explanation of conventional pipeline, wave pipeline and mesochronous pipeline is illustrated.

2.1 Conventional Pipeline

A single clock signal is implemented to enable data transmission between the registers stages within the pipeline as shown in Fig. 1.

The data speed will decrease from stage to stage due to clock skew which creates massive complications in this type of pipeline systems. The mathematical form of the clock signal is

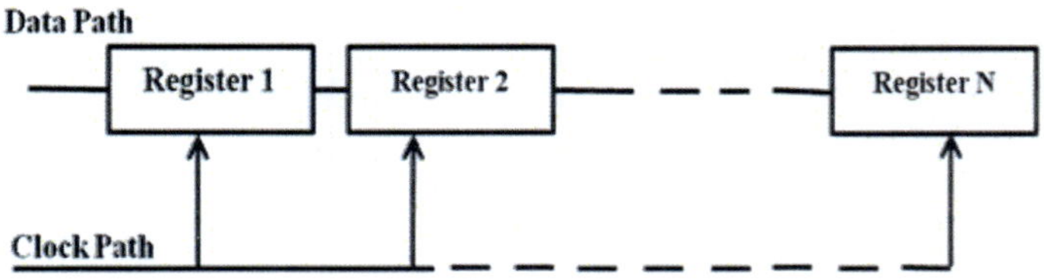

Fig. 1 Block diagram of the conventional pipeline system

$$T_{\mathrm{clk}_{\mathrm{conventional}}} \geq D\mathrm{max} + D_{\mathrm{r}} + T_{\mathrm{s}} + \Delta_{\mathrm{clk}} \quad (1)$$

where $D\mathrm{max}$: longest propagation delay, D_{r}: pipeline register delay, T_{s}: pipeline register setup time and Δ_{clk}: clock uncertainty. There are two types of conventional pipeline systems. The first type is a synchronous pipeline system [13] in which same clock is used in all stages, and the second type is an asynchronous pipeline [14] system which is used with asynchronous circuits and uses a request/acknowledge protocol, wherein the end (finish status) of each stage can be detected.

2.2 *Wave Pipeline*

The digital system clock frequency can be increased with wave pipeline methodology. Wave pipeline provides the pipelining maximum rate. The need of internal clock elements to increase the throughput is eliminated by the wave pipeline system. The difference between the longest and shortest path delays determines the rate at which the logic can propagate through the circuit. With the wave pipeline, the effective number of the pipeline stages is increased without changing the number of physical registers in the pipeline [15].

The output data should be clocked after the latest data has reached at the output and before the earliest data from the next cycle reaches the output, and in this manner the wave pipeline system will operate correctly. The clocking conditions of the earliest and the latest data propagation should be derived firstly in the circuit, and this will represent the register constraints [16, 17]. Let us consider parameter N to represent clock cycles in order to propagate a signal within the logic block before getting latched by the output register. The degree of the wave pipeline system is measured by the parameter N. The data should be clocked at a time (T_{L}):

$$T_{\mathrm{L}} = NT_{\mathrm{clk}} + \Delta \quad (2)$$

where Δ is the possible constructive skew, and it is of arbitrary value between the input and the output registers. The constraint of clocking of the latest data requires that during the Nth clock cycle, the latest possible signal arrives early enough to be clocked by the output register. This determines the lower bound on the above time as follows:

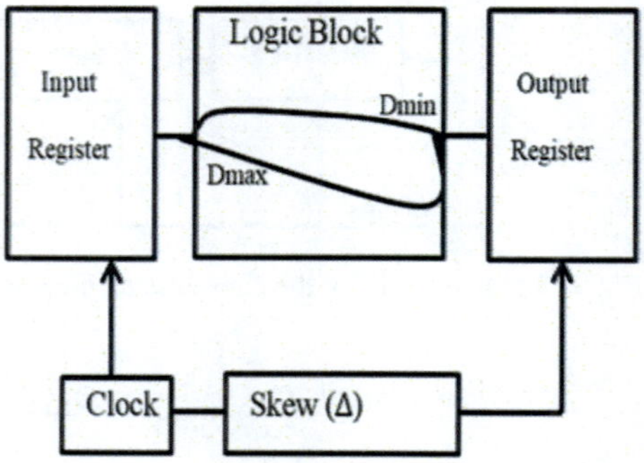

Fig. 2 Block diagram of the wave pipeline system

$$T_L > D_r + D_{max} + T_s + \Delta_{clk} \tag{3}$$

The constraint of clocking of the earliest data requires that there is no interference between the current wave and the arrival of the next wave $i + 1$. The earliest arrival of wave $i + 1$ is equal to

$$T_{clk} + D_r + D_{min} \tag{4}$$

The wave pipeline clock signal is expressed as follows:

$$T_{clk.w} \geq (D_{max} - D_{min}) + T_h + T_s + 2\Delta_{clk} \tag{5}$$

where $(D_{max} - D_{min})$: the difference between the longest and shortest path delays, T_h: hold time (must be allowed to remain steady), T_s: register setup time and Δ_{clk}: the skew element [18–20]. Reducing the delay $(D_{max} - D_{min})$ is an obvious way to improve the cycle time. The wave pipeline system is shown in Fig. 2.

2.3 Mesochronous Pipeline

The main idea of a mesochronous pipeline system is to introduce a delay element. This element should be introduced into the path of the clock signal. In this case, the clock synchronization will be controllable and the propagation delay will be reduced [21, 22] as shown in Fig. 3.

The delay value should be equal to the time needed when the pulse is passed from one pipeline stage to another. Simultaneously, more than one data signals are being operated by one pipeline stage and based on internal node physical properties the signals are separated [23]. The mathematical form of the clock signal is

$$T_{clk.m} \geq (D_{max}(f) - D_{min}(f)) + T_h + T_s + 2\Delta_{clk} \tag{6}$$

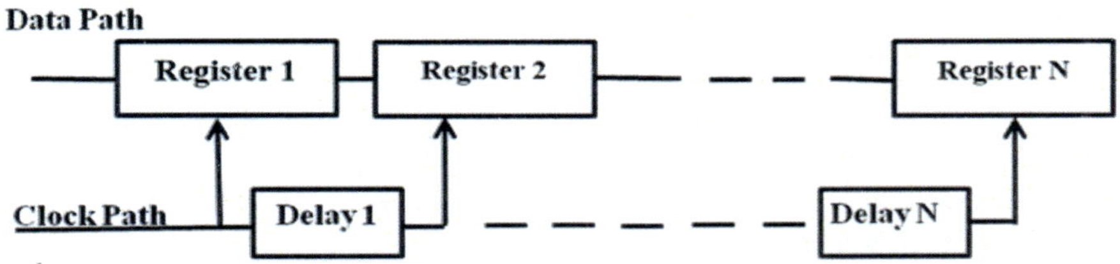

Fig. 3 Block diagram of the mesochronous pipeline system

3 Proposed Pipeline Scheme

In the digital systems, both setup and hold times have their effect on the output as the following: The output will reflect the new value of the input data if the input data varies before the setup time, and in contrast, the output will reflect the old value of the input data if the input data varies after the hold time. A halfway situation or unpredictable state may occur if the input data varies after the setup time and before the hold time. In this case, the output will be between low and high logic. So, the idea of detecting a pure logic, i.e., either high or low and avoiding halfway state comes into consideration in the proposed pipeline system. Simple logic gates are used to examine the binary information of the previous flip-flop. This will create a small delay before producing the clock signal to the next pipeline stage. In this case, the clock generator will not be allowed to pass the next clock signal until the next binary bit from the previous stages to the next stage is identified by the logic gates as shown in Fig. 4.

The binary checker block will verify the first stage data, and after ensuring valid logic the status will be forwarded to clock (2) gate to generate the next clock signal to the second stage and then this stage will receive the first stage data. A similar procedure will be repeated through rest stages.

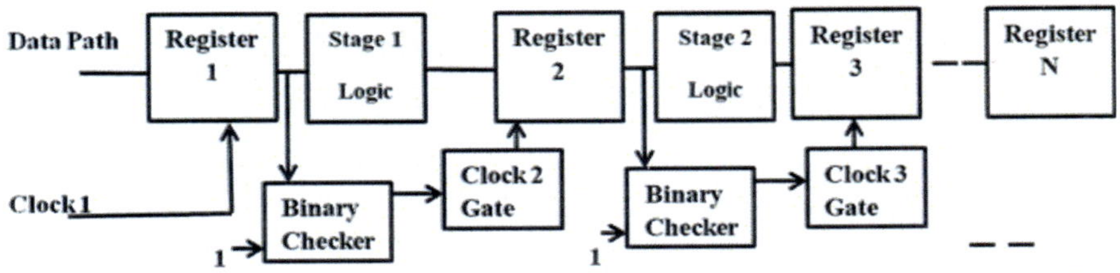

Fig. 4 Block diagram of the proposed pipeline system

4 Pipeline Schemes Comparison

This section illustrates a data propagation delay comparison between the above explained pipeline systems and the proposed pipeline system in a graph form. The logic case is considered. The proposed pipeline system gives the best propagation delay among the systems when the logic is introduced. The main drawback with the proposed pipeline system is its extra resources utilization, and therefore, it will require more area for its real-time implementation. However, the proposed pipeline system can be used practically when the area of the design is not the main constraint, and therefore, it will be suitable when the throughput of the design is the main concern.

Figure 5 shows the block diagram of the proposed pipeline system which is simulated using the Electronic Work Bench (EWB) software. In the following circuit, a function generator is used to supply the varying frequency input signal and a logic analyzer is used to calculate the propagation delay time from the input to the output. A similar procedure is repeated with the above explained pipeline system, and a comparison with respect to the data propagation delay is presented in both Figs. 6 and 7 for three and four pipeline stages, respectively. The comparison process is considered with input pulses in the frequency range of 5 Hz–999 MHz and for three and four pipeline stages.

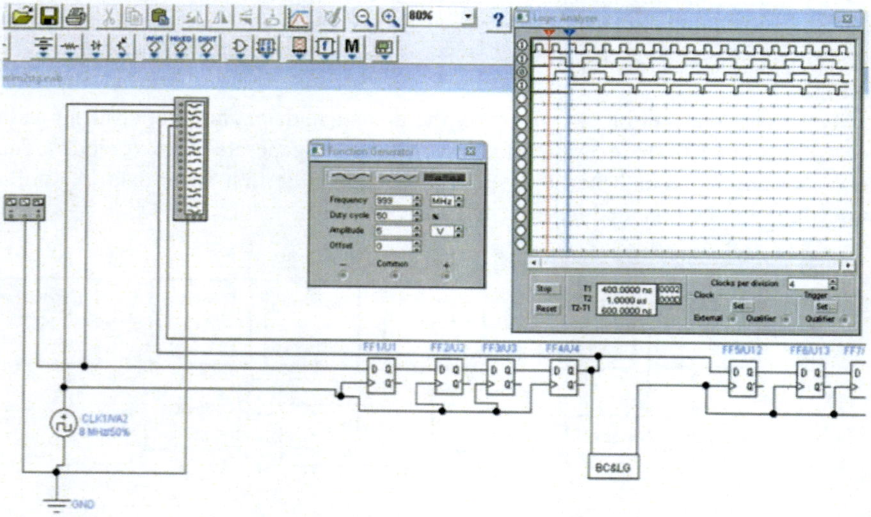

Fig. 5 Implementation of the proposed pipeline system

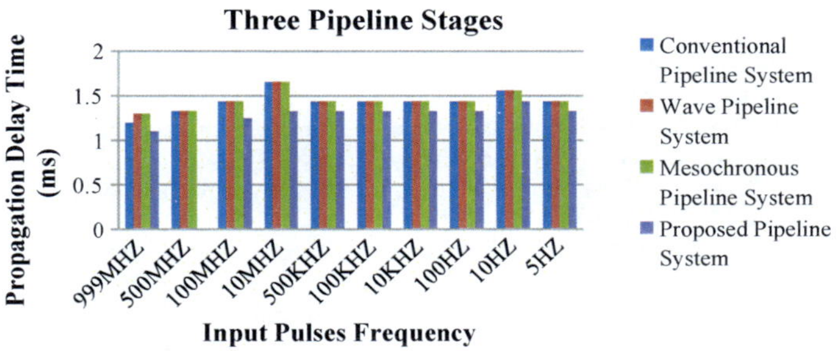

Fig. 6 Three pipeline stages propagation delay comparison

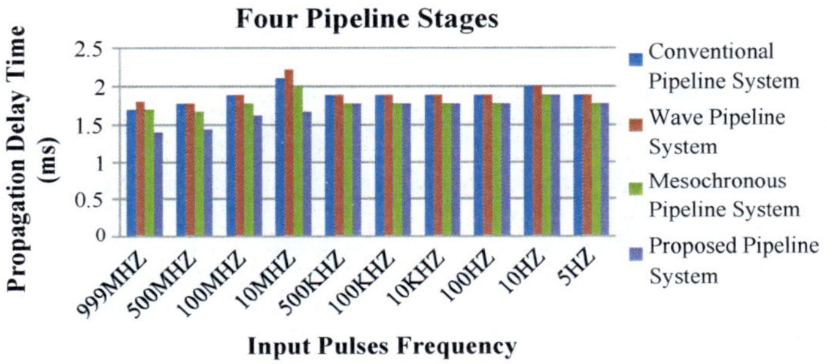

Fig. 7 Four pipeline stages propagation delay comparison

5 Conclusion

In the current work, the main focus is on data propagation delay minimization. With the proposed clocking scheme, the halfway or unpredictable logic state is avoided. The proposed pipeline system gives smaller data propagation delay in comparison with conventional pipeline, wave pipeline and mesochronous pipeline systems. The proposed pipeline system is recommended when the throughput of the design is the main concern.

References

1. Sung RJH et al (2007) Clock-logic domino circuits for high-speed and energy efficient microprocessor pipelines. IEEE Trans Circ Syst-II: Express Briefs 54(5):460–464
2. Yang S et al (2007) Surfing pipelines: theory and implementation. IEEE J Solid-State Circ 42(6):1405–1414
3. Lee H et al (2010) Pulse width allocation and clock skew scheduling: Optimization sequential circuits based on pulsed latches. IEEE Trans Comput Aided Des Integr Circ Syst 29(3)

4. Rantala A et al (2007) A DLL clock generator for a high speed A/D-converter with 1 ps jitter and skew calibrator with 1 ps precision in 0.35 μm CMOS. Springer Science, analog integrated circuit signal process, pp 69–79
5. Han Kihyuk et al (2011) Off-chip skew measurement and compensation module (SMCM) design for built-off test chip. Springer J Electron Test 27:429–439
6. Mango C T. Chao et al (2004) Static statistical timing analysis for latch based pipeline design. IEEE, pp 468–472. doi:0-7803-8702
7. Soliman WG, Reddy DVRK (2019) Microprocessor-based prototype design of a PMDC motor with its system identification and PI controller design. SN Appl Sci 1:549
8. Wasim S et al (2018) Microprocessor based permanent magnetic DC motor system identification and optimal PI controller design. In: Proceedings of the 12th INDIACom, IEEE conference ID: 42835, pp 1020–1026
9. Soliman WG, Reddy DVRK (2019) Microprocessor-based edge computing for an internet of things (IoT) enabled distributed motion control. In: Singh M, Gupta P, Tyagi V, Flusser J, Ören T, Kashyap R (eds) Advances in computing and data sciences. ICACDS 2019. Communications in computer and information science, vol 1046. Springer, Singapore
10. Soliman WG, Patel VP (2015) Modified PID implementation on FPGA using distributed arithmetic algorithm. Int J Adv Eng Res Dev 2(11):44–52
11. Harris D et al (1997) Skew-tolerant domino circuits. ISSCC digest of technical papers, pp 416–417
12. Unger S (1986) Clocking schemes for high-speed digital systems. IEEE Trans Comput c-35(10):880–895
13. Dash PK et al (2000) An extended complex Kalman filter for frequency measurement of distorted signals. IEEE, pp 1569–1574. doi: 0-7803-5935-6/00/2000
14. David H et al (1997) Skew-tolerant domino circuits. In: IEEE international solid-state circuits conference, session 25/processors and logic paper/sp 25.7, pp 422–423
15. Chuan PK et al (2000) Delay balancing using Latches. EECS department, University of Michigan
16. Daniel A et al (1994) Pipelining communication in large VLSI/ULSI systems. IEEE Trans Very Large Scale Integr (VLSI) Syst 2(1)
17. Joy DA et al (1993) Clock period minimization with wave pipelining. IEEE Trans Comput-Aided Des Integr Circ Syst 12(4):461–472
18. Boemo EI et al (1996) Wave pipelines via look-up tables. IEEE, pp 185–188. doi: 0-7803-3073-0/96/1996
19. McVey ES et al (1975) A digital error expansion circuit for measurement of frequency. IEEE Trans Instrum Measur 24(3):246–248
20. Seetharaman G, Venkataramane B, Lakshminarayanan G (2008) Automation techniques for implementation of Hybrid wave-pipelined 2D DWT. J Real-Time Image Process 3(3):217–229
21. Tatapudi SB, Delgadofrias JG (2006) A mesychronous high performance digital systems, 53(5)
22. Intel manuals, July, order Number: 003965-003 by Russian Military, http://doc.chipfind.ru/intel/8257.htm (1990)
23. Jordan HF (1984) Experience with pipelined multiple instruction steams. Proc IEEE 72(1):113–123 (Invited paper)

Design of Metamaterial Loaded Dipole Antenna for GPR

T. Pavani, A. Naga Jyothi, A. Ushasree, Y. Rajasree Rao, and Ch. Usha Kumari

Abstract A novel design of a dipole antenna for water detection is developed for ground-penetrating radar (GPR) system. The water decreasing day by day can increase the importance of the natural object water. Because of the degradation of surface water resources, the requirement for graphics of water resource is accumulated. GPR could be a promising machinery to find and establish formation of water. A dipole antenna incorporated with an inverted S-shaped metamaterial is proposed for GPR applications. The metamaterial-inspired antenna is designed on an FR4 substrate with overall dimensions of 100 × 300 mm. By placement of an inverted S-shaped metamaterial to induce additional resonance due to the occurrence of magnetic dipole moment, the antenna resonant frequency is changed from 1.88 to 1.71 GHz. The return loss and the VSWR plots have been studied along with the radiation patterns.

Keywords Metamaterial · Antenna · GPR

T. Pavani (✉) · A. Ushasree · Ch. Usha Kumari
Gokaraju Rangaraju Institute of Engineering & Technology, Hyderabad, India
e-mail: pav.tum@gmail.com

A. Ushasree
e-mail: ushasreemishra@gmail.com

Ch. Usha Kumari
e-mail: ushakumari.c@gmail.com

A. Naga Jyothi
Vignan's Institute of Information Technology (A), Visakhapatnam, Andhra Pradesh, India
e-mail: aggala.jyothi@gmail.com

Y. Rajasree Rao
LORDS Institute of Engineering Technology, Hyderabad, India
e-mail: rajasreerao@lords.ac.in

P. S. R. Chowdary et al. (eds.), *Microelectronics, Electromagnetics and Telecommunications*, Lecture Notes in Electrical Engineering 655,
https://doi.org/10.1007/978-981-15-3828-5_9

1 Introduction

Ground-penetrating radar (GPR) is realistic procedure for the detection of underground metalloids. A unique, diminished, and economical antenna, having trade-off between resolution and depth for different applications, is immensely needed. The important parameter to be considered when designing an antenna for GPR applications is wideband frequency of operation. GPR technology has been improving tremendously such that the present techniques can able to penetrate deeper into the ground for the various applications like groundwater exploration, minerals detection, etc. [1–3]. Depending upon the application requirements, the antenna will be designed to achieve the desirable performance. The mechanical and technical issues related to the design of antenna are discussed in [4–7]. Antenna's height, its direction, protective and subsurface properties all influence GPR responses which make planar antennas, like bowtie and frequency-independent structures, higher in demand. Basic antennas, like resistively loaded dipole, are easy to design and have linear polarization but have a main drawback of poor gain. Bowtie and Vivaldi antennas are used in impulse GPR applications considering their non-scattering effects and their relatively high gain, but still refinements were made in the designing of antennas to show better performance [8, 9].

A low-resolution rolled dipole antenna is used in GPR, which decreases the length of the antenna by a multiple of 4. By integrating rolled dipole with bowtie antenna, efficiency can be improved [10]. According to the laboratory prototype, antenna designed by adding extra radiating arms on a dipole antenna can significantly reduce antenna length by 55% compared to a conventional dipole of 100 MHz with a directivity gain of 2.06 dBi [11]. An improved dipole antenna is designed for the safety of excavation in soft ground regions using a stepped frequency cross-hole GPR system to detect enclosure structure defects [12]. A metamaterial has time consonant vital parameters that are uninterrupted in space though difficulties remain in terms of homogenization, bandwidth, loss, and realizations. Metamaterial has the permittivity and permeability both being less than the free space values most of all with twice negative metamaterials. The main objectives of using metamaterial in designing of antenna are reducing size of antenna, increasing gain and directivity, and reducing coupling effect [13]. By using metamaterial, the frequency range of operation can be extended up to optical range [14]. A wide impedance bandwidth of 58.5%, frequency ranging from 2.32 to 4.24 GHz, and antenna gain about 8 dB are the results obtained when dipole antenna is loaded with T-shaped array of metamaterials on the dielectric substrate using rectangular box-shaped reflector [15].

The proposed design in this paper is a dipole antenna loaded with inverted S-shaped metamaterial that is designed using CST. The results obtained are analyzed and interpretation was made which shows improved performance parameters.

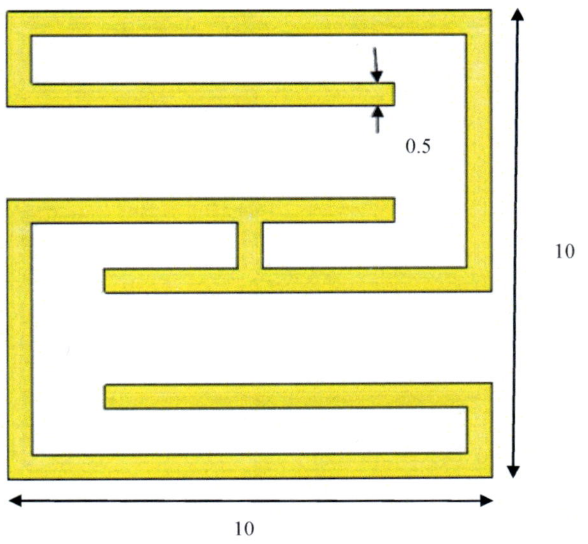

Fig. 1 Structure of inverted s-shaped metamaterial

1.1 Metamaterial Structure

The developed inverted S-shaped metamaterial structure is shown in Fig. 1. The developed inverted S-shaped structure consists of two split square resonators connected to each other. The proposed structure is constructed from a copper sheet of thickness 0.5 mm and the FR 4 glass substrate material of 10 × 10 mm.

2 Results

The designed metamaterial-loaded dipole antenna is shown in Fig. 2. The characteristics of metamaterial-loaded dipole are compared with that of conventional dipole. The plot of return loss is depicted in Fig. 3. The resonant frequency of antenna is shifted to low frequency when inverted S-shaped metamaterials are loaded (1.88–1.71 GHz). The % bandwidths of the antenna with and without SRR are 12.75 and 12.72. The bandwidth of the antenna is the range of frequencies over which the VSWR is less than 2. A VSWR specification commonly accepted is 2:1 VSWR. The impedance bandwidths of the dipole antenna with and without loading metamaterial are 256 MHz and 230 MHz, respectively. Measured return loss of dipole antenna loaded with SRR and without SRR is shown in Fig. 3. The far field patterns of dipole antenna are depicted in Figs. 4, 5, and 6. The gain of SRR-loaded dipole antenna and without SRR is 1.99 dB and 1.55 dB, respectively. The H-plane far field pattern of proposed antenna is given in Fig. 6. Measured VSWR of dipole antenna loaded with SRR and without SRR is shown in Fig. 7.

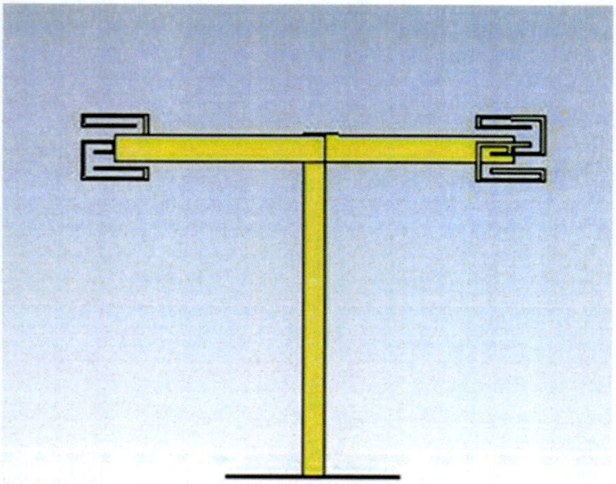

Fig. 2 Inverted S-shaped metamaterial-loaded dipole antenna

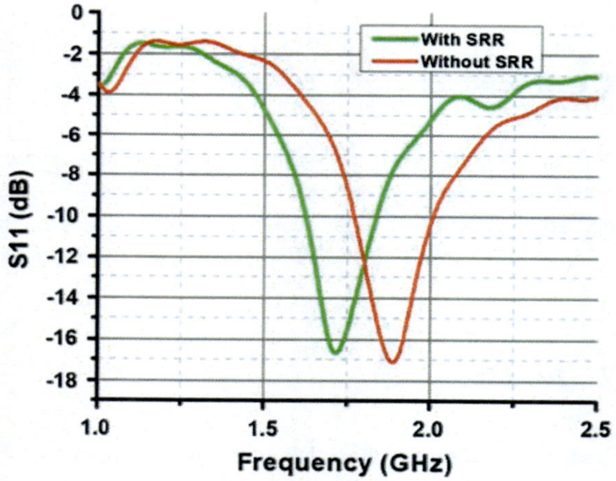

Fig. 3 Measured return loss of dipole antenna loaded with SRR and without SRR

3 Conclusion

In this paper, a design of wideband compact metamaterial-inspired dipole antenna for GPR applications is presented. The compact dimensions of the proposed antenna are 100 × 300 mm. The proposed antenna resonant frequency is shifted from 1.88 to 1.79 GHz by loading of SRR. A study on the proposed design suggests that by way of merely modifying the dimensions at the resonator, it is far viable to achieve significant versions in antenna resonance. The parameters of the antenna make the designed antenna an appropriate candidate for GPR applications.

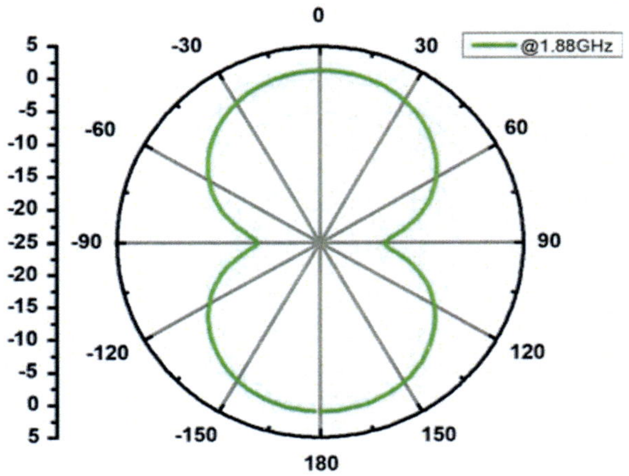

Fig. 4 Measured E-plane far field pattern at frequency of 1.88 GHz for conventional dipole antenna

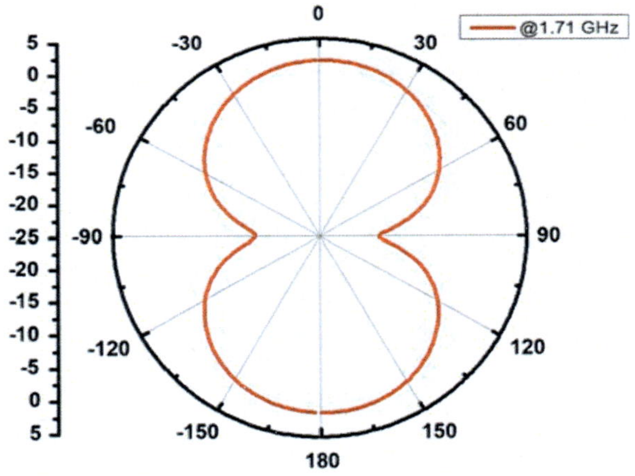

Fig. 5 Measured far field pattern at frequency of 1.71 GHz for metamaterial-loaded dipole antenna

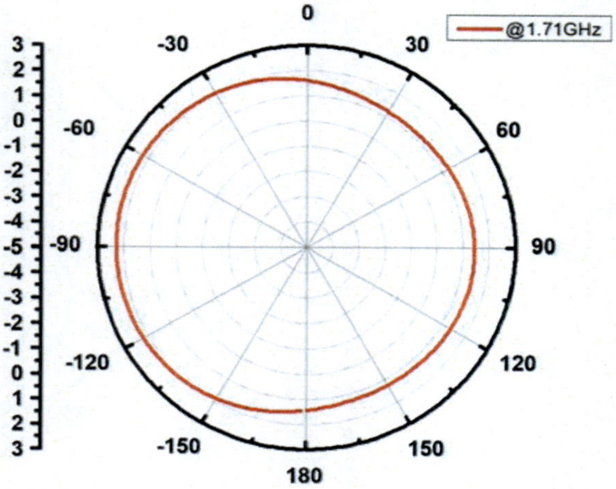

Fig. 6 Measured H-plane far field pattern at frequency of 1.71 GHz for metamaterial-loaded dipole antenna

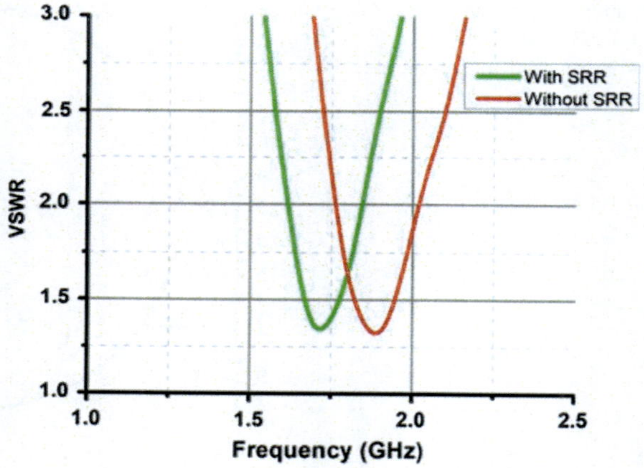

Fig. 7 Measured VSWR of dipole antenna loaded with SRR and without SRR

Acknowledgements This work is being supported by Collaborative Research Scheme, with the Grant No: JNTUH/TEQIP-III/CRS/2019/ECE/9.

References

1. Bernabini M, Pettinelli E, Pierdicca N, Piro S, Versino L (1995) Field experiments for characterization of GPR antenna and pulse propagation. J Appl Geophys 33(1–3):63–76
2. Terlapu SK, Jaya C, Srinivasa Raju GRLVN (2017) On the notch band characteristics of Koch fractal antenna for UWB applications. Int J Control Theor Appl 10(06):701–707

3. Terlapu SK, Chowdary PSR, Jaya C, Chakravarthy VVSSS, Satapathy SC (2018) On the design of fractal UWB wide slot antenna with notch band characteristics. In: Proceedings of 3rd international conference on micro-electronics, electromagnetics and telecommunications. Lecture Notes in Electrical Engineering, Springer, Singapore
4. Marwah N, Pandey GTV, Marwah S (2016) A novel wideband magneto-electric dipole antenna with improved feeding structure. Adv Electro Magnet 5(2):10–16
5. Travassos XL, Avila SL, Adriano RL, da S, Ida N (2018) A review of ground penetrating radar antenna design and optimization. J Micro Waves Optoelectron Electromagnet Appl 17(3):385–402
6. Pajewski L, Tosti F, Kusayanagi W (2015) Antennas for GPR systems. In: Benedetto A, Pajewski L (eds) Civil engineering applications of ground penetrating Radar. Springer transactions in civil and environmental engineering, 41–67
7. Kalbhor G, Vyas M, Patil BP (2013) Antenna design for ground penetrating radar system. Int J Adv Electr Electron Eng (IJAEEE) 2(6):1–5
8. Diamanti N, Annan AP (2013) Characterizing the energy distribution around GPR antennas. J Appl Geophys 99:83–90
9. Vijver E, DePue J, Cornelis W, Meirvenne M (2015) Comparison of air-launched and ground coupled configurations of SFCW GPR in time, frequency and wavelet domain. In: European Geosciences Union (EGU)—General assembly, Vienna, Austria, vol 17
10. Chen G, Richard C (2010) A 900 MHz shielded bow-tie antenna system for ground penetrating radar. In: 13th international conference on ground penetrating radar, Lecce, Italy
11. Ali J, Abdullah N, Ismail MT, Mohd E, Shah SM (2017) Ultra-wideband antenna design for GPR applications-a review. Int J Adv Comput Sci Appl (IJACSA) 8(7):392–400
12. Breinbjerg O (2015) Metamaterial antennas—the most successful metamaterial technology? In: 9th international congress on advanced electromagnetic materials in micro-waves and optics (METAMATERIALS). Oxford, pp 37–39
13. Lambot S, Giannopoulos A, Pajewski L, Slob E, Sato M (2016) Foreword to the special issue on advances in ground-penetrating radar research and applications. IEEE J Sel Top Appl Earth Obs Remote Sens 9(1):5–8
14. Smith DG, Jol HM (1995) Ground penetrating radar: antenna frequencies and maximum probable depths of penetration in quaternary sediments. J Appl Geophys 33:93–100
15. Lestari AA, Yulian D, Liarto Suksmono AB, Bharata E, Yarovoy AG, Ligthart LP (2017) Theoretical and experimental analysis of a rolled dipole antenna for low-resolution GPR. In: IEEE international conference on ultrawideband

Low-Power and High-Speed 2-4 and 4-16 Decoders Using Modified Gate Diffusion Input (M-GDI) Technique

Anusha Karumuri and Prema Kumar Medapati

Abstract This paper introduces the design of 2-4 and 4-16 line decoders using modified version of gate diffusion input (M-GDI) technique as it reduces the area that is the number of transistors and also it reduces the dissipation of power. The Combinational circuits like decoders which are used in the periphery circuitry of memory arrays like Static RAM are designed by using Modified Gate Diffusion Input (M-GDI) technique which eradicate the disadvantages of the pass transistor logic (PTL) and CMOS logic. The decoders which are designed using modified GDI technique offer better characteristics in terms of average power and delay, and also the transistor count is reduced compared to the decoders which are designed using mixed logic technology which are static CMOS, pass transistor logic and transmission gate logic. Finally, the decoders which are designed by using two techniques in 90 nm technology using Cadence Virtuoso compared in terms of transistor count, average power, delay and power delay product (PDP) show an improvement when compared to the decoders of mixed logic.

Keywords Decoders · Modified GDI · Mixed logic

1 Introduction

In modern age, the main considerations are area, speed and power which are the important parameters for increasing demand for low power and the design of electronic circuits needs to be compatible with minimized chip area which are handheld devices like mobiles, laptops, palmtops, wireless modems and remaining electronic devices.

A 2-4 decoder has two inputs A and B, and it generates its output $D_{0\text{-}3}$ based on its logic operation. Depending on the input combination A and B in non-inverting decoder, one of the four outputs (minterms) $D_{0\text{-}3}$ is selected and is set to '1' and rest are set to '0'. Similarly, inverting 2-4 decoder generates the complementary

A. Karumuri (✉) · P. K. Medapati
Shri Vishnu Engineering College for Women (Autonomous), Bhimavaram, India
e-mail: anushagupta557@gmail.com

P. S. R. Chowdary et al. (eds.), *Microelectronics, Electromagnetics and Telecommunications*, Lecture Notes in Electrical Engineering 655,
https://doi.org/10.1007/978-981-15-3828-5_10

outputs $I_{0\text{-}3}$; that is, selected output is set to '0' and the rest are set to '1'. Similarly, 4-16 decoders have 4 inputs and 16 outputs (minterms) generated based on the logic operation.

There are many efficient design styles suitable for the advancements in the VLSI technology; some of them are CMOS, PTL, gate diffusion input (GDI) techniques, etc. Among them, the gate diffusion input (GDI) designs the digital combinational circuits with less area, i.e., minimum transistor count and reduced power (low power), reduced propagation delay which overcomes the demerits of CMOS, PTL and other designing technologies [1]. It only uses two transistors for the implementation of logic gates.

GDI consists of three inputs G (common gate input of NMOS and PMOS), *P* (input to the source/drain of PMOS) and *N* (input to the source/drain of PMOS). The bulks of PMOS and NMOS are connected to *N* and *P*, respectively [2]; in the case of bulks, the fabrication of GDI cell is not possible in *P* well because when the bulk is connected to the drain, threshold voltage is increased and when in the case bulk is connected to the source the body effect is destroyed according to the threshold voltage equation so we have an alternate approach that is the modified GDI (M-GDI) in which bulks are not connected to respective drain/source [3] which is derived from GDI in order to eliminate the body effect and threshold voltage effects.

2 Decoder Design Using Mixed Logic

The combinational circuits like decoders can be designed by using different logic styles; the transistor count may defer in case of decoders designed by using the conventional CMOS; the 2-4 decoder takes 20 transistors; taking 2-4 decoder as pre-decoder, then we design the 4-16 decoder; it takes 104 transistors for the design using CMOS NAND and NOR gates. The circuits that are designed using mixed logic [4] exhibit good performance when compared to the circuits designed using the single logic. The existing decoders are designed by using the mixed logic style which is a combination of transmission gate logic (TGL) [5], static CMOS [6] and dual value logic (DVL) [7] which combines the advantages of the above three logic styles.

Transmission Gate Logic (TGL) is same as relay that conducts in both directions, and is also a CMOS based switch which is used to implement electronic switches and analog multiplexers. It reduces the noise margin and static power dissipation caused because of increased threshold voltage, but it requires control signal and its complement; generally, the clock signal is used as control signal. TGL implements logic gates, AND/OR gates; thus, the gates are used to design the decoders that gates are full swinging.

Static CMOS is logic circuit designs which have pull-up network which contains PMOS transistors and pull-down network which contains NMOS transistors which gives good performance and resistance to noise. In static CMOS, the operating voltages are low and small transistor sizes and inputs are connected to the gate terminal

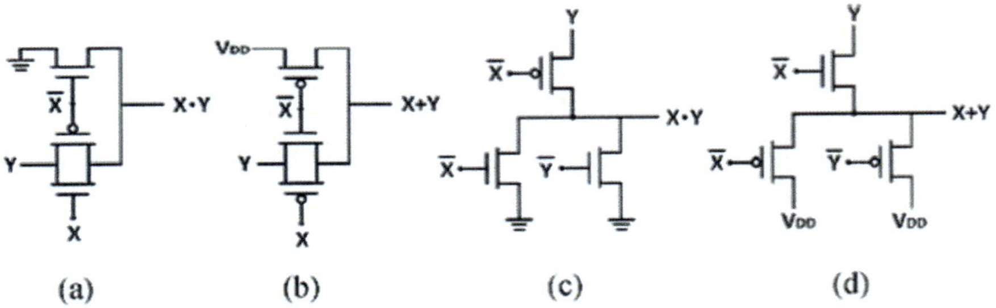

Fig. 1 **a** TGL AND gate. **b** TGL OR gate. **c** DVL AND gate. **d** DVL OR gate

of transistors; thereby, the designing becomes easy. It generates the output levels as long as the power supply is provided and consumes less power

Dual value logic (DVL) is one of the pass transistor logic (PTL) circuit techniques, and it is derived from the double pass transistor logic (DPL) with the main objective of speed improvement and power consumption [8]. Dual value is obtained from the double pass transistor logic by elimination of redundant branches present in the circuit, and there is rearrangement of signals in the circuit of DPL. The DVL gates have full swinging with less transistors, and it uses both NMOS and PMOS transistors with good speed of operation. Among the pass transistor logic design styles, DVL is more advantageous over conventional CMOS, double pass transistor logic (DPL), complementary pass transistor logic (CPL) and double pass transistor logic (DPL).

The above shown gates in Fig. 1 which are designed by using TGL and DVL both have the inputs *X* and *Y* which are applied at gate terminals. In TGL gates, input *X* is the control signal to the gates of three transistors and *Y* propagates output through the transmission gate. In DVL gates, input *X* is the control signal of two transistor gates and *Y* controls one transistor gate and propagates the output through the pass transistor [9, 10]. Hence, *X* and *Y* signals are considered as control and propagate signals, respectively. When we use inverted input as a propagate signal is not a good practice because it has some disadvantages, so an inverter is added to the propagation path which further increases the delay, respectively [4].

2.1 Transistor 2-4 Decoder

The 2-4 decoders can be designed by using TGL or DVL gates as it takes 16 transistors which includes 12 AND/OR gates and 2 inverters. By using some proper signal arrangement, we can eliminate one of the two inverters A or B; therefore, decoder can be designed by using 14 transistors shown in Fig. 2. The first minterm $D_0 = A'B'$ and the third minterm $D_2 = A'B$ can be implemented by using DVL logic, and the second and fourth minterms $D_1 = AB'$ and $D_3 = AB$ can be implemented by using TGL logic. For D_0 and D_2, A is the propagate signal, and for D_1 and D_3, B is

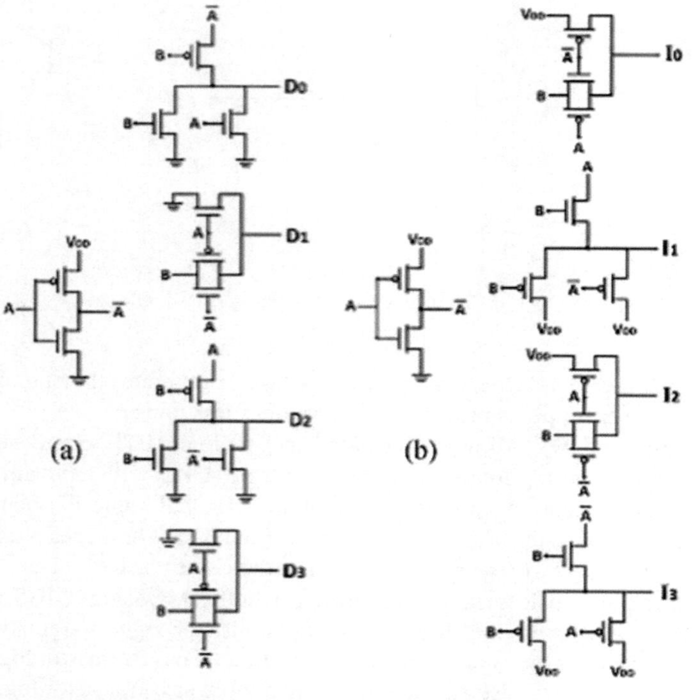

Fig. 2 14 transistor low-power 2-4 decoder. **a** 2-4 LP. **b** 2-4 LPI

the propagate signal; this eliminates the inverter B. Hence, the 14 transistor decoder consists of 9 NMOS and 5 PMOS for non-inverting decoder.

Similarly for the inverting decoder which is designed by using OR gates which is of 14 transistors topology consists of 5 NMOS and 9 PMOS. Same as non-inverting decoder, here also I_0 and I_2 minterms are implemented with TGL gates and I_1 and I_3 minterms are implemented with DVL gates. For I_0 and I_2, B is the propagate signal, and for I_1 and I_3, A is the propagate signal. By the elimination of inverter B, there is reduction of two transistors, logical effort, switching activity and also power dissipation. Here, we have two new topologies which are 2-4 LP and 2-4 LPI, I for invert.

2.2 Transistor 2-4 Decoder

In the 2-4 LP and LPI decoder, there is a drawback of worst-case delay because the complementary signal A is used as propagate signal for D_0 and I_3 minterms. So to overcome this, we can implement the minterms D_0 and I_3 using static CMOS gates. D_0 is implemented with CMOSNOR, and I_3 is implemented with CMOS NAND gates. Due to the addition of one transistor to the 14 transistor topology Decoder then this results new topology 15 transistor is obtained which has the advantage of delay but there is slightly there is increase in power.

Similarly as 14 transistor topology, 15 transistor topology also has two topologies called 2-4 HP and 2-4 HPI shown in Fig. 3. Hence, 15 transistor topology decoder is called high-performance decoder. Similarly, 4-16 decoders can be designed by using 2-4 decoders but when we cascade the PTL circuits it cannot give good performance due to poor driving capability so we can design 4-16 by using PTL and CMOS [8] combination instead of using mixed topology.

4-16 decoders are implemented with 2-4 decoder as pre-decoder combines with CMOS NOR or NAND gates which produce the decoded output. As 2-4 decoders have 4 new topologies, 4-16 decoders also have 4 new topologies as shown in Fig. 4, namely 4-16 LP, 4-16 LPI, 4-16 HP and 4-16 HPI. In case of 4-16, the transistor count decreases from 104 to 92 in case of low power (LP) and 94 in high performance (HP).

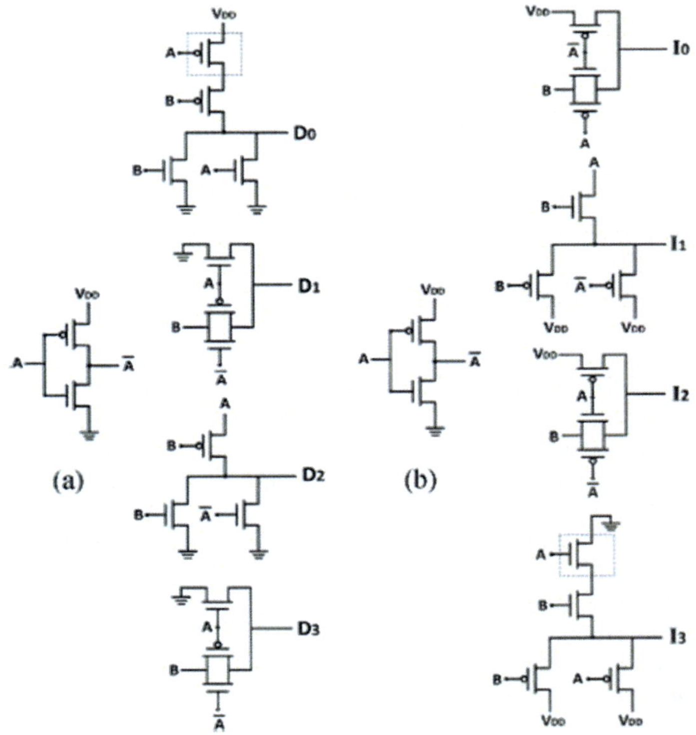

Fig. 3 15 transistor high-performance 2-4 decoder. **a** 2-4 HP. **b** 2-4 HPI

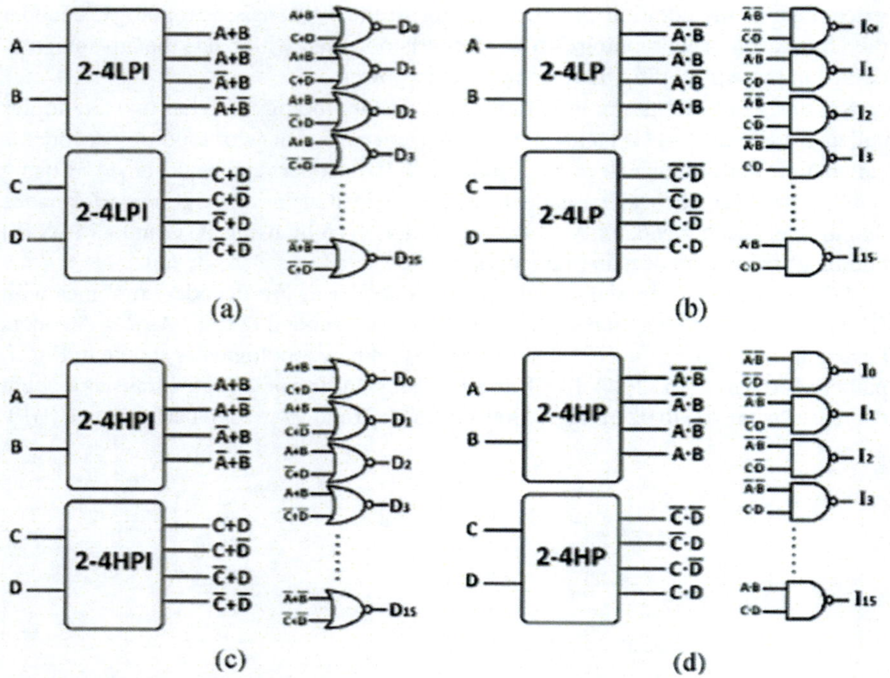

Fig. 4 4-16 decoders. **a** 4-16 LP. **b** 4-16 LPI. **c** 4-16 HP. **d**. 4-16 HPI

3 Decoder Design Using Modified GDI (M-GDI)

The proposed design is the modified GDI decoders which have minimum transistor count and consume less power and high speed when compared to the decoders designed using mixed logic. The GDI logic has the drawback of bulk issue in fabrication process which uses twin tub or silicon on insulator (SOI). This increases the complexity of the design as well as cost of fabrication then to overcome the above drawbacks the decoders designed using modified GDI designed using Modified GDI AND gate, Inverters and Modified GDI OR gates all these three gates can be designed using only two transistors.

The above gates in Fig. 5 are the modified GDI AND and modified GDI OR gates used in the design of the decoders. The operation of the modified AND gate is given as an example. In the AND gate, the drain terminal of PMOS is connected to input A and source terminal of NMOS is connected to input B. Both the gate terminals of PMOS and NMOS are connected to input A. When the two inputs are low (0), PMOS operates in linear region and NMOS is in cutoff region. When A = high (1) and B = low (0), the PMOS is in cutoff region and NMOS is in linear region. When A = low (0) and B = high (1), then PMOS is in linear region and NMOS is in cutoff

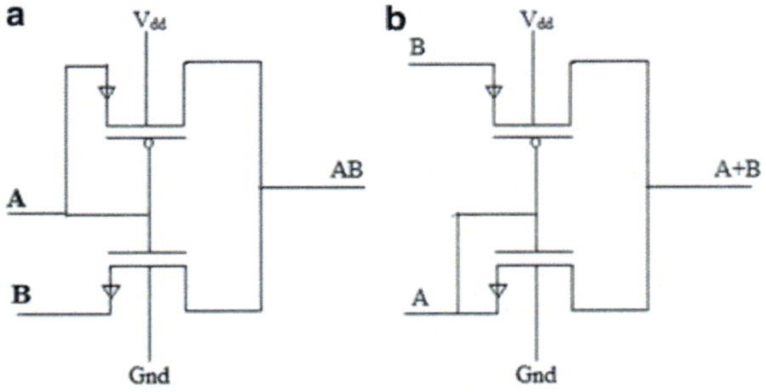

Fig. 5 **a** Modified GDI AND gate. **b** Modified GDI OR gate

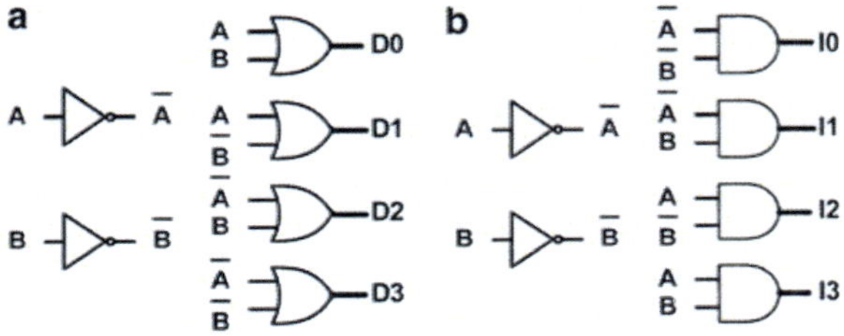

Fig. 6 **a** 2-4 modified GDI OR-based decoder. **b** 2-4 modified GDI AND-based decoder

region. When the two inputs are '1', then both PMOS and PMOS are in linear region and they produce the output as '1'.

Here Fig. 6 non-inverting decoder designed using four OR Gates and two inverters that are 12 transistors are required to design and also for inverting decoder requires Mod-GDI AND Gates and two inverters that is 12 transistors are required when compared to the existing decoders 3 transistors reduced in case of 2-4 HP and HPI, similarly 2 transistors are reduced in case of 2-4 LP and LPI. Taken 2-4 decoder as pre-decoders, 4-16 decoders are designed. Finally, the proposed decoders have the reduced transistor count, low power and high speed.

4 Results

The decoders of conventional CMOS, mixed logic and modified GDI are compared with transistor count, average power, delay and the power delay product (PDP). All the results are simulated using Cadence Virtuoso 90 nm CMOS technology. The power delay product (PDP) is defined as the product of average power and delay which is used to determine the performance of the digital circuits. Simulation results are performed in Cadence Virtuoso 90 nm CMOS technology.

Table 1 gives the information about the average power dissipation of 2-4 decoders and similarly Table 2 about the 4-16 decoders. Table 3 gives the information about the delay of all the decoders discussed above, Tables 4 and 5 about the power delay

Table 1 Average power dissipation of 2-4 decoders (in μw)

Time period (ns)	2			1			0.5		
Supply voltage (V_{dc})	0.8 v	1.0 v	1.2 v	0.8 v	1.0 v	1.2 v	0.8 v	1.0 v	1.2 v
NOR-based decoder	1.446	2.260	3.733	2.848	4.419	6.68	5.625	8.722	13.13
2-4LP (14T)	1.238	1.792	2.685	2.602	3.636	5.053	5.176	7.216	9.835
2-4 HP(15T)	1.398	1.970	2.942	2.756	3.886	5.486	5.467	7.695	10.63
2-4 M-GDI AND decoder	0.791	1.202	1.919	1.548	2.346	3.509	3.043	4.600	6.653
NAND-based decoder	1.727	2.612	4.215	3.416	5.133	7.795	6.763	10.18	15.00
2-4 LPI	0.846	1.346	2.522	1.671	2.653	4.49	3.353	5.273	8.426
2-4 HPI	1.016	1.520	2.701	2.005	2.991	4.834	4.014	5.925	9.089
2-4 M-GDI OR decoder	0.838	1.230	1.955	1.043	1.480	2.376	3.184	4.673	6.701

Table 2 Average power of 4-16 decoders (in μw)

Time period (ns)	2			1			0.5		
Supply voltage (V_{dc})	0.8 v	1.0 v	1.2 v	0.8 v	1.0 v	1.2 v	0.8 v	1.0 v	1.2 v
NOR-based decoder	6.001	9.339	14.51	11.85	18.38	27.56	23.46	36.40	53.56
4-16 LP	5.443	8.876	11.09	6.821	11.54	17.78	21.89	28.84	35.09
4-16 HP	6.906	11.22	17.95	8.829	13.33	22.14	30.98	35.39	42.66
4-16 M-GDI AND decoder	2.673	3.801	5.662	4.927	7.052	10.17	8.963	13.05	18.62
NAND-based decoder	6.088	9.586	14.99	11.99	18.81	28.44	23.16	36.95	55.22
4-16 LPI	5.380	6.644	10.71	5.821	11.09	13.18	21.18	22.83	35.08
4-16 HPI	8.974	11.00	21.31	12.33	19.46	22.76	23.33	28.38	44.79
4-16 M-GDI OR decoder	2.891	3.856	5.781	5.330	7.085	10.25	9.865	13.02	18.55

Table 3 Delay results (in ns)

2-4 decoder	Delay (ns)	4-16 decoder	Delay (ns)
NOR-based decoder	15.04	NOR-based decoder	38.12
2-4LP (14T)	8.015	4-16LP	20.05
2-4 HP(15T)	10.01	4-16 HP	25.09
NAND-based decoder	15.02	NAND-based decoder	34.89
2-4 LPI	11.017	4-16 LPI	19.56
2-4 HPI	13.02	4-16 HPI	24.87
M-GDI AND decoder	5.011	M-GDI AND decoder	13.09
M-GDI OR decoder	6.028	M-GDI OR decoder	15.02

product (PDP) of all the decoders Table 6 about the transistor count of the 2-4 and 4-16 decoders designed using the mixed logic, modified GDI and conventional CMOS. All the decoders are simulated at supply voltages 0.8, 1 and 1.2 v with time period 2, 1 and 0.5 ns.

5 Conclusion

This paper introduces an efficient and low-power design for decoder circuit. By using modified gate diffusion input (M-GDI) methodology, 2-4 and 4-16 decoders are designed with minimum transistor count, low power, high performance and minimum delay compared with mixed logic design and conventional CMOS decoders and simulation results are performed in Cadence Virtuoso 90 nm CMOS technology. These decoders are used in the applications where minimum area and less power are required and used as pre-decoders in designing the larger decoders, multiplexers and combinational circuits. The proposed modified GDI (M-GDI) decoder design excelled existing mixed logic decoder design in all aspects.

Table 4 Power delay product results (in fs) for 2-4 decoder

Time period (ns)	2			1			0.5		
Supply voltage (V_{dc})	0.8 v	1.0 v	1.2 v	0.8 v	1.0 v	1.2 v	0.8 v	1.0 v	1.2 v
2-4 NOR decoder	21.74	33.99	56.14	42.83	66.46	100.46	84.60	131.17	197.47
2-4LP (14T)	9.22	14.36	21.52	20.85	29.14	40.49	41.48	57.83	78.82
2-4 HP(15T)	13.99	19.71	29.44	27.58	38.89	54.91	54.72	77.02	106.40
2-4 M-GDI AND decoder	3.96	6.02	9.61	7.75	11.75	17.58	15.24	23.05	33.33
2-4 NAND decoder	25.93	39.23	63.30	51.30	77.09	117.08	101.58	152.90	225.30
2-4 LPI (14T)	9.32	14.82	27.78	18.40	29.22	49.46	36.94	58.09	92.82
2-4 HPI (15T)	13.22	19.79	35.16	26.10	38.94	62.93	52.26	77.14	118.33
2-4 M-GDI OR decoder	5.05	7.41	11.78	6.28	8.92	14.32	19.19	28.16	40.39

Table 5 Power delay product results (in fs) for 4-16 decoder

Time period (ns)	2			1			0.5		
Supply voltage (V_{dc})	0.8 v	1.0 v	1.2 v	0.8 v	1.0 v	1.2 v	0.8 v	1.0 v	1.2 v
4-16 NOR decoder	228.75	356.00	553.12	451.72	700.64	1050.58	894.29	1387.56	2041.70
4-16LP	109.13	177.96	222.35	136.76	231.37	356.48	438.89	578.24	703.55
4-16 HP	173.27	281.50	450.36	221.29	334.44	555.49	777.28	887.93	1070.33
4-16 M-GDI AND decoder	34.98	49.75	74.11	64.49	92.31	133.12	117.32	170.82	243.73
4-16 NAND decoder	212.41	334.24	523.00	418.33	656.28	992.27	808.05	1289.18	1926.62
4-16 LPI	105.23	129.87	209.48	113.83	216.92	257.80	414.28	446.55	686.16
4-16 HPI	223.18	273.57	529.97	306.64	483.97	566.04	580.21	705.81	1113.92
4-16 M-GDI OR decoder	43.42	57.91	86.83	80.05	106.41	153.95	148.17	195.56	278.62

Table 6 Transistor count of the following decoders

Types of decoders	2-4 decoders	4-16 decoders
CMOS decoders	20	104
2-4 LP/LPI	14	92
2-4 HP/HPI	15	94
M-GDI decoders	12	56

References

1. Morgenshtein A, Fish A, Wagner IA (2002) Gate diffusion input (GDI): a power–efficient method for digital combinational circuits. IEEE Trans VLSI Syst 10(5)
2. Morgenshtein A, Fish A, Wagner IA (2002) Gate diffusion input (GDI) A technique for low power design of digital circuits: analysis and characterization. In: IEEE international symposium on circuits and systems, vol 1
3. Soni D, Shah MV (2017) Review on modified gate diffusion input. Int Res J Eng Technol (IRJET) 4(04)
4. Balobas D, Konofaos N (2017) Design of low-power high-performance 2-4 & 4-16 mixed-logic decoders. IEEE Trans Circ Syst—II: Expr Briefs 64
5. Wu X (1992) Theory of transmission switches and its application of design of CMOS digital circuits. Int J Circ Theor Appl 20(4):349–356
6. Weste NHE, Harris DM (2011) CMOS VLSI design, a circuits and systems perspective, 4th ed. Addison-Wesley, Boston
7. Oklobdzija VG, Duchene B (1995) Pass-transistor dual value logic for low-power CMOS. In: Proceedings of international symposium VLSI technology, pp 341–344
8. Morgenshtein A, Schwartz I, Fish A (2010) Gate diffusion input logic in standard CMOS nanoscale process. In: IEEE 26th convention of electrical and engineers in Israel
9. Dake JL, Terlapu SK (2016) Implementation of high-throughput digit-serial redundant basis multiplier over finite field. IOSR J VLSI Sig Process (IOSR-JVSP) 6(4):35–45. Ver. I. e-ISSN; 2319-4200, ISSN No.: 2319-4197
10. Gali SJ, Terlapu SK (2018) On the implementation of VLSI architecture of FM0/Manchester encoding and differential manchester coding for short-range communications. In: Anguera J, Satapathy S, Bhateja V, Sunitha K (eds) Microelectronics, electromagnetics and telecommunications. Lecture notes in electrical engineering, vol 471. Springer, Singapore

A Sensitivity Based Approach for Optimal Allocation of OUPFC Under Single Line Contingencies

Srinivasa Rao Veeranki, Srinivasa Rao Rayapudi, and Ravindra Manam

Abstract In this paper, a sensitivity based approach is proposed for optimal allocation of optimal unified power flow controller (OUPFC) under single line contingency to eliminate overloads on transmission lines. The approach is formulated based on ranking index (RI) and performance index (PI). After outage of a branch element, a unitary variation of power flow (PF) in every transmission line is attained through RI. It is formulated to quantify loading level of network after a given outage. Contingencies are organized in descending order depending on the value of RI. Sensitivity factors are attained by differentiating real power flow performance index (RPFPI) subjected to system parameters of OUPFC. Optimal allocation of OUPFC is based on sensitivity factors obtained by considering line outages in the order of their severities which is given by RI. The proposed approach is programmed on 5-bus and IEEE 14-bus networks under MATLAB environment.

Keywords OUPFC · Contingency analysis · Sensitivity approach · Ranking index · Optimal placement

---

S. R. Veeranki (✉)
Department of Electrical and Electronics Engineering, Aditya Engineering College (A), ADB Road, Surampalem, Kakinada, India
e-mail: connectvsr@gmail.com

S. R. Rayapudi
Department of Electrical and Electronics Engineering, Jawaharlal Nehru Technological University Kakinada, Kakinada, India
e-mail: srinivas.jntueee@gmail.com

R. Manam
Department of Electrical and Electronics Engineering, Aditya college of Engineering, ADB Road, Surampalem, Kakinada, Andhra Pradesh, India
e-mail: ravieeejntu@gmail.com

P. S. R. Chowdary et al. (eds.), *Microelectronics, Electromagnetics and Telecommunications*, Lecture Notes in Electrical Engineering 655,
https://doi.org/10.1007/978-981-15-3828-5_11

1 Introduction

The sufficient generation has been put to satisfy the load which adequate transmission has been put into supply produced power to the load. Any piece of kit within the network will fail, either because of internal or external causes, like lightning strikes, due to objects hit transmission towers, due to errors in relay setting. It is extremely inefficient, if not difficult, to make a power network with most redundancy that inadequacy never leads load to be fallen down on the network. The system network is planned with the goal that load dropping is adequately small. Along these lines, maximum networks are planned to withstand many failures, and however, this does not ensure that the system will be 100% reliable [1, 2]. This paper focuses on probable consequences and corrective actions requisite by transmission line outage which is a major type of failure event. The failures in the transmission line create variations in PF and voltages on apparatus which are allied on transmission side to the system. Hence, the transmission failures analysis helps in developing techniques to retain PF and voltages within their respective parameters. Thus, a further newly attractive technique is to make full usage of the present transmission lines. Thus, the application of FACTS devices on the system is a choice to increase the system steady-state security. This led to the development of FACTS devices over which OUPFC is the effective device developed to improve power network reliability by controlling PF over lines. OUPFC consists of the combination of UPFC and phase-shifting transformer (PST) which exhibits their individual better features in controlling PF in transmission lines. UPFC can control real, reactive PF and voltage magnitude at the buses, and PST controls phase angle of the bus voltages [3, 4]. There are some strategies proposed in the literature for optimizing the location of OUPFC. A Lyapunov energy function based on appropriate angles in proper locations of OUPFC to improve transient stability of a power network has been proposed [5]. A current-based model of OUPFC with supporting mathematical derivation has been presented [6]. A fuzzy interactive based of OUPFC with minimum cost of OUPFC installation was presented [7]. The primary reasons of August 14, 2003, blackout in USA and Canada [8] are closely linked with the voltage collapse following line congestion. It was once observed that much consideration has been paid to the heuristic algorithm to solve the finest area of FACTS devices. In heuristic optimization approaches, the preliminary population is selected arbitrarily. Repeating of optimal results attained with equal primary condition is not definite with heuristic approaches. Though thinking about our current knowledge, no research work in this area has viewed the sensitivity approach for ideal region of OUPFC system under single line outage ($N - 1$ contingency). As indicated by particular of OUPFC, its use is to decrease the burden of the PF in intensely loaded energy transmission lines. Hence, the principle commitment of this paper is to ideally allocate OUPFC by utilizing the genuine PF record under a single line contingency. RPFPI sensitivity indices are attained by differentiating RPFPI subject to parameters of FACTS devices to be optimized. This technique is programmed on 5-bus and 14-bus network for allocation of OUPFC under single line contingency. The rest of the paper is ordered in a subsequent way. Section 2 consists of static modeling of OUPFC and methods for optimal allocation of OUPFC. Section 3 consists of simulation results, and Sect. 4 concludes the problem.

2 Static Modeling of OUPFC

2.1 OUPFC

The OUPFC is an arrangement of PST and UPFC which is connected by two tertiary transformers. PST is associated with auxiliary windings of energizing and inducing transformers that add voltage with a static stage to transmission line constrained by mechanical switches. Based on system conditions, added voltage alters the transmission angle. The injection or added voltage alters the transmission angle, subjected to network conditions. Two converters such as shunt and series converters are operated via DC connection along with storage capacitor (DC). The OUPFC schematic model is shown in Fig. 1.

The injection model can be attained by using circuit theory as shown in Fig. 2. The induced active power at bus B-i(P_{io}) and bus B-j(P_{jo}), and reactive powers Q_{io}

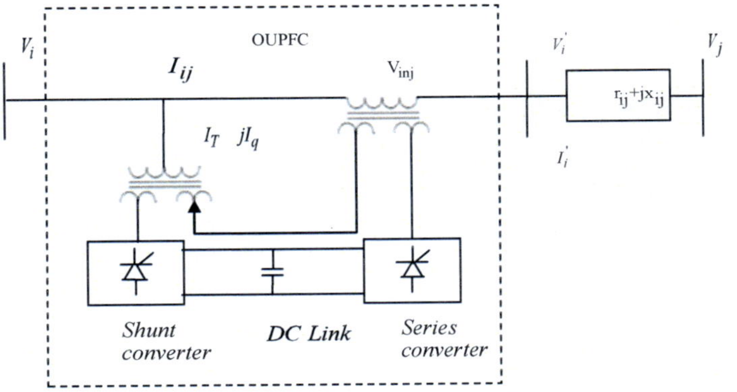

Fig. 1 Representation of OUPFC

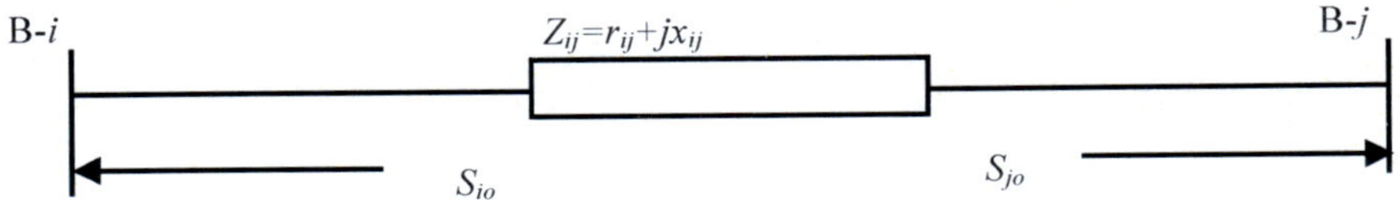

Fig. 2 Injection model of OUPFC

and Q_{jo} of line with OUPFC are:

$$\begin{aligned} P_{\mathrm{io}} = & -V_i^2K^2G_{ij} - V_iV_jK\left[G_{ij}\sin\delta_{ij} - \mathrm{B}_{ij}\cos\delta_{ij}\right] \\ & - V_T^2G_{ij} - 2V_iV_TG_{ij}\cos(\delta_T - \delta_i) \\ & + V_jV_T\left[G_{ij}\cos(\delta_T - \delta_j) + \mathrm{B}_{ij}\sin(\delta_T - \delta_j)\right] \end{aligned} \quad (1)$$

$$\begin{aligned} P_{\mathrm{jo}} = & -V_iV_jK\left[G_{ij}\sin\delta_{ij} + \mathrm{B}_{ij}\cos\delta_{ij}\right] \\ & + V_jV_T\left[G_{ij}\cos(\delta_T - \delta_j) - \mathrm{B}_{ij}\sin(\delta_T - \delta_j)\right] \end{aligned} \quad (2)$$

$$\begin{aligned} Q_{\mathrm{io}} = & V_i^2K^2B_{ij} + V_iV_jK\left[G_{ij}\cos\delta_{ij} + \mathrm{B}_{ij}\sin\delta_{ij}\right] \\ & + V_iI_q + V_iV_T\left[G_{ij}\sin(\delta_T - \delta_i) + \mathrm{B}_{ij}\cos(\delta_T - \delta_i)\right] \end{aligned} \quad (3)$$

$$\begin{aligned} Q_{\mathrm{jo}} = & -V_iV_jK\left[G_{ij}\cos\delta_{ij} - \mathrm{B}_{ij}\sin\delta_{ij}\right] \\ & - V_jV_T\left[G_{ij}\sin(\delta_T - \delta_j) + \mathrm{B}_{ij}\cos(\delta_T - \delta_j)\right] \end{aligned} \quad (4)$$

where $K = \tan(\varphi)$. This process is recurrent for outage of every line in turn. The RI indicates the average rate of flow of power in lines. The RI is arranged in order for all probable contingencies. The strategy begins with consideration of priori utmost critical contingency, going down in the list until the non-problematic contingency is evaluated.

2.2 *Sensitivity Indices*

The PI defined in Eq. (5) specifies the critical position of network loading under steady and contingency situations.

$$\mathrm{PI} = \sum_{m=1}^{N_l} \frac{W_m}{2n}\left(\frac{P_{lm}}{P_{lm}^{\max}}\right)^{2n} \quad (5)$$

where W_m is a weight coefficient, P_{lm} is RPF and $P_{lm}^{\max}$ is a measure of rated capacity of m-line N_l which is measured as total number of lines in the network. PI is considered to measure of critical condition of overloading of lines in network. Here, exponent value considered as 2.0, and weight coefficient is taken as 1.0. The OUPFC sensitivity factors c_1^{o}, c_2^{o}, c_3^{o} and c_4^{o} is formulated as partial derivative of PI subjected to induced voltage (V_T), induced voltage phase angle (δ_T), current (I_q) and phase-shifting transformer phase angle(φ) on *k*-line, respectively.

$$c_1^o = \left.\frac{\partial \mathrm{PI}}{\partial V_T}\right|_{V_T=0} = \text{PI Sensitivity subject to } V_T,$$

$$c_2^o = \left.\frac{\partial \mathrm{PI}}{V_T \partial \delta_T}\right|_{\delta_T=0} = \text{PI Sensitivity subject to } \delta_T,$$

$$c_3^o = \left.\frac{\partial \mathrm{PI}}{\partial I_q}\right|_{I_q=0} = \text{PI Sensitivity subject to } I_q,$$

$$c_4^o = \left.\frac{\partial \mathrm{PI}}{\partial \varphi}\right|_{\varphi=0} = \text{PI Sensitivity subject to } \varphi,$$

Using Eq. (5), the sensitivity of PI subject to OUPFC parameter $X_k(V_T, \delta_T, I_\mathrm{q}$ and $\varphi)$, linked among bus B-*i* and B-*j*, can be formulated as

$$\frac{\partial \mathrm{PI}}{\partial X_k} = \sum_{m=1}^{N_1} W_m\, P_{1m}^3 \left(\frac{1}{P_{1m}^{\max}}\right)^4 \frac{\partial P_{1m}}{\partial X_k} \tag{6}$$

Considering Equation [9], RPF over the *m*-line is formulated as follows. *s* is considered as slack bus.

$$P_{lm} = \begin{cases} \sum\limits_{n=1, n\neq s}^{N} S_{mn} P_n \; for\; m \neq k \\ \sum\limits_{n=1, n\neq s}^{N} S_{mn} P_n + P_j \; for\; m = k \end{cases} \tag{7}$$

where N is considered as number of buses in the network and S_{mn} is *mn*th component of matrix $[S_f]$ which transmits line PF with power injections [9] at buses without OUPFC. It is observed that in *k*-line which in between buses *i* to *j* has added flow of P_j at *j-bus*, by linking OUPFC as shown in Fig. 2. Using Eqs. (6) and (7), the following equations are formulated as,

$$\frac{\partial P_{1m}}{\partial X_k} = \begin{cases} \left(S_{mi} \frac{\partial P_i}{\partial X_k} + S_{mj} \frac{\partial P_j}{\partial X_k}\right) & \text{for } m \neq k \\ \left(S_{mi} \frac{\partial P_i}{\partial X_k} + S_{mj} \frac{\partial P_j}{\partial X_k}\right) + \frac{\partial P_j}{\partial X_k} & \text{for } m = k \end{cases} \tag{8}$$

The terms,

$$\left.\frac{\partial P_i}{\partial \varphi}\right|_{\varphi=0}, \left.\frac{\partial P_j}{\partial \varphi}\right|_{\varphi=0}, \left.\frac{\partial P_i}{\partial V_T}\right|_{V_T=0}, \left.\frac{\partial P_j}{\partial V_T}\right|_{V_T=0}, \left.\frac{\partial P_i}{V_T \partial \delta_T}\right|_{\delta_T=0}, \left.\frac{\partial P_j}{V_T \partial \delta_T}\right|_{\delta_T=0}, \left.\frac{\partial P_i}{\partial I_q}\right|_{I_q=0}$$

and $\left.\frac{\partial P_j}{\partial I_q}\right|_{I_q=0}$ can be derived using Eqs. (1–4), respectively.

The sensitivity factors c_1^o, c_2^o, c_3^o and c_4^o can be formulated by partially differentiating PI subject to V_T, δ_T, I_q and φ of *k*-line.

2.3 *Conditions for Optimal Allocation*

The conditions considered for optimal allocation of OUPFC is the same as in [9] which is outlined as:

- OUPFC is allocated in *k*-line having least sensitivity subject to injected voltage and current.
- OUPFC is allocated in *k*-line having biggest sensitivity subject to phase angle.
- OUPFC is not considered to be allocated in line enclosing PV buses even if sensitivity is high of these.

3 Simulation Analysis with Results

The proposed technique is tested on two different networks having 5 and 14 buses with OUPFC with line outages. The simulations are carried out in MATLAB programming environment, and the results are analyzed.

3.1 *5-Bus System*

The bus network [10] encloses three generators, two load buses and six transmission lines. Bus-5 is a reference bus and lines 1–2 and 1–4 have shunt susceptance 0.002 p.u each and impedance of $(0.002 + j0.01)$ p.u and remaining four lines have shunt susceptance 0.004 p.u each and impedance of $(0.004 + j0.02)$ p.u. Limit for line flow is considered as 800 MW. The PF without FACTS device with contingency is presented in Table 1. Table 1 presents sending end PF for base case and for all line outages. From Table (subcolumn 3, base case), the load flow results by NR method

Table 1 Line flows without any FACTS equipment (in p.u) of 5-bus network

k-Line		PF (in p.u)						
		No outage	Line outages					
No	*i*–*j*	Base case	1–2	1–3	1–4	2–5	3–4	4–5
£1	1–2	0.96	–	−0.53	−3.52	5.08	1.38	−1.2
£2	1–3	−6.99	−6.80	–	**−11.5**	−7.82	**−9.61**	7.43
£3	1–4	**−8.97**	**−8.20**	**−14.5**	–	**−12.3**	−6.77	**10.0**
£4	2–5	−4.04	−5.00	−5.53	**−8.59**	–	−3.62	3.10
£5	3–4	2.79	3.00	**10.0**	2.40	1.81	–	−2.5
£6	4–5	1.06	2.05	2.07	5.07	−3.41	0.55	–
Overloads		£3	£3	£3 and L5	£2 and £4	£3	£2	£3

Table 2 Sensitivities (c_1^o) of OUPFC subject to V_T (in p.u) of 5-bus network

k-Line		Base case	Outage of lines					
No	*i–j*		1–2	1–3	1–4	2–5	3–4	4–5
£1	1–2	**−0.6257**	–	**−3.1296**	**−0.5237**	**−0.7102**	**−0.2589**	−0.0000
£2	1–3	−0.0357	0.0536	–	1.3584	−0.3911	0.0529	−0.2644
£3	1–4	0.9474	0.4820	3.7404	–	3.1883	0.3269	0.4075
£4	2–5	−0.4095	0.0467	−1.9490	0.0499	–	−0.1704	0.0056
£5	3–4	−0.2321	**−0.1552**	0.0860	0.2266	−1.5997	–	**−0.4208**
£6	4–5	0.5042	−0.0001	2.668	0.6469	0.0019	0.2163	–
Factors		£1	£5	£1	£1	£1	£1	£5

Table 3 Sensitivities (c_2^o) of UPFC subject to δ_T (in p.u) of 5-bus network

k-Line		Base case	Outage of lines					
No	*i–j*		1–2	1–3	1–4	2–5	3–4	4–5
£1	1–2	−2.3363	–	−11.477	−2.3418	−0.5710	−1.0122	−0.0000
£2	1–3	−1.0725	−0.7024	–	**4.1147**	−5.0786	0.1469	**−2.9733**
£3	1–4	**4.0746**	**1.2759**	**15.1750**	–	**7.2144**	**1.1667**	**2.9571**
£4	2–5	−2.4411	0.0503	−12.784	−2.7928	–	−1.0478	0.0034
£5	3–4	−1.2537	−0.9828	1.9137	3.1907	−5.8821	–	−2.8008
£6	4–5	2.5421	0.0030	13.5326	3.4734	0.0047	1.0862	–
Factors		£3	£3	£3	£2	£3	£3	£2, £3

indicate that RPF in line-3 for base case is 8.97 p.u, which is high compared to line loading limit and in case of line outage, it can be observed that line-3 is getting overloaded in most of the cases except lines-3 and 5 outage case. In case, line-3 outage lines-2 and 4 get overloaded and in line-5 outage line-2 get overloaded. The lines which are overloaded are presented in bold. Last row of table clearly indicates overloaded line/lines for a given outage. Sensitivities are computed for every control parameter of OUPFC connected with every line one at a period for same operational constraints. The sensitivities of RPFPI considering OUPFC for various line outages and base case are shown in Tables 2, 3 and 4. The highest absolute or negative sensitivities are shown in bold case depending on device nature.

3.2 OUPFC Allocation

For OUPFC allocation, the RPFPI is differentiated subject to induced voltage magnitude, induced voltage phase angle, shunt current magnitude of UPFC and phase-shifting transformer phase angle.

Table 4 Sensitivities (c_4^o) with PST of OUPFC (in p.u) of 5-bus network

k-Line		Base case	Outage of lines					
No	*i–j*		1–2	1–3	1–4	2–5	3–4	4–5
£1	1–2	−2.3058	–	−4.3633	−0.7960	−0.2595	−0.4071	−0.0000
£2	1–3	−1.0444	−0.2794	–	1.4285	−1.9124	0.0593	**−1.1873**
£3	1–4	**4.0133**	**0.5146**	**5.7550**	**–**	**2.8923**	**0.4697**	1.0823
£4	2–5	−2.3999	0.0216	−4.9126	−0.9647	–	−0.4210	0.0014
£5	3–4	−1.2953	−0.4145	0.7923	1.3126	−2.5501	–	**−1.1778**
£6	4–5	2.6582	0.0013	5.5660	**1.4608**	0.0020	0.4662	–
Factors		£3	£3	£3	£6	£3	£3	£2, £5

OUPFC is allocated in a line with the least value of sensitivity factor (c_1^o). From Table 2, it seems that line-1 is having least sensitivity factor (c_1^o) for base case. In the case of line outages, line-1 is considered for four cases followed by line-5 in two cases. So, line-1 is only suitable for allocation of OUPFC. Last row of the table indicates line with the least sensitivity factor for base case and for a given outage. Table 3 presents sensitivity factors subjected to induced voltage phase angle (δ_T), and OUPFC is placed based on line with largest absolute value of sensitivity factor (c_2^o). From Table 3, it is presented that line-3 is the largest absolute sensitivity factor (c_2^o) for base case. In case of line outages, line-3 is suitable for five cases followed by line-2 which is suitable for one case (i.e., line-3 outage). So, line-3 is considered for allocation of OUPFC with negative phase shift because factors are positive values. Last row of the table indicates line with largest absolute sensitivity factor for base case and for a given outage. The sensitivity factor (c_3^o) corresponding to magnitude of shunt current (I_q), which is the third parameter of OUPFC, is always zero because it cannot control real RPF as it is in 90° phase shift with input voltage.

Table 4 presents sensitivity factors subject to phase-shifting transformer phase angle (δ_P), and OUPFC is allocated in line which has the highest value of sensitivity factor (c_4^o). From Table 4, it is perceived that line-3 is the highest absolute for base case. In case of line outages, line-3 is considered for four cases for line-5 and line-6 which are suitable for one case each. The line-5 and line-6 are connected between two generators. Therefore, 3-line is suitable for allocation of OUPFC with negative phase shift because factors are positive values. Last row of the table indicates line with largest absolute sensitivity factor for base case and for a given outage.

3.3 *Effect of Single Line Contingency on Optimal Allocation of OUPFC*

Table 5 summarizes the analysis. Line outages considered are present in row 1. Row 2 shows sending and receiving end buses of the corresponding line. Row 5 presents

Table 5 Results of each line outage with OUPFC in 5-bus test system

£.O-l	£NO	£2	£3	£4	£1	£6	£5
	i–j	1–3	1–4	2–5	1–2	4–5	3–4
Ranking index		0.671	0.631	0.624	0.521	0.512	0.459
Rank		1	2	3	4	5	6
OLUO (c_2^o and c_4^o)		**£3**	£2	**£3**	**£3**	£3	£3
OLUN (c_2^o and c_4^o)		**£3**					

*£.O-l is line outage in line-l, £. NO is line number, OLUO/OLUN is optimal location under outage/no outage

an optimal allocation of OUPFC one in each column under line outage. Phase angle control of series induced voltage can effectively control RPF in lines. Hence, the optimal allocation of OUPFC is considered based on sensitivity factor attained by differentiating RPFPI subject to phase angle. Last row of Table 5 presents the optimal allocation of OUPFC without any line outage. RI which represents the variation of PF in every transmission line after outage of branch element given in row 1 is given in row 3. RI indicates severity of fault. So, row 4 denotes the rank of line outages in descending order. Lines getting overloaded because of line outage are presented in Table 1. To limit PF in overloaded lines, the optimal allocation is determined based on sensitivity factors. One interesting observation from Table 5 is the optimal allocation for most severe fault coincides with the case of no contingency. The result for the allocation of OUPFC is shown in bold.

3.4 14-Bus Network

The IEEE 14-bus network [11] has 5 PV buses, 11 PQ buses and 20 transmission lines. The line flow constraint is considered to be 120 MW. Bus-1 is considered as reference bus. 100 MAVA is Base MVA considered. Detailed presentation is done for 5-bus network to validate the efficacy of the method proposed from Tables 1, 2, 3 and 4. Lines getting overloaded for every outage are observed on 14-bus network also. To limit PF in overloaded lines, OUPFC is allocated using sensitivity factors in the same way as it is done for 5-bus network. The summary of results is presented in Table 6. Table 6 organizes in the same fashion as that of Table 5. Line outages in Column 1 are considered in descending order of RI presented in Column 4. The actual rank is presented in Column 5. Column 3 presents the optimal allocation of OUPFC obtained by differentiation of RPFPI subjected to device parameters. Device parameter, induced voltage phase angle is considered for obtaining the sensitivity factor. Observations from Table 6 are optimal allocation for most severe fault does not coincide with the case of no contingency, and in several contingencies, line-2 and line-1 are suitable for optimal allocation, but line-1 is connected between two generators. Decision for allocation of OUPFC is presented in bold.

Table 6 Results of each line outage with OUPFC in IEEE 14-bus test system

£. O-*l*		OLUO	RI	Rank
£. NO	*i–j*	OUPFC (c_2^o and c_4^o)		
£10	5–6	£7, £5	0.3370	1
£3	2–3	£2, £1	0.3186	2
£2	1–5	£4	0.2885	3
£1	1–2	£7	0.2851	4
£13	6–13	£2, £1	0.2844	5
£9	4–9	£2, £1	0.2834	6
£11	6–11	£2, £1	0.2823	7
£4	2–4	£7	0.2815	8
£12	6–12	£2, £1	0.2803	9
£17	9–14	£2, £1	0.2800	10
£18	10–11	£2, £1	0.2787	11
£14	7–8	£2, £1	0.2783	12
£20	13–14	£2, £1	0.2783	13
£19	12–13	£2, £1	0.2777	14
£16	9–10	£2, £1	0.2773	15
£8	4–7	£2, £1	0.2701	16
£15	7–9	£2, £1	0.2696	17
£6	3–4	£2, £1	0.2655	18
£7	4–5	£2, £1	0.2584	19
£5	2–5	£4	0.2578	20
OLUN (c_2^o and c_4^o)		£2		

4 Conclusion

A new sensitivity based approach is proposed for optimal allocation of OUPFC under single line contingency in the network. The method is based on the ranking index and performance index. RI is used to quantify average loading level of all lines of the network after given outage. Contingencies are arranged in descending order depending on value of RI, identified lines getting overloaded in each contingency. For limiting PF in those overloaded lines, OUPFC is placed in optimal locations. OUPFC is allocated using sensitivity analysis. Sensitivity factors are obtained by differentiating RPFPI subjected to system parameters of OUPFC under normal as well as contingency cases. The proposed sensitivity based approach is programmed on 5-bus and 14-bus systems. Six contingencies are considered in 5-bus network, and 20 contingencies are considered in 14-bus network. It is observed in 5-bus network that optimal location determined for most severe fault coincides with optimal allocation obtained for base case which is not the same in case of 14-bus network.

But, the optimal location of OUPFC obtained for base case coincides with optimal location obtained in case of several contingencies. Similar observation is obtained with 5-bus networks also.

References

1. Ara AL, Aghaei J, Alaleh M (2013) Contingency-based optimal placement of optimal unified power flow controller (OUPFC) in electrical energy transmission systems. Sci Iran 20(3):778–785
2. Gyugyi L, Song YH, John AT (1999) Flexible AC transmission systems (FACT), 30 edn (ch. 1). Institution of Electrical Engineers, London. (Inst Elec Eng Power Energy Series)
3. Avaz Pour P, Lashkar Ara A, Nabavi Niaki SA (2017) Enhancing power systems transient stability using optimal unified power flow controller based on Lyapunov control strategy. Sci Iran 24(3):1458–1466
4. Srilatha D, Sivanagaraju S (2017) Analyzing power flow solution with optimal unified power flow controller. Int J Eng Technol 9(3):2278–2289
5. Lashkar Ara A, Shabani M, Nabavi Niaki SA (2015) Multi-objective optimal location of optimal unified power flow controller (OUPFC) through a fuzzy interactive method. Sci Iran 22(6):2432–2446
6. Andersson G, Donalek P, Farmer R, Hatziargyriou N, Kamwa I, Kundur P, Martins N, Paserba J, Pourbeik P, Sanchez-Gasca J, Schulz R, Stankovic A, Taylor C, Vittal V (2005) Causes of the 2003 major grid blackouts in North America and Europe, and recommended means to improve system dynamic performance. IEEE Trans Power Syst 20(4):1922–1928
7. Ejebe GC, Wollenberg BF (1979) Automatic contingency selection. IEEE Trans Power ApparUs Syst 98(1):92–104
8. Schauder C, Gernhardt M, Stacey E, Lemak T, Gyugyi L, Cease TW, Edris A (1995) Development of a ± 100 MVAr static condenser for voltage control of transmission systems. IEEE Trans Power Deliv 10(3):1486–1496
9. Srinivasa Rao R, Srinivasa Rao V (2015) A generalized approach for determination of optimal location and performance analysis of FACT devices. Int J Electr Power Energy Syst 73:711–724
10. Exposito AG, Conejo AJ, Canizares C (2009) Electric energy systems analysis and operation. CRC Press, New York
11. Freris LL, Sasson AM (1968) Investigation of the load flow problem. Proc IEEE 115(10):1459–1469

Impact Analysis of Black Hole, Flooding Attacks and Enhancements in MANET Using SHA-3 Keccak Algorithm

T. Sairam Vamsi, T. Sudheer Kumar, and M. Vamsi Krishna

Abstract In present-day wireless communication scenario, Mobile ad hoc network (MANET) plays a very important role, as it consists of many autonomous nodes which communicate together to form a proper communication network. Each node in a network will move in random path, so that nodes direction will change frequently. But, some of the nodes may misbehave which leads to many problems in that network. These nodes are called as malicious nodes which create severe data loss by dropping data packets, and network may loss its privacy due to these intruders. So, providing security is the major challenge, because the networks are more vulnerable to many attacks. Some major attacks like flooding and black hole affect the E2E delay, packet delivery ratio (PDR), and throughput of the network. So, this paper mainly explains on enhancements in security in AODV using efficient techniques called SHA-3 Keccak and dynamic threshold routing algorithm.

Keywords MANET · Black hole · Flooding · AODV · SHA-3 Keccak

1 Introduction

The Mobile ad hoc network (MANET) is a network of mobile nodes without any infrastructure which faces many attacks while they are communicating using any one of the routing mechanisms [1]. In this process, the cooperation between the devices plays a vital role. Some of the nodes deliberately or not deliberately do not cooperate each other, which leads to network damage. Active and passive attacks are the two different types of attacks in MANETS. The attack which alters the data and disturbs the network is called as active attack, and the attack which does not

T. Sairam Vamsi (✉) · T. Sudheer Kumar
Shri Vishnu Engineering College for Women, Bhimavaram, Andhra Pradesh, India
e-mail: vamsi.0438@gmail.com

T. Sudheer Kumar
e-mail: skterlapu@gmail.com

M. Vamsi Krishna
Centurion University of Technology and Management, Paralakhemundi, Odisha, India

P. S. R. Chowdary et al. (eds.), *Microelectronics, Electromagnetics and Telecommunications*, Lecture Notes in Electrical Engineering 655,
https://doi.org/10.1007/978-981-15-3828-5_12

transmit the data packet to neighboring nodes is called passive attack [2]. So, actions must be taken to identify and eliminate such nodes to maintain secrecy and integrity. Generally, attacks are observed in any of the layers in TCP/IP, but the attacks which are observed in network layers are because of routing.

1.1 Objective

The main aim of the paper is to investigate the performance of MANETs under many attacks like black hole and wormhole, adopting SHA-3 security algorithm and dynamic threshold algorithm to the network and observing changes in performance of MANET.

1.2 Methodology

The implementation starts with

1. Implementing of Mobile ad hoc network (MANET) and analyzing performance by employing different routing protocols like AODV, DSDV, etc.
2. Analyzing performance of MANET under different attacks like black hole and flooding.
3. Developing an algorithm to choose different route when attacker is introduced. (SHA-3 Keccak algorithms).

1.3 AODV Route Discovery

While packets are forwarding from source to destination node, the steps followed are illustrated below.

When a network node wants to forward the data from start to end node, it verifies the routing table to find whether there is a dedicated route from source to destination node [3].

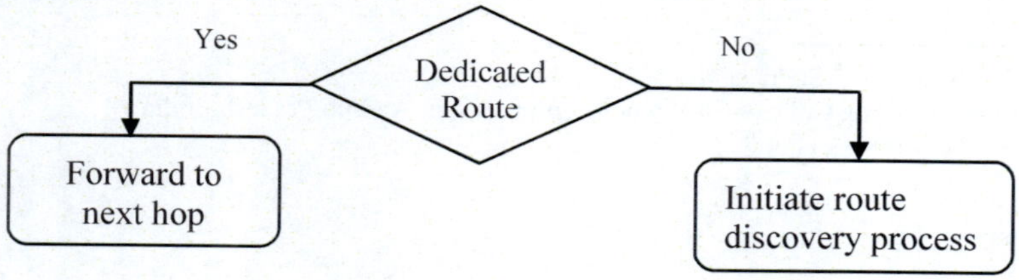

Route discovery process starts with the creating RRQ packet from the source node.

2 Performance Parameters, Security Attacks and Algorithm

Analyzing the performance of the network is very important while deciding the kind of protocol used and kind of security algorithm used for the implementation of original network. In this section discussed the common parameters used and different security attacks. Figure 1 shows the different performance parameters to be considered for analysis and Fig. 2 shows different security attacks to be considered.

Most of the common performance parameters are

See Fig. 1.

Security Attacks

See Fig. 2.

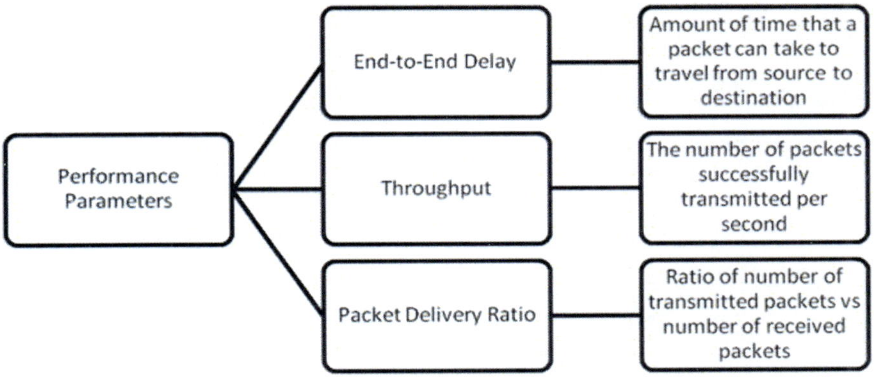

Fig. 1 Performance parameters

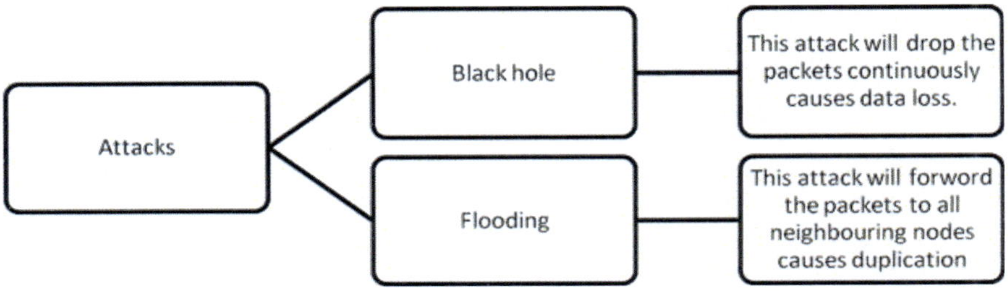

Fig. 2 Security attacks

3 Simulation Results and Comparative Analysis

The work starts from implementing a MANET scenario with N number of nodes, by defining start node agent as TCP0 and end node agent is sink, and employs file transfer protocol (FTP) for exchanging packets each other [4]. This is shown in Fig. 3.

While the TCL script is executed in the network simulator, two files are generated in the background. They are NAM file and trace file [5]. NAM file describes the network animation which shows the progression of the packets through the network. The simulation results are stored in trace file. This file contains information like no. of nodes, links between nodes, packet traces, etc. [6]. Figure 4 shows the trace file output.

3.1 *Performance Parameters Using AODV Protocol*

The performance parameters are calculated for various scenarios for with and without mobility. The three types of performance parameters like E2E delay, packet delivery ratio (PDR), and throughput have been considered and calculated with the

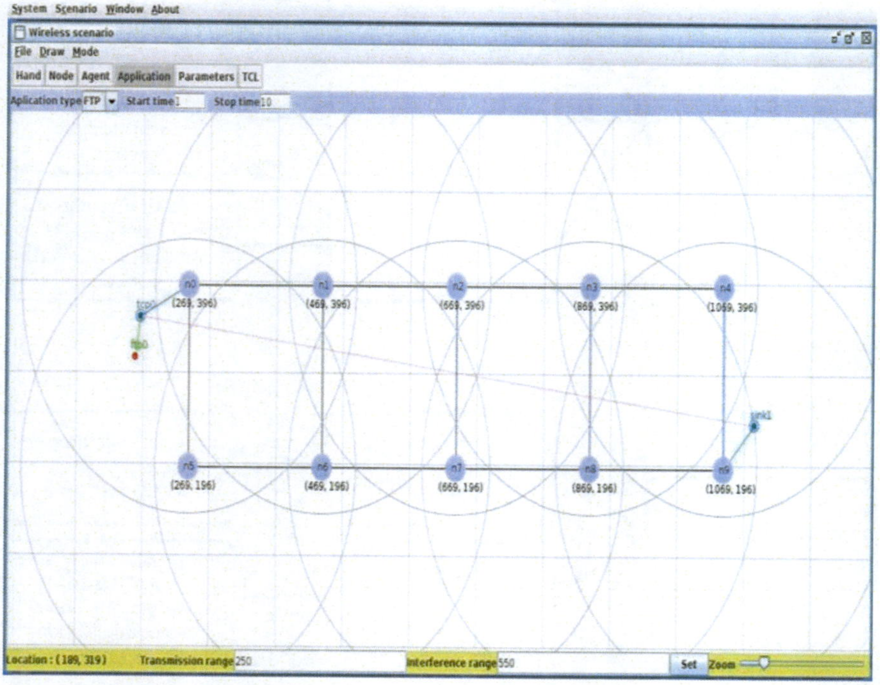

Fig. 3 Network scenario for ten nodes

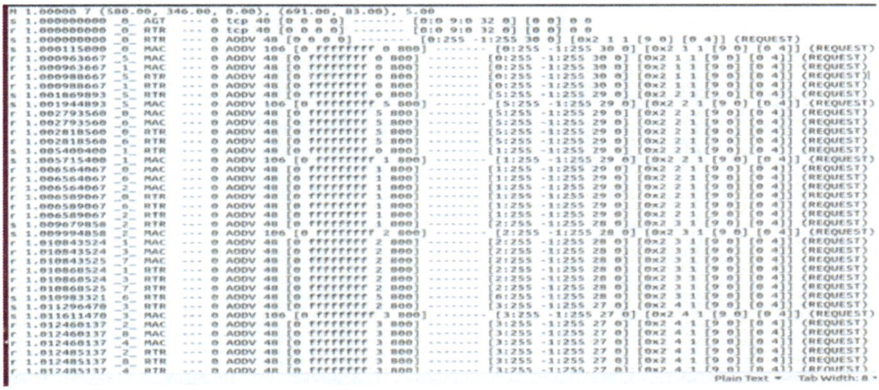

Fig. 4 Trace file

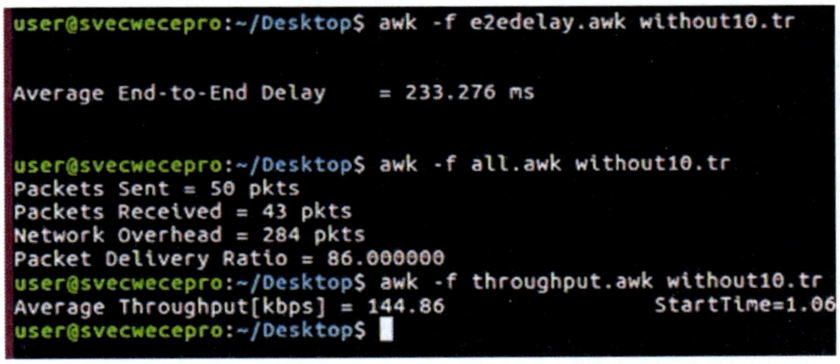

Fig. 5 Parameters without mobility

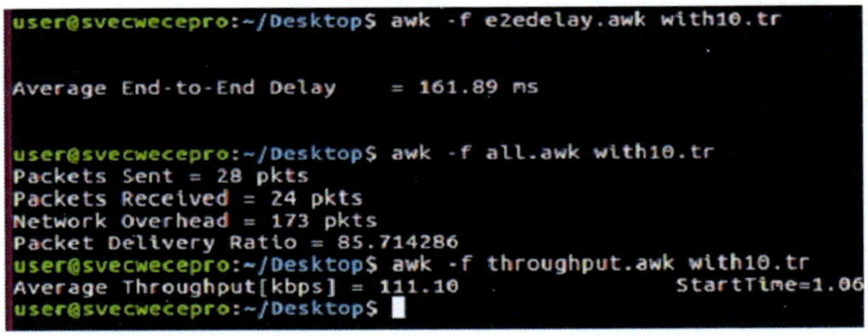

Fig. 6 Parameters with mobility

help of AWK files which are developed in C++ language for manipulating data and generating reports [7]. AWK file is used for processing test-based data.

Figures 5 and 6 show the performance parameters with and without mobility. The network scenario consists of one mobile node out of ten nodes. Hence, the average end-to-end delay decreases.

3.2 Attacks in MANET

The network scenario has been created by implementing a malicious node in the network [8]. The malicious node makes the packet dropping and does not allow the packets to reach destination [9].

By observing Figs. 7 and 8, data packets may drop continuously due to black hole

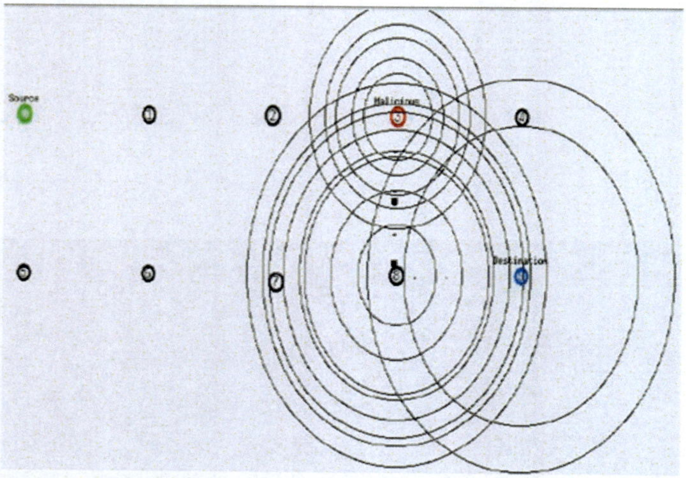

Fig. 7 Packet dropping due to black hole attack

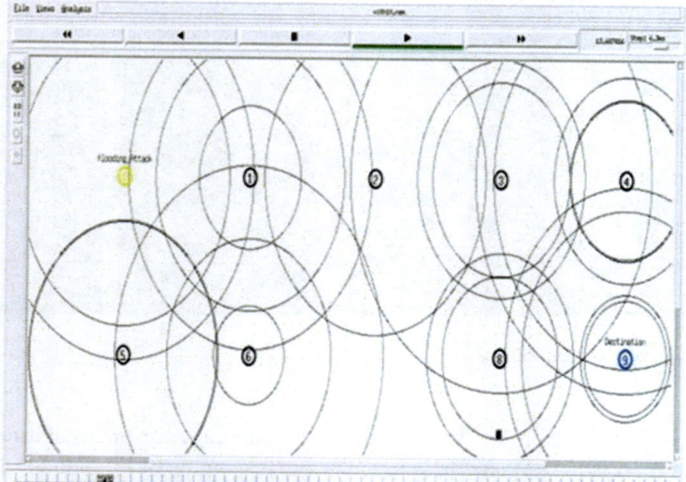

Fig. 8 Packet dropping due to flooding

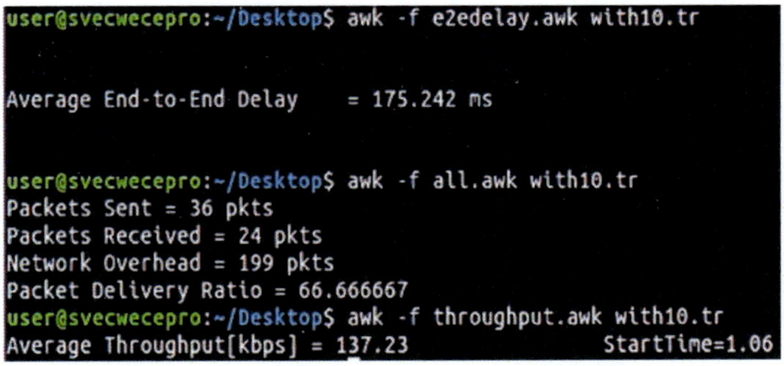

```
user@svecwecepro:~/Desktop$ awk -f e2edelay.awk with10.tr

Average End-to-End Delay    = 175.242 ms

user@svecwecepro:~/Desktop$ awk -f all.awk with10.tr
Packets Sent = 36 pkts
Packets Received = 24 pkts
Network Overhead = 199 pkts
Packet Delivery Ratio = 66.666667
user@svecwecepro:~/Desktop$ awk -f throughput.awk with10.tr
Average Throughput[kbps] = 137.23               StartTime=1.06
```

Fig. 9 Performance due to black hole attack

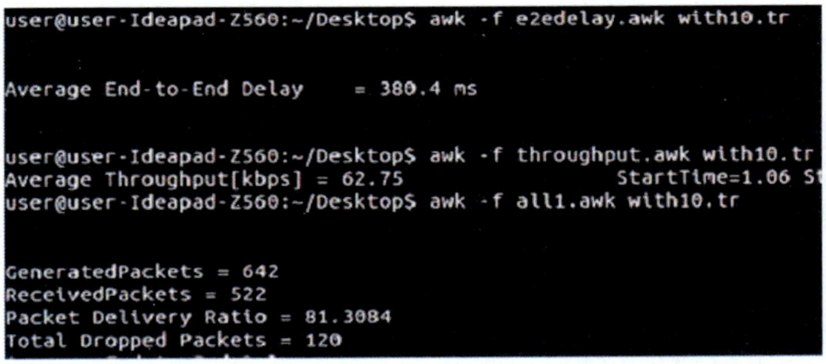

```
user@user-Ideapad-Z560:~/Desktop$ awk -f e2edelay.awk with10.tr

Average End-to-End Delay    = 380.4 ms

user@user-Ideapad-Z560:~/Desktop$ awk -f throughput.awk with10.tr
Average Throughput[kbps] = 62.75               StartTime=1.06 St
user@user-Ideapad-Z560:~/Desktop$ awk -f all1.awk with10.tr

GeneratedPackets = 642
ReceivedPackets = 522
Packet Delivery Ratio = 81.3084
Total Dropped Packets = 120
```

Fig. 10 Performance due to flooding attack

attack, and packets may reach to all the neighboring nodes due to flooding which causes data duplication.

Figures 9 and 10 show the performance parameters output from trace file, and by observation, we can say that the performance of network is degraded due to attackers.

The overall comparative analysis is shown in Fig. 11.

3.3 SHA-3 Keccak Algorithm

Securing hash algorithm (SHA-3), a Keccak security algorithm, has been used as a security algorithm to decrease the number of packets dropping in a MANET [10]. The trust-based authentication code is used as an HMAC code to achieve data integrity and authentication. The threshold-based algorithm is used to identify the malicious node and change its path in the short duration of time. Figure 12 shows the results after applying security algorithm.

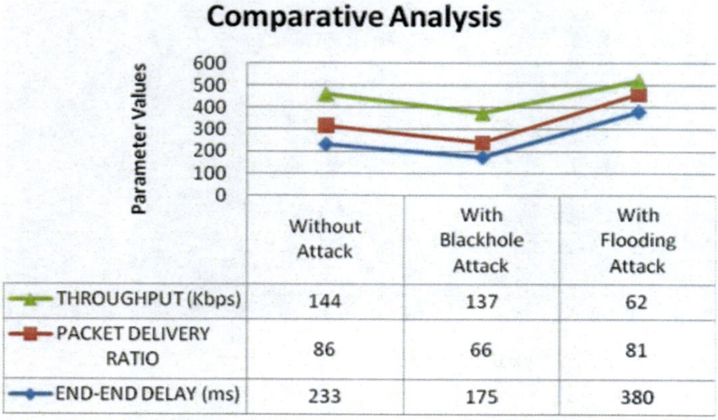

Fig. 11 Comparative analysis due to black hole and flooding attacks

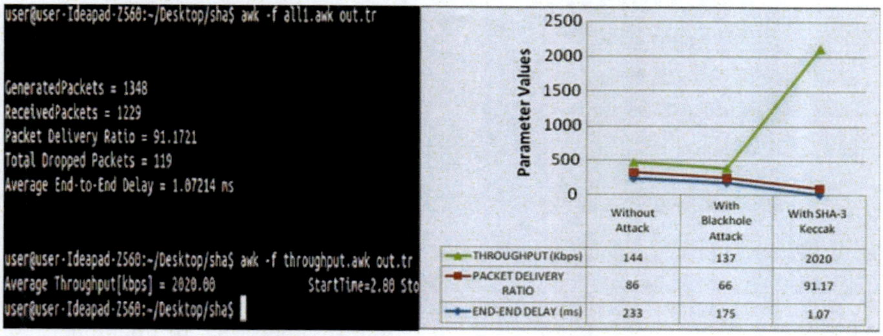

Fig. 12 SHA-3 results for black hole attacks

4 Conclusion and Future Scope

The impact analysis of black hole, flooding attack, and enhancement using security algorithm deals with all the network parameters and how they change for different scenarios. The network parameters like end-to-end delay, throughput, and packet delivery ratio change in accordance to the attack employed to the MANET. But, the main objective is to decrease the number of packets dropped when they are travelling from source node to destination node. To provide reliable communication after employing the attacks, SHA-3 Keccak algorithm has been employed to improve the performance of network parameters.

4.1 Future Scope

MANETs are more sensitive to attacks. So, to provide security enhancement, efficient security algorithm must be used. With the help of SHA-3 algorithm, some of the parameters' performance has been increased for black hole attack, but this algorithm cannot able to improve the performance of parameters for flooding attack. So, the researchers have to concentrate on this issue and try to investigate other security algorithms.

References

1. Moudni H, Er-rouidi M, Mouncif H, El Hadadi B (2016) Performance analysis of AODV routing protocol in MANET under the influence of routing attacks. In: International conference on electrical and information technologies (ICEIT), tangiers, pp 536–542
2. Ramya P, SairamVamsi T (2018) Impact analysis of blackhole, flooding, and grayhole attacks and security enhancements in mobile ad hoc networks using SHA3 algorithm. In: Anguera J, Satapathy S, Bhateja V, Sunitha K (eds) Microelectronics, electromagnetics and telecommunications. Lecture notes in electrical engineering, vol 471. Springer, Singapore; Fahmy AHH, Bahaa-Eldin A (2015) Agent-based trusted on-demand routing protocol for mobile ad-hoc networks. Wirel Netw 21(2):467–483
3. Mane Dadaso, Gothwal Deepali (2013) Improved security for attacks in MANET using AODV. Int J Innov Eng Technol (IJIET) 2(3):37–44
4. Mahmoud MMEA, Shen XS (2014) Secure routing protocols. Security for multi-hop wireless networks. Springer, New York, pp 63–93
5. Zhao Z, Hu H, Ahn GJ, Wu R (2010) Risk aware mitigation for MANET routing attacks. IEEE Trans Dependable Secur Comput 9(2):250–260
6. Tseng F-H, Chou L-D, Chao H-C (2011) A survey of black hole attacks in wireless mobile ad hoc networks. Hum-Centric Comput Inf Sci 1(1):1–16
7. Yi P et al (2005) A new routing attack in mobile ad hoc networks. Int'l J Info Tech 11(2)
8. Palanisamy V, Annadurai P (2009) Impact of rushing attack on multicast in mobile ad hoc network. Int J Comput Sci Inf Secur 4(1–2)
9. Shanmuganathan V, Anand T (2012) A survey on gray hole attack in MANET. IRACST–Int J Comput Netw Wirel Commun (IJCNWC) 2250–3501
10. Mahajan V, Natu M, Sethi A (2008) Analysis of wormhole intrusion attacks in MANETs. In: IEEE military communications conference, pp 1–7

Reconfigurable Rectangular Microstrip Patch Antenna for S-Band Applications

LalBabu Prasad, B. Ramesh, and K. P. Vinay

Abstract A novel reconfigurable rectangular microstrip patch antenna (RMPA) is presented for S-band applications which can be switchable at two different radiation patterns. The designed antenna consists of eight parasitic elements which are connected to the rectangular patch using eight radio frequency micro-electromechanical system (RF MEMS) switches, and by controlling the switches, antenna has omni-directional radiation pattern in OFF mode and monopole-like radiation pattern in ON mode. In OFF mode, the designed antenna resonates at 2.39 GHz in industrial, scientific and medical (ISM) band, and in ON mode, it resonates at 2.66 and 3.5 GHz. In both the modes, the return loss $S_{11} < -10$ dB and gain is more than 6 dBi. The antenna is simulated using CST MW studio and the simulated results are presented.

Keywords RMPA · RF micro-electromechanical system switches (RF MEMS) · Reconfigurable · ISM band · Parasitic elements

1 Introduction

Nowadays proficient and secure wireless communication demand leads to a design of various types of antennas having multifunctionality features with compact size and low cost. Microstrip patch antenna has attracted many researchers in wireless communications due to its low profile, simple fabrications and lightweight properties [1]. In the present trend, portable wireless devices operate at multiple frequency bands for different applications have increased the demand for reconfigurable microstrip antennas in the modern wireless communication systems. Reconfigurable feature in

L. Prasad (✉) · B. Ramesh · K. P. Vinay
ECE Department, Raghu Engineering College, Visakhapatnam 531162, India
e-mail: gulablal@gmail.com

B. Ramesh
e-mail: rameshboddu11@gmail.com

K. P. Vinay
e-mail: dr.kpvinay@ieee.org

P. S. R. Chowdary et al. (eds.), *Microelectronics, Electromagnetics and Telecommunications*, Lecture Notes in Electrical Engineering 655,
https://doi.org/10.1007/978-981-15-3828-5_13

microstrip antenna enhances the state of art in satellite and remote sensing applications. The reconfigurable microstrip antennas are able to adopt a certain frequency and radiation patterns in controlled manner which are useful for secure wireless communication network. The fundamental resonating frequency of the reconfigurable antenna can be change through mechanical, electrical and by other means [2, 3]. To achieve the reconfigurable property through electrical method is to vary the surface current distribution with altering the comparable regional structure, the effective radiated electrical length, and feeding position electrically. Similarly, to achieve a reconfigurable feature is by using lumped elements [4], optical switches [5], PIN diodes [6–8], RF micro-electromechanical switches [9–11] and tunable BST capacitors [12].

In this work, a reconfigurable rectangular microstrip patch antenna is proposed which can shift its operating frequency by controlling the RFMEMS switches. The RMPA antenna is designed to operate at ISM band and radio location applications in S-band region. To obtain reconfigurable feature, eight parasitic elements are switched with the RMPA and to understand the operation of the parasitic elements, we considered metal strip and vacuum gap instead of ON and OFF switches in the simulation design.

2 Antenna Design and Configuration

The reconfigurable rectangular microstrip patch antenna is designed using felt fabric material as a substrate of 3 mm thickness whose permittivity is 1.3 and loss tangent is 0.044. The conventional RMPA structure is modified by placing eight parasitic elements which are connected through eight switches as four on either side of the patch. The size of the dielectric substrate and the ground plane are $100 \times 100\ \text{mm}^2$ and the area of the patch is $46.5 \times 39\ \text{mm}^2$. Probe feeding method is used to excite the designed antenna. The design structure of the proposed work is depicted in Fig. 1 and their respective dimensions are depicted in Table 1.

There are eight parasitic stubs of size $2.25\ \text{mm} \times 6\ \text{mm}$ is embedded on both sides of the patch with four each are connected to the patch through switches of length 2.25 mm and width 1 mm along with these eight vias are connected as one via to each parasitic stub through ground plane in order to improve the radiation pattern of the antenna. The vias act as the shunt inductor and they will result in enhancing broadside radiation pattern at low frequencies. The vias are designed using copper and the diameter of the vias is 0.9375 mm. By using the switches antenna can be made to operate in two modes, i.e., ON and OFF mode as depicted in Fig. 2. In the ON mode, a copper strip is used instead of RF MEMS switch to connect the parasitic stub along vias with the radiating patch, and in the OFF mode, the vacuum gap is used to disconnect the stub with the patch.

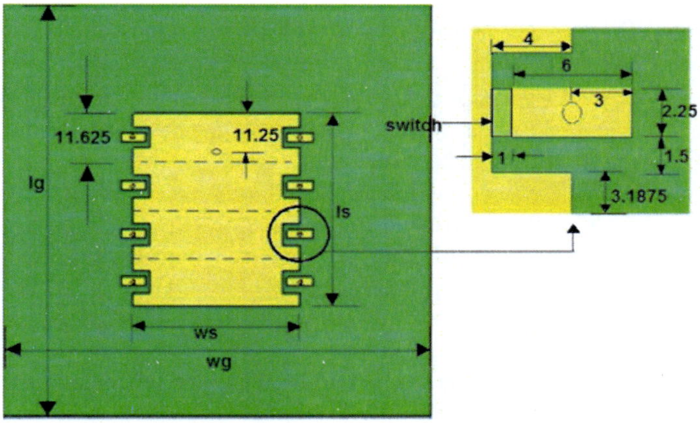

Fig. 1 Geometry of the designed RMPA

Table 1 Dimensions of the designed RMPA are as follows

Antenna parameters	Notation	Values (mm)
Length of the ground plane and substrate	L_g	100
Width of the ground plane and substrate	W_g	100
Length of patch	L_s	46.5
Width of patch	W_s	39
Diameter of vias	V_d	0.9375

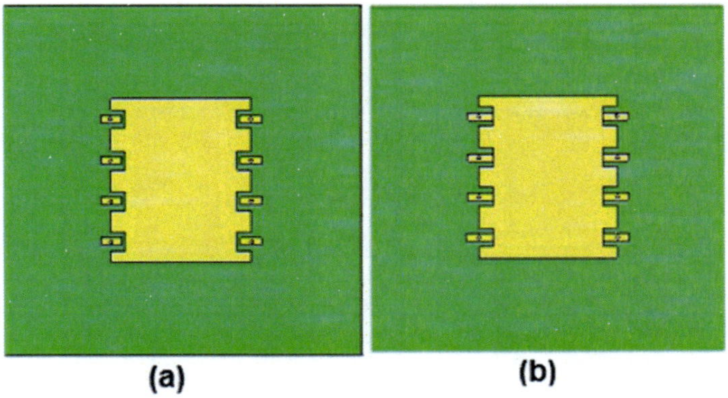

Fig. 2 Mode of operation the designed antenna, **a** OFF, **b** ON

3 Simulation Outcomes and Analysis

With the help of CST microwave studio tool, the performance of the designed antenna was simulated. Figure 3 depicts the S11 parameter of the designed antenna when all the parasitic stubs are separated by vacuum gaps from the RMPA, i.e., OFF mode, it resonant at 2.39 GHz frequency in the ISM band with return loss of −13.6 dB and impedance bandwidth of 111 MHz.

Similarly, when all the parasitic stubs are connected to the RMPA through copper strip, i.e., ON mode, the antenna resonates at 2.6 and 3.5 GHz with S11 is equal to −15.1 dB, −13.9 db and impedance bandwidth is 211 MHz, 227.4 MHz, respectively, which is depicted in Fig. 4.

From Fig. 5, the surface current distribution when the antenna is resonating at 2.39 GHz in the OFF mode is uniformly distributed throughout the radiating patch and it is found to be high near the feeding location that is the fact that the radiating

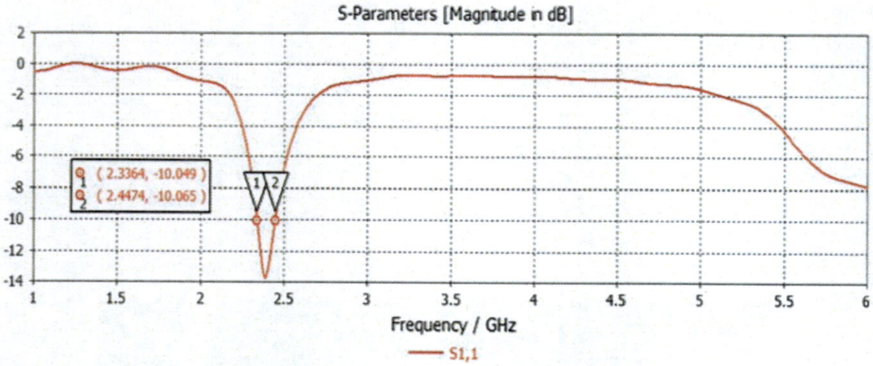

Fig. 3 S11 plot of the designed antenna in OFF mode

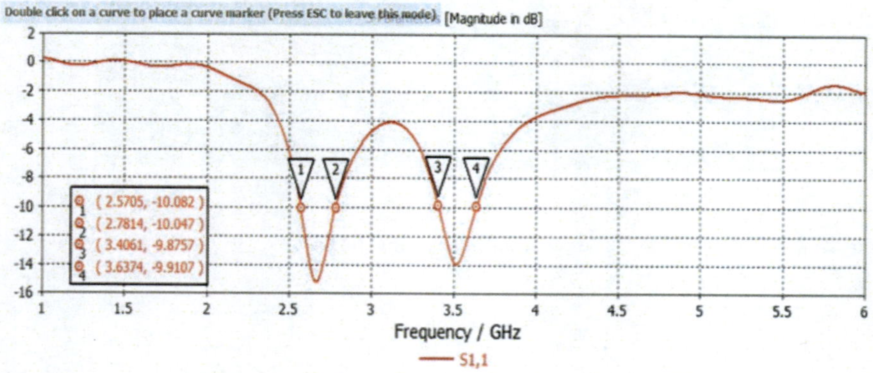

Fig. 4 S11 plot of the designed antenna in ON mode

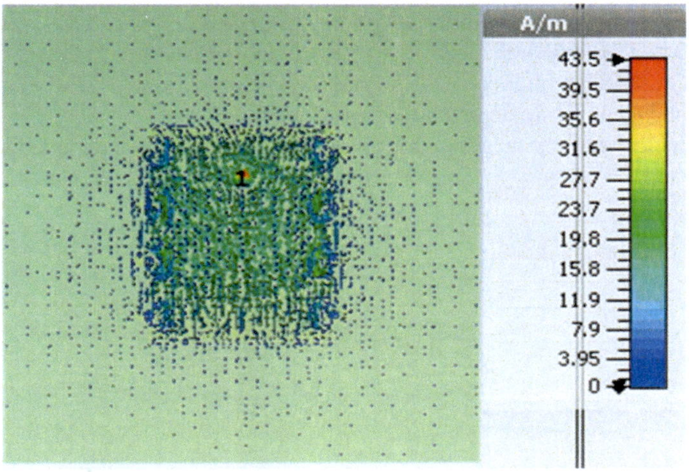

Fig. 5 Surface current distribution plot of the designed antenna in the OFF mode

patch surface current dominantly affecting and controlling the resonance frequency and radiating mechanism of the designed antenna which results in the omnidirectional radiation pattern with gain of 5.18 dB and directivity of 8.52 dBi. The 2D radiation pattern of the designed antenna in the OFF mode is depicted in Fig. 6.

In the ON mode, i.e., when all the parasitic stubs along with vias are connected to the RMPA through switch, the antenna resonates at two different frequencies, that is, 2.6 and 3.5 GHz. At 2.6 GHz, the surface current is mainly distributed on the parasitic stubs which are present on either sides of the radiating patch, which suggest that the parasitic stub portion of the radiating patch is act as a part of the radiating structure which results in the bidirectional radiation pattern at 2.6 GHz resonating frequency with gain 4.88 dB and directivity of 6.75 dBi. At 3.5 GHz, the surface current is distributed on both the radiating patch and the parasitic stubs which results in the new radiation pattern and it is combination of the radiation patterns at 2.39

Fig. 6 2D radiation pattern of the designed antenna in the OFF mode

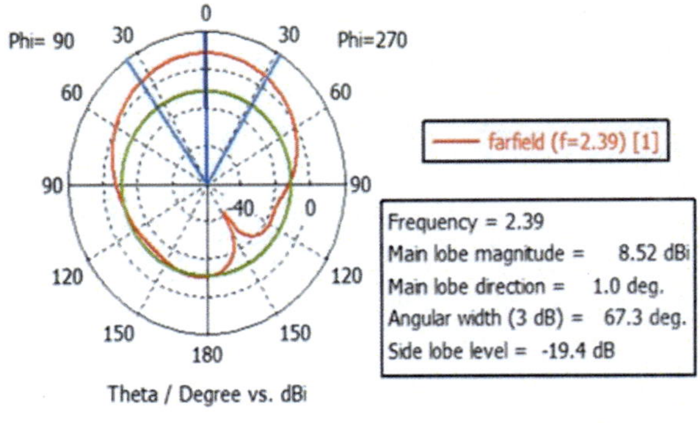

and 2.6 GHz. The gain and directivity of the designed antenna at 3.5 GHz frequency is 6.47 dB and 10.1 dBi, respectively, which is more than both gain and directivity at 2.39 and 2.6 GHz frequencies. The surface current and 2D radiation patterns of the designed antenna are depicted in Figs. 7 and 8.

The performance characteristics of the designed antenna are depicted in Table 2.

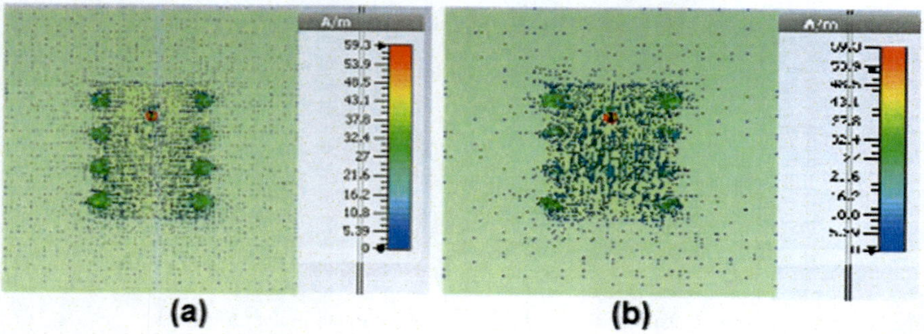

Fig. 7 Surface current distribution plot of the designed antenna in the ON mode, **a** 2.6 GHz, **b** 3.5 GHz

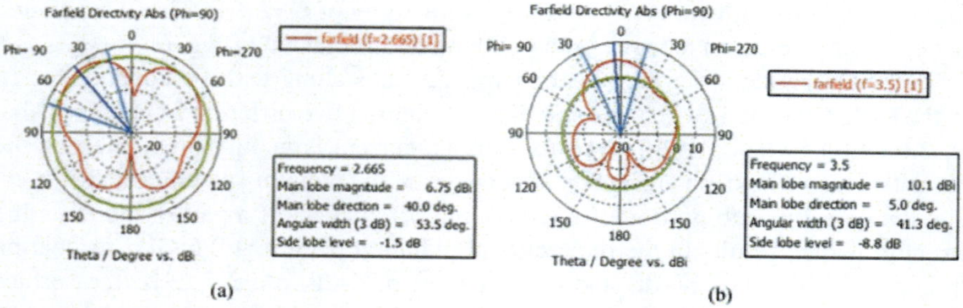

Fig. 8 2D radiation pattern of the designed antenna in the ON mode, **a** 2.6 GHz, **b** 3.5 GHz

Table 2 Performance characteristics of the designed antenna simulated results

Mode of operation	Resonant frequency (GHz)	S_{11} (dB)	VSWR	Gain (dB)	Directivity (dBi)	Bandwidth (MHz)
OFF	2.39	−13.6	1.53	5.18	8.52	111.0
ON	2.6	−15.1	1.42	4.88	6.75	211.0
	3.5	−13.9	1.5	6.47	10.1	227.4

4 Conclusion

A reconfigurable rectangular microstrip patch antenna which can operate in two different modes at three different frequencies is designed using CST MW studio software. The designed antenna is operating in ON mode and OFF mode by controlling the RF MEMS switches. In the simulation part, copper strips are used instead of RF MEMS switches. The antenna operates at 2.39 GHz with S11 < -10 dB, directivity of 8.52 dBi and impedance bandwidth of 111 MHz by yielding an omnidirectional radiation pattern in OFF mode and bidirectional radiation pattern at 2.6 GHz and 3.5 GHz in ON mode along with S11 < -10 dB, directivity >6 dBi and impedance bandwidth of 211 MHz, 227.4 MHz, respectively. The proposed antenna is suitable for ISM band and radiolocation applications in the wireless communication of S-band. For future simulation and testing process, different types of switches like PIN diode, optical switches along DGS are used to understand the performance of the designed antenna.

References

1. Balanis CA (2005) Antenna theory, analysis and design, 3rd edn, Wiley, New York
2. Wu W, Wang BZ, Sun S (2005) Pattern reconfigurable microstrip patch antenna. J Electromagn Waves Appl 19(1):107–113
3. Majid HA, Abd Rahim MK, Hamid MR, Ismail MF (2014) Frequency reconfigurable microstrip patch-slot antenna with directional radiation pattern. Prog Electromagn Res 144:319–328
4. Shah SA, Khan MF, Ullah S, Flint JA (2014) Design of a multi-band frequency reconfigurable planar monopole antenna using truncated ground plane for Wi-Fi, WLAN and WiMAX applications. In: 2014 international conference on open source systems and technologies (ICOSST). IEEE, pp 151–155
5. Roach TL, Huff GH, Bernhard JT (2007) On the applications for a radiation reconfigurable antenna. In: 2nd NASA/ESA conference on adaptive hardware and systems, Edinburgh, pp 7–13
6. Shah IA, Hayat S, Basir A, Zada M, Shah SAA, Ullah S (2018) Design and analysis of a hexa-band frequency reconfigurable antenna for wireless communication. Int J Electron Commun. doi:https://doi.org/10.1016/j.aeue.2018.10.012
7. Madhav BT, Rajiya S, Nadh BP, Kumar MS (2018) Frequency reconfigurable monopole antenna with DGS for ISM band applications. J Electr Eng 69(4):293–299
8. Anand H, Kumar A (2016) Design of frequency-reconfigurable microstrip patch antenna. Indian J Sci Technol 9(22). https://doi.org/10.17485/ijst/2016/v9i22/92812
9. Deng Z, Yao Y (2011) Ka band frequency reconfigurable microstrip antenna based on MEMS technology. In: Electrical power systems and computers, vol 99, LNEE, pp 535–541
10. Al-alaa MA, Elsadek HA, Abdallah EA, Hashish EA (2014) Pattern and frequency reconfigurable monopole disc antenna using pin diodes and mems switches. Microw Opt Technol Lett 56(1):187–195
11. Xu Y, Tian Y, Zhang B, Duan J, Yan L (2018) A novel RF MEMS switch on frequency reconfigurable antenna application. Microsystem Technol 24:1–9. https://doi.org/10.1007/s00542-018-3863-9
12. Bronckers LA, Roc'h A, Smolders AB (2019) A new design method for frequency-reconfigurable antennas using multiple tuning components. IEEE Trans Antennas Propag 1–1. https://doi.org/10.1109/tap.2019.2930204

Design and Development of Hindrance Application Using Vocal and Quick Responsible Code for Railway

Vadamodula Prasad and Kerru Jeevan Vamsi

Abstract According to the fast-growing world, people are completely dependent of technology and mobile applications. In this regard, a hindrance is required at all levels to provide the information about railways through the specific application. This application provides limited services related to railways like train details, passenger name record (PNR) status, live status, train routes, and ticket bookings. This application holds the basic vocal model using Google assistance and functions on analog involvement of the user. The authenticity required for this application is done through quick responsible (QR) code. Through this application, people can perform the operations as mentioned above. The user and detailed information of tickets were purchased, and inquiry is gathered through their Aadhaar card #number submission through QR. The output of this application would be in the form of voice generated by the application. This application is built using Android Studio.

Keywords Quick response code · Android Studio · Authenticity · Vocal model and hindrance

1 Introduction

Speech recognition (SR) is currently executed in different tongues around creation. Language is the greatest dispatch between humans and the device. The communication that is verified with the communication devices is changed from audio signal to a set of words in speech recognition. Conventional form of the name will act as output or implemented by the combination to articulate into noises that prove voice-to-voice statement.

V. Prasad (✉) · K. J. Vamsi
Department of Computer Science and Engineering, GMR Institute of Technology, Rajam, Andhra Pradesh, India
e-mail: prasad.v@gmrit.edu.in; prasad_v542@yahoo.co.in

K. J. Vamsi
e-mail: jeevanvamsi73@gmail.com

P. S. R. Chowdary et al. (eds.), *Microelectronics, Electromagnetics and Telecommunications*, Lecture Notes in Electrical Engineering 655,
https://doi.org/10.1007/978-981-15-3828-5_14

Survey [1] on voice recognition for the past 60 years briefed the well-known approaches used in different stages of voice credit structure with an objective to summarize and compare methods used in voice credit. The solicited [2] *works represent the* reduction of error rate with convolutional neural networks when compared with presently available techniques for speech recognition. This study [3] summarized the well-organized language credit system by means of changed methods like Mel-frequency cepstral coefficients (MFCC), vector quantization (VQ), and hidden Markov model (HMM). The experimentation [4] is to review the feature extraction methods which are used in language system. Linear predictive coding (LPC) [5], Mel-frequency cepstral coefficient (MFCC), RASTA filtering, and probabilistic linear discriminant analysis (PLDA) [6] explain the language-based speech verification system by using two main phases: feature excerption point and feature corresponding point. Matching phase [7, 8] uses time-wrapping algorithm to compute the wrapping distance between two sequences. This system made a safe arrangement that arranges the speech credit for a natural language by merging the digital and mathematical knowledge using MFCC.

Quick Response Code will be the symbol fixed for medium barcode, and this acts as mechanism—clear visual context that holds information describing the product to which it is involved. *Google Cloud Speech API* designs audio to script by involving influential neural network reproductions. These support in translating the script of users dictating to a feature recording device and record the audio files within various methods (Fig. 1).

Text to Speech: Text to speech [9] is an engine in android that converts the text written to speech. A text to speech instance can only be used to synthesize text on its completion of its initialization. Railway API serves a large and efficient group

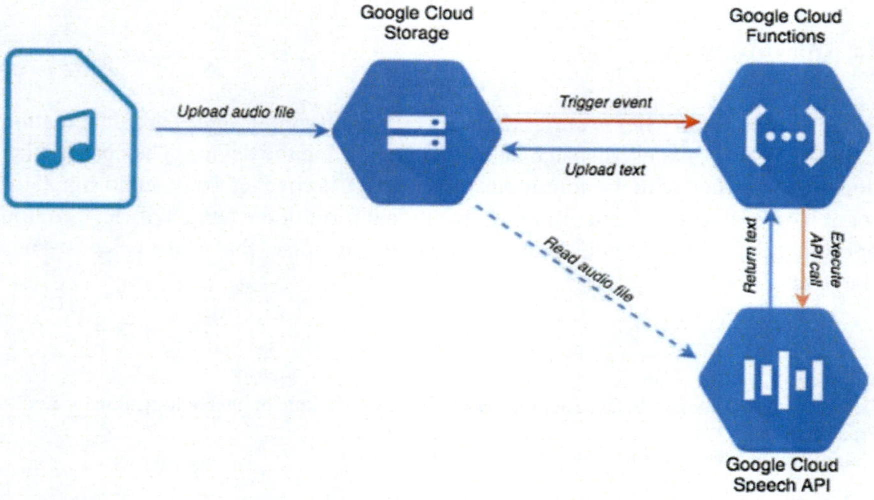

Fig. 1 Google cloud application (GCS) program interface

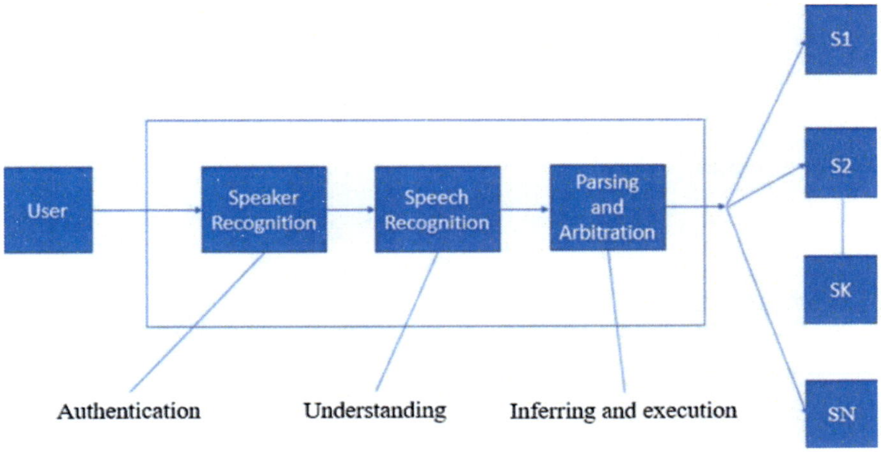

Fig. 2 Process of speech recognition

of trains' and stations' information along with structures like live position, PNR status, arrivals, and departures, and trains among stations. This information is made accessible through an influential and peaceful API in an organized JavaScript Object Notation (JSON) exposed library setup. The procedure is shortened for obtaining and opening railways data and helps power retrieval. *Strategy of the Application* is an explanation of the construction of the software to be executed, and *Architectural Plan* workings involve the construction of the scanner, commands, audio, search, and management. *Scanner* designs are used to scan and decode the human-readable data, with the assistance of an imaging device, like a camera or a scanner.

The methodology represents a unique key component of the present system which distinguishes the instructions assigned by the user as shown in Fig. 2. The model is answerable for identifying the instructions, and further, the *Search* mechanisms yield the input as the demand from the instruction component and regain the suitable result from the record. It bounces back to the display module and the speech module. Later, the *Language* module is used to carry the result in the form of the speech by means of Microsoft speech control which earns input form the search module. *Supervision* module cares employee's data and the instructions to the system as displayed.

2 System Design

See Fig. 3.

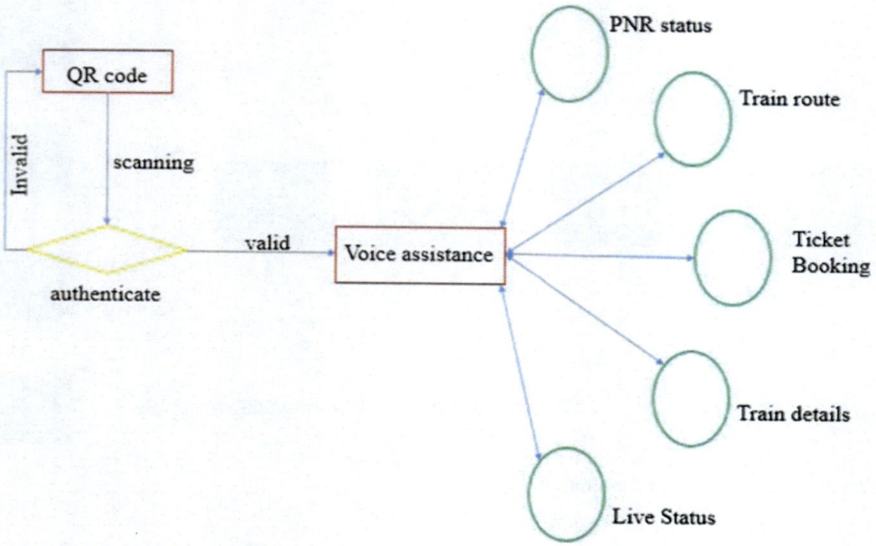

Fig. 3 Implementation modules and its design

3 Implementation Methodology

3.1 Data Encryption Method

In this data encoding, there are four modes of encrypting the data; these are kanji, numeric, alphanumeric, and bytes. It selects the method of encryption, and in data encoding process, it encrypts the data into 8 bits; long after these, the data will generate the error code words by the process called Reed–Solomon method of encryption. QR code reads both the information encryption words and the fault encryption confrontations and detects whether it is correct data or not, then after it arranges in the correct order in structure final message, then after it arranges in QR code matrix form and performs data masking and formats the information if necessary and it adds pixels to classify the error improvement stage and cover design being used in the QR encryption. Thus, QR code is encrypted (Figs. 4 and 5).

3.2 Data Decryption Method

In data decoding, the identifying units will identify dim and bright units by the way of an array of "0" and "1" bits by tracing and receiving a copy of the representation. This abstract the set-up data by decoding the arrangement data and releases the hiding design by putting the error improvement on the arrangement data units as necessary.

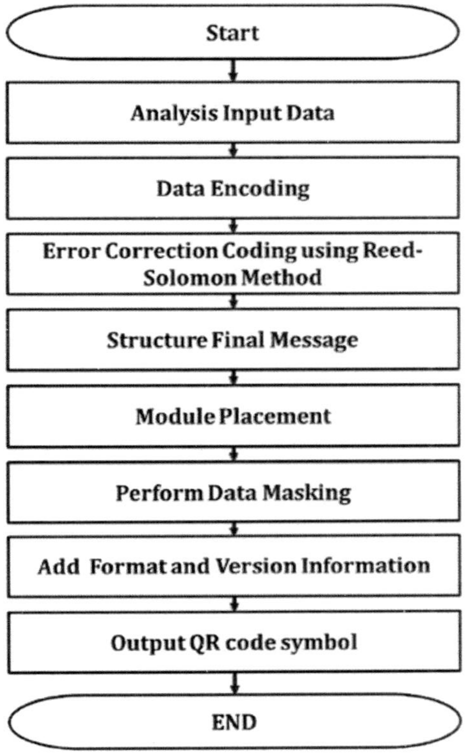

Fig. 4 QR code encoding

It acquires a cover design, and the form data will decode the form data area and determines the type of the QR encryption representation; it releases covering XOR, the programming area bit design with the cover design whose orientation has stayed removed from the setup data and bring back the information by understanding the representation fonts and in decode information code words it divides the data code words into parts rendering to the style pointers and character sum.

3.3 Audio-Phonetic Method

The audio-phonetic method is built on the model of audio phonetics. The phonetic elements are considered by usual belongings that are surrounded in the language sign or its field. Even though the audio belongings of phonetic elements are very adjustable; together with the talker and by the near phonetic units, it is expected that the instructions leading the unpredictability are traditional onward and can be willingly educated and useful in real-world circumstances.

Fig. 5 QR code decoding

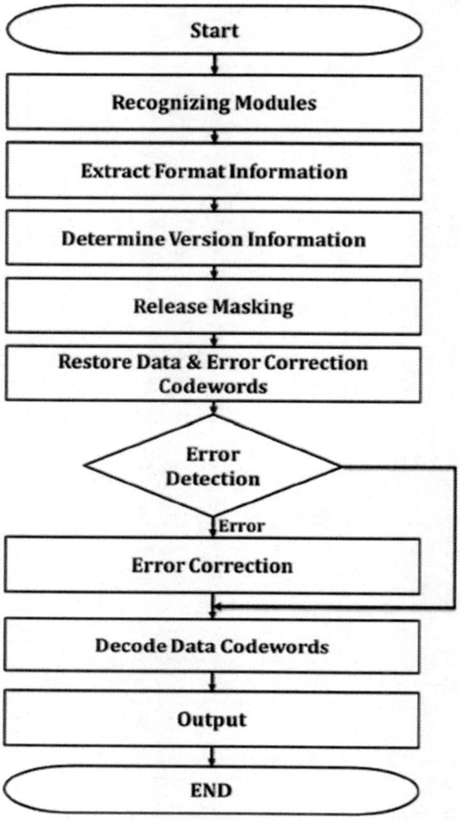

4 Resultant Application

Once the authentication is over by using the QR code scanner, it opens the voice assistance. In this paper, the user must give commands in the form of voice. On recognition of voice, a result is returned by combining upper- and lower-case versions of the recognized tokens.

The result may be empty; if some speech was detected, then it did not match the specified grammar. At the other end, if we give number as an input, then it simply returns number in the form of integer (e.g., 55 or 10); but it does not return string by matching the data in the knowledge base.

5 Conclusion

Language credit is an interesting problematic to an agreement with the review provided toward to show the progress in technology in the preceding ages. SR is unique of the greatest participating areas of engine cleverness; meanwhile, people do a regular action to SR. It is involved experts as significant correction and takes shaped a technical influence on culture in addition to, is likely to decoration more in part of people mechanism communication. In the detailed survey on the concept of voice/speech recognition system, it had provided wide information regarding the various methods/technologies available for ASR. This usage and the application of ASR related projects are nowadays more in public places which provide a great ease to the society and for this benefit. This study and development of this project had given me an idea which made my thought process to design an ASR with the help of JSON libraries and Android Studio.

References

1. Anusuya MA et al (2009) Speech recognition by machine: a review. Int J Comput Sci Inf Secur (IJCSIS) 6
2. Swamy S et al (2013) An efficient speech recognition system. Comput Sci Eng Int J (CSEIJ) 3(4)
3. Bhavesh SR et al (2015) A survey on speech recognition technique. Int J Res Appl Sci Eng Technol (IJRASET) 3(10):2321–9653
4. Bassam AQ, Al-Qatab et al (2010) Arabic speech recognition using Hidden Markov Model Toolkit(HTK). Int Symp Inf Technol
5. Junghare H et al (2016) Efficient methods and implementation of automatic speech recognition system. Int J Comput Sci Inf Technol 7(3):1116–1120
6. Arora SJ et al (2012) Automatic speech recognition: a review. Int J Comput Appl 60(9):34–44
7. Kavitha R et al (2014) Speech based voice recognition system for natural language processing. Int J Comput Sci Inf Technol 5(4):5301–530
8. Mansour AH et al (2015) Voice recognition using back propagation algorithm in neural networks. Int J Comput Trends Technol 23(3):132
9. Web Link. https://WWW.yosnalab.com/article?project=railway-enquiry-voice-recoganization

Estimation from Censored Sample: Size-Biased Lomax Distribution

A. Naga Durgamamba and Kanti Sahu

Abstract In this work, the scale parameter is derived with famous shape parameter from the censored sample victimization of the maximum likelihood technique for the size-biased Lomax distribution (SBLD). The predicting equations are changed to urge less-complicated and economical predictors. Two ways of modification are steered. The results are given.

Keywords Likelihood function · Size-biased Lomax distribution · Order statistics · Censored sample

1 Introduction

The probability density and distribution functions of the SBLD is given by

$$f(t) = \frac{\lambda(\lambda - 1)}{\theta}\frac{t}{\theta}\left(1 + \frac{t}{\theta}\right)^{-(\lambda+1)} \quad \text{for} \quad t \geq 0, \lambda > 1, \theta > 0 \tag{1}$$

$$F(t) = 1 - \left(1 + \frac{\lambda t}{\theta}\right)\left(1 + \frac{t}{\theta}\right)^{-\lambda}; \quad t \geq 0, \lambda > 1, \theta > 0 \tag{2}$$

where λ and θ are shape and scale parameters of the distribution.

The maximum likelihood technique of estimation of the parameters λ and θ from censored sample is established once, whereas one of the parameters is thought and the opposite is unknown. Some modifications to maximum likelihood method from censored sample are suggested as we solve estimating equations by numerical iterative techniques. Such studies primarily based on modified maximum likelihood estimation which includes from [1–11].

A. Naga Durgamamba (✉)
Raghu Institute of Technology, Dakamarri, Visakhapatnam 531162, India
e-mail: durgamamba@gmail.com

K. Sahu
DKS Post Graduate Institute and Research Center, Raipur, Chhattisgarh 492001, India

P. S. R. Chowdary et al. (eds.), *Microelectronics, Electromagnetics and Telecommunications*, Lecture Notes in Electrical Engineering 655,
https://doi.org/10.1007/978-981-15-3828-5_15

In this paper, we used two methods to estimate the scale parameter in size-biased Lomax distribution from censored sample. We present these in the following Sections. Since the method of estimation involves loads of numerical computations, all such results are given within the variety of numerical tables toward the tip of the paper with applicable identification and labels.

1.1 *Modified ML Estimation of a Scale Parameter from Censored Samples*

Censoring a given sample in life testing experiments now and then becomes essential to avoid delay and value of experimentation. One of the major, not unusual schemes of censoring may be a failure-censored sample, whereby prearranged n things are a place to checking out existence, and additionally, the test is terminated as rapidly as a predefined observation (say) 'r' is mentioned below ($r < n$). In such matters, we have a tendency to which are left with 'r' actual observations say $x_1 < x_2 < x_3 < \cdots < x_r$ and also the lifetimes of the remaining $(n - r)$ things are over x_r. This kind of sample is termed Type II right-censored sample.

Allow $x_1 < x_2 < x_3 < \cdots < x_r$ to be a Type II right-censored sample from a SBLD all through a deliberately stochastic pattern of length 'n'. The likelihood operation of the above-censored sample is

$$\prod_{i=1}^{r} f(t_i; \lambda, \theta).[1 - F(t_r; \lambda, \theta)]^{n-r} \tag{3}$$

here f (.) and F(.) severally represent the pdf and cdf of SBLD.

The log-likelihood operation to estimate θ from the given censored sample is given by

$$L \propto \left[\prod_{i=1}^{r} \frac{\lambda(\lambda - 1)}{\theta^2}.t_i.\left(1 + \frac{t_i}{\theta}\right)^{-(\lambda+1)}\right]\left[\left(1 + \frac{t_r}{\theta}\right)^{-\lambda}\left(1 + \frac{\lambda t_r}{\theta}\right)\right]^{n-r}$$

$$\begin{aligned}\text{Log } L = {} & \text{constant} - 2r \log\theta \\ & - (\lambda + 1)\sum_{i=1}^{r} \log\left(1 + \frac{t_i}{\theta}\right) - \lambda(n - r)\log\left(1 + \frac{t_r}{\theta}\right) \\ & + (n - r)\log\left(1 + \frac{\lambda \cdot t_r}{\theta}\right)\end{aligned}$$

where the constant is independent of the parameters to be estimated. Differentiating with respect to 'θ', we acquire the predicting equation for the parameter 'θ'

$$\frac{\partial \log L}{\partial \theta} = 0 \Rightarrow \frac{2r}{\theta} - \frac{\lambda+1}{\theta}\sum_{i=1}^{r}\frac{\frac{t_i}{\theta}}{1+\frac{t_i}{\theta}}$$
$$-\frac{\lambda(n-r)}{\theta}\frac{\frac{t_r}{\theta}}{1+\frac{t_r}{\theta}} + \frac{n-r}{\theta}\frac{\lambda\frac{t_r}{\theta}}{1+\lambda\frac{t_r}{\theta}} = 0$$

where $Z_i = \frac{t_i}{\theta}$ and $Z_r = \frac{t_r}{\theta}$

$$2r - (\lambda+1)\sum_{i=1}^{r}\frac{Z_i}{1+Z_i} - \lambda(n-r)\frac{Z_r}{1+Z_r} + (n-r)\frac{\lambda Z_r}{1+\lambda Z_r} = 0 \quad (4)$$

It may be seen that Eq. (4) cannot be solved analytically for θ. The MLE of θ has got to be obtained as an associate in nursing repetitious answer of (4). We have a tendency to approximate the expression

$$h(Z_i) = \frac{Z_i}{1+Z_i},\ h(Z_r) = \frac{Z_r}{1+Z_r},\ h^*(Z_r) = \frac{\lambda z_r}{1+\lambda z_r}$$

of the likelihood Eq. (4) for estimating θ by a linear expression as

$$h(z_i) = \gamma_i + \delta_i z_i,\ h(Z_r) = \gamma_r + \delta_r z_r,\ h^*(Z_r) = \gamma_r^* + \delta_r^* z_r \quad (5)$$

where γ_i and δ_i, γ_r and δ_r and γ_r^* and δ_r^* are to be fitly found, to induce a changed MLE of θ. As per the constant quantity specifications, we tend to take $\lambda = 3$.

$$\frac{\partial \log l}{\partial \theta} = 0 \Rightarrow 2r - 4\sum_{i=1}^{r-1}(\gamma_i + \delta_i Z_i) - (4 + 3(n-r))$$
$$(\gamma_r + \delta_r Z_r) + (n-r)\left(\gamma_r^* + \delta_r^* Z_r\right) = 0$$

$$\hat{\theta} = \frac{4\sum_{i=1}^{r-1}\delta_i t_i + [4+3(n-r)]\delta_r t_r - (n-r)\delta_r^* t_r}{2r - 4\sum_{i=1}^{r-1}\gamma_i - [4+3(n-r)]\gamma_r + (n-r)\gamma_r^*} \quad (6)$$

The ensuing MMLE's of θ from the censored sample can be obtained by using two ways namely Tiku (1967) and [1]. We propose two ways for finding γ_i and δ_i, γ_r and δ_r and γ_r^* and δ_r^* of Eq. (5).

Method I

Let $P_r = \frac{r}{n+1}$, $r = 1, 2, 3, \ldots n$ and $q_r = 1 - P_r$

Let Z_r^*, Z_r^{**} be the solutions of the following equations

$$F(Z_r^*) = p_r^* \text{ and } F(Z_r^{**}) = p_r^{**}$$

where $p_r^* = p_r - \sqrt{\frac{p_r q_r}{n}}$, $p_r^{**} = p_r$

The solutions of Z_r^*, Z_r^{**} in our size-biased Lomax model are

$$F(Z_r^*) = P_r^* \Rightarrow Z_r^* = F^{-1}(P_r^*) \quad F(Z_r^{**}) = P_r^{**} \Rightarrow Z_r^{**} = F^{-1}(P_r^{**})$$

$$h(Z_r^*) = \frac{Z_r^*}{1 + Z_r^*} \text{ and } h(Z_r^{**}) = \frac{Z_r^{**}}{1 + Z_r^{**}}$$

$$h^*(Z_r^*) = \frac{3Z_r^*}{1 + 3Z_r^*} \text{ and } h^*(Z_r^{**}) = \frac{3Z_r^{**}}{1 + 3Z_r^{**}}$$

The intercepts γ_r, γ_r^* and slopes δ_r, δ_r^* of Eq. (5) are given below

$$\delta_r = \frac{h(Z_r^{**}) - h(Z_r^*)}{Z_r^{**} - Z_r^*} \text{ and } \gamma_r = h(Z_r^*) - \delta_r Z_r^* \tag{7}$$

$$\delta_r^* = \frac{h^*(Z_r^{**}) - h^*(Z_r^*)}{Z_r^{**} - Z_r^*} \text{ and } \gamma_r^* = h^*(Z_r^*) - \delta_r^* Z_r^* \tag{8}$$

Method II

Consider Taylor's expansion of $h(\xi_r) = \frac{\xi_r}{1+\xi_r}$ within the neighborhood of the rth quantile of SBLD. We have a tendency to get another linear approximation for $h(\xi_r)$ which is given by

$$h(\xi_r) = \gamma_r + \delta_r \xi_r \tag{9}$$

where $\delta_r = h^1(\xi_r)$ with ξ_r as the rth quantile for the population is given by

$$F(\xi_r) = P_r \Rightarrow \xi_r = F^{-1}(P_r), \quad P_r = \frac{r}{n+1} \quad \gamma_r = h(\xi_r) - \delta_r \xi_r$$

$$h^*(\xi_r) = \frac{3\xi_r}{1 + 3\xi_r} \quad h^*(\xi_r) = \gamma_r^* + \delta_r^* \xi_r$$

where

$$\delta_r^* = h^{*^1}(\xi_r) \; \gamma_r^* = h^*(\xi_r) - \delta_r^* \xi_r \tag{10}$$

The relevant values of slopes and intercepts are calculated for $n = 5$, 10, 15 and 20 for 20% censored sample of r.

In the two ways referred higher than the fundamental principle, the expressions $\frac{Z_i}{1+Z_i}$, $\frac{Z_r}{1+Z_r}$ and $\frac{\lambda Z_r}{1+\lambda Z_r}$ are approximated by several linear functions in some neighborhood of the population. It may be seen that the development of the neighborhood over a definite performance is linearized based on the sample size additionally. Larger the size nearer is going to be the approximation, that is, the accuracy of the approximation becomes finer and finer for big values of n. Hence, the approximate log-likelihood equation and also the precise log-likelihood equation vary by tiny quantities for big n. Therefore, the solutions of the precise equation and approximate log-likelihood equations tend to every different as $n \to \infty$. Hence, the precise and changed MLEs are asymptotically identical (Tiku et al. 1986). However, identical cannot be aforementioned in little samples. At the identical time, the little sample variance of changed MLE is not mathematically tractable. We, therefore, compared these estimates in little samples through Monte Carlo simulation.

For mounted worth of $\lambda = 3$, the values of δ_r, γ_r and δ_r^*, γ_r^* for techniques I and II are bestowed within Tables 1 and 2. The bias, variance, and MSE of the estimates by the two ways of modification are obtained through simulation for $n = 5$ (5) 20 with all attainable issues of censored samples are given in Table 3. The subsequent conclusions are drawn from the simulated sampling characteristics supported 1000 Monte Carlo simulation runs.

2 Conclusion

The variances of modified MLEs from censored samples are determined to be over the analogous variances of modified MLEs from the complete sample as set up by means of the relationships among Tables 3 and 4. Because of censored samples, there is a loss of information leading to fallen efficiency and therefore raising variance.

Table 1 Slope ($\delta_{r,}$, δ_r^*) and intercept (γ_r, γ_r^*) in modification to MLE by technique I

r	n	$\lambda = 3$				r	n	$\lambda = 3$			
		$\delta_{r,}$	γ_r	δ_r^*	γ_r^*			$\delta_{r,}$	γ_r	δ_r^*	γ_r^*
1	5	0.454	0.1	0.506	0.335	11	15	0.128	0.392	0.075	0.692
2	5	0.368	0.116	0.373	0.352	12	15	0.095	0.455	0.052	0.739
3	5	0.226	0.226	0.174	0.522	13	15	0.064	0.528	0.032	0.787
4	5	0.116	0.368	0.07	0.665	14	15	0.035	0.62	0.016	0.839
1	10	0.593	0.051	0.839	0.215	1	20	0.703	0.025	1.207	0.13
2	10	0.54	0.055	0.72	0.218	2	20	0.669	0.026	1.104	0.129
3	10	0.431	0.101	0.471	0.33	3	20	0.588	0.047	0.838	0.201
4	10	0.344	0.151	0.321	0.423	4	20	0.523	0.069	0.663	0.262
5	10	0.271	0.208	0.22	0.505	5	20	0.467	0.092	0.535	0.317
6	10	0.208	0.271	0.148	0.58	6	20	0.418	0.116	0.438	0.366
7	10	0.151	0.344	0.095	0.651	7	20	0.374	0.142	0.361	0.412
8	10	0.101	0.431	0.056	0.72	8	20	0.333	0.169	0.298	0.455
9	10	0.055	0.54	0.027	0.793	9	20	0.296	0.198	0.247	0.495
1	15	0.66	0.034	1.053	0.161	10	20	0.261	0.228	0.203	0.534
2	15	0.62	0.035	0.943	0.162	11	20	0.228	0.261	0.167	0.572
3	15	0.528	0.064	0.68	0.249	12	20	0.198	0.296	0.136	0.608
4	15	0.455	0.095	0.513	0.321	13	20	0.169	0.333	0.109	0.643
5	15	0.392	0.128	0.394	0.386	14	20	0.142	0.374	0.086	0.678
6	15	0.337	0.163	0.306	0.444	15	20	0.116	0.418	0.066	0.713
7	15	0.287	0.201	0.237	0.499	16	20	0.092	0.467	0.049	0.748

(continued)

Table 1 (continued)

r	n	$\lambda = 3$				r	n	$\lambda = 3$			
		$\delta_{r,}$	γ_r	δ_r^*	γ_r^*			$\delta_{r,}$	γ_r	δ_r^*	γ_r^*
8	15	0.242	0.242	0.183	0.55	17	20	0.069	0.523	0.035	0.785
9	15	0.201	0.287	0.14	0.599	18	20	0.047	0.588	0.022	0.823
10	15	0.163	0.337	0.104	0.646	19	20	0.026	0.669	0.011	0.865

Table 2 Slope ($\delta_{r,}$, δ_r^*) and intercept (γ_r, γ_r^*) in modification to MLE by technique II

r	n	$\lambda = 3$				r	n	$\lambda = 3$			
		$\delta_{r,}$	γ_r	δ_r^*	γ_r^*			$\delta_{r,}$	γ_r	δ_r^*	γ_r^*
1	5	0.549	0.067	0.714	0.262	11	15	0.207	0.297	0.142	0.612
2	5	0.611	0.048	0.887	0.208	12	15	0.166	0.352	0.104	0.663
3	5	0.428	0.12	0.448	0.376	13	15	0.126	0.416	0.072	0.714
4	5	0.28	0.221	0.223	0.529	14	15	0.087	0.496	0.045	0.77
1	10	0.663	0.035	1.056	0.165	1	20	0.754	0.017	1.415	0.098
2	10	0.725	0.022	1.291	0.118	2	20	0.805	0.011	1.66	0.066
3	10	0.597	0.052	0.845	0.22	3	20	0.714	0.024	1.248	0.126
4	10	0.495	0.088	0.585	0.312	4	20	0.642	0.039	0.987	0.182
5	10	0.408	0.13	0.413	0.396	5	20	0.581	0.057	0.8	0.234
6	10	0.331	0.18	0.29	0.475	6	20	0.526	0.075	0.658	0.283
7	10	0.261	0.24	0.2	0.551	7	20	0.477	0.096	0.546	0.329
8	10	0.196	0.311	0.131	0.626	8	20	0.432	0.118	0.455	0.373
9	10	0.134	0.402	0.078	0.703	9	20	0.389	0.141	0.38	0.415
1	15	0.719	0.023	1.267	0.123	10	20	0.35	0.167	0.318	0.455
2	15	0.775	0.014	1.513	0.084	11	20	0.312	0.195	0.264	0.494
3	15	0.67	0.033	1.083	0.159	12	20	0.277	0.225	0.219	0.533
4	15	0.587	0.055	0.819	0.228	13	20	0.243	0.257	0.18	0.57
5	15	0.517	0.079	0.635	0.292	14	20	0.211	0.293	0.146	0.608
6	15	0.454	0.107	0.498	0.351	15	20	0.18	0.332	0.116	0.645
7	15	0.397	0.137	0.393	0.407	16	20	0.15	0.376	0.091	0.683

(continued)

Table 2 (continued)

r	n	$\lambda = 3$				r	n	$\lambda = 3$			
		$\delta_{r,}$	γ_r	δ_r^*	γ_r^*			$\delta_{r,}$	γ_r	δ_r^*	γ_r^*
8	15	0.345	0.171	0.31	0.461	17	20	0.121	0.426	0.068	0.721
9	15	0.296	0.208	0.243	0.512	18	20	0.092	0.484	0.049	0.762
10	15	0.25	0.25	0.188	0.563	19	20	0.065	0.556	0.031	0.807

Table 3 Sample characteristics of MMLE of θ from expurgated samples' techniques I and II

λ	*n*	Bias		Variance	MSE	
		MMLE-I	MMLE-II	MMLE-II	MMLE-I	MMLE-II
3	5	0.101	0.094	0.6858	0.705	0.6946
3	10	0.041	0.071	0.418	0.3917	0.4231
3	15	0.006	0.042	0.2975	0.2722	0.2992
3	20	0.021	0.053	0.1576	0.1487	0.1604

Table 4 Sample characteristics of MMLE of θ from expurgated samples' technique I and technique II

λ	*n*	Bias		Variance	MSE	
		MMLE-I	MMLE-II	MMLE-II	MMLE-I	MMLE-II
3	5	0.257	0.594	1.1352	0.8279	1.4876
3	10	0.089	0.297	0.3963	0.3007	0.4843
3	15	0.045	0.199	0.18	0.1379	0.2195
3	20	0.033	0.154	0.1058	0.0837	0.1295

References

1. Balakrishnan N (1990) Approximate maximum likelihood estimation for a generalized logistic distribution. J Stat Plan Inference 26:221–236
2. Dubey SD (1966) Transformations for estimation of parameters. J Indian Stat Assoc 4(3–4):109–124
3. Ebrahimi N (1984) Maximum likelihood estimation from doubly censored samples of the mean of a normal distribution with known coefficient of variation. Commun Stat-Theory Methods 13(5):651–659
4. Harter HL, Moore AH (1965) Maximum likelihood estimation of the parameters of gamma and Weibull populations from complete and censored samples. Technometrics 7:639–643
5. Kantam RRL, Priya MCh, Ravikumar MS (2013) Modified maximum likelihood estimation in linear failure rate distribution. Inter Stat, July, USA
6. Kantam RRL, Ramakrishna V, Ravi Kumar MS (2013) Estimation and testing in type-I generalized half logistic distribution. J Mod Appl Stat Methods 12(1):198–206
7. Kantam RRL, Ramakrishna V, Ravi Kumar MS (2014) Estimation and testing in type-II generalized half logistic distribution. J Mod Appl Stat Methods 13(1):267–277
8. Kantam RRL, Ravikumar MS (2015) Modified maximum likelihood estimation from censored samples in burr type X distribution. Pak J Stat Oper Res 11(4):601–629
9. Kantam RRL, Srinivasa Rao G (2002) Log-logistic distribution: modified maximum likelihood estimation. Gujarat Stat Rev 29(1):25–36
10. Kantam RRL, Sriram B (2003) Maximum likelihood estimation from censored samples- some modifications in length biased version of exponential model. Stat Methods 5(1):63–78
11. Subba Rao R, Kantam RRL, Prasad G (2015) Modified maximum likelihood estimation in pareto-rayleigh distribution. In: Conference proceedings of national seminar on recent developments in applied statistics—golden research thoughts, pp 140–152

Fuzzy Controlled High Gain Nonisolated DC to DC Hybrid Boost Converter

M. V. Sudarsan, Ch. Sai Babu, and S. Satyanarayana

Abstract DC to DC conversion at high voltage gain is an imperative feature for many applications particularly for photovoltaic grid-connected system. The voltage conversions at large gain in boost converter are restricted due to the diode reverse-recovery problem and the stress on the switch. In this paper, a hybrid boost converter (HBC) that operates at high voltage gain is analyzed. This converter topology has better features like large voltage conversion for the smaller duty cycles, reduced voltage and current stress on the active switches. Also, the dynamic performance of the HBC is analyzed in the closed-loop operation with fuzzy logic controller for the variations of supply voltages and load resistances. The simulation model of 400 V, 10 KW HBC is designed and implemented in MATLAB/Simulink, and the obtained results verify the better performance of HBC over boost converter.

Keywords Boost converter · Hybrid boost converter · Large voltage conversion ratio · Fuzzy logic controller

1 Introduction

Renewable energy resources can be a legitimate option for the fossil fuels due to their clean and cost-effective features. To use these renewable energy sources like wind,

M. V. Sudarsan (✉)
Electrical and Electronics Engineering, Vignan's LARA Institute of Technology and Science, Vadlamudi, India
e-mail: mvsudarsan.eee@gmail.com

Ch. Sai Babu
Electrical and Electronics Engineering, University College of Engineering Kakinada, Jawaharlal Nehru Technological University, Kakinada, Kakinada, India
e-mail: chs_eee@yahoo.co.in

S. Satyanarayana
Electrical and Electronics Engineering, Raghu Institute of Technology, Visakhapatnam, India
e-mail: ssn_professoreee@yahoo.com

P. S. R. Chowdary et al. (eds.), *Microelectronics, Electromagnetics and Telecommunications*, Lecture Notes in Electrical Engineering 655,
https://doi.org/10.1007/978-981-15-3828-5_16

solar, fuel cell as inputs for generation of electrical power, there is a necessity to step-up low-level voltage to the required voltage level to meet the AC utility. This high voltage conversion is possible with boost converter at large duty cycles. The operation at large duty cycles leads higher conduction losses, lower conversion efficiency and diode reverse-recovery problems; also, it uses large inductor to mitigate the current ripple produced due to the very short period of switch turn OFF time and requires a very fast, expensive comparator in the duty-cycle generation circuit. To overcome the above-mentioned drawbacks, there is a necessity of designing a high voltage gain converter [1].

In this paper, an HBC which operates in combination of active and passive switching elements boosts up the voltage at higher gain over a wide range and also mitigates the above-mentioned drawbacks. HBC is formed with two similar inductors connected with two power semiconductor switches on input side, diode and capacitor on load side. The large voltage conversion ratio in this HBC is possible with the inductors charging in parallel when the switches are in ON state and discharges their energy to the load by connecting in series when the switches turn OFF and the voltage stress and current stress of the switches are reduced. For achieving a controlled output voltage in the converter against line and load-side disturbances, there is a necessity of controlling the switches on time using the controllers. Here, to have a better performance of the controlling action, fuzzy logic controller is used which has advantages of simplicity, heuristic nature and adaptability for all linear and nonlinear systems. Also, the designing of fuzzy controller does not require the complete information of the system.

2 Boost Converter

Figure 1a explains the topological diagram of the boost converter, and its operation is analyzed in continuous mode of operation.

Stage-I (t_0–t_1): In this stage, the inductor charges linearly and stores energy for a period of DT_s. The turn ON of the switch isolates the load from the supply. The inductor voltage in this stage is

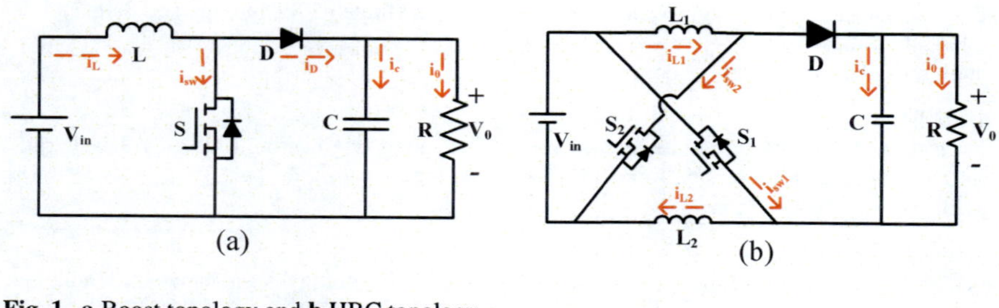

Fig. 1 **a** Boost topology and **b** HBC topology

$$v_{L}(t) = L\frac{\mathrm{d}i_{L}(t)}{\mathrm{d}t} = V_{\mathrm{in}} \tag{1}$$

Stage-II (t_1–t_2): In this stage, the turn OFF of the switch makes a path for charged inductor to discharge its stored energy and transfers its energy from supply to the load. The inductor voltage during OFF period (1–D) T_s is

$$v_{\mathrm{L}}(t) = L\frac{\mathrm{d}i_{\mathrm{L}}(t)}{\mathrm{d}t} = V_{\mathrm{in}} - V_0 \tag{2}$$

Thus, the voltage and current gains of the converter are obtained by averaging the inductor voltage to zero and are given as

$$\frac{V_0}{V_{\mathrm{in}}} = \frac{1}{1-D} \quad \text{and} \quad \frac{I_{\mathrm{in}}}{I_0} = \frac{1}{1-D} \tag{3}$$

The values of boosting inductor and the filtering capacitor are designed for an input ripple current ΔI and for an output ripple voltage ΔV_0 as

$$L = \frac{V_{\mathrm{in}}D}{\Delta I f_s} \quad \text{and} \quad C = \frac{V_o D}{R\Delta\, V_0 f_s} \tag{4}$$

3 Hybrid Boost Converter

The topology of HBC consists of two similar weighted inductors cross-connected with a pair switches S_1 and S_2 from the input supply to the load through the diode and capacitor and is shown in Fig. 1b. S_1 and S_2 operate for same duty cycles in general less when compared to the ratio of the boost converter. HBC operation is based on the switched inductor concept, where the two inductors of same inductance will charge in parallel during ON period of switches and will discharge during OFF period of switches by connecting the inductors in series [2]. This process of charging and discharging the inductors achieves the high voltage gain for increasing the voltage level with smaller duty cycles.

3.1 Continuous Mode of Operation

Stage-I (t_0–t_1): During this stage, the turn ON of the power switches S_1, S_2 connects the inductors L_1, L_2 in parallel to the supply voltage V_{in} and gets charged by increasing currents linearly from a minimum value to i_{Lpeak} at time t_1. The rectifying diode gets turned OFF and makes the load disconnecting from the input supply, whereas the current to the load is supplied by the charged capacitor. Thus, the inductor voltages and the blocking voltage of the diode during this interval are

$$v_D = V_0 + V_{in} \tag{5}$$

The input ripple current during this period of conduction is

$$I_{\text{in ripple}} = \frac{2V_{in}D}{f_s L} \tag{6}$$

Stage-II (t_1–t_2): In this stage, the turn OFF of the power switches S_1, S_2 connects the inductors L_1, L_2 in series and discharges their energy to load and filtering capacitor through the turned ON diode. The current thus decreases linearly from i_{Lpeak} to minimum value at time t_2. Thus, the inductor voltages during the discharging period are

$$v_{L1} = v_{L2} = \frac{V_{in} - V_o}{2} \tag{7}$$

As per the volt-second balance principle, making the voltage across inductor over one cycle of time period to zero, the voltage conversion gain of HBC [3] is

$$\frac{V_o}{V_{in}} = \frac{1+D}{1-D} \tag{8}$$

The blocking voltage of the power switches in stage-II during OFF period is given by

$$v_{S_1} = v_{S_2} = \frac{V_{in} + V_o}{2} \tag{9}$$

3.2 *Discontinuous Mode of Operation*

Stage-I (t_0–t_1): The operation in this stage is similar as in continuous mode of operation, and the inductors L_1, L_2 get charged in parallel from zero current to the peak value i_{Lpeak} within DT_s. The capacitor current in this interval is equal to the load current in opposite direction (Fig. 2).
Stage-II (t_1–t_2): The turn OFF of power switches S_1, S_2 in this stage connects the inductors L_1, L_2 to connect in series and discharges their energies to the capacitor and the load through the turned ON diode. The current in these inductors decreases from its peak value to zero at time instant t_2. The expression of peak currents with reference to the current profile in discharging period is

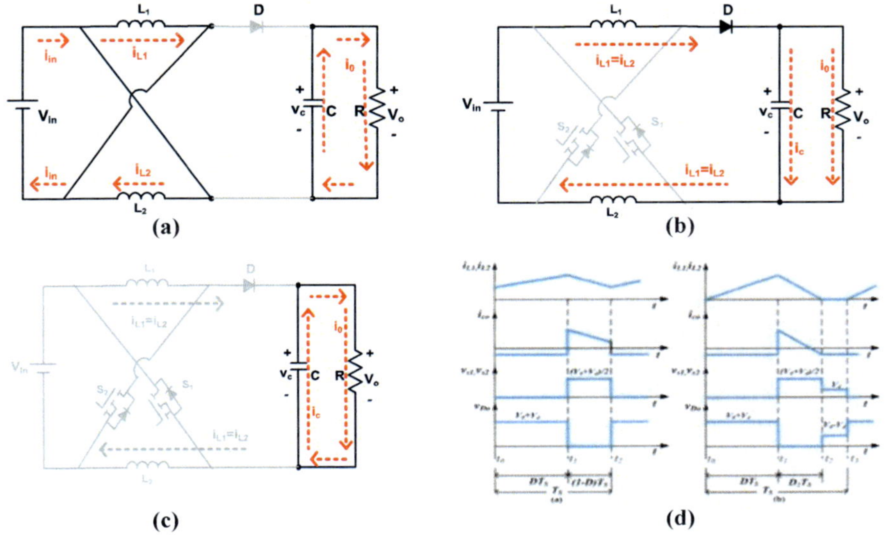

Fig. 2 Hybrid boost converter with switches in **a** ON state, **b** OFF state, **c** OFF state in DCM and **d** waveforms in CCM and DCM

$$i_{\mathrm{Lp1}} = i_{\mathrm{Lp2}} = \frac{V_0 - V_{\mathrm{in}}}{2Lf_s} D_2 \tag{10}$$

On equating the peak currents from stage-I and stage-II, the duty cycle D_2 is

$$D_2 = \frac{2DV_{\mathrm{in}}}{V_0 - V_{\mathrm{in}}} \tag{11}$$

Stage-III (t_2–t_3): The switches S_1, S_2 and diode D get turned OFF. The capacitor supplies the current to the load until the next cycle of operation. The zero current in these inductors makes the discontinuous operation. The steady-state analysis of the converter is done based on the capacitor current over a cycle of time period. The average current of the capacitor over a cycle of operation is [4].

$$I_{\mathrm{co}} = -\frac{V_0}{R} + \left(\frac{DV_{\mathrm{in}}}{V_0 - V_{\mathrm{in}}}\right)\left(\frac{V_{\mathrm{in}}D}{Lf_s}\right) = \frac{D^2V_{\mathrm{in}}^2}{(V_0 - V_{\mathrm{in}})Lf_s} - \frac{V_0}{R} \tag{12}$$

Defining the normalized time constant of inductor as τ_{L} equals to ratio of *L/R* to switching cycle period (T_s), and on substituting this in the above equation, the voltage gain in the discontinuous mode is $M_{\mathrm{DCM}} = \frac{1}{2} + \sqrt{\frac{1}{4} + \frac{D^2}{\tau_L}}$.

Fig. 3 Specifications of boost and HBC converters

Parameters	Boost	HBC
Input voltage V_{in}	100 V	100 V
Output voltage V_0	400 V	400 V
Output power P_0	10 KW	10 KW
Switch frequency f_s	20 kHz	20 kHz
Capacitor C_0	120 μF	96 μF
Inductors L_1 and L_2	0.75 mH	0.6 mH

The analysis of HBC at boundary conduction mode is

$$\tau_{LB} = \frac{(1-D)^2 D}{2(1+D)} \tag{13}$$

If time constant τ_L of the converter is smaller than the $\tau_{LB,}$ then the converter operates in DCM operation [5].

4 Closed-Loop Control of Active Network Converter

4.1 *Fuzzy Controller*

The Mamdani-type fuzzy logic controller in this work consists of three stages, namely fuzzification, fuzzy inference engine and defuzzification. In the first stage, the inputs error, change in error (e(t) and ce(t)) and output u(t) of the controller are fuzzified with triangular membership functions. Each input variable is defined with seven linguistic terms and output variable with eleven linguistic terms as shown in Fig. 4a, b. In the second stage, 49 if-then rules are formed with antecedent as e(t), ce(t) and consequent as u(t). An *and—or* operation is performed, and a fuzzy output is generated from the inference engine through the defined rule. In the third stage, the output in fuzzy terms is defuzzified to the control signal using centroid method.

5 Simulation Results

The MATLAB/Simulink results of converters during steady-state and their dynamic performance are analyzed in both open and closed-loop operations.

Figure 5a, b shows the output voltage and current responses of 10 KW boost converter boosting the voltage to 400 V with a duty ratio of 75% and delivers a load current of 25 A. The output voltage ripple and current ripple of boost converter are observed to be 8 V and 0.4 A, respectively.

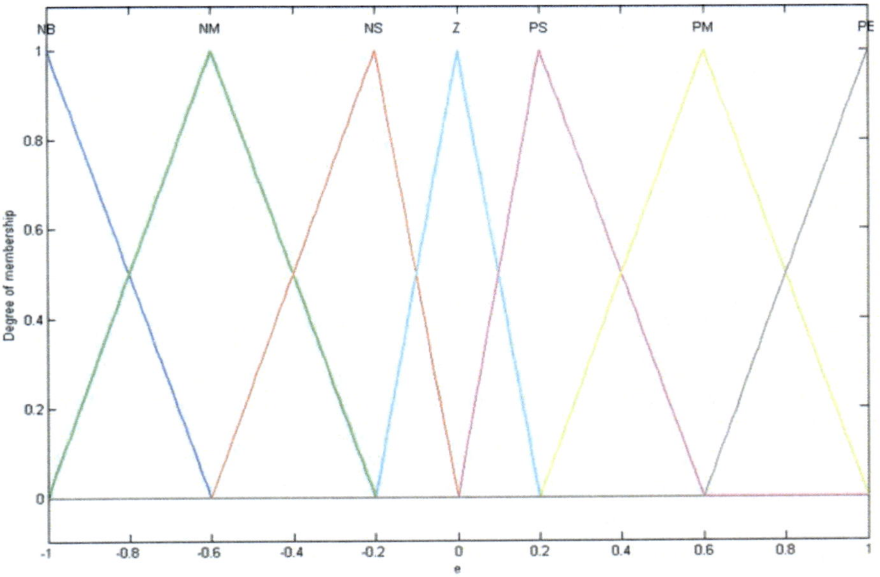

(a) Input variables e*(t)* and ce*(t)*

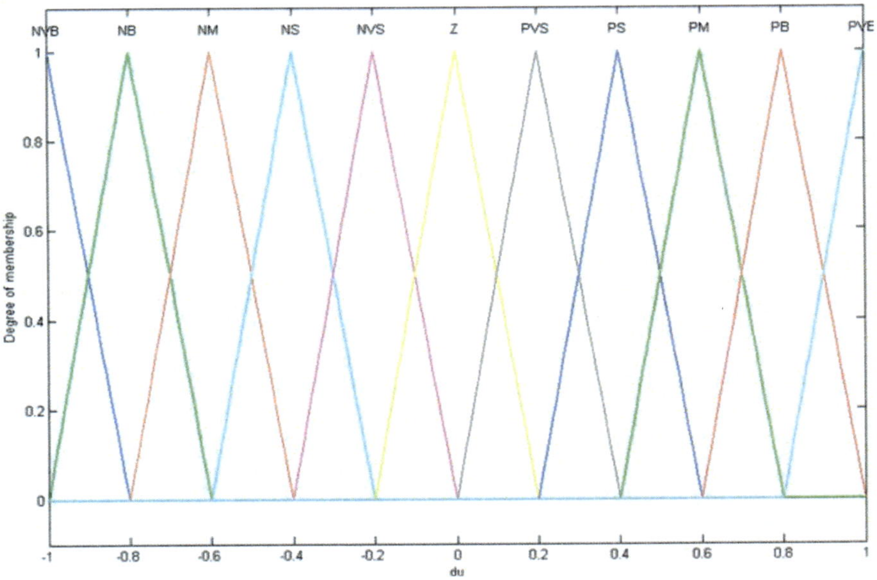

(b) Output Variable u*(t)*

Fig. 4 Membership functions of **a** inputs e(*t*) and ce(*t*) and **b** output u(*t*)

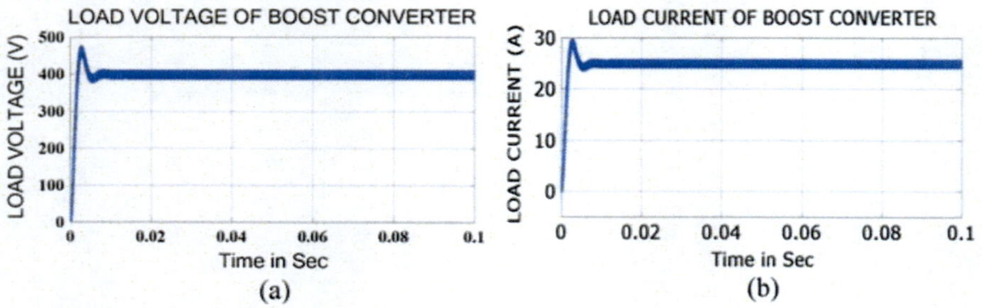

Fig. 5 **a** Output voltage and **b** output current responses of boost converter

Figure 6a depicts the inductor current response of 100 A with a ripple content of 5%, diode voltage and current responses which blocks −400 V and carries 100 A current, switch voltage and current responses which blocks 400 V during off state and carries 100 A current during on period.

Figure 7a, b shows the output voltage and current responses of 10 KW HBC boosting the voltage from 100 to 400 V with smaller duty ratio. Because of its parallel charging of inductors during ON state of switches and series discharging of inductors during OFF stage of switches, the boosting operation with lesser duty ratio of 60% only is possible.

Hence, the issue reverse-recovery problem in the diode and conduction loss of switches are minimized. The output voltage ripple and output current ripple of HBC are 6 V and 0.4 A, respectively. Figure 6b shows the responses of all elements in HBC as current response of inductors with a less current of 65 A designed for a ripple content of 5 A, diode voltage and current responses which blocks (−500 V) and carries only 65 A current and switch voltage and current responses which has

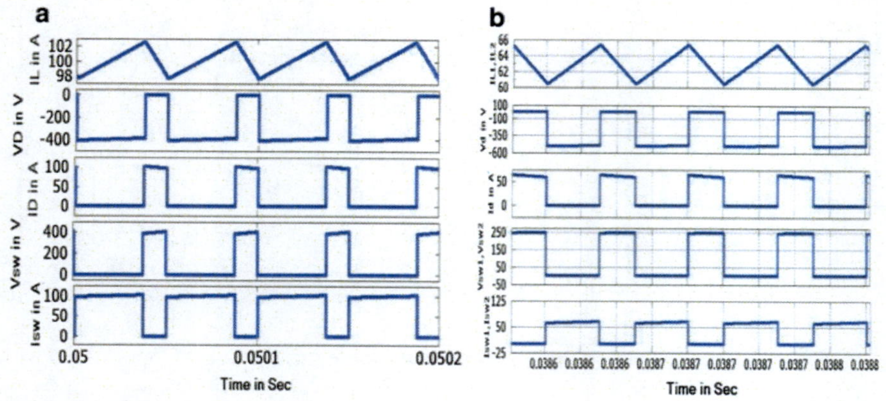

Fig. 6 Responses of inductor currents, diode voltage, diode current, switch voltages and switch currents of **a** boost converter and **b** HBC

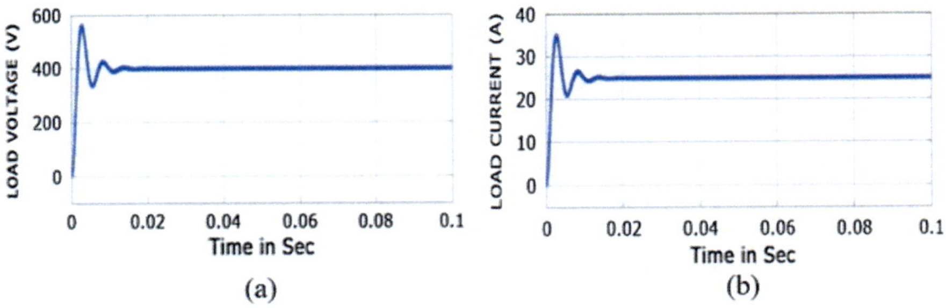

Fig. 7 **a** Output voltage and **b** output current responses of HBC

to block only 250 V during OFF state and carries less current of only 65 A current during on period.

Thus, from responses of Figs. 5, 6 and 7, the performance of HBC is superior than boost converter, and hence, the dynamic operation of HBC is analyzed in closed loop with PI and fuzzy logic controllers to regulate the voltage at 400 V under all conditions.

Figure 8 shows the voltage responses of HBC simulated in closed loop with PI and fuzzy logic controllers. The voltage response of PI-controlled HBC gives a peak overshoot of 75% and settles for a steady-state voltage of 400 V at 0.15 s, whereas the converter with fuzzy logic controller gives the better performance in terms of zero peak overshoot, less settling time of 0.05 s, small-ripple fast-rise time and zero steady-state error. Also, the robustness of fuzzy logic controller is observed for both load and supply disturbances.

Figure 9a shows the output voltage and current of HBC for a disturbance at input supply side for a decrement of voltage from 100 to 75 V at 0.5 s with a fixed load. Even under this voltage disturbance, the fuzzy controller delivers corresponding control signal which gives the required duty cycle of switches for getting a constant voltage of 400 V and 25 A at load side. Figure 9b shows waveforms of load voltage and load current of HBC with fuzzy controller for a disturbance on load side increasing

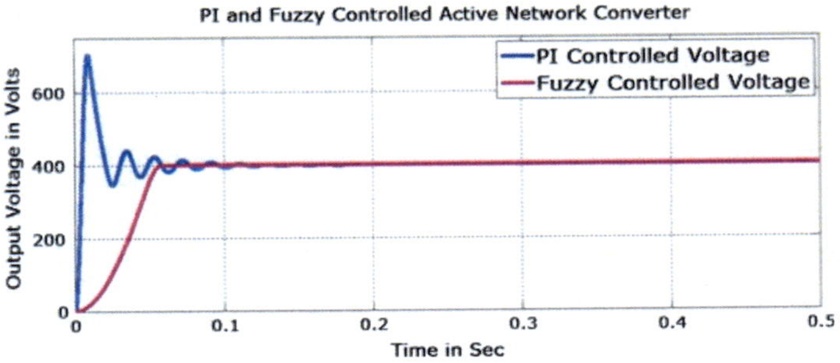

Fig. 8 Output voltage response comparison of HBC

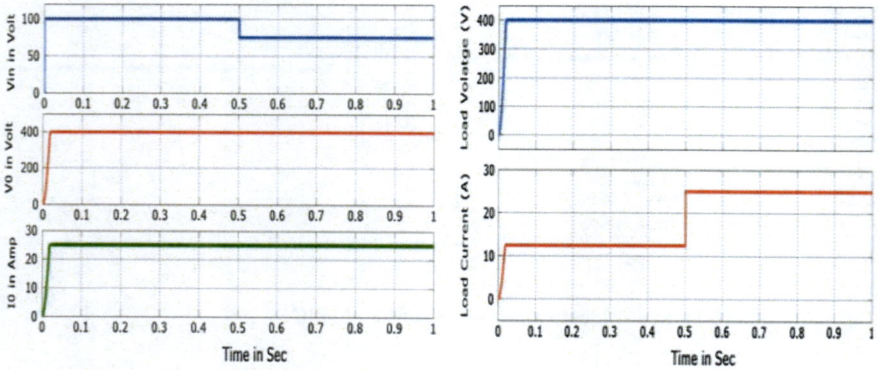

(a) (b)

Fig. 9 **a** Dynamic response of load voltage and load current of fuzzy-controlled HBC for decrease of supply voltage from 100 to 75 V at 0.5 s and **b** dynamic response of load voltage and load current of fuzzy-controlled HBC for increase of load by 50% at 0.5 s

the load from half to full load at 0.5 s for a fixed input voltage. Even under this disturbance also, the controller adjusts the duty cycle of the switches in such a way to generate a constant voltage of 400 V with a load current increasing from 12.5 to 25 A at 0.5 s.

6 Conclusion

The simulation study and analysis shows that the fuzzy-controlled HBC gives the better performance comparing with the PI-controlled in terms of its dynamic responses. Also, the HBC topology has better features like large voltage conversion for the smaller duty cycles, reduced voltage and current stress on the active switches.

References

1. Hossain MZ, Rahim NA, Jeyraj A, Selvaraj L (2018) Recent progress and development on power DC-DC converter topology, control, design and applications: a review. Renew Sustain Energy Rev 81(1):205–230
2. Forouzesh M, Yari K, Baghramian A, Hasanpour S (2017) Single switch high step-up converter based on coupled inductor and switched capacitor techniques with quasi-resonant operation. IET Power Electron 10(2):240–250
3. Lakshmi M, Hemamalini S (2018) Nonisolated high gain DC–DC converter for DC microgrids. IEEE Trans Ind Electron 65(2):1205–1212

4. Yang LS, Liang TJ, Chen JF (2009) Transformer less dc-dc converters with high step-up voltage gain. IEEE Trans Ind Electron 56(8):3144–3152
5. Axelrod B, Berkovich Y, Ioinovici A (2003) Transformerless DC–DC converters with a very high DC line-to-load voltage ratio. In: International symposium on circuits and systems 2003, IEEE, vol 3. IEEE, Thailand, pp III-435–III-438

Air Gap Coupled Microstrip Antenna for K/Ka Band Wireless Applications

K. S. Ravi Kumar, Yashpal Singh, and K. P. Vinay

Abstract A novel air gap coupled microstrip patch antenna is proposed for K/Ka band wireless applications. The antenna consists of defected ground structure and three-layer stacking of two FR4-epoxy and one air gap with two rectangular patches layered vertically with L-shaped slot. The dimensions of the antenna are 20 × 20 × 4 mm^3. The proposed antenna operates in the frequency ranges of 16.89–33.89 GHz which covers the total K-band (18–26.5 GHz) and partial Ka band (26.5–40 GHz). The designed antenna is simulated by using HFSS EM simulator. The simulation result shows the proposed antenna gain of 9.34 dB and radiation efficiency of 76%.

Keywords DGS · L-shaped patch · FR4-epoxy · Stacking · HFSS software

1 Introduction

Recent studies had provided the clear understanding of importance of antenna in the wireless communication system. So, choosing antenna based on its application is very important, and the other way to get good antenna is to design with compact size, high efficiency, high gain and bandwidth [1].

Today, most of us are using compact mobile devices in our daily life, so the proposed antenna design is to be too compact for fitting in that device. PCB antenna makes most handy devices as they include in the circuit. Patch antenna is one of the PCB antennas which can be used for different applications. The present 5G era operates in the range of 24–30 GHz which comes under K/Ka band, and this K/Ka

K. S. Ravi Kumar (✉) · K. P. Vinay
RAGHU Engineering College, Visakhapatnam 531162, India
e-mail: Srinivas.ravikumar260@gmail.com

K. P. Vinay
e-mail: phanikvinay@gmail.com

Y. Singh
Somany Institute of Technology and Management, Rewari, Haryana, India
e-mail: ypsingh10@rediffmail.com

P. S. R. Chowdary et al. (eds.), *Microelectronics, Electromagnetics and Telecommunications*, Lecture Notes in Electrical Engineering 655,
https://doi.org/10.1007/978-981-15-3828-5_17

band also has applications like multimedia transmission where ultra-wideband is necessary.

This design is focused mainly on ultra-wideband patch antenna operating in K/Ka band to achieve this. There are a lot of techniques used like stacking of patch vertically, introducing air layer between substrate [2], increasing height of substrate, slotted patch [3–5], increasing relative permittivity of substrate material [6], truncated ground plane [7], notches and slits in the ground plane [8] which are for making antenna to operate in K/Ka band, whereas defected ground structures and defected patch [9] are for controlling gain and resonant frequency, and decreasing ε_r of substrate is for increasing bandwidth. As the gain and bandwidth product remains constant for the antenna, any one of the parameters must be compromised to obtain optimized result in terms of gain and bandwidth by combining all the mentioned parametric changes. It is clearly observed from simulation results that the proposed antenna's bandwidth is increased from 8.5 to 17 dB, and gain and radiation efficiency remained satisfactory.

2 Antenna Design

The proposed compact UWB antenna with its dimensions is presented in Fig. 1, it consists of 4 figures where Fig. 1a is the upper patch which contains a rectangular patch with a rectangular slot total patch having width of 1 mm, Fig. 1b consists of lower patch of rectangular shape having L-shaped slot. This is a parasitical patch which is excited by the backward radiation of upper patch. As we are using co-axial feeding for the inner probe of feed that has to travel through lower patch to reach the upper patch, we will make circular slot of radius 0.6 mm at the feed point. Fig. 1c gives us the overview on layered pattern. It consists of three layers FR4-epoxy, air, FR4-epoxy on each FR4-epoxy a patch is fabricated, gives us information about co-axial feeding probe and also consists of two parts co-axial pin and inner probe. Inner probe passes through substrate to reach upper patch, and down the ground plane, co-axial pin consists of one inner and one outer probe, where inner probe is made up of pec having $\varepsilon_r = 1$ material and outer probe is made up of Teflon(tm) having $\varepsilon_r = 2.1$, and the height of inner probe must be taken as such that it just touches the upper patch so it is 4 mm. The radius of inner probe is 0.5 mm and outer probe is 1.66 mm. Fig. 1d gives us how ground has to be defected, and the total figure gives us the design considerations that we had considered.

3 Results and Discussion

The proposed antenna is simulated in HFSS EM simulator, and the following inferences are taken from the obtained results.

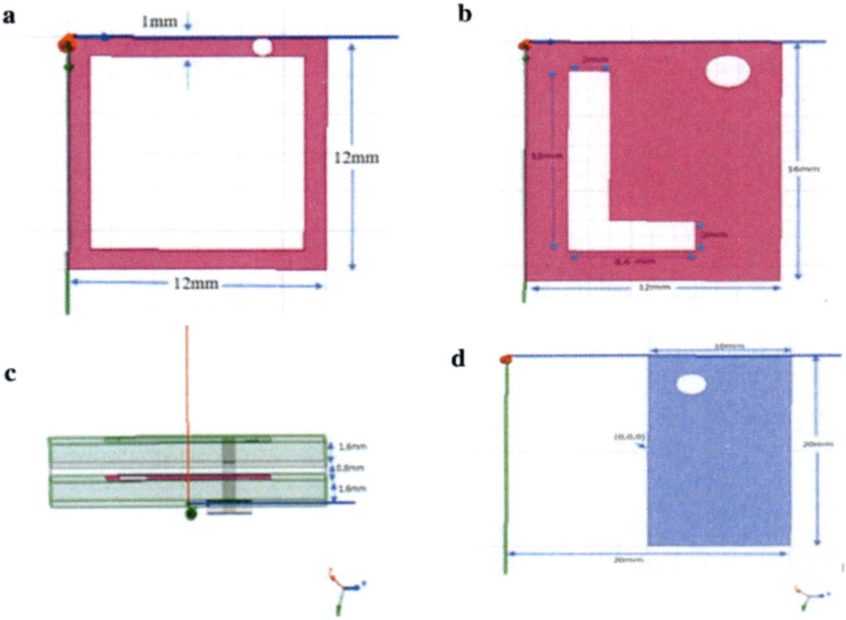

Fig. 1 **a** Upper patch with optimized dimensions layered on substrate with copper. **b** Lower parasitical patch with optimized dimensions. **c** Layered structure of sandwiched air between two substrates. **d** Defected ground structure with feed location (−3, 5.5, 0)

Figure 2a shows the S11 parameters where the range of frequencies that the return losses are less than −10 dB is 16.89–33.8 GHz, so the bandwidth of proposed antenna is 17 GHz which is 50% more than referenced antenna.

Figure 2b is about input impedance of patch antenna for perfect impedance matching with SMA 50 Ω co-axial probe, the input impedance must be 50 Ω in the simulation, and it is 49.98 Ω at 22.7 GHz and 50.19 Ω at 28.7 GHz, so that perfect impedance matching is acquired.

The simulated results of voltage standing wave ratio (VSWR) are shown in Fig. 2c where VSWR < 2 for frequency of operation.

Figure 2d contains the surface current distribution of antenna where the maximum current occurs at the edges of patch, the surface current density at 28.9 GHz is 88.4 A/m and at 22.9 GHz is 78 A/m, and the maximum density is observed at feed location.

Radiation efficiency is an important parameter to decide the antenna efficiency. Fig. 2e shows that the radiation efficiency of our antenna is 76% at 28 GHz which an acceptable value.

Gain, which is important consideration of proposed antenna, is shown in Fig. 2f where the gain is maximum at 31 GHz, i.e., 9.35 db. Gain at the resonating frequency is 8.03 db which is also very high for a patch antenna.

a

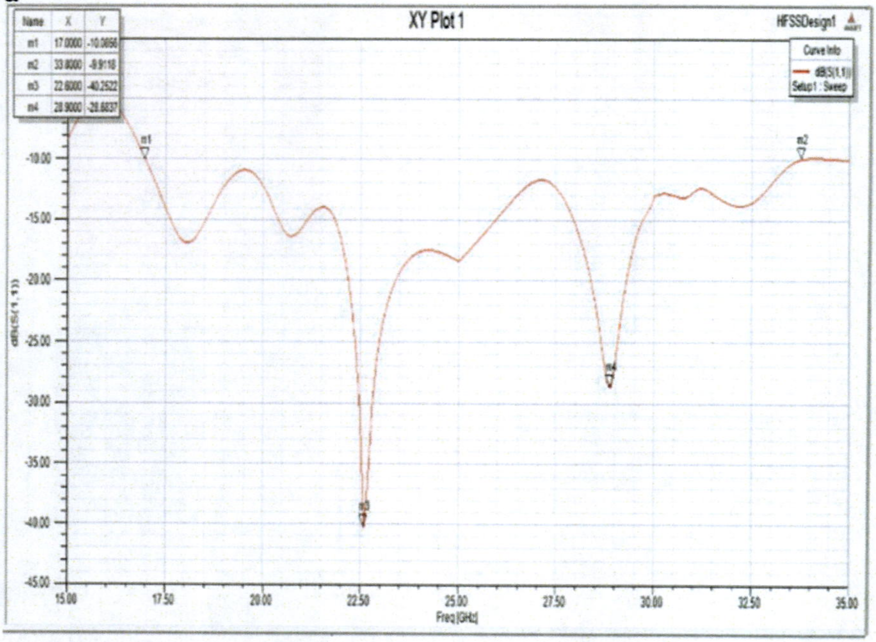

Fig. 2a Simulated return losses showing bandwidth as 17 GHz

b

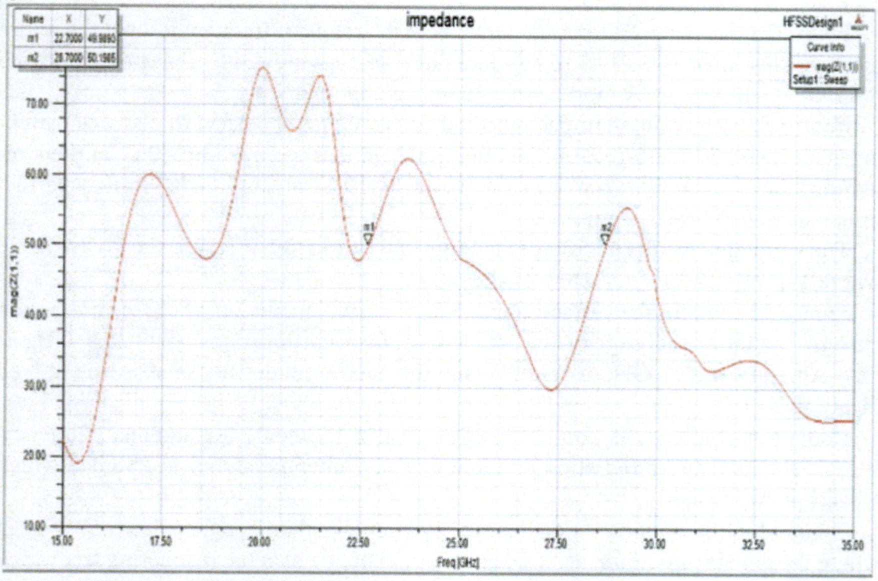

Fig. 2b Simulated result of input impedance (Z11)

c

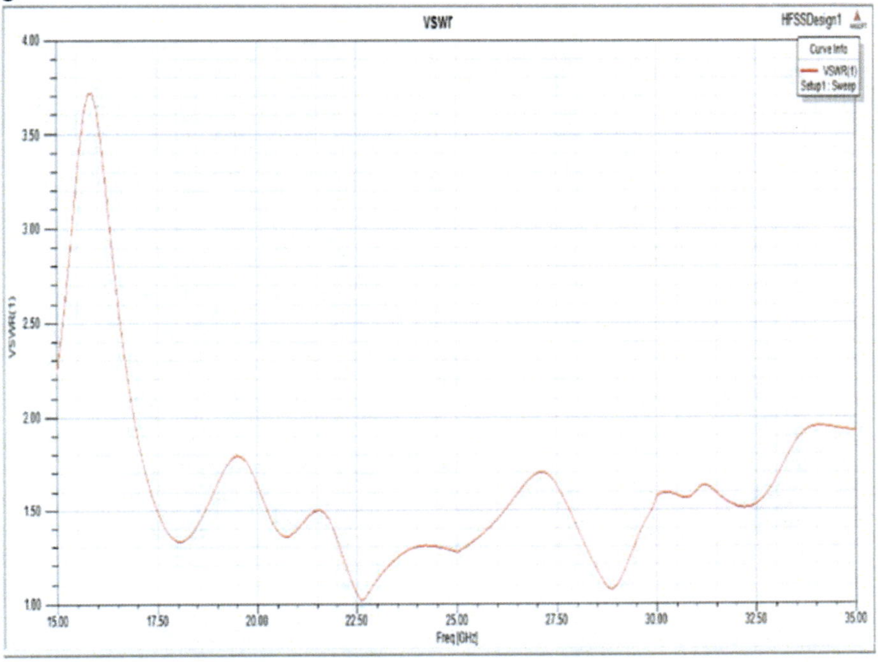

Fig. 2c Simulated graph of VSWR

Radiation pattern which is important to find the beamwidth and directivity is shown in Fig. 2g where only *E*-plane ($\phi = 0°$) and *H*-plane ($\phi = 90°$) at 22.6 GHz is considered. Here, vertical axis represents the gain in dB, and graph shows the beam that radiates with wide beam angle (Table 1).

4 Conclusions

The proposed antenna is having compact size and ultra-wideband characteristics (16.8–33.8 GHz) in K/Ka band. It has a 50% increase of bandwidth (8.5–17 GHz) than another antenna reported in table, and it also has a 1% increase in gain (8.5–9.35 dB). Therefore, the requirement of compact and ultra-wideband antenna has been satisfied. The proposed antenna operates mainly in K/Ka band which can be used for 5G and in the internal communication devices. As the bandwidth is also high, it is also used for in-house communication, RADAR and satellite communication.

d

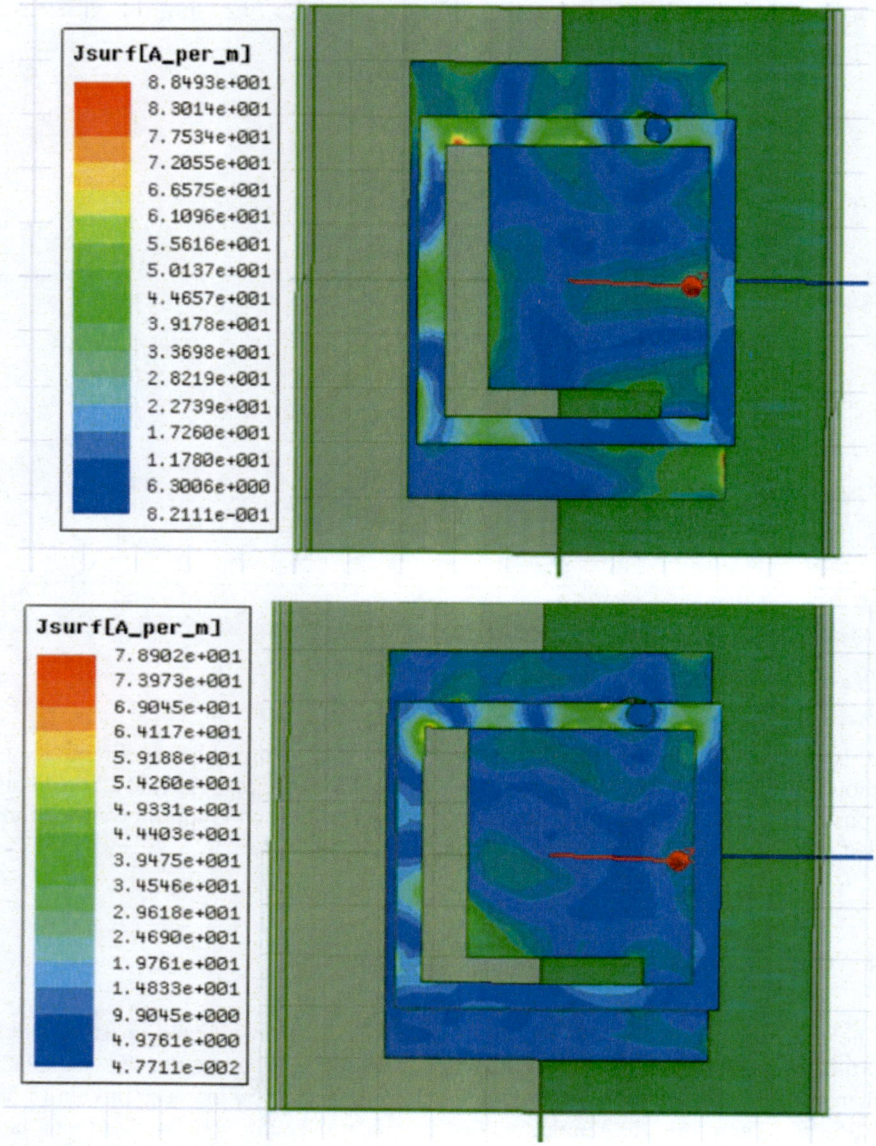

Fig. 2d Current distribution at 28.9 GHz and at 22.9 GHz

e

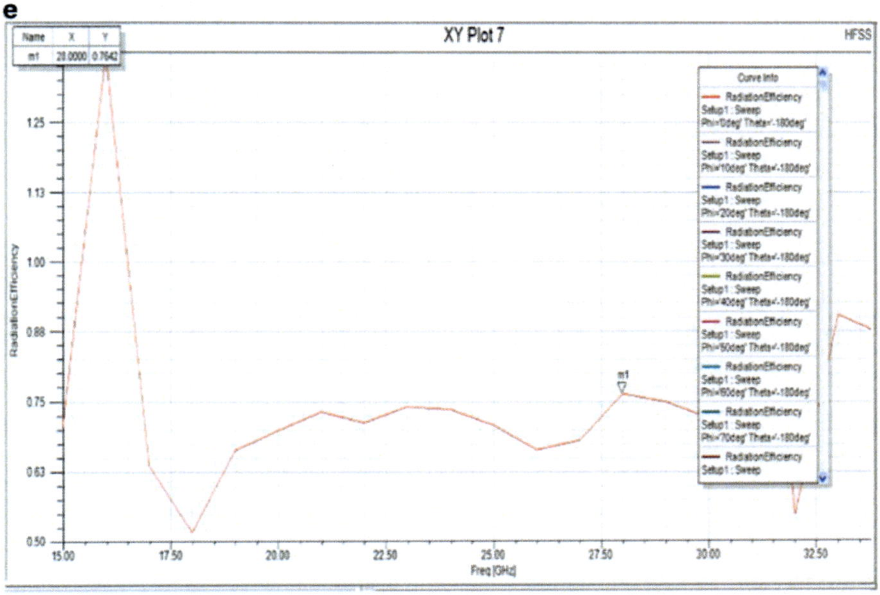

Fig. 2e Simulated radiation efficiency by considering a radiation around the patch antenna

f

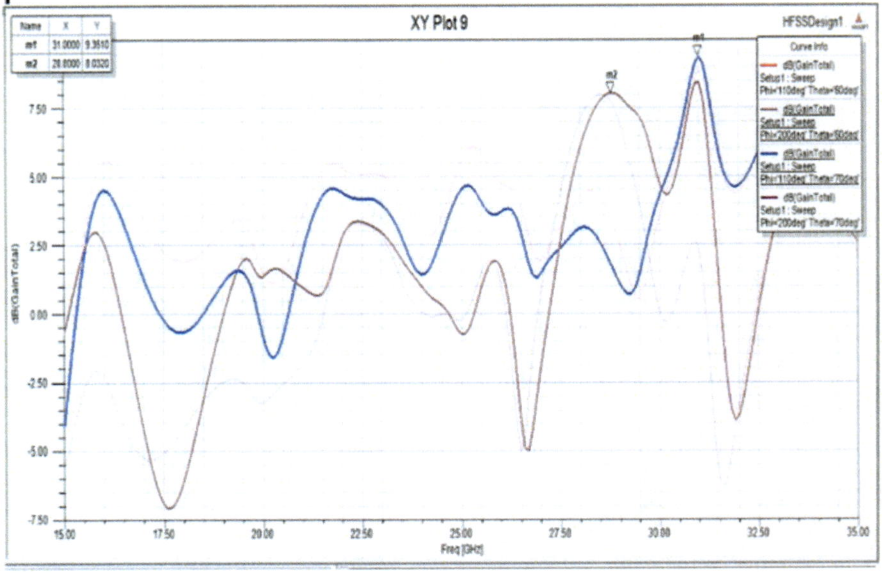

Fig. 2f Simulated result of gain

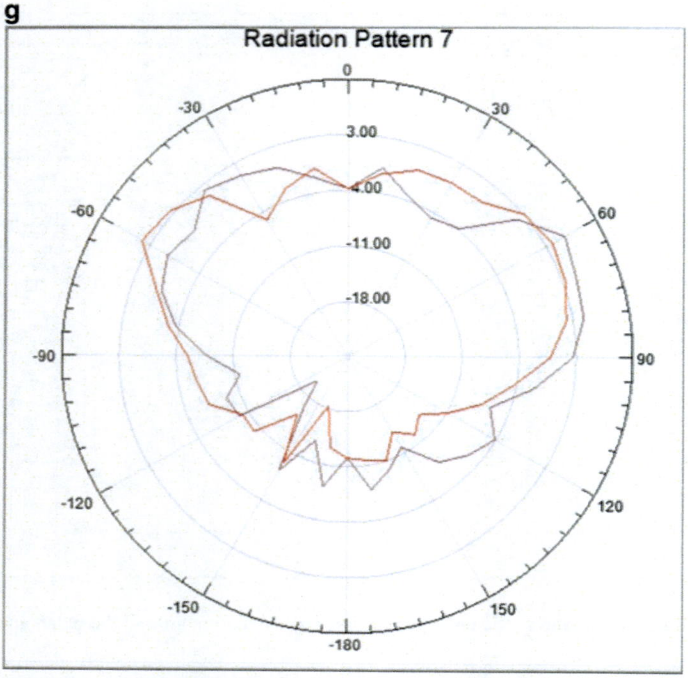

Fig. 2g Simulated radiation pattern for *E*-plane and *H*-plane

Table 1 Advancements in parameters

Antenna/parameters	Reference [2]	Proposed
Volume	20 * 20 * 4.8 mm^3	20 * 20 * 4 mm^3
Bandwidth	8.5 GHz (11.97–20.54 GHz)	16.89 GHz (16.89–33.8 GHz)
Gain	8.5 dB	9.35 dB
Radiation efficiency	88.5%	76%

These results are obtained by simulating proposed antenna in HFSS EM simulator

References

1. Kraus JD, Marhefka RJ, Khan AS, Antenna and wave propagation, 4th edn, pp 500–519
2. Mishra B, Singh V, Singh RK, Singh N, Singh R, A compact UWB patch antenna with defected ground for Ku/K band application. https://doi.org/10.1002/mop.30911
3. Bhadouria AS, Kumar M (2014) Wide Ku-band microstrip patch antenna using defected patch and ground. In: International conference on advanced engineering technology research (ICAETR—2014), IEEE, pp 1–5
4. Matin MA, Sharif BS, Tsimenidis CC (2011) Broadband stacked microstrip antennas with different radiating patch. Wirel Pers Commun 56:637–648

5. Patel AK, Jaiswal K, Pandey AK, Yadav S, Srivastava K, Singh R (2019) A compact inverted V-shaped slotted triple and wideband patch antenna for Ku, K, and Ka band applications. In: RTC2E-LNEE-524, pp 59–67
6. Khandelwal MK, Kanaujia BK, Dwari S, Kumar S, Gautam AK (2014) Bandwidth enhancement and cross-polarization suppression in ultra wideband microstrip antenna with defected ground plane. Microw Opt Technol Lett 56:2141–2146
7. Tan B-K, Withington S, Yassin G (2016) A compact microstrip-fed planar dual-dipole antenna for broadband applications. IEEE Antennas Wirel Propag Lett 15:593–596
8. Prasad PC, Chattoraj N (2013) Design of compact Ku band microstrip antenna for satellite communication. In: International conference on communicational signal process IEEE, pp 196–200
9. Khandelwal MK, Kanaujia BK, Dwari S, Kumar S (2013) Design and analysis of microstrip DGS patch antenna with enhanced bandwidth for Ku Band applications. In: International conference microwave photonics, IEEE, pp 1–4

Electromagnetic Analysis of MEMS-Based Tunable EBG Bandstop Filter Using RF MEMS Switch for Ku-Band Applications

G. Shanthi, K. Srinivasa Rao, and K. Girija Sravani

Abstract This paper presents an electromagnetic analysis of EBG Bandstop filter integrated with RF MEMS switch. The transmission line theory in microwave technology is used to analyze the proposed structure by studying isolation parameter of the filter. A fixed–fixed switch with low pull-in voltage of 4.05 V is designed and integrated on the signal line which is equidistant from the two EBG structures of filter. The filter resonates at 14.05 GHz without integrating the switches and is shifted to 15.85 GHz due to upstate capacitance of 41.56 μF when integrated with the switch at on-state. The frequency is tuned to 15.27, 14.5 and 11.4 GHz by actuating the beam to displace about 1, 2 and 3 μm. The switch produces a downstate capacitance of 19.11 pF during offstate and tunes the resonant frequency to 11.4 GHz. Thus, the tuning of the proposed EBG Bandstop filter is achieved by using RF MEMS switches and is efficiently used for Ku-band applications. The proposed electromagnetic analysis is carried out using HFSS 13.0v FEM tool.

Keywords Electromagnetic analysis · EBG Bandstop filter · RF MEMS switch · Microwave technology · Transmission line theory

G. Shanthi (✉) · K. Srinivasa Rao (✉)
Department of Electronics and Communication Engineering, KL Deemed to be University, Koneru Lakshmaiah Education Foundation, Vaddeswaram, Guntur 522502, India
e-mail: shanthi_g@vnrvjiet.in

K. Srinivasa Rao
e-mail: srinivasakarumuri@gmail.com

G. Shanthi
Department of Electronics and Communication Engineering, VNR Vignana Jyothi Institute of Engineering and Technology, Bachupally, Hyderabad, Telengana 500090, India

K. Girija Sravani
Department of Electronics and Communication Engineering, NIT Silchar, Assam, India

P. S. R. Chowdary et al. (eds.), *Microelectronics, Electromagnetics and Telecommunications*, Lecture Notes in Electrical Engineering 655,
https://doi.org/10.1007/978-981-15-3828-5_18

1 Introduction

Wireless communication has shown immense growth in satellite applications, millimeter-wave circuits and radio frequencies. As it is necessary to define the bandwidth/frequency of the signal in television broadcasting and radio systems, frequency tuning has become significant in avoiding the overlapping of signals with other bandwidths. The need of frequency tuning in receiving and transmitting systems and reconfigurable filters has been proved to be a solution. It provides solution to other challenges like mutual interference, interference with neighboring nodes and efficient usage of the spectrum. After development of RF MEMS Switches with excellent characteristics such as low losses, low power consumption, low distortion made these switches to be used in reconfigurable filters, tunable phase shifters, antennas and also in tuning of matching networks.

A two-pole tunable filter with distributed MEMS transmission line is designed with its resonant frequency tuned between 37.8 and 40.4 GHz and bandwidth ranges in between 1 and 1.6 GHz [1]. A MEMS tunable bandstop filter was investigated in which a tuning range of 17.3–19 GHz was observed with insertion loss varying from 1.3 to 2.4 dB [2]. The characteristics and RF behavior of a CSRR band reject filter were observed. The rejection of the stop bands for upstate is −20.18 dB and for upstate is −18.29 dB for center frequencies 38.80 GHz and 35.32 GHz, respectively [3]. The center frequency of a tunable bandstop filter is shifted by tuning the height of MEMS switch ranging from 6 to 14 GHz [4]. A CCSR-based CPW transmission line filter was proposed with the center frequency of 29.79 GHz, rejection of stop band is −19.87 dB, and at 33.17 GHz, it is observed to be 17.96 dB [5]. Single-ended absorptive bandstop filter using compact, low-Q resonators was designed with maximum attenuation of 63 dB and frequency range 2.9–4.3 GHz [6].

The tunable MEMS-based electromagnetic band gap filter for Ku-band applications with low insertion and high isolation has never been proposed till date. In this paper, an attempt to design Bandstop filter with electromagnetic band gap structure for Ku-band applications is carried out and achieved good RF characteristics. The tuning of the frequency is achieved by incorporating RF MEMS switch over the transmission line of the filter. The electromagnetic characteristics of EBG filter integrated with RF MEMS switch are clearly presented.

The rest of the paper is organized as follows: Sect. 2 illustrates the proposed design and working principle of proposed EBG Bandstop filter integrated with RF MEMS switches. Section 3 discusses the results and discussions based on integration of switch at different conditions and followed by conclusion.

Table 1 Dimensions of EBG Bandstop filter integrated with proposed fixed–fixed switch

S. no.	Component	Length (μm)	Width (μm)	Depth (μm)	Material
1	Substrate	560	760	600	Silicon
2	Substrate dielectric	560	760	2	SiO_2
3	CPW (G/S/G)	70/120/70		1	Gold
4	EBG radius	500			–
5	Switch cavity slots	80	400	1	–
6	Suspending beam	400	120	0.5	Gold
7	Signal line dielectric	120	140	0.5	HfO_2
8	Actuation electrodes	120	120	1	Gold

2 Device and Working Principle

2.1 EBG Bandstop Filter

The structure of proposed filter is consisting of CPW structure made up of gold over the silicon dioxide insulating layer which are both mounted on a silicon substrate with high dielectric constant of 11.9. The circular slots created on the ground planes of CPW serve as an electromagnetic band gap. The etched portion of the ground planes forms EBG, and these are used to develop good inductance for proposed filter. A shunt switch is placed over the signal line with a distance of λ/4 from the first EBG structure. The switch is made up of combination of dielectric layer and suspending beam in which a thin dielectric layer is placed over the signal line and beam is suspended over the dielectric layer with a gap of 3 μm. The dimensions of the proposed filter are described in Table 1 and are designed using FEM tool to understand its electromagnetic characteristics Fig. 1.

2.2 RF MEMS Switch

A fixed–fixed beam type RF MEMS switch is taken over the signal line to tune the frequency of the bandstop filter by creating a capacitive path between the signal and ground lines. A thin dielectric layer is placed over the signal line to generate high capacitance over the signal line suspending beam. The fixed–fixed beam is placed over the dielectric by leaving an air gap of 3 μm (Fig. 2).

The beam is fixed at both ends by using anchors where height of the anchors is equal to the height of the air gap. Two slots are created on the ground planes at the anchors to provide an inductance which is highly useful to obtain the resonant frequency of RF MEMS switch. The materials for EBG Bandstop filter and proposed switch are selected by using VIKOR method, and the dimensions along with materials are presented in Table 1.

Fig. 1 Proposed EBG Bandstop filter using RF MEMS switch

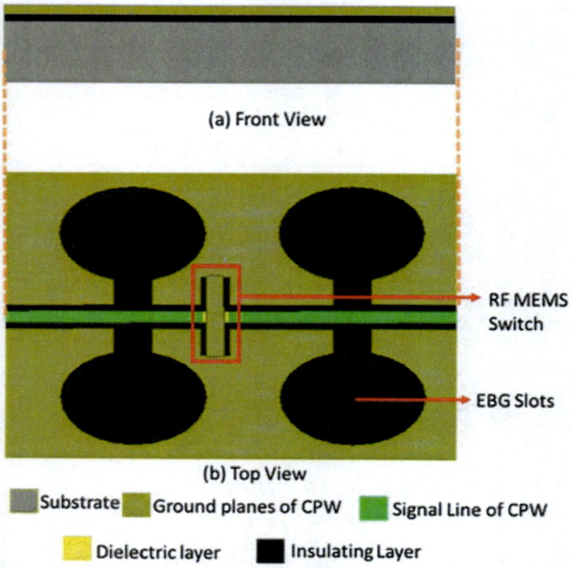

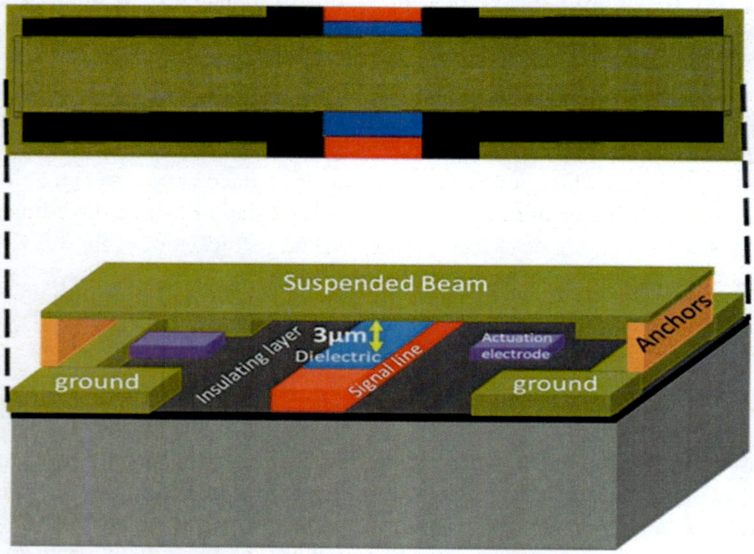

Fig. 2 Structure of proposed fixed–fixed switch

2.3 Working Principle

EBG structures are the periodic structures which are equivalent to the magnetic surface at the resonant frequency. These structures possess high surface impedance and are able to suppress the surface waves at certain operational frequencies. Each unit of EBG can be represented as the RLC tank circuit where the inductor is in series and resistor and capacitance are in parallel [7]. The single unit EBG is represented with RLC tank circuit configuration as shown in Fig. 3.

The proposed Bandstop filter attenuates the RF signal band of frequencies based upon the RLC values with central resonant frequency (f_0) which is obtained from Eq. (3) [8] is

$$f_0 = \frac{1}{2\pi\sqrt{LC}} \tag{1}$$

The proposed Bandstop filter consists of two EBG structures which are placed at a certain distance. The circuit configuration of proposed Bandstop filter containing two EBG unit structures with RF MEMS switch integrated between them is represented in Fig. 4. The R_m, L_m and C_m are the lumped parameter occurred due to switch configurations.

The RF MEMS switch is configured with RLC circuit which is in shunt configuration with the lumped parameter of EBG structures. The capacitance (C_m) developed by the switch can be varied by actuating the beam at different downward displacement. When the impedance due to inductance and impedance due to capacitance are equal, then the switch acts as a pure resistive path and the total signal is grounded by the switch without allowing to the output terminal. Hence, the proposed design acts as a Bandstop filter and can be tuned for different frequencies with the help variable capacitance during actuation. The total impedance occurred by the switch is given by Eq. (1) is

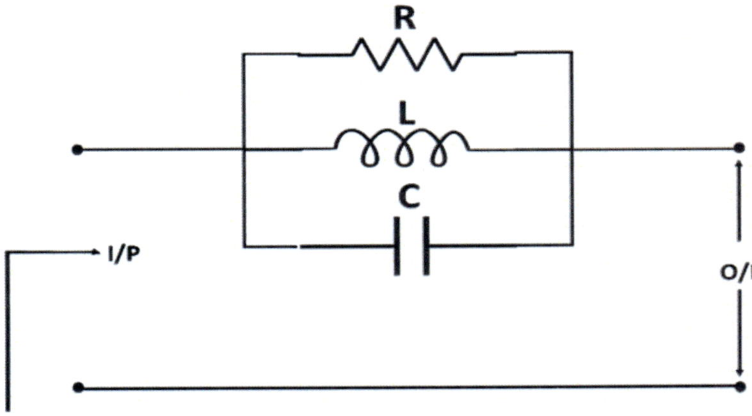

Fig. 3 RLC configuration of single unit EBG structure

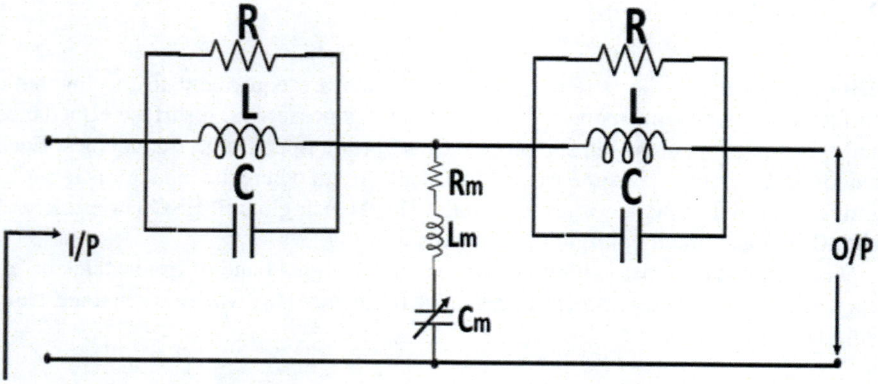

Fig. 4 RLC configuration of proposed Bandstop filter using RF MEMS switch

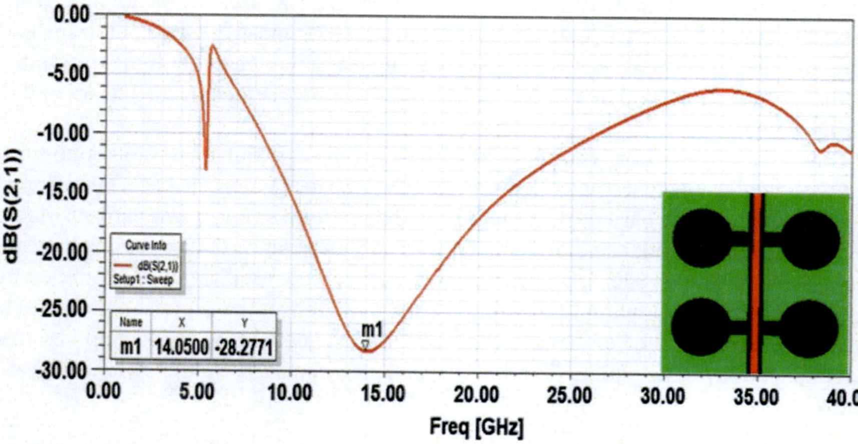

Fig. 5 Isolation of the proposed Bandstop filter without integration of RF MEMS switch

$$Z_m = R_m + X_L - X_m \tag{2}$$

where Z_m is the total impedance, R_m is the resistance, L_m is inductance and C_m is the capacitance developed by proposed switch. The impedance occurred during the resonance can be given as

$$Z_0 = \frac{j\omega L}{1 - \left(\frac{\omega}{\omega_0}\right)^2} \tag{3}$$

where the ω is the transmission signal frequency and the resonant frequency ω_0 can be determined by

$$\omega_0 = \frac{1}{\sqrt{L_m C_m}} \tag{4}$$

$$f_m = \frac{1}{2\pi\sqrt{L_m C_m}} \tag{5}$$

The signal which is transmitted through the filter is grounded when the transmission frequency is resonant to the resonance frequency of the switch. The resonant frequency of the switch can be varied by varying the capacitance (C_m), and the frequency of the Bandstop filter is reconfigured with the help of switch actuation.

3 Results and Discussions

The proposed structure is designed using HFSS 13.0v software, and electromagnetic characteristics have been analyzed for the proposed Bandstop filter with and without integration of the switch. The study of S—parameters—can illustrate the performance of the proposed EBG Bandstop filter and electromagnetic performance of RF MEMS switch.

3.1 EBG Bandstop Filter Without Switches

Initially, the proposed EBG Bandstop filter is not integrated with the RF MEMS Switch. Isolation of the filter is the parameter used to analyze the performance of proposed EBG Bandstop filter. The proposed filter does not transmit the selected band of frequencies as the name suggested it as Bandstop filter. The isolation of the filter over a range of Ku-band frequencies is observed by simulating the design and presented in Fig. 6 (Fig. 5).

3.2 EBG Bandstop Filter with Integration of Proposed RF MEMS Switch

The upstate of the capacitance of the switch occurs between the beam and signal line when the switch is not actuated. It is obtained by Eq. (6) when a small amplitude of RF signal is applied on the signal line of CPW [9].

$$C_u = \frac{\varepsilon_0 A_c}{g_0 + \frac{t_d}{\varepsilon_r}} \tag{6}$$

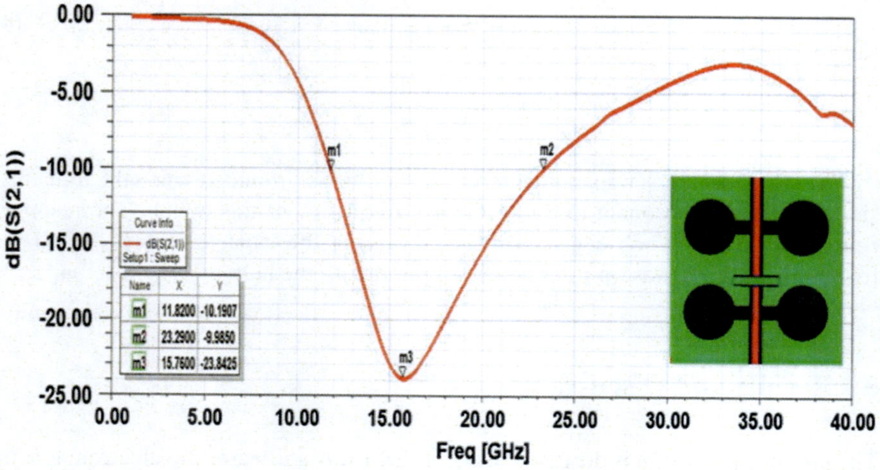

Fig. 6 Isolation of the proposed Bandstop filter with RF MEMS switch in onstate

where A_c is the overlapping area of beam and signal line, t_d is the dielectric thickness and g_0 is the air gap before actuation which is 3 μm. It is obtained by calculating Eq. (6) which is 41.56 μF. The isolation of the proposed EBG Bandstop filter is shifted from 14.05 to 15.76 GHz due to the upstate capacitance developed by the switch which is in shunt configuration. The filter gives −23.8 dB at 15.76 GHz which is the resonant frequency of filter when switch is in upstate. The filter gives a grat bandwidth of 11.2 GHz and does not allow the frequencies between 11.82 and 23.2 GHz.

The RF MEMS switch is actuated due to electrostatic forces between beam and biasing electrodes. These electrostatic forces are generated by creating a potential difference between beam and biasing electrodes. The beam is always connected to the ground planes, and thus, the voltage potential is applied on the biasing pads. The voltage required to pull the beam downwards and made it to completely fall on the dielectric layer present over the signal line is referred as the pull-in voltage, and it is given as [10]

$$V_p \sqrt{\frac{8Kg_0^3}{27\varepsilon_0 A}} \tag{7}$$

where K is the stiffness of the beam that can be calculated using conventional formulas as 0.525 N/m, g_0 is the gap between the beam and biasing electrodes and A is the overlapping area between electrodes. From the dimensions presented in Table 1, the pull-in voltage of the switch is calculated as 4.05 V. The switch beam completely falls on the dielectric layer over the signal line, and the capacitance increases during this condition. This capacitance can be referred as downstate capacitance and is given by [11].

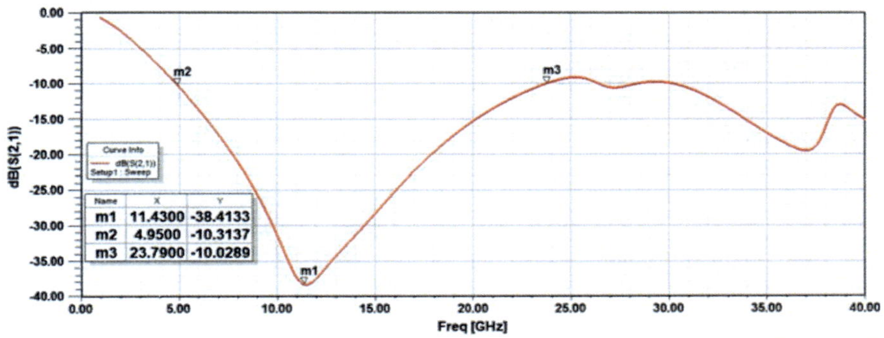

Fig. 7 Isolation of the proposed Bandstop filter with RF MEMS switch in offstate

$$C_{\text{down}} = \frac{\varepsilon_0 \varepsilon_r A_t}{t_d} \tag{8}$$

where A_t is the total overlapping area which is the summation of A_c and A, ε_r is the relativity permittivity of the dielectric medium which is 25 and t_d is the dielectric thickness. The total downstate capacitance obtained from Eq. (8) is 19.11 pF. This downstate capacitance which is in shunt configuration of the filter shifts the Bandstop filter frequency to 11.43 GHz as shown in Fig. 8. The filter shows a high isolation of −38.4 at 11.43 GHz and can be efficiently used in the frequency band from 4.9 to 23.7 GHz with a bandwidth of 12.3 GHz (Fig. 7).

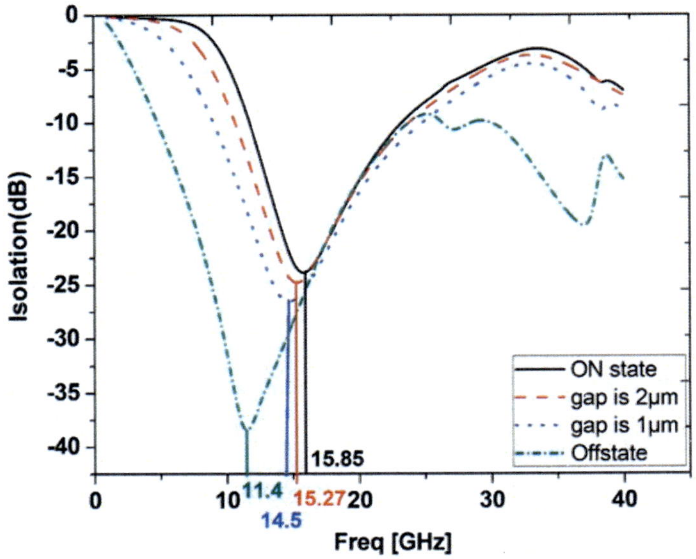

Fig. 8 Tuning of proposed EBG Bandstop filter

3.3 *Frequency Reconfiguration in EBG Bandstop Filter Integrated with RF Switches*

The center frequency (resonant frequency) of the proposed EBG Bandstop filter can be tuned by RF MEMS switch, and the tuning is shown in Fig. 8. The resonant frequency of the filter shifts from 15.85 to 15.27 GHz when the beam of the switch is displaced to 1 μm vertically downwards from onstate which is maintained with initial gap of 2 μm between beam and dielectric. It further shifts to 14.5 GHz when the beam is displaced to 2 μm maintaining a gap of 1 μm, and it again shifts to 11.4 GHz when the switch displaces 3 μm and falls on dielectric layer. No further shift in frequency occurs as the switch can not have downward displacement further.

4 Conclusion

In this paper, we have proposed a tuneable EBG Bandstop filter integrated with RF MEMS switch for Ku-band applications. The filter is designed at resonant frequency of 14.05 GHz without integrating of switch, and the isolation is observed to understand its performance. A fixed–fixed switch with low pull-in voltage of 4.05 V is proposed and integrated with proposed filter over the signal line. The fixed–fixed switch integrated with the filter shifts the filter frequency from 15.85 to 11.5 GHz, and the frequency can be varied in between the range which illustrates the fine tuning of the filter.

Acknowledgement The Authors would like to thank to NMDC supported by NPMASS, National Institute of Technology, Silchar for providing the necessary computational tools.

References

1. Mercier D, Blondy P, Cros D, Guillon P (2001) Distributed MEMS tunable filters. In: 31st European microwave conference, 2001. https://doi.org/10.1109/euma.2001.339123
2. Karim MF, Liu AQ, Yu AB, Alphones A (n.d.) MEMS-based tunable bandstop filter using electromagnetic bandgap (EBG) structures. In: Asia-pacific microwave conference proceedings. https://doi.org/10.1109/apmc.2005.1606719
3. Pradhan B, Gupta B (2014) RF MEMS tunable band reject filter using meta materials. In: International conference on electronics, communication and instrumentation (ICECI). https://doi.org/10.1109/iceci.2014.6767378
4. Zhang NB, Deng ZL, Zhao MY (2010) A novel tunable band-stop filter based on RF MEMS technology. In: 2010 international conference on computer, mechatronics, control and electronic engineering. https://doi.org/10.1109/cmce.2010.5609950
5. Pradhan B, Gupta B (2014) Novel tunable band reject filter using RF MEMS technology. In: Proceedings of the 2014 IEEE students' technology symposium. https://doi.org/10.1109/techsym.2014.6808071

6. Kim B, Lee J, Lee J, Jung B, Chappell WJ (2013) RF CMOS integrated on-chip tunable absorptive bandstop filter using Q-tunable resonators. IEEE Trans Electron Devices 60(5):1730–1737. https://doi.org/10.1109/ted.2013.2253557
7. Sinha T, Pandey AK, Chauhan RK (2016) A compact dualband bandpass-to-bandstop tunable filter for wireless applications. In: 2016 international conference on emerging trends in communication technologies (ETCT). https://doi.org/10.1109/etct.2016.7882998
8. Srinivasa RK, Girija Sravani K (2018) Design and performance analysis of uniform meandered structured RF MEMS capacitive shunt switch along with perforations. Microsyst Technol 24(2):901–908
9. Shah IA, Hayat S, Basir A, Zada M, Shah SAA, Ullah S (2019) Design and analysis of a hexa-band frequency reconfigurable antenna for wireless communication. AEU-Int J Electron Commun 98:80–88
10. Wright MD, Baron W, Miller J, Tuss J, Zeppettella D, Ali M (2018) MEMS reconfigurable broadband patch antenna for conformal applications. IEEE Trans Antennas Propag 66(6):2770–2778
11. Xu Y, Tian Y, Zhang B, Duan J, Yan L (2018) A novel RF MEMS switch on frequency reconfigurable antenna application. Microsyst Technol 1–9

Semi-circular MIMO Patch Antenna Using the Neutralization Line Technique for UWB Applications

Gorre Naga Jyothi Sree and Suman Nelaturi

Abstract In this paper, the semi-circular MIMO patch antenna is designed and reducing the mutual coupling between the MIMO patches by using neutralization line and bandwidth enhancement is done by monopole structure, and the proposed antenna is for UWB applications. The mutual coupling between the patches is reduced by inserting neutralization line. The dimensions of the semi-circular MIMO patch antenna are 50×25 mm^2. The envelop correlation coefficient (ECC) value is 0.002 and the diversity gain is 10dBi. The efficiency of the reported antenna is 98%. The designed structure has reflection coefficients and transmission coefficients of ≤ -10 dB and ≤ -20 dB, respectively. The presented antenna has stable patterns of radiation, gain, diversity gain, and group delay. The proposed antenna is simulated using computer simulation technology MW studio (CST) software.

Keywords Reflection coefficient · Neutralization line · Transmission coefficient · Mutual coupling · Envelope correlation coefficient (ECC)

1 Introduction

Nowadays, the wireless communication technology is transforming to 5G. To attain those 5G data, speed and quality of the system are improved by MIMO patch antennas. A smooth decoupling design is used to enhance port isolation. As well as, mutual coupling is reduced to more than −15 dB and less envelope correlation coefficient that is below 0.02. In [1], the Koch fractal-based MIMO UWB system for the rejection of WLAN band with low mutual coupling is described. With the help of shared antenna, compact UWB of notch–band characteristics whose ECC < 0.15 is maintained in [2]. A decoupling structure of H-type diversity array in [3] with high compact area is explained here. In [4], technique applied for the slitted ground

G. Naga Jyothi Sree (✉) · S. Nelaturi
Department of ECE, VFSTR (Deemed to Be University), Vadlamudi, Andhra Pradesh, India
e-mail: nagajyo53@gmail.com

S. Nelaturi
e-mail: nelaturi.suman4@gmail.com

P. S. R. Chowdary et al. (eds.), *Microelectronics, Electromagnetics and Telecommunications*, Lecture Notes in Electrical Engineering 655,
https://doi.org/10.1007/978-981-15-3828-5_19

Table 1 Comparison with existing techniques

Published literature	Antenna size (mm^2)	Frequency range (GHz)	Isolation (dB)	ECC	Efficiency (%)
[3]	80 × 48	3.1–10.6	≤−25	<0.15	75
[5]	55 × 120	2.4–5.2	≤−25	<0.15	73
[6]	90 × 55	2.0–3.0	≤−20	<0.18	82
[8]	100 × 55	1.85–11.9	≤−18	<0.20	81
[9]	70 × 37	4.0–9.0	≤−20	<0.28	84
[10]	91 × 38	2.8–10	≤−22	<0.05	74
Current design	50 × 25	3.0–10.0	≤−40	<0.02	98

plane for MIMO implementation system with isolation >18 dB is achieved. For two coupled-type antennas, a dual-type resonator is approached in [5] with higher ECC obtained within their structure. A high compact area 2-element antenna array is used for mobile terminals with the method of decoupling network [6] maintained the isolation >20 dB. In [7], 4-element compact MIMO UWB structure is for rejection of WLAN whose ECC is also observed as higher value. For the applications of mobile devices, a compact structured MIMO system ranged from 1.85 to 11.9 GHz in [8] had a large compact area. In reference [9] EBG structure explained and in [10] challenges of MIMO construction in the usage of UWB systems whose ECC < 0.05. Table 1 indicated the comparison with existing references. The novelty in the current design structure is a monopole antenna that is designed to obtain the UWB characteristics of the system. To improve the reflection coefficient, a small portion of the slots are cutting on the left and right sides of the circular patches. In order to reduce the mutual coupling among the MIMO analysis system neutralization technique is implemented in the current structure designed structure greatly improves the performance of the MIMO system compared to the literature mentioned systems. Placing the neutralization line between the circular patch with slots among the patch element also greatly improves the efficiency of the system and realized gain also improves the current design. The distribution of surface currents also observed the greater improvement by the parameter of surface current distributions.

2 Antenna Geometry and Analysis

The proposed patch is designed with dimensions of 50 mm × 25 mm with monopole ground plane. To obtain this structure, the width and lengths are considered in terms of WD and LT parameters. The next step is considered a circular ring with inner and outer radius designed and transformed on the substrate to produce MIMO structure. The top portion of the circle is cut to form a semi-circular patch. Then, at the center of a semi-circular patch, a rectangular patch is etched. Further, add a circle to the semi-circle rectangular etched surface to increase the refection coefficient of the proposed

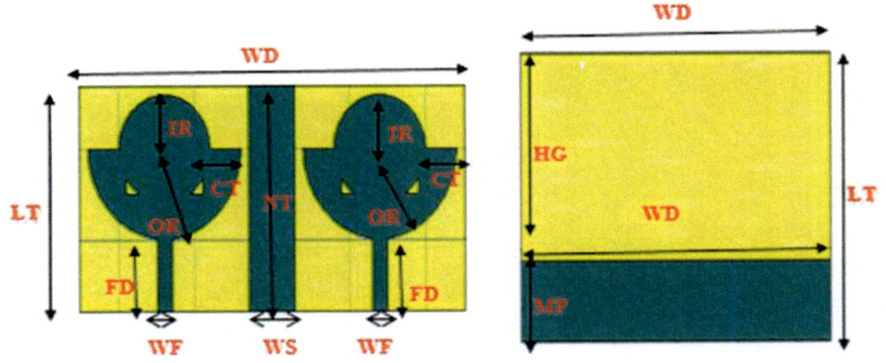

Front view Bottom View

Fig. 1 Geometry of proposed system

patch. Here, for exciting the patch, line feed mechanism is to be taken place. By this line feed mechanism, antenna is to be excited. This entire design can be obtained from the monopole ground to produce UWB characteristics. Figure 1 identifies the current design geometry of the system. Table 2 shows the parameters of the current design.

3 Results and Discussion

The UWB semi-circular MIMO antenna with the method of neutralization line is designed on FR-4 substrate with permittivity of 4.4 and loss tangent of 0.02. Figure 2 represents the comparison about scattering parameters indication resonating at tri-band of frequencies like 4.8, 6.8 and 8.5 GHz. At the tri-band resonant frequencies, the obtained S_{11} and S_{21} are −38 dB, −21 dB, and −29 dB and −40 dB, −60 dB, and −52 dB, respectively, which identified that a greater achievement in the all tri-band frequencies maintained is ≤−35 dB. From Fig. 2, it is observed that by the etched ground plane, we attain a <−10 dBi reflection coefficient for UWB and by adding the neutralization line, the mutual coupling is reduced to <−30 dBi. Figure 3 identifies the radiation efficiencies at the resonant frequencies that are 98, 87, and 78%.

Figure 4 resemblance the gain of the semi-circle UWB MIMO radiator which represents the values of 2.8, 5.8 and 4.4 dBi at the tri-band resonant frequencies which is shown in Fig. 4. Figure 5 shows the distribution of surface currents at tri-band frequencies. Figure 5a identifies that port 1 is excited and distribution of current is maximum in left side of the circular patch. Figure 5b indicates that port 2 excited and the maximum current flows at the feeding point. Figure 5c resemblance the port 1 is excited and the distribution of current is maximum right side of the circular patch and neutralization path. Figure 6 shows the envelope correlation coefficient

Table 2 Proposed antenna design dimensions

Design parameters	LT	WD	IR	OR	FD	NT	MP	HG	WF	CT	WS
Value(mm)	25	50	6	10	8	25	7	18	2	4	7

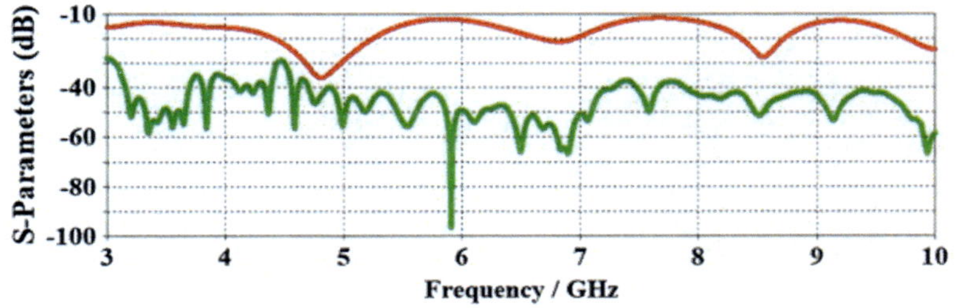

Fig. 2 Proposed MIMO antenna S-parameters

Fig. 3 Radiation efficiency of MIMO

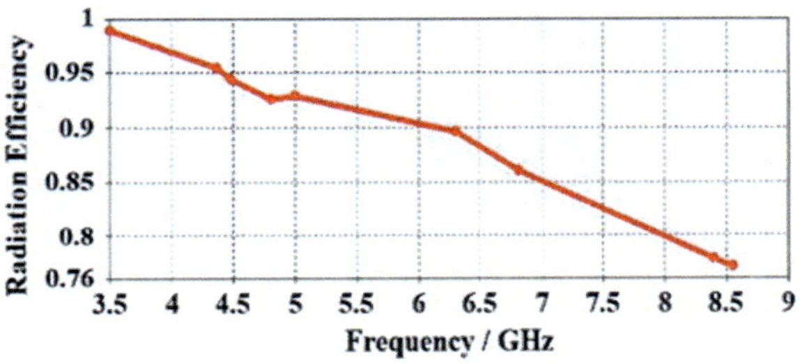

Fig. 4 Gain of MIMO radiator

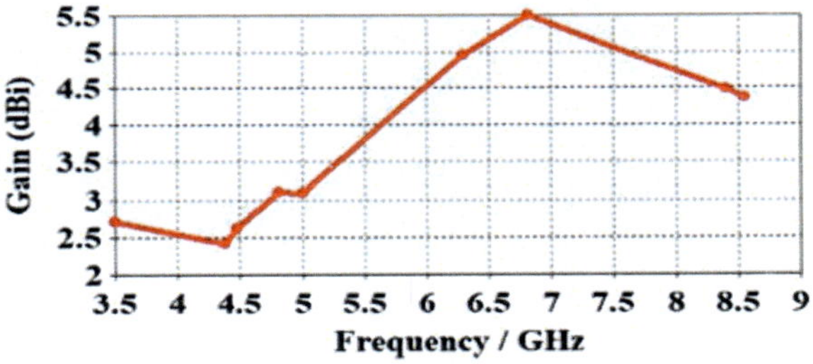

of the current designed structure which maintained the value ECC ≤ 0.002 at the resonating band of frequencies and also observed that the diversity gain of the MIMO system at the tri-band resonant frequencies is approximately equal to the 10 dBi of the current designed structure which is shown in Fig. 7 and produced the better antenna parameter values. The designed structure shows the better parameters for MIMO analysis which satisfied the MIMO parameters like ECC and diversity gain of the current design.

The radiation patterns of the current design are identified at the three resonant frequencies, and the distribution of fields in *xz* and *yz* planes indicates the distribution of E-field and H-field co-pole and cross-pole representations which always show that the co-pole is always larger than the cross-pole for every distribution radiation patterns at the resonant frequencies which are shown in Fig. 8.

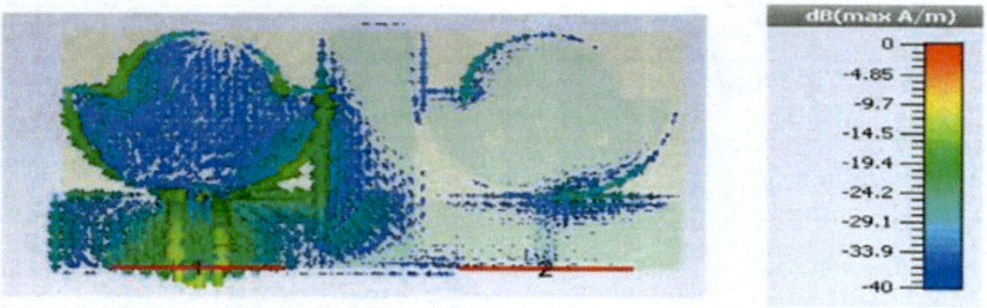

(a) Port 1 Excited @ 4.8 GHz

(b)
(c) Port 2 excited @ 6.8 GHz

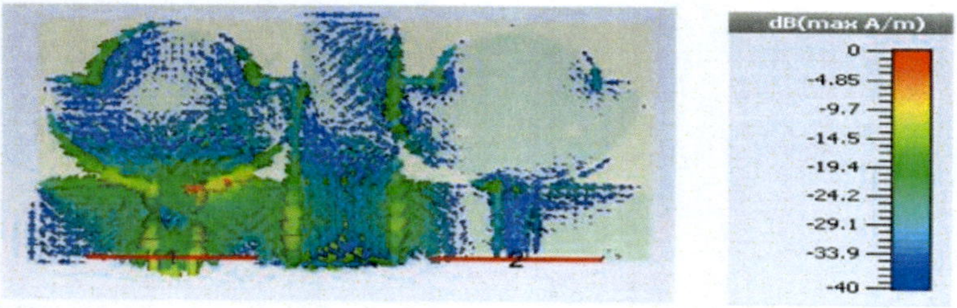

(d) Port 1 Excited @ 8.5 GHz

Fig. 5 Surface current distribution about tri-band UWB antenna

Fig. 6 ECC of the current design

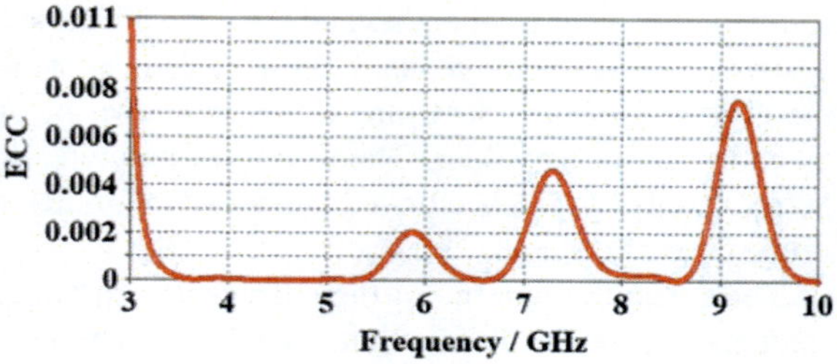

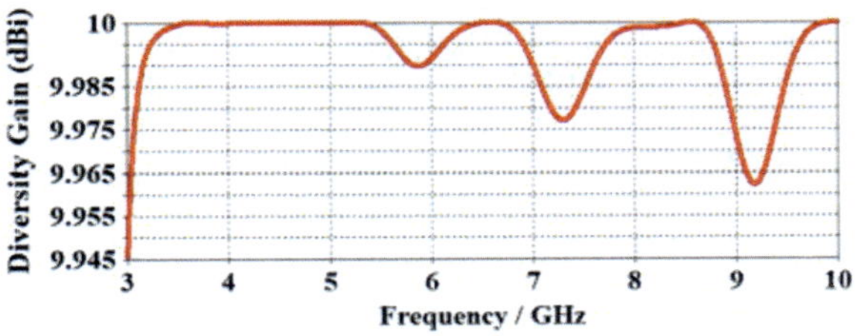

Fig. 7 Diversity gain current design

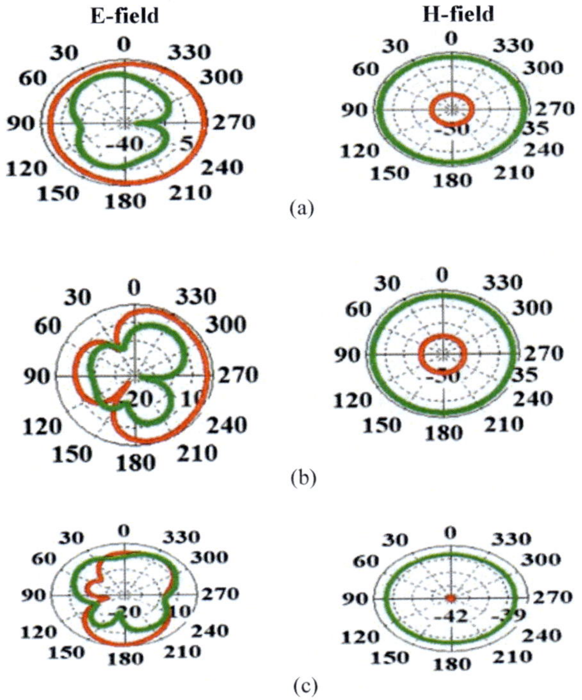

Fig. 8 Co- and cross-polarization at **a** 4.8 GHz, **b** 6.8 GHz, and **c** 8.5 GHz, for H-field: Co-pol: ----, Cross-pol: -----; for E-field: Co-pol: -----, Cross-pol: ----

4 Conclusions

In this article, a UWB semi-circular MIMO is designed for the applications of wireless systems. The resonant frequencies wide-band characteristics and maintained the transmission and reflection coefficients are ≤ -35 and ≤ -15 dB at the proposed designed structure. The semi-circular patches are separated by a neutralization line between the patches to reduce the mutual coupling between the patches here. So that, with the neutralization line technique, the mutual coupling is reduced to ≤ -60 dB. This shows that the system has a greater improvement in the proposed structure better applicable to MIMO applications. For bandwidth enhancement, here we have gone through monopole structure. This structure also has ECC ≤ 0.002 and diversity gain is approximately 10 dBi at the resonant band of frequencies for UWB applications of MIMO systems.

References

1. Tripathi S, Mohan A, Yadav S (2015) A compact Koch fractal UWB MIMO antenna with WLAN band-rejection. IEEE Antennas Wirel Propag Lett 14:1565–1568
2. Srivastava G, Kanuijia BK (2015) Compact dual band-notched UWB mimo antenna with shared radiator. Microw Opt Technol Lett 57(12):2886–2891
3. Gogosh Nayab et al (2013) An UWB diversity antenna array with a novel H-type decoupling structure. Microw Opt Technol Lett 55(11):2715–2720
4. Wu Yu-Ting, Chu Qing-Xin (2014) Dual-band multiple input multiple output antenna with slitted ground. IET MicrowS Antennas Propag 8(13):1007–1013
5. Zhao L, Wu K-L (2015) A dual-band coupled resonator decoupling network for two coupled antennas. IEEE Trans Antennas Propag 63(7):2843–2850
6. Zhao L, Yeung LK, Wu K-L (2014) A coupled resonator decoupling network for two-element compact antenna arrays in mobile terminals. IEEE Trans Antennas Propag 62(5):2767–2776
7. Khan SM et al (2016) A compact four elements UWB MIMO antenna with on-demand WLAN rejection. Microw Opt Technol Lett 58(2)
8. Lee J-M et al (2012) A compact ultrawideband MIMO antenna with WLAN band-rejected operation for mobile devices. IEEE Antennas Wirel Propag Lett 11:990–993
9. Yang F, Yahya R-S (2003) Microstrip antennas integrated with electromagnetic band-gap (EBG) structures: A low mutual coupling design for array applications. IEEE Trans Antennas Propag 51(10):2936–2946
10. Jusoh M et al (2012) A MIMO antenna design challenges for UWB application. Prog Electromagn Res 36:357–371

Multi-Beam Generation Using Quasi-Newton method and Teaching Learning Based Optimization algorithm

R. Krishna Chaitanya, P. Mallikarjuna Rao, and K. V. S. N. Raju

Abstract Satellite communication is extensively used in television broadcasting and mobile communications. Multi-beams are used in satellite communication to communicate different distinct locations with multiple users. Optimization methods like quasi-Newton method (QNM) and teaching–learning-based optimization (TLBO) are used to generate multi-beam pattern using linear antenna arrays. The desired amplitude and phase distributions are determined by using both QNM and TLBO algorithms. The desired multi-beam pattern is plotted in U domain where $u = \sin \theta$. QNM has converged much fast and with less number of iterations than TLBO algorithm in generating the multi-beam pattern. The convergence plots are generated for $n = 60, 80$ elements using QNM and TLBO algorithms.

Keywords Quasi-Newton method · Linear array antenna · Teaching–learning-based optimization (TLBO) · Multi-beam

1 Introduction

Linear array antennas are used to generate different shaped beams by changing the input parameters like amplitude and phase. These shaped beams are used for wide range of communication purposes for different wireless communications. To generate these shaped beams, different random optimization methods like particle swarm optimization, TLBO, and nonlinear optimization method like QNM are used. In [1], Yazhong has proposed a hybrid approach which combines Newton method and chaos optimization. The new optimization solves nonlinear equations. The new algorithm combines rapid convergence of Newton method and global search capability of chaos

R. Krishna Chaitanya (✉) · K. V. S. N. Raju
Department of ECE, S.R.K.R. Engineering College, WG District, Bhimavaram, India
e-mail: rkc639@gmail.com

P. Mallikarjuna Rao
Department of ECE, Andhra University, Visakhapatnam, India

P. S. R. Chowdary et al. (eds.), *Microelectronics, Electromagnetics and Telecommunications*, Lecture Notes in Electrical Engineering 655,
https://doi.org/10.1007/978-981-15-3828-5_20

optimization. Becker [2] has developed a new algorithm called variable metric (quasi-Newton) forward–backward splitting algorithm for solving non-smooth convex problems. Li [3] has developed a method for dynamic economic dispatch using QNM, in Lagrangian relaxation problem by updating the multipliers. The algorithm has provided better convergence, computational efficiency, and robustness. Adel Hamdi [4] has developed a reduced QNM for simultaneous design and optimization of space preconditioner for the approximated reduced gradient. Chakravarthy [5] developed a new social grouping algorithm to synthesis circular array antenna. Social grouping algorithm is used to find lowest side lobe level for a fixed beam width. Suresh Chandra Satapathy [6] has developed a new Improved Teaching-Learning Based Optimization (ITLBO) which is better than traditional TLBO and other optimization methods like Particle swarm optimization (PSO) and Differential Evolution algorithm (DE). ITLBO has better performance over different benchmark optimization problems than TLBO, PSO, and DE. Chakravarthy [7] has optimized non-uniform linear array using TLBO algorithm. The amplitude parameters are determined using TLBO algorithm. Recioui [8] has used TLBO for thinning of linear and planar arrays. TLBO is used to find the optimum configuration of thinned array with lower side lobe level and fixed half-power beamwidth. Venkata Rao [9] has reviewed all the applications of TLBO in solving constrained and unconstrained applications in all fields of engineering and sciences. The potential of TLBO algorithm in solving various optimization problems are presented. Krishna Chaitanya [10] used QNM, particle swarm optimization (PSO), and firefly algorithm (FF) to generate triangle pattern and stair-step pattern using linear array antenna. The performance of QNM over PSO, FF to generate the shaped beams has been presented.

In this paper, linear array is synthesized to obtain the desired multi-beam pattern using TLBO and QNM optimization algorithms. Generation of desired pattern is studied by TLBO and QNM algorithms. Convergence plots determine the time taken to generate the desired pattern in terms of number of iterations. The desired pattern is generated such that the side lobe level is completely minimized.

In this paper, Sect. 2 discusses the array factor of linear array antennas. Section 3 gives a brief introduction about QNM and TLBO algorithms. Section 4 stem plots of amplitude and phase values for 60-elements using TLBO and QNM are presented. Section 5 concludes the paper with the performance improvement in terms of convergence, speed, stability, and robustness of QNM over TLBO.

2 Array Factor for Linear Array

Linear array antenna that has array factor [11] for N number of elements is given in Eq. (1)

$$\left.\begin{array}{l}E_f(\theta) = \sum_{n_1=-N}^{-1} a_{n_1} e^{j(kd_{n_1}x+\varphi_{n_1})} + \sum_{n_2=1}^{N} a_{n_2} e^{j(kd_{n_2}x+\varphi_{n_2})} \\ x = \cos\theta \\ k = 2\pi/\lambda, \\ d_{n_1} = \frac{(2n_1+1)}{2} \times \frac{\lambda}{2} \text{ for } -N \le n_1 \le -1 \\ d_{n_2} = \frac{(2n_2-1)}{2} \times \frac{\lambda}{2} \text{ for } 1 \le n_2 \le N\end{array}\right\} \quad (1)$$

ϕ_{n_1}, ϕ_{n_2} = Array antenna element phase value.
ϕ = Angle of observation.
a_{n_1} a_{n_2} = Array antenna element amplitude value.
d_{n_1}, d_{n_2} = Separation between two elements.
N = Linear array antenna elements.
λ = Wave length.

2.1 Fitness Function for Generation of Shaped Beam

Fitness function (X) in Eq. (2) is an error value, which is minimization of difference between desired value and obtained value. TLBO and QNM optimization algorithms are used to minimize the fitness function and obtain the desired parameters of amplitude and phase for linear array antenna.

$$\mathrm{X} = \frac{1}{N_n} \sum |(\mathrm{EV})_n|^2 \quad (2)$$

$$\mathrm{EV} = \text{desired value} - \text{obtained value} \quad (3)$$

N_n = Represents total sample points in the region.
X = Mean square error value of sample point.

3 Optimization Algorithms

3.1 Quasi-Newton Method

QNM is an optimization technique used for solving nonlinear equations and also to find global minimum of a function. QNM has a particular advantage over Newton method in which the procedural steps do not need second derivative, and hence, its computation is very fast and also complexity gets reduced. It is an extension of Newton method. QNM does not require the Hessian matrix for computation. It makes use of secant method to find the roots of first derivative of multidimensional problem.

There are different types of QNM called the symmetry rank 1(SR1) and Broydens method.

3.2 Teaching–Learning-Based Optimization

In TLBO algorithm [12, 13], there are teaching phase and learning phase. In the first phase (teaching phase), the teacher explains the concept to the student. Where there will be maximum learning phase by the student. It is similar to global search in random optimizations like PSO and FF. After the teaching phase, the student tries to learn the subject by discussing with friends called the learning phase, and it is within the students, between the groups of students in the class. This method is similar to local search for finding the solution with in the students.

In TLBO, teaching phase is by Eq. (4)

$$L_N^{g_1} = L^{g_1} + \text{rand} * (L_{\text{teacher}}^{g_1} - K_{\text{factor}} * M^{g_1}) \tag{4}$$

where

$L =$ learner.
$g_1 =$ generation.
$K_{\text{factor}} =$ teaching factor range (1–2).

In learning phase, Eq. (5) is

$$L_{N_x}^{g_1} = \begin{cases} L_x^{g_1} + \text{rand} * \left(L_x^{g_1} - L_y^{g_1}\right) \\ L_x^{g_1} < L_y^{g_1} \\ L_x^{g_1} + \text{rand} * (L_y^{g_1} - L_x^{g_1}) \\ \text{Otherwise} \end{cases} \tag{5}$$

4 Results

In this paper, to generate a Multi-beam pattern with Linear array antenna, different optimization techniques like QNM and TLBO are used.

The linear array antenna has a different number of element configurations $n =$ 20, 40, 60, 80 that are used to generate the shaped beam. The error value reduced with minimum iterations for array antennas is shown in Table 1. The error value is reduced to 0.023 for minimum number of iteration of 41 in quasi-Newton method, and the same error value is obtained with 1968 iterations in TLBO algorithm for 20-element array. The error value is reduced to 0.007 for 17 iterations in quasi-Newton method, and the same error value is obtained with 597 iterations in TLBO algorithm

Table 1 Error values obtained for both QNM and TLBO algorithm for minimum iterations

Array dimension	QNM		TLBO	
	Minimum iterations	Error value	Minimum iterations	Error value
20 elements	41	0.023	1968	0.023
40 elements	17	0.007	597	0.007
60 elements	177	0.0001	265	0.0001
80 elements	82	0.0001	177	0.0001

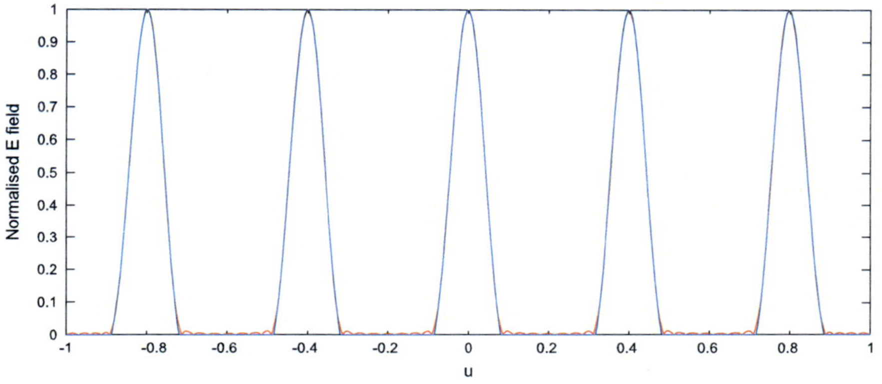

Fig. 1 Multi-beam pattern for 60-element array using QNM

for 40-element array. The error value is reduced to 0.0001 for 177 iterations in QNM, and the same error value is obtained with 265 iterations in TLBO algorithm for 60-element array. The error value is reduced to 0.0001 for 82 iterations using QNM, and the same error value is obtained with 177 iterations using TLBO algorithm for 80-element array. The results show that QNM has converged faster than the TLBO algorithm for different array configurations.

The excitation values for 60 elements for QNM are shown if Fig. 2a, b and for TLBO are shown in Fig. 4a, b. The obtained shaped beam pattern for 60 elements using QNM and TLBO is shown in Figs. 1 and 3.

5 Conclusion

A new method for the generation of multi-beam is presented in this paper using QNM and TLBO algorithm. A new algorithm implementation, amplitude, and phase distributions are presented. The convergence plot is presented for array length of 60- and 80-element arrays in Fig. 5. QNM has shown better results than TLBO

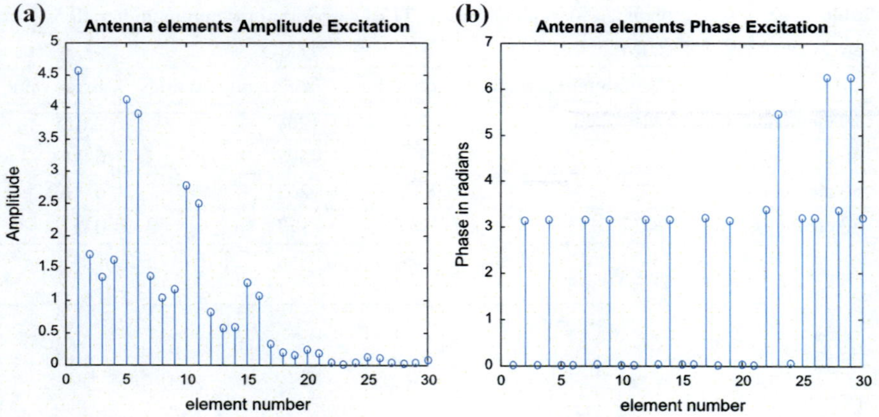

Fig. 2 **a, b** Stem plots of amplitude values and phase values for multi-beam pattern using QNM for 60-element array

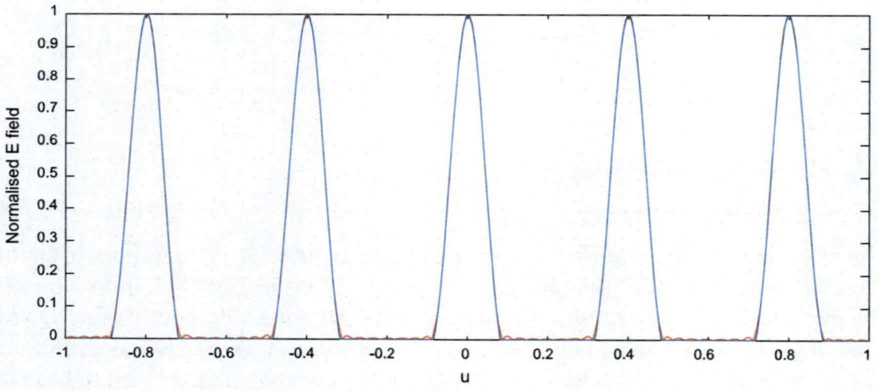

Fig. 3 Multi-beam pattern for 60-element array using TLBO

optimization method which has less algorithm-specific parameters. QNM has less number of iterations and faster than the random optimization methods. Further, QNM can be tested on all results obtained by random optimization methods.

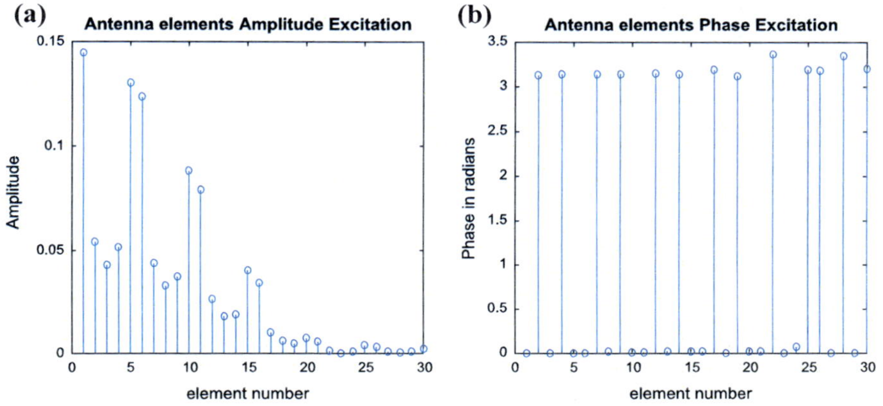

Fig. 4 **a, b** Stem plot for amplitude values and phase values for multi-beam pattern using TLBO for 60-element array

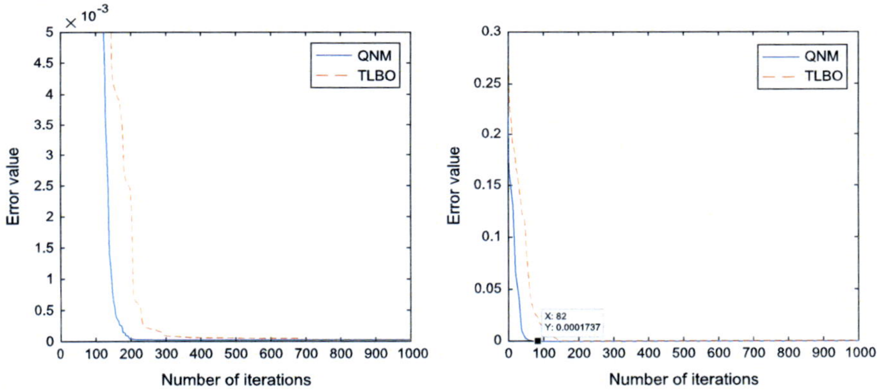

Fig. 5 Multi-beam pattern convergence plot using QNM and TLBO for 60- and 80-element array

References

1. Luo YZ, Tang GJ, Zhou LN (2008) Hybrid approach for solving systems of nonlinear equations using chaos optimization and quasi-Newton method. App Soft Comput 8(2):1068–1073
2. Becker S, Fadili J (2012) A quasi-Newton proximal splitting method. In: Advances in neural information processing systems. pp 2618–2626
3. Li Z, Wu W, Zhang B, Sun H, Guo Q (2013) Dynamic economic dispatch using Lagrangian relaxation with multiplier updates based on a quasi-Newton method. IEEE Trans Power Syst 28(4):4516–4527
4. Hamdi A, Griewank A (2011) Reduced quasi-Newton method for simultaneous design and optimization. Comput Optim Appl 49(3):521–548
5. Chakravarthy VS, Chowdary PS, Satpathy SC, Terlapu SK, Anguera J (2018) Antenna array synthesis using social group optimization. Microelectronics, electromagnetics and telecommunications. Springer, Singapore, pp 895–905

6. Satapathy S, Naik A (2013) Improved teaching learning based optimization for global function optimization. Decis Sci Lett 2(1):23–34
7. Chakravarthy VV, Babu KN, Suresh S, Devi PC, Rao PM (2015) Linear array optimization using teaching learning based optimization. In: Emerging ICT for bridging the future—proceedings of the 49th annual convention of the computer society of India CSI, vol 2. Springer, pp 183–190
8. Recioui A (2016) Design and thinning of linear and planar antenna arrays using a binary teaching learning optimizer. Acta Phys Pol, A 130:7–8
9. Rao R (2016) Review of applications of TLBO algorithm and a tutorial for beginners to solve the unconstrained and constrained optimization problems. Decis Sci Lett 5(1):1–30
10. Krishna Chaitanya R, Raju GSN, Raju KVSN, Mallikarjuna Rao P (2019) Antenna pattern synthesis using the Quasi-Newton method, firefly and particle swarm optimization techniques. IETE J Res, 1–9
11. Balanis CA (2005) Antenna theory analysis and design. Wiley, New Jersey
12. Bozorg-Haddad O (2018) Advanced optimization by nature inspired algorithms. Springer, Singapore
13. Zadehparizi F, Jam S (2019) An improved teaching-learning–based optimization for design of frequency reconfigurable antennas. Int J Commun Syst, e4030

Type E Fault Performance Improvement of DFIG Using Lookup Table-Based Control Scheme

D. V. N. Ananth, G. Joga Rao, and S. N. M. Venkatesh Gudela

Abstract The low and high voltage grid faults are to be taken seriously because of low converter ratings of doubly fed induction generator (DFIG). During low voltage fault, using the developed technique, the rotor flux reference value transforms to a new value from synchronous speed to a lesser value or even to zero based on the decrease in the stator voltage. This reference speed increases in case of high voltage faults. This changing in speed is for injecting rotor current at a rotor slip frequency based on the change in speed of rotor during the fault. So, the stator flux DC-offset component is mitigated and making it decomposes quickly by controlling the deviation of rotor speed from a reference value with controlled flux decomposition. The torque ripples, current surges, and overall system performance are improved when the system is analyzed for 50 and 130% of the rated voltage under fault. The system is studied under simulation using MATLAB/Simulink software.

Keywords Doubly fed induction generator · Field-oriented control · Low and high voltage fault ride through · Voltage mitigation

1 Introduction

Among all renewable electrical energy resources, wind and solar are getting popular due to controllability, location feasibility and are commercially available. In the wind energy systems, the doubly fed induction generator (DFIG) is considered better among wind energy conversion systems because of low-power rating converter, fault-tolerant behavior, available in large scale of 8 MW till date. Since low converter

D. V. N. Ananth (✉) · G. Joga Rao · S. N. M. Venkatesh Gudela
Department of Electrical and Electronics Engineering, Raghu Institute of Technology, Visakhapatnam, India
e-mail: nagaananth@gmail.com

G. Joga Rao
e-mail: jogarao.phd@gmail.com

S. N. M. Venkatesh Gudela
e-mail: venky0234@gmail.com

P. S. R. Chowdary et al. (eds.), *Microelectronics, Electromagnetics and Telecommunications*, Lecture Notes in Electrical Engineering 655,
https://doi.org/10.1007/978-981-15-3828-5_21

ratings, the DFIG is said to be sensitive if large inrush currents enter the rotor windings due to some external grid faults. These faults will produce large electromagnetic torque (EMT) and power oscillations in the DFIG, which simultaneously leads to damage of gearwheels, windings, and decreases the lifetime of the turbine-generator set. Hence, extensive research is required to manage the DFIG during low and high voltage grid faults. There are many strategies developed for rotor side converter (RSC) and grid side converter (GSC) to overcome grid faults and to improve the performance.

The voltage across the capacitor between the back to back converters is controlled using Grid Side Controller (GSC) to maintain a almost constant to improve the voltage profile at point of common coupling during the HVRT issues [1–3]. In most of these papers, rotor back emf is compensated at fault by limiting inrush fault current into the generator windings. In these methods, rotor and stator are controlled independently and the dynamic behavior is also independent. Therefore, such methods can improve performance during the fault, but deteriorate during a post-fault. Using FACTS devices like UPFC, static var compensator (SVC), etc. and different controllers was discussed in the review papers [4–6]. The authors used different control schemes and external devices to compensate voltage swell devoid of phase angle variation is investigated for HVRT studies. Battery- or SMES-type energy storage devices are used to meet transmission system operator constraints for a wind power conversion system for HVRT by considering different grid code deals [7, 8]. HVRT issue is more critical than LVRT was stated in [2, 7–14] and active resistance is used to restrain rotor current and electromagnetic torque oscillations for improved operation as described in [2, 9–12]. In most of these papers, stability of the overall system during grid disturbances is not explained clearly. Many papers [13, 14] did not explain the coordination between rotor voltage and current to work seamlessly during low and high voltage faults.

This paper presents a control technique for DFIG to withstand under low voltage or a high voltage grid fault, which uses offsetting of DC flux components in the stator windings. This suggested technique will

(1) Smoothen DFIG operation during and after the faults;
(2) Improved dynamic and transient response to grid faults;
(3) Appropriate for all types of grid faults;
(4) Damp oscillation in the torque and power and can regain to its pre-fault value even though the fault still existing and hence more sustainable and reliable operation;
(5) Maintains nearly constant speed or within limits not going to dangerous speeds;
(6) Limits surge currents and maintains a reliable and sustainable operation and further
(7) Preserve electromechanical energy exchange under interruptions.

In the Sect. 2, mathematical modelling of DFIG based converters using FOC is discussed to find the effect of stator and rotor voltages and currents on the real and reactive power flows from the DFIG. The Sect. 3 describe the control schemes of grid and rotor side converters. Later in Sect. 4, simulation results are analysed for low and high voltage faults for the same test system and the results are compared with earlier works to describe the effectiveness of proposed technique. The conclusion of the work is given in Sect. 5 followed by references.

2 Analysis of DFIG Back to Back Converters Using Field-Oriented Control (FOC) Scheme

The mathematical modeling for the internal control loop of RSC of DFIG and control parameters for rotor speed and reactive power are discussed in detail in this section. The equivalent circuit of DFIG is taken from [13–16] and is used in this paper in arbitrary rotating frame ω. In the analysis, the equations are referred and called using *dq* reference frame [6]:

The stator *dq*-axis currents can also be rewritten as shown in Eq. (1)

$$\left\{ \begin{aligned} I_{\text{ds}} &= \frac{1}{L_s}[\psi_{\text{ds}} - L_m I_{\text{dr}}] \quad \text{or} \quad \frac{V_s}{X_s} - \frac{X_m}{X_s} I_{\text{dr}} \\ I_{\text{qs}} &= -\frac{1}{L_s}[L_m I_{\text{qr}}] \end{aligned} \right\} \tag{1}$$

Now, the rotor voltage equations are represented in terms of stator and rotor currents by replacing the rotor flux parameters to extract a relationship with rotor currents and main stator voltage.

$$V_{\text{dr}} = (R_r + \rho L_r) i_{\text{dr}} - \omega_{\text{slip}} L_r i_{\text{qr}} + \omega_{\text{slip}} L_m i_{\text{qs}} - \rho L_m i_{\text{ds}} \tag{2a}$$

Substituting Eq. (2a) for stator currents in Eq. (1), the rotor d and q-axis voltage will be as in Eqs. (2b) and (2c).

$$V_{\text{dr}} = (R_r + \rho\sigma L_r) i_{\text{dr}} - \omega_{\text{slip}} \sigma L_r i_{\text{qr}} - \rho \frac{L_m}{L_s} \psi_{\text{ds}} \tag{2b}$$

$$V_{\text{dr}} = (R_r + \rho\sigma L_r) i_{\text{dr}} - \omega_{\text{slip}} \sigma L_r i_{\text{qr}} - \rho \frac{L_m}{\omega_s L_s} V_s \tag{2c}$$

The $1 + \frac{L_m^2}{L_s L_r}$ is defined as σL_r which is also called as rotor leakage inductance factor [6]. Similarly, deriving the equations for *q*-axis rotor voltage, the equations are

$$V_{\text{qr}} = (R_r + \rho L_r) i_{\text{qr}} + \omega_{\text{slip}} L_r i_{\text{dr}} - \omega_{\text{slip}} L_m i_{\text{ds}} - \rho L_m i_{\text{qs}} \tag{2d}$$

$$V_{\text{qr}} = (R_r + \rho\sigma L_r) i_{\text{qr}} + \omega_{\text{slip}} \sigma L_r i_{\text{dr}} - \omega_{\text{slip}} \frac{L_m}{L_s} \psi_{\text{ds}} \tag{2e}$$

and

$$V_{\mathrm{qr}} = (R_r + \rho\sigma L_r)i_{\mathrm{qr}} + \omega_{\mathrm{slip}}\sigma L_r i_{\mathrm{dr}} - \omega_{\mathrm{slip}}\frac{L_m}{\omega_s L_s}V_s \tag{2f}$$

Equations 2c and 2f are rotor *dq*-axis voltages in terms of rotor currents and main stator or grid voltage. The right side first two terms in these both equations are decoupled parameters and the last term is the disturbance observer term. These are controlled effectively in the inner control circuit of RSC to maintain the better rotor voltage as desired during the steady state.

The stator and rotor real, reactive and apparent powers are given in Eqs. (3a–3d) in terms of stator and rotor voltage and current components. For effective control of DFIG real and reactive power as per the grid requirement, the controllable RSC parameter knowledge plays a vital role. Hence, stator and rotor powers are represented in terms of rotor voltage and currents are derived [9].

$$P_r = \frac{3}{2}\left(V_{\mathrm{dr}}I_{\mathrm{dr}} + V_{\mathrm{qr}}I_{\mathrm{qr}}\right) \tag{3a}$$

$$Q_r = \frac{3}{2}\left(V_{\mathrm{qr}}I_{\mathrm{dr}} - V_{\mathrm{dr}}I_{\mathrm{qr}}\right) \tag{3b}$$

$$P_s = \frac{3}{2}V_s I_{\mathrm{qs}} = -\frac{3}{2}\frac{X_m}{X_s}V_s I_{\mathrm{qr}} \tag{3c}$$

$$Q_s = \frac{3}{2}V_s i_{\mathrm{ds}} = \frac{3}{2}V_s\left(\frac{V_s}{X_s} - \frac{X_m}{X_s}i_{\mathrm{dr}}\right) = \left(\frac{3V_s^2}{2X_s} - \frac{3X_m V_s}{2X_s}i_{\mathrm{dr}}\right) \tag{3d}$$

The rotor real power is expressed in terms of stator powers, stator voltage, and rotor voltages as

Or

$$P_r = \frac{3}{2}\left(V_{\mathrm{dr}}\frac{2}{3}\frac{X_s}{X_m}\frac{Q_s}{V_m} + \frac{V_{\mathrm{dr}}V_s}{X_m} + \frac{2V_{\mathrm{qr}}X_s}{3X_m V_s}P_s\right) \tag{4a}$$

Rearranging Eq. (4a), the rotor real and reactive powers will be as in Eqs. (4b) and (4c).

$$P_r = \frac{X_s}{X_m}\frac{V_{\mathrm{dr}}}{V_m}Q_s + \frac{3}{2}\frac{V_{\mathrm{dr}}V_s}{X_m} + \frac{X_s V_{\mathrm{qr}}}{X_m V_s}P_s \tag{4b}$$

$$Q_r = \frac{X_s}{X_m}\frac{V_{\mathrm{qr}}}{V_s}Q_s + \frac{3}{2}\frac{V_{\mathrm{qr}}V_s}{X_m} - \frac{X_s V_{\mathrm{dr}}}{X_m V_s}P_s \tag{4c}$$

The tuned PI (proportional and integral) controller helps the inner control loop of RSC to settle to the desired state. When stator or the grid voltage falls, both the rotor *d*- and *q*-axis voltages drop, respectively. If the decoupled rotor currents are controlled and increased, the voltage drop in the rotor can be mitigated and can be possible with fast-acting control action of RSC and maintenance of DC voltage across the capacitor by the GSC converter.

3 Design of Rotor and Grid Side Converters to Overcome Grid Voltage Faults

The grid and rotor side converters (RSC and GSC) control schemes are shown in Fig. 1a, b. In Fig. 1a, based on the desired grid reactive power requirement, the stator real power is varied. This is developed based on the characteristic lookup table technique. This reference real power to be developed from the stator of DFIG is compared with the actual real power, and then GSC power is obtained. This is the power to be extracted from the GSC converter. Based on the PI control action of DC-link voltages, the actual GSC real power is extracted. Hence, real and reactive powers are controlled from stator and GSC terminals, and based on PCC terminal voltage, stator voltage is developed as shown in Fig. 1a.

The RSC control scheme is shown in Fig. 1b which plays a vital role in curtailing the fault current entering into the RSC terminal and also helps in maintaining the rotor speed by controlling the mutual flux between stator and the rotor. Here, based on the wind speed, the rotor speed reference is estimated using a lookup table technique which is based on characteristics on the environment and the wind turbine profile. The actual rotor speed is compared with the reference rotor speed and the error is to be maintained to zero and as low as possible. The output from this PI controller is d-axis rotor reference current. The actual and reference d-axis rotor currents are compared and designed to maintain error to a low value. The output of the PI is reference d-axis voltage. Like d-axis reference voltage generation, the q-axis voltage generation is also obtained using desired and actual reactive power and q-axis currents. Based on the stator voltage and current values, synchronous speed values are developed using the proposed IFOC control scheme. If considerable decrease in the voltage is found, stator angular speed decreases relatively smaller value. If the voltage decreases to zero or beyond 0.1 pu, the angular speed will also go to zero. Similarly, if voltage and current increases beyond nominal value, the angular speed also increases, thereby flux also increases. This will force the rotor speed and current to be within the limits.

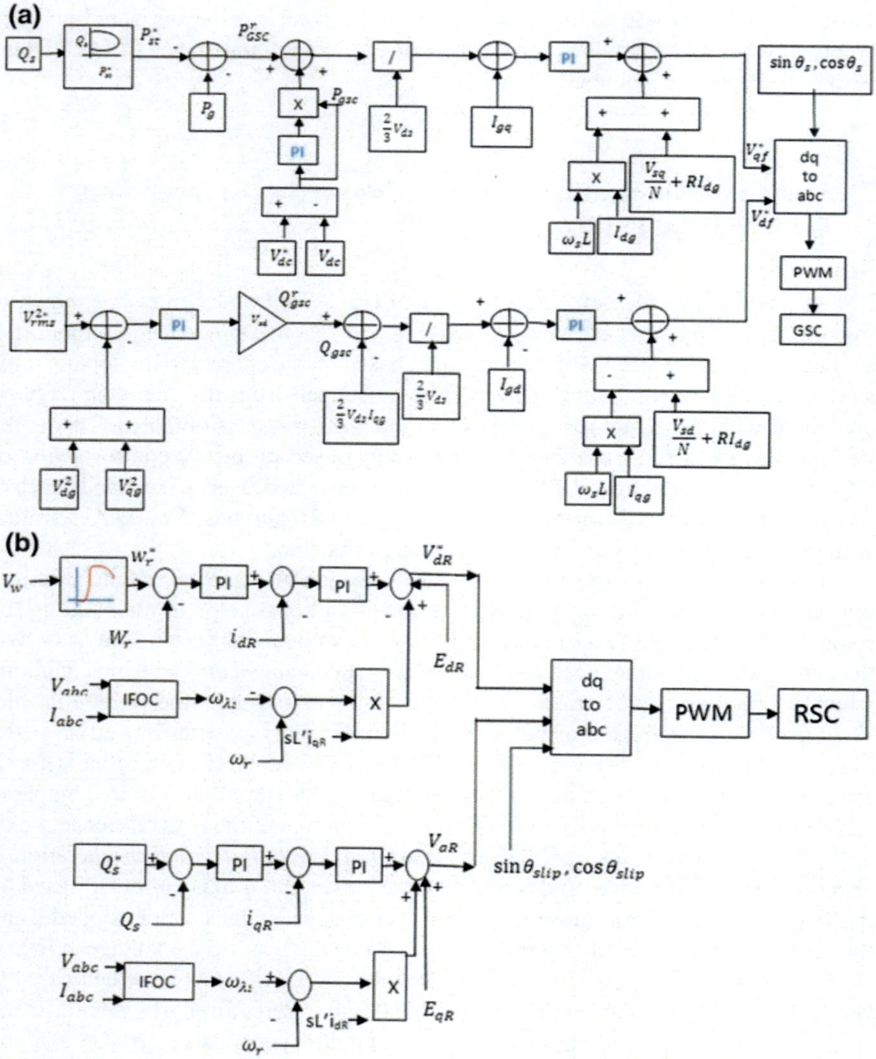

Fig. 1 **a** GSC controller to overcome any symmetrical low or high voltage faults. **b** RSC controller to overcome any symmetrical low or high voltage faults

4 Result Analysis

In this analysis, the conventional DFIG testbed system is considered based on [15]. It takes the definitions for different types of faults from the review paper. Voltage sags from 270 to 150 V and swells of 1380 V during 1–1.2 s. The rotor, stator voltage and current, rotor speed and electromagnetic torque (EMT) of the DFIG are studied for the grid fault. Now the voltage transients are observed and the stator steady state voltage decreased from 270 V to 150 V in the first case and this voltage increased from same 270 V to 380 V in the second case.

4.1 Case 1: DLG Fault Near PCC or Type E Type of Fault

Type E fault refers to double phase to ground and here A and B phases to ground fault are considered for the same system under this case study. A fault resistance of 0.001 Ω is placed at PCC for 0.7–0.8 s, and the system behavior is analyzed. During fault instants as shown in Fig. 2a, b, the rotor current changed from 0.3 to 0.45 pu and finally attained to its normal value. The waveform distortions are observed, but with lower-order harmonics which is found to be better to a standard FOC method.

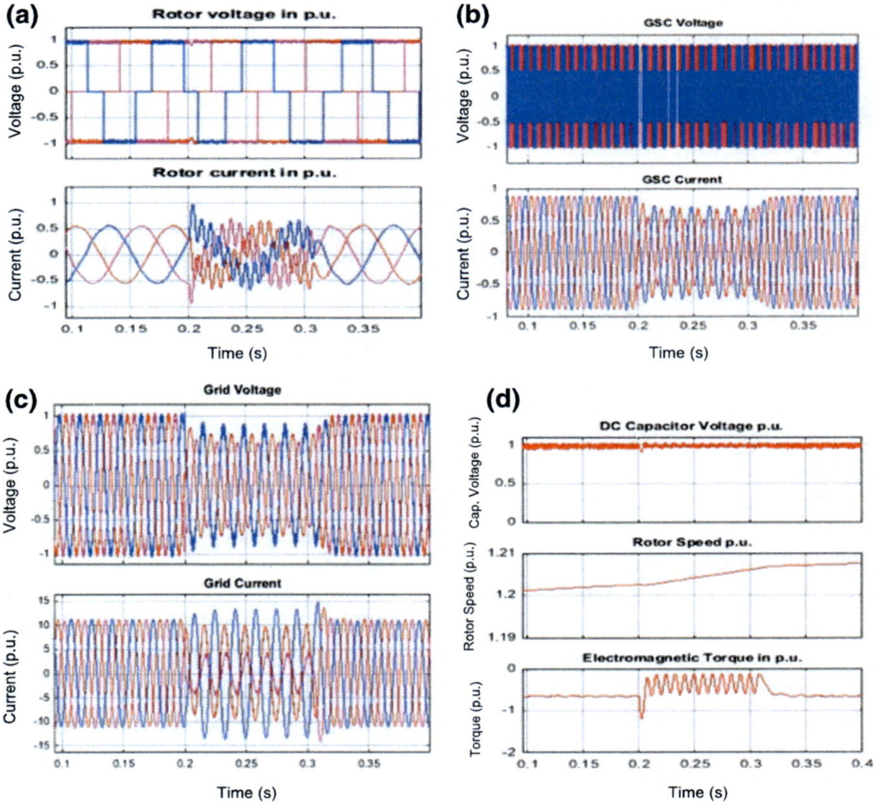

Fig. 2 **a** Rotor voltage and current versus time in sec, **b** GSC terminal voltage and current, **c** grid terminal voltage and current, **d** DC capacitor voltage, rotor speed, and EMT

The GSC voltage is almost constant but the faulty current in these two phases dipped to 0.3 pu from 0.45 pu at the fault period, but the healthy phase is now only 0 pu. The GSC performance without any surges or spikes is observed and is curtailed with the proposed scheme. The AB voltage decreased to 0.1 pu whereas the current is now 0.3 pu in these AB phases and 0.65 pu in C phase as in Fig. 2c. The grid current in AB phases became 0.5 and 1.2 pu current is observed in healthy C phase.

The voltage across the capacitor and the rotor speed along with EMT in pu is shown in Fig. 2d. The capacitor voltage is mostly constant even under the fault conditions. The rotor speed is almost constant at 1.201 pu with a negligible deviation to 1.208. The EMT which is initially at −0.8 pu reached to −1.05 pu at fault occurring instant and later during the fault reacted with oscillations at 0.2 pu. The oscillations and deviation in the torque are observed due to mismatch in the mechanical and electrical powers. If during faults surge current of amplitude greater than 2 pu for more than 3 seconds is produced in the DFIG windings, it will damage stator and rotor windings. Hence, our developed IFOC is better in performance and is recommended for system to have safe operation for any double phase to ground voltage fault.

5 Conclusion

The DFIG wind turbine system performance with grid disturbance like voltage dip and swell is analyzed in this paper. It is observed that with proposed FOC technique generator, torque oscillations are damped effectively and are maintained to pre-fault state magnitude during the faults. Better control over inrush current to rotor windings can be achieved by limiting the rapid decay in rotor flux and making the rotor current to be controlled by decomposing DC-offset component of stator flux using FOC. This method has limited rapid increase or decrease in rotor speed during sags or swells and is maintained at a permissible limit without losing synchronism with the grid. The limit for a permissible increase in rotor speed is decided based on $\omega_{\Phi s}$ as explained in the flowchart.

The system behavior is observed for case raising the voltage to 1.3 times for 1–1.2 s. The behavior seems like the fault occurs at PCC. The DFIG parameters like rotor, stator voltage and currents, rotor speed and electromagnetic torque (EMT) are observed for these cases. The developed FOC method is observed and found that it can improve performance during LVRT as well as HVRT. Here, the current and voltage waveforms and magnitude at stator and rotor terminals at the time of disturbances are improved considerably using the developed rotor side converter (RSC) control scheme. During severe grid under voltage fault, the oscillations in stator terminal voltage decreased, stator and rotor winding currents are controlled within two times increase, the capacitor voltage is almost constant, there is small increase in rotor speed, but in control and EMT, oscillations damped and also reached to normal during fault with the proposed scheme compared with conventional controller techniques.

Further, the main advantages of this method are (i) minimization of the DC-offset current components in the DFIG to a small quantity and (ii) quick and effective control of oscillations in the torque and rotor speed. The stator and rotor currents are nonzero or huge values and recover to pre-fault value yet the grid fault continues. Therefore, the work holds better for any type of fault.

References

1. Zheng Z, Yang G, Geng H (2013) High voltage ride-through control strategy of grid-side converter for DFIG-based WECS. In: IECON 2013-39th annual conference of the IEEE industrial electronics society, pp 5282–5287. IEEE
2. Safaei A, Hosseinian SH, Abyaneh HA (2017) Enhancing the HVRT and LVRT capabilities of DFIG-based wind turbine in an islanded microgrid. Eng, Technol Appl Sci Res 7(6):2118–2123
3. Xiuhe L, Jianxue M (2018) HVRT optimization strategy of DFIG with dynamic shift virtual resistor based on fuzzy control. Electr Meas Instrum 10:14
4. Jerin ARA, Kaliannan P, Subramaniam U, El Moursi MS (2018) Review on FRT solutions for improving transient stability in DFIG-WTs. IET Renew Power Gener 12(15):1786–1799
5. Taher SA, Arani ZD, Rahimi M, Shahidehpour M (2018) Model predictive fuzzy control for enhancing FRT capability of DFIG-based WT in real-time simulation environment. Energy Syst 9(4):899–919
6. Duggirala VNA, Gundavarapu VNK (2016) Improved LVRT for grid connected DFIG using enhanced field oriented control technique with super capacitor as external energy storage system. Eng Sci Technol, Int J 19(4):1742–1752
7. Justo JJ, Mwasilu FA (2019) Low voltage ride through enhancement for wind turbines equipped with DFIG under symmetrical grid faults. Tanzan J Eng Technol 37(2)
8. Bluhm R, Fortmann F (2020) Enhanced features of DFIG wind energy converters regarding new grid code requirements
9. Srinivasan P, Samiappan D (2018) Modeling and simulation of HVRT and LVRT enhancement capability for doubly fed induction generator based wind energy conversion system. Int J Eng Technol 7(2.33):405–408
10. Falehi AD, Rafiee M (2018) LVRT/HVRT capability enhancement of DFIG wind turbine using optimal design and control of novel $PI^{\lambda}D^{\mu}$-AMLI based DVR. Sustain Energy, Grids Netw 16:111–125
11. Ahmad A, Loganathan R (2010) Development of LVRT and HVRT control strategy for DFIG based wind turbine system. In: 2010 IEEE international energy conference, pp 316–321. IEEE
12. Liasi SG, Afshar Z, Harandi MJ, Kojori SS (2018) An improved control strategy for DVR in order to achieve both LVRT and HVRT in DFIG wind turbine. In: 2018 international conference and exposition on electrical and power engineering (EPE), pp 0724–0730. IEEE
13. Ruiqi LI, Hua G, Geng Y (2016) Fault ride-through of renewable energy conversion systems during voltage recovery. J Mod Power Syst Clean Energy 4(1):28–39
14. Wang Y, Qiuwei W, Honghua X, Guo Q, Sun H (2014) Fast coordinated control of DFIG wind turbine generators for low and high voltage ride-through. Energies 7(7):4140–4156
15. Rolan A, Corcoles F, Pedra J (2011) Doubly fed induction generator subject to symmetrical voltage sags. IEEE Trans Energy Convers 26(4):1219–1229
16. Ling Y, Cai X, Wang N (2013) Rotor current transient analysis of DFIG-based wind turbines during symmetrical voltage faults. Energy Convers Manag 76:910–917

Positioning Strategies: Implementation and Applications of Major Source Localization and Positioning Approaches Over Indian Subcontinent

Ganesh Laveti, D. Eswara Chaitanya, P. Chaya Devi, and T. Vinodh Kumar

Abstract The Indian subcontinent comprising of a number of countries has a long border and a large coastline. Hence, it is essential to have a strong surveillance and tracking system to locate and monitor the movements of unknown objects and sources from across the border. Global positioning (GP) of an unknown object and source localization (SL) of an unknown radiating source have thus become very critical from a defence point of view. There are many measurement techniques available for the purpose of global positioning and source localization such as time of arrival (TOA), time difference of arrival (TDOA), received signal strength (RSS) and angle of arrival (AOA). However, for any given application, the precision in the estimated position varies with the measurement uncertainty and measurement technique chosen. This brings about the problem of associating applications specific to Indian geographic conditions with the existing measurement techniques. Hence, this paper discusses the choice of the relevant technique for specific applications concerning diverse fields like defence, medicine, etc. To aid better understanding, this paper uses real-world data collected from GPS receiver located at Andhra University, Visakhapatnam, for validation of TOA and TDOA measurement techniques. The RSS and AOA techniques are validated using simulated data.

Keywords Angle of arrival · Global positioning system · Received signal strength · Source localization · Time of arrival · Time difference of arrival

---

G. Laveti (✉)
Gayatri Vidya Parishad College of Engineering for Women, Madhurawada, Visakhapatnam, India
e-mail: ganeshlaveti2010@gvpcew.ac.in

D. Eswara Chaitanya
RVR and JC College of Engineering, Guntur, India

P. Chaya Devi
Anil Neerukonda Institute of Technology and Sciences, Chittivalasa, Visakhapatnam, Andhra Pradesh, India

T. Vinodh Kumar
Sai Ganapati Polytechnic College, Visakhapatnam, India

P. S. R. Chowdary et al. (eds.), *Microelectronics, Electromagnetics and Telecommunications*, Lecture Notes in Electrical Engineering 655,
https://doi.org/10.1007/978-981-15-3828-5_22

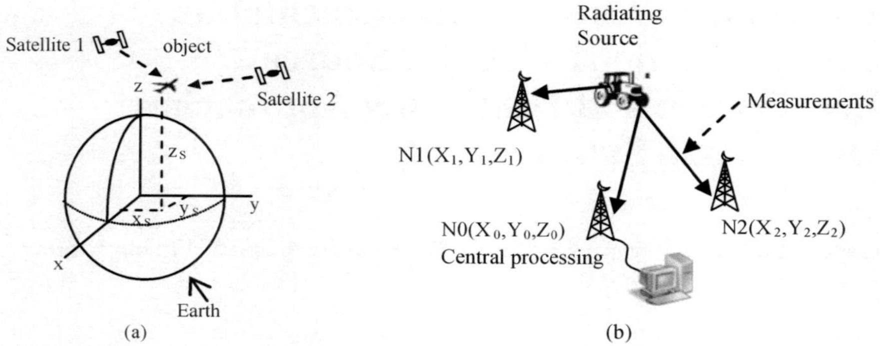

Fig. 1 Positioning **a** global positioning of source **b** source localization

1 Introduction

The demand for services that perform the task of determining the position of an unknown object and radiating source has seen a rapid increase in the recent years. The words global positioning and localization are often used interchangeably and spread uncertainty in their perception. GP refers to determining the coordinates of an object with reference to a global coordinate system like the Earth-Centred, Earth-Fixed (ECEF) system, whereas SL is the process that determines the position of a radiating source with reference to the origin (master station or centralized system) of a local coordinate system [1] as shown in Fig. 1.

The essential aspect of all the GP and SL systems is to maintain precision and accuracy of their output, which is influenced by various factors like measurement technique employed, geographic location, positioning algorithm, etc. Of all the factors that affect system performance, the type of measurement technique employed is considered important because the proper use of a measurement technique for a class of applications results in increased precision by compensating the effects due to error sources and system bias.

2 Positioning with TOA Measurements

This measurement is one popular technique often used in positioning systems and is defined as the time a signal takes to arrive at the destination from its origin. Though the measurement represents the travel time of a signal, most of the positioning systems use range information for position determination which is calculated from TOA as given by Eq. 1.

$$\text{Range}_{\text{SN}} = \text{TOA}_{\text{SN}} \times v_c$$

$$N = 1, 2, 3, \ldots \text{ no. of nodes} \tag{1}$$

where Range_{SN} is the range between node N and the source S, TOA_{SN} is the travel time of the signal to node N and V_c is the velocity of the signal. It is evident from Fig. 2 that a minimum of three TOA measurements are required for SL with nodes placed in coplanar configuration, and the same would need four TOA measurements for a non-coplanar node arrangement [2] (i.e. spheres form instead of circles).

GPS is a constellation of satellites that uses trilateration for determining its receiver position (i.e. unknown object) as shown in Fig. 3, and the pseudo-range (PR_i) measurements between satellite and the receiver can be calculated from Eq. 2.

$$\begin{aligned} \text{PR}_i &= \sqrt{(\text{SAT}_{Xi} - R_X)^2 + (\text{SAT}_{Yi} - R_Y)^2 + (\text{SAT}_{Zi} - R_Z)^2} \\ i &= 1, 2, 3, 4 \end{aligned} \tag{2}$$

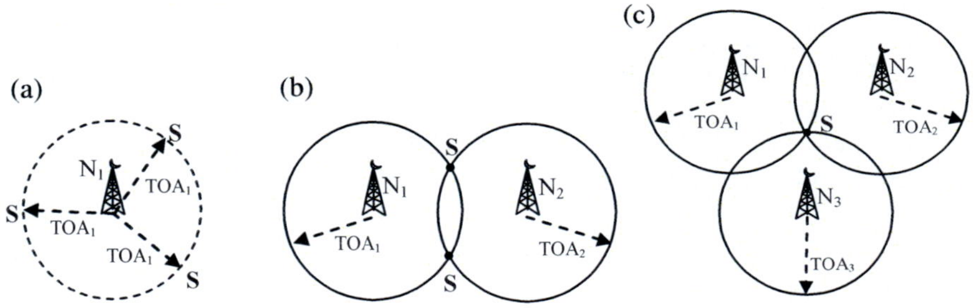

Fig. 2 Source localization with TOA measurements using trilateration. **a** Infinite possible locations **b** two possible locations **c** resolved ambiguity in source location/position

Fig. 3 Global positioning of an unknown object with four satellites using trilateration

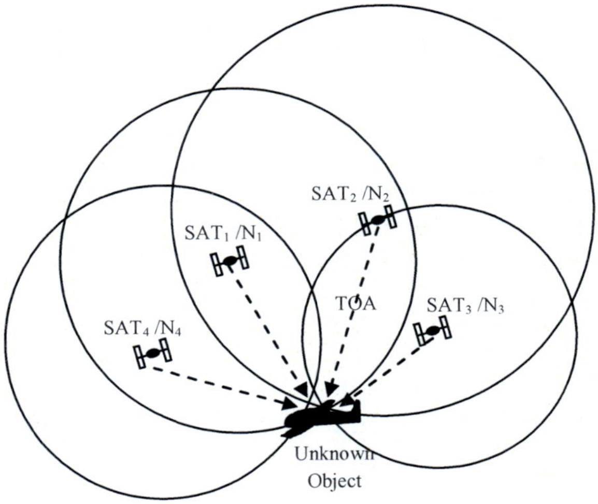

where PR_i is the observed pseudo-range measurement between ith satellite and the receiver R, SAT_{Xi}, SAT_{Yi}, SAT_{Zi} are the ith satellite coordinates and R_X, R_Y, R_Z represent unknown receiver coordinates. Solving Eq. 2 for the unknown receiver coordinates is complex as it is a nonlinear equation and hence is linearized with Taylor series and approximated to first-order series expansion which is given in Eq. 3.

$$\begin{bmatrix} \psi_{X1} & \psi_{Y1} & \psi_{Z1} \\ \vdots & \vdots & \vdots \\ \psi_{X4} & \psi_{Y4} & \psi_{Z4} \end{bmatrix} \begin{bmatrix} \delta R_X \\ \delta R_Y \\ \delta R_Z \end{bmatrix} = \begin{bmatrix} \delta \mathrm{PRD}_1 \\ \delta \mathrm{PRD}_2 \\ \delta \mathrm{PRD}_3 \end{bmatrix} \tag{3}$$

where $\Psi_{Xi} = \frac{\tilde{R}_X - \mathrm{SAT}_{Xi}}{\tilde{P}\tilde{R}_i}$, $\Psi_{Yi} = \frac{\tilde{R}_Y - \mathrm{SAT}_{Yi}}{\tilde{P}\tilde{R}_i}$, $\Psi_{\mathrm{Zi}} = \frac{\tilde{R}_Z - \mathrm{SAT}_{Zi}}{\tilde{P}\tilde{R}_i}$ and $\tilde{P}\tilde{R}_i = \sqrt{\left(\mathrm{SAT}_{Xi} - \tilde{R}_X\right)^2 + \left(\mathrm{SAT}_{Yi} - \tilde{R}_Y\right)^2 + \left(\mathrm{SAT}_{Zi} - \tilde{R}_Z\right)^2}$ $i = 1, 2, 3, 4$.

Here, $\delta \mathrm{PR}_i = 1, 2, 3, 4$ is the difference between observed (PR_i) and estimated ($\tilde{P}\tilde{R}_i$) range, (δR_X, δR_Y, δR_Z) are the changes in position estimate, the variables ($\tilde{R}_X$, $\tilde{R}_Y$, $\tilde{R}_Z$) represent the receiver position estimate and $\tilde{P}\tilde{R}$ is the computed/estimated range from estimated receiver position. For most of the time, GPS system receives more TOA range measurements than the unknowns, and hence, the solution δR is estimated using Eqs. 4 and 5.

$$\delta R = \left(\Psi^{\mathrm{T}}\Psi\right)^{-1}\Psi^{\mathrm{T}}\delta \mathrm{PR} \tag{4}$$

$$\left[\tilde{R}_X \tilde{R}_Y \tilde{R}_Z\right]^{\mathrm{T}} = \left[\tilde{R}_X \tilde{R}_Y \tilde{R}_Z\right]^{\mathrm{T}} + [\delta R_X \delta R_Y \delta R_Z]^{\mathrm{T}} \tag{5}$$

3 Positioning with TDOA Measurements

This technique enables SL with the measurements developed from the arrival time difference of the signal between pairs of nodes [3]. In general, the node which receives the signal first is considered as the reference node, and TDOA measurement to other nodes is computed from it. Similar to the TOA positioning systems, some TDOA systems need range information which is computed from Eq. (6).

$$\begin{aligned} \mathrm{RDOA}_{\mathrm{RN}} &= (\mathrm{AT}_R - \mathrm{AT}_N) \times v_c \\ N &= 1, 2, 3, \ldots \text{ No. of nodes } \;\forall N \neq R \end{aligned} \tag{6}$$

where $\mathrm{RDOA}_{\mathrm{RN}}$ is the range difference of arrival between reference node R and the other node N and AT is the signal arrival time. Similar to the TOA measurement,

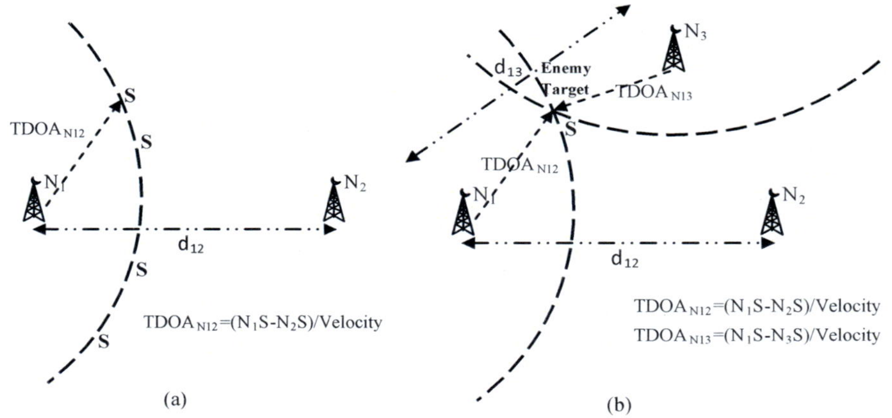

Fig. 4 Source localization with TDOA measurements using hyperbolic multilateration **a** infinite possible locations **b** resolved ambiguity in source position

TDOA measurement between a pair of nodes is represented graphically with a hyperbola [3]. This principle of a hyperbola is extended to SL with TDOA measurements and is illustrated in Fig. 4.

This method of source localization with hyperbolas is well known as multilateration [1]. With the overwhelming advantages of TDOA technique such as signal initiation-time independent measurements, reduced measurement error and system cost, this technique is widely used in many real-world applications [4]. For more quantitative performance analysis of TDOA technique, real-world RDOA measurements obtained from GPS receiver located at Andhra University, Visakhapatnam, are used in determining its position, and the error in estimation is deliberated upon in the results section. The satellite/node with maximum elevation or with the smaller range is considered as the reference satellite and the RDOAs to other satellites are computed with respect to this reference satellite as given by Eq. 7.

$$\mathrm{RDOA}_{\mathrm{sref},\,i} = \mathrm{PR}_{\mathrm{sref}} - \mathrm{PR}_i \tag{7}$$

$$\begin{aligned}\mathrm{RDOA}_{\mathrm{sref},\,i} = &\sqrt{(\mathrm{SAT}_{X\mathrm{ref}} - R_X)^2 + (\mathrm{SAT}_{Y\mathrm{ref}} - R_Y)^2 + (\mathrm{SAT}_{Z\mathrm{ref}} - R_Z)^2} \\ &- \sqrt{(\mathrm{SAT}_{X\mathrm{i}} - R_X)^2 + (\mathrm{SAT}_{Y\mathrm{i}} - R_Y)^2 + (\mathrm{SAT}_{Z\mathrm{i}} - R_Z)^2} \\ &i = 1, 2, 3, \ldots \text{ and } i \neq \mathrm{sref}\end{aligned}$$

where $\mathrm{RDOA}_{\mathrm{sref},\,i}$ is the observed pseudo-range difference of arrival at the receiver R, between satellite i and the reference satellite sref, $(\mathrm{SAT}_{X\mathrm{ref}}, \mathrm{SAT}_{Y\mathrm{ref}}, \mathrm{SAT}_{Z\mathrm{ref}})$ are the reference satellite coordinates, $(\mathrm{SAT}_{Xi}, \mathrm{SAT}_{Yi}, \mathrm{SAT}_{Zi})$ are the ith satellite coordinates and R_X, R_Y, R_Z represent unknown receiver coordinates.

The approximated linearized equations for the four satellites in matrix form ($\Gamma \times \delta R = \delta\text{PRD}$) is given in Eq. 8.

$$\begin{bmatrix} \Gamma_{X1} & \Gamma_{Y1} & \Gamma_{Z1} \\ \Gamma_{X2} & \Gamma_{Y2} & \Gamma_{Z2} \\ \Gamma_{X3} & \Gamma_{Y3} & \Gamma_{Z3} \end{bmatrix} \begin{bmatrix} \delta R_X \\ \delta R_Y \\ \delta R_Z \end{bmatrix} = \begin{bmatrix} \delta\text{PRD}_1 \\ \delta\text{PRD}_2 \\ \delta\text{PRD}_3 \end{bmatrix} \tag{8}$$

where $\Gamma_{Xi} = \frac{\tilde{R}_X - \text{SAT}_{Xi}}{\tilde{PR}_i} - \frac{\tilde{R}_X - \text{SAT}_{X\text{ref}}}{\tilde{PR}_{\text{ref}}}$, $\Gamma_{Yi} = \frac{\tilde{R}_Y - \text{SAT}_{Yi}}{\tilde{PR}_i} - \frac{\tilde{R}_Y - \text{SAT}_{Y\text{ref}}}{\tilde{PR}_{\text{ref}}}$, $\Gamma_{Zi} = \frac{\tilde{R}_Z - \text{SAT}_{Zi}}{\tilde{PR}_i} - \frac{\tilde{R}_Z - \text{SAT}_{Z\text{ref}}}{\tilde{PR}_{\text{ref}}}$, and $\delta\text{PRD}_i = \text{RDOA}_{\text{sref, i}} - \tilde{\text{RDOA}}_{\text{sref, i}}\ \forall i = 1, 2, 3, ..$ with $\text{i} \neq \text{sref}$.

$$\begin{aligned} \tilde{\text{RDOA}}_{\text{sref,i}} &= \sqrt{\left(\text{SAT}_{X\text{ref}} - \tilde{R}_X\right)^2 + \left(\text{SAT}_{Y\text{ref}} - \tilde{R}_Y\right)^2 + \left(\text{SAT}_{Z\text{ref}} - \tilde{R}_Z\right)^2} \\ &\quad - \sqrt{\left(\text{SAT}_{Xi} - \tilde{R}_X\right)^2 + \left(\text{SAT}_{Yi} - \tilde{R}_Y\right)^2 + \left(\text{SAT}_{Zi} - \tilde{R}_Z\right)^2} \\ &\quad i = 1, 2, 3, \text{ and } i \neq \text{sref} \end{aligned}$$

Here, $\delta\text{PRD}_i = 1, 2, 3, \ldots$ is the difference between observed ($\text{RDOA}_{\text{sref}, i}$) and estimated ($\tilde{\text{RDOA}}_{\text{sref}, i}$) range differences, ($\delta R_X, \delta R_Y, \delta R_Z$) are the changes in position estimate, the variables ($\tilde{R}_X,\ \tilde{R}_Y,\ \tilde{R}_Z$) represent the receiver position estimate and $\tilde{\text{RDOA}}_{\text{sref}, i}$ is the computed/estimated range difference from estimated position. For most of the time, GPS system receives more TOA range measurements than the unknowns, and hence, the solution δR is estimated using Eqs. 9 and 10.

$$\delta R = \left(\Gamma^{\text{T}}\Gamma\right)^{-1}\Gamma^{\text{T}}\delta\text{PRD} \tag{9}$$

$$\left[\tilde{R}_X \tilde{R}_Y \tilde{R}_Z\right]^{\text{T}} = \left[\tilde{R}_X \tilde{R}_Y \tilde{R}_Z\right]^{\text{T}} + [\delta R_X \delta R_Y \delta R_Z]^{\text{T}} \tag{10}$$

4 Positioning with RSS Measurements

The positioning techniques discussed in the previous section use travel or/and arrival time of the signal in determining the unknown position, and it is also clear from Eqs. 1 to 6 that with velocity as constant, a small deviation in time leads to a huge error in range measurement, resulting in a low accuracy position estimate [5]. Though most of the errors are reduced by proper maintenance and modelling of the system, the effect due to multipath is inevitable and unpredictable [6–8]. Hence, TOA and TDOA techniques do not provide precise position estimate for the systems used especially in indoor and urban canyons.

The SL technique discussed in this section relies on the strength/power of the received signal in localizing the source position. The strength of the signal propagated over long distances undergoes signal strength changes in a random and erratic manner and needs to be characterized statistically as given in Eq. 11.

$$\mathrm{RSS}^{d}_{\mathrm{NS}} = \mathrm{SS}_S - (P_L + \text{other effects})\ N = 1, 2, 3, \ldots \text{ No. of node} \tag{11}$$

where $\mathrm{RSS}^{d}_{\mathrm{NS}}$ is the received signal strength in dBm between source S and node N at a distance d, SS_S is the source transmitted signal strength and P_L is the path loss. Thus, by relating the RSS to a P_L model, the range between S and N is estimated [5]. Various empirical path loss expressions used in determining the range measurements are given in Eqs. 12–14. Equations 12 and 13 represent calculation of path loss for free-space environment at a reference distance, do, and required distance d, respectively.

$$\mathrm{PL}^{\mathrm{do}}_{\text{free Space}}(\mathrm{dB}) = 20\log\left(\frac{4\Pi\,\mathrm{do}}{\lambda_c}\right)(\mathrm{dB}) \tag{12}$$

$$\mathrm{PL}^{d}_{\text{free space}}(\mathrm{dB}) = \mathrm{PL}^{\mathrm{do}}_{\text{free space}} + 10 * \mathrm{n} * \log\left(\frac{\mathrm{do}}{d}\right) \tag{13}$$

where $\mathrm{PL}^{\mathrm{do}}_{\text{free space}}$ is the free-space path loss at a distance do, λ_c is the wavelength of operating signal, $\mathrm{PL}^{d}_{\text{free space}}$ is the free-space path loss at the distance d and n is the path loss exponent [9]. The predicted path loss model employed in generating the results shown in Sect. 7 is Okumura–Hata [9] urban area model and is given in Eq. 14.

$$\begin{aligned}\mathrm{PL}^{d}_{\mathrm{Hata}}\mathrm{Urban}(\mathrm{dB}) &= 69.55 + 29.16\log f_c - 13.82\log h_s - a(h_N)\\ &\quad + (44.9 - 6.55\log h_s)\log d\end{aligned} \tag{14}$$

Here, $a(h_N) = (1.1\log f_c - 0.7)h_N - (1.56\log f_c - 0.8)$, and $\mathrm{PL}^{d}_{\mathrm{Hata}}$ is the Hata path loss in urban environment at a distance d, f_c is the operating carrier frequency, h_S is the antenna height of the source and h_N is the node antenna height. As discussed previously, upon collection of RSS measurements at the nodes, the range, d_{NS}, between node and source is computed from Eq. 15 on substitution of Eq. 14 in Eq. 11 and neglecting the other propagation effects.

$$\begin{aligned}d_{\mathrm{NS}} &= \left|\log^{-1}\left(\frac{\mathrm{RSP}^{d}_{\mathrm{NS}} - \mathrm{SS}_s + 69.55 + 29.1\log f_c - 13.8\log h_s - a(h_N)}{44.9 - 6.5\log h_s}\right)\right|\\ N &= 1, 2, 3, \ldots \text{ No. of nodes}\end{aligned} \tag{15}$$

The range equations are then solved for optimal solution with genetic algorithm on employing the objective function in Eq. 16.

$$0.5\sum\nolimits_{N=1}^{\text{No. of nodes}}\left(d_{\mathrm{NS}} - \sqrt{\left((X_N - X_S)^2 + (Y_N - Y_S)^2 + (Z_N - Z_S)^2\right)}\right)^2 \tag{16}$$

5 Positioning with AOA Measurements

The SL techniques discussed earlier offer definite and precise position estimates when operating in a homogeneous medium (e.g. air) under explicit system conditions. These methods are not suitable for a non-homogeneous medium [5, 10]. Thus, the measurement techniques that rely directly on RSS or travel time of the signal have poor performance in SL under non-homogeneous medium. Figure 5 depicts the process of SL with AOA measurement technique. It is observed from Fig. 5 that the source S can be anywhere on the line of bearing formed at an angle, θ_{N1}, between node, $N1$, and source S.

The final position of source is obtained on considering one more AOA measurement, θ_{N2}, as shown in Fig. 5. The calculation of AOA, θ_{Ni}, is given in Eq. 17, which represents slope of the bearing line between source and node.

$$\theta_{Ni} = \tan^{-1}\left(\frac{Y_S - Y_{Ni}}{X_S - X_{Ni}}\right) \tag{17}$$

where θ_{Ni} is the observed angle of signal arrival at node Ni, (Y_S, X_S) are the vertical and horizontal coordinates of source S and (Y_{Ni}, X_{Ni}) are the vertical and horizontal coordinates of node Ni. This paper makes use of GA to solve Eq. 17 for the source coordinates, and the objective function is given in Eq. 18.

$$\sum_{i=1}^{\text{No. of nodes}} 0.5\left(\tan(\theta_{Ni}) - \left(\frac{Y_S - Y_{Ni}}{X_S - X_{Ni}}\right)\right)^2 \tag{18}$$

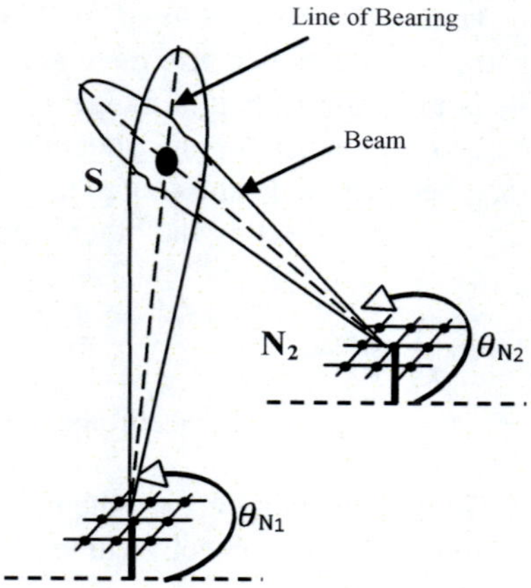

Fig. 5 Two-dimensional source localization with bearing angle of arrival measurements

6 Results and Discussion

The performance analysis of the previously discussed measurement techniques is carried out in this section. The software used for this purpose is MATLAB R2010a. Real-time data is used for generating results of TOA and TDOA techniques. This data is collected from DFGPS receiver located at Andhra University, Visakhapatnam, over a period of 23 h 56 min. The least-squares estimator is used for solving the system of linear range equations for TOA and range difference equations for TDOA [11]. The variation of receiver position error with time for TOA measurements is shown in Fig. 6a. The mean error in receiver x, y and z positions is 54.2 m, 147.7 m and 47.4 m, respectively. The variation of receiver position error with time for TDOA measurements is shown in Fig. 6b. The mean error in receiver x, y and z positions is 51.7 m, 104.8 m and 35.5 m, respectively. It is evident from the above results that the mean error in the case of TDOA is less when compared to that of TOA due to cancellation of common errors (Ionospheric, tropospheric errors, etc.) while calculating range differences.

The results of the other two techniques (RSS and AOA) are simulated after making various assumptions. Genetic algorithm is used for solving nonlinear system of equations of RSS and AOA technique. Table 1 details the simulation assumptions used for evaluation of RSS measurement technique. The Hata model is used to characterize the path loss aspect of the channel. The error in estimated position of the source is depicted for each Monte Carlo simulation in Fig. 7 The random noise samples added to the measurements follow uniform distribution.

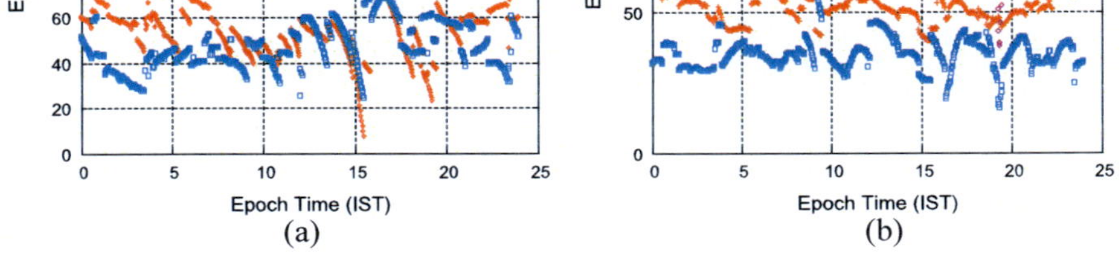

Fig. 6 AU DFGPS receiver position estimate error for the data collected on 12 January 2017, for 23 h 56 min with **a** TOA measurements **b** TDOA measurements

Table 1 Simulation assumptions for RSS and AOA technique

Parameter (RSS)	Value	Parameter (AOA)	Value
Operating frequency (f_c)	200 MHz	Node/sensor 1 position	(0, 0) km
Transmitter height (h_t)	70 m	Node/sensor 2 position	(100, 0) km
Receiver height (h_r)	6 m	True source position	(50, −86.6) km
Power transmitted (P_t)	6 W	No. of monte carlo simulations	100
Node/sensor 1 position	(0, 0, 0.006) km	Measurement noise range	0°–10°
Node/sensor 2 position	(10, 0, 0.006) km		
Node/sensor 3 position	(20, 10, 0.006) km		
True source position	(8, 15, 0.07) km		
No. of monte carlo simulations	100		
Measurement noise bound	(0–0.01) dB		

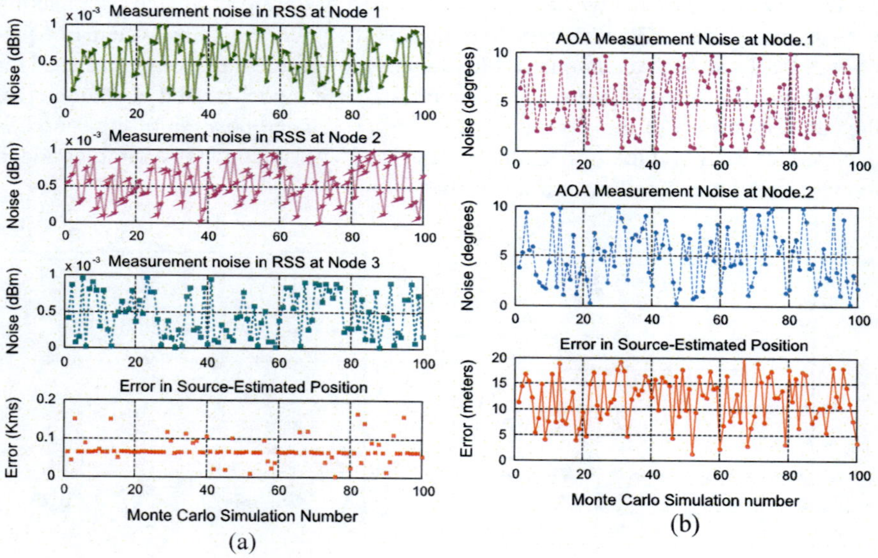

Fig. 7 Error in estimated position with **a** RSS measurements **b** AOA measurements

7 Conclusions

The dilemma in choosing a measurement technique that meets the demands of a specific positioning application over Indian geographic conditions was tackled in this paper. To achieve this end, this paper presented the concepts for implementing pivotal measurement techniques and discussed the motive for choosing a particular technique for a class of applications. It is obvious from the results and discussion that for long-range GP and SL under homogeneous and low multipath environments, time-based measurements TOA and TDOA perform well. The pseudo-range measurements collected by the GPS receiver for 5744 epochs over a period of 23 h 56 min were used for evaluation of both techniques. These two measurement techniques are often used for long and medium-range positioning applications but fail to provide accurate system performance in non-homogeneous and multipath environment. To tackle this problem, two other techniques RSS and AOA were discussed and analysed with simulated data of 100 Monte Carlo simulations. Though results show degradation in the RSS performance, it can be improved with proper path loss modelling and is best suited for short distance positioning systems.

References

1. Chaitanya DE, Kumar MNVSS, Rao GS, Goswami R (2014) Convergence issues of Taylor series method in determining unknown target location using hyperbolic multilateration. IEEE Int Conf Sci Eng Manag Res, Chennai
2. Seyed A, Zekavat R, Buehrer RM (2012) Hand book of position location-theory, practice, and advances. IEEE Press, Wiley Inc. Publications, New Jersey-54, Ch. 1 and 2
3. Xu B, Sun G, Yu R, Yang Z (2013) High-accuracy TDOA-based localization without time synchronization. IEEE Trans Parallel Distrib Syst 1567–1576, 24 Aug 2013
4. Kabir H, Kanesan J, Reza AW, Ramiah H (2015) A mathematical algorithm of locomotive source localization based on hyperbolic technique. Int J Distrib Sens Netw (Oct 2015)
5. Cheng L, Wu C, Zhang Y, Wu H, Li M, Maple C (2012) A survey of localization in wireless sensor network. Int J Distrib Sens Netw
6. Liu H, Zhang Y, Su X, Li X, Xu N (2015) Mobile localization based on received signal strength and Pearson's correlation coefficient. Int J Distrib Sens Netw (Aug 2015)
7. Bidikar B, Goswami R, Rao GS (2018) Sagnac and orbital eccentricity-based pseudo-range modeling for GPS, navigation. In: Proceedings of the fourth ICMEET, pp 767–775
8. Bidikar B, Rao GS, Ganesh L, Santosh Kumar MNVS (2016) GPS C/A code multipath error estimation for surveying applications in urban canyon. In: Microelectronics, electromagnetics and telecommunications. Lecture notes in electrical engineering, vol 372, pp 135–142
9. Rao GS (2013) Mobile cellular communication. Pearson, India
10. Goh ST, Seyed A, Zekavat R, Pahlavan K (2014) DOA-based endoscopy capsule localization and orientation estimation via unscented Kalman filter. IEEE Sens J 14(11):3819–3829
11. Rao GS (2010) Global navigation satellite systems. McGraw Hill Education Pvt. Ltd. New Delhi-08, India

Fast Nonlinear Filter-Based Local Phase Quantization for Texture Classification

Sonali Dash, Ayyagari Sai Ramya, and B. Priyanka

Abstract Texture divides an image into subparts called regions. Texture analysis characterizes region using texture content. Local phase quantization is a conventional method, which gives highest classification accuracy. Bilateral filter is also a good approach to obtain smoothness of a digital image while preserving the edges. Also, fast bilateral filter is another approach for high dynamic range images. In this study, new approach is proposed by integrating local phase quantization and fast bilateral filter, which in turn results in good classification accuracy. Using different filter domain parameter and filter range parameter texture, features such as mean, standard deviation, entropy, skewness, and kurtosis are extracted. Finally, for classification, these features are given to *k*-nearest neighbor (K-NN) classifier. The new hybrid technique is for testing and training images from Brodatz database. The results of proposed technique are compared with conventional local phase quantization results. Best classification accuracy is obtained by hybrid method for different values of domain and range parameters with different window sizes.

Keywords Texture feature · Texture classification · Local phase quantization · Fast bilateral filter

1 Introduction

1.1 A Subsection Sample

Texture analysis is one of the most important aspects in computer vision so that one can understand the process and can simulate in computer technologies. Texture is a feature, which can divide an image into its subparts called regions. It can be classified as the portioned region. The regions of an image can be characterized using texture analysis with the help of texture content. Several suggested methods are available

S. Dash (✉) · A. Sai Ramya · B. Priyanka
Department of Electronics and Communication Engineering, Raghu Institute of Technology, Visakhapatnam, Andhra Pradesh 531162, India
e-mail: sanali.isan@gmail.com

P. S. R. Chowdary et al. (eds.), *Microelectronics, Electromagnetics and Telecommunications*, Lecture Notes in Electrical Engineering 655,
https://doi.org/10.1007/978-981-15-3828-5_23

in the literature by which texture features can be extracted, such as co-occurrence matrix method [1], Gabor filter [2], DWT [3], local binary patterns [4], Law Mask [5], and local phase quantization (LPQ) [6]. Later, few more hybridized methods are proposed to improve classification accuracy further [7]. One of the best efforts is LPQ with which higher accuracies can be achieved. When LPQ is used, low frequencies are not included which may not be a major problem in the analysis. Later on, in due course of time filtering methods came into existence that gained attention of users because of their simulation properties. Some of the filtering methods, such as Global filter, bilateral filter, adaptive filter, computational bilateral filter, are proposed for extraction of texture properties [6]. Bilateral filter is basically exhibiting nonlinear properties [8]. One of the simplest and most important properties of an image is range, which is the difference between maxima and minima. The bilateral filter does in the range what a conventional filter performs in its domain. With the help of filtering technique, even low frequencies can be included in the analysis without concerning the window matrix. With the help of bilateral filter, the images can be smoothed and preserved the edges [9]. Image smoothing can be achieved by properly choosing the spatial values of bilateral filter range and bilateral filter domain [10]. However, bilateral filter smoothens the edges it suffers from introduction of false edges. Thus, to liberalize, this filter an algorithm that computes FFT in addition to bilateral filter is implemented called fast bilateral filter.

In this paper, the LPQ and fast bilateral filter(FBF) are hybridized to enhance the classification rate than the traditional LPQ approach. The proposed method is assessed on Brodatz database. The paper is organized as follows. Section 2 represents the literature survey on traditional approaches, Sect. 3 delivers the proposed method in detail, and Sect. 4 is results and discussions followed by conclusion as Sect. 5.

2 Review of LPQ and FBF

2.1 *Local Phase Quantization*

It is a texture descriptor, using which, texture features like mean, standard deviation, entropy, skewness, and kurtosis can be extracted. In this method, every pixel is computed by local frequency coefficient at some frequency. The classification accuracy is obtained for different window sizes like 3×3, 5×5, 7×7. Convolution of the image intensity and a point spread function yields to spatial blurring. In the frequency domain representation, these results are generated as a multiplication of $c = a * \mathrm{b}$. Where *c, a,* and *b* are the discrete Fourier transforms (DFT) of the blurred image, LPQ transformed image, and the PSF, respectively. If the only phase is considered, then the equation becomes as $\angle c = \angle a + \angle b$. Here, *b* is always real-valued, $\angle b \in \{0, \pi\}$. Furthermore, the shape of b resembles the sin*c* function, because the rectangular value of *b* is having low frequencies, which are positive. In LPQ, phase value is represented in local neighborhood V_p at each pixel position $P = [p_1\ p_2]^{\mathrm{T}}$. Then the local spectra can be computed as follows.

$$f(z,p) = \sum_{y} f(y) q_R(y-p) e^{-j2\pi z T y} \quad (1)$$

Where z is the frequency, q is the window function, q_R is the rectangular function which is given as $q_R = 1$, p is the pixel position, $f(y)$ is the input image, and T represents the transpose of z. In LPQ, phase can be obtained at an individual pixel point y of the range.

2.2 *Fast Bilateral Filter*

The FBF is considered as a high dynamic range filter. With this filter, bilateral filter can be linearized, and when used with FFT, it is one of the fast filtering alternatives.

$$\text{Complexity} = O(1 + G/(\sigma_s \sigma_r)) \quad (2)$$

G Number of grids of intensity
σ_s Spatial domain parameter
σ_r Range domain parameter.

The discrete form of bilateral filter with arbitrary spatial and range kernels can be

$$S^{BF}(z) = \frac{\sum y \varepsilon M(z) K_s(z-y) K_r(O(Z) - F(y)F(y))}{\sum y \varepsilon M(z) K_s(z-y) K_r(O(Z) - F(y))} \quad (3)$$

where K_s is spatial kernels, K_r is range kernel, $O(z)$ is original image, $M(z)$ is median values, and $F(y)$ is filtered image.

3 Proposed Methodology

In this work, the key objective is to develop an integrated LPQ and FBF that is used to enhance the classification accuracy. The procedural steps are depicted in the block diagram shown in Fig. 1.

Algorithm for the recommended approach for enhancing the classification accuracy is done in different stages, which are described below.

Stage 1: Initially, image preprocessing is done by splitting the original images.
Stage 2: In this stage, the original images are passed through the traditional LPQ and transformed images are generated.
Stage 3: The LPQ transformed images are applied to the FBF with various values of σ_s and σ_r. By keeping σ_s constant, different values of σ_r are applied to the LPQ transformed images for different window sizes such as 3×3, 5×5, 7×7.

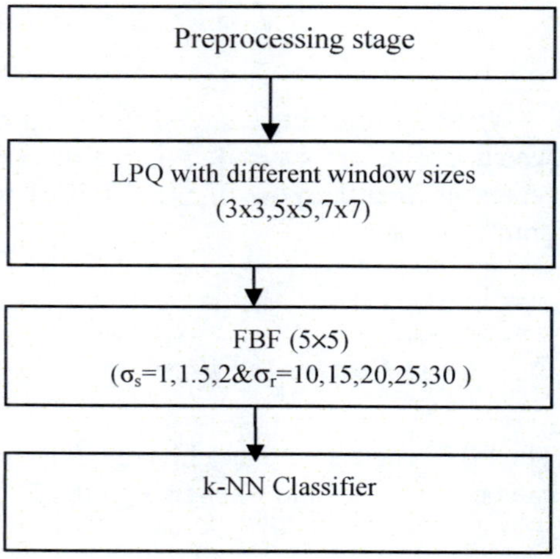

Fig. 1 Procedural steps of the proposed approach

Stage 4: Texture features like mean, standard deviation, entropy, skewness, and kurtosis are extracted by the recommended approach
Stage 5: Finally, the *k*-NN classifier is used to classify the texture features.

4 Experiments and Validation

The validation of proposed method is discussed here with the help of following experimentation. The suggested method is verified on the Brodatz database. Different 24 classes of images are taken from the Bodatz database. The original image of size 640 × 640 pixels is divided into sub-images of size 128 × 128 pixels. Hence, altogether 600 (24 × 25) sub-images are created, out of which 312 and 288 are utilized for testing and training, respectively (Fig. 2).

Initially in the first experiment, the traditional LPQ is considered for the texture feature extraction. LPQ is applied on the original images to generate the LPQ transformed images. Features are extracted for three different window sizes 3 × 3, 5 × 5, 7 × 7 of LPQ. The traditional LPQ texture descriptor gives classification accuracy for different window sizes. For window size 3 × 3, the classification accuracy is of 57.25%, and for window size 5 × 5, the classification accuracy is of 58.33%,

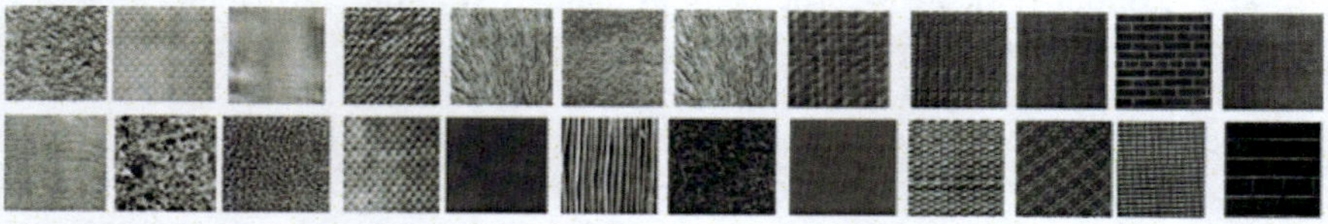

Fig. 2 Examples of Brodatz images

and for window size 7 × 7, the classification accuracy is of 50.96% are achieved. Table 1 shows the classification accuracies obtained from original LPQ technique for different window sizes.

Bilateral filter depends on two parameters that indicate the size and contrast of the features to be preserved. The two parameters are represented as space domain (σ_s) and range domain (σ_r). The space domain is the set of possible positions in an image and related to resolution. The range domain is the set of possible pixel values. Thus, the parameters σ_s and σ_r measure the amount of filtering. In this work, space domain parameter values are selected as: 1, 1.5, and 2, and range domain parameter values are selected as: 10, 15, 20, 25, and 30. For a particular fixed value of σ_s, different σ_r are applied on texture images.

The classification success rates of the suggested approach for various window sizes and various values of FBF are shown from Tables 2, 3, and 4. From the tables, it

Table 1 Results of the original LPQ approach

Different window sizes	Classification accuracy (%)
3 × 3	57.25
5 × 5	58.33
7 × 7	50.96

Table 2 Results of the suggested fusion model FBF and LPQ with $\sigma_s = 1$

Different window sizes with various σ_r values	Classification accuracy (%)
Window size (3 × 3)	
$\sigma_r = 10$	53.85
$\sigma_r = 15$	55.77
$\sigma_r = 20$	58.65
$\sigma_r = 25$	56.41
$\sigma_r = 30$	58.65
Window size (5 × 5)	
$\sigma_r = 10$	58.33
$\sigma_r = 15$	57.37
$\sigma_r = 20$	58.33
$\sigma_r = 25$	54.49
$\sigma_r = 30$	56.41
Window size (7 × 7)	
$\sigma_r = 10$	56.73
$\sigma_r = 15$	58.65
$\sigma_r = 20$	57.33
$\sigma_r = 25$	55.45
$\sigma_r = 30$	58.01

Table 3 Results of the suggested fusion model FBF and LPQ with $\sigma_s = 1.5$

Window size (3 × 3)	
$\sigma_r = 10$	51.60
$\sigma_r = 15$	52.65
$\sigma_r = 20$	58.65
$\sigma_r = 25$	55.13
$\sigma_r = 30$	58.65
Window size (5 × 5)	
$\sigma_r = 10$	55.13
$\sigma_r = 15$	57.69
$\sigma_r = 20$	56.41
$\sigma_r = 25$	57.37
$\sigma_r = 30$	55.45
Window size (7 × 7)	
$\sigma_r = 10$	57.69
$\sigma_r = 15$	57.69
$\sigma_r = 20$	57.37
$\sigma_r = 25$	60.26
$\sigma_r = 30$	57.69

Table 4 Results of the suggested fusion model FBF and LPQ with $\sigma_s = 2$

Different window sizes with various σ_r values	Classification accuracy (%)
Window size (3 × 3)	
$\sigma_r = 10$	49.36
$\sigma_r = 15$	50.64
$\sigma_r = 20$	52.27
$\sigma_r = 25$	54.81
$\sigma_r = 30$	54.49
Window size (5 × 5)	
$\sigma_r = 10$	52.27
$\sigma_r = 15$	54.17
$\sigma_r = 20$	53.53
$\sigma_r = 25$	54.49
$\sigma_r = 30$	54.81
Window size (7 × 7)	
$\sigma_r = 10$	55.77
$\sigma_r = 15$	59.29
$\sigma_r = 20$	56.41
$\sigma_r = 25$	56.73
$\sigma_r = 30$	56.73

is observed that highest classification accuracy of 60.26% is obtained for window size 7 × 7 at $\sigma_s = 1.5$ and $\sigma_r = 25$. The best classification accuracy of 59.29% is obtained for window size 7 × 7 at $\sigma_s = 2$ and $\sigma_r = 15$. The third-best classification accuracy of 58.65% is obtained for five different values of filter range parameter and filter domain parameter. Further, the proposed technique presented higher classification accuracy of 60.26% at some values of σ_s and σ_r.

Hence, it is observed that the suggested approaches deliver better classification accuracies than the original LPQ method. The increment in the classification accuracies is 2.44% for window size 3 × 3 at $\sigma_s = 1$ and $\sigma_r = 20, 30$, and also for window size 7 × 7, the incremental in classification accuracy is 9.3% at $\sigma_s = 1.5$ and $\sigma_r = 25$.

5 Conclusion

In this paper, a hybrid method for better classification accuracy is proposed. In this approach, LPQ is integrated with FBF. The classification accuracy is obtained for different window sizes for different domain and range filter parameters. The proposed technique has achieved best classifications accuracies of 60.26% for window size 7 × 7 at $\sigma_d = 1.5$ and $\sigma_r = 25$. The second-best classification accuracy of 59.29% is obtained for window size 7 × 7 at $\sigma_d = 2$ and $\sigma_r = 15$. The third-best classification accuracy of 58.65% is obtained for two different window sizes at five different values of filter range parameter and filter domain parameter by using Brodatz database.

References

1. Haralick RM (1979) Statistical and structural approaches to texture. Proc IEEE 67(5):786–8041979
2. Derin H, Elliot H (1987) Modelling and segmentation of noisy and textured images using Gibbs random fields. IEEE Trans Pat Ana Mach Int 9(1):39–55
3. Mandelbrot BB (1968) The fractal geometry of nature. Freeman, New York
4. Pietikainen M, Hadid A, Zhao G, Ahonen T (2011) Computer vision using local binary patterns, vol 40. Springer
5. Laws KI (1979) Texture energy measures. Image understanding workshop
6. Ojansivu V, Rahtu E, Heikkila J (2008) Rotation invariant local phase quantization for blur insensitive texture analysis. In: IEEE 19th international conference on pattern recognition
7. Roux SG, Clausel M, Vedel B, Jaffard S, Abry P (2013) Self similar anisotropic texture analysis: the hyperbolic wavelet transform contribution. IEEE Trans Image Process 22(11):4353–4363
8. Tomasi C, Manduchi R (1998) Bilateral filtering for gray and color images. In: Proceedings of the IEEE international conference on computer vision, pp 839–846
9. Cho H, Lee H, Kang H, Lee S (2014) Bilateral texture filtering. ACM Trans Graph 33(4):1–82014 (SIGGRAPH 2014)
10. Al-Hinnawi AR, Daear M (2013) Image texture descriptors to quantify bilateral filter on low dose computerized tomography. Int J Signal Process Image Process Pattern Recognit 5(3):123–136

Escalation of Energy Performance in Many User-Several Inputs and Several Output System with Spectral Ability Compulsion

Kommisetti Murthy Raju and Vemu Srinivas Rao

Abstract During the system design, achieving energy performance (EP) has been a performance task for the operators, and spectral ability (SA) has always been a predominant metric of choice. In many user-several input and several output system, multiple users have received multiple streams from the base station that has major concerns and an idea of accomplishing both SA and EP. The two important major factors (EP and SA) contention with each other and attentive examine their trade-off which is compulsory. Commenced a resource allowance algorithm to examine benefit between EP and SA in DL-MU-SISO system, which is accepted by standard organization (3GPP). The SA obligation is stiff development trouble wherever the consideration renamed with cubical disparity is defined to exploit EP. Next to achieve maximum energy performance, unique reserve allowance algorithm has been designed. The capability of the planned scheme and the principle trade-offs among energy performance and spectral ability transmissions are validated and depicted by the simulation results.

Keywords MU-SISO · EP-SA · 3GPP LTE-advanced

1 Introduction

In the MU-SISO downlink systems, observe the compensation between energy performance and spectral ability [1]. Energy utilization is a major task to consider system design, as administrators' aspect to maximize the energy ability usually measured in *b/j*. The aim of next generation mobile networks across TCP/IP uses the 3GPP LTE standard because it is a new energy-efficient way [2]. LTE-advanced has high spectral efficiency, ropes up to 3 Gigabits/second data rate in downlink. MU-SISO downlink methodology is analyzed for attaining high SA MU-SISO systems, accomplishing spatial multiplexing gain to attain at the BS [3]. To overcome the faults of single

K. M. Raju · V. Srinivas Rao (✉)
Department of Electronics and Communication Engineering, Shri Vishnu Engineering College for Women, Andhra Pradesh, India
e-mail: venkateswara.106@gmail.com

P. S. R. Chowdary et al. (eds.), *Microelectronics, Electromagnetics and Telecommunications*, Lecture Notes in Electrical Engineering 655,
https://doi.org/10.1007/978-981-15-3828-5_24

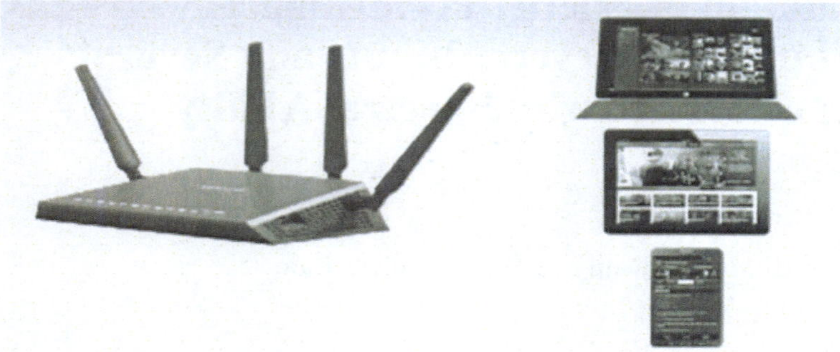

Fig. 1 MU-SISO system

user-several inputs and several output system, new long-term evolution standards defined a significant standard of the many user-several inputs and several output systems which add high capacity gain [3].

Energy ability of the orthogonal FDMA method has been initially inscribed in the examination with the consumption of the circuit. For single cell orthogonal FDMA networks [4], energy performance is a vital parameter to estimate. It uses both modulation order change and power for analysis. Energy ability is proportional fair organized in many user-several input and several output system. There is no work discussing energy performance with spectral ability constraints in the downlink MU-SISO system [5]. In this paper, our aim is to boost energy performance while using spectral ability obligation. First determine the energy consumption related to transmitted bits per unit, which is exceptionally advisable for framing renew broadcast systems. Initially, outline the development issue where restrictions are exhibited as a cubicle disparity and then suggest a new reserve distribution procedure to attain extreme energy performance in a given spectral ability [6, 7].

This paper is formulated as follows: Chap. 1 includes introduction, Chap. 2 deals with related work, Chap. 3 depicts proposed work, Chap. 4 includes implementation, Chap. 5 depicts outcomes, and Chap. 6 shows the conclusion (Fig. 1).

2 Related Work

2.1 Many User-Several Input and Several Output System

Many user-several input and several output system is an allowance of space-division multiple access (SDMA). This knowledge guides several influences on a distinct predictable channel where different users are acknowledged by spatial signatures. Space-division multiple access uses SM and permits for higher data rates [7] that can be accomplished by using many paths as altered channels for carrying data.

One more advantage of expending the space-division multiple access technique in mobile cellular networks is to alleviate the outcome nosiness coming from neighboring cells [8–11].

Many user-several input and several output system is with cooperation where assistance is recognized among the base stations (Figs. 2 and 3).

Base station is equipped with antennas $T_{x1}, T_{x2}, \ldots, T_{xN}$ and offers communication with user armed with M receiving antennas R_{x1}, R_{x2}, …, R_{xM}. The given many

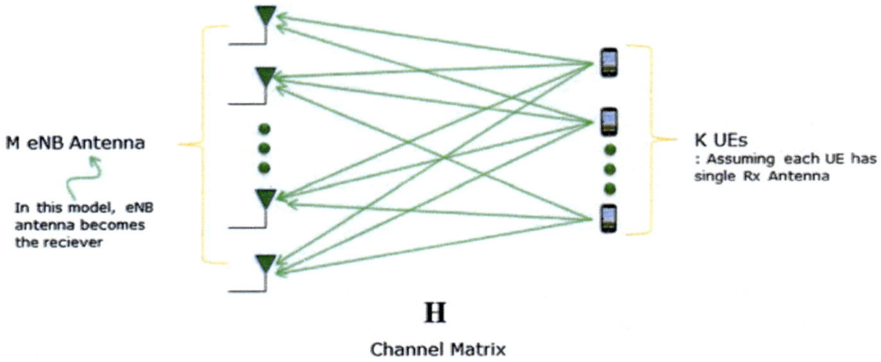

Fig. 2 Channel matrix of MU-SISO system

Fig. 3 MU-SISO structure

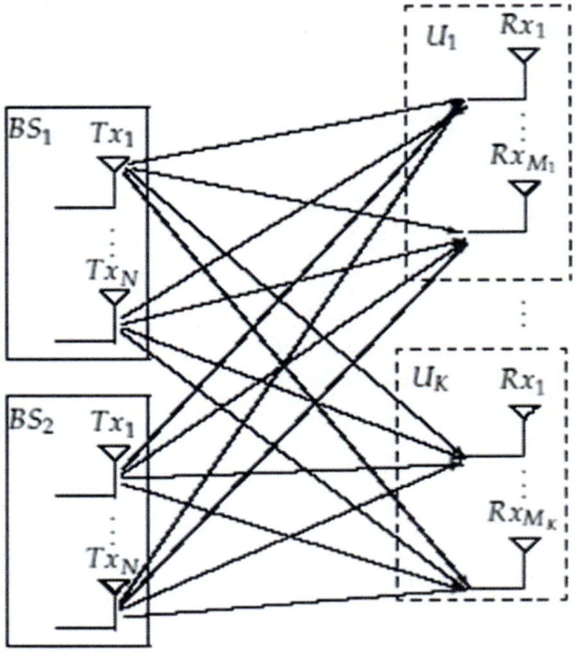

user-several input and several output system contains two base stations BS_1 and BS_2, and every one is armed with N antennas. Comprehensive many user-several input and several output systems may entail many base stations where the number of antennas could be changed. K users U_1, U_2, ..., U_k at the receive side with, respectively, $M_1, M_2, \ldots, M_k$ antennas connect with the transmit base stations. The same communication model is performed for the many user-several input and several output system with cooperation where collaboration is established between BS_1 and BS_2 [6].

2.2 Many User-Several Input and Several Output System Versus Single User-Several Input and Several Output System

In extension, many user-several input and several output systems are planned to involve with several receivers to enhance the degree of communication while protecting the same level of accuracy. These systems are talented to accomplish the total multiplexing gain acquired as the minimum value between the number of antennas at users and the number of antennas at base stations. Verify that several users keep concurrently connect over the same spectrum progresses the system execution. Still, MU-SISO networks are unprotected to firm co-channel nosiness which is not the case for SU-SISO ones. Considerable approaches are proposed to resolve the problem of nosiness in MU-SISO systems. A few of these accesses are established on beam forming technology (Figs. 4, 5, 6 and 7).

Fig. 4 SU-SISO structure

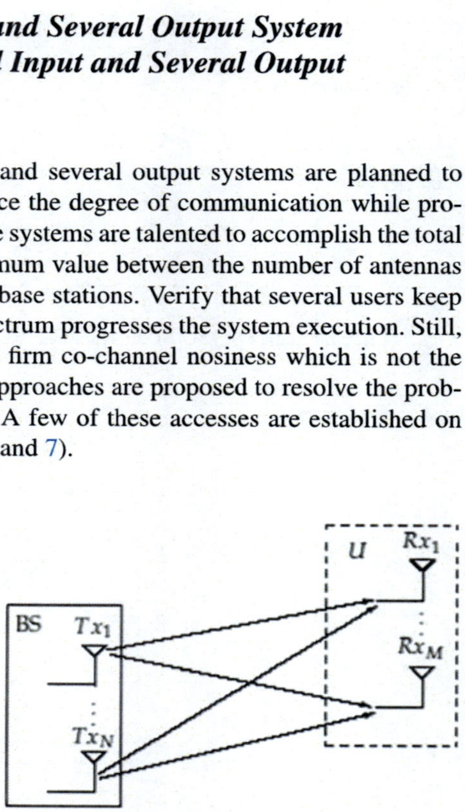

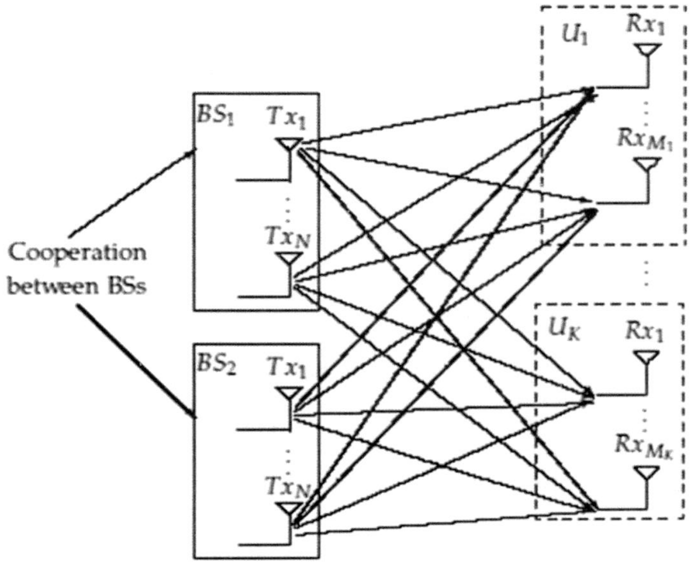

Fig. 5 MU-SISO with cooperation structure

Fig. 6 Uplink MU-SISO

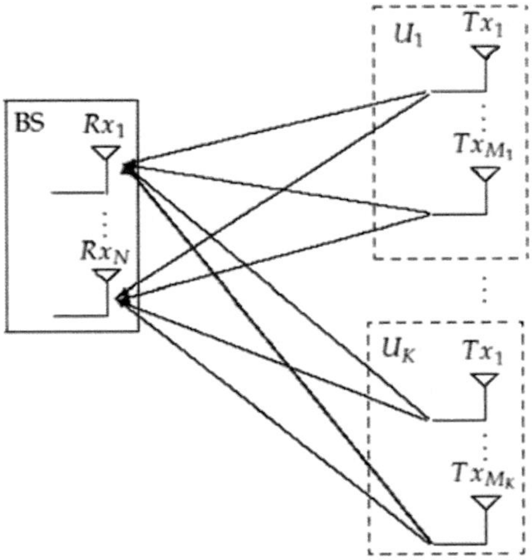

Fig. 7 Downlink MU-SISO

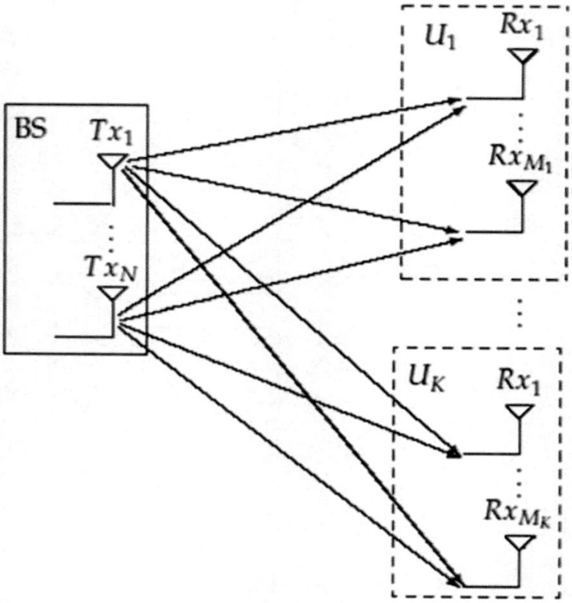

2.3 *Many User-Several Input and Several Output Systems Communication Schemes*

There are two types of communication schemes offered for many user-several input and several output systems which include UL-MU-SISO and downlink DL-MU-SISO. Two representations schemes are shown in below figures. The BS is assembled with N antennas. In downlink MU-SISO, the BS seeks to convey signals to K users U_1, U_2, ..., U_k, and these are commonly supplied with antennas of facts M_1, M_2, ..., M_k.

An antenna acts like a transmitting antenna, and it is represented by T_{xk} or else it is represented by R_{xk}.

3 Proposed Architecture

3.1 *Many User-Several Input and Several Output System Hexagonal Cell Design for the Single Cell*

In a single cell hexagonal cell layout of MU-SISO system, nosiness is confined as intracell intervention that is many user intervention in case of many user-several input and several output intracell interventions are commenced from the user equipments

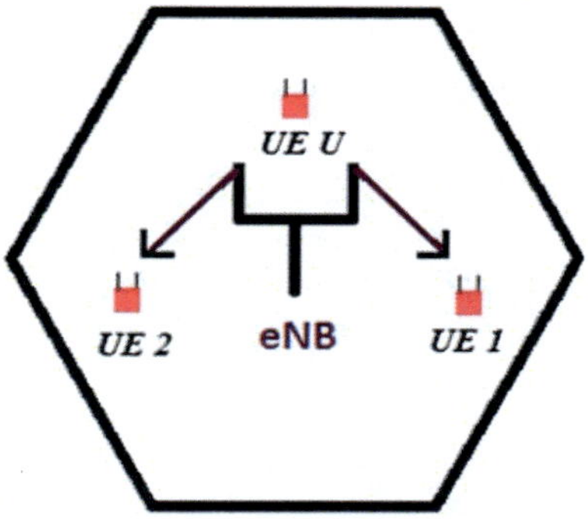

Fig. 8 Hexagonal cell layout

of the similar cells. To alleviate this nosiness, right precoding methods are generally restrained by coming wireless systems due to its clearness and benefit the enactment of the proposed system of the future system (Fig. 8).

In the above figure, eNB is assembled with TA and UEs are equipped with RA. MU-SISO system has one enhanced Node B by a consistent hexagon, serving user equipments (UEs) which are equally distributed over its coverage area. In MU-SISO systems, observed both multiuser intrusion and intracell intrusion. Disturbance is commenced from user equipments of that cell. Coding techniques are used to suppress the disturbances in future wireless communication systems. Zero-forcing (ZF) precoding is a low-complexity technique which improves the sum rate. In that technique, the transmitter transmits the signals close to the fixed user direction, and no other information to other users exists in M physical resource blocks assigned to one or more user equipments.

Robust power distribution using the channel condition between PRBs is studied according to transmission medium provision of the user equipments, and the overall transmit power P presented on each HC is allocated between the PRBs. The main transmission media dimension of the user equipment in the system is signal-to-interference-plus-noise ratio. Opt whether a transmission has been successful, and the SINR restrained for a given path is connected to measure the packet error rate (PER) for that chunk of data sent on each MU-SISO cell that can be expressed as

$$\Gamma_u^m = \frac{p_u^m \left| H_u^m W_u^m \right|^2}{\sum_{u' \neq u}^{U} p_{u'}^m \left| H_u^m W_{u'}^m \right|^2 + \sigma^2} \tag{1}$$

Associate shadowing, path loss and fast fading, which represents the link between the user equipment and evolved Node B of the cell C in the channel matrix elements. The precoding matrix for the connection between the evolved Node B and the UE u is CTA_RA. The intracell intrusion is alleviated by expending zero-forcing precoding strategy, pleasing the pseudo-inverse of the channel matrix. σ^2 is the Additive White Gaussian Noise power as supposed by the user equipment. The entire bandwidth B is correspondingly separated into physical resource block, each of bandwidth is W.

The spectral ability (SE) attained by Shannon theorem, of user u on physical resource block m is

$$\mathrm{SE}_u^m = \log_2\left(1 + \Gamma_u^m\right) \tag{2}$$

The top attainable data rate $\bar{s}_u^m$ of handler u physical resource block m is

$$\bar{s}_u^m = W \cdot \mathrm{SE}_u^m \tag{3}$$

The data rate for user u on physical resource block at any apt is s_u^m. The complete structure throughput S and P are the total transmitter power of evolved Node B for user u on physical resource block m. Transmission power is also significant on the power amplifier ability denoted as α where its value lies in between 0 and 1 and depends on transmitter ideal and exploitation. The power ingestion is defined as

$$S = \sum_{u=1}^{U} \sum_{m=1}^{M} s_u^m$$
$$P = \sum_{u=1}^{U} \sum_{m=1}^{M} p_u^m \tag{4}$$

Isolated from the transmission power, consider circuit power also. Circuit energy ingesting P_c could be divided into fixed and active parts that lean on parameters of dynamic links, and P_{st} is the fixed circuit components power in the convey mode and a persistent δ denoting active power utilization per unit throughput.

$$P_c = P_{\mathrm{st}} + \delta \cdot S$$

Cell Site Generation

For the cell generation, we consider the radius of the cell as 0.5 units. The enhanced Node NB will be at the center of the cell, i.e., (0, 0). The user equipment (UE) is fixed at random locations as shown in the figure. The boundaries of the cell site can be determined by using the below equations.

Cell radius (CR) = 0.5

$$x = \mathrm{CR} * [\cos(0 : \pi/3 : 2\pi)]$$
$$y = \mathrm{CR} * [\sin(0 : \pi/3 : 2\pi)]$$

(eNBx, eNBy) = (0, 0)

3.2 Coordinates of Cell Generation

X	0.5	0.25	−0.25	−0.5	−0.25	0.25	0.5
Y	0	0.433	0.433	0	−0.433	0.433	0

In cellular communication, a cell site is accommodated with electronic communication equipment and antenna that are placed, the cellular network consists of many cells that are distributed over areas, and each cell served at least one fixed location transceiver. Normally, three cell sites used to assign the physical location of radio equipment that provides coverage within a cell. The cell site proposed system is as follows (Figs. 9, 10 and 11).

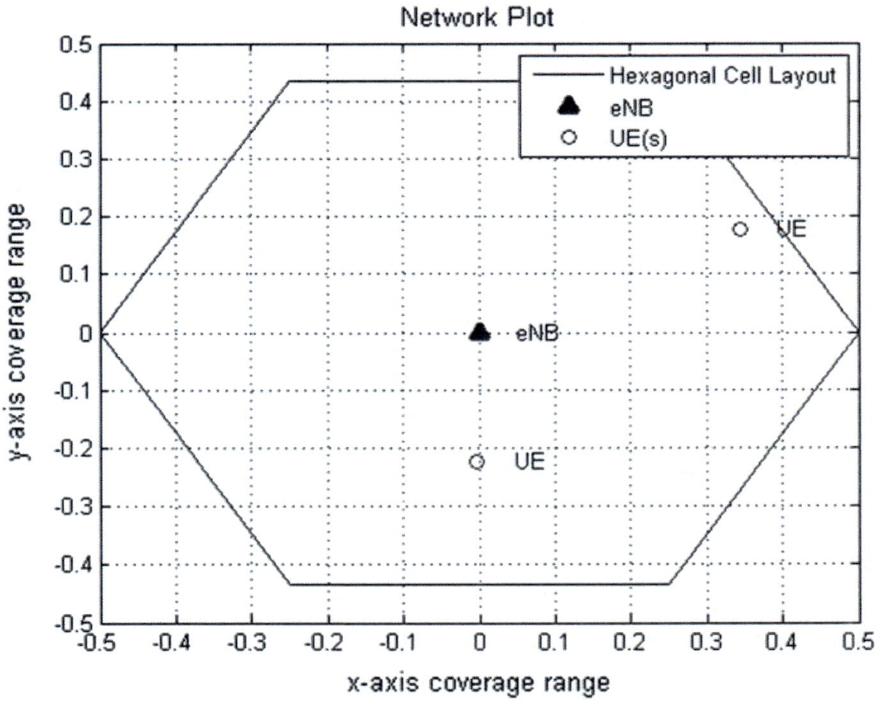

Fig. 9 Cell sites with eNB and UEs with 2X2 antenna configuration

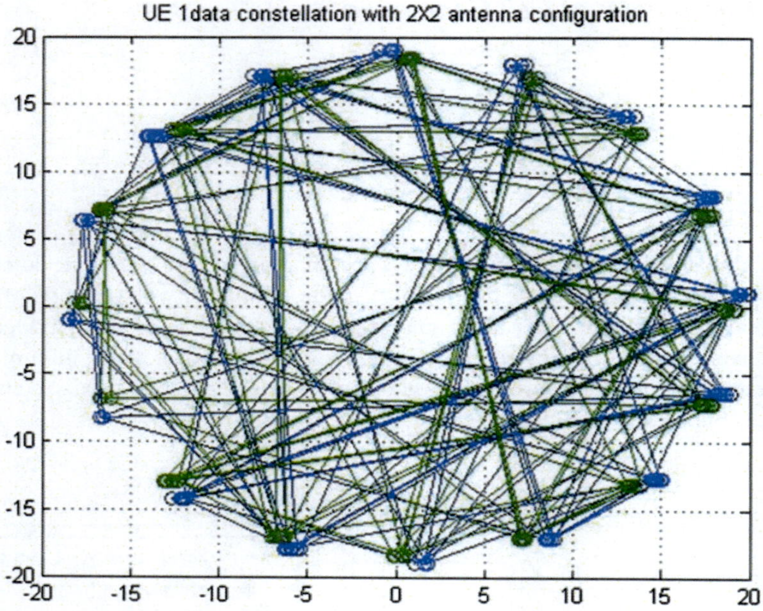

Fig. 10 UE1 data constellation

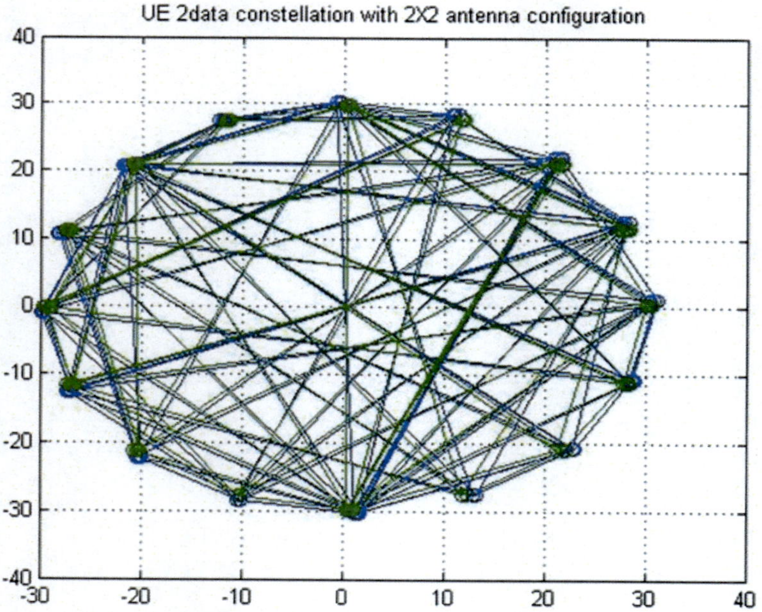

Fig. 11 UE2 data constellation with 2X2 antenna configuration

4 Implementation

4.1 Suggested the Ideal Ability Provision Procedure for Attaining Extreme in the DL-MU-SISO in Steps

Provision for Attaining Maximum EP:

Step 1: Initialization: Set two values are $s^{(0)}$ and $\lambda u^{\mathrm{m}(0)}$ for users and resource blocks;
Step 2: **Execute** modified method: Execute modified technique using equations;
Step 3: Limits design for analytical examination: Calculate the higher limit (Ω_{UL}) and the lower limit (Ω_{LL});
Step 4: s_u^m Calculation: Take $\Omega_u^m = \Omega_{\mathrm{UL}} + \Omega_{\mathrm{LL}}/2$ and calculate s_u^m;
Step 5: If $s_u^m > \bar{s}_u^m$, then $\Omega_{\mathrm{UL}} = \Omega_u^m$; otherwise $\Omega_{\mathrm{LL}} = \Omega_u^m$;
Step 6: Iterate until the execution meets the optimality.

5 Outcomes

5.1 Concurrence Attainment of the Optimization Model

Plotted spectral efficiency (SA) versus number of iterations. Examine that the increasing value of iterations spectral efficiency also increases. But the value of SE became constant after five iterations allowing no further increment in the value of SE. This shows that the optimized value of SA is obtained at five iterations.

In the above figure increasing number of iterations, spectral ability (SE) also increases linearly. Spectral efficiency value is constant after five iterations; this value does not increase after five iterations. Better value of spectral ability is achieved at five iterations. Figure 12 shows the concurrence action of the advanced algorithm. Energy efficiency value is constant after five iterations. Energy performance (EE) value does not increase after five iterations. With the help of these results, develop correct energy consumption (Energy performance is maximized) based on network if the power is constant (Fig. 13).

5.2 Evaluation of Merging of EP with Different Physical Resource Blocks

It aids the result of altered physical resource blocks loads with the proposed procedure and shows that the increase in the number of physical resource blocks enhances the energy efficiency of that design. Less number of PRBs means less iteration to

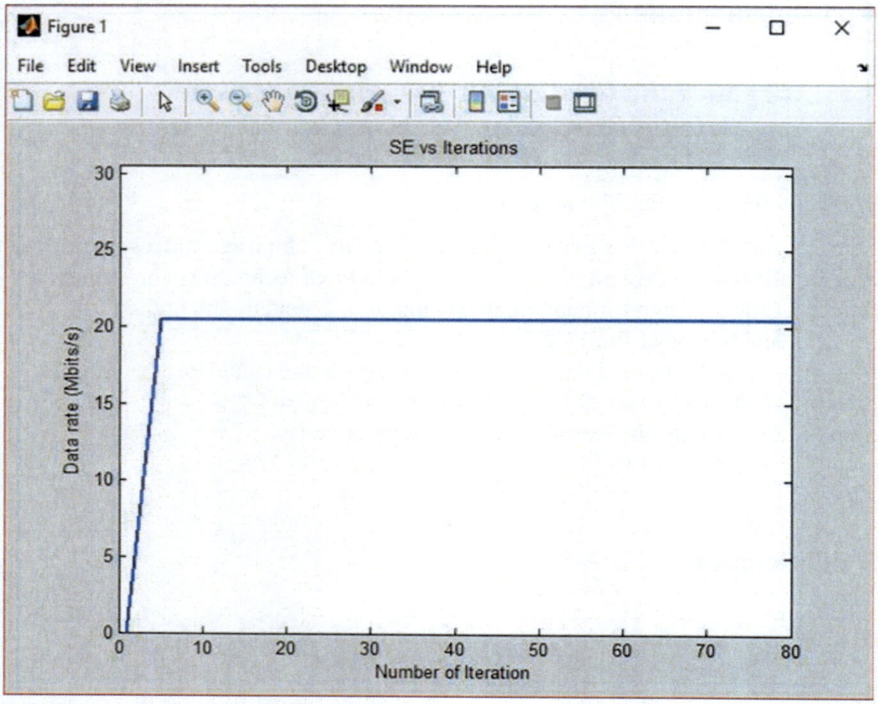

Fig. 12 SE versus iterations

obtain concurrence. More number of PRBs means much iteration required to obtain concurrence (Figs. 14 and 15).

5.3 *Concurrence of Diverse Active Circuit Power*

Delta value influence on energy efficiency. It proves the influence of the delta factor in the upgraded process. The energy efficiency improves as the value of delta decreases. Delta value increases, more iteration required to concurrence (Figs. 16 and 17).

5.4 *SA Versus EP Trade-off Relation*

Figure 18 shows energy performance as the function of the spectral ability. Observed that Energy performance growths monotonically with Spectral ability up to some point and then decreases. It observes the effect of power amplifier ability on energy ability. It shows that higher power amplifier performance leads to higher energy

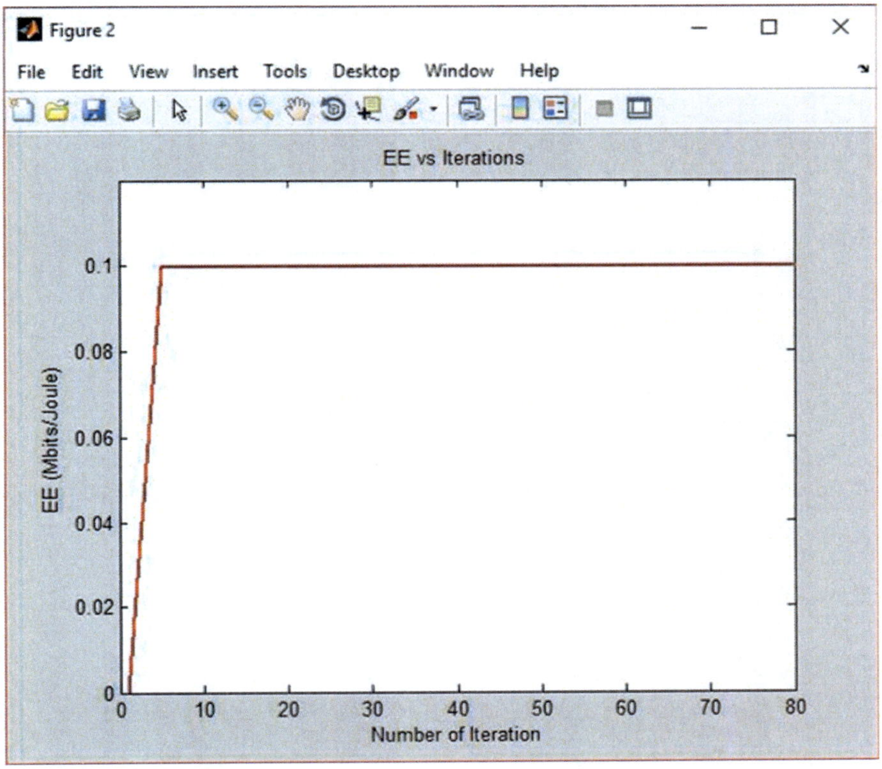

Fig. 13 EE versus iterations

performance. Conclude that the design of better energy utilization networks based on EP-oriented methods for MU-SISO can be done even if the system SA is not limited (Fig. 19).

6 Conclusion

To maximize EP, a new resource distribution algorithm has been proposed. This paper presents a systematic way for better energy performance of the MU-SISO system with a cost to limitations of target SA. Different network parameters are analyzed in this method has been used. Upper and lower limits of the proposed algorithm can be obtained by using a numerical search. Recreation results show the concert of the future effort and indicate the fundamental transaction among spectral ability transmission and energy performance. Supplementary research work to be prolonged to many cell situations with assistance among the cells is also required.

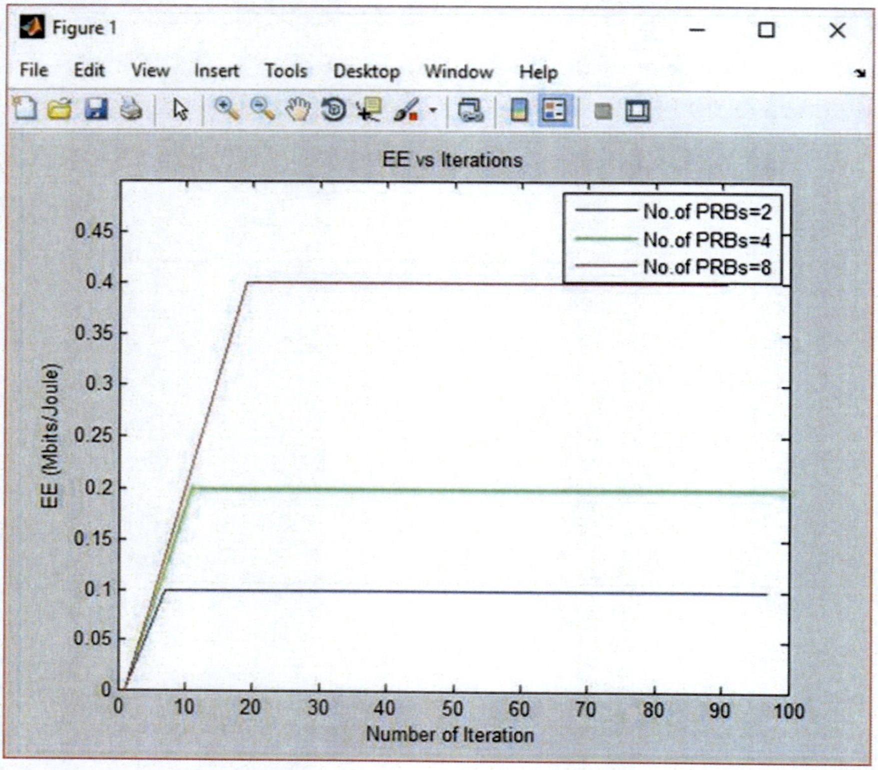

Fig. 14 EE versus iterations

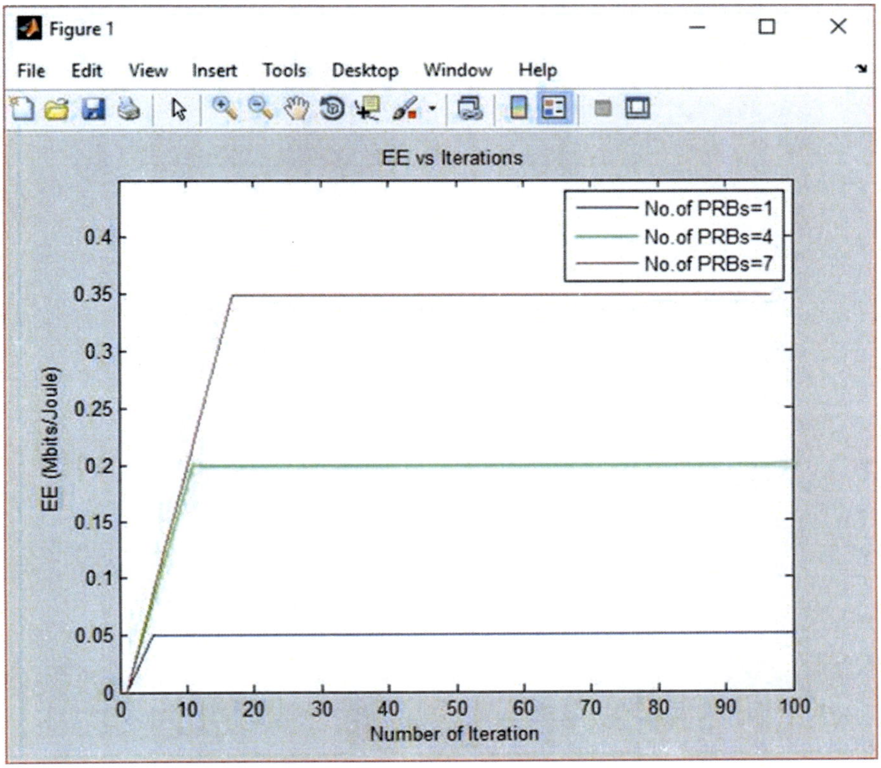

Fig. 15 EE versus iterations

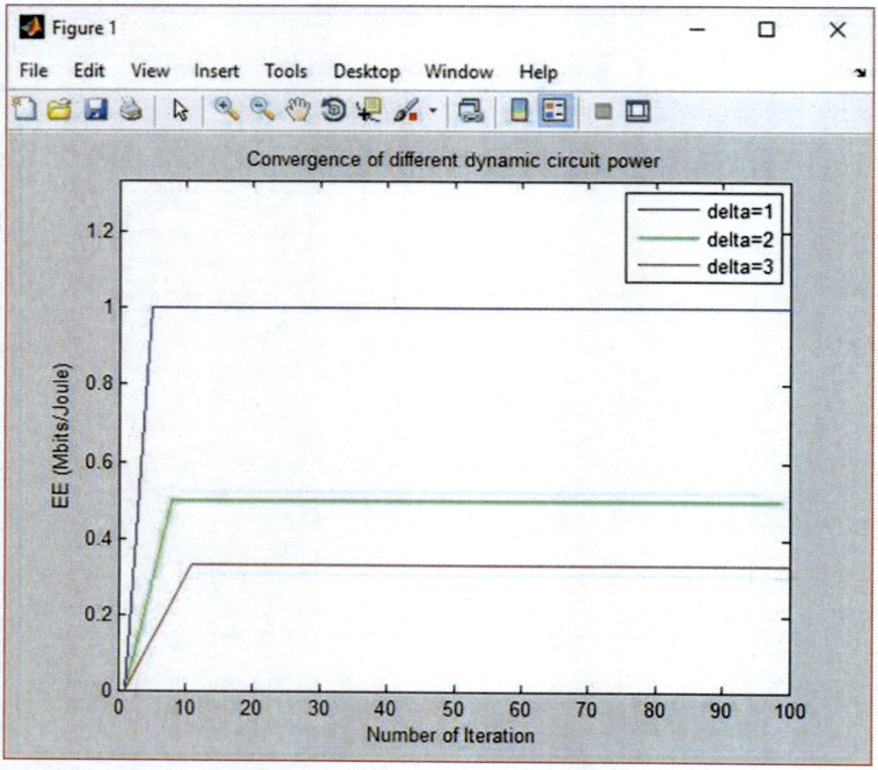

Fig. 16 Concurrence of diverse active circuit influence representation with different values of delta

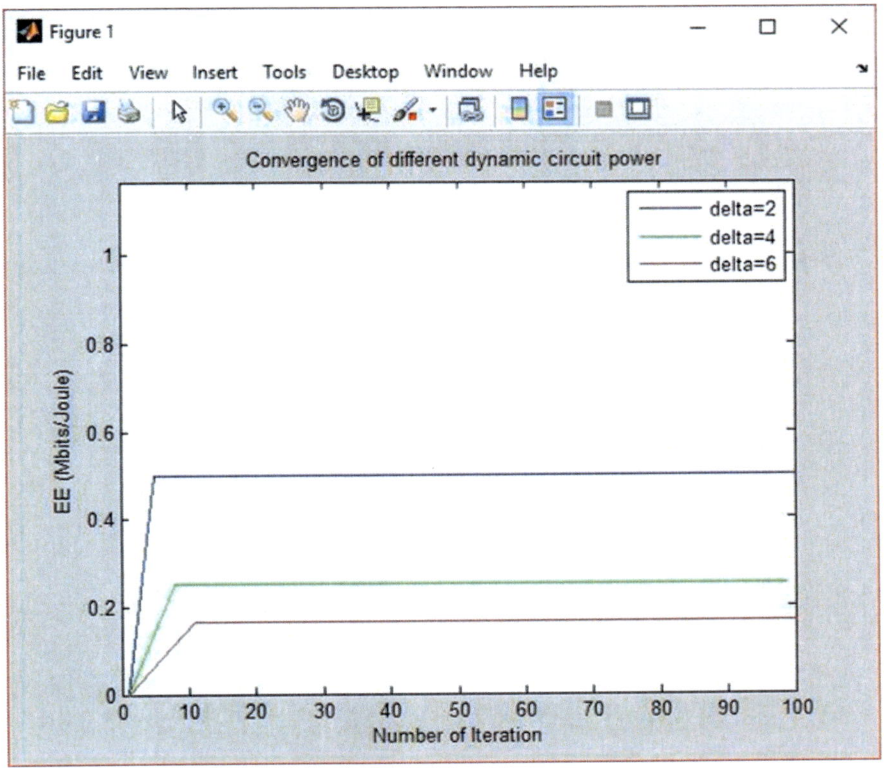

Fig. 17 Concurrence of diverse active circuit power with different values of delta

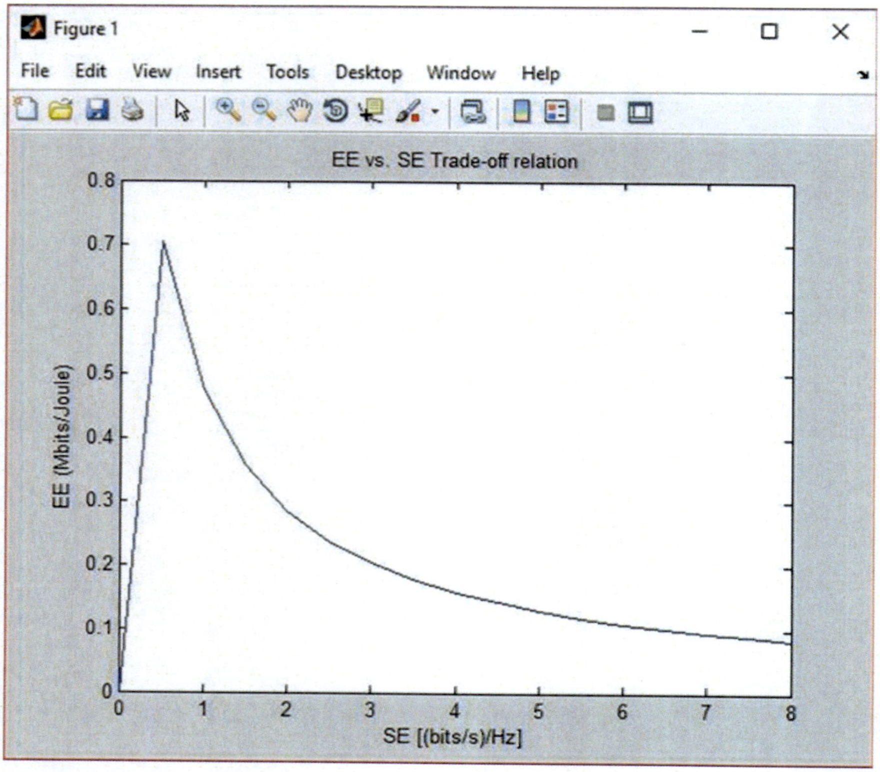

Fig. 18 SA versus EP trade-off relation

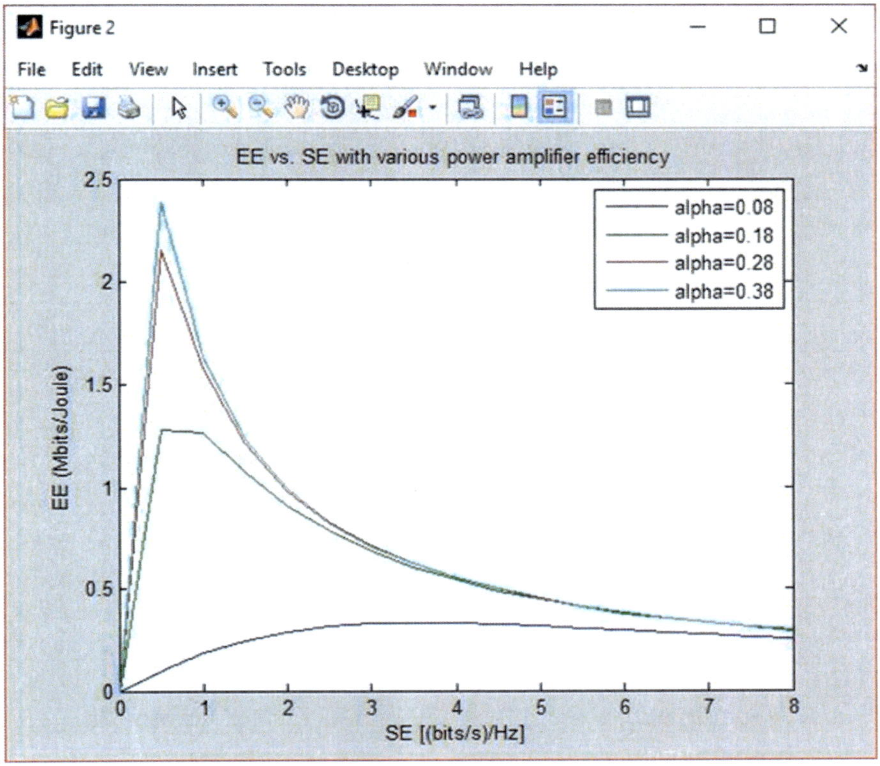

Fig. 19 SA versus EP trade-off relation

References

1. De Sanctis M, Cianca E, Joshi V (2011) Energy efficient wireless networks towards green communications. Wirel Pers Commun 59(3):537–552. Available: http://www.springerlink.com.miman.bib.bth.se/content/e764hq2h2h83081t/
2. Sundaram RK (1996) A first course in optimization theory. Cambridge University Press (June 1996)
3. Miao G, Himayat N, Li Y, Bormann D (2008) Energy efficient design in wireless OFDMA, in Communications, ICC'08. In: IEEE international conference on, May 2008, pp 3307–3312
4. GPP: Spatial channel model for mimo simulations, TR 25.996, V6.1.0, Tech. Rep. (September 2003)
5. Miao G, Himayat N, Li Y, Bormann D (2008) Energy efficient design in wireless OFDMA, in Communications, ICC '08. In: IEEE international conference on, May 2008, pp 3307–3312
6. Xiong C, Li GY, Zhang S, Chen Y, Xu S (2011) Energy- and spectral efficiency tradeoff in downlink OFDMA networks. IEEE Trans Wirel Commun 10(11):3874–3886
7. He C, Sheng B, Zhu P, You X (2012) Energy efficiency and spectral efficiency tradeoff in downlink distributed antenna systems. IEEE Wirel Commun Lett 1(3):153–156
8. Venkata Subbarao M, Samundiswary P (2018) Spectrum sensing in cognitive radio networks using time–frequency analysis and modulation recognition. In: Springer lecture notes in electrical engineering, vol 471, pp 827–837

9. Venkata Subbarao M, Samundiswary P (2016) An intelligent cognitive radio receiver for future trend wireless applications. Int J Comput Sci Inf Secur (IJCSIS) 14:7–12
10. Venkata Subbarao M, Samundiswary P (2018) Automatic modulation recognition in cognitive radio receivers using multi-order cumulants and decision trees. Int J Recent Technol Eng (IJRTE) 7(4):61–69
11. Liu L, Miao G, (Charlie) Zhang J (2012) Energy-efficient scheduling for downlink multi-user MIMO. In: 2012 IEEE international conference on communications (ICC), p 4394
12. Spencer Q, Swindlehurst A, Haardt M (2004) Zero-forcing methods for downlink spatial multiplexing in multiuser MIMO channels. IEEE Trans Signal Process 52(2):461–471

Nature-Inspired Biogeography-Based Optimization for Estimation of GPS Receiver Position in Low Latitude Regions of the Indian Subcontinent

N. Ashok Kumar and G. Sasibhushana Rao

Abstract The concept of biological species distribution in a geographical area laid a path to develop biogeography-based optimization (BBO) in order to solve real-time optimization problems. BBO is implemented on GPS receiver data that belongs to low latitude region of the Indian subcontinent in order to estimate the receiver position. 3D positional coordinates of the receiver obtained by using BBO algorithm are compared with the surveyed location coordinates to analyze the performance of BBO algorithm. From the estimated mean position errors, BBO is observed to be a better evolutionary computing algorithm in the estimation of any GPS receiver position which is located in low latitude regions of the Indian subcontinent. This algorithm can also be used for QZSS and IRNSS satellite systems.

Keywords Biogeography · Distribution · GPS · Evolutionary algorithm

1 Introduction

Biogeography-based optimization (BBO) is a nature-inspired technique which stochastically and iteratively optimizes a function to determine an accurate solution for a given fitness function. BBO's mathematical equations were developed during 1960s [1]. The concept of biogeography can be seen in the works of Charles Darwin [2] and Alfred Wallace [3]. The biogeography describes the migration and immigration nature of various species in nearby geographical locations (like nearby islands) [4, 5]. This concept has been mathematically formulated to develop BBO algorithm. In this article, the BBO algorithm is used in order to achieve a precise estimation of a GPS receiver location in low latitude regions of Indian subcontinent.

N. Ashok Kumar (✉)
Department of Electronics and Communication Engineering, Raghu Engineering College (A), Dakamarri, Visakhapatnam 531162, Andhra Pradesh, India
e-mail: ashok0709@gmail.com

G. Sasibhushana Rao
Department of Electronics and Communication Engineering, Andhra University College of Engineering (A), Andhra University, Visakhapatnam 530003, Andhra Pradesh, India

P. S. R. Chowdary et al. (eds.), *Microelectronics, Electromagnetics and Telecommunications*, Lecture Notes in Electrical Engineering 655,
https://doi.org/10.1007/978-981-15-3828-5_25

The real-time implementation is done on the GPS data that has been collected from the GPS receiver located at 17.73° N/83.31° E. From the observation of estimated 3D position obtained by using BBO, it is evident that the BBO can be used for the estimation of any GPS receiver 3D position in low latitude regions of India. Section 2 briefly introduces basics of GPS for receiver position estimation. Section 3 explains the BBO algorithm and its implementation procedure. Section 4 presents results and discussion. Section 5 presents final conclusions.

2 Global Positioning System

Global positioning system (GPS), a popular navigational satellite system with global coverage (GNSS) [6–8] that provides positioning, timing, and navigation services worldwide continuously. GPS provides continuous global coverage using 24 operational satellites in space segment. These satellites are placed in medium earth orbit (MEO) in 6 orbits and from the earth, these orbits are at an altitude of 20,200 km. And the orbits are inclined to the earth's equator with an inclination of 55°. The user location, who is having a GPS receiver, is determined by using these satellites. Trilateration principle is used for GPS in order to estimate the receiver position or user. To achieve this, pseudo-range measurements from minimum of four satellites are required. More than four satellites are also visible in nowadays. Therefore, pseudo-range equations are formed from these observed measurements. These equations are required to be solved in order to find the receiver location. Equation (1) represents the basic pseudo-range equation, that is,

$$P_{\text{sudo}} = \rho_{\text{geo}} + c(\Delta t - \Delta T) + \Delta_{\text{trop}} + \Delta_{\text{iono}} + \xi \tag{1}$$

where ρ_{geo} is the distance between satellite and receiver, P_{sudo} is observed pseudo-range measurement, Δt and ΔT are receiver clock and satellite clock offsets. Δ_{trop} is the error due to troposphere, and Δ_{iono} is error introduced by propagation through ionosphere. ξ is multipath error. In (1), receiver clock bias $\text{Clk}^{\text{Re}} = c \cdot \Delta t$ is kept as it is and the other correctable error factors are ignored for the simplified presentation. Therefore, the pseudo-range equation is given by,

$$P_{\text{sudo}} = \sqrt{(x_{\text{sv}} - x_u)^2 + (y_{\text{sv}} - y_u)^2 + (z_{\text{sv}} - z_u)^2} + \text{Clk}^{\text{Re}} \tag{2}$$

Based on number of visible satellites, (2) forms four or more equations which are nonlinear equations and they are solved by using any evolutionary algorithm to estimate the unknown parameters, i.e., receiver position. The four unknown parameters to be estimated are 3D coordinates of the receiver, i.e., x_u, y_u, z_u, and Clk^{Re} is receiver clock bias.

3 Biogeography-Based Optimization

Biogeography is a scientific concept that describes how various species are geographically distributed in nearby geographical areas, for example, distribution of different species among nearby islands based on various conditions [4]. In this paper, the term 'island' is replaced with 'habitat' as the island indicates geographical isolation of one habitat from the other habitats. Geographical areas are classified based on suitability of residence for biological species and are indicated by 'habitat suitable index' (HSI). High HSI represents well suitability for residence. The features like land area, type of vegetation, rain fall, etc., are the variables that characterize and correlate with HSI of species which are suitability index variables (SIVs). Therefore, HSI is considered as dependent variable and SIV is considered as independent variable. Species migrate from high HSI location to low HSI location as the low HSI location is able to accommodate more number of other species. But, high HSI location provides limited acceptance for other species to its location as it is already crowded.

Let us consider two habitats in two nearby locations. The emigration rate of species is represented by 'E' and immigration rate is represented by 'I'. The habitat with high HSI has more number of species with less or no possibility for other species to reside and hence it has high emigration rate $\left(E^{h}\right)$ and less immigration rate $\left(I^{h}\right)$, i.e., $E^{h} > I^{h}$. Similarly, low HSI habitat has less number of species that provides high possibility for other species and already residing species to reside. Hence, it has high immigration rate $\left(E^{l}\right)$ and less emigration rate $\left(I^{l}\right)$, i.e., $E^{l} < I^{l}$.

When the species of neighboring habitats are considered, the emigration rate of high HSI habitat is more than that of low HSI habitat, i.e., $E^{h} > E^{l}$. Similarly, the immigration rate to high HSI habitat is less than that of low HSI habitat, i.e., $I^{h} > I^{l}$. Figure 1 shows the ideal single habitat species model in which different emigration and immigration rates are due to various reasons such as geographical area, temperature, food availability, etc. The equilibrium number of species 'S_E'

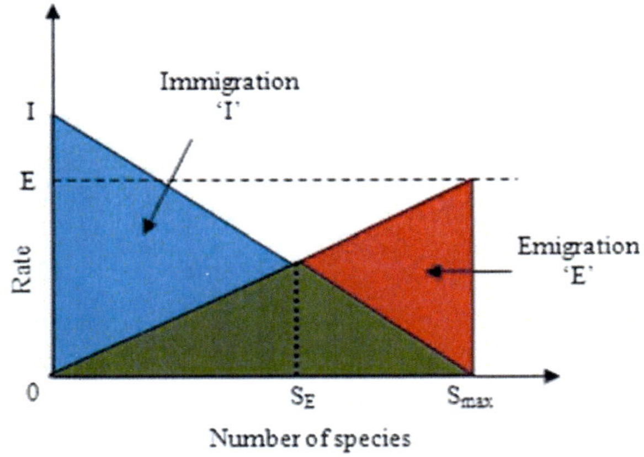

Fig. 1 Single habitat species model

occurs at the point of intersection of immigration and emigration curves. However, there may be change in equilibrium point due to temporal effects. Consider 'P_s' as the probability of a habitat having 'S' number of species. This probability changes from time 't' to 'δt' and is given by,

$$P_s(t + \delta t) = P_s(t)(1 - I_s\delta t - E_s\delta t) + P_{s-1}E_{s-1}\delta t + P_{s+1}E_{s+1}\delta t \tag{3}$$

The application of biogeography theory for solving the optimization problems is processed in two stages. They are (i) Migration and (ii) Mutation.

3.1 Migration

Consider, there is an optimization problem exists with a population of solutions which are represented as parameters of a vector. Each solution in the vector is considered as SIV. These solutions are obtained by using a fitness function. Good solutions are considered to be habitats with high HSI which provides good condition for many species to reside here. Whereas poor solutions are low HSI habitats, which provides unfavorable condition for species to reside and only some species can able to reside here.

Figure 2 illustrates the two solutions for some optimization problem in which S_1 is a poor solution and S_2 is a good solution. For simplicity, immigration and emigration curves are taken as equally distributed, i.e., $E = I$ which is shown in Fig. 2. The solution 'S' value depends on its HSI. In Fig. 2, S_1 represents a low HSI solution which indicates that the S_1 is a habitat has few species. S_2 represents a high HSI solution which indicates that the S_2 is a habitat has more species. Therefore, the emigration rate $\left(E^h\right)$ for S_1 is less than the emigration rate $\left(E^l\right)$ for S_2. Similarly, the immigration rate $\left(I^h\right)$ for S_1 is more than the immigration rate $\left(I^l\right)$ for S_2. The immigration and

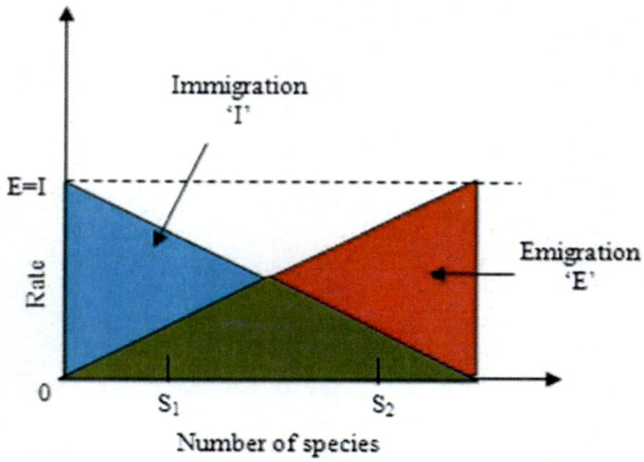

Fig. 2 Two habitat species model

emigration rates of the solutions S_1 and S_2 share the information between habitats based on probability. The probability P_{modify} of each solution is modified based on the other solutions. SIV is being decided by immigration rate whether the selected solution is to be modify or not. If a given SIV in a given solution S_i is to be modified, then, the emigration rate of other solutions are used probabilistically to decide to migrate the selected SIVs of the solution S_i. The migration process in BBO is adaptive and is used to modify the existing solutions, whereas in evolutionary algorithms like genetic algorithm (GA), new solutions are created by using global recombination approach.

3.2 Mutation

The HSI of a natural habitat is drastically changed due to cataclysmic events which also cause the change in species count from its equilibrium point value. Therefore, the BBO is modeled as SIV mutation for which species count probabilities are used to compute mutation rates. In Fig. 2, very low probabilities are observed for the high species and low species count. Significant probability is observed at medium species count. That means, if a given solution 'S' with low probability P_s is mutated to other solution. Conversely, high probability solution is less likely to mutate to other solution. The mutation rate 'M' of a solution is given by,

$$M(S) = M_{\max}\left[\frac{1 - P_s}{P_{\max}}\right] \tag{4}$$

where $M_{\max}$ is a user-defined value. The mutation process increases the diversity among the population otherwise high probable solution will be dominant that leads to inaccurate estimation.

The mutation process gives a chance to low HSI solutions to improve further. Pseudo code of BBO algorithm is presented in Table 1. The BBO algorithm, in Table 1, is presented in simple steps in order to implement it for any optimization problem; the same is implemented in the next section for the estimation of GPS receiver 3D position.

4 Results and Discussion

The suitability of the BBO algorithm, in order to achieve precise estimation of a GPS receiver location in low latitude regions of Indian subcontinent, is tested on real-time GPS ephemeris data. This data has been recorded by using a dual frequency GPS receiver located at Lat. 17.73° N/Long. 83.31° E, i.e., at Andhra University College of Engineering (A) in the Dept. of ECE, Visakhapatnam, India. The objective function is given by,

Table 1 Pseudo code of BBO algorithm

- Specify initialization, maximum number of species count ($S_{\max}$), immigration, emigration rates ($E_{\max}, I_{\max}$), and maximum mutation rate $M_{\max}$
- Generate randomly, initial set of habitats for which each habitat represents a potential solution for the given optimization problem
- for loop = 1, 2, 3
- Species count S is being mapped by HSI for each habitat with the emigration rate E and the immigration rate I
- Use emigration and immigration rates probabilistically to modify each non-elite habitat that is given by,

$P_s(t + \delta t) = P_s(t)(1 - I_s\delta t - E_s\delta t) + P_{s-1}E_{s-1}\delta t + P_{s+1}E_{s+1}\delta t$

- Update the probability of the species count for each habitat
- Then, each non-elite habitat is being mutated based on the probability, that is given by,

$M(S) = M_{\max}\left[\frac{1-P_s}{P_{\max}}\right]$

- End
- The above loop is terminated based on predefined condition for the given optimization problem

$$\text{Objective Function} = P_{\text{sudo}} - \sqrt{(x_{\text{sv}} - x_u)^2 + (y_{\text{sv}} - y_u)^2 + (z_{\text{sv}} - z_u)^2} + \text{Clk}^{\text{Re}} \quad (5)$$

For the implementation, the assigned BBO algorithm initial parameters are as follows: the number of habitats is 50, the number of iterations is 1000, emigration rate is $E = [0 \quad 1]$, immigration rate is $I = 1 - E$, and mutation rate is 0.9. The estimated receiver coordinates by using BBO algorithm are compared with the receivers survey position (706,970.90, 6,035,941.02, 1,930,009.58 m). Figure 3 shows the estimated three-dimensional coordinate errors in x-, y-, and z-directions. And the estimated receiver clock bias is shown in Fig. 4. These results are also presented in Table 2.

From Table 2, it is observed that the x-directional position error is maximum between 20:00 and 22:00 h, i.e., 52.4 m and the minimum position error is observed between 04:00 and 06:00 h, i.e., 29.8 m. The y-directional position error is observed to be maximum between 22:00 and 24:00 h, i.e., 109.5 m and the minimum position error is observed between 04:00 and 06:00 h, i.e., 47.0 m. The z-directional position error is observed to be maximum between 22:00 and 24:00 h, i.e., 39.6 m and the minimum position error is observed between 04:00 and 06:00 h, i.e., 14.9 m.

The mean position errors are obtained by eliminating other correctable error terms from (1). Hence, with due consideration of these error terms, the position error can be reduced further. That is, by considering atmospheric error, multipath errors, and satellite clock offsets into (1), improved position accuracy can be achieved. From the observation of Table 2, it is an evident that the BBO is a suitable and a better evolutionary algorithm in order to achieve precise estimation of a GPS receiver location in low latitude regions of Indian subcontinent.

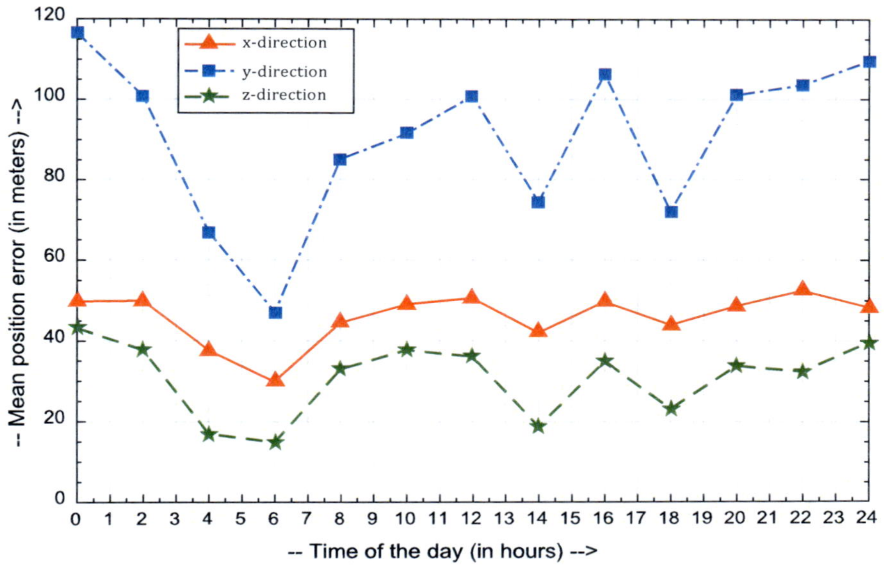

Fig. 3 Estimated position errors

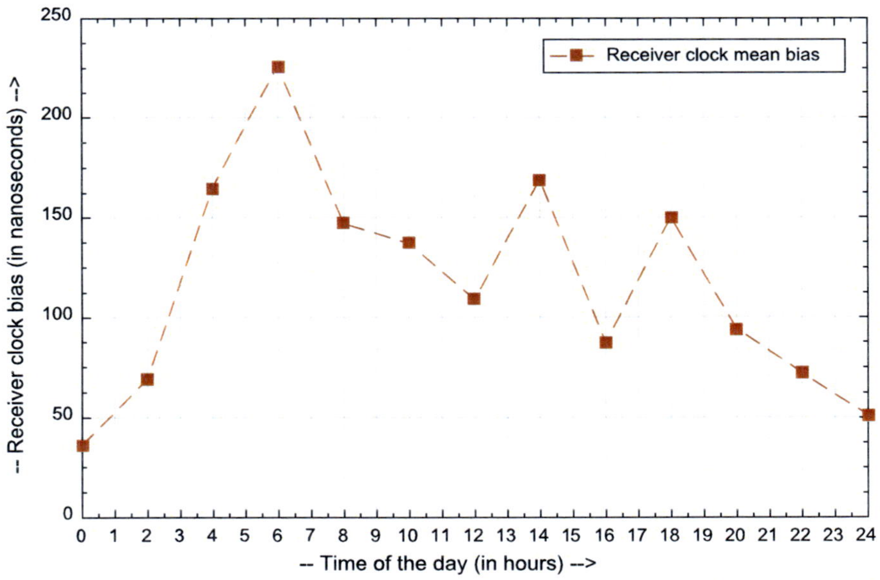

Fig. 4 Estimated receiver clock mean bias

Table 2 Estimated position errors by using biogeography-based optimization

Time (in h)	Estimated error in position			Receiver clock bias (in ns)
	x-direction (m)	y-direction (m)	z-direction (m)	
00–02	49.8	100.9	37.7	69
02–04	37.5	66.8	16.9	165
04–06	29.8	47.0	14.9	226
06–08	44.4	85.1	32.9	147
08–10	49.0	91.5	37.7	137
10–12	50.4	100.8	36.0	109
12–14	41.9	74.4	18.8	169
14–16	49.7	106.4	34.9	87
16–18	43.9	71.9	23.2	150
18–20	48.6	101.1	33.8	90
20–22	52.4	103.5	32.3	72
22–24	48.2	109.5	39.6	51
Mean value	45.4	88.2	29.9	123

5 Conclusions

The implementation of BBO algorithm is carried out on the real-time GPS data that is collected from the GPS receiver located in low latitude regions of Indian subcontinent. The initial parameters of BBO are considered as follows: the number of habitats is 50, the number of iterations is 1000, emigration rate is $E = [0 \quad 1]$, immigration rate is $I = 1 - E$, and mutation rate is 0.9. The estimated mean position errors by using BBO are 45.4, 88.2, and 29.9 m in x-, y-, and z-directions, and the estimated receiver clock bias is 123 ns. By considering all the ignored correctable error terms makes the further improvement in the position accuracy. Therefore, it is concluded that the BBO is suitable and a better nature-inspired algorithm in order to achieve precise estimation of any GPS receiver location in low latitude regions of Indian subcontinent.

References

1. Simon D (2008) Biogeography-based optimization. IEEE Trans Evol Comput 12(6):702–713
2. Darwin C (1995) The origin of species. Gramercy, New York
3. Wallace A (2005) The geographical distribution of animals (two volumes). Adamant Media Corporation, Boston, MA
4. MacArthur R, Wilson E (1967) The theory of biogeography. Princeton University Press, Princeton, NJ
5. Hanski I, Gilpin M (1997) Metapopulation biology. Academic, New York
6. Rao GS (2010) Global navigation satellite systems. McGraw Hill Education Private limited

7. Wellenhof BH, Wasle E, Lichtenegger H (2008) GNSS—global navigation satellite systems: GPS, GLONASS, Galileo and more. Springer, pp 365–430
8. Teunissen P, Montenbruck O (2017) Springer handbook of global navigational satellite systems, pp 561–563. https://doi.org/10.1007/978-3-319-42928-1

Network-on-Chip Xilinx Implementation of WBCDMA System and Its AWGN Performance Analysis

S. Rama Devi, T. Vedavyas, and M. Satya Anuradha

Abstract Communication is a solution to achieve data transfer between various system components. In this paper, an efficient Walsh-code-based code division multiple access (WBCDMA) was implemented. The encoder and decoder blocks are developed using Walsh-code-based orthogonal codes. An eight-node WBCDMA network was developed and synthesized. WBCDMA Network-on-chip communication is realized by using Xilinx ISE 14.7 edition. The area requirements are obtained from the synthesis report and using Vivado estimation tool. The power requirements are analyzed. This paper also concentrates on the real-time applicability of the system that is possible with the help of mixed-signal processing. Xilinx system generator is used to analyze BER performance of the above system under AWGN channel condition. The simulation reveals that the 8-bit data can be simultaneously transferred to the corresponding nodes without the loss of information. The WBCDMA NoC data transfers latency of 20. The power requirements are also minimized (Ahmed et al. in IEEE Trans Very Large Scale Integr (VLSI) Syst 25(6), 2017). Noise impact on the data transfer between different nodes is almost negligible when SNR crosses above 20 dB. These systems can be used to validate the real-time implementation.

Keywords Network-on-Chip · Latency · AWGN · BER · SNR · Mixed-signal synthesis

1 Introduction

According to technology development, the number of components to be integrated on the single silicon chip increases, so communication/data transfer between the different components becomes complicated with an increase in the number of system components. Network-on-Chip (NoC) is proposed to solve the on-chip communication problems. Circuit-switched network, packet switching network, is the most

S. Rama Devi (✉) · T. Vedavyas · M. Satya Anuradha
Andhra University College of Engineering (A), Andhra University, Visakhapatnam, Andhra Pradesh, India
e-mail: Ramadevi493@gmail.com

P. S. R. Chowdary et al. (eds.), *Microelectronics, Electromagnetics and Telecommunications*, Lecture Notes in Electrical Engineering 655,
https://doi.org/10.1007/978-981-15-3828-5_26

commonly used data transfer mechanisms in NoC structures. CDMA communication is nothing but spread-spectrum multiple access techniques. Each user is assigned with a unique orthogonal code to distinguish from each other in a shared communication resource. User's data spreads with these chip codes, which adds redundancy to the data, and hence, much wider bandwidth is required to transmit the data among different users [1]. Ahmed et al. proposed CDMA Network-on-Chip with serial and parallel overloaded CDMA interconnect(OCI) [2]. Wang et al. realized on-chip CDMA as a new standard-basis based encoding/decoding method with concentration on area, power assumption and network throughput [3]. A new CDMA approach, aggregated CDMA (ACDMA), is proposed according to [4]. Karthikeyan et al. discussed 3D NOC router architecture [5]. Study on the real-time applicability of WBCDMA NoC system in terms of its noise performance is not reported earlier, so that this motivated to work on NoC implementation of WBCDMA system. In this work, initially the NoC is implemented with Xilinx ISE 14.7 edition, and communication is established to transfer data using the star topology. Complete gate-level implementation is carried out to study output data latency, area and power. In the second part, the noise analysis is carried out to test the real-time applicability. Xilinx system generator-based implementation is used to analyze the BER behavior using MATLAB 2013a. This paper bridges the gap between research and the real-life.

The rest of the paper is arranged as follows. Section 2 discusses NoC CDMA architecture. Section 3 discusses the WB encoding/decoding and in Sect. 4 AWGN noise analysis. The simulation and synthesis results of CDMA NoC are addressed in Sect. 5. Finally, Sect. 6 concludes the paper.

2 NoC CDMA Architecture

A CDMA Network-on-Chip communication having star topology structure is considered for this work as shown in Fig. 1, where P2S is a parallel to serial and S2P is serial to parallel data converters. E and D are the encoding and decoding modules, A is the summation module, and NI is the network interface module. The primary parameter affecting the latency and throughput in any network is the path length. In star topology, central hub is the main part; it has the advantage of new nodes/stations which

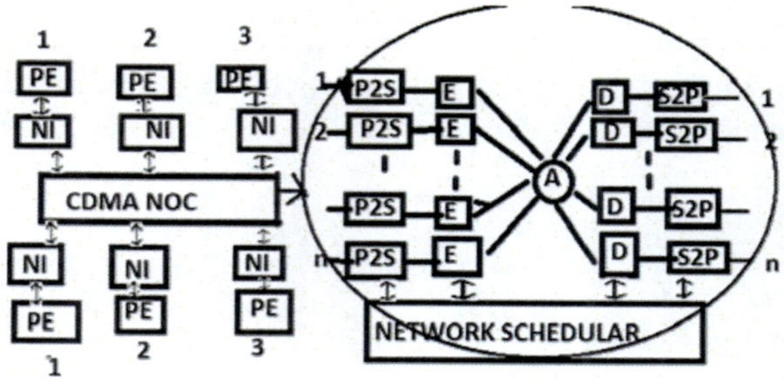

Fig. 1 Structure of CDMA NoC

can be added without changing the rest of the network. In this network, transmitted signal reaches the intended destination without passing through different nodes, and hence, minimum latency can be obtained. In this work, specific importance is given to Xilinx implementation of encoding and decoding blocks. WBCDMA encoding and decoding scheme is considered for simulation purpose.

3 WBCDMA Encoding and Decoding Method

3.1 WBCDMA Encoding

Walsh-code-based orthogonal codes are used to distinguish different users and prevent the users from interfering with each other and also provide integrity to data. Walsh codes are unique orthogonal codes (chips) assigned to each user, and these are also known as spreading sequences. The user's symbols are XORed with chips assigned to the individual user. The encoded data from each node is summed up to multi-bit sum by using arithmetic addition. It can be represented by the following Eqs. 1 and 2 (Fig. 2).

$$E_i^j = D_i^j \oplus W_i^j \tag{1}$$

$$Y^j = E_1^j + E_2^j + \cdots + E_n^j \tag{2}$$

Here,

E is encoded data
W is Walsh-code data
D is input data.

Fig. 2 Block diagram of WBCDMA encoding scheme

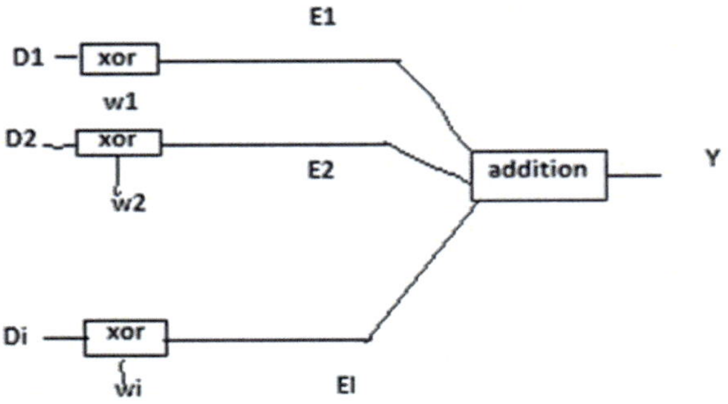

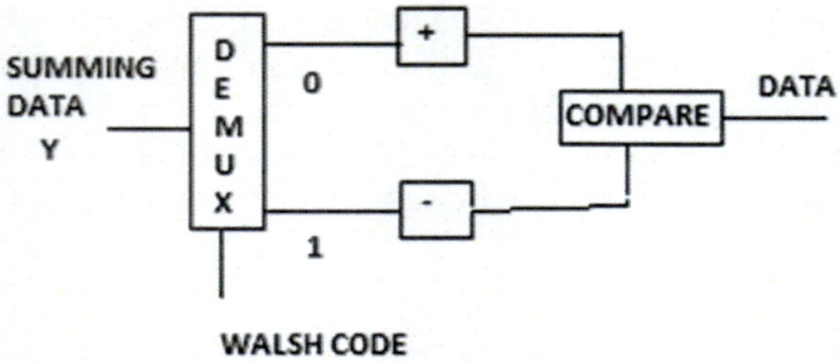

Fig. 3 Block diagram of WBCDMA decoding scheme

3.2 WBCDMA Decoding

In the Walsh-code-based decoding method, the received data from the encoder is a multi-bit sum. This sum is applied to the demultiplexer/decoder block. The multi-bit sum is transferred to the negative accumulator when the chip value is '0,' else it is directed to the positive accumulator. The output of the comparator is '1' when positive accumulator value exceeds the value in the negative accumulator (Fig. 3).

4 Noise Analysis

In real-time applications, the noise analysis plays an important role; it is used to study the performance of the system and is analyzed with bit error rate (BER).

$$\text{BER} = \text{Number of bits with error/total number of bits sent}$$

Additive white Gaussian noise (AWGN) comes from many natural causes. This AWGN channel is suited for both wired and wireless communications. This AWGN channel adds white Gaussian noise [6]. In order to perform noise analysis in the wireless channel, the data is to be converted to analog format. MATLAB 2013a Simulink is used to analyze in mixed-signal processing. For noise analysis, the data from the modulator is transformed into the analog domain using Phase-shift keying (PSK). The phase of the carrier is varied according to modulating signal [7]. The instantaneous value of the digital signal can be written as

$$x(t) = A\sin(2\pi f_c t) \tag{3}$$

when

A 1 binary value is '1'
A -1 binary value is '0'.

WBCDMA NoC for AWGN analysis is represented in Fig. 4. Here, 16-PSK is used for modulation and demodulation purposes.

WBCDMA NoC for AWGN analysis is represented in Fig. 5. The workspace in MATLAB environment is used for BER analysis.

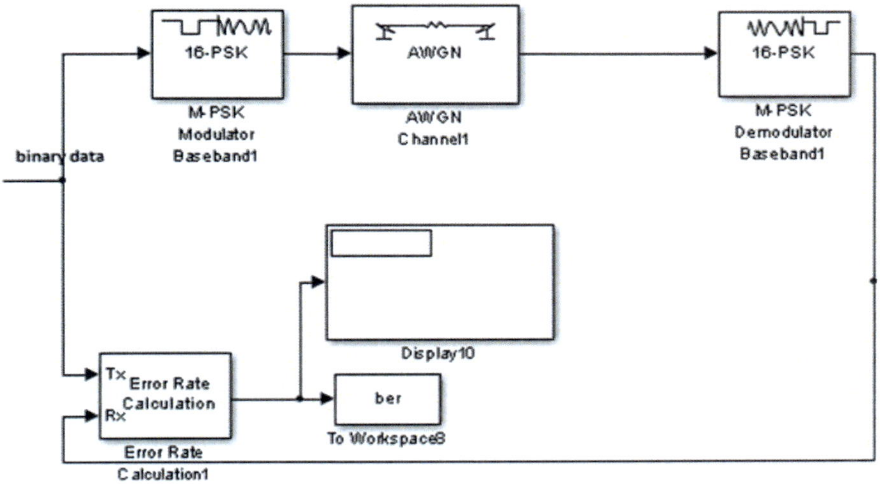

Fig. 4 AWGN analysis for WBCDMA system

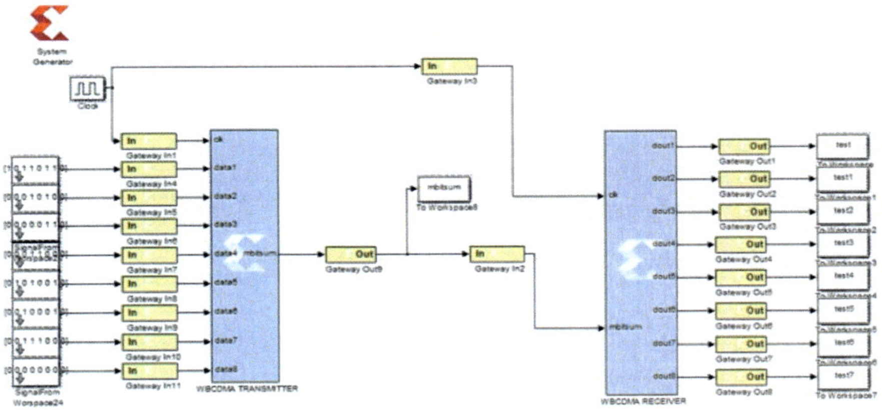

Fig. 5 WBCDMA NOC using Xilinx system generator

5 Simulation Results

Transmission and reception of one byte of data with eight different users (nodes) are observed in Xilinx issue 14.7 edition software with system configuration Windows 10 with 4 GB RAM. The system's performance is analyzed in terms of latency based on timing diagrams. LUTs and number of flip-flop requirements are obtained from the synthesis report and also power analysis by using Vivado software. The noise performance is analyzed in Xilinx system generator.

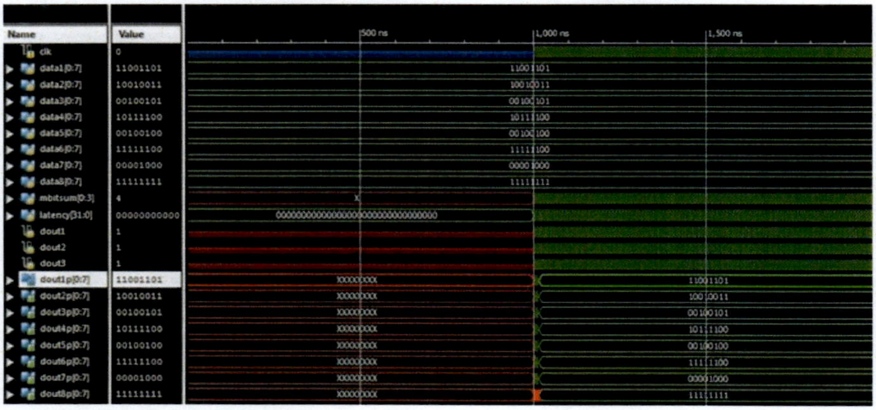

Fig. 6 WBCDMA transmitted and recovered data from eight nodes

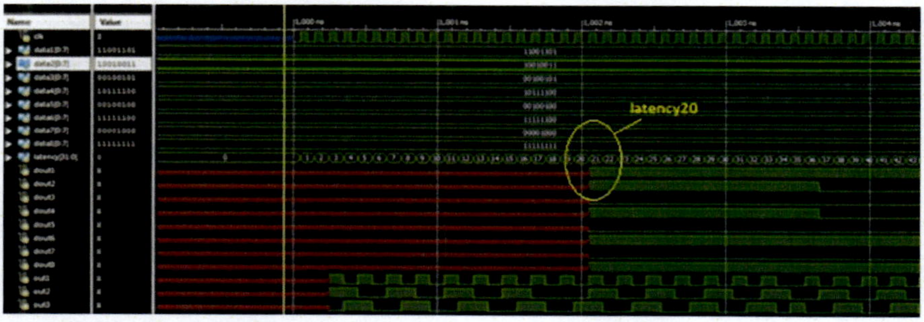

Fig. 7 WBCDMA transmitted and recovered data from eight nodes with latency

Figures 6 and 7 represent the transmitted data and recovered data using WBCDMA Network-on-Chip systems. Timing diagrams ensure that data is encoded and decoded. It is successful in recovering the transmitted data at the corresponding node. Latency refers to delays in transmitting or processing data. The timing diagrams reveal that WBCDMA receiver has constant data latency of 20 at each node. From the simulations, the synthesis report is generated. Device utilization for NoC implementation, WBCDMA system (encoder and decoder) is presented in Fig. 8.

Device Utilization Summary (estimated values)			
Logic Utilization	Used	Available	Utilization
Number of Slice Registers	258	126800	0%
Number of Slice LUTs	299	63400	0%
Number of fully used LUT-FF pairs	144	413	34%
Number of bonded IOBs	129	210	61%
Number of BUFG/BUFGCTRLs	1	32	3%

Fig. 8 WBCDMA synthesis report

Total on-chip power is measured using Xilinx Vivado software. It belongs to Artix7 and device XC7A100T with package CSG324. From Fig. 9, the total on-chip power utilized for encoding and decoding schemes with eight simultaneous users is 1.842 W.

The BER performance of WBCDMA decoder for variations in the SNR values is shown in Fig. 10. Bit errors are reduced as SNR increase. More error rate is observed at each node when SNR is less than −20 dB, but when it crosses −20 dB, BER starts to reduce. When BER becomes 20 dB, the recovered bits replicate bits transmitted at each node, indicating noise does not influence data transmission.

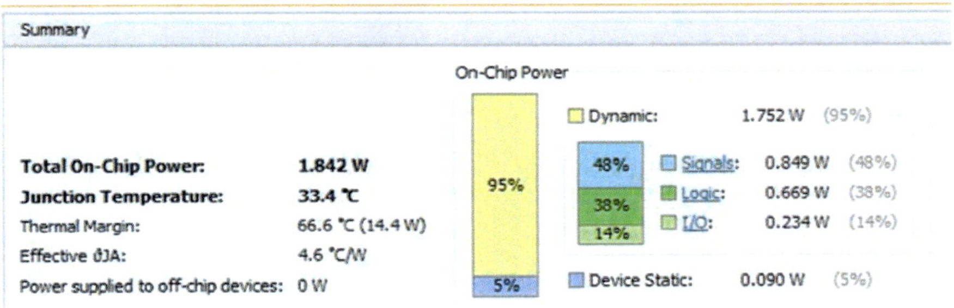

Fig. 9 WBCDMA power report

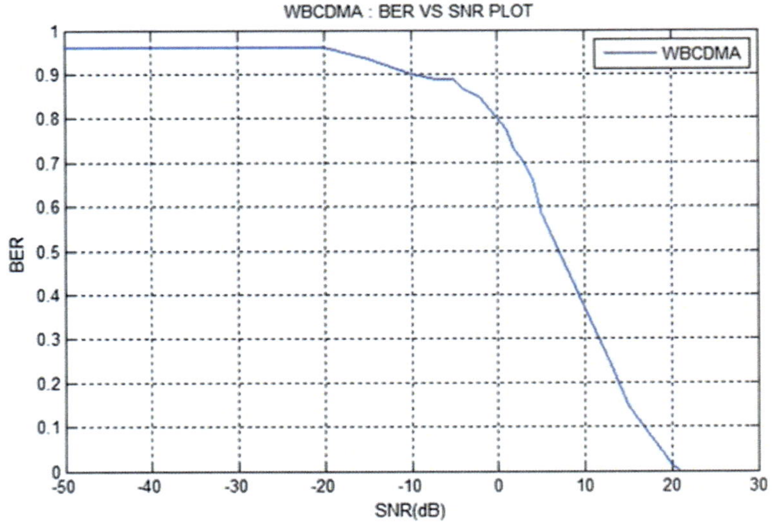

Fig. 10 WBCDMA power report

6 Conclusion and Future Work

An eight-node WCDMA network was realized using Xilinx ISE 14.7 version, and WCDMA on-chip communication network (encoders and decoders) is implemented with digital circuits. The simulation reveals that the 8t-bit data can be transferred to each node without the loss of information. From the simulation results, the WCDMA Network-on-Chip at each node achieves a constant data transfer latency of 20, irrespective of the communication media in the code domain. Based on the synthesis report, the number of LUT utilization is 0.47% in XC7A100T device. The above system uses 1.824 W of total on-chip power. Noise analysis is a requirement for NoC in real-time applications. Mixed-signal analysis reveals a 20 dB SNR which is sufficient to the simultaneous recovery of data from all eight receiving nodes under AWGN noise environment. In the future course of work, it can be further extended to different data rates.

References

1. Dinan EH, Jabbari B (1998) Spreading codes for direct sequence CDMA and wideband CDMA cellular networks. IEEE Commun Mag 48–54
2. Ahmed KE, Rizk MR, Farag MM (2017) Overloaded CDMA crossbar for network-on-chip. IEEE Trans Very Large Scale Integr (VLSI) Syst 25(6), June 2017
3. Wang J, Lu Z, Li Y (2015) A new CDMA encoding/decoding method for on-chip communication network. IEEE Trans Very Large Scale Integr (VLSI) Syst 1063–8210 ©2015 (IEEE)
4. Ahmed KE, Rizk MR, Farag MM (2016) Aggregated CDMA crossbar for network-on-chip, 978-1-5090-5721, IEEE
5. Karthikeyan A, Senthil Kumar P (2017) GALS implementation of randomly prioritized bufferless routing architecture for 3D NoC. Springer, 20 June 2017
6. Latha HN, Palachandra MV, Raoc M (2012) Real time implementation and performance evaluation of WCDMA system over AWGN channel on TMS320C6713DSK. Procedia Technol 4:82–86
7. Parthibun P, Prabu T (2013) FPGA implementation of DSSS-wideband WCDMA transmitter and receiver using QPSK. Int J VLSI Embed Syst (IJVES) 04, Article 09155

Comparison of Conformal and Planar CPW-Fed Circularly Polarized UWB Square Slot Antennas for WLAN, WiMAX, and 5G Applications

Sateesh Virothu and M. Satya Anuradha

Abstract In this article, a compact circular-polarized CPW-fed square slot antenna conformed on the cylindrical structure is designed to suit for WiFi, WiMAX, and 5G applications. The antenna is designed to resonate in the frequency band of 2–4.5 GHz, thus covering all major frequency bands of IEEE 802.11 (2.4–2.48 GHz), IEEE 802.16 (2.5–2.69 GHz/3.4–3.69 GHz), and 5G (3.3–3.69 GHz), with acceptable gain and radiation patterns for the above applications. Obtained 3-dB axial ratio bandwidth is in the range of 2.95–4.20 GHz. The antenna is designed using semi-flexible RT/duroid substrate having a dielectric constant $\varepsilon_r = 2.2$ with a loss tangent of $\tan\delta = 0.0009$. The proposed antennas consist of F-shaped feeding structure and L-shaped strip line connected to the square-ring ground plane. Initially, planar version of antenna is simulated with FR4 and RT/duroid substrate materials, and compared the performance of these two antennas in terms of impedance bandwidth, axial ratio, radiation patterns, and gain. Secondly, RT/duroid material is considered for conformal structure, and its performance is compared for various cylindrical curvatures. From the simulation results, the conformal antenna with different radii has achieved almost similar performance metrics, except a slight variation in the axial ratio and radiation patterns. The proposed antennas are designed and simulated on ANSYS HFSS 16.1 software.

Keywords CPW feed · Circular polarization · Conformal antenna · UWB

S. Virothu (✉) · M. Satya Anuradha
Department of Electronics and Communication, Andhra University College of Engineering (A), Andhra University, Visakhapatnam, Andhra Pradesh, India
e-mail: sateeshvirothu@gmail.com

M. Satya Anuradha
e-mail: radhamsa@gmail.com

P. S. R. Chowdary et al. (eds.), *Microelectronics, Electromagnetics and Telecommunications*, Lecture Notes in Electrical Engineering 655,
https://doi.org/10.1007/978-981-15-3828-5_27

1 Introduction

Printed monopole antennas fed by a coplanar waveguide (CPW) exhibit several advantages over microstrip antennas, features like lower dispersion, radiation loss, broader bandwidth, integration of active and passive components, conformability and ease of fabrication using MMIC made CPW-fed antennas most popular. Circular-polarized (CP) antennas play a vital role in various communication systems like RFID, WLAN, GPS, and WiMAX, which avoids polarization mismatch for different orientation of equipment, also provides advantages such as the reduction in multipath losses and resistance to adverse weather conditions [1–3].

Various techniques have been incorporated in CPW-fed square slot antennas to enhance the -10 dB impedance and CP bandwidths, reported in the literature [4–10]. Flexible monopole antennas have been experimentally demonstrated for wireless applications with narrow impedance bandwidths [11–15].

However, contributions on conformal antennas for wireless applications with wider impedance and 3-dB axial ratio bandwidths are limited in the open literature. Therefore, in this paper, a simple conformal CPW-fed square slot antenna with L-shape strip-line embedded into the square slot has been proposed to obtain desired bandwidth for the WiFi (IEEE 802.11, 2.4–2.48 GHz), WiMAX (IEEE 802.16, 2.5–2.69 GHz/3.4–3.69 GHz), and 5G (3.4–3.69 GHz) applications. The effect of curvature on the antenna parameters like S_{11}, axial ratio, gain, and radiation patterns are presented.

The rest of the paper is organized as follows: Sect. 2 discussed the antenna design considerations. In Sect. 3, simulation results, and finally, conclusions are made in Sect. 4.

2 Antenna Design

The structure and dimensions of the proposed rigid and semi-flexible antennas are shown in Fig. 1. To design the firm antenna, FR4_epoxy is chosen as a substrate with a dielectric constant of $\varepsilon_r = 4.4$ with a loss tangent of $\tan\delta = 0.02$. The same configuration is used in the semi-flexible antenna, Rogers RT/duroid 5880, with dielectric constant and loss tangent of $\varepsilon_r = 2.2$, $\tan\delta = 0.0009$, respectively. Table 1 presents the overall dimensions of both the antennas.

To excite the antennas, an F-shaped feeding structure is used and is illustrated in Fig. 1. Strip-line lengths L_1 and L_2 contribute TE_{10} and TE_{01} modes, respectively. Square slot dimensions are calculated based on Eq. (1).

$$f_c = \frac{c}{2\pi\sqrt{\varepsilon_{\text{reff}}}} \times \sqrt{\left[\left(\frac{m\pi}{a}\right)^2 + \left(\frac{n\pi}{b}\right)^2\right]} \tag{1}$$

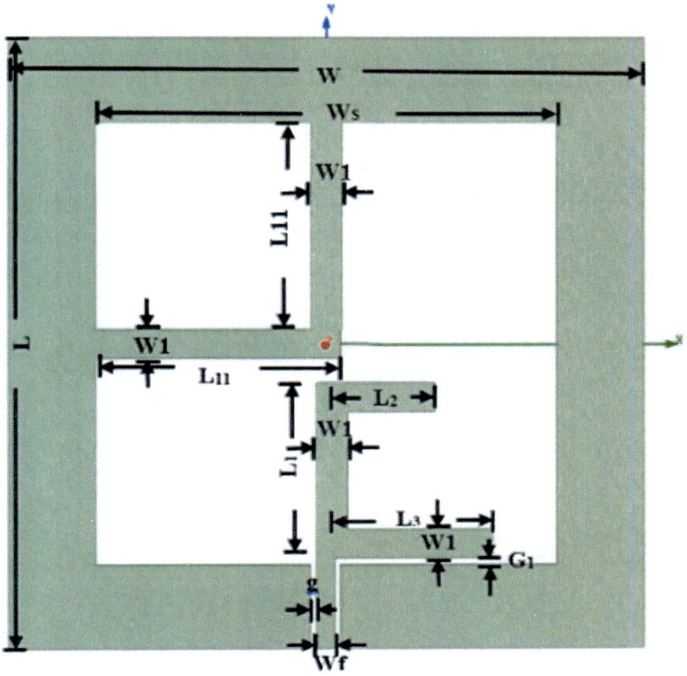

Fig. 1 Configuration of the proposed antenna

The inverted L-shaped strip located at the top left corner of the square slot is used to produce TE_{10} and TE_{01} radiating modes with 90° phase difference and equal amplitudes to generate circularly polarized waves. L_3 is used to set the required input impedance to get the large impedance bandwidth.

3 Simulation Results and Discussions

In this section, simulation results of the broadband CPW-fed circular-polarized antennas with rigid (FR4) and semi-flexible (RT/duroid) substrates are analyzed. These antennas are designed and simulated with the help of ANSYS HFSS 16.1 software. Figure 2 shows the geometry of flat and conformal antennas for different curvatures. Curved surface with radii ∞, 60, 45, and 30 mm are considered.

Variation in return loss, axial ratio, and gain as a function of frequency for antennas under consideration is illustrated in Fig. 3. From return loss plot, all the five antennas have achieved a broader impedance bandwidth covering the frequency band from 2.01

Table 1 Antenna dimensions for FR4 and RT duroid

Antenna with FR4	Parameter	W	L	h	W_s	W_f	g	L_1	L_2	L_3	G_1	W_1	L_{11}
	Value (mm)	34	34	0.8	25.5	1	0.4	9.5	7.4	10	0.5	2	13.75
Antenna with RT/duroid	Value (mm)	42	42	0.787	30.5	1.2	0.4	12.6	7.2	11	0.5	2	16.25

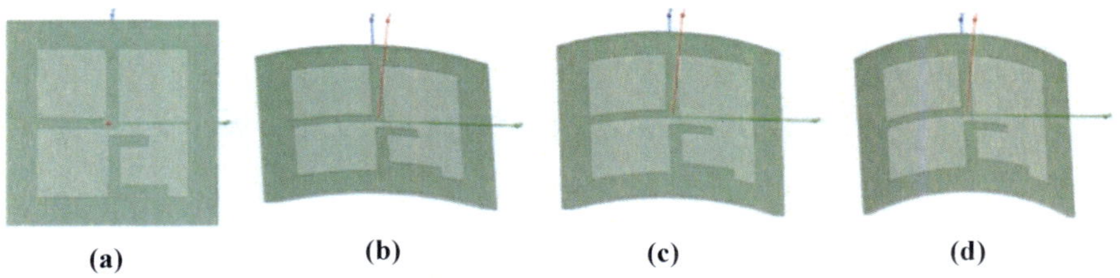

Fig. 2 Geometry of a FR4 (or) RT/duroid planar antenna ($R = \infty$), b RT/duroid conformal antenna with $R = 60$ mm, c $R = 45$ mm, and d $R = 30$ mm

to 4.56 GHz. Inflexible RT/duroid and FR4 antenna configurations have obtained an impedance bandwidth of 75% and 62%, respectively. From this observation, an antenna with RT/duroid substrate material gives better impedance bandwidth, by a factor 13.32% compared to FR4 substrate. In conformal antenna structures with different cylindrical radii ($R = 60$, 45 and 30 mm), there is not much a noticeable variation in the resonant frequency and impedance bandwidth is observed.

2D radiation plots of all the antenna designs considered in this work are shown in Figs. 4 and 5. These plots are presented for 2.4 and 3.5 GHz frequencies. From these radiation plots, flat antennas radiate symmetrical bidirectional radiation patterns and produce RHCP and LHCP waves in the $+z$- and $-z$-directions.

By comparing conformal antennas with planar antenna, there is a slight change in symmetry of radiation patterns in $+z$- and $-z$-directions. From Table 2, as curvature radii decrease, the antenna 3-dB beamwidth of LHCP increases in XOZ-plane and decreases in YOZ-plane, whereas 3-dB beamwidth increases for RHCP pattern in both the planes. Almost no change in peak gain is observed in curved structures with different radii.

Table 3 presents a comparison of the proposed antenna configurations with earlier works to cover WLAN (2.4 GHz), WiMAX (2.5 and 3.5 GHz), and 5G (3.3–3.69 GHz) wireless applications. The proposed rigid antennas are compact, radiate with better gain, and maintain relatively comparable bandwidths [1, 2, 6, 7]. The proposed conformal antenna exhibits good performance in terms of gain, axial ratio, and impedance bandwidth over the earlier published works. Hence, it is suitable for multiple wireless applications (WiFi, WiMAX, and 5G technology) in the 2–4.5 GHz band.

4 Conclusion

A CPW-fed broad impedance bandwidth circular-polarized square slot antenna with flat and conformal structures has been presented in the frequency band of 2–4.5 GHz. Initially , planar versions of rigid FR4 and semi-flexible RT/duroid substrate antenna

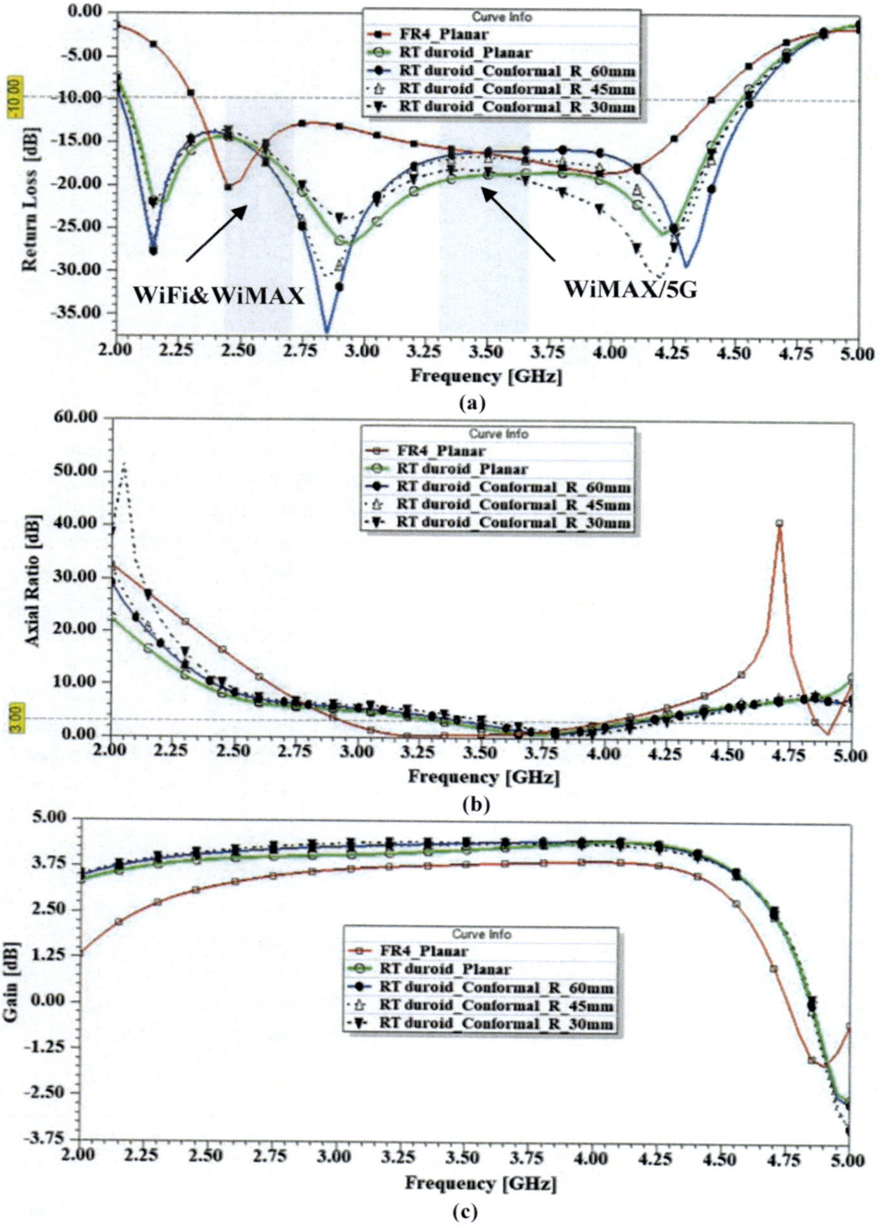

Fig. 3 Plots for **a** return loss versus frequency, **b** axial ratio versus frequency, and **c** gain versus frequency

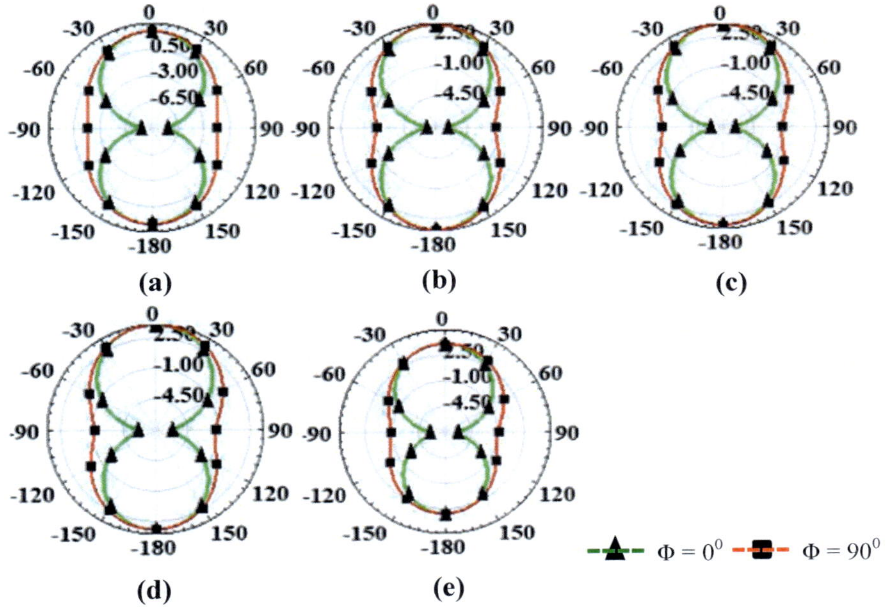

Fig. 4 2D radiation patterns of antenna structures for *XOZ*-plane ($\Phi = 0°$) and *YOZ*-plane ($\Phi = 90°$) at 2.4 GHz. **a** FR4 planar antenna. **b** RT/duroid planar antenna, **c** RT/duroid conformal antenna with $R = 60$ mm, **d** $R = 45$ mm, and **e** $R = 30$ mm

performance metrics (S_{11}, axial ratio, gain, and radiation plots) are compared. A considerable improvement in the gain in the range of 3.4–4.44 dBi is achieved for all three structures. Further, the curvature effect on conformal antennas is analyzed. From this analysis, all the antenna parameters are almost similar, except, a small noticeable change in antenna beamwidth is observed for both flat and conformal structures. Hence, this conformal design is a strong candidate for WLAN, WiMAX, and 5G applications, where flexibility and miniaturization are prime concern.

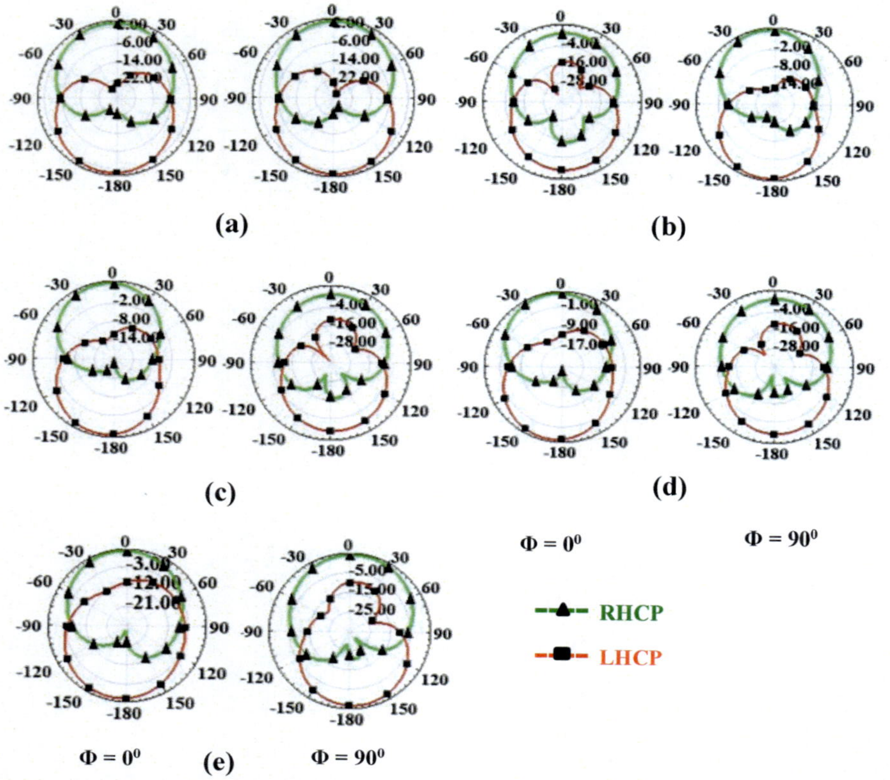

Fig. 5 2D radiation plots of antenna structures for *XOZ*-plane ($\Phi = 0°$) and *YOZ*-plane ($\Phi = 90°$) at 3.5 GHz. **a** FR4 planar antenna. **b** RT/duroid planar antenna. **c** RT/duroid conformal antenna with $R = 60$ mm, **d** $R = 45$ mm, and **e** $R = 30$ mm

Table 2 Comparison of overall simulation results

Antenna/parameter	f_c, $S_{11} \leq -10$ dB frequency range (GHz)	$S_{11} \leq -10$ dB BW (GHz)/% BW	3-dB A.R bandwidth/AR (GHz)	Peak gain (dBi)	3 dB beamwidth (degrees) LHCP	RHCP
FR4 planar	3.35, (2.31–4.39)	2.08/62.09	1.02/(2.95–3.97)	3.88	*91.8* (*Φ*= 0°)	*91.89* (*Φ*= 0°)
					88.87 (*Φ*= 90°)	*87.89* (*Φ*= 90°)
RT/duroid planar	3.28, (2.04–4.51)	2.47/75.41	0.67/(3.37–4.04)	4.44	81.48	80.84
					79.89	80.17
RT/duroid (R = 60 mm)	3.29, (2.01–4.56)	2.55/77.63	0.64/(3.46–4.10)	4.46	82.43	82.49
					76.12	82.26
RT/duroid (R = 45 mm)	3.28, (2.02–4.53)	2.51/76.64	0.64/(3.45–4.09)	4.42	84.22	84.03
					75.78	83.04
RT/duroid (R = 30 mm)	3.28, (2.03–4.54)	2.51/76.40	0.64/(3.56–4.20)	4.41	89.90	85.75
					74.30	85.77

Table 3 Comparison of the proposed work with earlier CPW-fed antennas

Ref. No.	Substrate/type	f_c (GHz)	$S_{11} \leq -10$ dB BW (GHz)/% BW	3 dB A.R BW (GHz)	Peak gain (dBi)	Dimensions (mm^3)
[1]	FR4 (rigid)	2.76	2.76/96.5	1.9	3.5	50 × 55 × 1
[2]	FR4 (rigid)	2.82	2.02/71.63	0.7, 0.25	2.5	63 × 75 × 1.6
[6]	FR4 (rigid)	2.17	2.32/106.9	0.55, 0.14	3.81	70 × 70 × 1.6
[7]	FR4 (rigid)	2.48	2.00/8.06	0.32	3.5	50 × 48 × 1.6
[12]	Polyimide (flexible)	3.55	0.58/16.34		2.06	18 × 20 × 0.1
Proposed Planar	FR4 (rigid)	3.35	2.08/62.09	1.02	3.88	34 × 34 × 0.8
Proposed Planar	RT/duroid 5880 (semi-flexible)	3.28	2.47/75.41	0.67	4.44	42 × 42 × 0.8
Proposed conformal (R = 60 mm)	RT/duroid 5880 (semi-flexible)	3.29	2.55/77.63	0.64	4.46	42 × 42 × 0.8

References

1. Ding K, Guo YX, Senior Member, IEEE, Gao C, Member, IEEE (2017) CPW-fed wideband circularly polarized printed monopole antenna with open loop and asymmetric ground plane. IEEE Antennas Wirel Propaga Lett 16
2. Saini RK, Student Member, IEEE, Dwari S, Member, IEEE, Mandal MK, Senior Member, IEEE (2017) CPW-fed dual-band dual-sense circularly polarized monopole antenna. IEEE Antennas Wirel Propag Lett 16
3. Simons RN (2001) Coplanar waveguide circuits, components, and systems. Copyright 2001 Wiley, Inc. ISBNs: 0-471-16121-7 (Hardback); 0-471-22475-8 (Electronic)
4. Chen HD (2003) Broadband CPW-fed square slot antennas with a widened tuning stub. IEEE Trans Antennas Propag 51
5. Chainmool S, Kerdsumang S, Prayoot A, Vivek V (2004) A broadband CPW-fed square slot antenna using loading metallic strips and a widened tuning stub. ISCIT
6. Chen YY, Jiao YC, Zhao G, Zhang F, Liao ZL, Tian Y (2011) Dual-band dual-sense circularly polarized slot antenna with a C-shaped grounded strip. IEEE Antennas Wirel Propag 10
7. Deng IC, Lin RJ, Chang KM, Chen JB (2006) Study of a circularly polarized CPW-fed inductive square slot antenna. Microw Opt Technol Lett 48(8):1665–1667
8. Kim SW, Kim GS, Choi DY (2017) CPW-fed wideband circular polarized antenna for UHF RFID applications. Hindawi Int J Antennas Propag 2017. Article ID 3987263, p 7
9. Zhang S, Huang H, Yin Y (2014) A broadband CPW-fed circularly polarized square slot antenna for UHF RFID applications. Prog Electromagn Res C 50:39–46
10. Lu JH, Wang SF (2013) Planar broadband circularly polarized antenna with square slot for UHF RFID reader. IEEE Trans Antennas Propag 61(1):45–53
11. Hachi A, Lebbar H, Himidi M (2017) Flexible and conformal printed monopole antennas. Prog Electromagn Res Lett 67:89–95
12. Kumar Naik K, Gopi D (2018) Flexible CPW-fed split-triangular shaped patch antenna for WiMAX applications. Prog Electromagn Res M 70:157–166
13. Kim DO, Kim CY, Yang DG (2012) Flexible Hilbert curve loop antenna having a triple-band and omnidirectional pattern for WLAN/WiMAX applications. Hindawi Int J Antennas Propaga 2012, Article ID 687256, p 9
14. Sahoo R, Vakula D, Sarma NVSN (2017) A wideband conformal slot antenna for GPS application. In: Progress in electromagnetics research symposium-fall (PIERS-FALL), Singapore, pp 19–22
15. Sahoo R, Vakula D (2019) Bow-tie-shaped wideband conformal antenna with wide-slot for GPS application. Turk J Electr Eng Comput Sci 27: 80–93

Design of a Quad-Band Annular Ring-Loaded Circular Patch Antenna with Meander Line Slot and DGS for Wireless Applications

Mahesh Babu Kota, T. V. Rama Krishna, Ketavath Kumar Naik, E. Eswar Sai Yaswanth, G. Hanimi Reddy, and K. Gowtam Chowdary

Abstract In this paper, a quad-band antenna with circular patch antenna loaded with annular ring with CSRR DGS is presented. To improve the impedance bandwidth, a meander line structure is etched on the patch and complementary split ring resonator (CSRR) structure on ground plane of the antenna. The proposed antenna has resonated at quad-band for wireless applications. The operating frequencies are observed at 2.31 GHz, 3.05 GHz, 4.69 GHz and 5.6 GHz with reflection coefficients −9.47 dB, −22.80 dB, −18.87 dB and −25.27 dB, respectively. The gains at the resonating frequencies are 2.67, 3.74, 4.5 and 4.31 dB. The radiation patterns are also presented. The obtained frequency bands are suitable for wireless applications like WiMAX, WiBro and WLAN.

Keywords Circular patch · Loaded annular ring CSRR · Meander line · WiMAX · WLAN

1 Introduction

In day-to-day life, the advancements in the wireless communication technology increased rapidly probing the designers to design multiband antennas with the features like compact size, high efficiency, low profile, lightweight and ease of fabrication [1]. The microstrip patch antennas became the right choice to fulfil the requirements of modern wireless communication. However, at low frequencies, the patch antennas cannot attain the compact size. To overcome this limitation, the structures like meander line [2–4] and CSRR [5, 6] are introduced either in patch or ground of microstrip patch antenna (MSPA). The operating frequency shifts occur due to changes in effective inductance and capacitance values of the patch obtained by CSRR DGS structure and Meander line slots which enhance the performance of the

M. B. Kota (✉) · T. V. Rama Krishna · K. Kumar Naik · E. Eswar Sai Yaswanth · G. Hanimi Reddy · K. Gowtam Chowdary
Department of Electronics and Communication Engineering, KL Deemed to be University, Koneru Lakshmaiah Education Foundation, Guntur, Andhra Pradesh, India
e-mail: kmaheshbabu@kluniversity.in

P. S. R. Chowdary et al. (eds.), *Microelectronics, Electromagnetics and Telecommunications*, Lecture Notes in Electrical Engineering 655,
https://doi.org/10.1007/978-981-15-3828-5_28

Patch antenna without altering the design specifications [7]. Most of the patch antennas operating for wireless applications consist of CSRR structures. A circular slot patch etched with vertical meander line [8] is designed for dual-band applications operating under X-band. CSRR was incorporated in ground plane for conformal antennas [9] for ISM band applications. An elliptical ring flexible patch [10] was designed with dual bands for wireless applications. The loading of ring structure to a circular patch for performance enhancement through mutual coupling was analysed in [11]. In this paper, an annular ring-loaded circular patch antenna is designed by etching the meander line structure on the patch and CSRR structure on the ground to improve the return loss and multiband operation. The paper presents a detailed description of the antenna design geometry and design analysis on the parameters like return loss, radiation pattern and gain. The parametric analysis was performed for different widths and various substrate materials like Rogers Duroid and FR-4.

2 Antenna Geometry

The geometrical development of the proposed antenna is shown in Fig. 1. The basic structure consists of probe-fed circular patch (Fig. 1a) on a FR-4 dielectric material with substrate permittivity of $\varepsilon_r = 4.3$. A meander line structure with four meander sections is etched on the circular patch (Fig. 1b) to enhance the impedance matching. The width of each meander section is 0.03 cm, length of 0.5 cm and the separation between each section is 0.1 cm. The overall dimensions of the meander line are $0.94 \times 0.5\ \text{cm}^2$. An annular ring is loaded to the structure (Fig. 1c) to suppress the fringing effect of the patch. A dual contact is made with the ring using metallic strips of 0.05 cm width and 0.03 cm length (Fig. 1d) for proper impedance matching of the patch at the resonant frequencies.

Figure 2 shows the ground structure of the antenna with dimensions of $1.7 \times 1.7\ \text{cm}^2$. Firstly, a planar grounded was constructed (Fig. 2a) which provided a single resonant frequency. To obtain multiband operation, complementary split ring resonator (CSRR) defective ground structure was considered (Fig. 2b). The CSRR constructed is of rectangular shape with each ring having width of 0.05 cm and the separation between each ring is of 0.05 cm. The total dimensions of CSRR structure

a. Circular patch b. Etchedmeanderline c. Loaded annularring d. Final Design

Fig. 1 Design steps of the proposed model patch

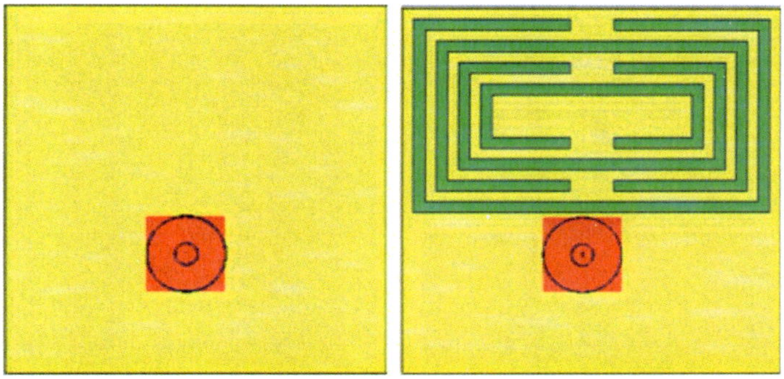

a. Without CSRR b. With CSRR

Fig. 2 Ground structure of proposed model

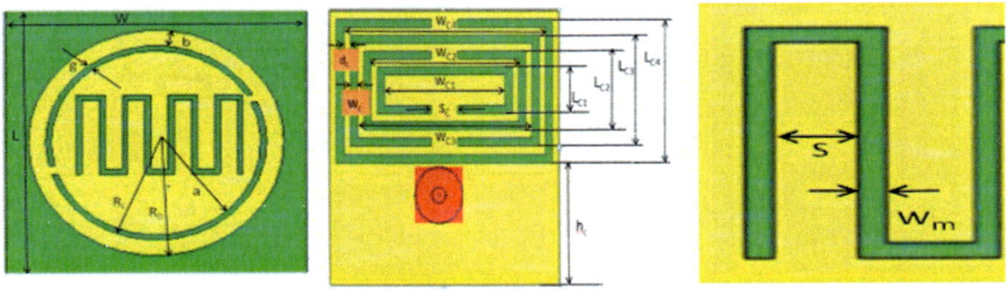

a. Patch Geometry b. CSRR Geometry c. Meander-Line

Fig. 3 Geometrical variables of the proposed model

are $1.60 \times 0.9\ \text{cm}^2$. The geometric variables of the proposed patch antenna are shown and tabulated in Fig. 3 and Table 1, respectively.

3 Results and Discussions

The proposed model was designed and simulated using CST Microwave Studio. The various parameters considered from the simulation are S_{11} (return loss), VSWR, gain, E-field, H-field and surface currents. The parametric analysis was performed and simulation results were compared for substrate materials FR4 with permittivity value 4.3 and Roger RT Duroid 5880 with permittivity 2.2.

Table 1 Geometry variables of the model (all dimensions are in cm)

Variable	Value	Description	Variable	Value	Description
L	1.7	Length of ground/substrate	W_{C1}	0.9	CSRR ring1 width
W	1.7	Width of ground/substrate	W_{C2}	1.1	CSRR ring2 width
a	0.595	Radius of circular patch	W_{C3}	1.3	CSRR ring3 width
R_o	0.725	Ring outer radius	W_{C4}	1.5	CSRR ring4 width
R_i	0.625	Ring inner radius	L_{C1}	0.3	CSRR ring1 length
g	0.03	Circle and ring separation	L_{C2}	0.5	CSRR ring2 length
$b = R_o - R_i$	0.1	Width of the ring	L_{C3}	0.7	CSRR ring3 length
h_C	0.75	Position of CSRR	L_{C4}	0.9	CSRR ring4 length
S	0.1	meander sections space	d_c	0.05	Separation of rings
W_m	0.03	Width of meander section	W_C	0.05	Thickness of CSRR
f_p	0.5	Probe feed position	S_C	0.2	CSRR slot gap

3.1 Return Loss S_{11} (DB)

Figure 4 shows the return loss S_{11} curve for a basic circular patch with and without CSRR. The patch provides only one resonant frequency 5.415 GHz having return loss

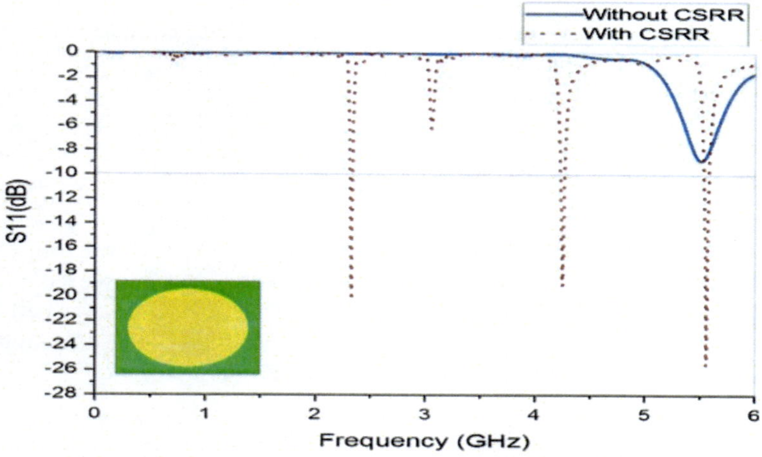

Fig. 4 Return loss response curve for circular patch with and without CSRR

value $S_{11} = -8.743$ dB without loading of CSRR and multiband phenomenon providing three resonant frequencies at 5.556 GHz, 4.428 GHz and 2.328 GHz with S_{11} values −25.59 dB, −19.11 dB and −20.007 dB, respectively. The loading of CSRR on the ground improved the patch performance providing multiband resonance.

Figure 5 shows the S_{11} response curve for the final patch. The patch has provided a significant improvement in the impedance characteristics with loading of CSRR on the ground. Figure 6 represents the response curve of S_{11} for modelled patch antenna for FR-4 and Duroid substrate materials.

The patch antenna with Duroid substrate provides dual-band resonant frequencies at 3.876 GHz with $S_{11} = -16.621$ dB and 2.952 GHz with $S_{11} = -18.061$ dB. The return loss response of patch with FR-4 shown a quad-band performance resonating at frequencies 5.598 GHz, 4.692 GHz, 3.054 GHz and 2.316 GHz with return loss values −25.276 dB, −18.871 dB, −22.803 dB and −9.4733 dB, respectively.

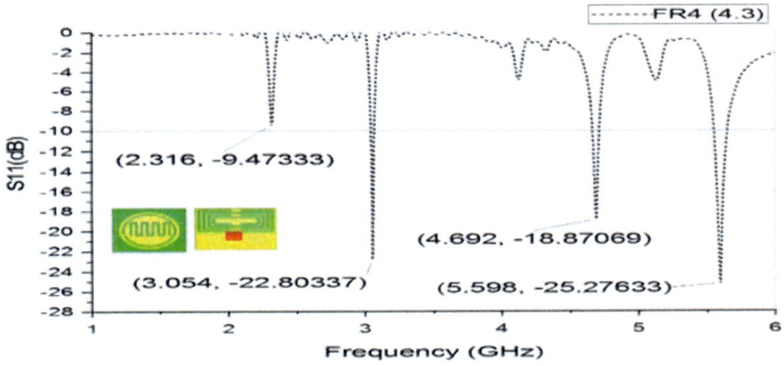

Fig. 5 Return loss response curve of proposed patch

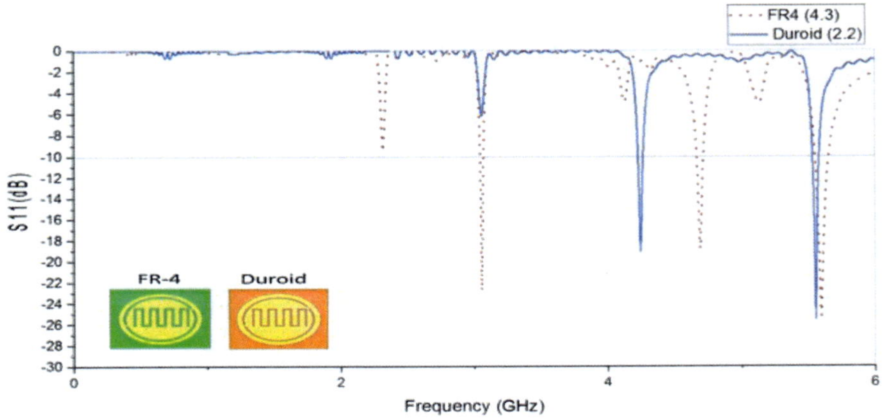

Fig. 6 Return loss response curve of patch for Duroid and FR-4 dielectric substrate

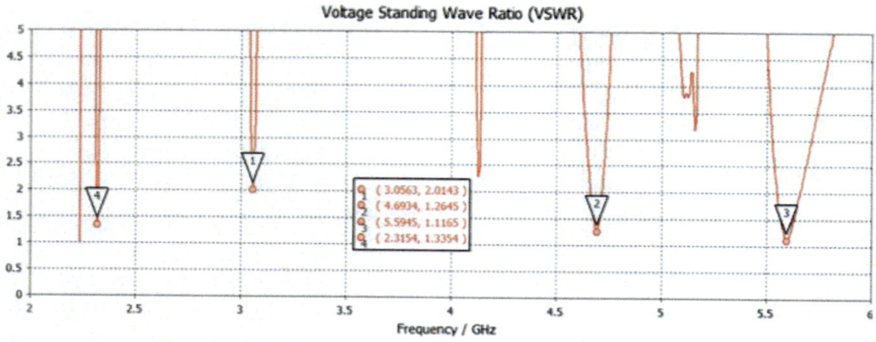

Fig. 7 VSWR response curve for FR-4 substrate

3.2 *VSWR*

The VSWR of the proposed ring coupled patch with etched meander line and CSRR-loaded DGS has been considered for different substrate materials FR-4 and RT Duroid. Figure 7 shows the response curve of VSWR for the proposed antenna with FR-4 dielectric material. The VSWR values at the corresponding resonating frequencies 2.316 GHz, 3.054 GHz, 4.692 GHz and 5.598 GHz are 1.33, 2.01, 1.26 and 1.11, respectively.

3.3 *Radiation Pattern*

Figure 8 provides the radiation patterns of the operating bands for the proposed model. The patterns obtained are of bidirectional. The main lobe of the radiation provides around (10°–15°). The radiation patterns had an advantage of circular polarisation due to the meander line and CSRR structures as the surface currents move around the slots. Although the patterns are of bidirectional, the radiation intensity was much higher when compared at the ground suppressing the cross polarisation.

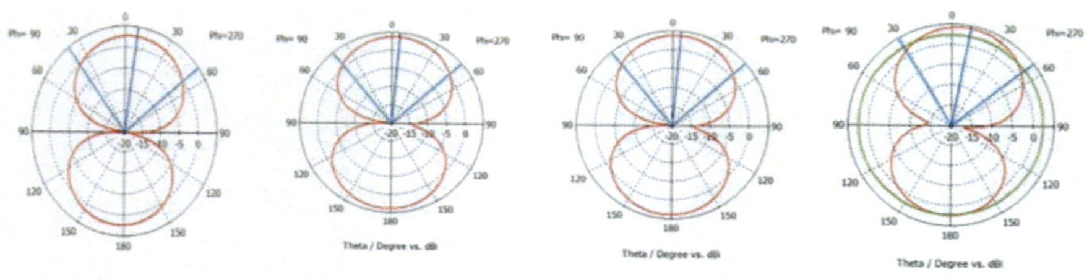

a. 2.316 GHz b. 3.054 GHz c. 4.692 GHz d. 5.598 GHz

Fig. 8 Polar plot patterns of the proposed model at the resonating frequencies

3.4 Gain

The polar as well as 3D gain plots of the proposed model are as shown in Fig. 9. Due to the coupling annular ring in the structure, the radiation losses have been reduced as the current takes long path to flow which increases the radiation efficiency of the path especially at 4.692 and 5.598 GHz resonating frequencies. The gains obtained at the resonating frequencies are 2.67, 3.74, 4.5 and 4.31 dBi. All the radiations patterns obtained are of bidirectional providing a minimum level of side lobes.

The following Table 2 shows the comparison of the proposed patch antenna for FR-4 and RT Duroid substrate materials.

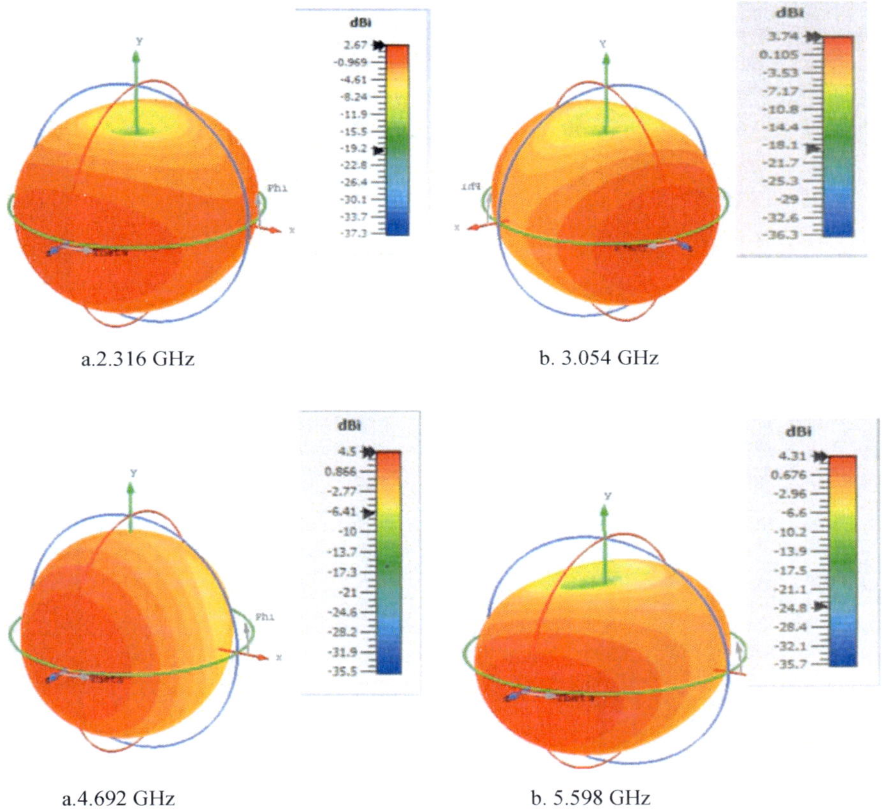

Fig. 9 3D radiations patterns at resonating frequencies

Table 2 Comparison of the proposed model for FR-4 and Duroid substrate materials

Material	Operating bands	Resonant frequencies	Return loss (S_{11})	VSWR	Gain	Efficiency (%)
Duroid	2	2.952 3.876	−18.06 −16.62	1.28 1.34	3.06 3.74	85.98
FR-4	4	2.316 3.054 4.692 5.598	−9.47 −22.80 −18.87 −25.27	1.33 2.01 1.26 1.11	2.67 3.74 4.5 4.31	96.49

4 Conclusion

The proposed antenna operates for wireless applications providing good resonance at quad-bands namely 2.31, 3.05, 4.69 and 5.59 GHz providing better impedance bandwidths, VSWR, gain and radiation patterns. The simulation results shown greater than 2.5 dB for all the resonating bands with reflection coefficients −9.47 dB, −22.8 dB, −18.87 dB and −25.27 dB, respectively, suitable for WLAN, WiBro and WIMAX applications.

References

1. Stutzman WL, Thiele GAT (1998) Antenna Theory and design, 2nd edn. Wiley, New York
2. Pasakawee S, Hu Z (2011) Electrical small meander line patch antenna. In: 6th European conference on antennas and propagation (EUCAP), IEEE
3. Alibakhshi Kenari M, Naser-Moghadasi M, Sadeghzadeh RA, Bal S (2016) A new planar broad-band antenna based on meandered line loops for portable wireless communication devices. Radio Sci, AGU Publications
4. Tirkey SR, Jha N, Pandeeswari R, Raghavan S (2016) Design of flexible meandered loop antennas loaded with CSRR and SRR for implantable applications. IEEE WISPNET
5. Khan MS, Asif SM, Shubaira RM (2017) Compact CSRR-enabled UWB diversity antenna. IEEE Antennas Wirel Propag Lett 16
6. Lu ZL, Yang XX (2017) A multidirectional pattern-reconfigurable patch antenna with CSRR on the ground. IEEE Antennas Wirel Propag Lett 16
7. Lee U, Hao Y (2008) Characterization of microstrip patch antennas on metamaterial substrates loaded with complementary split-ring resonators. Microw Opt Technol Lett 50(8)
8. Naik KK (2018) Design of circular slot on rectangular patch with meander line antenna for satellite communications. Int Conf Invėn Commun Comput Technol (ICICCT 2018). IEEE 9781538619742
9. Naik KK, Saibabu, Surendra C (2018) Design of conformal antenna with slots on path and CSRR on ground plane for ISM band applications. Int Conf Commun Signal Process. IEEE, 9781538635216, (April 2018)
10. Dattatreya G, Naik KK (2019) A low volume flexible CPW-fed elliptical-ring with split tri-angular patch dual-band antenna. Int J RF Microw Comput Aided Eng. Wiley Publications (March 2019)
11. Kota MB, Krishna TVR (2019) Design analysis and performance evaluation of annulus patch antennas. Int J Recent Technol Eng (IJRTE) 8(3):2277–3878. Blue Eyes Intelligence Engineering and Sciences Publication (Sept 2019)

CPW Fed Hexa-to-Hexa Fractal Antenna for Multiband Applications

K. Yogaprasad and V. R. Anitha

Abstract In this paper, a hexa-to-hexa-shaped fractal antenna with four iterations is presented for multiband applications. The substrate material employed for the design is FR-4 substrate with a dielectric constant of 4.4. The corresponding thickness is 1.6 mm and the other dimensions are 35 mm. The proposed antenna is designed and simulated with the help of electromagnetic tools. The observed parameters like reflection coefficient, VSWR, far-field patterns, fabricated prototype, and validated with vector network analyzer result up to 15 GHz. The antenna resonates at five different frequencies like 5.87 GHz, 9.83 GHz, 13.43 GHz, 14.63 GHz, and 18.6 GHz. The operating bands are referred as C-band/X-band/Ku-band applications.

Keywords Hexa-to-hexa fractal · Multiband antenna · Vector network analyzer (VNA) · Coplanar waveguide (CPW)

1 Introduction

In wireless applications, multiple resonance-based antennas are preferred and some of the bands of the proposed antenna under microwave frequencies are L-band (operating frequency 1–2 GHz), S-band (operating frequency 2–4 GHz), C-band (operating frequency 4–8 GHz), X-band (operating frequency 8–12 GHz), and Ku-band (operating frequency 12–18 GHz). The biggest task of the antenna designers is to design an antenna which can be used to operate for two or more bands, and these can be replaced with many antennas that can help in cost reduction and also reduction of

K. Yogaprasad (✉)
Research Scholar, Department of Electronics and Communication Engineering, Rayalaseema University, Kurnool, India
e-mail: kyogaprasad@gmail.com

V. R. Anitha
Antenna Research Lab, Department of Electronics and Communication Engineering, Sree Vidyanikethan Engineering College, Rangampeta, Thirupathi, India
e-mail: anithavr@gmail.com

P. S. R. Chowdary et al. (eds.), *Microelectronics, Electromagnetics and Telecommunications*, Lecture Notes in Electrical Engineering 655,
https://doi.org/10.1007/978-981-15-3828-5_29

space as well as time to design. The different scenarios are projected to implement dual or multiband applications, and some techniques are slots, shorting pin, patches, etc.

In the earlier days, because of cost-effectiveness in constructions, the microstrip antennas were used as a patch above the substrate and bottom acts as a ground and had used for wireless communication applications due to less size and weight [1–4]. During several decades, these microstrip patch antennas [5–8] have been developed to overcome many restrictions. Generally, shapes like triangular, rectangular and circular are used for designing these antennas.

In the study presented in this paper, initially an antenna is designed for multiple resonances as well as multiband applications and excited with coplanar waveguide (CPW) feed. The design enables multiband operations with a single antenna [9–13]. The design of antennas for WLAN applications has been developed with a much-customized decorum and structures [14, 15]. The wire antennas in the due course are replaced with wireless with the developments in the planar structure fabrication technology which is widely used to design the antenna. In the last decade, fractal designs have been proposed and started replacing with normal antenna's merits like size reduction, cost reduction, and multiband or multi-frequency applications.

This research paper, hexa-to-hexa fractal-based antenna with four iterations is given in Sect. 2. The discussion of generalized parameters of the antennas like reflection coefficient, VSWR, and radiation patterns, and comparison between simulated and measured results are given in Sect. 3. Finally, the conclusions are given in Sect. 4.

2 Proposed Structure

The hexa-to-hexa fractal with four iterative models is shown in Fig. 1. The $_1, $_2, $_3, $_4 represents first, second, third, and fourth iterative hexa-to-hexa fractal models. Figure 2 describes the parameter representation of the proposed antenna and that parameter values are shown in Table 1. The fabricated prototype of the proposed design is shown in Fig. 3, and their dimensions are 24 mm × 30 mm × 1.6 mm.

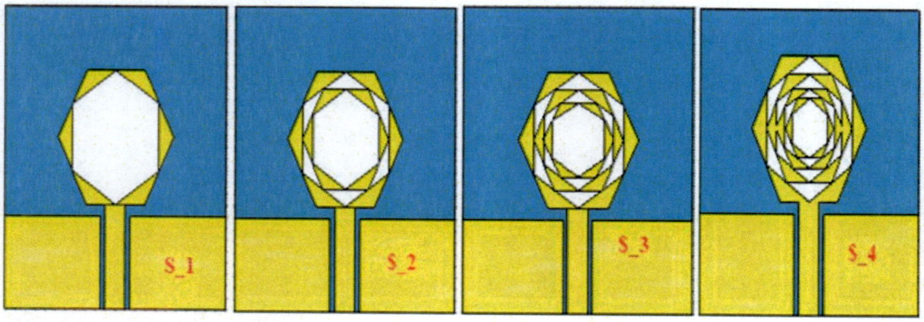

Fig. 1 Iterated antenna structure

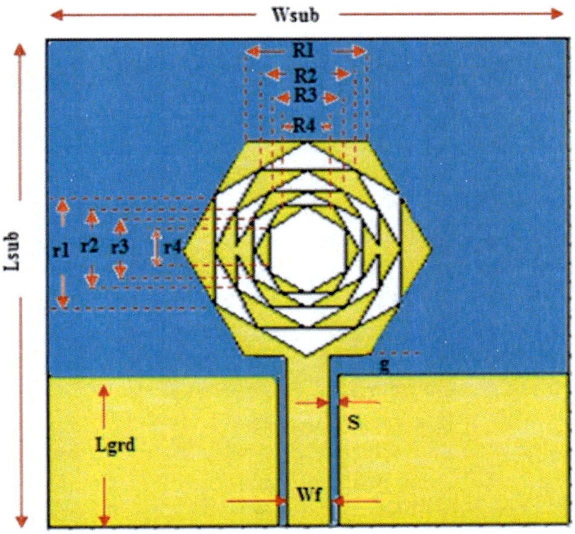

Fig. 2 Proposed antenna with parameter representation

Table 1 Parameters used in this antenna

Parameter	Dimension (mm)	Parameter	Dimension (mm)
*W*2	8.5	*Gr*	0.5
*W*sub/*L*sub	24/26	*Wf*	2
*L*grd	8	*g*	1
*R*1/*R*2/*R*3/*R*4	5.75/4.3/3.4/2.4	*r*1	4.3
*r*3	2.5	*r*2	3.3
*r*4	1.8	*S*	0.5

Fig. 3 Fabricated prototype

3 Results and Discussion

The reflection coefficient of hexa-to-hexa fractal antenna with four iterative structures is shown in Fig. 4. The red indicates the result of the first iteration which produces three resonant frequencies. Those are 2.9, 5.6, 18.2 GHz, and their reflection coefficients are −13, −14, −12 dB. The blue indicates the result of second iteration which resonates at 18 GHz; the reflection coefficient is −10.2 dB. The green indicates the result for third iteration model which resonates at four frequencies, and those are 4.8 GHz (4–6.2 GHz), 9.5 GHz, 14 GHz (8–16 GHz), 20 GHz (19–20); their reflection coefficient values are −22, −20, −19, −12 dB. The yellow represents the fourth and the final iteration models that resonate six frequencies. The proposed design simulation result is shown in Fig. 5 and in that reflection coefficient is shown in Fig. 5a and VSWR is shown in Fig. 5b.

This five resonant frequencies are 5.872, 9.832, 13.43, 14.63, 18.64 GHz and occupy a bandwidth of 2.976 GHz with a resonant frequencies of 5.872, 2.19 GHz bandwidth, with resonant frequencies of 9.832, 4.08 GHz bandwidth with resonant frequencies 13.43, 14.63, and 2.43 GHz bandwidth and with resonant frequency 18.64 GHz with respect to −10 dB reference line and VSWR 2:1 ratio. The reflection coefficient and VSWR values are −22.11, −12.952, −26.51, −14.59, −19.2 dB and 1.17, 1.515, 1.136, 1.45, 1.246 at the frequencies of 5.872, 9.832, 13.43, 14.63, and 18.64 GHz. The comparison of measured and simulated results of an antenna in terms of S_{11} and VSWR is shown in Fig. 6. The measurement results are reported up to 15 GHz with little bit fluctuations which were observed in reflection coefficient (S_{11}) due to connector losses and dispersion losses, and also, slight deviation is observed at fractal edges in a fabricated prototype.

The far-field radiation pattern is shown in Fig. 7, and diagrams on the right and left sides represent H-plane and E-plane at different frequencies and also observed that bidirectional radiation in both planes.

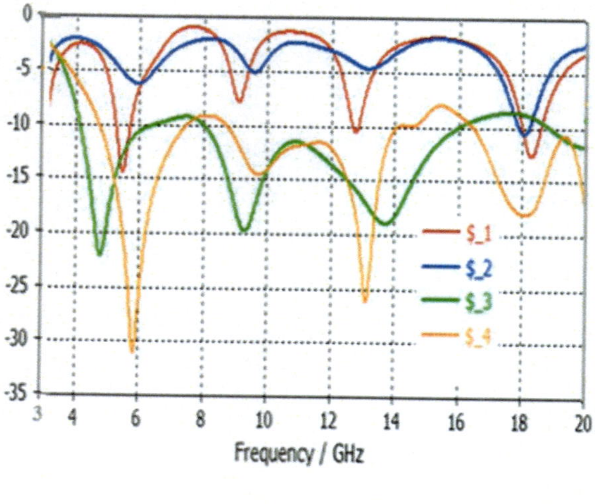

Fig. 4 Reflection coefficient for four iterated structures

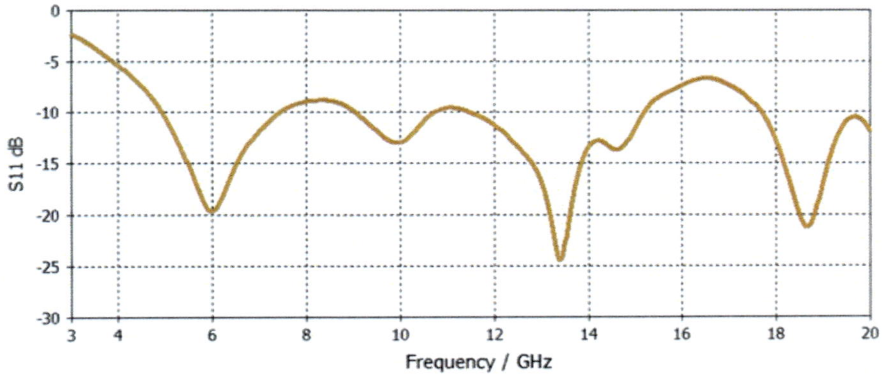

a). Refelection Coefficient S_{11}

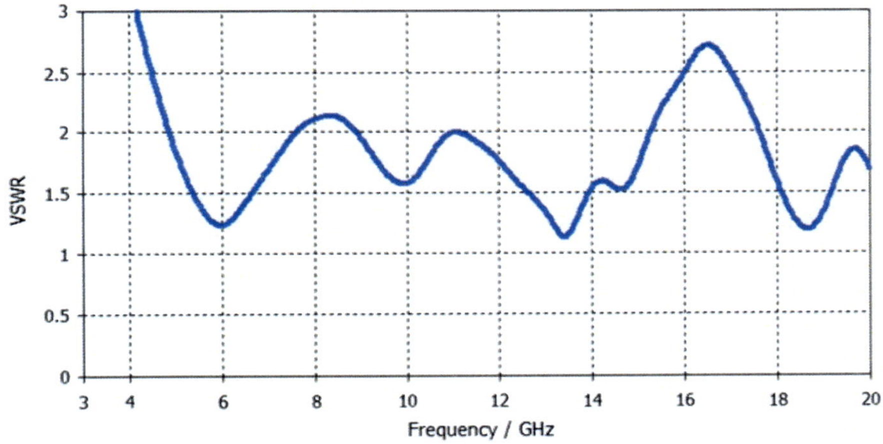

b). VSWR

Fig. 5 Simulation result

4 Conclusion

In this paper, the multiband fractal antenna has been invented. The shape used in this design is hexa-to-hexa with four iterations fed by coplanar waveguide. The antenna has been simulated, fabricated, tested, and validated with VNA results with the dimensions of 24×26 mm^2. The proposed antenna is useful to operate four bands, and those are C, X, Ku, K band applications. The antenna is tested with vector network analyzer to validate the parameters of the reflection coefficient and VSWR up to 15 GHz. The measurement results are shifted prior to 2 GHz frequency due to connector losses, and small gaps are also presented in the edges of the fractals. The measured antenna resonates at 2.8 GHz, 3.6 GHz, 4.2 GHz, 6 GHz, 10.2 GHz, and 15 GHz, respectively.

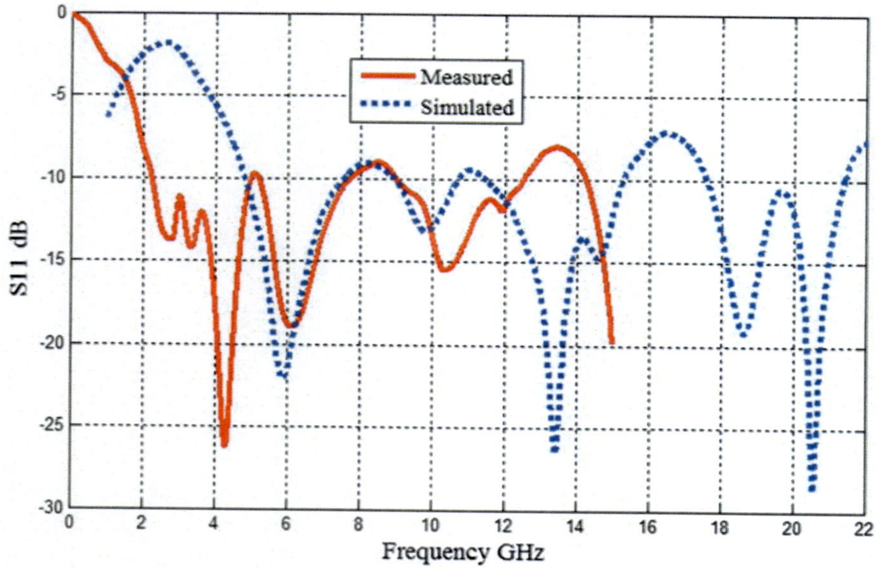

a) S_{11}

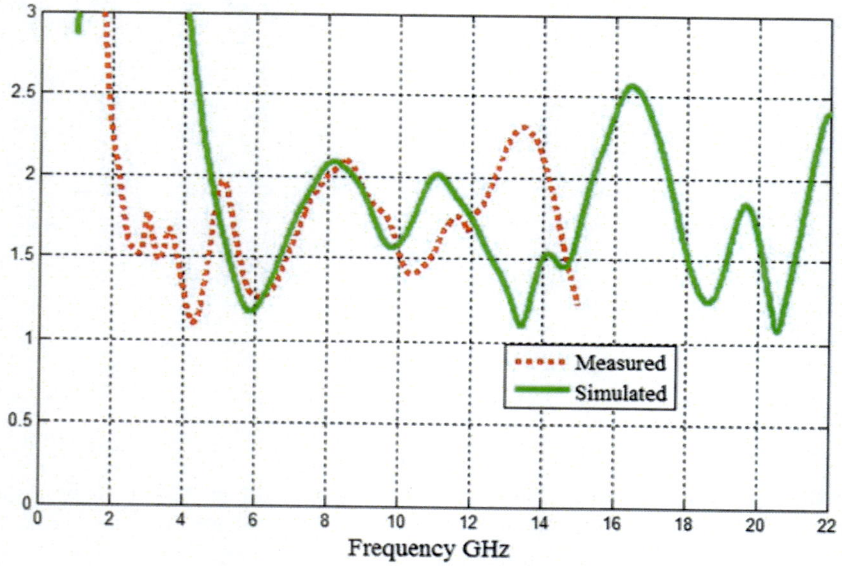

b) VSWR

Fig. 6 Simulation and measured comparison results

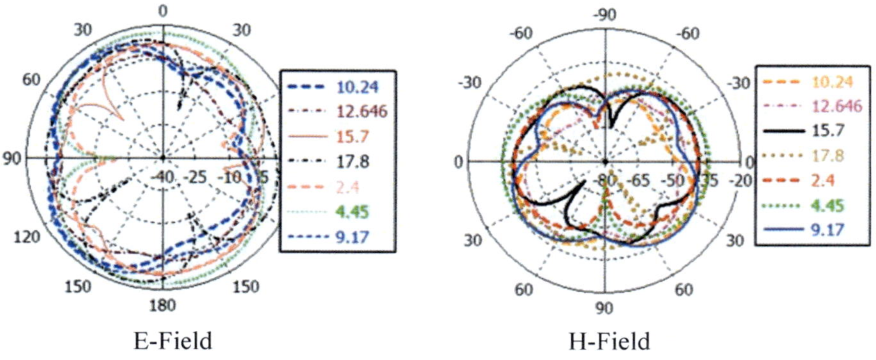

Fig. 7 Far-field result

References

1. Bashar B (2016) Qas Elias, design of broadband circular patch microstrip antenna for KU-band satellite communication applications. Int J Microw optic Technol 11:362–368 (2016)
2. Kumar MN, Shanmugnantham T (2016) Current and future trends in SIW antennas-an overview. In: IEEE international conference on advanced computing, Bhimavaram, Andhra Pradesh
3. Kwaha BJ, Inyang ON, Amalu P (2016) The circular microstrip patch antenna—design and implementation. IJRRSS 8:88–95
4. Kumar MN, Shanmugnantham T (2018) Microstrip Fed SIW venus shaped slot antenna for millimeter wireless communication applications. IEEE ICSCS-2018, Kollam
5. Kumar MN, Shanmugnantham T (2018) Division shaped SIW slot antenna for millimeter wireless/automotive radar applications. Comput Electr Eng—J (Elsevier) 71:667–675
6. Kumar MN, Shanmuganantham T (2019) Back to back Pi shaped slot with SIW cavity for millimeter wireless applications. Int J Microw Opt Technol (IJMOT) 14(6):402–408
7. Kumar MN, Shanmugnantham T (2018) Microstrip fed SIW venus shaped slot antenna for millimeter wireless communication applications. Int J Eng Technol 7(2.33):878–881
8. Kumar MN, Yogaprasad K, Anitha VR (2018) Meandered shaped microstrip feed antenna for Ku-band applications. Int J Mech Prod Eng Res Dev (IJMPERD) 8(2):276–280
9. Kumar MN, Yogaprasad K, Anitha VR (2020) A quad band Sierpenski based fractal Antenna fed by CPW. Microw Opt Technol Lett (MOTL-Wiley) 2(2):893–898
10. Dinesh M, Nandakumar M, Balachandra K (2018) Micro-strip feed reconfigurable antenna for wideband applications. Journal: Lecturer Notes in Electrical Engineering. Springer, pp 665–671
11. Kizhekke J, Kanagasabai M (2015) Circularly polarized broad band ant deploying fractal slot geometry. IEEE Antennas Wirel Propag Lett 14:1286–1289
12. Andrenko AS, Ivanchenko IV, Ivanchenko DI (2006) Active broad X-band circular patch antenna. IEEE Antennas Wirel Propag Lett 5:529–533
13. Reddy VV, Sharma NVSN (2015) Compact circularly polarized asymmetrical fractal boundary micro-strip ant for wireless applications. IEEE Antennas Wirel Propag Lett 13:118–120
14. Subbarao A, Raghavan S (2013) Compact coplanar waveguide fed planar antenna for ultra-wideband and WLAN applications. Wirel Pers Commun 7:2849–2862
15. Kumar MN, Shanmuganantham T (2019) Broad-band I-shaped SIW slot antenna for V-band applications. Appl Comput Electromagn Soc (ACES) 34(11):1719–1724

GA Tuned Kalman Filter for Precise Positioning

Nalineekumari Arasavali, G. Sasibhushana Rao, and N. Ashok Kumar

Abstract The Global Positioning System (GPS) plays a predominant role in various navigation applications; their precise position values are required. Kalman filter (KF) is a navigation solution, used to predict and estimate the anonymous state by suppressing the noise existing in aviation control systems. However, the effect is more susceptible to parameters of KF, whose choice purely based on previous experience of an operator. In this paper, the genetic algorithm (GA)-based KF approach is presented. GA technique is used to optimize the covariance values of error in initial state, measurement noise and process noise. For experimental validation, the data collected at Andhra University, Visakhapatnam, is used which is located at (706,970.9093, 6,035,941.0226 and 1,930,009.5821) (m).

Keywords GPS · Kalman filter · Genetic algorithm

1 Introduction

The Global Positioning System (GPS) [1] is a satellite-based navigation system consists of 24 satellites and developed by the US Department of Defense. To find the position in terms of latitude, longitude and altitude, three or more satellites are required, and in order to determine 3D position, four or more satellites are required. The accuracy of GPS depends on the quality of the two basic measurements, namely pseudorange and carrier phase, and the satellite ephemeris data [2]. GPS receiver location is considered as a reference location. Here, the receiver predicts the position using signal received from minimum of four satellites and compares the predicted

N. Arasavali (✉) · G. Sasibhushana Rao
Department of Electronics and Communication Engineering, Andhra University College of Engineering (A), Andhra University, Visakhapatnam, Andhra Pradesh 530003, India
e-mail: naliniarasavali@gmail.com

N. Ashok Kumar
Department of Electronics and Communication Engineering, Raghu Engineering College (A), Visakhapatnam, Andhra Pradesh 531162, India

P. S. R. Chowdary et al. (eds.), *Microelectronics, Electromagnetics and Telecommunications*, Lecture Notes in Electrical Engineering 655,
https://doi.org/10.1007/978-981-15-3828-5_30

location with known location. The accurate position of the receiver depends on precise range measurements [3].

2 Kalman Filter

Kalman filter [4] is the predictor and a corrector type, navigation solution, which consists of an array of mathematical equations. It has profound research applications, especially in the field of navigation. It has an array of mathematical expressions (see equations from 1 to 7).

2.1 Time Update Equations of Kalman Filter Algorithm

Predicted state estimate:

$$s_{\xi/\xi-1} = \varphi_{\mathrm{tr}} s_{\xi-1/\xi-1} \tag{1}$$

Predicted covariance estimate:

$$p_{\xi/\xi-1} = \varphi_{\mathrm{tr}} P_0 \varphi_{\mathrm{tr}}^{\mathrm{T}} + Q_v \tag{2}$$

The time update equations give the state and covariance estimation values from time step $\xi - 1$ to step ξ.

2.2 Measurements Update Equations of Kalman Filter Algorithm

Measurement residual

$$y_{\xi/\xi} = y_{\xi/\xi-1} - H s_{\xi/\xi} \tag{3}$$

Residual covariance

$$p_{\xi/\xi} = H_\xi p_{\xi/\xi-1} H_\xi^{\mathrm{T}} + R_v \tag{4}$$

Near-optimal linear quadratic estimator gain

$$k = p_{\xi/\xi-1} H_\xi^{\mathrm{T}} (H_\xi p_{\xi/\xi} H_\xi^{\mathrm{T}} + R_v)^{-1} \tag{5}$$

Updated state estimate

$$x_{\xi/\xi} = x_{\xi/\xi-1} + ky_{\xi/\xi} \quad (6)$$

Updated covariance estimate

$$p_{\xi/\xi} = (1 - kH_{\xi})p_{\xi/\xi-1} \quad (7)$$

where H, observation matrix, k is Kalman gain, P_0 is the initial covariance, Q_v is the process noise covariance and R_v is the measurement noise covariance. In this algorithm, the gain, K, is computed based on the initial covariance, process and the measurement noise, P_0, Q_v and R_v, respectively. R_v is taken as $E[ee^{\mathrm{T}}]$.

3 Genetic Algorithm

Genetic algorithm [5] is an adaptive computational metaheuristic technique, which is also a random search technique, used to provide best feasible solutions based on previously collected data. GA principle depends on the structure of genes and etiquette of chromosomes in the population. Figure 1 depicts the basic steps of genetic algorithm. Selection, crossover and mutation are three important operators of GA. GA is available in two different forms as binary-coded GA and real-coded GA. In this paper, real-coded GA is used.

Selection: This is the process that determines which solutions are best to preserve and allowed to produce new offsprings. The main aim of the selection operator is to

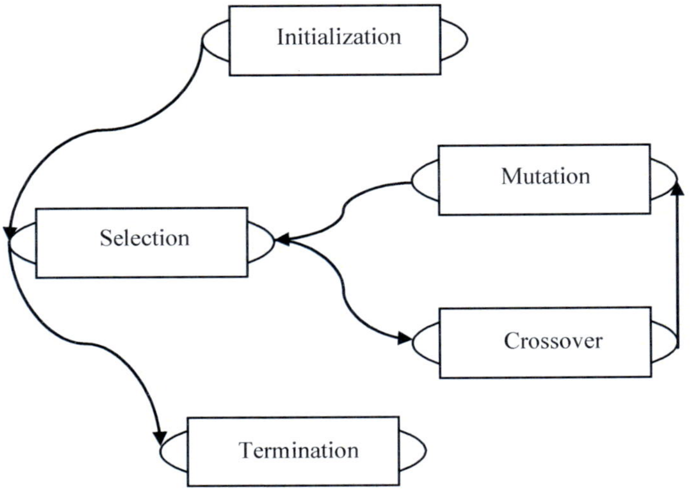

Fig. 1 Basic steps in genetic algorithm (GA)

select the best solutions and discard the rest of the solution in a population by keeping the population size constant. After identifying the best solution, selection operator will make multiple copies of these solutions and is placed in the places of discarded population. To identify the best population, fitness or cost function will be used. The optimality of the solution is quantified by the fitness function value. Fitness value is assigned to each and every solution. Based on this value, good solutions will be preserved. Tournament selection, roulette wheel selection, rank selection, etc. are the different techniques to implement selection in GA.

Crossover: It is used to create new offsprings from the existing population after selection. Crossover operator is used to exchange the genes between the existing population. In the case of real crossover, two parents will be selected randomly and created new children by combining them.

$$b = \begin{cases} (2 * \delta)^{\frac{1}{\mu+1}} & \text{if } \delta \leq 0.5 \\ \left(\frac{1}{(2*(1-\delta)}\right)^{\frac{1}{\mu+1}} & \text{if } \delta > 0.5 \end{cases} \tag{8}$$

where δ is a random number and μ is a crossover operator. From two parents P1 and P2, two children will be produced (Eqs. 9 and 10).

$$\text{child}_1 = \frac{1}{2}((1+b) * \text{parent}_1 + (1-b) * \text{parent}_2) \tag{9}$$

$$\text{child}_2 = \frac{1}{2}((1-b) * \text{parent}_1 + (1+b) * \text{parent}_2) \tag{10}$$

Mutation: After selection and crossover, the old population is replaced with new offsprings which are produced by crossover, and the rest of individuals are directly copied. Mutation is used to ensure the genetic diversity within the population. How often the parts of a chromosome will be mutated is indicated by mutation probability. In real-coded GA, the mutation process consists of below Eqs. (11) and (12). η is the mutation operator.

$$d = \begin{cases} ((2 * \delta)^{\frac{1}{\eta+1}} - 1)\ldots & \text{if } \delta \leq 0.5 \\ \left(1 - ((2 * (1-\delta))^{\frac{1}{\eta+1)}}\right)\ldots & \text{if } \delta > 0.5 \end{cases} \tag{11}$$

$$\text{child}_1 = \text{parent}_1 + d \tag{12}$$

The parameters of GA is as shown in Table 1. The used parameter values for real-coded GA are listed in Table 1. In this paper, optimization of parameters of KF is proposed to achieve more accurate navigation solution, and the parameters are initial state covariance (P_0), process noise covariance (Q_v) and measurement noise covariance (R_v). The unsurpassed optimization technique, i.e., genetic algorithm is used to optimize these three parameters [6]. Chosen best fitness function as the

Table 1 Parameters of real-coded GA

Parameter	Value
Size of population	50
Crossover probability	0.8
Mutation probability	0.2
Factor of mutation	20
Crossover ratio	20

objective function which reaches its best value to improve the performance of existing algorithms.

4 Results and Discussion

Parameters of Kalman filter are optimized with GA technique. In the case of GA, a number of individuals are taken as 50. Figure 2a shows how the parameter values are going close to their best values with the variation in the fitness value, and the

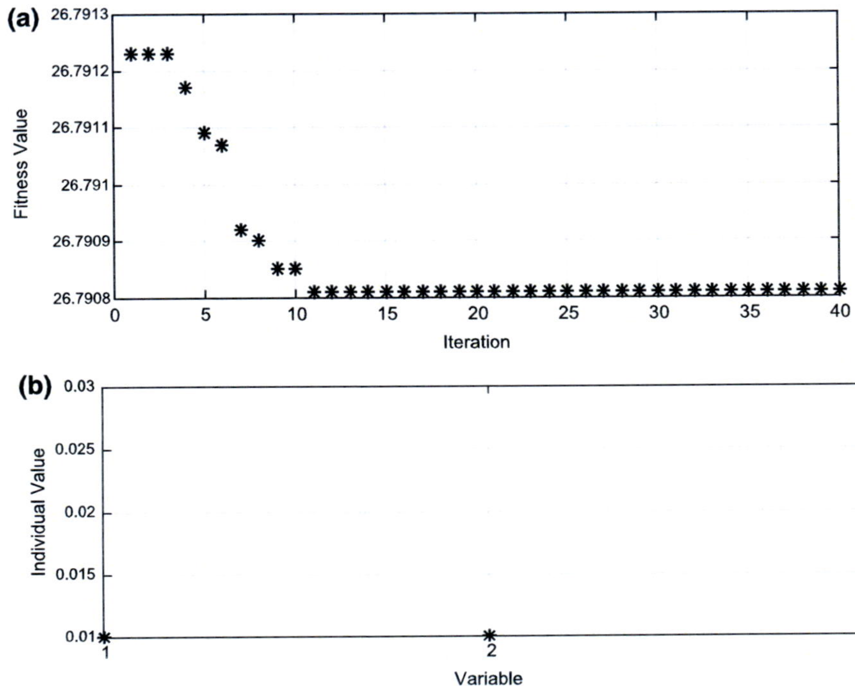

Fig. 2 **a** Fitness value using GA. **b** Best values of individuals

status of current individuals at the best fitness value are shown in Fig. 2b. In this analysis, R, Q_v, P_0 are considered as identity matrices multiplied with constants. Those multiplier constants are optimized with well-known optimization technique, genetic algorithm (GA).

Result of Kalman Filter tuned with GA

The optimal values of KF parameters are substituted into KF equations. The position error values are reduced in X-, Y- and Z-directions as shown in figures below.

KF tuned with GA

The parameter values are obtained through GA technique reduced error values as X-error is reduced about 2 m, Y-error is reduced about 3 m and Z-error is reduced about 1 m. These results are better depicted as shown in Figs. 3, 4 and 5, how the error values are reduced with optimal values of KF parameters.

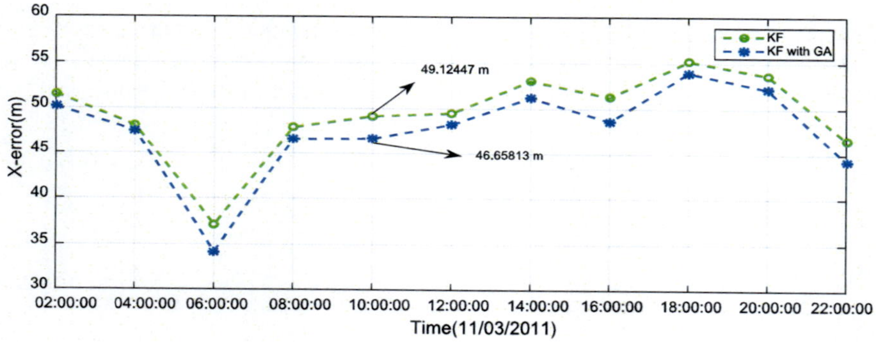

Fig. 3 Positioning performance in X-direction (receiver: Visakhapatnam) (KF-Kalman filter, GA-genetic algorithm)

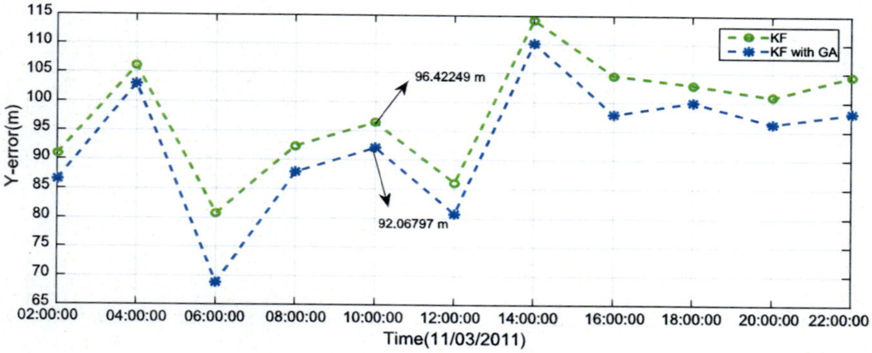

Fig. 4 Positioning performance in Y-direction (receiver: Visakhapatnam) (KF-Kalman filter, GA-genetic algorithm)

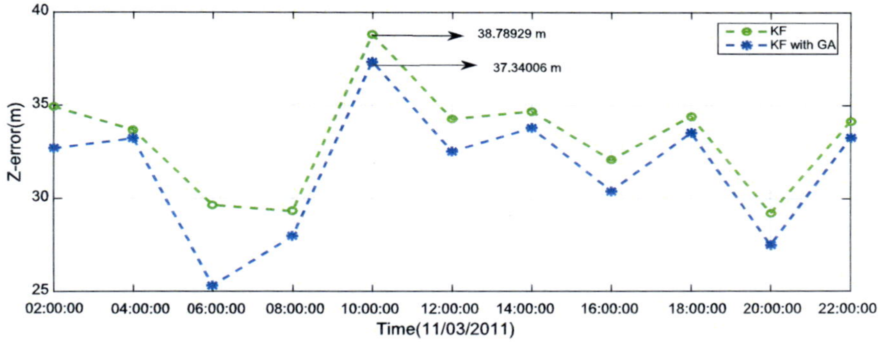

Fig. 5 Positioning performance in *Z*-direction (receiver: Visakhapatnam) (KF-Kalman filter, GA-genetic algorithm)

Table 2 Position error difference during 10:00:00 on 11/03/2011 at Visakhapatnam

Direction	Difference between KF and adaptive KF with GA in m
X-error	2.46
Y-error	4.35
Z-error	1.45

As shown in Fig. 3, during 10:00:00 (Hours:Minutes:Seconds) on March 11, 2011, observed *X*-error value with KF is 49.12447 and 46.65813 m is with KF-GA.

From Fig. 4, it is observed that during 10:00:00 (Hours:Minutes:Seconds) on March 11, 2011, the *Y*-error values are with KF and KF-GA as 96.42249 m and 92.06797 m, respectively. From Fig. 5, it is observed that during 10:00:00 (Hours:Minutes:Seconds) on March 11, 2011, observed *Z*-error value with KF is 38.78929 and 37.34006 m is with KF-GA. The positional difference in *X*-, *Y*- and *Z*-directions are given in Table 2.

5 Conclusion

In this paper, the way of reducing the position error with random search technique was discussed. And if needed cultural algorithm (CA) can be used to optimize the initial user position. This is verified with the data collected at GPS receiver located at (706,970.9093, 6,035,941.0226 and 1,930,009.5821) m. The results are shown that the optimized parameter values improved the position accuracy. Accuracy achieved is 2.46 m, 4.35 m and 1.45 m in *X*-, *Y*- and *Z*-directions, respectively.

References

1. Kaplan ED (1996) Understanding GPS, principles and applications. Artech House
2. Rao GS (2010) Global navigation satellite systems. McGraw Hill Education Private Limited
3. Farrell J (1998) The global positioning system and inertial navigation. McGraw-Hill Professional, New York
4. Wong RVC (1981) A Kalman filter/smoother for the Ferranti inertial land surveyor system. Proceedings of the Second International Symposium on Inertial Technology for Surveying, Canada
5. Mosavi MR, Sadeghian M, Saeidi S (2011) Increasing DGPS navigation accuracy using Kalman filter tuned by genetic algorithm. Int J Comput Sci Issues (IJCSI) 8(6), no 3
6. Yan J, Yuan D, Xing X, Jia Q (2008) Kalman filtering parameter optimization techniques based on genetic algorithm. In: 2008 IEEE conference on college of automation department northwestern polytechnical university Xi'an. Shaanxi Province, China

An Optimized Path Loss Model for Urban Wireless Channels

Sreevardhan Cheerla, D. Venkata Ratnam, and J. R. K. Kumar Dabbakuti

Abstract The mobile network planner relies on a signal propagation path loss model to enhance the wireless communication system in order to avail an acceptable limit of quality of service for the mobile users. Hence, it is very crucial to find a robust propagation model suitable for a range of environmental conditions which may be implemented as guidelines for planning of cell in wireless communication systems. Path loss is the regulating factor in limiting the performance of the system in urban areas. It is essential to develop an appropriate path loss model which predicts the path loss values depending on the received signal strength. In the present paper, the COST 231 propagation model has been optimized by making use of Newton's method. The statistical measures like absolute average error and root-mean-square error were calculated for the frequencies 800 and 1800 MHz. From the simulation results, it is found that the optimized model best acclimatizes with a smaller mean relative error. The lesser value of mean error supports successful implementation of the optimization technique and therefore suggested that the present optimized model can be useful for telecommunication providers to improve the service for mobile user satisfaction.

Keywords Path loss · GSM · CWI model · Optimization · Newton's method

1 Introduction

The reduction in the signal strength obstructions in between transmitter and receiver is termed as path loss. Path loss can be determined from numerous elements of communication system such as transmitter power and antenna height, location and gain parameters. Path loss parameters play a key role in the design of link power budget of cellular communication system. Hence, an accurate path loss model is essential for better prediction of path loss in the urban areas ensuring an improved received signal

---

S. Cheerla (✉) · D. Venkata Ratnam · J. R. K. K. Dabbakuti
Department of Electronics and Communication Engineering, KL Deemed to be University, Koneru Lakshmaiah Education Foundation, Vaddeswaram, Guntur, Andhra Pradesh 522502, India
e-mail: sreevardhancheerla@kluniversity.in

P. S. R. Chowdary et al. (eds.), *Microelectronics, Electromagnetics and Telecommunications*, Lecture Notes in Electrical Engineering 655,
https://doi.org/10.1007/978-981-15-3828-5_31

strength (RSS). Propagation models are used to predict the received level of power at different locations under a given coverage area. Based on approaches, the existing models can be segregated into different types: empirical, semi-deterministic and deterministic models [1]. Empirical models are developed by statistical modelling of large measurement data sets such as Okumura–Hata model and COST 231 Hata model. However, for the prediction of radio waves being propagated in space more accurate information is necessary. Meanwhile, the results obtained through empirical models are not that much accurate. The shortcoming in these models is that they are approximated and do not predict phenomena accordingly [2]. The empirical methods deal by reconfiguring certain parameters of the model such as wall and floors between receiver and transmitter in a statistical way through the performed measurements in a realistic environment [3]. The deterministic models are derived from the principles of physics like ray tracing [4], radiosity [5], Maxwell's equation [6], uniform theory of diffraction [7], without affecting the accuracy, and require high computational power and more site parameters. However, suitable implementation of deterministic models may produce relatively accurate predictions than the empirical models [8]. A P-norm-based adaptive penalized least mean square algorithm is proposed for channel estimation in [9]. In recent studies, the prediction of path loss is carried out by using neural network and some other regression methods. An adaptive neuro-fuzzy interference system-based path loss prediction system is proposed in [10]. By considering the site-specific parameters as the input entities in the neural network, the values of path loss are predicted for an urban environment in [11]. An artificial neural network-based hybrid model is used to estimate the path loss values at 800 and 1800 MHz frequency bands for an urban environment over Vijayawada, India, in [12]. From the literature review, it is convinced that the path loss is modelled depending on the site-specific conditions. The problems associated with site-dependent modelling can be avoided by tuning the parameters for the appropriate environments through which path loss can be optimized [13]. A statistical tuning of COST 231 WI (CWI) model was proposed using particle swarm optimization (PSO) in [14] where the coefficients are calculated based on the statistical tuning of empirical parameters like roof height and separation of buildings. An optimized Hata model based on existing Hata model and measured path loss for 400–1800 MHz band of frequency over Putrajaya, Malaysia, is described in [15]. Similarly, a framework has proposed the downlink of a multi-tier HetNet by considering the joint user, power and subcarrier allocations [16]. The proposed framework retains the users' quality of service needs, such as user minimum rate and the base stations' maximum transmission power [16]. An optimization method for ultra-dense network balancing mobility and densification is proposed in [17]. The proposed method is proved to be efficient and practical to meet the service demands with the consumption of least bandwidth resources. In this work, optimization of path loss of CWI model for 800 and 1800 MHz band of frequency over Vijayawada, India, is presented with respect to real-time measured data.

2 Methodology

2.1 COST 231 Walfisch Ikegami Model

For a frequency over a range of 800–2000 MHz, CWI model is developed as an extended version to the Ikegami model [18]. CWI model is chosen to efficiently estimate the path loss by considering the environmental characteristics such as height of buildings, gap between the buildings, width of road, frequency and street orientation. A free-space path loss model can be considered to effectively estimate the signal strength in an existed line-of-sight (LOS) path. For a non-LOS (NLOS) path, CWI-NLOS model is superior in estimating the path loss when compared to other empirical models [19]. This model adopts the theoretical Walfisch–Bertoni model [20] and consists of three parts: the first and second components are taken from the group led by Bertoni [20–22], while the third is obtained from Ikegami [15]. The free-space loss (*Lf*), loss due to scattering and diffraction from top of roof to street (*Lr*) and loss due to the multi-screen diffraction (*L*m) are considered for modelling the CWI model for NLOS conditions [14]. The modified NLOS path loss of CWI model is expressed as [14]

$$\mathrm{PL}\,(\mathrm{dB}) = 32.4 + 20\log d + 30\log f - 8.8 - 10\log(1 + \Delta h_\mathrm{t}) + k_\mathrm{a} + k_\mathrm{d}\log d \tag{1}$$

In the optimization process, the entities E_o and E_sys are considered as initial offset and system design parameters whereas β_sys is taken as the established slope of the model curve. The parameters are expressed from (1) as

$$E_\mathrm{o} = 32.4 \tag{2}$$

$$E_\mathrm{sys} = 20\log d + 30\log f - 8.8 - 10\log(1 + \Delta h_\mathrm{t}) \tag{3}$$

$$\beta_\mathrm{sys} = k_\mathrm{a} + k_\mathrm{d} * \log d \tag{4}$$

2.2 Measurement Data

As part of the verification of the optimized path loss model, a real-time data set of global systems for mobile communications (GSM) consisting of cellular communication path losses at 800 and 1800 MHz is measured considered in urban environment at Vijayawada location, Andhra Pradesh. The data collection process was started at Benz Circle, Vijayawada, India (16.49 N, 80.65 E), and ended at Patamata (16.49 N, 80.66 E) via Gurunanak Colony (16.5 N, 80.67 E) over a 4.6 km long distance as

Fig. 1 Measurement location (Vijayawada urban) in Google Maps

shown in Fig. 1. Notable variations on signal strengths are due to high dense buildings and vegetation throughout the observation campaign.

The parameters considered for path loss evaluation in [12] are considered for this analysis also.

3 Optimization Method

Optimization process of CWI model is illustrated in Fig. 2. Newton's iterative method

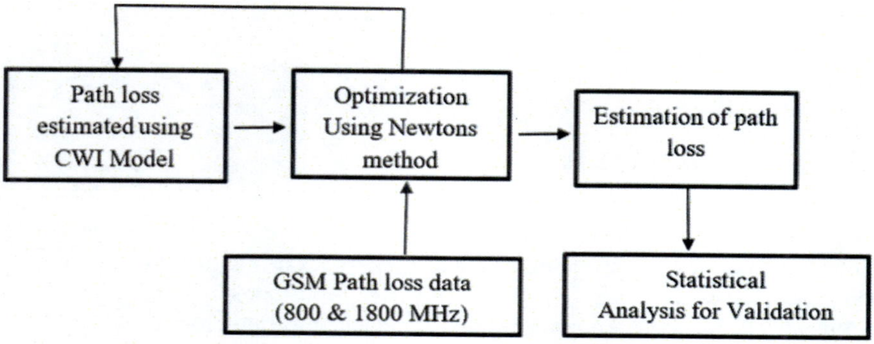

Fig. 2 Schematic representation of optimization technique

is used during the optimization of path loss. Standard deviation, variance and relative error are estimated in the statistical analysis for accessing the performance of the optimized model. Further, the optimized path loss is validated with measured and CWI model to extract the inference in this study.

In this study, an innovative approach is followed to optimize the measured data by CWI model. With a given set of experiment data, the best fit of the theoretical model curve can be obtained if a minimum limit is satisfied by the function of sum of deviation squares:

$$P(a, b) = \sum_{i=1}^{n} [y_i - E_{\mathrm{R}}(x_i, a, b)]^2 = \min \tag{5}$$

where y_i is the measured value at a given distance x_i;
$E_{\mathrm{R}}(x_i, a, b)$ is the modelling result at a given distance x_i through optimization;
a, b are the parameters of the optimization model;

$$a = E_0 + E_{\mathrm{sys}}$$
$$b = \beta_{\mathrm{sys}}$$

The expression of the COST model in Eq. (1) transformed into:

$$E_{\mathrm{R}} = a + b$$

E_{R} is the path loss (dB).

Both factors a and b remain fixed for a given set of calculation. The results from Eq. (5) can be expressed as:

$$\frac{\partial E_{\mathrm{R}}}{\partial a} = \sum_{i=1}^{n} ((y_i - E_{\mathrm{R}}(x_i, a, b)) * 1) = 0 \tag{6}$$

$$\frac{\partial E_{\mathrm{R}}}{\partial b} = \sum_{i=1}^{n} ((y_i - E_{\mathrm{R}}(x_i, a, b)) * x_i) = 0 \tag{7}$$

The refined coefficients are calculated by solving Eqs. (6) and (7) via the iterative numerical methods. The coefficients obtained through the optimization process are listed in Table 1.

Table 1 Updated coefficients for CWI model

S. No.	Parameter	800 MHz	1800 MHz
1	E_{sys}	0.79	0.63
2	β_{sys}	10.1	27.33

4 Results

The real-time measured RSS of GSM signal at 800 and 1800 MHz frequency bands is taken for estimating the path loss values using optimization process method. Later, the obtained path loss values from CWI model, measured data as well as optimized CWI model are considered in the analysis. The accurate results were obtained when Newton's method is used for optimization process. Figure 3 shows the changes in path loss estimated through CWI model, optimized CWI and real-time measured data in an urban environment at 800 MHz. The results show that the path loss obtained through the optimization technique is following the measured data than the CWI model for 800 MHz frequency band. In Fig. 3 (lower panel), we analyse the errors in path loss values in terms of absolute error distribution. The absolute error of CWI is ~15 dB, whereas for optimized model it is 0.5 dB.

Figure 4 depicts the discrepancies between the path loss values from CWI model, optimized model and real-time measurement in an urbanized area at 1800 MHz. It is evident from the figure that the optimized model concurred well with the measured path loss values. Figure 4 (lower panel) shows absolute error between predictions of optimized and real-time measurement. The maximum error is ~17 dB for CWI model, whereas for optimized model it is ~1.5 dB.

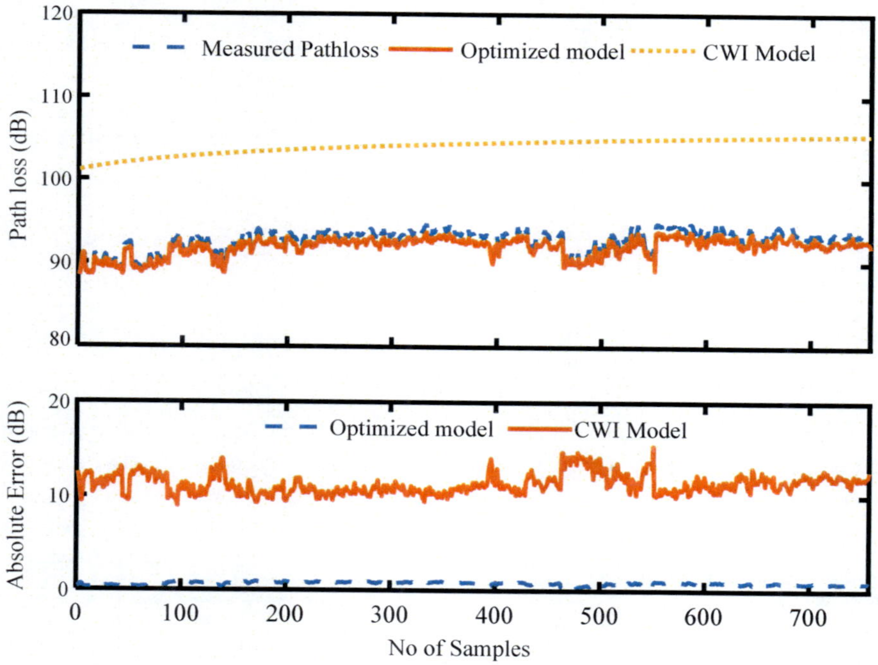

Fig. 3 Comparison of path loss obtained from optimized model and CWI model with the measured path loss at 800 MHz

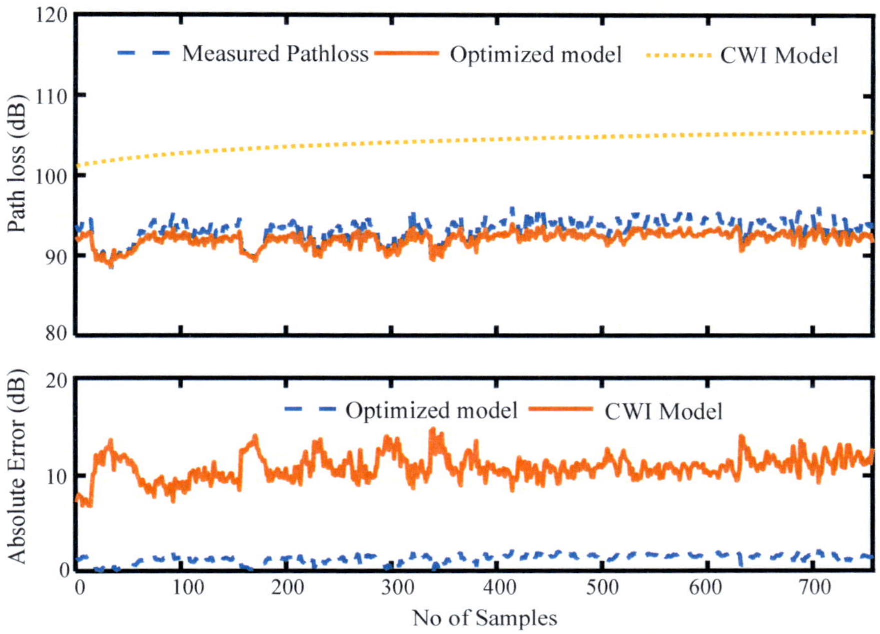

Fig. 4 Comparison of path losses from optimization and CWI model with the path loss measured at 1800 MHz

Table 2 Performance evaluation of CWI and optimized models for 800 and 1800 MHz bands

Frequency	Model	AAE (dB)	RMSE (dB)
800 MHz	CWI model	11.39	11.44
	Optimized model	0.78	0.80
1800 MHz	CWI model	10.81	10.89
	Optimized model	1.26	1.33

The comparison of errors obtained through both CWI model and optimized model is summarized in Table 2.

5 Conclusions

The prediction of path loss is the main objective of all the propagation models. In this study, an optimized path loss model is proposed using Newton's method. The simulation results obtained by using optimization are validated using real-time measured data. It is clear from the analysis that the optimized path loss values do agree with the measured data than the empirical CWI model. The path loss values are calculated through the CWI model in urban area conditions. It is clear that the error metrics calculated by using the optimized model are much accurate when compared with the CWI model values. The RMSE values measured with optimized model (0.80 dB for 800 MHz and 1.33 dB for 1800 MHz) show that the prediction accuracy is very much high over that of CWI model. The important extracts from this study would complement towards the planning frequency reuse and better system designing of wireless communication systems.

Acknowledgements This work was supported by the Department of Science and Technology (DST), New Delhi, India, SR/FST/ESI-130/2013(C), under DST-FIST.

References

1. Alqudah YA (2013) On the performance of Cost 231 Walfisch Ikegami model in deployed 3.5 GHz network. In: 2013 international conference on technological advances in electrical, electronics and computer engineering (TAEECE). IEEE, pp 524–527
2. Lopez-Barrantes AJ, Gutierrez O, Saez de Adana F, Kronberger R (2012) Comparison of empirical models and deterministic models for the analysis of interference in indoor environments. In: 2012 Asia-Pacific symposium on electromagnetic compatibility (APEMC). IEEE, pp 509–512
3. Seidel SY, Rappaport TS (1992) 914 MHz path loss prediction models for indoor wireless communications in multifloored buildings. IEEE Trans Antennas Propag 40(2):207–217
4. Rappaport TS, Seidel SY, Schaubach KR (1993) Site-specific propagation prediction for PCS system design. In: Wireless personal communications. Springer, Boston, MA, pp 281–315
5. Tsang K-F, Chan W-S, Jing D, Kang K, Yuen S-Y, Zhang W-X (1998) Radiosity method: a new propagation model for microcellular communication. In: IEEE antennas and propagation society international symposium, 1998, vol 4. IEEE, pp 2228–2231
6. Hrovat A, Kandus G, Javornik T (2014) A survey of radio propagation modeling for tunnels. IEEE Commun Surv Tutor 16(2):658–669
7. Tan SY, Tan HS (1996) A microcellular communications propagation model based on the uniform theory of diffraction and multiple image theory. IEEE Trans Antennas Propag 44(10):1317–1326
8. Popescu I, Nafornita I, Constantinou P (2005) Comparison of neural network models for path loss prediction. In: IEEE international conference on wireless and mobile computing, networking and communications, 2005 (WiMob'2005), vol 1. IEEE, pp 44–49
9. Wang Y, Jiang T (2016) Norm adaption penalized least mean square/fourth algorithm for sparse channel estimation. Sig Process 128:243–251
10. Alotaibi FD, Abdennour A, Ali AA (2008) A robust prediction model using ANFIS based on recent TETRA outdoor RF measurements conducted in Riyadh city–Saudi Arabia. AEU-Int J Electron Commun 62(9):674–682
11. Sotiroudis SP, Siakavara K (2015) Mobile radio propagation path loss prediction using artificial neural networks with optimal input information for urban environments. AEU-Int J Electron Commun 69(10):1453–1463
12. Cheerla S, Venkata Ratnam D, Borra HS (2018) Neural network-based path loss model for cellular mobile networks at 800 and 1800 MHz bands. AEU-Int J Electron Commun 94:179–186
13. Mollel MS, Kisangiri M (2014) Optimization of Hata model based on measurements data using least square method: a case study in Dar-es-Salaam—Tanzania. Int J Comput Appl 102(4)

14. Tahat A, Taha M (2012) Statistical tuning of Walfisch-Ikegami propagation model using particle swarm optimization. In: 2012 IEEE 19th symposium on communications and vehicular technology in the Benelux (SCVT). IEEE
15. Roslee MB, Kwan KF (2010) Optimization of Hata propagation prediction model in suburban area in Malaysia. Prog Electromagn Res 13:91–106
16. Munir H, Hassan SA, Pervaiz H, Ni Q, Musavian L (2017) Resource optimization in multi-tier HetNets exploiting multi-slope path loss model. IEEE Access 5:8714–8726
17. Zhu J, Zhao M, Zhou S (2018) An optimization design of ultra dense networks balancing mobility and densification. IEEE Access 6
18. Damosso E, Correia LM (1999) COST action 231: digital mobile radio towards future generation systems: final report. European Commission
19. Alqudah YA (2013) On the performance of COST 231 Walfisch Ikegami model in deployed 3.5 GHz network. IEEE. ISBN: 978-1-4673-5613-8
20. Walfisch J, Bertoni HL (1988) A theoretical model of UHF propagation in urban environments. IEEE Trans Antennas Propag 36(12):1788–1796
21. Xia HH, Bertoni HL (1992) Diffraction of cylindrical and plane waves by an array of absorbing half-screens. IEEE Trans Antennas Propag 40(2):170–177
22. Maciel LR, Bertoni HL, Xia HN (1993) Unified approach to prediction of propagation over buildings for all ranges of base station antenna height

A Comprehensive Review on Effect of Band Selection on the Recital of Hyper-spectral Image Classification

Kalidindi Kishore Raju, G. P. Saradhi Varma, and Davuluri Rajyalakshmi

Abstract In every research field algorithms have been realized by various authors. These algorithms like geo-spatial-based land cover and land use research are to reassess in day by day. The recital of land cover and land use (LCLU) nomenclature of hyper-spectral image chiefly depends on two principal concerns listed, namely (i) huge number of predictive pixels with hundreds of spectral bands as dimensionality and (ii) noisy and redundant bands that may mislead the classification accuracy. When compared with a number of spectral bands due to less training sample instances, they have sceptical collision on the accuracy of supervised classifiers which is called as the Hughes effect. This paper is to study the result of reduction in dimensionality by selecting relevant bands and eliminating irrelevant and redundant ones by varied feature selection techniques. Once the bands are selected, they will be supplied to unlike classifiers namely support vector machine (SVM), Bayes and decision tree classifiers to examine the effect on classification accuracy and also estimate the decency of fit. In this regard, we employ two hyper-spectral image data sets, namely Indian Pines and Botswana, which are used. With a minimal spectral bands subset, it can achieve maximum classifier accuracy with the help of support vector machine-REF, joint mutual information (JMMI) and high-dimensional model representation (HDMR), and the feature selection methods are proposed.

Keywords Classifier accuracy · Dimensionality reduction · Feature selection · Hyper-spectral image · Land cover land use classification

---

K. K. Raju (✉)
Department of Computer Science and Engineering, Jawaharlal Nehru Technological University Kakinada, Kakinada, Andhra Pradesh, India
e-mail: kkishoreraju78@gmail.com

G. P. Saradhi Varma
Department of Computer Science and Engineering, KL Deemed to be University, Guntur, Andhra Pradesh, India
e-mail: drvarmagps@gmail.com

D. Rajyalakshmi
Department of Computer Science and Engineering, Jawaharlal Nehru Technological University Kakinada, University College of Engineering Narasaraopet, Narasaraopet, Andhra Pradesh, India
e-mail: drrajyalakshmid@gmail.com

P. S. R. Chowdary et al. (eds.), *Microelectronics, Electromagnetics and Telecommunications*, Lecture Notes in Electrical Engineering 655,
https://doi.org/10.1007/978-981-15-3828-5_33

1 Introduction

Nomenclature of hyper-spectral image data is the process of labelling pixels into multiple classes which has arriving emergent research concern in a wide diversity of applications [1–5]. Recently, hyper-spectral image taxonomy has become an out of the ordinary field of study in different image processing applications, especially in remote sensing data analysis [6–12]. Hyper-spectral image is a 3D data cube, which contains two-dimensional spatial information (image feature) and one-dimensional spectral information (spectral bands). Especially, the spectral bands contain values at very fine wavelengths from an independent pixel (x_i, y_j), while the image features such as land cover features and shape features disclose the disparity and association among adjacent pixels from different directions at a confident wavelength [13, 14]. Each instance (pixel) in a hyper-spectral image data set is a bundle of numerous spectral values (spectral bands) with different frequencies and is termed as high spectral resolution. Two major concerns make the hyper-spectral image classification very sensitive such as (i) huge number of spectral bands and predictive pixels and (ii) noisy and redundant bands [15]. In other cases, computational complexity of classification and its accuracy mainly depend on subset of bands and also subset of pixels for training.

Spectral band selection (i.e. feature selection) approach is the best option to make the classifier effective in terms of computational time without distressing the classifier. The main role of feature selection groundwork is to dig up the fewest possible subset of spectral bands from the actual set without forfeit the accuracy of taxonomy [16, 17]. Feature selection methods are basically categorized into (i) filter methods (called filters), (ii) wrapper methods (called wrappers), (iii) embedded methods and (iv) hybrid approach (filter and then wrapper). For any classifier, filter approaches are the well-established heuristic-based feature selection methods applied to assess the eligibility of features by considering each feature independently [9]. In this method, a highly relevant subset of bands for the specified class is selected based on rank derived from relevance score for each feature, computed by learning algorithm. Usually, highest ranked features are considered due to their contribution in classification accuracy. Obviously, low-ranked features (redundant or noisy features) are eliminated from data set [18]. Feature selection by means of filters is based on two issues: Feature relevancy and feature redundancy. Feature relevancy is a measure of predictability of respective class, whereas feature redundancy is the similarity between the most relevant features. Based on these two issues, filters are characterized into univariate-based feature selection filters in which feature scoring computation is based on feature relevancy and multivariate-based feature combination filters in which feature scoring is based on both relevancy (should be maximum) and redundancy (should be minimum).

On the other hand, wrapper approaches predict the feature subset by assessing the merit of feature using a classifier [19]. Filter approaches are computationally faster on high-dimensional data when compared to wrapper approaches, but they failed to handle interdependence between the spectral features. Wrappers outperform high classification accuracy than filters, but computationally expensive due to repeated execution of classifier [20]. In embedded methods, feature selection algorithm is embedded in predictive model [21–23] and it has precise learning algorithm for feature selection like wrappers. It utilizes the complete data set instead of splitting of data set into training and validation. For each individual feature or subset of features, retraining is not required for the predictor leads to faster decision-making. Due to faster execution on high-dimensional data, wrapper methods are replaced by embedded methods [24].

Hybrid methods are two-stage implementation feature selection procedure for further improvement in classification of high-dimensional data sets in terms of both computational speed and classification accuracy [25]. In stage one, suitable filter is implemented to extract most relevant subset of bands by eliminating irrelevant ones. In stage two, predictive model-specific wrapper method is implemented for further possible reduction in previous feature subset [26].

2 Literature Review

Pearson correlation coefficient (PCC) proposed by Guyon and Elisseeff [27] is a linear correlation between features with regard to specified target based on covariance and variance. Correlation value '−1' represents perfect negative correlation, whereas '+1' represents perfect positive correlation between the features. Feature and the target are independent when the correlation is '0'. This is sensitive to linear dependency between the features but insensitive to nonlinear dependencies. Fisher score (FSCR) and ANOVA are the filters proposed by Duda et al. [28] that extract the high score feature in a class relating to mean and variance. The main drawback with these techniques is redundant features that may appear due to suboptimal subset of features. Statistical-based chi-square (CHSQ) proposed by Guo et al. [2] measures the dependency of two features. Large chi-square-valued features are considered as maximum contributing features to predict the class. Information theoretic-based feature selection technique like information gain and Gini index, proposed by Sui [29] is based on entropy. Feature with higher IG is a member of feature subset for strong prediction capability. Due to univariant approach, correlation between features will not be considered. Relief (RELF) is a machine learning technique proposed by Kira and Rendell [30]. In this approach weight based feature selection is done using nearest hit and nearest miss. It is mainly introduced for two-class problems. Relief is failed in handling incomplete data sets with huge redundancy. To address these problems, Relief is extended to Relief-F in Kononenko et al. [31] to handle multiclass data sets with redundancy and noise. It is a multivariate algorithm in which features are assigned with rank based on their relevance. Minimum Information Maximization (MIM) proposed by Lewis [32] estimates the each feature's importance in class representation using Information Theory. To improve the feature selection penalty is calculated for the less correlated features in mutual information feature selection

(MIFS) by Battiti [33]. Based on joint random variable, a new feature is complementary to existing features in joint mutual information (JMI) by Yang et al. [7]. Maximum relevance minimum redundancy (mRMR) proposed by Peng et al. [34] integrates redundancy and relevance of each feature to find out the most relevant feature subset to improve classifier accuracy.

Wrapper models are having an advantage of a predefined classifier to evaluate the pre-eminence of features, and emblematic biases of the classifier are outwitting by the FS process. Wrapper techniques are broadly classified in Chandrashekar and Sahin [35] into sequential selection algorithms and heuristic search algorithms. Sequential forward selection algorithm (SFS) appends a feature into feature subset if it optimizes (maximizes) the objective function, and it is also done with sequential backward selection (SBS). But these techniques are failed in handling correlation and mutual effectiveness between the features. An extended adaptation of SS algorithm by including backtracking is called sequential floating forward selection (SFFS). Still, redundant features may exist in the final feature subset because of the nesting effect of SFS and SFFS. Instead of sequential search, heuristic methods generate the new subsets and then evaluate using objective function Chandrashekar and Sahin [35]. The Genetic Algorithm is proposed in Oh et al. [36], describes about arbitrary mutations subsequent to each crossover to find the optimal feature subset. In addition to this, GA also tunes the abstract parameters in support vector machine (SVM). Guyon and Elisseeff [27] proposed a linear SVM-based elimination of irreverent features recursively from the predictor training model as a pruning-based embedded model.

SLogReg is a sparse logistic regression proposed by Shevade and Keerthi [37] aimed at sparsely using Laplace prior. In this approach, model parameter tuning is expensive and will leads to univariate. BLogReg is Bayesian logistic regression proposed by Cawley and Talbot [38], to overcome the tuning problem by maximization of likelihood objective function of SLogReg. Still, BLogReg technique is insensitive to multiple class problems. Sparse multinomial logistic regression method (SBMLR) is an extended adaptation of BLogReg to handle multiclass problems and multinomial data sets. Like BLogReg initial stage (model selection) is not required in SBMLR and training is primed using component-wise training algorithm as discussed in Shevade and Keerthi [37].

Finally, the contributions of various authors on a variety of feature selection (FS) techniques are empirically evaluated based on the principle, parameters to handle, applicability, capabilities and performance requirements in FS. Sixteen algorithms listed in Table 1 are evaluated based on various accepts mentioned above. As shown in Table 1, among sixteen FS algorithms (i) only one is unsupervised and all other are supervised. (ii) Multivariate analysis will handle effectively six algorithms and all remaining are capable to handle only univariate analysis. (iii) Ten algorithms are capable to handle redundant features effectively in the process of FS.

Table 1 Evaluation of different FS techniques based on principle, applicability and other capabilities

S. No.	Feature selection technique	Proposed authors	Filter/wrapper/embedded	Learning type	Analysis type	Based on feature set (FS)/feature weighting (FW)	Can handle feature redundancy
1	BLogReg	Cawley and Talbot [38]	Embedded	Supervisor	Multivariate	FS	No
2	CFS	Hall and Smith [40]	Filter	Supervisor	Multivariate	FS	No
3	Chi-square	Tang and Liu et al. [24, 41]	Filter	Supervisor	Univariate	FW	Yes
4	FCBF	Tang and Liu et al. [24, 41]	Filter	Supervisor	Multivariate	FS	No
5	Fisher score	Duda et al. [28]	Filter	Supervisor	Univariate	FW	Yes
6	Gini index	Gini [42]	Filter	Supervisor	Univariate	FW	Yes
7	Information gain	Cover and Thomas [43]	Filter	Supervisor	Univariate	FW	Yes
8	Kruskal–Wallis	Wei [44]	Embedded	Supervisor	Univariate	FW	Yes
9	mRMR	Ding et al. [45]	Filter	Supervisor	Multivariate	FS	No
10	Relief	Tang et al. [24]	Filter	Supervisor	Univariate	FS	Yes
11	Relief-F	Tang et al. [24]	Filter	Supervisor	Univariate	FW	Yes
12	SBMLR	Cawley and Talbot [38]	Embedded	Supervisor	Multivariate	FS	No
13	*T*-test	Runger et al. [46]	Filter	Supervisor	Univariate	FW	Yes

(continued)

Table 1 (continued)

S. No.	Feature selection technique	Proposed authors	Filter/wrapper/embedded	Learning type	Analysis type	Based on feature set (FS)/feature weighting (FW)	Can handle feature redundancy
14	Spectrum	Liu and Motoda [41]	Filter	Unsupervised	Univariate	FW	Yes
15	SFFS	Chandrashekar and Sahin [35]	Wrapper	Supervised	Univariate	FS	No
16	GA–SVM	Oh et al. [36], Guyon and Elisseeff [27], Parmar et al. [47] and Raju et al. [48]	Wrapper	Supervised	Multivariate	FS	Yes

3 Band Selection in Hyper-spectral Imagery

Land use and land cover classification using hyper-spectral imagery is a sensitive task for the classifiers because of rich in dimensionality that includes spatial (more number of pixels per unit area) and spectral bands (more number of bands with small wavelengths in each pixel) information. Generally, data is represented in terms of three-dimensional hyper-cubes (3DHC) as shown in Fig. 1. Two dimensions $\{X, Y\}$ of the hyper-cube represent the spatial information in terms of latitude and longitude, and third dimension $\{\lambda\}$ represents the spectral information (i.e. spectral bands). Generally, multispectral images' maximum number of spectral bands is 10–15 with maximum band widths. But in case of hyper-spectral images, spectral information is stored with very small wavelength bands (called high spectral resolution). To predict the label of each unlabelled pixel, hundreds of bands information by the classifier are need to be considered. There are three foremost concerns. (i) Classifiers may not able to handle high-dimensional data. (ii) All the bands are not needed to predict the pixel class. (iii) Some bands may not carry unique information and some may noisy. The literature suggested some feature selection techniques to address the above three concerns. For the Given $\lambda = \{\lambda_1, \lambda_2, \lambda_3, \ldots, \lambda_m\}$ are the 'm' number of fine spectral bands for the set of instances (pixels) $P = \{p_1, p_2, \ldots, p_n\} \in R_{m \times n}$. Each instance is assimilate to any of the class labels of $C = \{C_1, C_2, C_3 \ldots, C_k\}$ which is a 'k' number of Class label set. The minimal subset i.e. $\lambda_s \subseteq \lambda$ that attains maximum classifier performance to represent target classification labels C by using particular classifier and classification metrics.

(a) *Chi-square (Chi2)*: In statistics, interdependency between two features and the episode of class can be measured by using chi-square as explained in '*Algorithm-1*'. For the given features λ and target set of classes C, chi-square χ^2 can be calculated using Eq. (1).

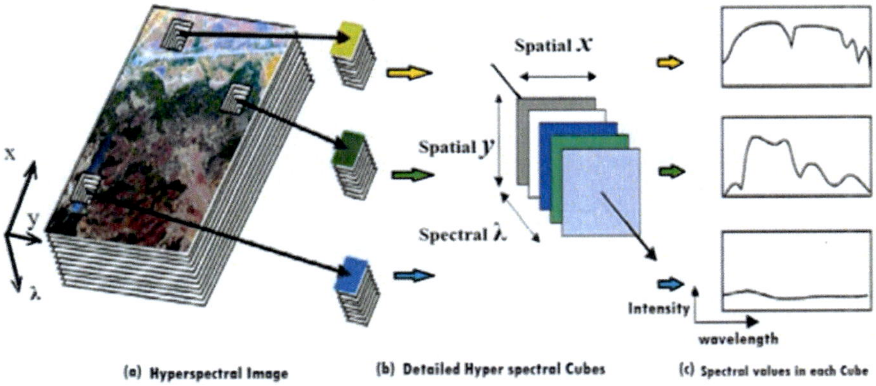

Fig. 1 Hyper-spatial and hyper-spectral cube representation of hyper-spectral image

$$\chi^2(\lambda, c) = \frac{N * (\gamma_{\lambda,c}\gamma'_{\lambda,c} - \gamma_\lambda\gamma_c)}{(\gamma_{\lambda,c} + \gamma_c)*(\gamma'_{\lambda,c} + \gamma_\lambda)*(\gamma_{\lambda,c} + \gamma_\lambda)*(\gamma'_{\lambda,c} + \gamma_c)} \tag{1}$$

where $\gamma_{\lambda,c}$ = the frequency of co-occurrence of feature λ and class c, $\gamma'_{\lambda,c}$ = the frequency of neither feature λ and class c occurs, γ_λ = frequency of only λ occurs without c and γ_c = frequency of only c occurs without λ

Algorithm-1: Chi-Square

```
Input :  λ – Feature set of size m
         C – Class labels set of size k
         S⊆I – sample Instances of s
           ρ – Threshold value
Output
         λ'⊆ λ
Set initially feature subset λs = {φ}
For each i = 1 to m do
Begin
 Measure γλ,c , γ'λ,c , γλ , γc
  χ2 (λ , c) = N*( γλ,c γ'λ,c − γλ γc) / (γλ,c + γc)*(γ'λ,c + γ□ )*( γλ,c + γλ )*(γ'λ,c + γc)
   if χ2 (λ , c) > ρ
   then λs {}← λi
end
```

Algorithm-2: Feature Ranking $\pi(C\backslash\lambda)$

```
Initialization: S = 0; C = set of a Class labels;
λ = Set of Features
For each ci ϵ C do:
   calculate π(c[i]);
   πc = S + π(c[i]) * log2(π(c[i]));
   S← πc
End for
For each λj ϵ λ do:
   calculate π(λ [j]);
   sum = S + π(λ [j]) * log2(π(λ [j]));
  S← sum
End For
For each ci ϵ C do:
   For each λj ϵ λ do:
     calculate π(c[i] λ [j]);
     πm = S + π(c[i] λ [j]) * log2(π(c[i] λ [j]));
    S← πm
End for
   π(C\ λ) = (−1) * Sum * (−1) πm
   IG(λ) = π(c) − π(C\ λ)
Return IG(λ)
```

(b) *Fisher score technique*: *F*-score technique selects high-scored features by evaluating the features independently and is basically derived from Laplacian score. Let $R_{\lambda \times n}$ is a feature data set with m features to be reduced to $R_{\lambda' \times n}$ high fisher score feature subset where $\lambda' \subseteq \lambda$. The features are picked by using principle that high coherence among the features in the identical class and low coupling among the features located in diverse classes. Fisher score for the given feature is measured by using Eq. (2):

$$F(X_i) = \frac{\sum_k^K n_j (\mu_{ij} - \mu_i)^2}{\sigma^2} \tag{2}$$

where $\sigma^2 = \sum_{k=1}^K n_j \sigma_{ij}^2$, μ_{ij} = mean of jth class related to ith feature, μ_i = mean of set of ith features and σ_{ij}^2 = variance of jth class related to ith features.

F-score technique needs labelled data to train and then will be applied to unlabelled data as it is an efficient supervised method. Each feature instance is independent of training and prediction phase that leads to inability in picking redundant features.

(c) *Information gain (IG)*: In this technique, information gain will be measured by computing class entropy before considering a sensitive feature that is represented as $\pi(c)$ and after considering feature, λ is denoted as $\pi(C\backslash\lambda)$ and finally IG is calculated by using Eq. (3).

$$\mathrm{IG}(\lambda) = \pi(c) - \pi(C\backslash\lambda) \tag{3}$$

IG measures the likeliness between the class labels and its respective features. Value of IG varies between 0 and 1. A feature is relevant to represent a class when it is having high IG value. Each feature in a given set IG value is calculated using '*Algorithm-2*' individually and selects top-*k* high gain value features that will be selected for effective classification. But, IG values will not help in handling redundant features.

(d) *Relief-F technique*: Relief-F is a comprehensive version of Relief because later one is not able to determine multiclass problems and incomplete data sets. In this, each feature (spectral band) is tagged with weights $W[\lambda]$ to approximate how well this feature make a distinction among instances that are neighbours and the detailed approach is explained in '*Algorithm-3*'. Relief-F considers both K-nearest hits (correlated bands) and nearest misses (distinct bands) and takes the average of both for all given sample instances to measure locality of assessment.

Algorithm-3: Relief-F Algorithm

set for all bands, weights W[λ] = 0.0
For each i = 1 to m do:
begin
  randomly select an instance I
  find *k* nearest hits Hj
end
For each class c ≠ class(I) do:
begin
  find *k* nearest misses M_j(c)
end
For each λ = 1 to #features do:
begin

$$W[\lambda] = W[\lambda] - \sum_{j=1}^{k} \frac{\mathrm{diff}(\lambda, I, H_j)}{m*k} + \sum_{c \neq \mathrm{class}(I)} \frac{\left[\frac{P(c)}{1-P(\mathrm{class}(I))} \sum_{j=1}^{k} \frac{\mathrm{diff}(\lambda, I, H_j)}{m*k}\right]}{m*k}$$

End

Algorithm-4: Forward Search in JMI Algorithm

Initialisation : set λ ← feature set of size m
S ← empty feature subset
For each $\lambda_i \in \lambda$ compute $I(C, \lambda_i)$
Find the max $I(C, \lambda_i)$
  Set λ ← $\lambda \backslash \{\lambda_i\}$
  Set S ← $\{\lambda_i\}$
Repeat until |S| = k
Choose the feature $\lambda_i = \underset{S=1,2,\ldots k}{Min}\ I(\lambda_i, \lambda_s, C)$
  Set λ ← $\lambda \backslash \{\lambda_i\}$
  Set S ← S ∪ $\{\lambda_i\}$
Return the set S

(e) *Maximal joint mutual information (mJMI)*: λ be the complete feature set and *S* be the effective feature subset. Let $\lambda_s \in \lambda - S$ and $\lambda_s \in S$ then the lowest joint mutual information value of candidate feature λ_s predicts the class label *C*, joined with each feature in *S*,

$$\text{Hence, } \underset{S=1,2,\ldots k}{\mathrm{Min}}\ I(\lambda_i, \lambda_s, C) \tag{4}$$

$$\text{where, } I(\lambda_i, \lambda_s, C) = I(\lambda_s, C) + I(\lambda_i, C/\lambda_s)$$

JMI algorithms shown in '*Algorithm-4*' are classifier-independent having high computation efficiency and also high scalability. Usually, the number of features increases, and both relevancy and redundancy will grow simultaneously. These algorithms are inefficient in handling interaction between the features that may weaken the identification of redundant features. Sometimes algorithm may select feature which is not unambiguously concern to target class.

(f) *Maximum relevance minimum redundancy (mRMR)*: Founded on the importance and similarity of the features in predicting the class of an instance maximum relevance minimum redundancy (mRMR) is an optimization technique that is explained in '*Algorithm-5*' shown below. Maximum relevance is measured in terms of maximum dependency. Compact superior features subset is extracted in two stages at very low cost. Each pixel in Hyperspectral image is having number of features. But it is not required to select all features to classfy those pixels. Because some features are not relevelt and some features are almost similar to decide the pixel class. So, in this sentence i am trying to estimate the importance of each feature by measuring maximal relavence and minimal redundancy. To increase class discriminative power with minimal features is denoted as min $R(\lambda)$ and is defined in Eq. (6).

$$\text{Max } D(\lambda, C), \ \ D = \frac{1}{|\lambda|} \sum_{\lambda_i \in \lambda} I(\lambda_i, C) \tag{5}$$

$$\text{Min } R(\lambda), \ \ R = \frac{1}{|\lambda|^2} \sum_{\lambda_i, \lambda_j \in \lambda} I(\lambda_i, \lambda_j) \tag{6}$$

Final objective function to achieve minimum redundancy and maximum relevance with regard to the above two limitations is defined as

$$\text{Max } \varphi(D, R), \quad \varphi = D - R \tag{7}$$

(g) *High-dimensional model representation (HDMR)*: Choosing the subset of meaningful distinct features from the given huge number of features to get high classification accuracy is a costly afire because of vast search space. High-dimensional model representation proposed by Taşkın et al. [39] is sensitivity analysis approach that is explained in '*Algorithm-6*', in which sensitivity score will be computed for each feature based on its impact on classifier. High-sensitivity score features will be selected as best features.

Algorithm-5: mRMR	**Algorithm-6: Sensitivity Score Calculation**

λ_s is a selected feature subset for the given $\boldsymbol{\lambda}$ feature set
Initially $\lambda_s = \{\varphi\}$
For each i = 1 to (m − 1) do:
begin
calculate the importance $D(\lambda_i, C)$ of the feature
calculate the similarity $R(\lambda_i, \lambda_{i+1})$ of ith and $(i + 1)^{th}$ features
End For
For each i = 1 to (m − 1) do:
$\lambda_i \leftarrow$ Max $\boldsymbol{\varphi(D, R)}$
$\lambda_s \leftarrow \lambda_i$
End For
Return λ_s
	Input
	Training pixels: $P_t = \{P_1, P_2, \ldots, P_r\}$, $R^n \in [0,1]^n$
	Class Labels: $C = \{c_1, c_2, \ldots, c_r\}$, $c_i \in [-1,+1]$
	Output
	Sensitivity Score: $S = \{s_1, s_2, s_3 \ldots s_n\}$
	Parameters
	Samples = M, k = Neighbours, p = polynomials
	Sobol sequence samples: $P_t' = \{P_1', P_2', \ldots, P_r'\}$
	C' = k-NN (P_t, C, P_t', k)
	$\Sigma_g \leftarrow$ Compute global variance
	$\Sigma_i \leftarrow$ Compute Individual variance
	$S_i \leftarrow \frac{\Sigma_i}{\Sigma_g}$

Feature selection by means of HDMR is implemented as, let $P_t = \{P_1, P_2, \ldots, P_r\}$ are the r training instances and their class tags are $C = \{c_1, c_2, \ldots, c_r\}$ where each instance lie in n-dimensional cube $P_i = \{P_{i1}, P_{i2}, \ldots, P_{in}\}$, $R^n \in [0, 1]^n$, and $c_i \in [-1, +1]$. To speed up the process auxiliary training sample $P'_t = \{P'_1, P'_2, \ldots, P'_r\}$ is using Sobol sequence used by Taşkın et al. [39] and class labels of new pixels are predicted using KNN-classifier.

Then, sensitivity score for each feature is computed using algorithm. HDMR feature selection approach is mainly designed for two-class problems labelled with [−1, +1]. By converting multiple class problems into sequence of two-class problems, we can apply this technique.

(h) *Support vector machine using recursive feature elimination (SVM-RFE) technique*: The major intention of SVM-RFE is to obtain non-redundant highly noteworthy features from the given feature set by eliminating irrelevant features using backward elimination approach and is explained in '*Algorithm-7*'. Overfitting can be avoided when the data set is having high dimensionality. The outcome behaviour of the technique may deviate when there is high correlation between features.

```
Algorithm-7: Backward Elimination in SVM-RFE
Initialize:
R := {Training DAtaset}
λ := {All input features}
f := filter_out_factor
s := size of the final informative band feature subset
Begin
While the size of λ > s do:
Train the SVM on R in the feature space defined by F
Rank the each feature in λ by w_i^2 in the descending order
If f < 0
            λ2 := λ – {f low ranked features in λ}
elseif
      λ2 := λ – {λ2 size is closest smaller number
                 of power of 2 by removing highest
                 ranked features}
else
   λ2 := λ – {λ * 100% of features in λ with highest rank}
end
if the size of λ2 < s
   adjust λ2 to be composed of s top ranked features in λ
end
     λ← λ2
end
return λ
end
```

4 Experimental Set-Up

Performance evaluation of different feature selection techniques discussed in previous section is done on multiclass high-dimensional hyper-spectral data sets. In our experiment, two hyper-spectral remote sensing image data sets, namely Indian Pines shown in Fig. 2 and Botswana shown in Fig. 4 are considered. AVIRIS sensor gath-

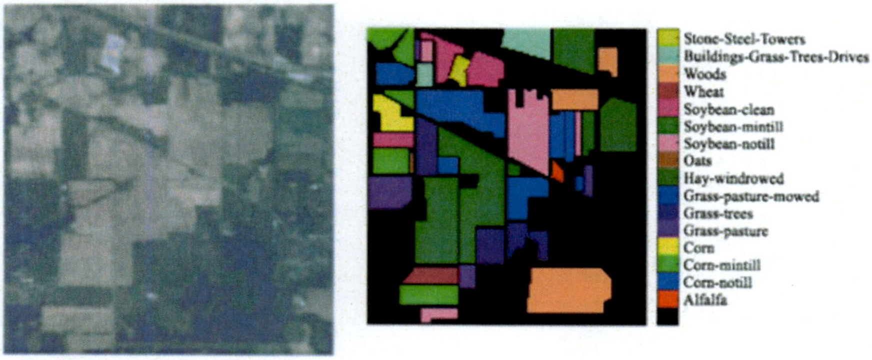

Fig. 2 Indian Pines hyper-spectral image data set and ground truth with 15 classes

ered this Indian Pines sight from North-Western Indiana with [145 × 145] pixels and 224 spectral bands. It contains 15 assorted land cover classes like different agriculture crops, vegetation, housing and other built structures (see Fig. 2). It also contains the ground reality of the same sight which shows all the 15 classes represented by different colures, respectively. Each land cover object is having spectral reflectance that is captured in hyper-spectral image. After pre-processing of hyper-spectral image, out of 224 spectral reflectance bands, 200 are included in the data set and Alfalfa (Class-1) and Hay-Windrowed (Class-15) spectral bands details (see Fig. 3).

Botswana scene is captured by NASAEO-1 satellite over the Okavango Delta. This is captured with 242 features (spectral bands) and 30 m pixel resolution. After rectification of noisy bands, 145 bands are considered as features to represent 14 varieties of land cover classes like water bodies, vegetation, different types of lands, etc. (see Fig. 4). It also contains the ground reality of the same area which shows all the 14 classes represented by different colures, respectively. 145 spectral band values of Water (Class-1) and also Exposes soils (Class-14) in Botswana dataset (see Fig. 5).

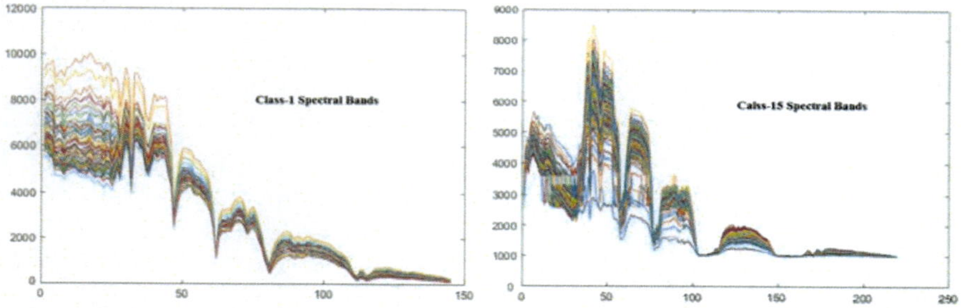

Fig. 3 Graph representing spectral value on Y-axis and number of spectral features on X-axis (224 spectral band values of the classes 1 and 15 in Indian Pines data set)

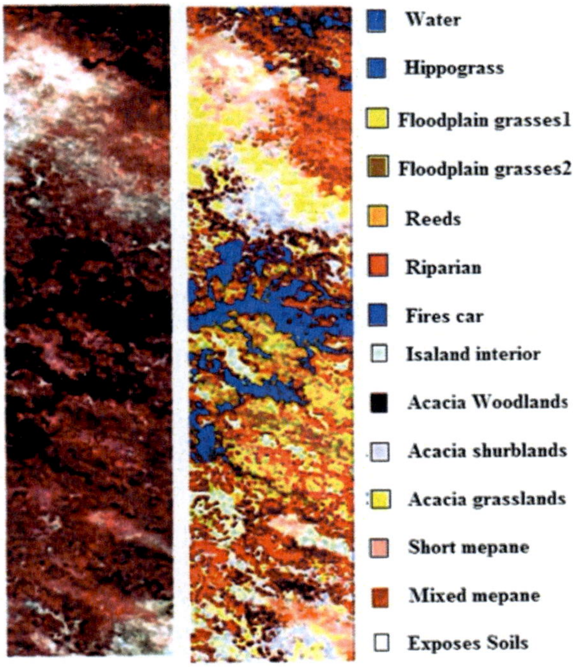

Fig. 4 Botswana hyper-spectral image data set and ground truth with 14 classes

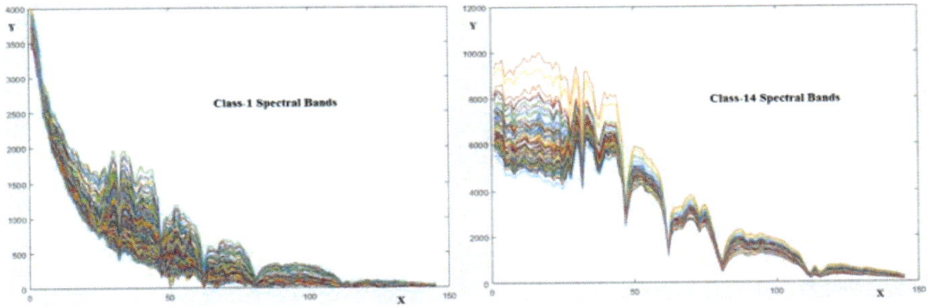

Fig. 5 Graph representing spectral value on *Y*-axis and number of spectral features on *X*-axis (145 spectral feature values of classes 1 and 14 in Botswana data set)

To assess the feature selection algorithms (i.e. chi-square (Chi2), fisher score (*F*-score), information gain (IG), support vector machine-based reverse feature elimination (SVM-RFE), Relief-F, mRMR, JMMI and HDMR using Sobol sequence), experiments are performed on both the data sets with respect to their ground truth. As a part of design of experiment, all the band values are transformed into {0, 1} to avoid larger values for the computational flexibility. Training, validation and testing experiments are designed by taking data from the two data sets randomly. We investigated the selected bands after band selection and its consequence on classifier accuracy. To evaluate the feature selection methods on two hyper-spectral data sets, two main criteria are considered: classification accuracy and the feature selection stability suggested by Taşkın et al. [39]. Major intention of this schoolwork is to analyse the effect of bands and band selection methods on different classifiers, when they work at default parameters. Improvement of classifier efficiency is out of prospect of this study. To carry out classification accuracy study, three classifiers are used: linear support vector machine (L-SVM) classifier, Naive Bayes classifier and decision tree classifier. To study the stability of methods in band selection, top 20% high-ranked bands out of 50 features selected by feature selection method is considered. If the feature selection method calculates same rank for very similar features in top 20% high-ranked features, then the feature selection is highly stable. If not, relevant features may sometimes be ignored by feature selection method.

5 Results and Discussion

In our experiments, each data set is separated into ten disjoint sub sets (i.e. 10-fold cross-validation) in view of sinking the training and testing times by keeping maximum accuracy. Out of 10 subsets of data set, one set is for testing and remaining sets are for training and will be continued for all subsets. Finally, all the ten testing and training outputs are averaged. The purpose of selecting different classifiers is to investigate how these are prejudiced by diverse feature selection methods. At what subset of bands out of full band set, classifier reaches to its maximum accuracy (see Fig. 6a, b) visualizing the learning curves of respective classifiers and data sets which represents how the classifiers are learning the fact from the bands supplied by different band selection algorithms that we deduce in our study. The effect of bands on Indian Pines data set visualized in Fig. 6a depicts that the three classifiers are at its maximum accuracy when all bands are considered.

In all the three graphs shown in Fig. 6a indicating that up to 140 features, there is an increment or some variation in classifier accuracy, and after that, there is no considerable increment in accuracy. In case of selected bands, both linear SVM and decision tree classifiers improve their accuracy when the feature selection method supplies more bands, but in the case of Bayes classifier, other than HDMR and SVM, remaining are having negative effect. On the other hand, learning curves of three classifiers using different band selection methods on Botswana data set visualized in Fig. 6b depicts that the three classifiers are at its maximum accuracy when all 145 bands are considered. In case of selected bands, both linear SVM and decision tree classifiers improve their accuracy when the band selection method supplies more bands, but in case of Bayes classifier, other than HDMR, SVM and JMI, remaining are having negative effect. Accuracy against number of bands selected by the FS-methods for the three classifiers (see Fig. 6b) indicates that up to 100 features

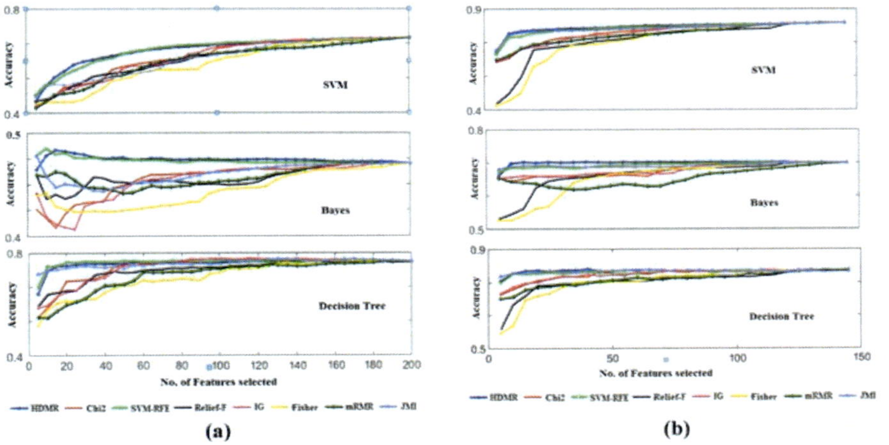

Fig. 6 Effect of band selection by applying seven band selection techniques on three classifiers' accuracy using **a** Indian Pines data set and **b** Botswana data set

there is an increment or some variation in classifier accuracy, after that, there is no considerable increment in accuracy. With the help of learning curve 'goodness-of-fit' can be estimated by taking area below the curve at maximum classifier accuracy.

With the help of bar chart (see Fig. 7a, b), 'goodness-of-fit' is depicted for each classifier using different feature selection techniques on two data sets, respectively. The height of the bar is inversely proportional to the 'goodness-of-fit'. In Indian Pines data set, both HDMR and linear SVM are having smaller height rectangles. In case of Botswana data set along with HDMR and SVM, JMI is also having almost same 'goodness-of-fit'.

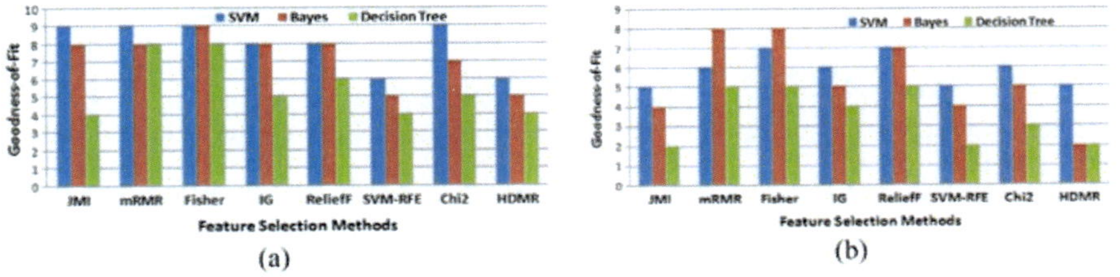

Fig. 7 'Goodness-of-fit' of classifiers when feature selection techniques are applied on **a** Indian Pines data set and **b** Botswana data set

6 Conclusion

This review paper covers land cover land use (LCLU) categorization of hyper-spectral image that actually depends on two issues namely dimensionality reduction by eliminating the redundant and noisy bands, as compared with number of spectral bands have pessimistic impact on the accuracy of supervised classifiers namely the Hughes effect. To observe the effect of reduction in dimensionality by selecting pertinent bands and eliminating non-correlated and superfluous ones using a mixture of feature selection methods like chi-square, information gain, fisher score, SVM-RFE, Relief-F, mRMR, JMMI and HDMR. Bands are selected will be supplied to different classifiers like SVM, Bayes and decision tree classifiers to examine the effect on classification accuracy and also estimate the goodness-of-fit. Two hyper-spectral image data sets like Indian Pines and Botswana are used. Finally, experimentation has been done to extract minimal spectral bands subset that achieved maximum classifier accuracy in this task, and SVM-REF, JMMI and HDMR are better feature selection methods.

References

1. Landgrebe D (2002) Hyper-spectral image data analysis. IEEE Signal Process Mag 19(1):17–28
2. Guo B, Gunn S, Damper R, Nelson J (2008) Customizing kernel functions for SVM-based hyper-spectral image classification. IEEE Trans Image Process 17(4):622–629
3. Jiao H, Zhong Y, Zhang L, Li P (2011) Unsupervised remote sensing image classification using an artificial DNA computing. In: Proceedings of the international conference on computing, networking and communications, July 2011, vol 3, pp 1341–1345
4. Leng Q, Yang H, Jiang J (2019) Label noise cleansing with sparse graph for hyperspectral image classification, MDPI. Remote Sens 11:1116. https://doi.org/10.3390/rs11091116
5. Liu T, Zhang L, Li P et al (2012) Remotely sensed image retrieval based on region-level semantic mining. J Image Video Process. https://doi.org/10.1186/1687-5281-2012-4
6. Zhu L, Chen Y, Ghamisi P, Benediktsson JA (2018) Generative adversarial networks for hyperspectral image classification. IEEE Trans Geosci Remote Sens 56:5046–5063
7. Yang L, Yang S, Jin P, Zhang R (2014) Semi-supervised hyper-spectral image classification using spatio-spectral Laplacian support vector machine. IEEE Geosci Remote Sens Lett 11(3):651–655
8. Yang S, Qiao Y, Yang L, Jin P, Jiao L (2014) Hyper-spectral image classification based on relaxed clustering assumption and spatial Laplace regularizer. IEEE Geosci Remote Sens Lett 11(5):901–905
9. Yang CSL, Chuang L, Ke CH, Yang CH (2008) A hybrid feature selection method for microarray classification. IAENG Int J Comput Sci 35(3)
10. Zhang X, Pan Z, Lu X, Hu B, Zheng X (2018) Hyperspectral image classification based on joint spectrum of spatial space and spectral space. Multimed Tools Appl 77(22):29759–29777
11. Jiang J, Ma J, Chen C, Wang Z, Cai Z, Wang L (2018) SuperPCA: a superpixelwise PCA approach for unsupervised feature extraction of hyperspectral imagery. IEEE Trans Geosci Remote Sens 56:4581–4593
12. Medjahed SA, Saadi TA, Benyettou A, Ouali M (2016) A new post-classification and band selection frameworks for hyperspectral image classification. Egypt J Remote Sens Space Sci 19:163–173

13. Dai Q, Cheng J-H, Sun D-W, Zeng X-A (2015) Advances in feature selection methods for hyper-spectral image processing in food industry applications: a review. Crit Rev Food Sci Nutr 55(10):1368–1382. https://doi.org/10.1080/10408398.2013.871692
14. ElMasry G, Sun D-W, Allen P (2012) Near-infrared hyper-spectral imaging for predicting colour, pH and tenderness of fresh beef. J Food Eng 110(1):127–140
15. Medjahed SA, Ouali M (2018) Selection based on optimization approach for hyper-spectral image classification. Egypt J Remote Sens Space Sci
16. Vaddi R, Prabukumar M (2018) Comparative study of feature extraction techniques for hyper spectral remote sensing image classification: a survey. In: International conference on intelligent computing and control systems (ICICCS), vol 11. IEEE. https://doi.org/10.1109/iccons.2017.8250521
17. Jiang SY, Wang LX (2016) Efficient feature selection based on correlation measure between continuous and discrete features. Inf Process Lett 116(2):203–2015
18. Lazar C, Taminau J, Meganck S, Steenhoff D, Coletta A, Molter C, Nowe A (2012) A survey on filter techniques for feature selection in gene expression microarray analysis. IEEE/ACM Trans Comput Biol Bioinform (TCBB) 9(4):1106–1119
19. Saeys Y, Inza I, Pedro L (2007) A review of feature selection techniques in bioinformatics. Bioinform Adv 23(13):2507–2517
20. Lin P, Thapa N, Omer I, Zhang J (2011) Feature selection: a pre-process for data perturbation. IAENG Int J Comput Sci 38(2):168–175
21. Qi M, Fu Z, Chen F (2016) Research on a feature selection method based on median impact value for modeling in thermal power plants. Appl Therm Eng 94:472–477
22. Kursa MB (2016) Embedded all relevant feature selection with random ferns. arXiv preprint arXiv:1604.06133
23. Ma S, Huang J (2008) Penalized feature selection and classification in bioinformatics. Brief Bioinform 9(5):392–403
24. Tang J, Alelyani S, Liu H (2014) Feature selection for classification: a review. In: Data classification: algorithms and applications, p 37
25. Silvestre C, Cardoso MG, Figueiredo M (2015) Feature selection for clustering categorical data with an embedded modelling approach. Expert Syst 32(3):444–453
26. Bermejo P, de la Ossa L, Gámez JA, Puerta JM (2012) Fast wrapper feature subset selection in high-dimensional datasets by means of filter re-ranking. Knowl-Based Syst 25(1):35–44
27. Guyon I, Elisseeff A (2003) An introduction to variable and feature selection. J Mach Learn Res 3:1157–1182
28. Duda OR, Hart EP, Stork GD (2012) Pattern classification
29. Sui B (2013) Information gain feature selection based on feature interactions. Doctoral dissertation, University of Houston
30. Kira K, Rendell LA (1992) The feature selection problem: traditional methods and a new algorithm. AAAI 2:129–134
31. Kononenko I, Šimec E, Robnik-Šikonja M (1997) Overcoming the myopia of inductive learning algorithms with RELIEFF. Appl Intell 7(1):39–55
32. Lewis DD (1992) Feature selection and feature extraction for text categorization. In: Proceedings of the workshop on speech and natural language. Association for Computational Linguistics, Morristown, NJ, USA, pp 2012–2017
33. Battiti R (1994) Using mutual information for selecting features in supervised neural net learning. IEEE Trans Neural Netw 5(4):537–550
34. Peng H, Long F, Ding C (2005) Feature selection based on mutual information: criteria of max dependency, max-relevance, and min-redundancy. IEEE Trans Pattern Anal Mach Intell 27(8):1226–1238
35. Chandrashekar G, Sahin F (2014) A survey on feature selection methods. Comput Electr Eng 40:16–28
36. Oh I, Lee J, Moon B (2004) Hybrid genetic algorithms for feature selection. IEEE Trans Pattern Anal Mach Intell 26(11):1424–1437

37. Shevade SK, Keerthi SS (2003) A simple and efficient algorithm for gene selection using sparse logistic regression. Bioinformatics 19:2246–2253
38. Cawley GC, Talbot NLC (2006) Gene selection in cancer classification using sparse logistic regression with bayesian regularization. Bioinformatics 22(19):2348–2355
39. Taşkın G, Kaya H, Bruzzone L (2016) Feature selection based on high dimensional model representation for hyper-spectral images. IEEE Trans Image Process 1057–7149 (c). https://doi.org/10.1109/tip.2017.2687128
40. Hall MA, Smith LA (1999) Feature selection for machine learning: comparing a correlation based filter approach to the wrapper. In: Proceedings of the twelfth international Florida artificial intelligence research society conference, pp 235–239. ISBN: 1-57735-080-4
41. Liu H, Motoda H (2008) Computational methods of feature selection. Chapman & Hall
42. Gini C (1912) Variabilitae mutabilita. In: Memori di metodologia statistica
43. Cover TM, Thomas JA (1991) Elements of information theory. Wiley
44. Wei LJ (1981) Asymptotic conservativeness and efficiency of Kruskal-Wallis test for k dependent samples. J Am Stat Assoc 76(376):1006–1009
45. Ding F, Peng C, Long H (2015) Feature selection based on mutual information: criteria of max-dependency, max-relevance, and min-redundancy. IEEE Trans Pattern Anal Mach Intell 27(8):1226–1238
46. Runger GC, Montgomery DC, Hubele NF (2007) Engineering statistics. Wiley, Hoboken, NJ
47. Parmar C, Grossmann P, Bussink J, Lambin P, Aerts HJ (2015) Machine learning methods for quantitative radiomic biomarkers. Sci Rep 5:13087. https://doi.org/10.1038/srep13087
48. Raju KK, Varma GPS, Rajyalakshmi D, Alluri S (2017) An effective semi supervised classification of hyper spectral remote sensing images with spatially neighbour hoods. J Adv Res Dyn Control Syst 15:701–711

Study of Changes in Land Use and Land Covers Using Temporal Landsat-8 Images

Parminder Kaur Birdi and Varsha Ajith

Abstract Changes in land use and land cover play an important role in understanding the interactions of human activities with the environment. This makes it necessary to monitor and detect the changes happening over a period of time to maintain a sustainable environment. In this paper, an attempt has been made to study the changes in land use and land cover of Aurangabad district in Maharashtra, India. The study is carried out using multitemporal satellite images from Landsat-8 sensors for the period from 2013 to 2019. The results show changes occurring in different covers including green vegetation cover, settlement, bare land and water bodies. Different classification algorithms are applied to observe and record changes happening in different land covers. The classified maps provide information which can be used by district authorities and other government agencies to take future decisions for development and sustaining environment.

Keywords Land use · Land cover · Change detection · Landsat-8 · Remote sensing

1 Introduction

In the last few decades, uncontrolled sprawl of urbanization has posed a lot of challenges for sustaining environment. In fact, this is a global environmental issue which needs to be discussed to find suitable solutions. To address this issue, remote sensing can be used to make a decision support tool for evaluating land use changes and provide an insight to find solutions for proper planning. Large amount of remote sensed data is generated periodically by satellites covering entire earth surface, thus providing images which can be used to compute changes happening over various land covers. Images captured by sensors of the same earth region over different time

P. K. Birdi (✉) · V. Ajith
MGM's Jawaharlal Nehru Engineering College, N-6, CIDCO, Aurangabad, Maharashtra 431003, India
e-mail: dhingra.param@gmail.com

V. Ajith
e-mail: varsha469@gmail.com

P. S. R. Chowdary et al. (eds.), *Microelectronics, Electromagnetics and Telecommunications*, Lecture Notes in Electrical Engineering 655,
https://doi.org/10.1007/978-981-15-3828-5_34

periods are helpful in change detection. Temporal images covering same area can be downloaded from various agencies like US Geological Survey (USGS) Earth Explorer [1]. Most widely used remote sensed classification algorithms like Maximum Likelihood Classifier (MLC), Decision Trees (DT) and Neural Networks (NN) have been applied in applications of land use and land cover (LULC) identification and change detection [2].

Aung et al. have presented an analysis of changes in land cover of a township in Myanmar using Landsat-7 and 8 images. Random Forest classification method is used to detect urban sprawl. As per the results reported a large area is detected to be urbanized [3]. Radhika et al. have proposed a classification technique based on Neural Network for change detection. Ensemble-based classifier, i.e. results from different classifiers, is used for assigning class to every pixel during classification process. Support Vector Machine (SVM), Maximum Likelihood and K-Nearest Neighbour Classifiers are used. For test image, an accuracy of 91% is attained [4]. Grivei et al. have presented a SVM-based learning technique for change detection by extracting evolution classes from satellite image time series data. The experiments are performed on Landsat-4 and 5 images [5]. Adam et al. have assessed accuracy of different classification algorithms for LULC classification. ASTER remote sensed images are used for experimentation. The higher classification accuracy of 72.92% was recorded using MLC algorithm [6]. Another study carried by Birdi et al. used MLC, Parallelepiped Classifier, Minimum Distance classifier, Spectral Angle Mapper and Neural Network Method to apply classification algorithm to fused images. The data set used is from Landsat-8 sensor with spatial resolution on 30 m. Higher overall accuracies were recorded using MLC method [7, 8].

From all the literature studied related to this work, it has been found that for effective change detection, supervised classification methods are more beneficial. In this paper, remote sensed images from Landsat-8 sensor have been acquired for a period of six years from 2013 to 2019 and classified using Maximum Likelihood Classifier and Neural Network to measure the changes occurred in different land covers of the study area. Data is provided as output in the format of land area used in the form of maps and tables where quantitative data is present to represent the change detection. This study evaluates the rate of land use change from 2013 to 2019. It also computes the regions of rapid change and magnitude of change for assessing the past and present condition of land covers which help to understand the dynamics and trend of change.

The paper is organized with details of study area and data set used in Sects. 2 and 3. Section 4 describes methodology used and the results and discussions are explained in Sect. 5 and final conclusion is drawn in Sect. 6.

Table 1 Landsat-8 OLI sensor data set characteristics

Mode	Band	Spectral region	Spectral resolution (μm)	Spatial resolution (m)
Multispectral	Band 1	Visible	0.43–0.45	30
	Band 2	Visible	0.450–0.51	
	Band 3	Visible	0.53–0.59	
	Band 4	Red	0.64–0.67	
	Band 5	Near-infrared (NIR)	0.85–0.88	
	Band 6	SWIR 1	1.57–1.65	
	Band 7	SWIR 2	2.11–2.29	

2 Study Area

The study area is part of Aurangabad district from Maharashtra, India, located at 19.8762° N, 75.3433° E covering an area of approximately 139 km^2. The population of the area is 11.8 lakhs as per census carried out in 2011. The study area covers a variety of land covers including vegetation, settlement, trees, water bodies and bare land.

3 Data Set Used

Satellite sensors are used to collect remote sensed data with different spectral, spatial, radiometric and temporal resolutions. Remote sensed images are downloaded from US Geological Survey (USGS), captured by Landsat-8 Operational Land Image (OLI) sensor. The OLI sensor captures eight multispectral band images at 30 m spatial resolution and one panchromatic band image at 15 m spatial resolution. The sensor revisit time is 16 days. The swath width is 185 km. Table 1 gives details of Landsat-8 sensor data set [9]. Figure 1 shows four images of different dates (a: 11 April 2013, b: 5 May 2016, c: 8 May 2017, d: 28 April 2019) used for experiments. A spatial subset of the image is cropped from the data set.

4 Methodology Used for Experiments

Images are pre-processed to remove atmospheric and radiometric errors before applying classification method. Classified images are used for computing changes occured in different land covers present in the study area. For experiments, layer-stacked images are prepared using band 1 to band 7. Landsat-8 images are downloaded in L1T product format which is radiometrically and geometrically corrected. Images are pre-processed, and DN values are rescaled to reflectance values by applying radiometric calibration in ENVI software [9].

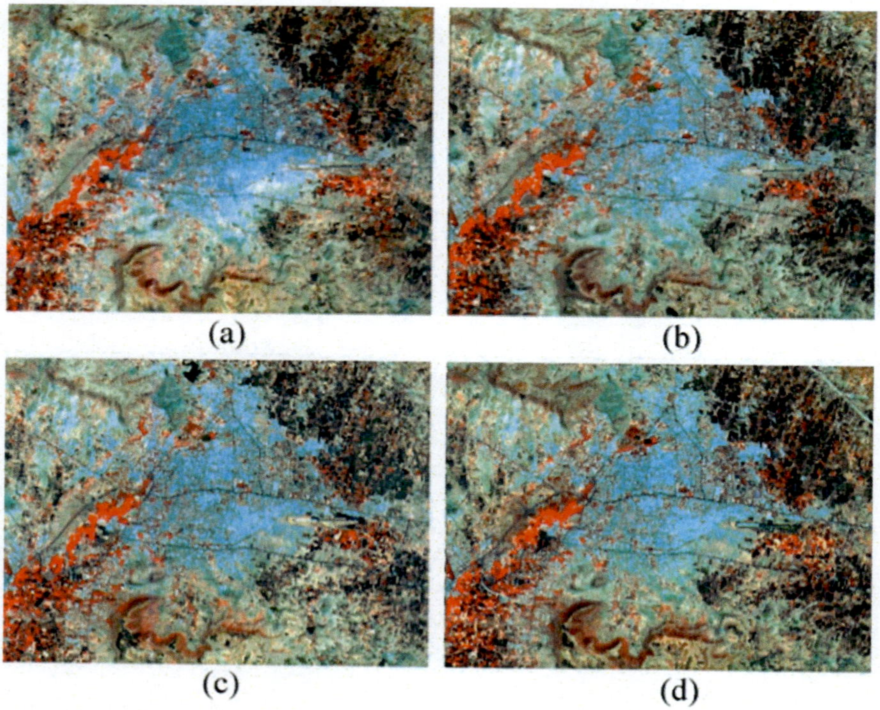

Fig. 1 Multispectral Landsat-8 images of different dates used for experiments

4.1 Methodology

To make the change analysis of the study area, QGIS and ENVI software are used. Details of the area are collected from various resources including maps from government office. Map of the study area is acquired and digitized using QGIS software. Figure 2 shows the proposed methodology for image classification for Landsat-8 images.

4.2 Field Work

The study site is visited for field data collection, and GPS-enabled smartphone is used to record coordinates of different land covers. The training and test data set is created using the field data where 60% data is used for training the classifier and remaining 40% is used for testing purpose.

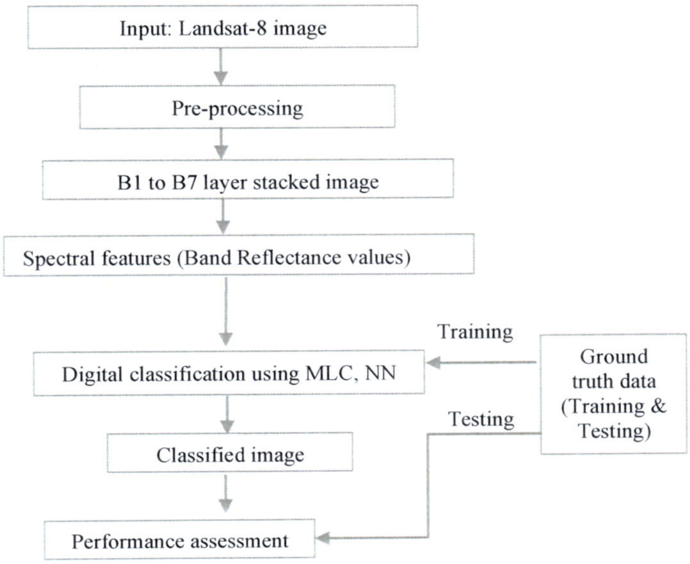

Fig. 2 Proposed methodology for image classification using Landsat-8 images

4.3 Image Classification

The proposed methodology takes Landsat-8 image as input. The image is pre-processed, and layer-stacked images are created using band 1 to band 7. The image classification methods make use of spectral features for classification. Supervised classification methods, Maximum Likelihood Classifier (MLC) and Neural Network (NN) are applied on the layer-stacked image [10]. MLC is a parametric supervised classification method to compute probability value for every pixel. The pixel is assigned the class having highest probability value. Second method used is NN, a nonparametric classifier with layered feed-forward architecture. For learning, back-propagation learning algorithm is applied. The architecture is finalized after evaluating the performance for different parameters including number of hidden layers, initialization of weight values, learning rate and error value. The number of hidden layers is changed from two to ten, and the results are recorded. For this study area, the images are classified into five classes, i.e. water body, vegetation, barren land, settlement and unclassified area. The classified images generated using MLC and NN classifier for different dates: (a) 11 April 2013, (b) 5 May 2016, (c) 8 May 2017, (d) 28 April 2019 are as shown in Figs. 3 and 4, respectively.

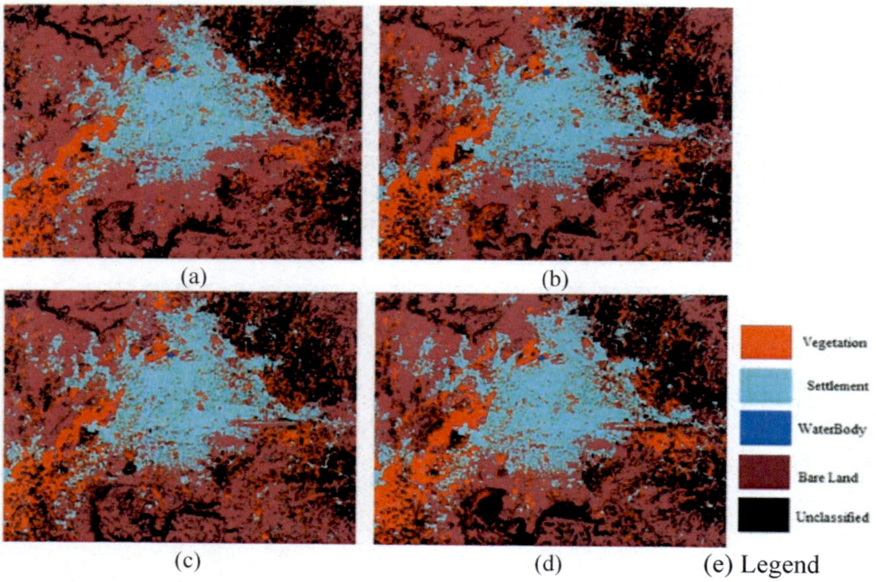

Fig. 3 MLC classified multispectral Landsat-8 images

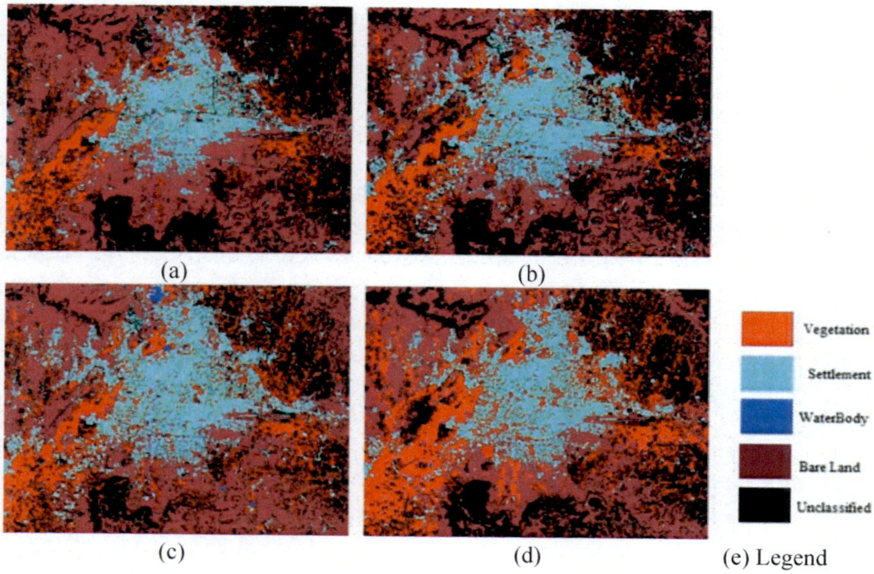

Fig. 4 NN classified multispectral Landsat-8 images

5 Results and Discussion

The accuracy of classification methods is measured using confusion matrix, which represents correspondence between field/ground truth values and predicted values. It presents a summary of predicted values produced by the classification method. Using confusion matrix, overall accuracy (OA) and Cohen's kappa coefficient are computed. Kappa is the amount of agreement between predicted and field values, 0 indicating no agreement and 1 indicating complete agreement [11]. Kappa value is computed for performance comparison of two classifiers using Eq. (1).

$$k = \frac{N\sum_{i=1}^{n} m_{ii} - \sum_{i=1}^{n}(G_iC_i)}{N^2 - \sum_{i=1}^{n}(G_iC_i)} \tag{1}$$

where i represents class number, N represents total number of predicted values, m_{ii} is overall accuracy value, C_i is total number of predicted values belonging to ith class and G_i is total number of ground truth values belonging to class i [12, 13]. Commission Error (CE) and Omission Error (OE) are also calculated. Commission Error is defined as fraction of predicted values assigned to a class, which actually do not belong to that class. Omission Error is fraction of ground truth values omitted from the actual class during classification process. User's accuracy (UA) and producer's accuracy (PA) for each class are also computed. UA is the probability that a randomly chosen point from the classified image belongs to the same class as present on ground truth image. PA is the probability that a randomly chosen point from the reference image has the same predicted class in the classified image. Tables 2 and 3 show

Table 2 OA, kappa, UA, PA, CE and OE values for classified images using MLC

	Apr-2013		May-2016		May-2017		Apr-2019	
	PA %	UA %	PA %	UA %	PA %	UA %	PA %	UA %
Settlement	94.16	91.84	95.14	89.56	96.11	90.15	95.72	87.86
Vegetation	99.67	99.52	91.93	99.18	91.26	99.02	88.44	98.3
Water body	100	100	100	100	100	100	97.92	98.95
Bare land	90.58	88.36	90.98	86.84	86.6	84.26	87.4	78.27
Unclassified	91.38	95.5	89.53	81.05	89.16	78.61	90.02	82.23
OA (%)	95.83		91.87		90.81		89.64	
Kappa	0.938		0.88		0.866		0.85	
	CE %	OE %	CE %	OE %	CE %	OE %	CE %	OE %
Settlement	8.16	5.84	10.44	4.86	9.85	3.89	12.14	4.28
Vegetation	0.48	0.33	0.82	8.07	0.98	8.74	1.7	11.56
Water body	0	0	0	0	0	0	1.05	2.08
Bare land	11.64	9.42	13.16	9.02	15.74	13.4	21.73	12.6
Unclassified	4.5	8.62	18.95	10.47	21.39	10.84	17.77	9.98

Table 3 OA, kappa, UA, PA, CE and OE values for classified images using NN

	Apr-2013		May-2016		May-2017		Apr-2019	
	PA %	UA %	PA %	UA %	PA %	UA %	PA %	UA %
Settlement	93.39	93.75	95.14	91.74	95.53	88.31	92.02	92.38
Vegetation	99.67	99.81	92.74	98.58	94.36	97.34	92.98	94.24
Water body	100	100	100	100	100	100	86.46	93.26
Bare land	88.99	90.07	85.94	90.88	89.26	83.5	85.94	84.71
Unclassified	94.21	92.62	91.01	77.14	83.25	86.45	87.56	84.84
OA (%)	95.99%		91.66%		91.61%		90.45%	
Kappa	0.94		0.878		0.877		0.86	
	CE %	OE %	CE %	OE %	CE %	OE %	CE %	OE %
Settlement	6.25	6.61	8.26	4.86	11.69	4.47	7.62	7.98
Vegetation	0.19	0.33	1.42	7.26	2.66	5.64	5.76	7.02
Water body	0	0	0	0	0	0	6.74	13.54
Bare land	9.93	11.01	9.12	14.06	16.5	10.74	15.29	14.06
Unclassified	7.38	5.79	22.86	8.99	13.55	16.75	15.16	12.44

OA, kappa, UA, PA values and CE and OE recorded by applying MLC and NN to temporal Landsat-8 images.

Overall classification accuracy values are recorded to be more than equal to 89% for all four images, indicating that classification is carried out with higher correction rate. This is supported by the kappa coefficient values of 0.85 and higher. The PA for settlement class from Apr-19 image shows a value of 95.72%, whereas UA is 87.86%, indicating that even though 95.72% of reference settlement area is correctly identified as settlement, only 87.86% of that area identified as "Settlement" in the classification was actually settlement. Figure 5 shows the plot of the kappa coefficient recorded using MLC and NN classifiers, indicating that the accuracies are higher using NN method.

The main objective of this study was to record changes occurred in different land covers for a period from April 2013 till May 2019 for the identified study site. From classification results, the area classified under different land covers is recorded as shown in Table 4 using two classifiers used, i.e. MLC and NN, respectively.

Fig. 5 Kappa coefficient plot for classified images using MLC and NN

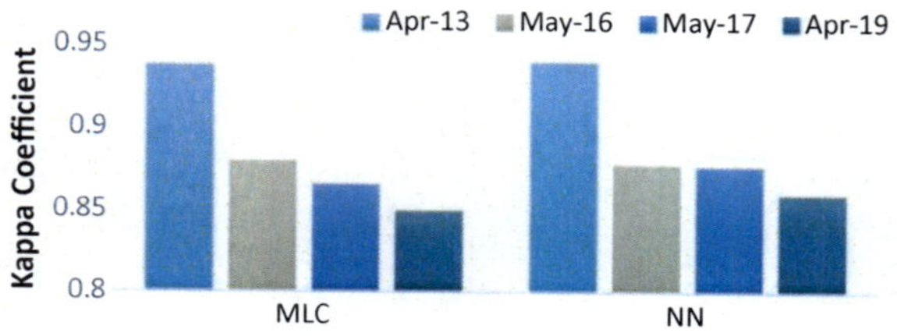

Table 4 Area classified under different classes using MLC and NN classifier

Date of imagery	Water	Settlement	Vegetation	Bare land	Unclassified
Area classified under different classes (in %) using MLC					
Apr-2013	0.039	17.95	12.76	44.51	24.74
May-2016	0.039	18.79	15.83	39.86	25.48
May-2017	0.038	19.28	15.97	38.26	26.54
Apr-2019	0.041	19.98	18.38	35.10	26.56
Area classified under different classes (in %) using NN					
Apr-2013	0.078	13.33	12.51	37.51	36.61
May-2016	0.27	17.81	16.2	29.44	36.34
May-2017	0.311	20.01	19.36	28.74	29.58
Apr-2019	0.062	20.19	22.34	28.91	28.51

From Table 4 showing area covered, it is clearly visible that vegetation land cover has increased by approximately more than 7%. Settlement land cover also is on the rise, whereas bare land cover is reducing. From the classified images, we can see that bare land cover is converted either into vegetation land or settlement. These results are supported by classified images using both the classifiers.

6 Conclusions and Future Scope

In this paper, we have studied the change in land covers of Aurangabad district area from the year 2013 to 2019. It has been observed that due to large efforts carried out regarding global warming and sustainable environment, people are working towards improving green cover. The study conducted using Landsat-8 remote sensor images show this. The images are classified using supervised classification methods, MLC and NN.

The results can be further used by government agencies in decision making and planning for a better environment. The settlements like new sites for urban development can also be planned accordingly. This study focused on change detection and can be further enhanced to suggest locations for city development.

References

1. http://earthexplorer.usgs.gov. Accessed on 22 Oct 2018
2. Mather PM (2004) Computer processing of remotely-sensed images: an introduction, 3rd edn. Wiley, pp 149–169, 203–245
3. Aung HPP, Aung ST (2019) Analysis of land cover change detection using satellite images in Patheingyi township. In: Zin T, Lin JW (eds) Big data analysis and deep learning applications.

ICBDL 2018. Advances in intelligent systems and computing, vol 744. Springer, Singapore
4. Radhika K, Varadarajan S, Li Z (2018) A neural network based classification of satellite images for change detection applications. Cogent Eng 5:1. https://doi.org/10.1080/23311916.2018.1484587
5. Grivei A, Radoi A, Datcu M (2017) Land cover change detection in satellite image time series using an active learning method. In: 9th international workshop on the analysis of multitemporal remote sensing images (MultiTemp), Brugge, pp 1–4
6. Adam HE, Csaplovics E, Elhaja ME (2016) A comparison of pixel-based and object-based approaches for land use land cover classification in semi-arid areas, Sudan. In: 8th IGRSM international conference and exhibition on remote sensing & GIS (IGRSM), pp 1–10
7. Birdi PK, Kale KV (2019) Accuracy assessment of classification on Landsat-8 data for land cover and land use of an urban area by applying different image fusion techniques and varying training samples. In: LNEE, vol 521. Springer, Singapore, pp 189–197
8. Birdi PK, Kale KV (2018) Enhancement of land cover and land use classification accuracy using spectral and textural features of fused images. In: CCIS, vol 876. Springer, Singapore, pp 317–325
9. http://usgs.gov/Landsat8DataUserhandbook.pdf
10. Richards JA, Jia X (2006) Remote sensing digital image analysis: an introduction. Springer-Verlag, Heidelberg, pp 193–238
11. Cohen J (1960) A coefficient of agreement for nominal scales. Educ Psychol Measur 20:213–220
12. Congalton RG, Green K (1999) Assessing the accuracy of remotely sensed data: principles and practices. CRC, Lewis Publishers, pp 47–65
13. Congalton RG (1991) A review of assessing the accuracy of classifications of remotely sensed data. Remote Sens Environ 37(1):35–46

Maintenance Scheduling of Heavy Machinery Using IoT for Wide Range of Real-Time Applications

Jasti Lavanya, P. Kusuma Vani, and N. Srinivas Gupta

Abstract This proposed system presents the design of machinery maintenance scheduling system using Twitter feed with the use of ARM processor FRDMKL25Z, cloud service "ThingSpeak" and ESP8266 Wi-Fi module. The sensors in the system intensively monitor temperature, vibrations and smoke in the machinery. Measured parameters from sensors are sent to cloud for analysis through Wi-Fi module by the processor to monitor the data continuously. When the sensor data crosses certain safety level in the machinery, then an alert notification is sent to the Twitter feed. The sensor data is continuously monitored by the processor and is stored in the cloud service "ThingSpeak" which analyzes the data. When the data crosses certain safety levels, then a live Twitter feed notification is updated to the linked organizational Twitter account. We can also set a reminder for machinery service by giving date and time to ThingSpeak.

Keywords Internet of things · Scheduling · Machine maintenance · Smoke sensor · Vibration sensor · Temperature sensor · KL25Z · ARM Cortex

1 Introduction

Machinery plays a vital role in any factory [1, 2] and maintaining them in pristine order is of paramount importance. If a problem which could have been solved by a simple maintenance is left unattended, it could lead to severe damage of that machinery and costing the company both time and money. One of the most practical approaches for maintenance is the combination of corrective and preventive maintenance [3]. Maintenance can be scheduled by getting the best information regarding equipment by using continuous monitoring, testing and historical data of the same.

J. Lavanya (✉) · P. K. Vani · N. S. Gupta
Raghu Institute of Technology (RIT), Dakamarri, Bheemunipatnam, Visakhapatnam, Andhra Pradesh, India
e-mail: lavanyaj@vjit.ac.in

P. S. R. Chowdary et al. (eds.), *Microelectronics, Electromagnetics and Telecommunications*, Lecture Notes in Electrical Engineering 655,
https://doi.org/10.1007/978-981-15-3828-5_35

Sometimes, earlier systems give unreliable results [4]. What maintenance was really necessary can be determined, when it is torn down for maintenance by using accurate records of the "as found" condition of equipment [5].

1.1 Existing Methods

From past history, three categories are there in maintenance, predictive maintenance, periodic maintenance and corrective maintenance [6]. By using condition monitoring techniques, predictive maintenance concentrates on forecasting of failures followed by early detection [7]. Recurrent checks and schedules come under periodic maintenance to avoid failures. Corrective maintenance is a reactive action for action, whereas action is failure of equipment and reaction is repairing the assets after the failure. Optimization of the maintenance period has been the key aspect in most of the earlier proposed systems. In this paper, condition-based maintenance [5] plays a key role in maintenance scheduling. Maintenance scheduling in early studies has been reviewed in Ref. [8]. According to Ref. [9], both advantages and sharper drawbacks are also there in earlier proposed systems. The most widely used systems are as follows:

- *Periodic Maintenance*: These systems are used to have regular or periodic equipment checks [11, 13]. This means the malfunction in a system is analyzed for in a certain period of time, say, every 3 months or so.
- *Corrective Maintenance (CM)*: This system aims to schedule the maintenance after the malfunction in the system has occurred [12, 13]. It aims to have immediate corrective action to restore the machinery/equipment back to its original state.
- *Predictive Maintenance (PM)*: This system is a special case which aims to do both of the above stated methods and more so with higher efficiency [13]. It uses various data from the past and performs its actions.

1.2 Advantages

This merit bunch from earlier various designs and implementations [9] is as follows: Periodic maintenance aims to solve errors by checking regularly over a span of time. This means, if any errors creep into the system just after a major issue was solved, it can still be avoided due to the periodic check scheduled to occur at that time. Corrective maintenance is the simplest technique to implement as it only aims to solve the problem at hand without any analysis and computations. Predictive maintenance is by far the biggest improvement in terms of maintenance systems. It aims at integrating periodic maintenance as well as analyzing the past failure data to come up with a probabilistic solution to avoid future failures or to avoid them before they turn into a problem [10].

1.3 Disadvantages

Same as like other systems in the world, these above-mentioned systems also have some drawbacks. We have to consider those to rectify them in a better way in the proposed system. The key disadvantage of existing systems is facing a tough time during maintenance. Sometimes, earlier systems give unreliable results [4], though unlikely.

Performance is limited by very high ratio of cost to function. Lesser the improvement, lesser the functionality of the system [14]. As can be clearly seen, the periodic and corrective maintenance systems are just a temporary solution that tend to solve the problem either before or after it has occurred [15]. Even though predictive maintenance is the best solution but it is quite expensive, requires huge computation and difficult to implement [16].

This paper is best described in Sects. 2, 3, 4 given below.

Section 2 briefs this system along with the modeled block diagram.

Section 3 composes the reconciliation of the proposed structure. The details about the components and their specifications are included too.

Section 4 has techniques of testing in real time. It also includes results in different scenarios.

2 Proposed Methodology

The basic objective of this proposed method is to fabricate a predictive maintenance scheduling system so as to have a safe workplace/environment. It helps integrate maintenance with the Internet of things in reporting possible incidents of malfunction or damage to the machinery (condition-based maintenance). The availability of various cloud platforms helps make things easier. The whole process contains the following steps:

Design a system which alerts about the abnormality noticed at the machinery or equipment. Use Wi-Fi technology in sending notifications to the system administrator's Twitter feed notifying them of a possible malfunction to the equipment. As measured, it takes much less time and cost compared to that involved in investigating the tragedy that might occur.

2.1 Block Diagram

Figure 1 shows the block diagram of proposed system. This diagram gives complete picture of individual blocks, how they are connected to each other and simple understanding of functionality. It consists of all the major hardware components including the sensors also.

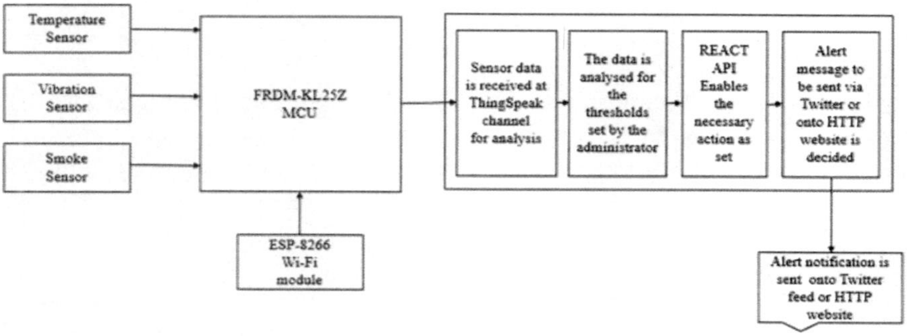

Fig. 1 Block diagram of proposed system

2.2 Functionality

Onboard microcontroller unit FRDM-KL25Z is used to operate and completely control the total devised system. All of the sensors can eternally watch by this ARM controller.

The sensor outputs are continually monitored by the FRDM MCU. These values are sent out to the ThingSpeak cloud platform via the Wi-Fi module.

These values are analyzed by ThingTweet service for the respective set thresholds. If the sensor values are below set thresholds, no alert is sent out.

When the set thresholds are reached or crossed, an alert is sent out on the Twitter feed of the administrator updating the state of the machinery.

As soon as the device is powered on, it turns the Wi-Fi module ON. Then, the values of the sensors are continually monitored and further actions are succeeded.

2.3 Flowchart

Figure 2 shows the execution flow of the system. It sharply concentrates on running of the while loop and verification method for the status of detected malfunction. The other responsibilities taking care by this flowchart are collecting of data from sensors and sending it to the administrator in the form of notification.

The flowchart consists of an inner loop in a whole single while loop which is responsible for verification and processing of sensor data with maximum possible less delay. To give a simple brief, the flowchart has a main loop that runs the system, one sensors' check loop and two functions that analyze and end the notification alert when initiated, respectively.

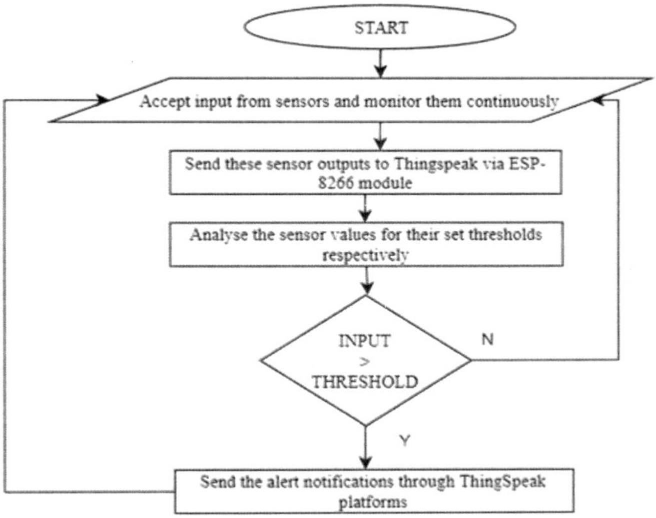

Fig. 2 Flow chart of the proposed methodology

3 Synthesis

3.1 Hardware Implementation

Details in brief about components, devices and other physical aspects of the proposed design consist in Hardware Implementation. Implementation of proposed system has been done by using FRDM-KL25Z, Wi-Fi module (ESP8266), smoke, temperature and vibration sensors. Pre-calibration of sensors has been done to get operating range and accuracy according to system requirement.

A. FRDM-KL25Z Board: Full form of FRDM is Freescale Freedom Development board. ARM microcontroller is embedded as the core in the KL25Z kit which has been used in this proposed system. It is an L series controller which is built on the Cortex M0+ core. Many key factors like operating clock frequency of 48 megahertz, a full USB controller to transfer data with wired and wireless connection, many analog/digital peripherals with inbuilt converters and 128 kilobytes flash memory lead to select this optimized board as the test device.

The KL25Z with its all available shields is compatible for hardware placement same as that of an Arduino Uno. Provision of interfaces like a single touch capacitive slider, RGB LED and a three-axis accelerometer is also there in the KL25Z. Other significant features like debugging, flashing and serial communication make it as an optimal board for embedded projects. By using two incorporated OpenSDA ports, this board can be connected to any system for the purpose of development or debugging.

B. ESP-8266 Wi-Fi Module: Highly integrated ESP8266 chip is designed to reach the requirements of wireless connected world. This tiny chip furnishes inbuilt and

100% complete Wi-Fi networking solution, allowing it to either host the application or isolate another application processor's wireless networking functions. The sole purpose of this chip is to receive the sensor values connected to the FRDM board and upload it to the cloud analysis platform for further action.

C. Temperature Sensor: The sensor used in this project to detect temperature variations is the LM35. It uses two inbuilt transistors and calculates the output temperature using what is known as the delta VBE architecture. By default, the output is calibrated linearly in degrees Celsius. It so happens that, for every 1° increase in temperature, there is a linear increase of 10 mV in the output voltage. We can also make suitable adjustments in the code to output the temperature in Fahrenheit, i.e., by conversion from deg C to deg F. The threshold set to test the project was 35 °C.

D. Vibration Sensor: It is a sensor having the propensity to perceive mechanical vibrations in a given area which is useful in a different variety of fields. Nowadays, one of the significant fields is security systems in which these sensors are used to alert someone of trouble with a system. They have quite a few different uses. The one used in this project is the SW-420, which has a piezoelectric sensing element and an onboard comparator to give the digital output. Usually the vibration sensors are calibrated to sense very mild to very high vibrations. The SW-420 module used in this project just uses a pre-determined threshold to output a digital signal when it detects a vibration irrespective of its magnitude. So, when a vibration is sensed, it induces a voltage proportional to its magnitude. Now, this value is compared against the set threshold which is intentionally low, and as a result outputs a digital pulse.

E. MQ2 (Smoke) Sensor: The MQ2 sensor is a multi-purpose sensor used to detect various types of gases. It can detect anything from CO to alcohol including butane, LPG, methane, etc. It has the ability to measure the presence of a combustible gas in air having concentration from 300 to 10 K ppm. So, in pure air, its readings are around 180–210. But when it is placed in the presence of a match or fire, the smoke reading output goes from 700 to 1500. Hence, this is the threshold that would be set to detect the presence of smoke or fire in the machinery.

3.2 Functionality of Proposed System

The system is proposed with a systematic working behavior. At the start, setup code for the Wi-Fi module runs in this system. It takes nearly 15–20 s which is mostly called as primary startup includes prevention of local echo in the serial line. At a constant baud rate, three sensors' output is measured all the time as a sub-loop in main loop after the initial setup. These sensors give their response to the respective parameters in the function of voltage. The onboard ADC and DAC of the FRDM board perform the respective conversions and give output to the Wi-Fi module which sends this data to ThingSpeak. If any abnormality is detected, as per the set thresholds for each of the parameters, the cloud platform ThingSpeak activates ThingTweet service which updates this malfunction on to the administrator's Twitter feed. If no thresholds are reached, the system continues to monitor further. There are various functions such as the ones listed below. Each of these is associated with a sensor and to output the parameter value to the FRDM board.

temp.read()* 100 gives the output temperature.
smo.read() and vib.read() give the output values of smoke and vibration sensors, respectively.
wifi.initialize(), wifi.reset(), etc., are some functions related to setting up the Wi-Fi module.

4 Results

According to the recorded procedure, results are observed in three stages, whereas just starting up and initialization come under the first stage. When the sensor outputs are detected as invalid, then it can be considered as the second stage. Finally, the last stage verifies the malfunction and initiates the alert to be sent on to the Twitter feed.

Case (i) From Fig. 3, Initializing: This is the initial setup where the parameters of the Wi-Fi module are set up. These include setting the IP address, selecting the SSID and authenticating using password.
Case (ii): From Fig. 4, invalid output detected: No alert will be given even though output is detected in the second case. Because these values are below set thresholds, no alert must be sent out.
Case (iii): From Fig. 5, malfunction detected: In this case, the output of the sensors is measured, and if any one of the sensors reach their set thresholds as shown in Fig. 6 (alert by glowing LED) ThingSpeak activates the ThingTweet service so as to send out an alert notification on the administrator's Twitter feed.

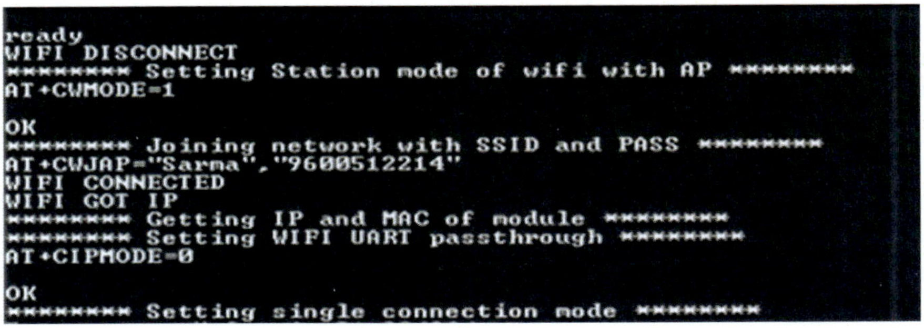

Fig. 3 Initialization of Wi-Fi module

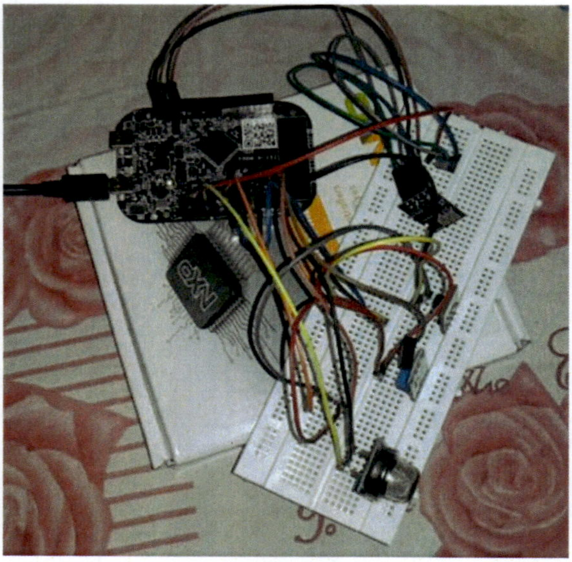

Fig. 4 Result when there is no malfunction

Fig. 5 Result when malfunction is detected

5 Conclusion

The proposed system design with a predictive and fail-safe mechanism could successfully overcome drawbacks like disjunct checks in periodic maintenance, failure detection after occurrence in CM and exorbitant, computational complexity and huge processing time in PM. The basic model of the more advanced maintenance scheduling is designed by us in this proposed system which is affordable to all with its simple integrated KL25Z board as hardware tool and open software.

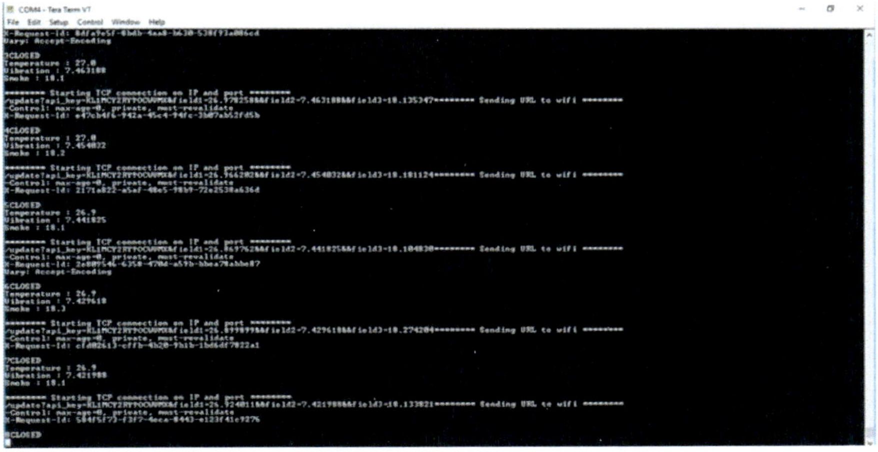

Fig. 6 Terminal output in the event of a malfunction

References

1. Camci F (2015) Maintenance scheduling of geographically distributed assets with prognostics information. Eur J Oper Res
2. Froger A, Gendreau M, Mendoza JE, Pinson E, Rousseau L-M (2014) Maintenance scheduling in the electricity industry: a literature review. CIRRELT
3. Niu G, Yang B-S, Pecht M (2010) Development of an optimized condition-based maintenance system by data fusion and reliability-centered maintenance. Eur Saf Reliab Assoc Saf Eng Risk Anal Div
4. Ribrant J (2005/2006) Reliability performance and maintenance—a survey of failures in wind power systems. KTH School of Electrical Engineering
5. Ahmed R, Kamaruddin S (2012) An overview of time-based and condition-based maintenance in industrial application
6. Canfield RV (1986) Cost optimization of periodic preventive maintenance. IEEE Trans Reliab 35(1):78–81
7. Grall A, Dieulle L, Berenguer C, Roussignol M (2002) Continuous-time predictive-maintenance scheduling for a deteriorating system. IEEE Trans Reliab 51(2)
8. Jayabalan V, Chaudhuri D (1992) Cost optimization of maintenance scheduling for a system with assured reliability. IEEE Trans Reliab 41(1):21–25
9. Dekker R (1996) Applications of maintenance optimization models: a review and analysis. Reliab Eng Syst Saf
10. Zhou X, Xi L, Lee J (2005) Reliability centered predictive maintenance scheduling for a continuously monitored system subject to degradation
11. Lee J, Ni J, Djurdjanovic D, Qiu H, Liao H (2006) Intelligent prognostics tools and e-maintenance. Comput Ind 57(6):476–489
12. Labib AW (2004) A decision analysis model for maintenance policy selection using a CMMS. J Qual Maint Eng 10(3):191–202
13. Yam RCM, Tse PW, Li L, Tu P (2001) Intelligent predictive decision support system for condition-based maintenance. Int J Adv Manuf Technol 17:383
14. Löfsten H (1999) Management of industrial maintenance—economic evaluation of maintenance policies. Int J Oper Prod Manag 19(7):716–737

15. Usher JS, Kamal AH, Syed WH (1998) Cost optimal preventive maintenance and replacement scheduling. IIE Trans
16. Tsang AHC (1995) Condition-based maintenance: tools and decision making. J Qual Maint Eng 1(3):3–17

A Tapered Microstrip-Fed Steering-Shaped Super-Wideband Printed Monopole Antenna

Srinivasarao Alluri and Nakkeeran Rangaswamy

Abstract A super-wideband (SWB) steering-shaped printed monopole antenna (SPMA) having a ratio bandwidth (RBW) close to 11:1 is presented in this paper. Partial ground plane with rounded top corner and a rectangular notch are used to get wide bandwidth. Good impedance matching is achieved by a microstrip feed line with linear tapering. A 22.78 GHz bandwidth for $S_{11} \leq -10$ dB is achieved from 2.22 GHz to 25 GHz. A 4 dB average peak gain and a variation from 96% to 57% in radiation efficiency are observed. The simulated far-field patterns in the two principal planes at different frequencies are presented. The proposed radiator can be used for different civilian and military applications.

Keywords Super-wideband · Printed monopole · Tapered microstrip feed line · Impedance bandwidth · Partial ground plane · Ratio bandwidth

1 Introduction

Over the past two decades, in wireless communication technology (WCT) use of a single radio device over a large frequency range is a common practice to provide different communication services. Multiple antennas with each one covering a specific frequency band are installed in the radio device to provide several communication services. However, a large space requirement for these antennas in the present-day portable wireless communication devices is a major problem. Most importantly, the installation of multiple antennas in a single device will cause electromagnetic interference/compatibility (EMI/EMC) problems and the system complexity also increases. Therefore, a single antenna with wideband response and stable radiation patterns throughout the band of interest is required. Several ultra-wideband (UWB) antennas [1–5] were proposed in the literature for short-range communication for 3.1–10.6 GHz frequency range.

S. Alluri (✉) · N. Rangaswamy
Department of Electronics Engineering, School of Engineering and Technology, Pondicherry University, Puducherry 605014, India
e-mail: asrao.81@gmail.com

P. S. R. Chowdary et al. (eds.), *Microelectronics, Electromagnetics and Telecommunications*, Lecture Notes in Electrical Engineering 655,
https://doi.org/10.1007/978-981-15-3828-5_36

In the present day wireless communication systems (WCS), users were demanding for high data rate over an extremely wideband of frequencies to accommodate both short- and long-range communications. A super-wideband (SWB) antenna is a good candidate to mitigate all these problems in WCS. Rumsey first proposed the concept of SWB technology in early 1960 [6]. Few printed monopole SWB antennas are reported in the literature [7–17]. In [7], an impedance bandwidth (IBW) of 1.02–24.1 GHz with a ratio bandwidth (RBW) of 23.63:1 was achieved by a modified tapered CPW-fed elliptical monopole antenna. A rectangular monopole with T-shaped tapered CPW ground plane was proposed in [8] for enhanced IBW from 2.4 GHz to 24.3 GHz for the reflection coefficient ≤ -10 dB. More than 50:1 RBW was achieved using a Mickey Mouse-shaped 45 mm × 42 mm × 1.5 mm size SWB monopole antenna with a rounded top-corner partial ground and a slot in ground plane in [9]. A 3–35 GHz printed propeller-shaped monopole with a RBW of 11.6:1 was proposed in [10] with a gain variation from 4 dBi to 5.2 dBi. A triangularly tapered feed line and exponential feed region were employed in [11] to achieve a simulated IBW of 2.5–80 GHz with $S_{11} \leq -10$ dB. In [12], a novel complementary Sierpinski triangular fractal structure surrounded by two semicircular sectors and slotted partial ground for improved bandwidth and coupling was proposed. A Chebyshev tapered antipodal Vivaldi antenna having SWB characteristics from 1 GHz to 35 GHz was presented in [13] with overall dimension of 95 mm × 97 mm × 1.6 mm. A phi-shaped CPW-fed monopole antenna was proposed in [14] with 3.5–37.2 GHz IBW for $S_{11} \leq -10$ dB. A fractional bandwidth (FBW) of nearly 167% and a 11:1 RBW was achieved in [15] by a partially segmented and elliptical slot-loaded circular patch antenna. Some other SWB antennas were also proposed in [16–23] with different geometrical shapes to achieve RBW of more than 10:1.

In open literature, some SWB antennas covering lower frequencies below 3 GHz are bulk in size, whereas SWB antennas which are compact in size are not covering frequencies below 3 GHz. A steering-shaped compact SWB antenna is proposed in this paper with overall size of $35 \times 35 \times 1.6$ mm^3 which provides an IBW from 2.22 GHz to 25 GHz.

2 Antenna Design

The proposed steering-shaped SWB radiator and geometrical dimensions are shown in Fig. 1. The antenna structure is printed on 1.6 mm height FR-4 material which is having $\varepsilon_r = 4.4$ and a $\tan\delta = 0.02$ with a size of $35 \times 35 \times 1.6$ mm^3. The optimal values of dimensions for the designed radiator structure are shown in Table 1. A steering-shaped circular radiator with a tapered microstrip feed line is printed on top of the substrate and a notch-loaded partial rectangular ground with rounded top corners on the back. The capacitive effect between the partial ground plane and radiating patch decreases with a notch in ground, and hence, IBW is improved. The bottom feed line width (W_f) is 2.95 mm corresponding to a characteristic impedance $Z_0 = 50\ \Omega$, approximately, and a top feed line width (W_{f1}) is 1.6 mm corresponding

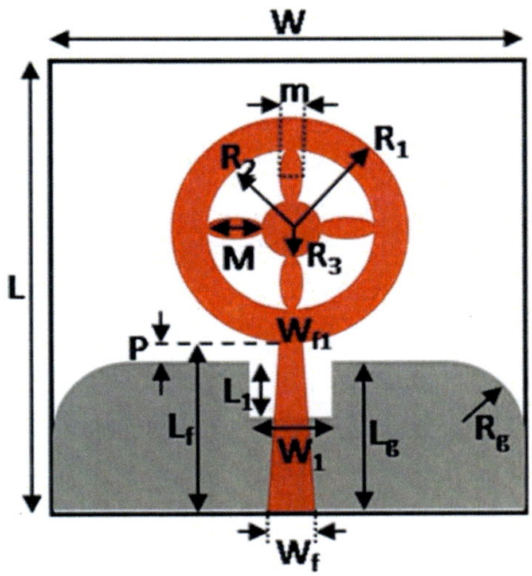

Fig. 1 Structure of the designed antenna

Table 1 Antenna structural parameters with their value

Parameter	Value (mm)	Parameter	Value (mm)
L	35	W	35
L_1	3	W_1	2
L_f	12	W_f	2.95
L_g	11	W_{f1}	1.6
R_1	9	R_2	7
R_3	2	R_g	9
M	5	m	2.5
P	1	h	1.6

to nearly 71 Ω. The gradual transformation of Z_0 from 50 to 71 Ω by tapered feed line provides a good impedance matching and further improves IBW.

3 Results and Discussion

The simulation of the proposed steering-shaped antenna is done using finite element method (FEM)-based ANSYS HFSS 3D electromagnetic (EM) simulator. Simulated S_{11} (in dB) plot is shown in Fig. 2. IBW of at least 22.78 GHz for reflection coefficient ≤ -10 dB is observed from 2.22 GHz to 25 GHz. The variation in group delay is shown in Fig. 3. The frequency dependency of time delay of an antenna is characterized by group delay $\tau_g(\omega)$, and in frequency domain, it is defined as [23]

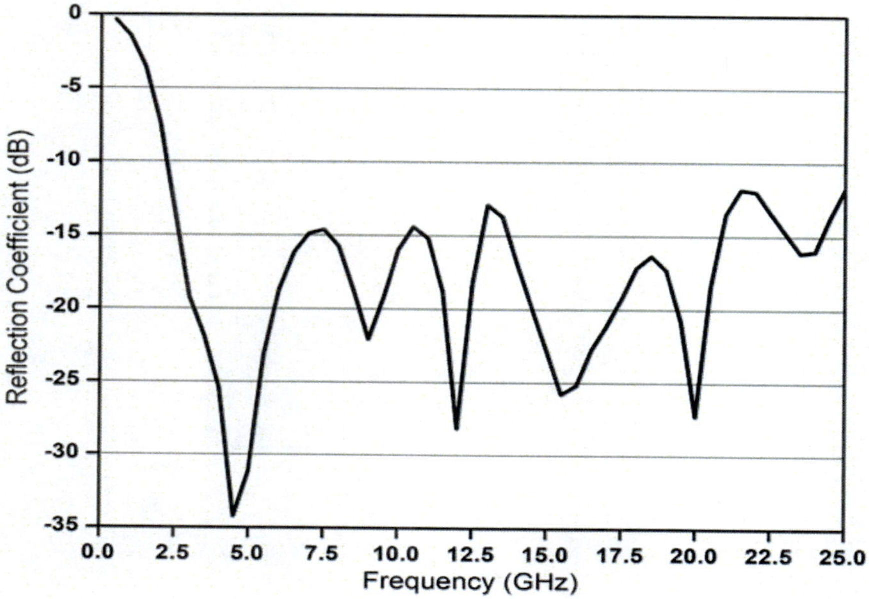

Fig. 2 Simulated reflection coefficient versus frequency characteristics

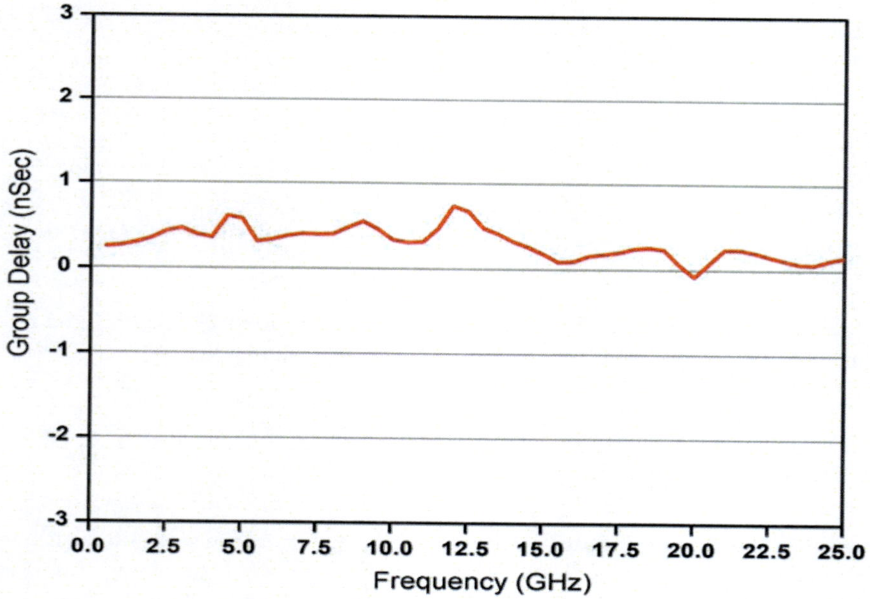

Fig. 3 Group delay versus frequency characteristics

$$\tau_g(\omega) = -\frac{d\varphi(\omega)}{d\omega} = -\frac{d\varphi(f)}{2\pi df} \tag{1}$$

where $\varphi(f)$ is a frequency-dependent phase angle of radiated signal. It is observed that variation in $\tau_g(\omega) < 1$ ns throughout the operating bandwidth. A $\tau_g(\omega) > 1$ ns results in pulse distortion in far-field region due to nonlinear phases of transfer function of the radiator.

The far-field patterns at four different frequencies in $\Phi = 0°$ and $\Phi = 90°$ planes for the proposed radiator are shown in Fig. 4. It is observed from Fig. 5 that the peak gain increases with frequency which is due to large size of radiating patch compared with corresponding wavelengths at higher frequencies. A 4 dB average peak gain is achieved using proposed antenna. Due to losses in substrate material at higher frequencies, a decrement from 96% to 57% in radiation efficiency is observed with increasing frequency as in Fig. 5.

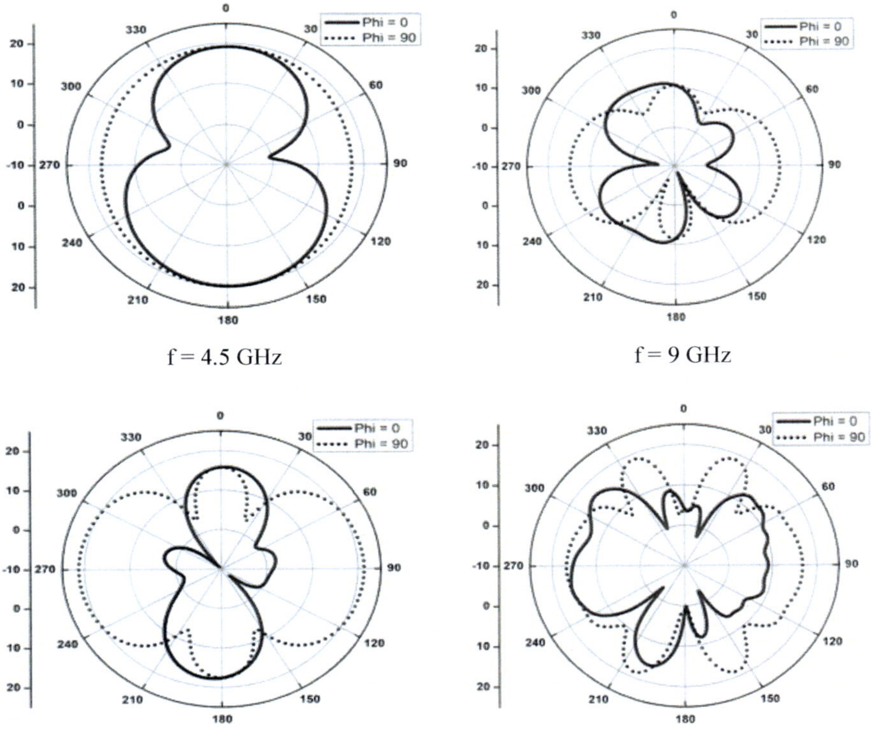

f = 4.5 GHz

f = 9 GHz

f = 12 GHz

f = 20 GHz

Fig. 4 Simulated far-field patterns in $\Phi = 0°$ and $\Phi = 90°$ planes for different frequencies

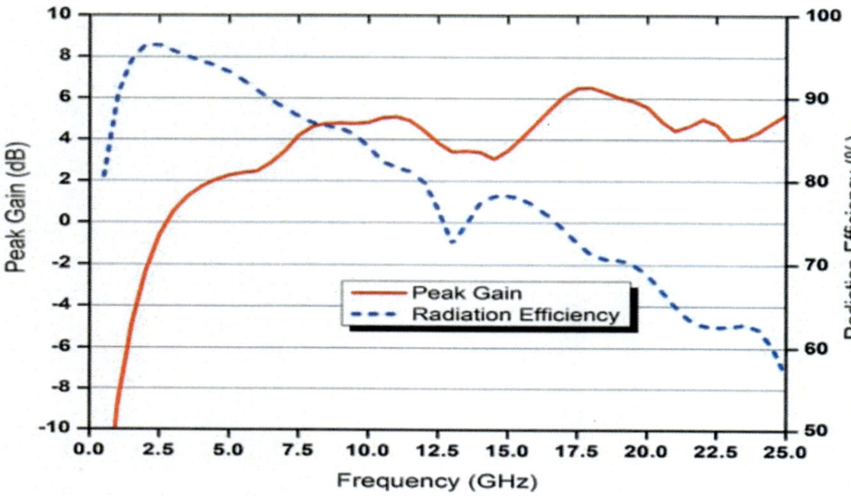

Fig. 5 Simulated peak gain and radiation efficiency characteristics

The surface current distribution for four different frequencies is shown in Fig. 6. It is noticed that the current distribution is uniform along with the patch, feed line, and ground plane.

4 Conclusion

A printed steering-shaped monopole (PSM) having a ratio bandwidth of 11:1 for SWB applications was discussed. The presented PSM provides a bandwidth of 22.78 GHz from 2.22 GHz to 25 GHz at a reflection coefficient ≤ -10 dB. A good impedance matching was witnessed by a tapered microstrip feed line, and a rectangular notch-loaded partial ground plane with rounded top corner was used to get wide bandwidth. An average peak gain of 4 dB and varying radiation efficiency from 57% to 96% over the operating bandwidth was noticed. The radiation patterns both in $\Phi = 0°$ and $\Phi = 90°$ planes were simulated for different frequencies. A figure of eight-shaped pattern in $\Phi = 0°$ and an omnidirectional pattern in $\Phi = 90°$ planes were observed at low frequencies, and as the frequency increases, patterns turn out to be directional in both the planes. The designed radiator is a wonderful candidate for ISM, Wi-Fi, WLAN, WiMAX, UWB, S, C, X, and Ku bands.

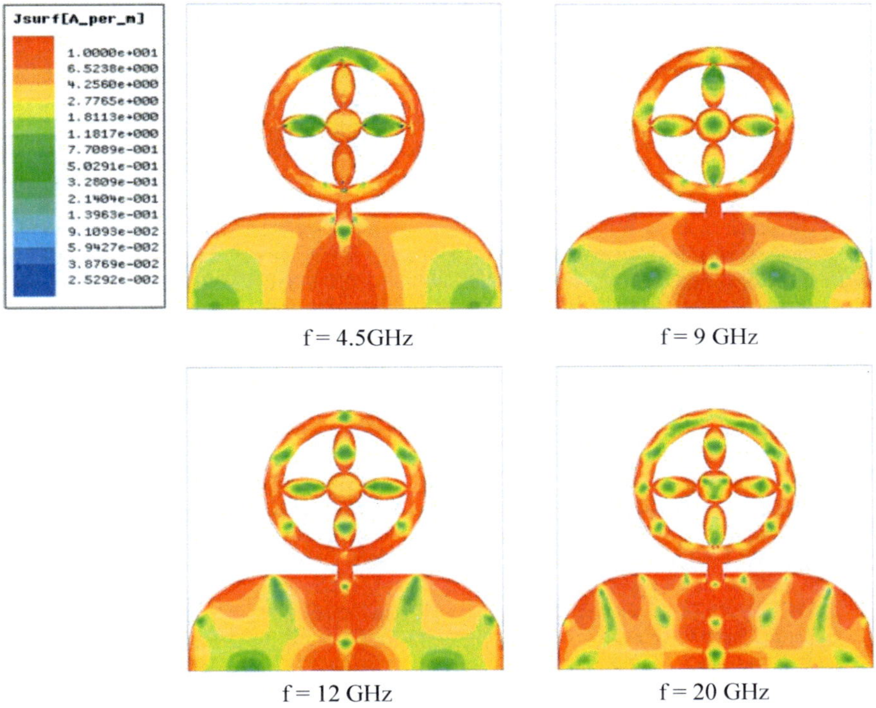

Fig. 6 Surface current distribution at different frequencies

References

1. Anveshkumar N, Gandhi AS (2018) A five-port integrated UWB and narrowband antennas system design for CR applications. IEEE Trans Antennas Propag 66(4):1669–1676
2. Ray KP (2008) Design aspects of printed monopole antennas for ultra-wide band applications. Int J Antennas Propag 2008:1–8. https://doi.org/10.1155/2008/713858
3. Sudheer Kumar T, Jaya Ch, Srinivasa Raju GRLVN (2017) On the notch band characteristics of Koch fractal antenna for UWB applications. Int J Control Theory Appl 10(6):701–707
4. Anveshkumar N, Gandhi AS (2018) A survey on planar antenna designs for cognitive radio applications. Wirel Pers Commun 98:541–569
5. Terlapu SK, Chowdary PSR, Jaya Ch, Chakravarthy VVSSS, Satapathy SC (2018) On the design of fractal UWB wide slot antenna with notch band characteristics. In: Satapathy S, Bhateja V, Anguera J (eds) Proceedings of 3rd international conference on micro-electronics, electromagnetics and telecommunications, Jan 2018. Lecture notes in electrical engineering, vol 471. Springer, Singapore, pp 907–912. https://doi.org/10.1007/978-981-10-7329-8_94
6. Rumsey V (1966) Frequency independent antennas. Academic Press, New York
7. Liu J, Zhong S, Esselle KP (2011) A printed elliptical monopole antenna with modified feeding structure for bandwidth enhancement. IEEE Trans Antennas Propag 59(2):667–670
8. Deng C, Xie Y, Li P (2009) CPW-fed planar printed monopole antenna with impedance bandwidth enhanced. IEEE Antennas Wirel Propag Lett 8:1394–1397
9. Cao P, Huang Y, Zhang J, Alrawashdeh R (2013) A compact super wideband monopole antenna. In: 7th European conference on antennas and propagation, pp 3107–3110

10. Gorai A, Karmakar A, Pal M, Ghatak R (2013) A CPW-fed propeller shaped monopole antenna with super wideband characteristics. Prog Electromagn Res C 45:125–135
11. Manohar M, Kshetrimayum RS, Gogoi AK (2013) Printed monopole antenna with tapered feed line, feed region and patch for super wideband applications. IET Microw Antennas Propag 1–7. https://doi.org/10.1049/iet-map.2013.0094
12. Torres CAF, Monroy JLM, Morales HL, Perez RAC, Tellez AC (2017) A novel fractal antenna based on the Sierpinski structure for super wide-band applications. Microw Opt Technol Lett 59(5):1148–1153
13. Gorai A, Karmakar A, Pal M, Ghatak R (2015) A super wideband Chebyshev tapered antipodal Vivaldi antenna. Int J Electron Commun (AEÜ) 69:1328–1333
14. Singhal S, Singh AK (2016) CPW-fed phi-shaped monopole antenna for super-wideband applications. Prog Electromagn Res C 64:105–116
15. Okas P, Sharma A, Das G, Gangwar RK (2018) Elliptical slot loaded partially segmented circular monopole antenna for super wideband application. Int J Electron Commun (AEÜ) 88:63–69
16. Akbari M, Koohestani M, Ghobadi C, Nourinia J (2011) Compact CPW-fed printed monopole antenna with super-wideband performance. Microw Opt Technol Lett 53(7):1481–1483
17. Manohar M, Kshetrimayum RS, Gogoi AK (2017) A compact dual band-notched circular ring printed monopole antenna for super-wideband applications. Radio Eng 26(1):64–70
18. Barbarino S, Consoli F (2010) Study on super-wideband planar asymmetrical dipole antennas of circular shape. IEEE Trans Antennas Propag 58(12):4074–4078
19. Singhal S (2018) Octagonal Sierpinski band-notched super-wideband antenna with defected ground structure and symmetrical feeding. J Comput Electron 17(3):1071–1081
20. Okan Y, Smith D, Michael E (2013) Printed slot loaded bow-tie antenna with super wideband radiation characteristics for imaging applications. IEEE Trans Antennas Propag 61(12):6206–6210
21. Singhal S, Singh AK (2017) CPW-fed octagonal super-wideband fractal antenna with defected ground structure. IET Microw Antennas Propag 11(3):370–377
22. Rafique U, Din SU (2018) Beveled-shaped super-wideband planar antenna. Turk J Electr Comput Sci 26:2417–2425
23. Weisbeck W, Adamiuk G, Sturm C (2009) Basic properties and design principles of UWB antennas. Proc IEEE 97(2):372–385

High-Throughput VLSI Architectures for VLSI Signal Processing

R. Ashok Chaitanya Varma, M. Venkata Subbarao, D. Ramesh Varma, and G. R. L. V. N. S. Raju

Abstract The purpose of this work is to develop VLSI DSP architectures for CRC-32 generator polynomial equation to improve better throughput with less number of clock pulses. In this paper, IIR filter-based design method is proposed. Different levels of architectures are proposed to achieve the requirement. LFSR is used in developing VLSI DSP architectures. These architectures had been implemented in Xilinx tool.

Keywords CRC · LFSR · IIR filter · Look-ahead technique

1 Introduction

In the present day scenario, high-speed communication and signal processing requirements are rapidly increasing day by day continuously to achieve high throughput and low latency requirement at a desired frequency level. To achieve these requirements, cyclic redundancy check (CRC) parallel-level architectures [1–3], with linear feedback shift register (LFSR), is required. LFSR architecture is simple and can run at desired frequency level, but with low-throughput limitation [4–7]. Designing multilevel parallel architecture does not double the throughput because of feedback loops. The full speedup can be achieved by pipelining the architecture with feed-forward paths [8]. To achieve these speedup requirements, different techniques, based on combined pipelining and parallelism [9, 10], look-ahead technique computations, are proposed to reduce the critical path [11–17] for IIR filter design. The limitations with combined pipelining and parallelism are number of delay elements; XOR elements will be increased; and hardware complexity increases. The rest of the paper is structured as follows. The analysis of VLSI architectures is presented in Sect. 2, Sect. 3 contains the simulations of CRC-32 architectures and comparison of various levels. Lastly, Sect. 4 ends with conclusion of the paper.

R. A. C. Varma · M. V. Subbarao (✉) · D. R. Varma · G. R. L. V. N. S. Raju
Department of Electronics and Communication Engineering, Shri Vishnu Engineering College for Women (A), Bhimavaram, India
e-mail: mvsubbarao@ieee.org

P. S. R. Chowdary et al. (eds.), *Microelectronics, Electromagnetics and Telecommunications*, Lecture Notes in Electrical Engineering 655,
https://doi.org/10.1007/978-981-15-3828-5_37

2 Analysis for CRC-32 Polynomial Equation

2.1 CRC-32 Sequential and One-Level Parallel Design

Generator polynomial of CRC-32 is

$$z(l) = z32 + z31 + z26 + z23 + z22 + z16 + z12 + z11 \\ + z10 + z8 + z7 + z5 + z4 + z2 + z + 1 \tag{1}$$

Figure 1 shows CRC-32 serial architecture. $v(l)$ represents input bits, $w(l)$ is back response data, and $z(l)$ is lastly generated output bits. One bit after the other will be processed for one clock cycle. In one-level parallel architecture after proposed formulation $v(l)$ represents input bits, $f(l)$ is back response data and $z(l)$ is overall generated output data stream. In these levels of parallel architectures, a new technique with feed forward paths, feedback paths and multiplexer with sel as select signal is provided. When sel is '0', input is '0'. When sel is '1', input is $z(l)$.

The output generated equation for serial architecture is represented as

$$z(l) = +z(-32+l) + z(-31+l) + z(-30+l) + z(-28+l) \\ + z(-27+l) + z(-25+l) + z(-24+l) + z(-22+l) \\ + z(-21+l) + z(-20+l) + z(-16+l) + z(-10+l) \\ + z(-9+l) + z(-6+l) + z(-1+l) + f(l) \tag{2}$$

$$f(l) = v(-32+l) + v(-31+l) + v(-30+l) + v(-28+l) \\ + v(-27+l) + v(-25+l) + v(-24+l) + v(-22+l) \\ + v(-21+l) + v(-20+l) + v(-16+l) + v(-10+l) \\ + v(-9+l) + v(-6+l) + v(-1+l) \tag{3}$$

Figure 2 shows CRC-32 one-level parallel architecture. The output generated equation for single-level parallel architecture is represented as Eqs. (2) and (3). Hence, a number of clock pulses are more with a reduced amount of throughput

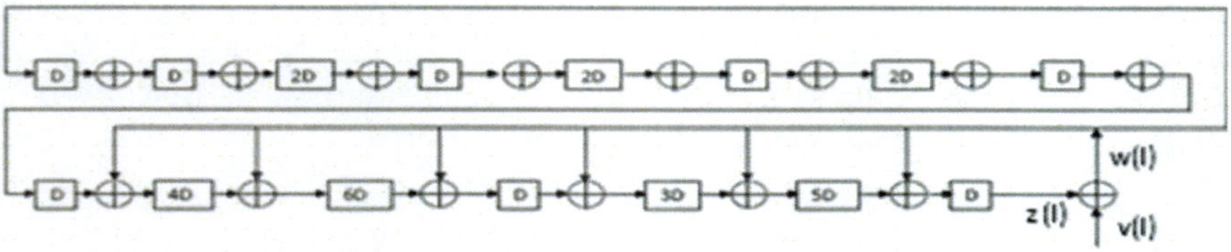

Fig. 1 CRC-32 serial-level architecture

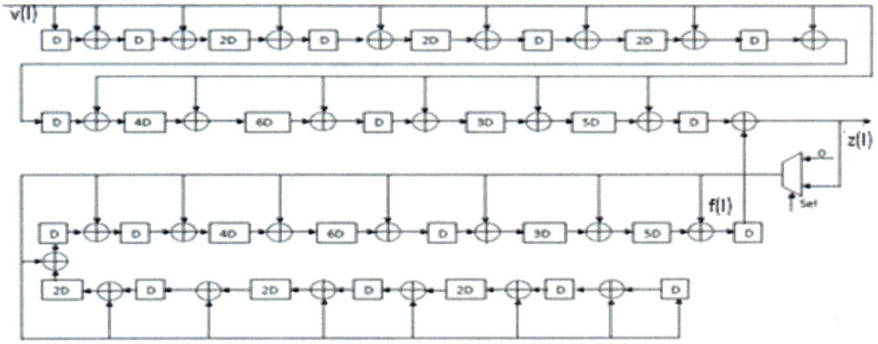

Fig. 2 CRC-32 one-level parallel architecture after proposed formulation

rate. These architectures have limitation in VLSI signal processing. To overcome limitation, combined pipelining and parallelism is projected.

2.2 *CRC-32 Two-Level Combined Pipelining and Parallelism*

In this two-level combined pipelining and parallelism, two bits will be processed per clock cycle [9]; hence, the throughput rate is two and number of clock cycles required to process the data is $N/2$ [10]. In Figs. 3 and 4, every delay element is delayed by two units. In three-level combined pipelining and parallelism, three bits will be processed per clock cycle; hence, the throughput rate is three and number of

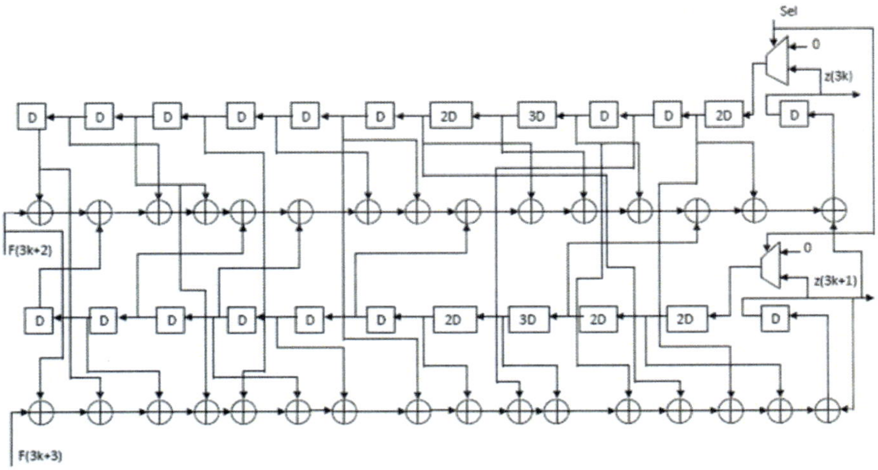

Fig. 3 Two-level parallel architecture *z* structure

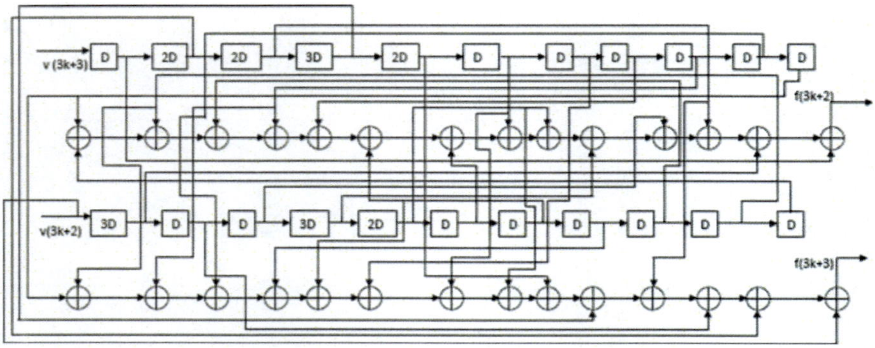

Fig. 4 Two-level parallel architecture f structure

clock cycles required to process the data is $N/3$. 'N' represents the number of bits. Taking unit delay for the $z(l)$ and after solving, the solved equations are

$$\begin{aligned} z(l-1) = {} & z(l-33) + z(l-32) + z(l-31) + z(l-29) + z(l-28) + z(l-26) \\ & + z(l-25) + z(l-23) + z(l-22) + z(l-21) + z(l-17) \\ & + z(l-11) + z(l-10) + z(l-7) + z(l-2) + f(l-1) \end{aligned} \tag{4}$$

After solving Eqs. (1) and (3), the final equations are

$$\begin{aligned} z(l) = {} & z(l-33) + z(l-30) + z(l-29) + z(l-27) + z(l-26) + z(l-24) \\ & + z(l-23) + z(l-20) + z(l-17) + z(l-16) + z(l-11) \\ & + z(l-9) + z(l-7) + z(l-6) + z(l-2) + f(l-1) + f(l) \end{aligned} \tag{5}$$

let $x = 3k$

$$\begin{aligned} z(2+x) = {} & z(-30+x) + z(-29+x) + z(-28+x) + z(-26+x) \\ & + z(-25+x) + z(-23+x) + z(-22+x) + z(-20+x) \\ & + z(-19+x) + z(-18+x) + z(-14+x) + z(-8+x) \\ & + z(-7+x) + z(-4+x) + z(1+x) + f(+2) \end{aligned} \tag{6}$$

$$\begin{aligned} z(3+x) = {} & z(-30+x) + z(-27+x) + z(-26+x) + z(-24+x) + z(-23+x) \\ & + z(-21+x) + z(-20+x) + z(-17+x) + z(-14+x) \\ & + z(-13+x) + z(-8+x) + z(-6+x) + z(-4+x) \\ & + z(-3+x) + z(1+x) + f(2+x) + f(3+x) \end{aligned} \tag{7}$$

$$\begin{aligned} f(x+2) = {} & v(x-30) + v(x-29) + v(-28+x) + v(-26+x) + v(-25+x) \\ & + v(-23+x) + v(-22+x) + v(-20+x) + v(-19+x) \end{aligned}$$

$$+ v(-18 + x) + v(-14 + x) + v(-8 + x) + v(-7 + x)$$
$$+ v(-4 + x) + v(1 + x) \quad (8)$$

$$f(x + 3) = v(-29 + x) + v(-28 + x) + v(-27 + x) + v(-25 + x)$$
$$+ v(-24 + x) + v(-22 + x) + v(-21 + x) + v(-19 + x)$$
$$+ v(-18 + x) + v(-17 + x) + v(-13 + x) + v(-7 + x)$$
$$+ v(-6 + x) + v(-3 + x) + v(2 + x) \quad (9)$$

2.3 CRC-32 Three-Level Combined Pipelining and Parallelism

In Figs. 5 and 6, every delay element is delayed by three units [9, 10]. Taking unit delay and two delays for the $z(l)$ and after solving, the solved equations are

$$z(l) = z(-33 + l) + z(-30 + l) + z(-29 + l) + z(-27 + l) + z(-26 + l)$$
$$+ z(-24 + l) + z(-23 + l) + z(-20 + l) + z(-17 + l)$$
$$+ z(-16 + l) + z(-11 + l) + z(-9 + l) + z(-7 + l)$$
$$+ z(-6 + l) + z(-2 + l) + f(-1 + l) + f(l) \quad (10)$$

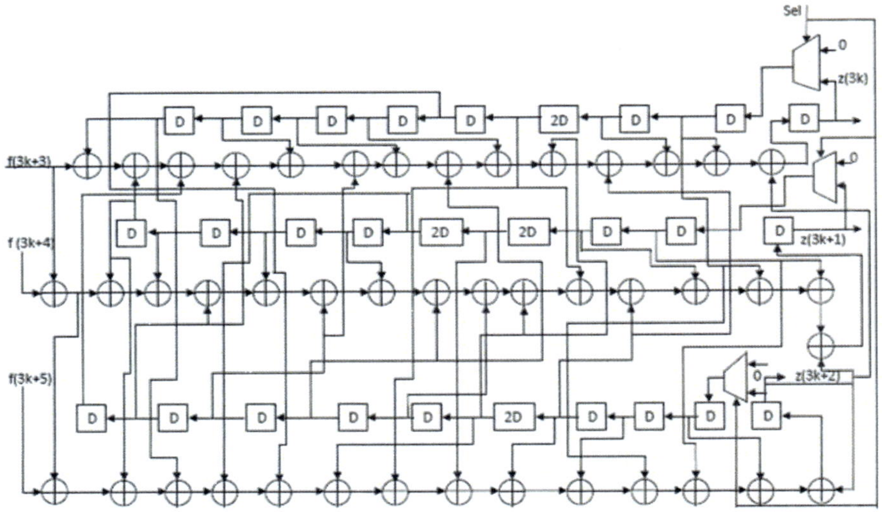

Fig. 5 Three-level parallel architecture z structure

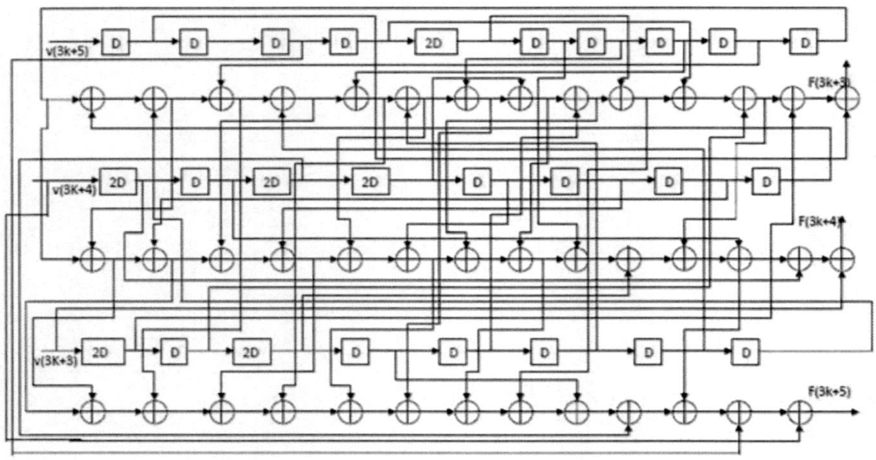

Fig. 6 Three-level parallel architecture f structure

$$\begin{aligned} z(l) = {} & z(-34+l) + z(-32+l) + z(-22+l) + z(-20+l) + z(-18+l) \\ & + z(-17+l) + z(-16+l) + z(-12+l) + z(-9+l) \\ & + z(-8+l) + z(-7+l) + z(-6+l) + z(-3+l) \\ & + f(-2+l) + f(-1+l) + f(l) \end{aligned} \tag{11}$$

$$\begin{aligned} z(x+3) = {} & z(-29+x) + z(-28+x) + z(-27+x) + z(-25+x) + z(-24+x) \\ & + z(-22+x) + z(x-21) + z(-19) + z(-18+x) \\ & + z(-17+x) + z(-13+x) + z(-7+x) + z(-6+x) \\ & + z(-3+x) + z(2+x) + f(3+x) \end{aligned} \tag{12}$$

$$\begin{aligned} z(4+x) = {} & z(-29+x) + z(-26+x) + z(-25+x) + z(-23+x) + z(-22+x) \\ & + z(-20+x) + z(-19+x) + z(-16+x) + z(-13+x) \\ & + z(-12+x) + z(-7+x) + z(-5+x) + z(-3+x) \\ & + z(-2+x) + z(2+x) + f(3+x) + f(4+x) \end{aligned} \tag{13}$$

$$\begin{aligned} z(5+x) = {} & z(-29+x) + z(-27+x) + z(-17+x) + z(-15+x) + z(-13+x) \\ & + z(-12+x) + z(-11+x) + z(-7+x) + z(-4+x) \\ & + z(-3+x) + z(-2+x) + z(-1+x) + z(2+x) \\ & + f(3+x) + f(4+x) + f(5+x) \end{aligned} \tag{14}$$

$$\begin{aligned} f(x+3) = {} & v(-29+x) + v(-28+x) + v(-27+x) + v(-25+x) \\ & + v(-24+x) + v(-22+x) + v(-21+x) + v(-19+x) \end{aligned}$$

$$+ v(-18 + x) + v(-17 + x) + v(-13 + x) + v(-7 + x) + v(-6 + x) + v(-3 + x) + v(+2 + x) \quad (15)$$

$$f(x + 4) = v(-28 + x) + v(-27 + x) + v(-26 + x) + v(-24 + x) + v(-23 + x) + v(-21 + x) + v(-20 + x) + v(-18 + x) + v(-17 + x) + v(-16 + x) + v(-12 + x) + v(-6 + x) + v(-5 + x) + v(-2 + x) + v(3 + x) \quad (16)$$

$$f(x + 5) = v(-27 + x) + v(-26 + x) + v(-25 + x) + v(-23 + x) + v(-22 + x) + v(-20 + x) + v(-19 + x) + v(-17 + x) + v(-16 + x) + v(-15 + x) + v(-11 + x) + v(-5 + x) + v(-4 + x) + v(-1 + x) + v(4 + x) \quad (17)$$

3 Simulation Results

The simulation results shown are simulated using Xilinx, and the comparison is shown in Table 1, for which the results had been simulated.

The input data provided is 10000000000000000000000000000011. The output observed is 01111010000101100101000001111110. Figures 7 and 8 show the CRC-32 serial architecture simulated waveform and one-level combined parallelism and pipelining architecture for which the output generated after 32 clock pulses.

Figure 9 shows the CRC-32 two-level combined parallelism and pipelining architecture for which the output generated after 16 clock pulses.

Figure 10 shows the CRC-32 three-level combined parallelism and pipelining architecture for which the output generated after 11 clock pulses.

Table 1 describes the performance comparison among various levels of architectures. The proposed three-level parallel architectures have improved the throughput with less number of clock cycles compared to the level of other architectures.

Table 1 Various levels of CRC-32 generator polynomial

Architecture levels	Checksum bits	XOR elements	Delay elements	Critical path	Throughput rate	Clock pulses required
Serial	33	15	32	2	1	32
One level	33	15	32	2	1	32
Two level	33	59	64	30	2	16
Three level	33	85	64	30	3	11

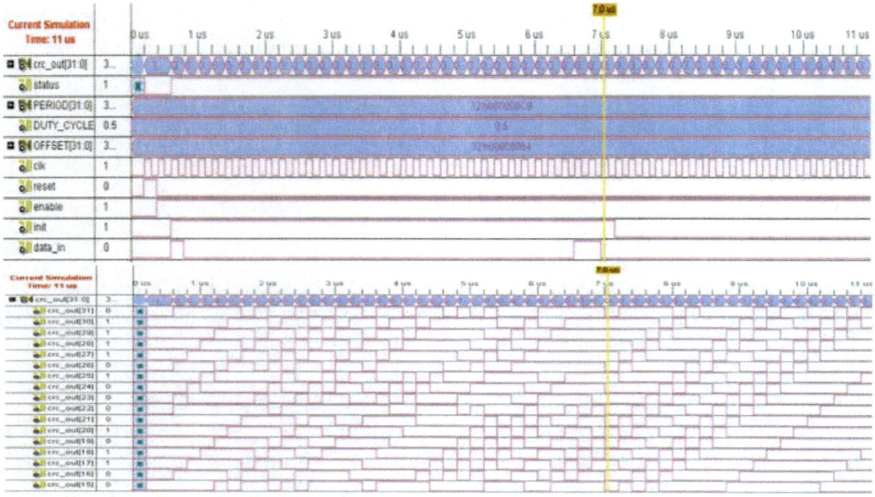

Fig. 7 CRC-32 serial architecture simulated waveform

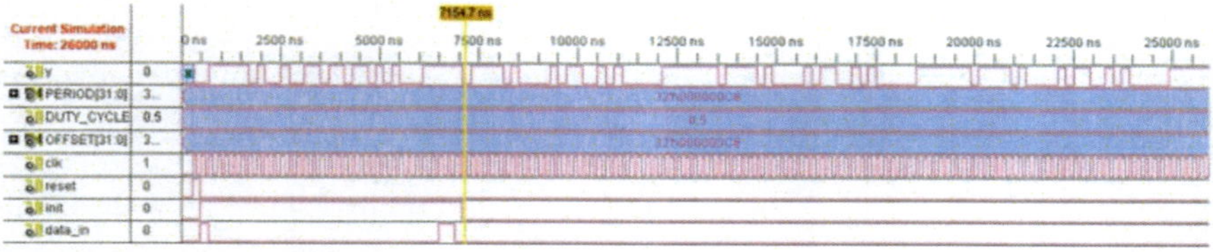

Fig. 8 CRC-32 one-level combined parallelism and pipelining architecture simulated waveform

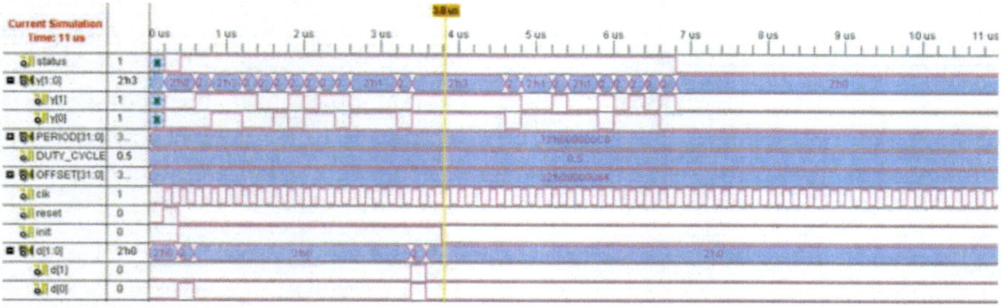

Fig. 9 CRC-32 two-level combined parallelism and pipelining architecture simulated waveform

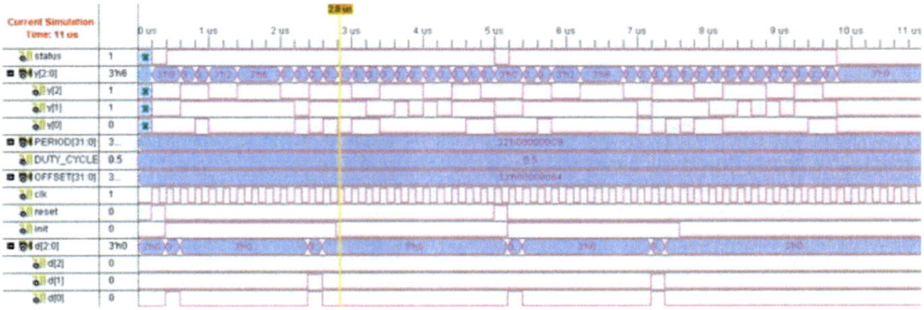

Fig. 10 CRC-32 three-level combined parallel and pipelining architecture simulated waveform

4 Conclusion

This paper concludes with a new approach for increasing the throughput rate and decreasing the number of clock pulses in parallel architectures from serial architectures. As the parallelism levels increase, the bit processing per a clock pulse also raises from one-level parallel architecture to three-level parallel architecture. In upcoming work, various DSP architectures like folding transformations are used for further improvement in throughput rate, critical path reduction, reduction in number of clock pulses and reduction in hardware complexity.

References

1. Campobello G, Patane G, Russo M (2003) Parallel CRC realization. IEEE Trans Comput 52(10):1312–1319
2. Zhang X, Parhi KK (2004) High-speed architectures for parallel long BCH encoders. In: Proceedings of the ACM great lakes symposium on VLSI, Boston, MA, Apr 2004, pp 1–6
3. Derby JH (2001) High speed CRC computation using state-space transformation. In: Proceedings of the global telecommunications conference 2001, GLOBECOM'01, vol 1, pp 166–170
4. Cheng C, Parhi KK (2009) High speed VLSI architecture for general linear feedback shift register (LFSR) structures. In: Proceedings of the 43rd Asilomar conference on signals, systems and computers, Monterey, CA, Nov 2009, pp 713–717
5. Jung J, Yoo H, Lee Y, Park I (2015) Efficient parallel architecture for linear feedback shift registers. IEEE Trans Circuits Syst II Express Briefs 62(11):1068–1072
6. Huo Y, Li X, Wang W, Liu D (2015) High performance table-based architecture for parallel CRC calculation. In: The 21st IEEE international workshop on local and metropolitan area networks, Beijing, pp 1–6
7. Ayinala M, Parhi KK (2011) High-speed parallel architectures for linear feedback shift registers. IEEE Trans Signal Process 59(9):4459–4469
8. Ayinala M, Parhi KK (2010) Efficient parallel VLSI architecture for linear feedback shift registers. In: IEEE workshop on SiPS, Oct 2010, pp 52–57
9. Varma RAC, Subbarao MV, Raju GRLVNS (2019) High throughput VLSI architectures for CRC-12 computation. In: Satapathy SC et al (eds) ICETE 2019. LAIS, vol 3. Springer Nature Switzerland AG 2020, pp 704–711

10. Varma RAC, Apparao YV (2018) High throughput VLSI architectures for CRC-16 computation in VLSI signal processing. In: Anguera J et al (eds) Microelectronics, electromagnetics and telecommunications. Lecture notes in electrical engineering, vol 471. Springer Nature Singapore Pte Ltd
11. Ayinala M, Brown MJ, Parhi KK (2012) Pipelined parallel FFT architectures via folding transformation. IEEE Trans VLSI Syst 20(6):1068–1081
12. Garrido M, Parhi KK, Grajal J (2009) A pipelined FFT architecture for real-valued signals. IEEE Trans Circuits Syst I Regul Pap 56(12):2634–2643
13. Cheng C, Parhi KK (2007) High-throughput VLSI architecture for FFT computation. IEEE Trans Circuits Syst II Express Briefs 54(10):863–867
14. Cheng C, Parhi KK (2008) Hardware-efficient low-latency architecture for high-throughput rate Viterbi decoders. IEEE Trans Circuits Syst II Express Briefs 55(12):1254–1258
15. Liu Y, Parhi KK (2016) Architectures for recursive digital filters using stochastic computing. IEEE Trans Signal Process 64(14):3705–3718
16. Prakash MS, Shaik RA (2013) Low-area and high-throughput architecture for an adaptive filter using distributed arithmetic. IEEE Trans Circuits Syst II 60(11):781–785
17. Dake JL, Terlapu SK (2016) Implementation of high-throughput digit-serial redundant basis multiplier over finite field. IOSR J VLSI Signal Process (IOSR-JVSP) 6(4):35–45. Ver. I, e-ISSN; 2319-4200, ISSN No.: 2319-4197

Reform Initiatives for Electrical Distribution Utilities in Jharkhand, India

Palacherla Srinivas, Rajagopal Peesapati, Muddana Harsha Vardhan, and Katchala Appala Naidu

Abstract In order to lower the cost of power, many countries have started reforming their power sectors with the participation of private companies. The Jharkhand State Electricity Board (JSEB) has been restructured into four different companies in the year 2014. In this paper, the relative efficiencies of electrical distribution utilities (EDUs) of the state are evaluated for the period 2008–2011 through the application of data envelopment analysis (DEA). The analysis of the relative efficiencies reveals the need of efficient reform initiatives for the EDUs of JSEB. In this regard, the present work proposes few initiatives that are additionally useful for electrical distribution sector in the state Jharkhand, India. Grouping of similar types of EDUs and change of circle based on geographical nature are proposed as two efficient initiatives for the distribution sector of the state. The mean efficiency score is evaluated before and after the implementation of proposed initiatives to verify the effectiveness. The findings of the research show the improvement of efficiencies after the application of the proposed initiatives.

Keywords Reform initiatives · Data envelopment analysis · K-means cluster

1 Introduction

In order to reach the economic benefits effectively to the end consumer, the power sector should play an important role by considering all the utilities. Decentralized management of distribution system has been playing a great role in both ongoing and upcoming power sector reforms [1]. The obstacles to optimal network pricing, relying on real world, topical examples are discussed in [2]. A clear description regarding the flaws in the unbundling of the European electricity sector is also presented. Finally, it is concluded that flawed coordination in electricity unbundling brings additional stress to the system. The drawbacks of energy reforms on grid investments and government coordination in Mexico energy sector are presented in

P. Srinivas (✉) · R. Peesapati · M. H. Vardhan · K. A. Naidu
Raghu Engineering College (A), Visakhapatnam 531162, India
e-mail: palacherlasrinu@gmail.com

P. S. R. Chowdary et al. (eds.), *Microelectronics, Electromagnetics and Telecommunications*, Lecture Notes in Electrical Engineering 655,
https://doi.org/10.1007/978-981-15-3828-5_38

[3]. The various aspects that influence the consumers are availability of electricity and its price [4–6]. Further, zone and circles are the subclassifications of MSEDCL. Each circle is subdivided into divisions and similarly each division into subdivisions [7]. Benchmarking measures are also proposed to characterize efficient ones and to find out the finest EDDs [8]. Reorganization model is presented to improve the operating performances of several EDDs [9] in later research work [10]. In the past researches, DEA has been applied to evaluate the three efficiencies, namely overall efficiency (OE), unified efficiency under natural disposability (UEN) and unified efficiency under natural and managerial disposability (UENM) [11]. It is incorporated with discriminant analysis (DA) to rank the Japanese electric power industry and with index decomposition analysis–artificial neural network for an appraisal of energy consumption in South African industrial sectors [12, 13]. Assemblage of similar type of companies/agencies is done on the basis of Fuzzy c-means (FCM) algorithm. FCM minimizes an objective function to attain good classifications [14]. Many cluster validity indices have been proposed, such as the Xie-Beni index (XB) [15], Dunn's index (DI) [16], partition coefficient (PC) and classification entropy (CE). An integrated algorithm is presented to measure the efficiencies of Iranian electric distribution companies (IDCs). DEA is applied to measure the efficiencies of each IDC. The outliers are also presented to overcome the sensitivity in the DEA results. Slack analysis provides the direction for improving the efficiencies [17]. At first, the performances of EDUs are evaluated by the application of data envelopment analysis. K-means clustering algorithm is proposed to group the similar EDUs in order to initiate reform by dividing the entire sector into an efficient number of groups. Change of circles based on geographical nature is proposed as a second initiative.

2 Methodologies

2.1 Data Envelopment Analysis (DEA)

DEA is a linear optimization methodology to generate the efficiency graph for analogous DMUs. DEA considers multiple inputs and produces multiple outputs with optimal weight vectors [18]. The relative efficiency of a firm consisting of multiple inputs and multiple outputs is defined as

$$\text{Overall Efficiency} = \frac{\text{weighted sum of outputs}}{\text{weighted sum of inputs}} \tag{1}$$

The efficiency scores of n DMUs taking m inputs and producing s outputs can be evaluated by a solution to the linear programming (LP) problem. The introduction to this additional restriction will produce a scalar variable, which is subtracted in the model presented in Eq. (2) [19]. The canonical form of the input-oriented model of DMU_0 is given as:

θ_o is overall efficiency score of DMU_0, y_{sn} is amount of output s produced DMU_n, x_{mn} is amount of m utilized by DMU_n, λ_n is weights for inputs and outputs of the DMU_n.

$$\text{Min}\,\theta_o \tag{2}$$

Subject to:

$$\begin{bmatrix} x_{11} & x_{12} & .. & x_{1n} \\ x_{21} & x_{22} & & x_{2n} \\ . & . & & \\ . & . & & \\ x_{m1} & & & x_{mn} \end{bmatrix} \begin{bmatrix} \lambda_1 \\ \lambda_2 \\ . \\ . \\ \lambda_n \end{bmatrix} - \theta_o \begin{bmatrix} x_{1o} \\ x_{2o} \\ . \\ . \\ x_{mo} \end{bmatrix} \leq 0 \tag{3}$$

$$\begin{bmatrix} y_{11} & y_{12} & .. & y_{1n} \\ y_{21} & y_{22} & .. & y_{2n} \\ . & . & & \\ . & . & & \\ y_{s1} & & & y_{sn} \end{bmatrix} \begin{bmatrix} \lambda_1 \\ \lambda_2 \\ . \\ . \\ \lambda_n \end{bmatrix} - \begin{bmatrix} y_{1o} \\ y_{2o} \\ . \\ . \\ y_{mo} \end{bmatrix} \geq 0 \tag{4}$$

$$\begin{bmatrix} \lambda_1 & \lambda_2 & .. & \lambda_n \end{bmatrix}^{\mathrm{T}} \geq 0 \tag{5}$$

The scale efficiency of the DMU_0 is given as:

$$\text{Scale Efficiency (SE)} = \frac{\text{OE}}{\text{TE}} \tag{6}$$

2.2 K-Means Clustering

Partitioning a set of entities into a number of homogeneous clusters is the task of clustering. K-means is one of the simplest unsupervised learning algorithms that clusters the data based on the locations and distances [20]. The objective function is as shown below.

$$J = \sum_{j=1}^{k} \sum_{i=1}^{n} \left| b_i^j - c_j \right|^2 \tag{7}$$

where $\left| b_i^j - c_j \right|^2$ is the distance measure between the data points b_i^j and cluster center c_j.

3 Input and Output Selection

The several parameters for efficiency analysis are chosen depending on the insight given by [17]. The total amount of ES during that financial year influences the operation of distribution utility. So, this was considered as an output. The better performance of distribution utility can be engraved out from the NC on respective utility, so the parameter NC is also considered as an output parameter. *O&M* cost is the important parameter for maintaining the EDU under specific financial constraints. *Number of employees* is also considered as one of the significant inputs as the works under the EDU are carried out by them. The positive correlation results for the considered period of study are shown in Table 1.

4 Results

4.1 Judgment on Relative Efficacies Over the Period of 2008–2011

The details of EDUs and the overall efficiencies of individual EDUs for the period 2008–2011 are presented in Table 2. The mean overall efficiencies of the EDUs during the considered period are 65.33% (2008–2009), 58.95% (2009–2010) and 63.94% (2010–2011). The overall efficiency score has shown wide fluctuation among the EDUs throughout the period of study. Efficiency scores reveal that sufficient margin exists in enhancing the efficiency of inefficient EDUs. The utilities Sahibganj and Dumka are found to be operating on the efficiency locus throughout the considered period of the study. The overall efficiencies of 36% utilities have been deteriorated throughout the period. For example, Ranchi (Doranda) utility scores 54.61%, 43.6% and 42.27% in the years 2008–09, 2009–10 and 2010–11, respectively. Similarly, the Deoghar utility worsens the overall efficiency from 69.28 to 54.19%. Around 26% utilities improved their overall efficiencies throughout the period, like Ranchi (Central) improved its overall efficiency from 48.7 (2008–09) to 52.27% (2010–2011). There is an unclear description on overall efficiencies of 33% utilities like Adityapur, as their overall efficiencies throughout the period are neither increasing nor decreasing. Inefficient utilities are required to reduce their inputs or increase their outputs (keeping the other as a constant) to become efficient. Only 41% utilities are operating at efficiency score more than mean efficiency score, and the rest are operating under the mean efficiency score. The efficiency score of utility Khunti is 21.33%, which indicates relatively bad performance and poor resource utilization. The technical and scale efficiency scores that are produced by VRS scale provide the information related to the overall inefficiencies. This reveals that the EDUs like Chatra, Adityapur, Loyabad, Pakur, Garhwa-2, Giridih (North) and Latehar are found to be scale inefficient but technically efficiency. As the scale is IRS, they may increase their scale sizes to improve overall efficiencies. The EDUs like Giridih (South) and

Table 1 Correlation results for the considered years

	2008–2009			2009–2010			2010–2011		
	DLL	TC	O&M	DLL	TC	O&M	DLL	TC	O&M
DLL	1	0.3214	0.0914	1	0.0644	0.0712	1	0.7770	0.6204
TC		1	0.0308		1	0.1949		1	0.8516
O&M			1			1			1

Table 2 Efficiencies and grouping results of 39 EDUs throughout the period 2008–10

Name of the circle	Name of the utility	2008–2009	2009–2010	2010–2011				Group
		OE	OE	OE	TE	SE	RTS	
Ranchi	Ranchi (East)	0.2865	0.4893	0.4538	0.4618	0.9826	IRS	1
	Ranchi (Kokar)	0.4984	0.5642	0.5379	0.5799	0.9276	IRS	1
	Ranchi (New Capital)	0.3564	0.4316	0.4026	0.4670	0.8620	IRS	1
	Ranchi (Central)	0.487	0.5207	0.5222	0.6579	0.7938	IRS	1
	Ranchi (Doranda)	0.5461	0.436	0.4227	0.5881	0.7187	IRS	1
	Ranchi (West)	0.3275	0.3605	0.4093	0.5437	0.7527	IRS	1
	Khunti	0.1186	0.1378	0.2133	0.4418	0.4828	IRS	2
Hazaribagh	Hazaribagh	0.6196	0.4995	1	1	1	CRS	2
	Chatra	0.6873	0.7413	0.6511	1	0.6511	IRS	2
	Koderma	0.5478	0.4864	0.4470	0.4893	0.9135	IRS	1
	Ramgarh	1	0.7163	1	1	1	CRS	1
Dhanbad	Dhanbad	0.4280	0.3708	0.3781	0.4388	0.8616	IRS	1
	Govindpur	0.4675	0.2711	0.4808	0.6369	0.7549	IRS	1
	Nirsa	0.7459	0.3302	0.7807	0.8888	0.8783	IRS	1
	Jharia	0.6026	0.4304	0.5953	0.8639	0.6891	IRS	1
Jamshedpur	Jamshedpur	1	0.7562	1	1	1	CRS	1
	Adityapur	0.9285	0.945	0.9156	1	0.9156	IRS	1
	Ghatsila	0.4507	0.4786	0.5395	0.7607	0.7092	IRS	1
Chaibasa	Chaibasa	0.5432	0.4541	0.3941	0.4380	0.8997	IRS	1
	Chakradhapur	0.5619	0.4898	0.5196	0.7069	0.7351	IRS	2
	Sarikela	0.573	0.6088	0.6718	0.7512	0.8942	IRS	1
Chas	Chas	0.7901	0.6992	0.5804	0.5843	0.9933	IRS	1
	Loyabad	0.9874	0.4514	0.4619	1	0.4619	IRS	2
	Tenughat	0.7879	0.6122	0.5052	0.7442	0.6788	IRS	2
Sahibganj	Sahibganj	1	1	1	1	1	CRS	1
	Pakur	1	0.8829	0.8462	1	0.8462	IRS	2
Gumla	Gumla	0.3087	0.3655	1	1	1	CRS	2
	Simdega	0.3125	0.2838	0.2387	0.6382	0.3740	IRS	2
	Lohardaga	0.2432	0.3516	0.2585	0.9029	0.2863	IRS	2
Deoghar	Deoghar	0.6928	0.6048	0.5419	1	0.5419	DRS	1
	Godda	1	0.9759	0.9594	0.9621	0.9973	IRS	1
Dumka	Dumka	1	1	1	1	1	CRS	1
	Jamtara	0.9613	1	1	1	1	CRS	1
Garhwa	Garhwa-1	0.5294	0.5227	0.4028	0.7588	0.5308	IRS	2
	Garhwa-2	0.2949	0.3405	0.3940	1	0.3940	IRS	2

(continued)

Table 2 (continued)

Name of the circle	Name of the utility	2008–2009	2009–2010	2010–2011				Group
		OE	OE	OE	TE	SE	RTS	
Giridih	Giridih (South)	0.9858	0.9481	0.9261	0.9935	0.9321	DRS	1
	Giridih (North)	1	0.7508	0.9796	1	0.9796	IRS	2
Daltonganj	Daltonganj	0.8099	0.686	0.6544	0.8319	0.7866	DRS	1
	Latehar	1	1	0.8525	1	0.8525	IRS	2
	Mean efficiency	0.6533	0.5895	0.6394	–	–	–	

Daltonganj are neither technically efficient nor scale efficient. To enhance the overall efficiency of these utilities, the scale size should be decreased as their scale is DRS. The utility Deoghar is technically efficient but scale inefficient and showing the decreasing returns to scale.

It is evident from the previous section that the distribution sector reforms are needed for improving the performance and efficient operation of EDUs which JSEB has been yet to initiate. Therefore, precise reform initiatives are proposed in the following sections for accelerated turnaround of the distribution sector. The first initiative to be taken in this regard is the disintegration of the distribution sector into several numbers of groups. The K-means algorithm proposed in Sect. 2 is aided to disintegrate the EDUs, forming them into several groups. The similar types of EDUs are grouped based on the 2010–2011 data. Figure 1 shows the mean silhouette value for different number of clusters. The plot determines that the optimal numbers of

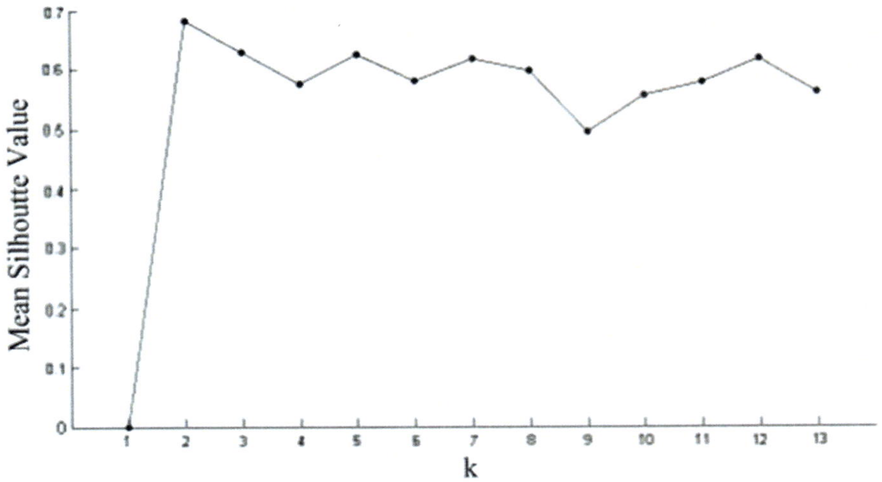

Fig. 1 Mean silhouette value for different number of clusters

clusters are to be two. Hence, the 39 EDUs are divided into two groups containing 25 and 14, respectively. The entire distribution sector is divided into two main groups, namely Jharkhand Southern Power Distribution Company Limited (JSPDCL) and Jharkhand Northern Power Distribution Company Limited (JNPDCL) which will manage the 25 and 14 EDUs, respectively. The last column of Table 3 reveals the grouping results of 39 EDUs. It is clear that many utilities which belong to the same circle are grouped differently. For example, the utilities Daltonganj and Latehar belong to the same circle but grouped differently. Therefore, in order to heal this difficulty, change of a circle based on geographical nature is proposed as a second initiative for the distribution sector reforms in the state. In this, the utilities are subject to change their circle with respect to grouping and geographical nature. The geographical map of state Jharkhand is shown in Fig. 2. The Khunti utility is under Ranchi circle but grouped differently from the remaining utilities of Ranchi. According to the map, Khunti is surrounded by Gumla, Simdega, Chaibasa, Chakradharpur (situated near to Chaibasa, not shown on map) and Ranchi. Out of these Gumla and Simdega are of the same circle and belongs to the same group. The utility Chakradharpur is also associated with the same group as Khunti. Therefore, Khunti utility can be amalgamated with the circle/utility of the same group for improving its operational efficiency. Here, the utility Khunti is merged with Chakradharpur, and both of them created as a new circle. Similarly, the remaining utilities are also merged with same group circles to enhance the efficiency of each utility. An enhancement in efficiency of the individual EDUs is observed by implementing the proposed two initiatives in the distribution sector of the state Jharkhand. Hence, it is revealed that the proposed reforms are additionally required to improve the operating efficiencies of the EDUs in state Jharkhand, India.

5 Conclusions

In this paper, three different reform initiatives are proposed for the accelerated development of the electricity distribution sector of the state Jharkhand, India. The work begins with the efficiency evaluation of the EDUs over the period 2008–2011, post-implementation of R-APDRP. The VRS outcomes reveal the unbalanced utilization of the resources of the EDUs. To improve the performance, two efficient reform initiatives are proposed for the utilities of the JSEB. Division of distribution sector and change of the circle are proposed as two additional reform initiatives, respectively. The results reveal that the proposed reform initiatives boost up the efficiencies of each EDU. Two new companies, namely JNPDCL and JSPDCL, are proposed by the implementation of the first reform initiative. Many of the utilities have changed their EDCs during the process of the second reform initiative. Finally, an increment in mean overall efficiency from 63.94 to 69.15% is observed during the entire process of the reforms. It is concluded that the proposed reform initiatives will help in accelerating the economic growth and achieving higher standards of living due to availability of reliable power at lower cost.

Table 3 Results of EDUs circles, overall, technical and scale efficiencies

Sl. No.	Utility	Group	OE	TE	SE	RTS	New circle
1	Ranchi (East)	1	0.505	0.590	0.856	IRS	Ranchi
2	Ranchi (Kokar)	1	0.54	0.777	0.704	IRS	
3	Ranchi (New Capital)	1	0.423	0.686	0.616	IRS	
4	Ranchi (Central)	1	0.503	1	0.5034	IRS	
5	Ranchi (Doranda)	1	0.423	1	0.423	IRS	
6	Ranchi (West)	1	0.310	0.947	0.327	IRS	
7	Koderma	1	0.559	0.609	0.918	IRS	
8	Ramgarh	1	1	1	1	IRS	
9	Dhanbad	1	0.600	0.754	0.796	IRS	Dhanbad
10	Govindpur	1	0.481	0.798	0.603	IRS	
11	Nirsa	1	0.787	1	0.787	IRS	
12	Jharia	1	0.616	0.996	0.618	IRS	
13	Jamshedpur	1	1	1	1	CRS	Jamshedpur
14	Adityapur	1	0.915	1	0.915	DRS	
15	Ghatsila	1	0.540	1	0.540	IRS	
16	Chaibasa	1	0.590	0.741	0.796	IRS	Chaibasa
17	Sarikela	1	0.672	0.882	0.763	IRS	
18	Chas	1	0.788	0.811	0.972	IRS	Chas
19	Giridih (South)	1	1	1	1	CRS	
20	Dumka	1	1	1	1	CRS	Dumka
21	Jamtara	1	1	1	1	CRS	
22	Daltonganj	1	0.680	0.867	0.784	DRS	Daltonganj
23	Sahibganj	1	1	1	1	CRS	Deoghar
24	Deoghar	1	0.723	1	0.723	DRS	
25	Godda	1	1	1	1	CRS	
26	Pakur	2	1	1	1	CRS	
27	Hazaribagh	2	1	1	1	CRS	Hazaribagh
28	Chatra	2	1	1	1	CRS	
29	Chakradhapur	2	0.540	0.711	0.760	IRS	Chakradharpur
30	Khunti	2	0.237	0.442	0.537	IRS	
31	Gumla	2	1	1	1	CRS	Gumla
32	Simdega	2	0.247	0.247	1	IRS	
33	Lohardaga	2	0.260	0.903	0.288	IRS	
34	Garhwa-1	2	0.480	0.759	0.633	IRS	Garhwa
35	Garhwa-2	2	0.453	1	0.453	IRS	

(continued)

Table 3 (continued)

Sl. No.	Utility	Group	OE	TE	SE	RTS	New circle
36	Latehar	2	0.853	1	0.853	IRS	
37	Giridih (North)	2	1	1	1	CRS	Giridih
38	Loyabad	2	0.582	1	0.582	IRS	
39	Tenughat	2	0.662	0.764	0.866	IRS	

Fig. 2 Jharkhand geographical map

References

1. Dyner I, Larsen ER (2001) From planning to strategy in the electricity industry. Energy Policy 29:1145–1154
2. Brunekreeft G (2015) Network unbundling and flawed coordination: experience from electricity sector. Util Policy 34:11–18
3. Ibarra-Yunez A (2015) Energy reform in Mexico: imperfect unbundling in the electricity sector. Util Policy 35:19–27
4. Thillai Rajan A (2000) Power sector reforms in Orissa: an ex-post analysis of the casual factors. Energy Policy 28:657–669
5. Kundu GK, Mishra BB (2011) Impact of reform and privatization on consumers: a case study of power sector reform in Orissa, India. Energy Policy 39:3537–3549
6. Kundu GK, Mishra BB (2012) Impact of reform and privatization on employees: a case study of power sector reform in Orissa, India. Energy Policy 45:252–262

7. Totare NP, Pandit S (2010) Power sector reform in Maharashtra, India. Energy Policy 38:7082–7092
8. Yadav VK, Padhy NP, Gupta HO (2011) Performance evaluation and improvement directions for an Indian electric utility. Energy Policy 39:7112–7120
9. Yadav VK, Chauhan YK, Padhy NP, Gupta HO (2013) A novel power sector restructuring model based on data envelopment analysis (DEA). Electr Power Energy Syst 44:629–637
10. Badelt G, Yehia M (2000) The way to restructure the Lebanese electric power sector: a challenge for the transitional management. Energy Policy 28:39–47
11. Goto M, Ofsuka A, Sueyoshi T (2014) DEA (data envelopment analysis) assessment of operational and environmental efficiencies on Japanese regional industries. Energy 66:535–549
12. Sueyoshi T, Goto M (2012) Efficiency-based rank assessment for electric power industry: a combined use of data envelopment analysis (DEA) and DEA discriminant analysis (DA). Energy Econ 34(3):634–644
13. Olanrewaju OA, Jimoh AA, Kholpane PA (2013) Assessing the energy potential in the South African industry: a combined IDA-ANN-DEA (index decomposition analysis-artificial neural network-data envelopment analysis) model. Energy 63(15):225–232
14. Bezdek JC (1981) Pattern recognition with fuzzy objective function algorithms. Plenum, New York
15. Xie Y, Raghavan VV, Zhao X (2002) 3M algorithm: finding an optimal fuzzy cluster scheme for proximity data. In: International proceedings of IEEE world congress on computational intelligence, Honolulu, HI, vol 1, pp 627–632
16. Dunn JC (1973) A fuzzy relative of the ISODATA process and its use in detecting compact well-separated clusters. J Cybern 3:32–57
17. Simab M, Haghifam M-R (2010) Using integrated model to assess the efficiency of electric distribution companies. IEEE Trans Power Delivery 25(4):1806–1814
18. Charnes A, Cooper WW, Rhodes E (1978) Measuring the efficiency of decision making units. Eur J Oper Res 2:429–444
19. Abbott M (2015) Reform and efficiency of New Zealand's airport. Util Policy 36:1–9
20. Ghosh S, Dubey SK (2013) Comparative analysis of k means and fuzzy c means clustering algorithms. Int J Adv Comput Sci Appl 4(4):35–39

FPGA Performance Evaluation of Present Cipher Using LCC Key Generation for IoT Sensor Nodes

Srikanth Parikibandla and Alluri Sreenivas

Abstract IoT which enables data transmission through different kind's interrelated networks, mostly the data is exchanged between wireless networks, and chances of hacking. Security is the most important aspect, and the data should be confidential to avoiding hacking. The cryptography solutions are utilized as an answer for improving security and the customary calculations because their limitation setting is not perfect for IoT gadgets. It is therefore possible to use the lightweight cryptographic algorithm as a solution to IoT security problems. However, there are a number of algorithms to choose from the distinctive execution criteria and conditions; the PRESENT cipher template in this paper is the encryption method using the 64-bit key for 64-bit data for hardware-level data protection input. To improve the security, Lorenz Chaotic Circuit with Dual-port Read Only Memory-based Present Algorithm (LCC-DROM-PA) architecture is proposed in this work, and for generating the key value, LCC is an essential design, and DROM is used for S-box design and P-layer design. After designing this architecture, FPGA performances are evaluated by the count of LUTs, flip-flops, slices, and frequency.

Keywords PRESENT cipher · IoT · Wireless sensor nodes · Lightweight cryptography · LCC · FPGA

1 Introduction

In coming future generation, technology is trying to grab the human thought process by designing the smart objects like smart home appliances to smart emotional decision in the design of robotics. In this technology development, Internet of Things

S. Parikibandla (✉) · A. Sreenivas
Department of Electronics and Communication Engineering, GITAM (Deemed to be University), Visakhapatnam, India
e-mail: psrikanth.bt@gmail.com

A. Sreenivas
e-mail: drallurisreenivas@gmail.com

P. S. R. Chowdary et al. (eds.), *Microelectronics, Electromagnetics and Telecommunications*, Lecture Notes in Electrical Engineering 655,
https://doi.org/10.1007/978-981-15-3828-5_39

(IoT) plays vital role. The IoT paradigm connects all industrial and consumer electronics manufacturers to worldwide communication networks which aim to make everything intelligent [1, 2].

The advancement in wireless technology and tactile devices has contributed to IoT being implemented [3]. The networks of wireless sensors accumulate thousands of tiny devices called sensor nodes, which have the capacity of detecting, processing, and transmitting data in the network. Sensor systems allude to a heterogeneous framework that consolidates small sensors and actuators with figuring components for explicit purposes. Sensor node is a smart, lightweight, and self-organizing system equipped with battery, radio, microcontroller, and sensor [4]. Among IoT devices; most of the data is transmitted through wireless networks, in which the system includes hardware, middleware, and presentation layers. The hardware layer provides actuators and sensors; middle layer gives computation with storage; and presentation layer comprises interpretation tools for taking information from various base forms. Via IoT applications using wireless sensor nodes, research including equipment framework structure, information the executives, security, and social variables is developed as present-day innovation. Most of the popular novel attacks are executed on physical layer as they are the sources of the raw information [5]. Due to computational restraints, IoT schemes cannot afford for realizing the security in middle and presentation levels, in the whole system security implementing at lowest level that is physical layer [4, 6, 7].

It is correlated with so many considerations such as data security, design area, operating speed, expense, reliability, and additional network sensor constraints, and the main consideration is to increase security over existing attacks without impacting the quality and complexity of the overall wireless sensor network [8–10]. Different keys use asymmetric-key cryptosystems where the same key is used as the symmetric-key cryptosystems. This paper is organized as follows: Sects. 1 and 2 review introduction to cryptography, Sect. 3 represents present algorithm, Sect. 4 represents key generation method, Sects. 5 and 6 proposed method, and Sect. 7 concludes with a conclusion.

2 Cryptography and Lightweight Cryptography

Cryptography is the science of security that converts the information from a readable state (plaintext) to unreadable form (ciphertext), which enables the confidentiality to insecure communication channel. It uses cryptographic algorithms to transform a plaintext into a ciphertext, using most of the time a key and the process called encryption [11] and which converts ciphertext to plaintext is called decryption. There are two forms of encryptions, one of which is that symmetric-key cryptosystems which utilize a similar key to encode and unscramble a message, while asymmetric-key cryptosystems utilize various keys. In this paper, we proposed a symmetric encryption.

The conventional algorithms provides ciphers that are having higher design area, power consumption, lower speed, and efficiency which make them unsuitable for applications like IoT due to its constrained environment. Most omnipresent computing devices do not meet these demands, so there is a need for lightweight ciphers that fulfill security requirements when considering low-resource devices [8, 9, 12]. Lightweight cryptography requires extremely low specifications for important target system resources, and area and energy consumptions are crucial metrics for evaluating LWC properties [13]. This work centers on the structure, execution, and examination for the lightweight symmetric cipher PRESENT.

3 PRESENT Algorithm

PRESENT cipher is one of the most popular encryption algorithms which is used in the applications using the ubiquitous computing, and power and cost constraints were given high priority [14, 15]. PRESENT is ISO/IEC 29192-2:2012 standard, as "one of the symmetric ultra-lightweight cipher which suits for lightweight cryptomodules. It is customized for usage in compelled situations" [16]. The process relies upon a round-based substitution permutation network (SPN). SPN utilizes an S-box of four bits and permutation layer of bit shifts.

The cipher has three basic operations to maintain confidentiality over input data:

- AddRoundKey applies the state using finite-field arithmetic to 64 bits from the round key.
- SBoxLayer uses sixteen substitution boxes (S-box) to perform a four-bit-to-four-bit substitution.
- PLayer applies bit-level shifts over the data.

The key usage for the decoding procedure is done similarly as in the encryption procedure. The distinction lies in the key development decision job. Blocks from the key list are taken from the final values to the original ones in the decryption process, which is the private key of the user. It means the last Ki sub-key used to encrypt will be the first to decrypt. Therefore, to start from this last key Ki, the encryption process must include all key generation rounds, executing the reverse process before reaching the original key. Therefore, the round counter must be decreased and mixed with the previously indicated nibble in each round [14, 17–19] (Fig. 1).

Lara-Nino et al. proposed two novel designs of lightweight hardware architecture for the PRESENT algorithm, minimizing implementation energy consumption and size and considering the key generation mechanism in design for FPGA [20–22]. The creators assembled the calculation with a short key to be utilized to such an extent that it may relinquish security to decrease usage size. For this situation, the key schedule is executed utilizing four 20-bit registers which store the underlying key and update each rounds its worth. The 16-bit key consists of the four four-bit main registers; with each round moving to the rightmost, four bits get the right shift to get the key in each round to process the 16-bit state word. Toward the finish of

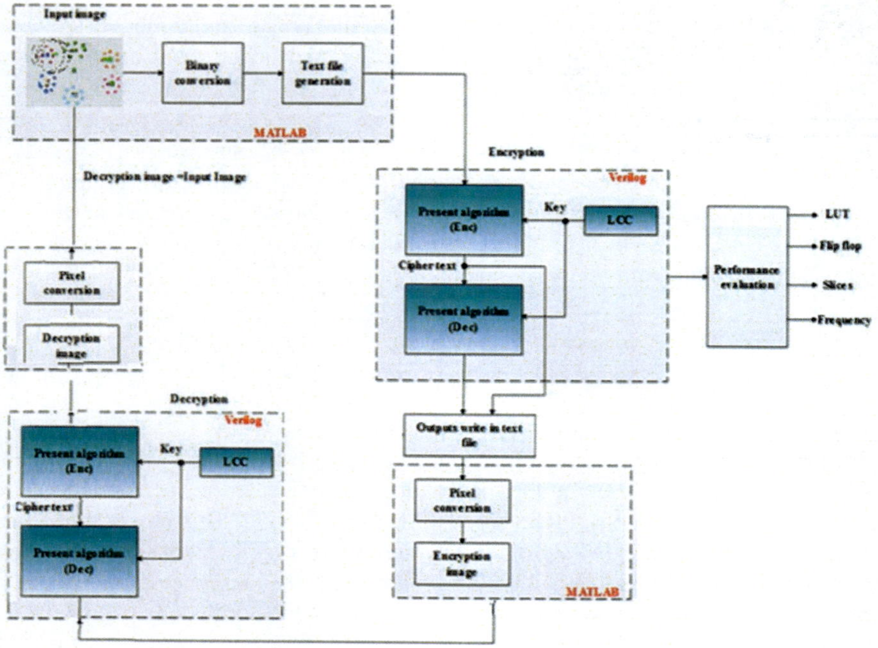

Fig. 1 Block diagram of the design

each round, the round key is changed parallel to the information handling, so no extra cycles are required. However, in order to break the input key into 16-bit letters, five cycles are required as shown in Fig. 2.

4 The Key Generation

The basic design proposed in this paper for the two architectures follows the strategies stated in [20] regarding data path construction. Thoughts are taken from [21] utilizing four 16-bit registers and from [22] utilizing both parallel and moving right or left access to such registers to present the state space. The 64-bit data is separated into four 16-bit words to include the information. At that point, each word is disseminated as four-bit words in the four registers. During each round of calculation, the piece of state to be handled comprises the four rightmost bits in each state register. These 16-bits are XORed with the relating 16-bit round catchphrase and after that handled by S-box and P layers. These 16-bit state words are then stored in state registers as four-bit words, whose value is then moved to the right. For this situation, the extra stage ought not to be designed; nonetheless, the register content should be updated toward the end of each round. The change allows initial permutations to be sequentially

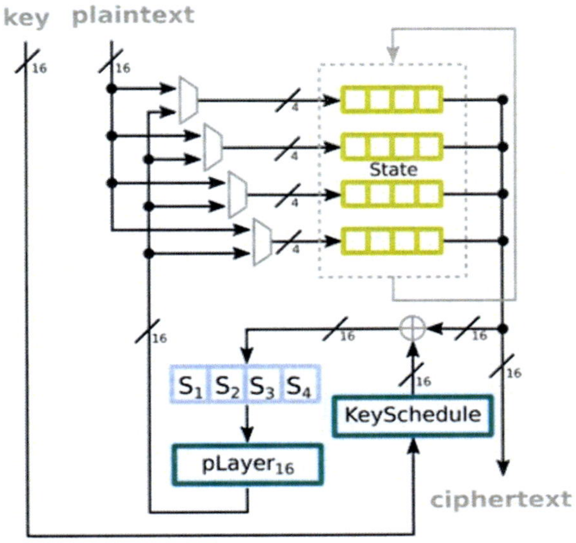

Fig. 2 Data path for proposed PRESENT architecture

executed and the corresponding permutation to be executed simultaneously. As a consequence, one loop decreases the latency per round.

5 The Proposed Design

Since many cryptosystems concentrate on generating pseudo-random sequences to conceal clear messages, the use of chaotic systems as a generator of these sequences has for many years become an important study subject. Chaos equation is derived from nonlinear dynamic system; it refers to aperiodic, unpredictable behavior that is sensitive to minor changes in initial values and control parameters [21].

In the proposed method to design and implement of the Lorenz Chaotic Circuit with Dual-port Read Only Memory-based PRESENT Algorithm (LCC-DROM-PA), LCC is an essential design for generating the key value, and DROM is used for S-box and P-layer designs. After designing this architecture in FPGA, its performances are evaluated by the count of LUTs, flip-flops, slices, and frequency. The sensor node image is read in MATLAB which will convert the image into the file containing binary values. In MATLAB, pixel-to-binary conversion is completed. This binary file is given to Verilog as the source file to protect the data.

The Lorenz model is an example of a complex and dynamic nonlinear system that fits the long-term action of the Lorenz system. The Lorenz system is a dynamic system that depicts chaotic motion in three dimensions. Such a system's graphical representation demonstrates how the state of a dynamic system (three-dimensional models' three variables) evolves over time in a complex, unrepeatable pattern [19, 21, 23].

6 Chaos-Based Image Encryption Algorithm

A new image authentication scheme is implemented that combines the position shuffling with the gray pixel values to confuse the relationship between the cipher image and the plain image.

One of the contrasts between the PRESENT models proposed is the key generation method, handling 16-bit information keys and utilizing standard logic to supply each key round. The LCC method is used for generating the key for the encryption and decryption operation shown in Fig. 3. In a cryptographic, key management is a challenge task, and the size of the key is very important in the restricted cryptosystem area. A large key size guarantees randomness, but the load increases with complexity proportionally. The LCC-based key used for PRESENT architecture is introduced to resolve this problem by generating the soft key [19, 21].

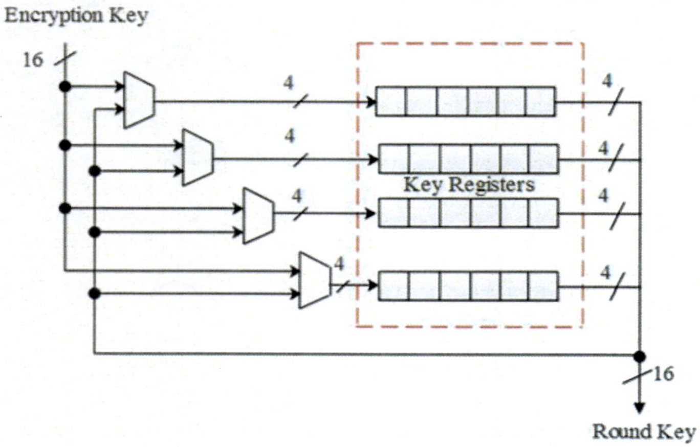

Fig. 3 16-bit key schedule for proposed architecture

First stage:

$$\begin{aligned} R_1[n-4] &= X[n] + Y[n] \\ \text{Cmul}_1[n-3] &= R_1[n-4] \times M_1 \\ \text{Cmul}_2[n-2] &= \text{Cmul}_1[n-3] \times M_2 \\ R_2[n-1] &= \text{Cmul}_2[n-2] + X[n] \\ \text{Register}_1 &= R_2[n-1] \end{aligned} \tag{1}$$

$$X[n] = \text{Register}_1 \tag{2}$$

Second stage:

$$\begin{aligned}
\text{Mul}_1[n-6] &= X[n] \times Y[n] \\
\text{Cmul}_3[n-5] &= X[n] \times M_3 \\
R_3[n-4] &= \text{Mul}_1[n-6] + \text{Cmul}_3[n-5] \\
R_4[n-3] &= R_3[n-4] + Y[n] \\
\text{Cmul}_4[n-2] &= R_4[n-3] + M_4 \\
R_5[n-1] &= \text{Cmul}_4[n-2] + Y[n] \\
\text{Register}_2 &= R_5[n-1]
\end{aligned} \tag{3}$$

$$Y[n] = \text{Register}_2 \tag{4}$$

Final stage:

$$\begin{aligned}
\text{Mul}_2[n-5] &= X[n] \times Y[n] \\
\text{Cmul}_5[n-4] &= Z[n] \times M_5 \\
R_6[n-3] &= \text{Mul}_2[n-5] + \text{Cmul}_5[n-4] \\
\text{Cmul}_6[n-2] &= R_6[n-3] \times M_6 \\
R_7[n-1] &= \text{Cmul}_6[n-2] + Z[n] \\
\text{Register}_3 &= R_7[n-1]
\end{aligned} \tag{5}$$

$$Z[n] = \text{Register}_3 \tag{6}$$

Figure 4 shows three different stages, depending on each stage. Each stage has two inputs like p and q which are stored in different registers like Register_1, Register_2, and Register_3. The outputs of the LCS are $X[n]$, $Y[n]$, and $Z[n]$. The random number that is possible in digital designs was successfully created by LCS. The architecture of PRESENT Algorithm mentioned in the earlier section was modeled using Verilog HDL using Xilinx version of 14.2, and the FPGA device Virtex-6XC6VCX75t was used for synthesis.

7 Conclusion

In this research, proposed architecture performances can be improved in LCC-DROM-PA architecture compared to conventional methods. The proposed design will be written in the Verilog and was synthesized in the Xilinx Virtex-6XC6VCX75t FPGA system with the intention of making efficient use of the FPGA platform. The suggested architectures could increase computer resources and provide less power consumption compared to other current implementations, making it ideal for use in applications of lightweight cryptography.

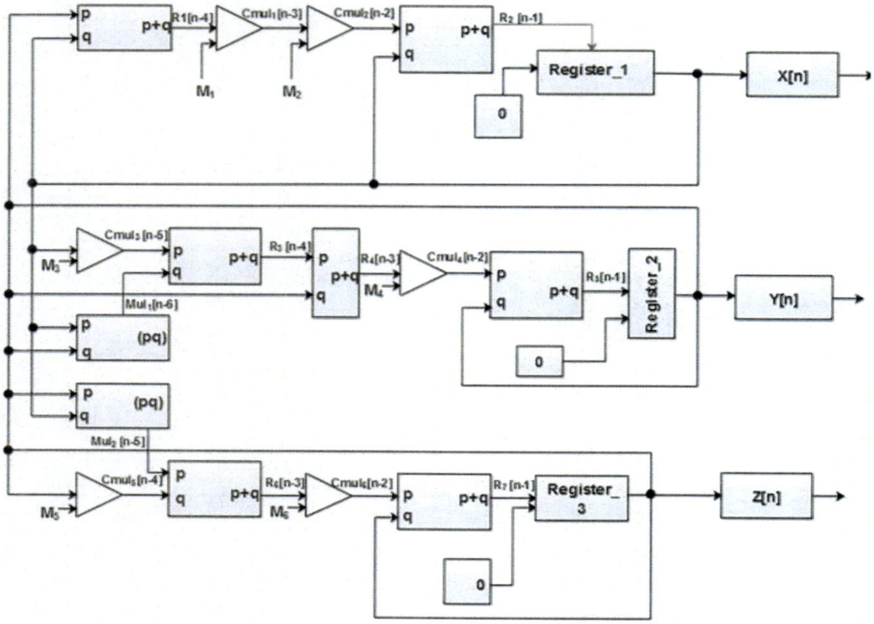

Fig. 4 Block diagram of the Lorenz chaotic circuit

References

1. Okabe T (2016) Efficient FPGA implementations of print cipher. JETIR 3(4). ISSN-2349-5162
2. Akash D, Shanthi P (2016) Lightweight security algorithm for wireless node connected with IoT. Indian J Sci Technol 9(30). https://doi.org/10.17485/ijst/2016/v9i30/99035
3. Karri R, Kuznetsov G, Goessel M (2003) Parity-based concurrent error detection of substitution-permutation network block ciphers. In: International workshop on cryptographic hardware and embedded systems. Springer, Berlin, Heidelberg
4. McKay KA, Bassham L, Turan MS, Mouha N (2017) Report on lightweight cryptography. US Department of Commerce, National Institute of Standards and Technology
5. Yalla P, Kaps J-P (2009) Lightweight cryptography for FPGAs. In: An international conference on reconfigurable computing and FPGAs—ReConFig'09, Dec 2009. IEEE, pp 225–230. https://doi.org/10.1109/ReConFig.2009.54
6. Banik S, Bogdanov A, Isobe T, Shibutani K, Hiwatari H, Akishita T, Regazzoni F (2014) Midori: a block cipher for low energy. In: International conference on the theory and application of cryptology and information security. Springer, Berlin, Heidelberg
7. Venugopal M, Doraipandian M (2017) Lightweight cryptographic solution for IoT—an assessment. Int J Pure Appl Math 117(16):511–516
8. Książak P, Farrelly W, Curran K (2014) A lightweight authentication protocol for secure communications between resource-limited devices and wireless sensor networks. Int J Inf Secur Privacy 8(4):62–102
9. Jyothirmayi G, Madhu GC (2018) Present cipher architecture implementation on Xilinx 14.3. IJEECS 7(4). ISSN: 2348-117x
10. Reddy PP, Thrimurthulu V, Kumar KJ (2014) Implementation of multi mode AES algorithm using Verilog. Int J Eng Res 3(12):780–785. ISSN: 2319-6890

11. Panasenko S, Smagin S (2011) Lightweight cryptography: underlying principles and approaches. Int J Comput Theory Eng 3(4)
12. Chaitra B, Kiran Kumar VG, Shatharama Rai C (2017) A survey on various lightweight cryptographic algorithms on FPGA. IOSR J Electron Commun Eng 12(1):54–59, Ver. II. E-ISSN: 2278-2834, P-ISSN: 2278-8735
13. Bogdanov A, Leander G, Knudsen LR, Paar C, Poschmann A, Robshaw MJ, Seurin Y, Vikkelsoe C (2007) Present—an ultra-lightweight block cipher. In: Proceedings of CHES 2007. Lecture notes in computer science, vol 4727. Springer-Verlag, pp 450–466 [Online]. Available: https://doi.org/10.1007/978-3-540-74735-2_31
14. Anurupam K (2018) Dynamic S-box implementation in present cipher. Int J Comput Sci Eng 6(9). E-ISSN: 2347-2693
15. Suresh H, Vignesh Chandrasekhar R (2018) Lightweight hardware architectures for present cipher in FPGA. IJEDR 6(1). ISSN: 2321-9939
16. Azari HD, Joshi PV (2018) An efficient implementation of present cipher model with 80 bit and 128 bit key over FPGA based hardware architecture. Int J Pure Appl Math 119(4):1825–1832
17. Gomez E, Hernández C, Martinez F (2017) Performance evaluation of the present cryptographic algorithm over FPGA. Contemp Eng Sci 10(12), 555–567. https://doi.org/10.12988/ces.2017.7653
18. Guan Z-H, Huang F, Guan W (2005) Chaos-based image encryption algorithm. Elsevier B.V. https://doi.org/10.1016/j.physleta.2005.08.006
19. Lara-Nino CA, Morales-Sandoval M, Diaz-Perez A (2016) Novel FPGA-based low-cost hardware architecture for the present block cipher. In: 2016 Euromicro conference on digital system design. IEEE. ISBN: 978-1-5090-2817-7/16. https://doi.org/10.1109/dsd.2016.46
20. Hanley N, O'Neill M (2012) Hardware comparison of the ISO/IEC 29192-2 block ciphers. In: Proceedings of IEEE computer society annual symposium on VLSI (ISVLSI), pp 57–62
21. Lara-Nino CA, Diaz-Perez A, Morales-Sandoval M (2017) Lightweight hardware architectures for the present cipher in FPGA. IEEE. ISSN: 1549-8328
22. Ali-Pacha A, Hadj-Said N, M'Hamed A, Belgoraf A (2007) Lorenz's attractor applied to the stream cipher (Ali-Pacha generator). Chaos Soliton Fract 33(5):1762–1766
23. Merah L, Ali-Pacha A, Said NH, Mamat M (2013) Design and FPGA implementation of Lorenz chaotic system for information security issues. Appl Math Sci 7(5):237–246

Estimation of FSO Link Availability for Visakhapatnam Coastal Region

Mogadala Vinod Kumar, G. Sasibhushana Rao, D. Amani, and Ch. Babji Prasad

Abstract Free-space optical (FSO) communications provide high data rates, secure, license-free transmission and are immune to electromagnetic interference. However, the reliability of FSO link depends on given geographic location and atmospheric conditions. In this paper, FSO link availability is estimated for Visakhapatnam region using visibility data collected from IMD Visakhapatnam. Atmospheric attenuation is calculated by using Kim's model. From the simulation results, it is observed that the atmospheric attenuation is as high as 30.26 dB/km, 22.77 dB/km and 20.24 dB/km for the transmission windows 850 nm, 1300 nm and 1550 nm, respectively.

Keywords Free-space optics · Visibility · Link availability · Kim's model

1 Introduction

The next-generation cellular systems are expected to support thousand times more backhaul traffic than the current cellular generation. Design of an efficient backhaul system for the next-generation cellular system is a major challenge for engineers and researchers. In order to achieve this, the network operators require new infrastructure for wireless networks based on high-speed links between the base stations and the backbone networks. The existing backhaul links have limited capacity, and hence, it is necessary to upgrade backhaul links in order to accommodate increased data rates [1]. Therefore, choosing a suitable technology in the design of the backhaul network

M. Vinod Kumar (✉) · G. Sasibhushana Rao · D. Amani · Ch. Babji Prasad
Department of Electronics and Communication Engineering, AU College of Engineering (A), Andhra University, Visakhapatnam, Andhra Pradesh 530003, India
e-mail: mogadala.vinod@gmail.com

G. Sasibhushana Rao
e-mail: sasigps@gmail.com

D. Amani
e-mail: amani927dogga@gmail.com

Ch. Babji Prasad
e-mail: babjiprasad.ch@gmail.com

P. S. R. Chowdary et al. (eds.), *Microelectronics, Electromagnetics and Telecommunications*, Lecture Notes in Electrical Engineering 655,
https://doi.org/10.1007/978-981-15-3828-5_40

architecture plays a vital role in the performance of the next-generation cellular networks [2]. The major existing backhauling techniques are wired and wireless backhaul networks. Wired backhaul includes copper and fiber. When compared to copper, fiber always supports high data rates, but the installation of fiber cables for the small cell base station may not be a good solution due to the high cost of installation, whereas RF-based wireless backhaul links have limited data rates and also not secure. Millimeter wave (mmW) RF communications operate over unlicensed spectrum, but its range is limited to very short distance due to attenuation. FSO communication system has attracted significant attention due to virtually unlimited bandwidth as it operates in the unlicensed terahertz spectrum band. The other advantages of this system include ease of deployment, interference-free transmission and suitability for the backhaul connections in the next-generation cellular networks [3]. Compared to fiber communication, FSO is also cost effective as it does not require digging. However, FSO signals are prone to atmospheric conditions (fog, clouds, snow, etc.). Hence, these signals experience random fluctuations in intensity, while propagating through the turbulent atmosphere. The received signal intensity will fluctuate due to interaction of the wave front with gas molecules and suspended particles present in the atmosphere. Absorption, scattering and turbulence are the atmospheric phenomena, which affect the FSO beam propagation [4]. Hence, FSO link availability is dependent on weather conditions in a given geographic location, where it is supposed to be installed. In this paper, attenuation due to scattering is calculated from the visibility data collected from IMD Visakhapatnam. Further, FSO link availability is calculated based on the results obtained from the attenuation due to scattering.

The rest of the paper is organized as follows. Section 2 discusses the effect of atmosphere conditions on the FSO signal. Results and discussions are presented in Sect. 3. Conclusions are given in Sect. 4.

2 Effect of Atmosphere Conditions on the FSO Signal

The received signal power in terms transmitted signal power is given by the following relation [5],

$$P_{\mathrm{r}} = P_{\mathrm{t}} \frac{a_2^2}{[a_1^2 + \theta L]^2} \mathrm{e}^{-\beta(\lambda)L} \tag{1}$$

where P_{r} is the power received, P_{t} is the transmitted power, a_1 is the diameter of transmitter aperture, a_2 is the diameter of receiver aperture, L is the link distance, θ is the beam divergence angle and $\beta(\lambda)$ is the attenuation coefficient.

2.1 Attenuation Due to Scattering and Absorption

Scattering and absorption lead to attenuation of the optical signal. The attenuation of optical signal due to scattering and absorption can be estimated using Beer–Lambert's law [6] which is given by,

$$\tau(\lambda) = \exp[-\beta(\lambda)L] \tag{2}$$

where $\tau(\lambda)$ is the transmittance, $\beta(\lambda)$ represents the total attenuation/extinction coefficient of the atmosphere at wavelength λ defined as,

$$\beta(\lambda) = A_a + S_a \tag{3}$$

where A_a and S_a are the absorption and scattering coefficient, respectively. FSO signal wavelength can be chosen such that it will have low absorption. Therefore, the total attenuation coefficient is approximately due to scattering effect. For clear and foggy weather conditions, the attenuation coefficient can be determined from the visibility data through Kim's model as [7]:

$$\beta(\lambda) = \frac{3.91}{V}\left(\frac{\lambda}{550}\right)^{-q} \tag{4}$$

where V is the visibility in km and q being the particle size distribution coefficient, which is defined as

$$q = \begin{cases} 1.6 & V > 50\,\text{km} \\ 1.3 & 6\,\text{km} < V < 50\,\text{km} \\ 0.16V + 0.34 & 1\,\text{km} < V < 6\,\text{km} \\ V - 0.5 & 0.5\,\text{km} < V < 1\,\text{km} \\ 0 & V < 0.5\,\text{km} \end{cases} \tag{5}$$

Link availability (L_a) can be obtained from cumulative distribution function (CDF) of visibility [8] and is defined as

$$L_a = \text{Prob}[V \geq V_{\min}(L)] = 1 - F[V_{\min}(L)] \tag{6}$$

where $F[V_{\min}(L)]$ is the CDF of visibility.

3 Results and Discussion

Atmospheric attenuation is calculated using Eqs. (4) and (5) based on visibility data (2016 and 2017) for three wavelengths (i.e., 850, 1300 and 1550 nm). The results are shown in Figs. 1, 2, 3, 4, 5 and 6 and are further summarized in Tables 1 and 2.

The minimum, maximum and average values of attenuation are 1.58, 20.24, 7.747 dB/km at 1550 nm; 3.45, 30.26, 12.91 dB/km at 850 nm; 1.986, 22.77, 8.994 dB/km at 1300 nm wavelengths, respectively, for the year 2016. The effect of wavelength on attenuation can be observed from the values and can be concluded that 1550 nm transmission window has less atmospheric attenuation when compared to other two transmission windows, whereas the minimum, maximum and average values of attenuation for the year 2017 are 1.58, 20.24, 7.138 dB/km at 1550 nm; 3.45,

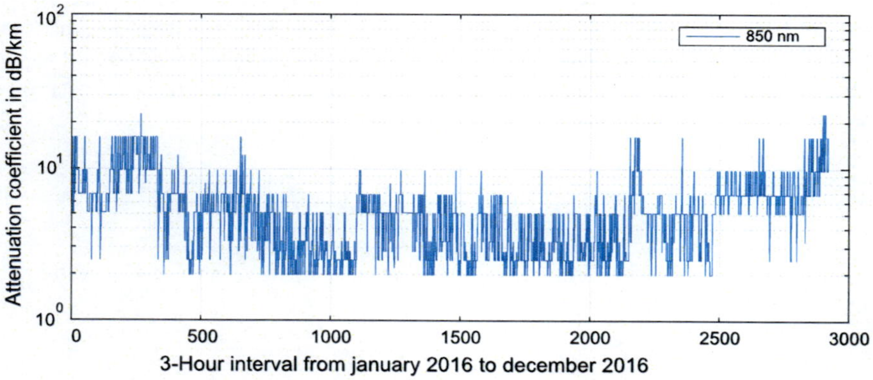

Fig. 1 Atmospheric attenuation plot at 850 nm for the year 2016

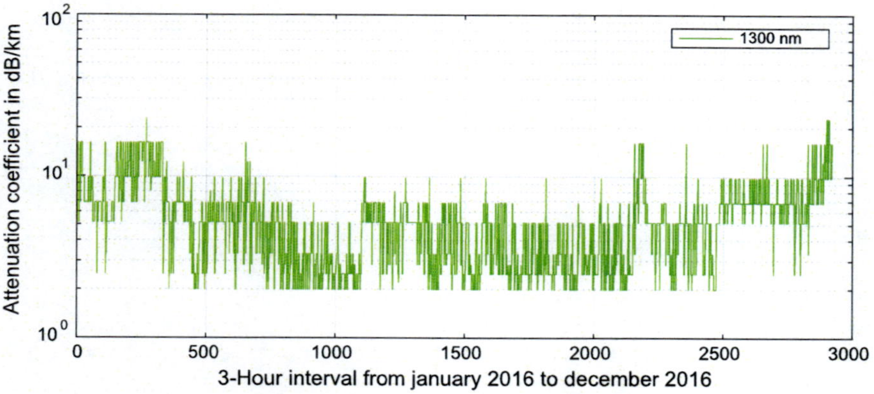

Fig. 2 Atmospheric attenuation plot at 1300 nm for the year 2016

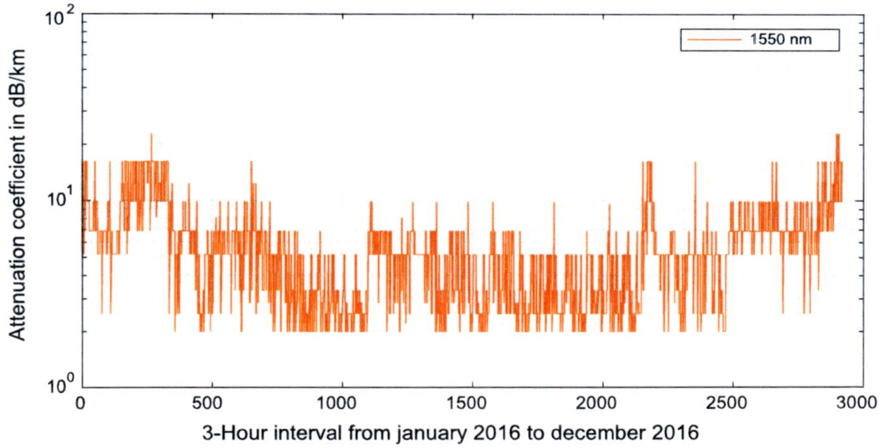

Fig. 3 Atmospheric attenuation plot at 1550 nm for the year 2016

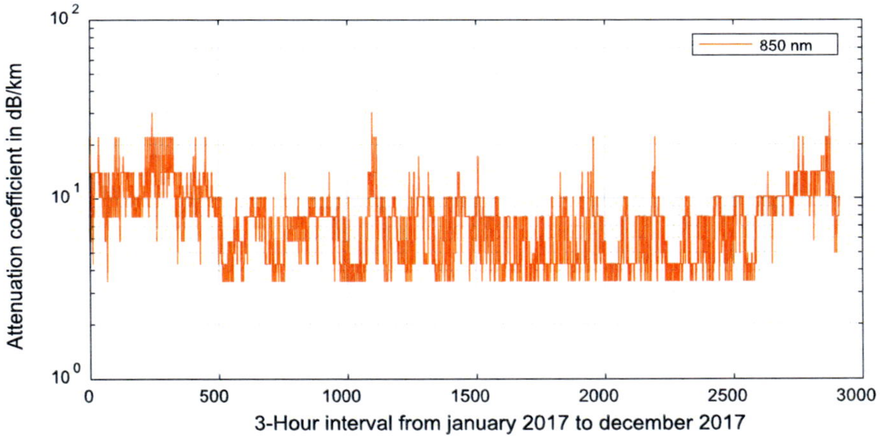

Fig. 4 Atmospheric attenuation plot at 850 nm for the year 2017

30.26, 12.07 dB/km at 850 nm; 1.986, 22.77, 8.321 dB/km at 1300 nm wavelengths, respectively.

Table 3 shows the CDF and PDF values of visibility. Percentage of FSO link availability is obtained from the CDF values of visibility, and it is shown in Table 4. It can be observed that percentage of link availability reduces if the visibility value required is more.

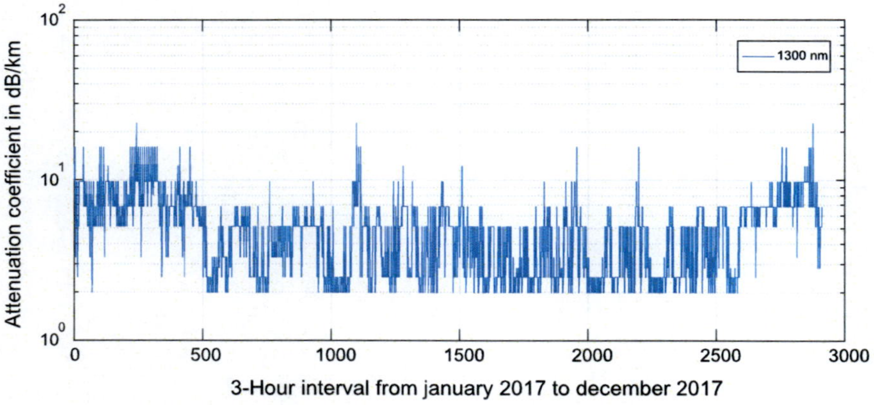

Fig. 5 Atmospheric attenuation plot at 1300 nm for the year 2017

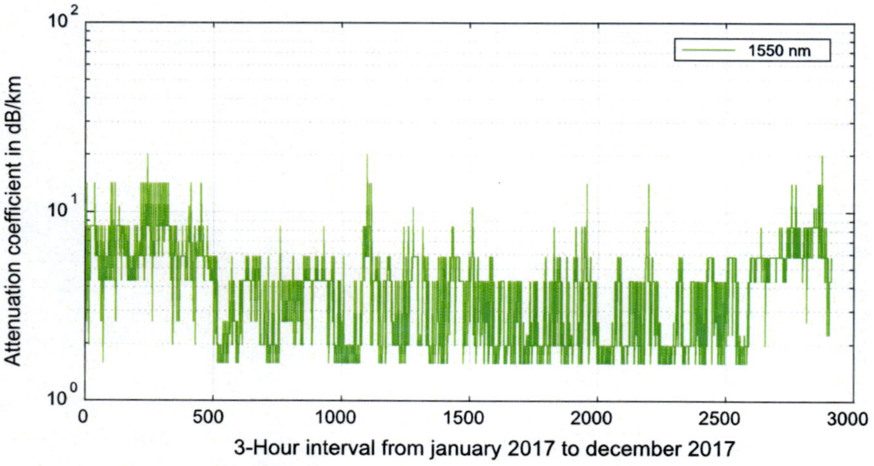

Fig. 6 Atmospheric attenuation plot at 1550 nm for the year 2017

4 Conclusions

FSO communication meets the requirements of the next-generation cellular networks such as high data rates, secure, cost effective and energy efficient. However, FSO link performance is affected by weather conditions which severely reduce the FSO link reliability. In this paper, FSO link availability is calculated by using visibility data collected for Visakhapatnam coastal region from IMD Visakhapatnam. From the simulation results, it is observed that 1550 nm wavelength has less atmospheric attenuation when compared to 850 and 1300 nm wavelengths.

Table 1 Summary of the atmospheric attenuation for the year 2016

Month	Attenuation 1550 nm (dB km^{-1})			Attenuation 850 nm (dB km^{-1})			Attenuation 1300 nm (dB km^{-1})		
	Min	Max	Average	Min	Max	Average	Min	Max	Average
Jan	1.97	14.15	7.74	4.31	22.04	12.91	2.48	16.11	8.99
Feb	1.58	20.24	7.15	3.45	30.26	12.02	1.98	22.77	8.32
Mar	1.58	14.15	4.69	3.45	22.04	8.47	1.98	16.11	5.57
Apr	1.58	15.40	5.60	3.45	18.52	8.47	1.98	13.25	5.62
May	1.58	8.44	3.28	3.45	14.03	6.31	1.98	9.80	3.96
Jun	1.58	8.44	3.77	3.45	14.03	7.08	1.98	9.80	4.53
Jul	1.58	8.44	3.77	3.45	14.03	7.08	1.98	9.80	4.53
Aug	1.58	8.44	2.59	3.45	14.03	5.29	1.98	9.80	3.19
Sep	1.58	14.15	3.60	3.45	22.04	6.79	1.98	16.11	4.32
Oct	1.58	14.15	4.08	3.45	22.04	7.52	1.98	16.11	4.88
Nov	1.97	20.24	6.87	4.31	30.26	11.67	2.48	22.77	8.02
Dec	1.97	20.24	6.99	4.31	30.26	11.84	2.48	22.77	8.16

Table 2 Summary of the atmospheric attenuation for the year 2017

Month	Attenuation 1550 nm (dB km^{-1})			Attenuation 850 nm (dB km^{-1})			Attenuation 1300 nm (dB km^{-1})		
	Min	Max	Average	Min	Max	Average	Min	Max	Average
Jan	1.58	20.24	6.78	3.45	30.26	11.54	1.98	22.77	7.92
Feb	1.97	14.15	7.13	4.31	22.04	12.07	2.48	16.11	8.32
Mar	1.58	16.24	3.52	3.45	25.22	6.99	1.98	17.24	5.24
Apr	1.58	8.44	3.73	3.45	14.03	7.03	1.98	9.08	4.49
May	1.58	20.24	3.63	3.45	30.26	6.77	1.98	22.77	4.35
Jun	1.58	15.34	3.52	3.45	25.47	6.99	1.98	18.45	5.24
July	1.58	10.68	3.37	3.45	17.20	6.44	1.98	12.17	4.07
Aug	1.58	8.44	2.75	3.45	14.03	5.48	1.98	9.08	3.36
Sep	1.58	14.15	2.92	3.45	22.04	5.71	1.98	16.11	3.55
Oct	1.58	8.44	3.73	3.45	14.03	7.52	1.98	9.80	4.47
Nov	1.58	8.44	4.08	3.45	14.03	7.52	1.98	9.80	4.88
Dec	1.98	9.80	3.36	4.92	30.26	11.55	2.83	22.77	7.92

Table 3 CDF and PDF values of visibility

Year		Visibility (V in km) out of 365 days				
		1	2	3	4	5
2016	PDF	0.00137	0.05068	0.08836	0.2082	0.2771
	CDF [$F(V)$]	0.00137	0.05205	0.1404	0.3486	0.6257
2017	PDF	0.00103	0.01786	0.1058	0.2023	0.2624
	CDF [$F(V)$]	0.00103	0.01890	0.1247	0.3269	0.5893

Table 4 Link availability

Year	% Link availability				
	$1 - F$ [1]	$1 - F$ [2]	$1 - F$ [3]	$1 - F$ [4]	$1 - F$ [5]
2016	99.89	94.79	85.96	65.14	37.43
2017	99.89	98.11	87.53	67.31	41.07

Acknowledgements The corresponding author would like to acknowledge the Ministry of Electronics and Information Technology (MeitY), Government of India, for providing financial assistance to this research work through "Visvesvaraya PhD scheme for Electronics and IT."

References

1. Rao GS (2013) Mobile cellular communication, 1st edn. Pearson Education, New Delhi
2. Schulz D et al (2016) Robust optical wireless link for the backhaul and fronthaul of small radio cells. J Lightwave Technol 34(6):1523–1532
3. Alzenad M, Shakir MZ, Yanikomeroglu H, Alouini M (2018) FSO-based vertical backhaul/fronthaul framework for 5G+ wireless networks. IEEE Commun Mag 56(1):218–224
4. Popoola W, Ghassemlooy Z, Awan MS, Eitgeb EL (2008) Atmospheric channel effects on terrestrial free space optical communication links. In: 3rd international conference on electronics, computers and artificial intelligence, Romania, pp 17–23
5. Kaushal H, Jain VK, Kar S (2017) Free space optical communication, 1st edn. Springer
6. Ghassemlooy Z, Popoola W, Rajbhandari S (2013) Optical wireless communications: system and channel modelling with MATLAB. CRC Press, Boca Raton
7. Majumdar A, Ricklin JC (2008) Free-space laser communications: principles and advances. Springer, New York
8. Kshatriya AJ, Acharya YB, Aggarwal AK et al (2016) Estimation of FSO link availability using climatic data. J Optics 45(4):324–330

CPW-Fed Monopole Antenna for RCS Reduction in X-Band Applications

V. Suryanarayana, N. Uzwala, V. Sateesh, M. Satya Anuradha, and S. Paul Douglas

Abstract This paper presents the design of CPW-fed monopole antenna for RCS reduction in the X-band. A basic CPW-fed monopole antenna is designed using FR-4 substrate and parameters such as return Loss, VSWR, gain, and monostatic and bistatic RCS are analyzed. A single artificial magnetic conductor (AMC) cell of size 50 mm × 15 mm is introduced on the basic structure and its RCS is studied. The resultant RCS was found to be more than that of the basic antenna. The size of the AMC cell is diminished to 4 mm × 2 mm and the periodicity of the cells is increased on the surface of basic antenna. With the introduction of AMC cells, reactance on the surface of the antenna is modified resulting in the alteration of scattering properties and hence RCS is reduced. The simulation study has been done by incorporating HFSS 2016.1 platform. From the simulation results, monopole antenna with AMC cells exhibits more than 10 dB RCS (monostatic and bistatic) reduction when compared with the other two antenna designs and the remaining parameters are maintained at the same level (for all the three designs).

Keywords CPW feed · AMC cells · RCS

V. Suryanarayana (✉) · N. Uzwala · V. Sateesh · M. Satya Anuradha
Department of Electronics and Communication Engineering, AU College of Engineering (A), Andhra University, Visakhapatnam, Andhra Pradesh, India
e-mail: surya.velamala@gmail.com

N. Uzwala
e-mail: naravauzwala1507@gmail.com

V. Sateesh
e-mail: sateeshvirothu@gmail.com

M. Satya Anuradha
e-mail: radhamsa@gmail.com

S. Paul Douglas
Department of Chemical Engineering, AU College of Engineering (A), Andhra University, Visakhapatnam, Andhra Pradesh, India
e-mail: pauldouglas12@gmail.com

P. S. R. Chowdary et al. (eds.), *Microelectronics, Electromagnetics and Telecommunications*, Lecture Notes in Electrical Engineering 655,
https://doi.org/10.1007/978-981-15-3828-5_41

1 Introduction

RCS reduction has become crucial in many applications. There are few methods which provide RCS reduction, such as radar absorbing material (RAM), ground cut slots, miniaturization, wireless distributed loading, frequency selective surfaces (FSS), artificial magnetic conductors (AMC), and electronic bandgap structures (EBG). The selection of these methods is based on our requirement or application. As an exceptional scattering device, antenna plays a crucial role as a source for stealth objects. The RCS reduction of an antenna must conserve the basic radiation properties. Thus, it is very problematic to accomplish the RCS reduction of the antenna without degrading the radiation properties [1–5]. Artificial magnetic conductors (AMCs) are such surfaces that reflect the incident waves with zero degrees reflection phase. Generally, the reflection phase of AMCs touches zero degree phase at one frequency known as the resonant frequency. The useful bandwidth of an AMC is in general defined as $\pm 90°$ to on either side of the resonant frequency as these phase values do not cause destructive interference between reflected and directed waves [6]. By loading AMC, wideband radar cross section and RCS reduction larger than 10 dB can be achieved for a conventional patch antenna [7, 8]. AMC metasurface also achieves an ultra-wideband RCS reduction using the polarization conversion technique [9]. A novel low-scattering monopole antenna based on the artificial magnetic conductor covered a band of 3.4–12.4 GHz, and RCS reduction is reached within 8.9–11.2 GHz for about 26 dB [10, 11]. Apart from previous results by the authors, the work on CPW-fed monopole with AMC cells is limited and thus encouraged us to work on this. In this paper Sect. 2 gives brief discussion about Antenna Design Methodology, Sect. 3 gives Results and Discussion and Sect. 4 gives Conclusion and Reference.

2 Antenna Design Methodology

2.1 CPW-Fed Monopole Antenna Design

Initially, a CPW-fed monopole antenna of dimensions 50 mm × 50 mm × 1.6 mm is designed and simulated over an FR4 substrate. Results such as return loss, VSWR, gain, monostatic and bistatic RCS are studied. Figure 1 shows monopole antenna design.

In order to reduce RCS, a single AMC cell of size 50 mm × 15 mm is introduced on the basic structure and simulated. Results are analyzed and were found that RCS of the monopole antenna with a single AMC cell is more when compared with the monopole antenna. The size of the AMC cell is been diminished in order to reduce RCS. Figure 2 shows the monopole antenna with a single AMC cell design.

Fig. 1 Monopole antenna

Fig. 2 Monopole antenna with single AMC cell

2.2 *Diminished AMC Cell Design*

A unit length of 5 mm is considered, and AMC cells of different dimensions with the ground beneath are introduced over this unit length. The reflection phase of these cells is calculated using floquet port and boundary conditions. AMC cell of size, 4 mm × 2 mm, gave good performance in the desired frequency range. The dimensions of the AMC cells are shown in Fig. 3, and Fig. 4 shows AMC cell with the master–slave boundary and floquet port.

These AMC cells are arranged periodically on the structure and simulated. Figure 5 shows the final design of the monopole antenna with AMC cells. The dimensions of the designed antenna are tabulated in Table 1, and the information about the comparison of material properties for designed and reference antenna is tabulated in Table 2.

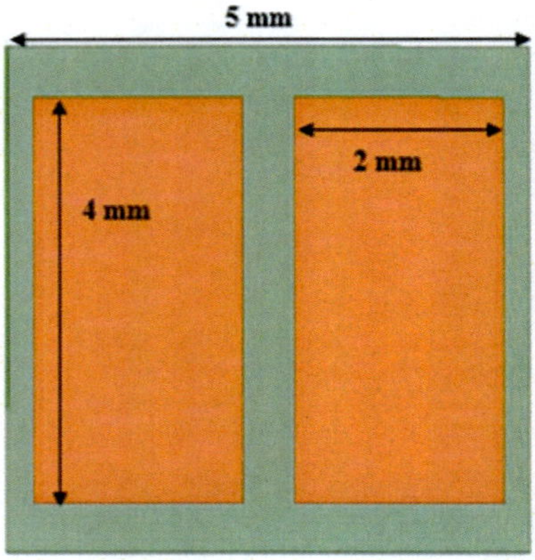

Fig. 3 AMC cells dimensions

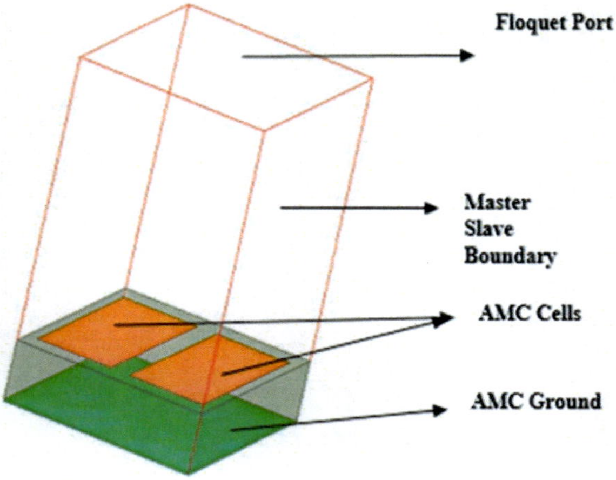

Fig. 4 AMC cells with master–slave boundary and floquet port

3 Results and Discussion

3.1 *Reflection Phase*

Figure 6 shows the reflection phase of AMC cells. At 0° phase, AMC cells exhibit resonant frequency of 10.43 GHz and desired frequency range, i.e., 8.4–12.2 GHz is observed between phases of ±90°.

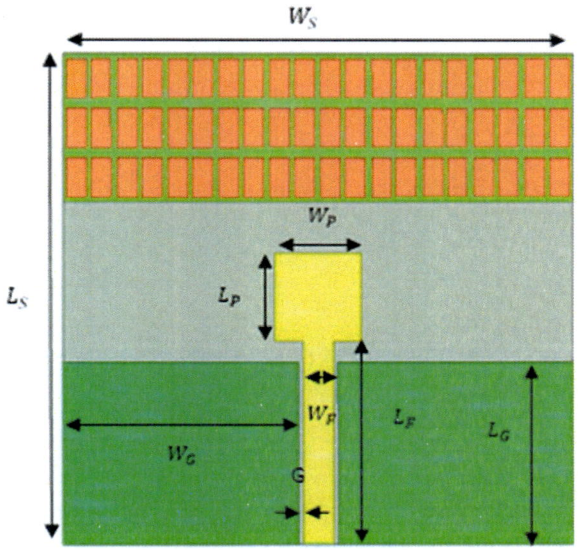

Fig. 5 Monopole antenna with AMC cells

Table 1 Comparison of dimensions for designed antenna and reference antenna

Dimensions (in mm)	Designed antenna	Ref antenna [11]
L_S	50	50
W_S	50	50
H_S	1.6	2
L_P	9	8
W_P	8.52	–
L_F	20.8	20.8
W_F	2.96	–
L_G	18.6	18.1
W_G	23.02	–
G	0.5	0.5
AMC shape and size	4 × 2 rectangle cell	4.6 × 4.6 square cell

Table 2 Comparison of material properties for designed antenna and reference antenna

Property	Accessible frequency range (GHz)	Dielectric constant	Thermal conductivity [W/(m K)]	Loss tangent
FR-4	1–15	4.4	0.81	0.02
Poly-tef Ref. [11]	3–16	2.65	–	0.003

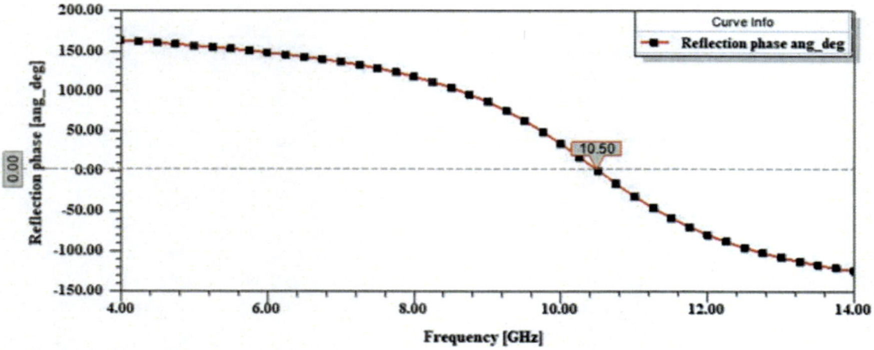

Fig. 6 Reflection phase of AMC cells

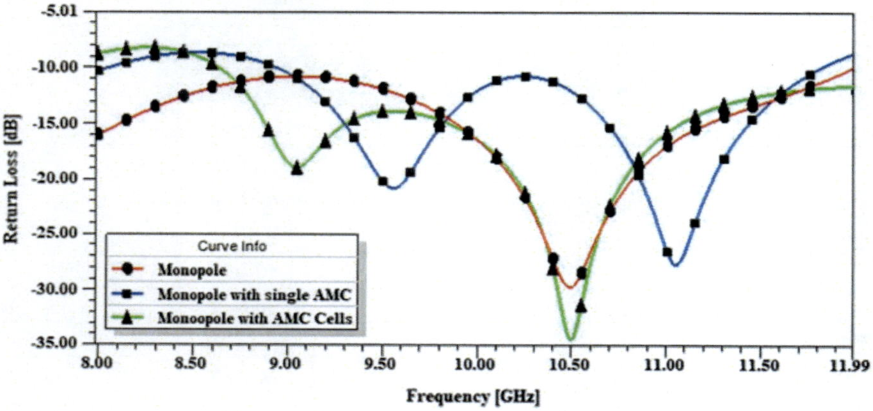

Fig. 7 Return loss comparison of all the three designs

3.2 Return Loss

Figure 7 shows the frequency versus return loss curves of all the three designs. It is found that S_{11} of all the three designs is below -10 dB and exhibit good performance in the desired frequency range. In the designed antenna, a wideband of approximately 8.6–12 GHz is observed with a return loss of -34.4 dB at 10.5 GHz.

3.3 VSWR

Figure 8 shows frequency versus VSWR curves of all the three designs, and it is found that VSWR is maintained less than 2 for the designed antenna.

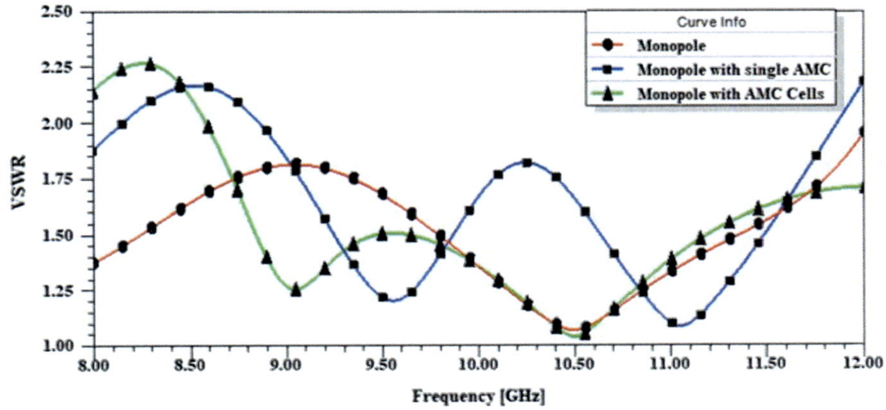

Fig. 8 VSWR comparison of all the three designs

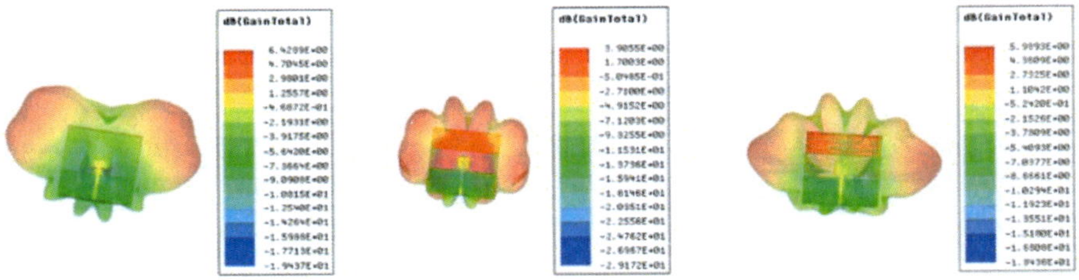

Fig. 9 Gain of all the three designs

3.4 Gain

Figure 9 shows gain plots of all the three designs. It is observed that the gain of a single monopole antenna is 6.42 dB; for monopole with single AMC, the gain is found to be 3.90 dB; and for monopole with AMC cells, the gain is 5.98 dB. Thus, gain is maintained good for the designed antenna.

3.5 Monostatic RCS

Figure 10 shows plot of frequency versus monostatic RCS of three designs. The RCS value of monopole antenna with AMC cells reached up to −23 dB at 10.05 GHz, which is approximately 10 dB less compared to the basic design, and 13 dB less when compared with monopole with single AMC.

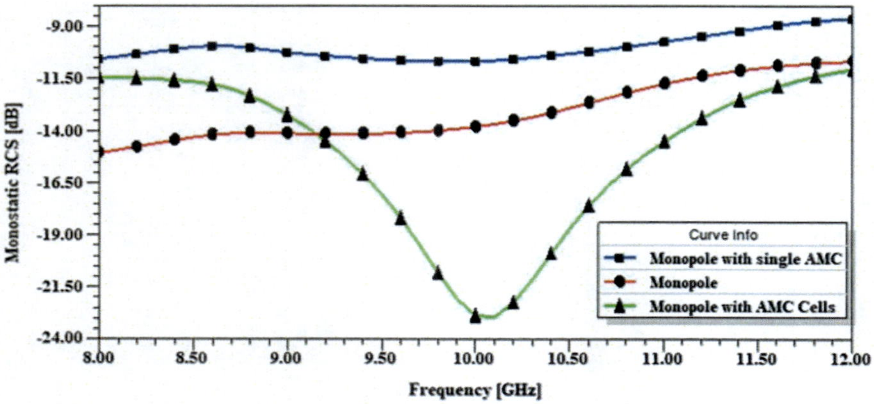

Fig. 10 Monostatic RCS comparison of all the three designs

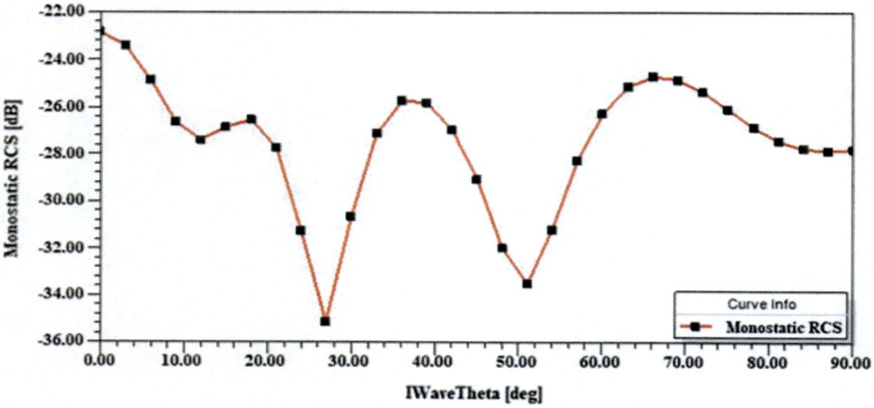

Fig. 11 Monostatic RCS

Figure 11 shows the plot of IWave Theta versus monostatic RCS for monopole antenna with AMC Cells. For all the angles of the incident wave, antenna exhibits performance below −10 dB.

3.6 Bistatic RCS

Figure 12 shows the plot of frequency versus bistatic RCS $\theta = 60°$ for all the three designs. It is clear from the plot that bistatic RCS is below −10 dBsm for all the three designs, and for monopole with AMC cells, the value reached up to −36 dBsm in the desired frequency range.

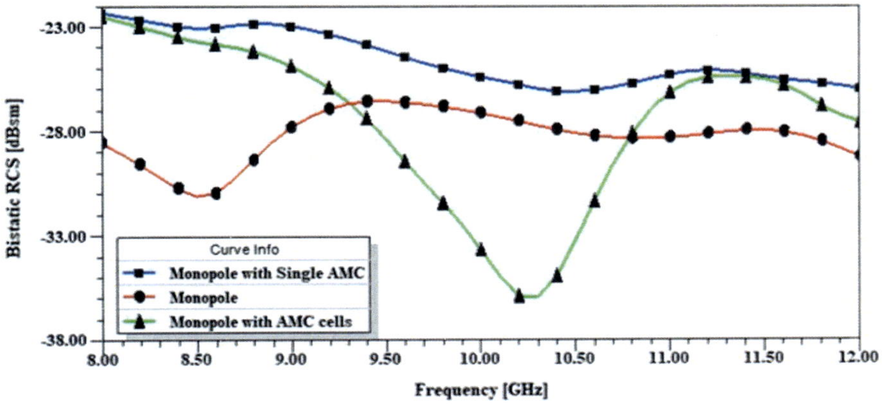

Fig. 12 Bistatic RCS at $\theta = 60°$

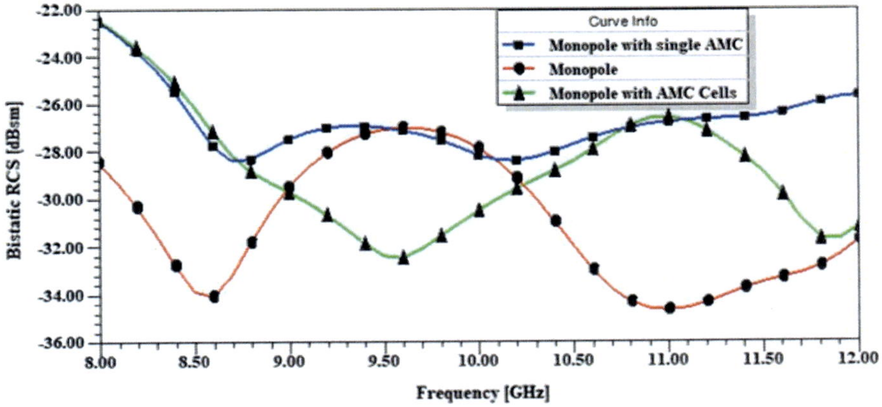

Fig. 13 Bistatic RCS at $\theta = 90°$

Figure 13 shows the plot of frequency versus bistatic RCS at $\theta = 90°$ for all the three designs. It is clear from the plot that bistatic RCS is below −10 dBsm (Table 3).

4 Conclusion

The monopole antenna with AMC cells is designed and simulated using ANSYS HFSS 2016.1. The reactance on the surface of the antenna is modified with the introduction of AMC cells resulting in the alteration of scattering properties. The designed rectangular AMC cells monopole antenna exhibits a wideband of approximately 8.6–12 GHz is observed which is in the desired frequency range. The monostatic

Table 3 Comparison of results between designed and reference antenna

Property	Monopole antenna with AMC cells	Reference antenna
Reflection phase (between ±90°)	8.4–12.2 GHz	7.3–9.7 GHz
Bandwidth (GHz)	8.6–12	3.3–12.2
Return loss (dB)	−34.4	−18 and −17.5
Highest resonant frequency (GHz)	10.05	4 and 11.8
VSWR	<2	–
Gain (dB)	5.98	–
Monostatic RCS (dB)	−23 at 10.50 GHz	−26.2 at 10.1 GHz
Bistatic RCS (dBsm)	−35.93 at 10.23 GHz	–

RCS is reduced to an extent of −23 dB at 10.05 GHz and bistatic RCS value reduced up to an extent of −36 dB. The antenna designed with FR4 substrate makes it cost-efficient, and design of monopole antenna with rectangular AMC cells is easy to fabricate. For further reduction in monostatic RCS, a combination of different AMC cells can be used.

References

1. Knott EF, Shaeffer JF, Tuley MT (2004) Radar cross section, 2nd edn. SciTech Publishing, Inc. ISBN 1-891121-25-1
2. Prabhakaran S, Jacob C, Sabith K, Bhaskaran N (2015) A study of radar cross section (RCS) reduction techniques. J Netw Commun Emerg Technol (JNCET) 3(1). ISSN: 2395-5317. www.jncet.org
3. Edalati A, Sarabandi K (2014) Wideband, wide angle, polarization independent RCS reduction using nonabsorptive miniaturized-element frequency selective surfaces. IEEE Trans Antennas Propag 62(2):747–754
4. Wang F, Jiang W, Hong T, Xue H, Gong S, Zhang Y (2014) Radar cross section reduction of wideband antenna with a novel wideband radar absorbing materials. IET Microwav Antennas Propag 8(7):491–497
5. Jiang W, Hong T, Gong SX (2013) Research on the scattering characteristics and the RCS reduction of circularly polarized microstrip antenna. Int J Antenna Propag 2013. Article ID 735847, 9 p
6. Wang F, Hong T, Gong S (2016) Left-handed material superstrate applied to the RCS reduction of microstrip antenna. J Electromagn Waves Appl 30(11):1428–1439
7. Dewan R, Rahim MKA, Hamid MR, Yusoff MFM, Samsuri NA, Murad NA, Kamardin K, Artificial magnetic conductor for various antenna applications: an overview e21105. https://doi.org/10.1002/mmce.21105,wileyonlinelibrary.com/journal/mmceVC2017WileyPeriodicals,Inc
8. Zheng Y-J, Gao J, Zhou Y-L, Cao X-Y, Xu L-M, Li S-J, Yang H-H (2017) Metamaterial-based patch antenna with wideband RCS reduction and gain enhancement using improved loading method. ISSN 1751-8725

9. Galarregui JCI, Pereda AT, de Falcón JLM, Ederra I, Gonzalo R, Member, IEEE, de Maagt P, Fellow, IEEE (2013) Broadband radar cross-section reduction using AMC technology. 0018-926X © (2013). IEEE
10. Ameri E, Esmaeli SH, Sedighy SH (2019) Ultra-wideband radar cross section reduction by using polarization conversion metasurfaces. Sci Rep 9:478. https://doi.org/10.1038/s41598-018-36542-6
11. Yao P, Zhang B, Duan J, Bai Q (2017) A novel low-scattering and wideband monopole antenna based on artificial magnetic conductor. IOP Conf Ser J Phys 887:012037. https://doi.org/10.1088/1742-6596/887/1/012037
12. Knott EF, Shaeffer JF, Tuley MT (2012) Radar cross section measurements, Springer Science & Business Media 2012. ISBN 1468499041, 9781468499049

Performance Improvement of Solar PV Maximum Power Point Tracking Using Sliding Mode Control Algorithm

M. Ravi Kumar, S. Satyanarayana, and V. Ganesh

Abstract The extraction of maximum power from solar photovoltaic (PV) is important for the performance improvement of the system. Maximum power point tracking (MPPT) algorithms are used to meet the above objective. In recent years, some control techniques along with MPPT are being investigated to further enhance the performance of the system. Among these control methods, a sliding mode control (SMC) technique is integrated to the power converter along with classical MPPT methods like perturb and observe (PO). In this paper, we compare the voltage level, power output and voltage tracking of the system with and without SMC. The system is modelled and simulated in the MATLAB/Simulink. The SMC gives fast convergence to the MPPT to track the reference voltage and gives maximum power output without much oscillation.

Keywords Solar photovoltaic system · MPPT control · Sliding mode control · Perturb and observe · Voltage tracking · Maximum power

1 Introduction

The gap between increasing electrical power demand and generation can be well managed with the usage of renewable energy resources. The major resources of renewable energy are solar irradiation, wind and biomass. Solar energy is gaining the importance because of its wide availability. This leads to the focus on research of solar photovoltaic systems, extraction of maximum power from the PV panels, battery management, and solar PV integration to the grid and control techniques for different solar converter topologies [1, 2]. The maximisation of power extraction

M. Ravi Kumar (✉)
PVP Siddhartha Institute of Technology, Vijayawada, India
e-mail: Ravikumar.melimi@gmail.com

S. Satyanarayana
Raghu Institute of Technology, Visakhapatnam, India

V. Ganesh
JNTUA College of Engineering, Pulivendula, India

P. S. R. Chowdary et al. (eds.), *Microelectronics, Electromagnetics and Telecommunications*, Lecture Notes in Electrical Engineering 655,
https://doi.org/10.1007/978-981-15-3828-5_42

from PV panel is possible by adjusting the duty cycle of the converter connected to the PV panel [3, 4].

The change in duty cycle in accordance with the parameter variation like solar irradiance is governed by the maximum power point tracker (MPPT) algorithm. Studies are being conducted on control techniques of converter along with MPPT to further enhance the performance of the system in terms of tracking reference voltage and extracting maximum power without much oscillation. In this paper, we considered the classical method such as perturb and observe along with sliding mode control (SMC). P&O is applied to DC–DC converter and SMC is applied to DC–AC converter. The paper is organised as follows: in Sect. 2, photovoltaic system connected to grid and the mechanism of MPPT is discussed. In Sect. 3, different methods of MPPT are highlighted followed by Sect. 4 with proposed sliding mode control technique. Section 5 consists of simulation results and discussion. In Sect. 6, conclusions are presented.

2 Photovoltaic System Connected to Grid and Maximum Power Point Tracker Mechanism

Solar PV system can be connected to transmission level or distribution level depending on the scale of generation of solar power. When a photovoltaic system is connected to distribution grid power, electronic-based converters are required as shown in Fig. 1.

2.1 Maximum Power Point Tracker Mechanism

The *V-I* characteristics of solar PV panel are dependent on many parameters mainly like solar irradiance and operating temperature. As these parameters change, the input to the boost converter changes. The DC-to-DC boost converter consists of an

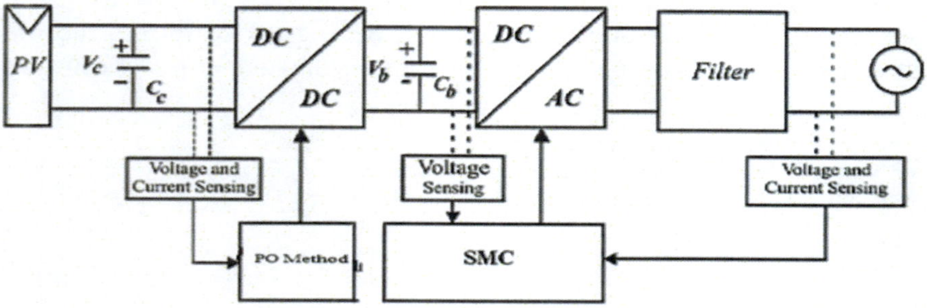

Fig. 1 Block diagram of solar PV system connected to grid

inductor in series to the source, MOSFET, and capacitor in parallel to the output. The boost converter is controlled by varying the duty ratio to MOSFET. The right setting of duty ratio to get maximum power from the boost converter is done with the help of MPPT method [5].

3 Different Methods of PPT

There are different methods that can be distinguished in many aspects such as complexity, economical, speed of convergence, correct tracking of voltage based on change in irradiance and operating temperature [6]. Some most popular MPPT techniques are perturb and observe (PO), incremental conductance (IC), fractional short-circuit current, fractional open-circuit voltage, fuzzy logic, neural networks, etc. [7].

3.1 Perturb and Observe Method (P&O)

The P&O algorithm is simple algorithm from all of the MPPT algorithms [8]. By using this algorithm, find the voltage of operating point at which maximum power can be taken out and it is found by changing solar system's operating voltage by a small different value in both positive and negative direction, which can be considered as ΔV and called as perturbation by maintaining sampling rate constant, to track MPP, with considering operating conditions. By the end, the system will become settled at some point which is nothing but the operating point which corresponds to maximum output power (Fig. 2) [9].

4 Sliding Mode Control

Controllers are used to decide the delivery of firing pulse to the converters. Tuning of controller parameters under uncertain conditions is very difficult with proportional–integral controllers. To overcome this, sliding mode control is used. The trajectory is brought to sliding surface and then to equilibrium point. Sliding states are obtained from instantaneous states with the help of control input. Integration of equivalent control with switching control forms sliding mode control.

$$u(t) = u_{\mathrm{equi}}(t) + u_{\mathrm{swi}}(t) = u_{\mathrm{equi}}(t) + \rho * \mathrm{signum}(\sigma) \tag{1}$$

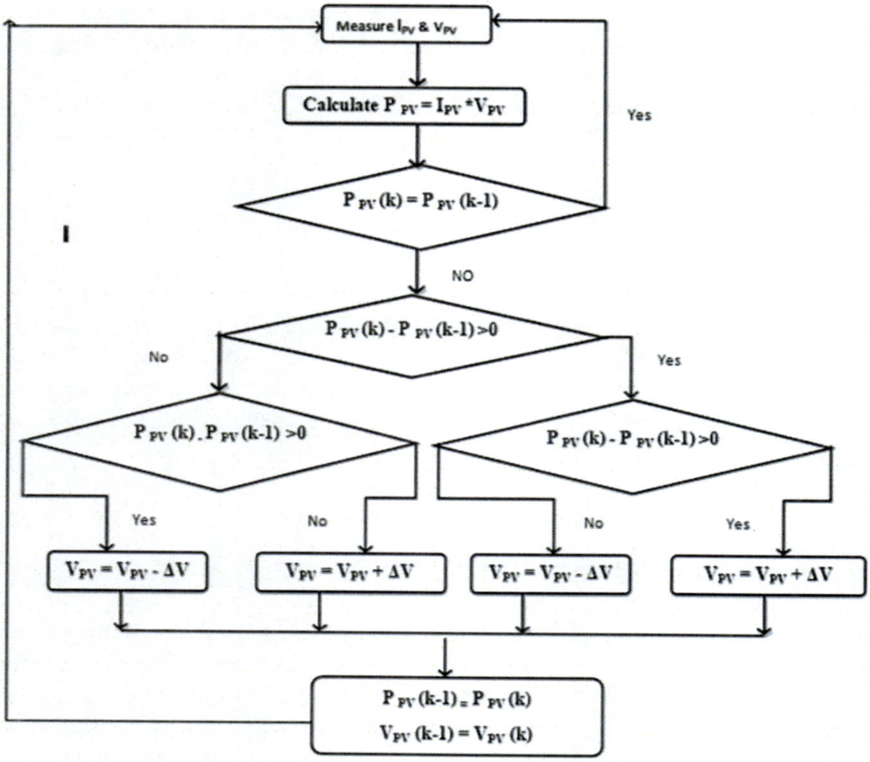

Fig. 2 P& O algorithm

where ρ is a positive constant, $u_{\text{equi}}(t)$ is an equivalent function, $u_{\text{sw}i}(t)$ is a switching function and σ is the error. Signum is a hard switching function so controller performance is highly chattering, and this can be reduced by using hyperbolic tangent function [10].

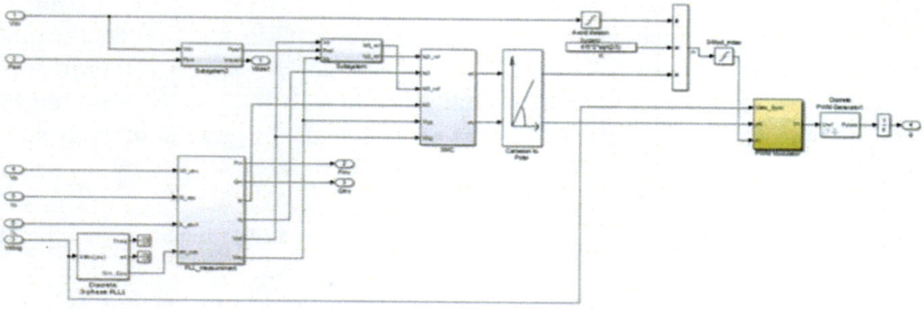

Fig. 3 Simulink diagram of implementation of SMC

5 Simulation Results and Discussion

Total PV system power rating considered is 11.9 kW. Panels in series are 13 so 13 * 54.7 V = 711.1 V is PV system output voltage at maximum power tracking. Panels in parallel are 3 so 3 * 5.58 = 16.74 A is PV output current at maximum power point tracking. Total power = $V_{\mathrm{mpp}} * I_{\mathrm{mpp}} = 711.1 * 16.74 = 11.9$ kW. In all the different cases for different irradiance, i.e. 0, 1000, 1356 W/m^2, maximum power output that can be generated by PV panel will change. Voltages will be more or less same but current will change, so overall power will change. If algorithm is perfect, we must get 11.9 kW at 1000 irradiance at 25 centigrade temperature.

Different methods will produce different amount of maximum power for same irradiance. The reference voltage to be tracked is set as $V_{\mathrm{dc}} = 720$ V. If reference voltage, i.e. V_{dc} is set below V_{mpp}, we cannot get maximum power. The grid considered is Indian system distribution level 50 Hz three phase 33/11–11 kV/400 V. Figure 4 shows the PV system output voltage from PO method without SMC. At 0.2 s, it has reached its maximum value which is less than the V_{mpp}. Figure 5 shows the PV system output voltage with SMC. At 0.1 s, because of the integration of SMC, it has reached the maximum value which is very nearer to the V_{mpp}.

Figure 6 shows the PV system output power at 1000 W/m^2 solar irradiance without SMC maximum value obtained is 10.1 kW. Moreover, the power is fluctuating within 0–0.2 s. Figure 7 shows the output power with SMC. It is observed that the maximum value obtained is above 10.1 kW with SMC. The power has reached a maximum value in less than 0.1 s with SMC. Figure 8 shows the reference voltage tracked by the DC-to-DC boost converter without SMC. It has taken nearly 0.2 s to reach the steady state. Figure 9 shows the reference voltage tracked by the DC-to-DC boost converter

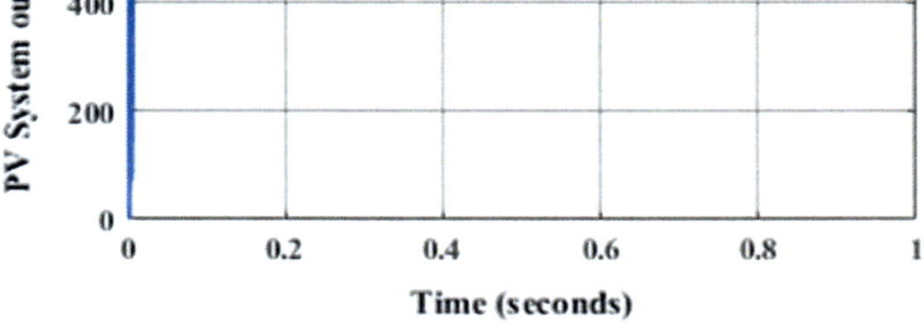

Fig. 4 PV system output voltage from only PO method

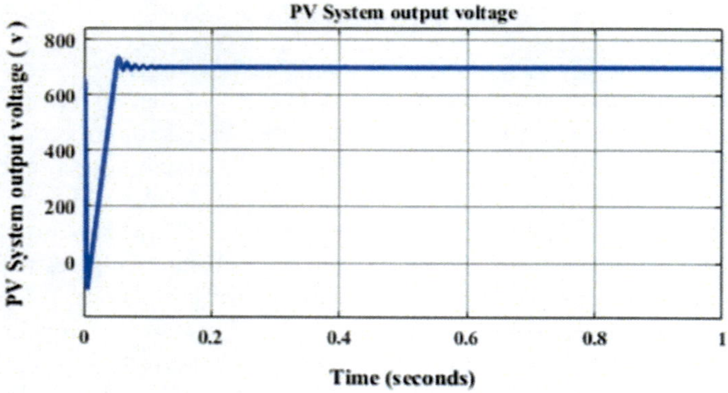

Fig. 5 PV system voltage from PO and SMC method

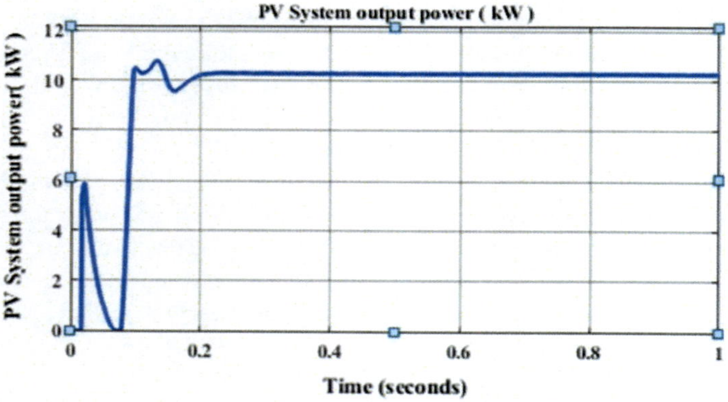

Fig. 6 PV system output power from only PO method

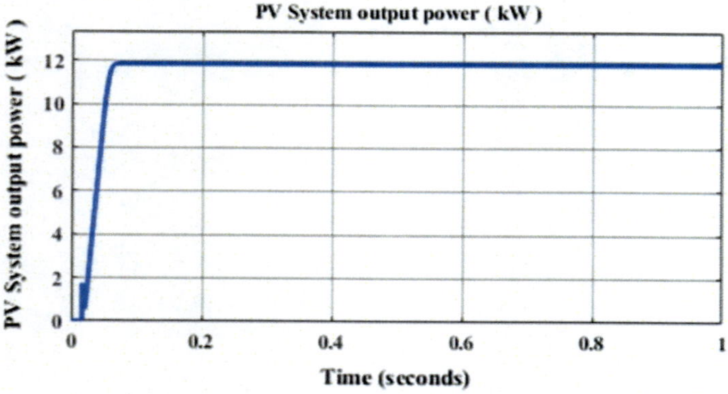

Fig. 7 PV system output power from PO and SMC method

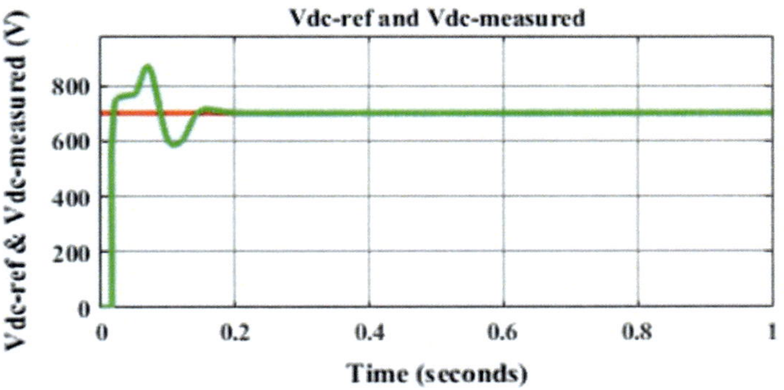

Fig. 8 Voltage measured and reference value of a converter from only PO method

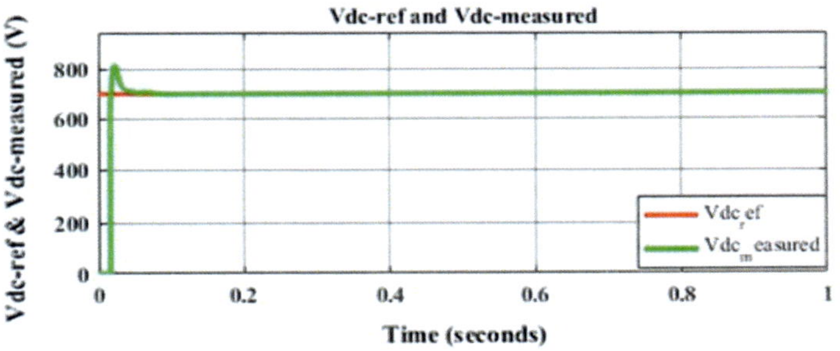

Fig. 9 Voltage measured and reference value of a converter from PO and SMC method

with SMC. At 0.09 s, the steady state is obtained which is very fast when compared without SMC.

6 Conclusion

PV system connected to grid is implemented with PO type MPPT method and it is observed that output power from PV system is unstable, having more oscillations and taking more time for tracking of the reference voltage in order to get maximum power at the MPP operating point. So, a controller is integrated to reduce these problem and to improve the performance of the system. Integrating PO along with SMC technique into the PV system, the voltage reference is tracked very fast compared to only PO without using SMC. Output power from PV system is more for same irradiance level and also very stable with SMC. Results are compared with and without SMC by implementing the simulation model in MATLAB/Simulink environment.

References

1. Rajput SK, Singh OV (2017) Reduction in CO_2 emission through photovoltaic system: a case study. In: 3rd IEEE international conference on nanotechnology for instrumentation and measurement, GBU, India, 16–17 July 2017
2. Devi VK, Premkumar K, Bisharathu Beevi A, Ramaiyer S (2017) A modified perturb & observe MPPT technique to tackle steady state and rapidly varying atmospheric conditions. Sol Energy 157:419–426
3. Attou A, Massoum A, Saidi M (2014) Photovoltaic power control using MPPT and boost converter. Balk J Electr Comput Eng 2(1):23–27 (2014)
4. Bernardo MCP, Peixoto ZMA, Machado Neto, LVB.: A high efficient micro-controlled buck converter with maximum power point tracking for photovoltaic systems. In: International conference on renewable energies and power quality (ICREPQ'09), Valencia (Spain), 15–17 Apr 2009
5. Sivagamasundari MS, Mary PM, Velvizhi VK (2013) Maximum power point tracking for photovoltaic system by perturb and observe method using buck boost converter. Int J Adv Res Electr Electron Instrum Eng 2(6):2433–2439
6. Wu G, Li X (2015) Application of Newton interpolation method in maximum power tracking of photovoltaic power generation. Power Technol 39(7):1432–1434
7. Lyden S, Haque ME, Mahmud MA (2016) Maximum power point tracking methods for PV systems. In: Advances in solar photovoltaic power plants, green energy and technology, pp 79–105, June 2016
8. Femia N, Petrone G, Spagnolo G, Vitelli M (2004) Optimizing duty-cycle perturbation of P&O MPPT technique. In: IEEE 35th annual power specialists conference, Aachen, Germany
9. Amrouche B, Belhamel M, Guessoum A (2007) Artificial intelligence based P&O MPPT method for photovoltaic systems. Revue des Energies Renouvelables ICRESD-07 Tlemcen, pp 11–16 (2007)
10. Khalil HK (2002) Nonlinear systems, 3rd edn. Prentice Hall, Upper Saddle River. ISBN 978-0-13-067389-3

Surfactant Effect on Bandgap and Crystallite Size of $ZnO–TiO_2–CeO_2$ Nanocomposites

H. Srinivasa Varaprasad, P. V. Sridevi, and M. Satya Anuradha

Abstract Multiphase, mixed metal oxide material composites at nanosize with wide bandgap are capable of good electronic functionality at higher temperature compared with existing semiconducting materials. In this work, $ZnO–TiO_2$ semiconducting metal oxide doped with 0.1 M CeO_2 ternary composite is prepared by solid-state method and surfactants like hydroxypropyl cellulose (HPC) and Sodium dodecylbenzene sulfonate (SDBS) are used as reacting agents. Prepared nanocomposites were characterized by X-ray diffraction (XRD), FESEM, and UV–Visible spectroscopy for the observation of crystallite size, surface morphology, and bandgap calculation sequentially. Crystallite size was calculated by using Debye–Scherrer method and was reconfirmed with Williamson–Hall and size–strain plot methods. The results revealed that mixed metal oxide composite generates a new energy level at 2.6 eV. Nanocomposite modified with hydroxypropyl cellulose (HPC) exhibits good optical absorption, around 16–20 nm crystallite size is obtained and also homogeneous distribution of particles is observed.

Keywords Solid state · HPC · SDBS · Wide bandgap

1 Introduction

Nowadays, most of the researchers are focused on mixed metal oxide and its composites to enhance properties like optical, thermal, structural, morphological, electrical, physical, chemical, mechanical, bio-medical, and interdisciplinary applications [1, 2]. It is more secure to use two or more material combinations in the place of one material by carefully choosing and tuning combinations of metal oxides [3, 4]. ZnO and TiO_2 are such kind of materials which can form $ZnTiO_3$, $ZnTiO_4$ perovskites which attracts the attention toward dye degradation, photocatalysis, pigments, sensors, etc. [5–7]. The addition of metals and metal oxide to the binary composite may further

H. Srinivasa Varaprasad (✉) · P. V. Sridevi · M. Satya Anuradha
Department of Electronics and Communication Engineering, Andhra University College of Engineering (A), Andhra University, Visakhapatnam, Andhra Pradesh, India
e-mail: hsv.prasad017@gmail.com

P. S. R. Chowdary et al. (eds.), *Microelectronics, Electromagnetics and Telecommunications*, Lecture Notes in Electrical Engineering 655,
https://doi.org/10.1007/978-981-15-3828-5_43

enhance the results. CeO_2 (3.5 eV) and SnO_2 (3.7 eV) are the metaloxides which have similar bandgaps of ZnO (3.37 eV) and TiO_2 (3.2 eV). The wide bandgap materials are capable of electronic functionality at higher temperatures compared with Si, Ge [8, 9].

All the materials at nanosize exhibit different properties compared with micron size. But the problem with nanomaterials is that they are highly reactive, oxidized very fast, and they get contaminated in the open atmosphere [10]. The surfactants are compounds that modify surface tension, material properties like particle shape, size, phase, surface morphology, and homogenies distribution of particles [11].

The properties of materials are varied with particle size, phase of the materials, calcination temperature, weight ratio, doping material, amount of dopants, catalysts, surface morphology, and preparation methods. The amount and phase distribution, inhomogeneity in the composition, size and shape distribution of crystallite and crystallographic orientations can be determined by diffraction lines of crystalline materials. The different methods to determine crystallite size and lattice strain from the diffraction lines are (1) Variance method, (2) Integral breadth method, (3) Warren–Averbach method, (4) Scherrer method, (5) Williamson–Hall method, (6) Size–strain plot method, etc. [12, 13].

The organization of the paper is as follows, materials used, preparation of CeO_2–ZnO–TiO_2 Semiconducting ternary Nanocomposite, and characterization methods are given in Sect. 2. Section 3 describes the results and analysis of crystallite size and bandgap of the composite by using X-ray diffraction and UV–visible spectroscopy. Surface morphology observed from FESEM. Finally, Sect. 4 concludes.

2 Composite Material Preparation

2.1 Materials

Commercially procured nanopowder materials ZnO (99.9% pure, 30 nm), TiO_2 (Anatase 99.9% pure, 25 nm), CeO_2 (99.9% pure, 35 nm) are used. Isopropanol (C_3H_8O), deionized water (H_2O), ethanol (C_2H_6O), and acetone (C_3H_6O) are used as medium and cleaning agents. Sodium hydroxide (NaOH) is used to adjust the value of pH. Hydroxypropyl cellulose (HPC), sodium dodecylbenzene sulfonate (SDBS) are used as reactive agents. All chemicals are used without any modifications.

2.2 Preparation for Nanocomposite

ZnO–TiO_2 (1:1) Nanocomposite is doped with 0.1 Mol of CeO_2 has been prepared by a solid slate method. The high purity commercially available nanopowders ZnO (99.9% pure) and TiO_2 anatase (99.9% pure) were equally weighted and mixed according to stoichiometry. Then, 0.1 Mol percentage of CeO_2 (99.9% pure) nanopowder was mixed to prepare 1:1 ZnO–TiO_2 nanocomposite. The mixed nanopowder was milled for 45 min at 600 rpm in a 125 ml jar with 1:10 ZrO_2 balls with isopropanol as a medium by using high energy ball mill. The obtained wet powder was collected and magnetically stirred until xerogel is formed at 50 rpm by maintaining 55 °C temperature. Further 50 ml deionized water, 20 m Mol NaOH, 1 Mol hydroxypropyl cellulose (HPC)/1 Mol sodium dodecylbenzene sulfonate (SDBS)/1 Mol hydroxypropyl cellulose (HPC), and sodium dodecylbenzene sulfonate (SDBS) were added and magnetically stirred for 3 h at 250 rpm. Some of the formed gels were collected. The remaining gel was dried at 85 °C for 1 h on heat mount. White-colored fine nanosized powders were obtained. Then these nanopowders were sintered up to 450 °C for 2 h at the increment of 5 °C/min in the muffle furnace. 0.1 M CeO_2–ZnO–TiO_2 composites are named as C1A0 (Pure), C1B0 (HPC), C1C0 (SDBS), and C1D0 (Modified by HPC and SDBS).

2.3 Characterization

The crystallite size and phase structures of the ZnO–TiO_2 (1:1) doped with 0.1 Mol of CeO_2 nanocomposite powder were observed by X-ray diffraction peak analysis (PAN ANALYTICAL, XPERT PRO) at a scanning rate 2°/min in 2θ range from 10° to 90° within CuKα radiation with $\lambda = 1.54060$ Å at 45 kV and 40 mA. The optical properties were calculated by UV–visible spectraphotometer (FILMETRICS F20, UV–VIS spectrophotometer). The micrographs are obtained by FESEM-EDS (JEOL Asia PTE Ltd) operated at 12 kV with 100 nm resolution.

3 Results and Discussion

3.1 X-ray Diffraction

The X-ray diffraction of the prepared sample modified by surfactants is as shown in Fig. 1 and it is analyzed with Philips Xpert high score plus software package. All the diffraction peaks exhibit ZnO, TiO_2, CeO_2 mixed phases structure. Crystallite size is calculated using Debye–Scherrer method, W–H method, and SSP method. The X-ray diffraction peaks of 0.1 M CeO_2 is doped to ZnO–TiO_2 modified by HPC are as shown in Fig. 2. The structure parameters of ZnO–TiO_2 doped with 0.1 M CeO_2 is represented in Table 1.

Debye–Scherrer method: The crystallite size can be calculated by the Scherrer method using Scherrer equation

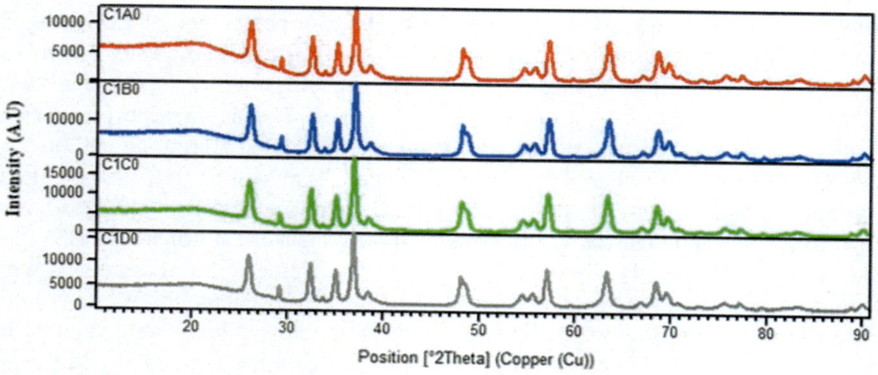

Fig. 1 XRD pattern of $ZnO–TiO_2$ doped with 0.1 M CeO_2 modified by HPC, SDBS, HPC, and SDBS

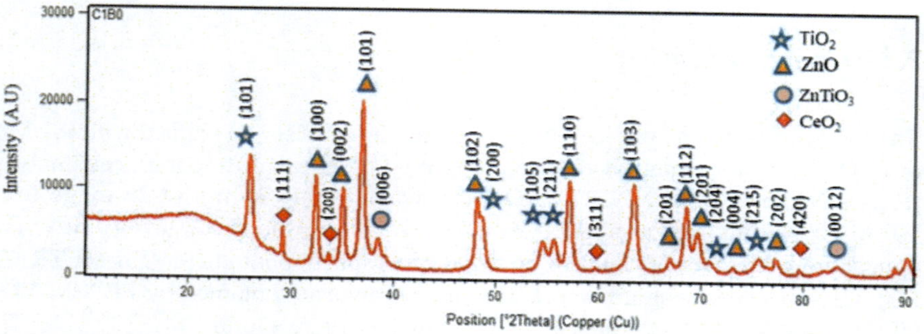

Fig. 2 XRD pattern of $ZnO–TiO_2$ doped with 0.1 M CeO_2 modified by HPC

$$D = \frac{K\lambda}{\beta_{hkl}\cos\theta} \tag{1}$$

Where D is crystallite size in nanometers, $k = 0.94$. The wavelength of radiation $\lambda = 1.5406$ Å (CuKα radiation), β_{hkl} = full width at half maximum (FWHM) of the peak in radians, and θ = Bragg angle.

Lattice strain of the material particles can be calculated by the relation

$$\varepsilon = \frac{\beta_s}{4\tan\theta} \tag{2}$$

Figure 3 shows the $\cos\theta$ verses $1/\beta$. The obtained crystallite sizes of 0.1 M CeO_2 are added $ZnO–TiO_2$ modified composite is as shown in Table 2.

W-H Plot Method. The crystallite size can be calculated by Williamson–Hall method. According to the Williamson–Hall method, the reflections can be determined from the relation

Table 1 X-ray diffraction parameters of ZnO–TiO_2 doped with 0.1 M CeO_2 modified by HPC

S.No.	Compound name and phase	2θ	Hkl	Unit cell	Crystal structure	ICOD
1	ZnO (zincite)	32.40, 35.03, 36.86, 48.06, 57.13, 63.40, 66.93, 68.48, 69.61, 73.10, 77.27	(100), (002), (101), (102), (110), (103), (201), (112), (201), (004), (202)	$a = b =$ 3.22 Å $c = 5.17$ Å $\alpha = \beta =$ 90° $\Upsilon = 120°$ Volume = 46.67 Å^3	Hexagonal	01-075-1526
2	TiO_2 (anatase)	25.84, 48.57, 54.35, 55.48, 63.32, 75.52	(101), (200), (105), (211), (204), (215)	$a = b =$ 3.75 Å $c = 9.42$ Å $\alpha = \beta = \Upsilon$ = 90° Volume = 132.8 Å^3	Tetragonal	00-002-0406
3	$ZnTiO_3$ (ecandrewsite)	38.32, 83.83	(006), (00 12)	$a = b =$ 0.355 Å $c =$ 0.140 Å $\alpha = \beta =$ 90° $\Upsilon = 120°$ Volume = 0.28 Å^3	Trigonal (hexagonal axis)	00-026-1500
4	CeO_2 (cerium oxide)	29.08, 33.60, 57.07, 79.53	(111), (200), (311), (420)	$a = b = c$ = 5.41 Å $\alpha = \beta = \Upsilon$ = 90° Volume = 158.34 Å^3	Cubic	00-001-0800

$$\beta_{hkl}\cos\theta = \frac{K\lambda}{D} + 4\varepsilon\sin\theta \tag{3}$$

where D = crystallite size, ε = lattice strain.

Figure 4 shows the $\beta\cos\theta$ verses 4sin θ. The obtained crystallite sizes of 0.1 M CeO_2 is doped with ZnO–TiO_2 modified composite is as shown in Table 2.

Size–Strain Plot Method. The crystallite size can be calculated by size–strain method. The equation related to the SSP Method is

$$(d_{hkl}\beta_{hkl}\cos\theta)^2 = \frac{K}{D}\left(d_{hkl}^2\beta_{hkl}\cos\theta\right) + \left(\frac{\varepsilon}{2}\right)^2 \tag{4}$$

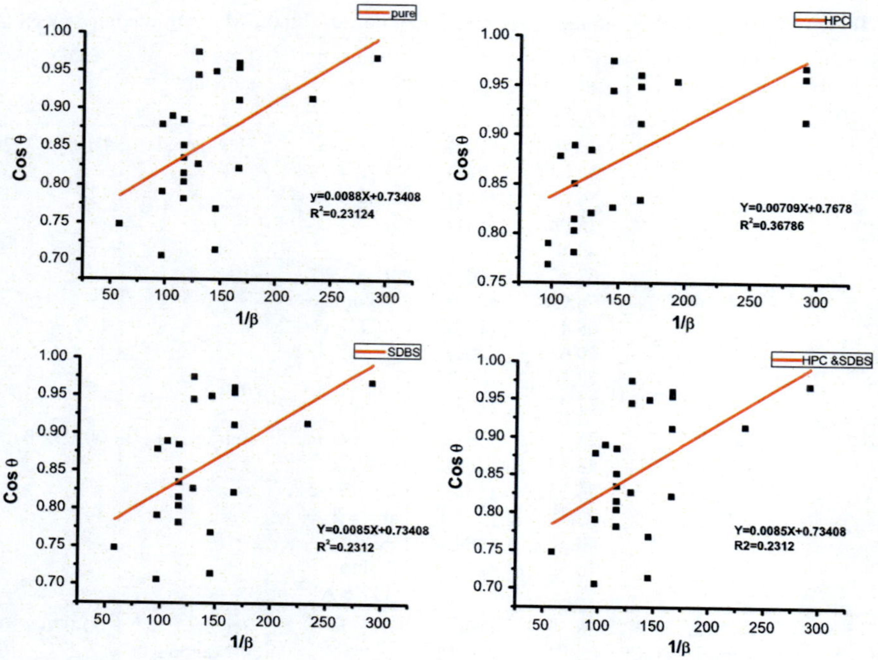

Fig. 3 Scherrer plot of $ZnO–TiO_2$ doped with 0.1 M CeO_2 modified by HPC, SDBS, HPC, and SDBS

Table 2 Crystallite size of $ZnO–TiO_2$ doped with 0.1 M CeO_2 modified composite in different methods

0.1 M CeO_2 added $ZnO–TiO_2$	Debye–Scherrer method (nm)	W–H plot method (nm)	Size–strain plot method (nm)
Pure	16.46	17.13	15.04
+HPC	20.43	21.91	19.06
+SDBS	17.04	18.91	16.23
+HPC and SDBS mixed	17.04	18.13	16.04

Figure 5 shows the $(d\beta \cos\theta)^2$ versus $d^2\beta \cos\theta$. The obtained crystallite sizes of 0.1 M CeO_2 is added to $ZnO–TiO_2$ modified composite is shown in Table 2.

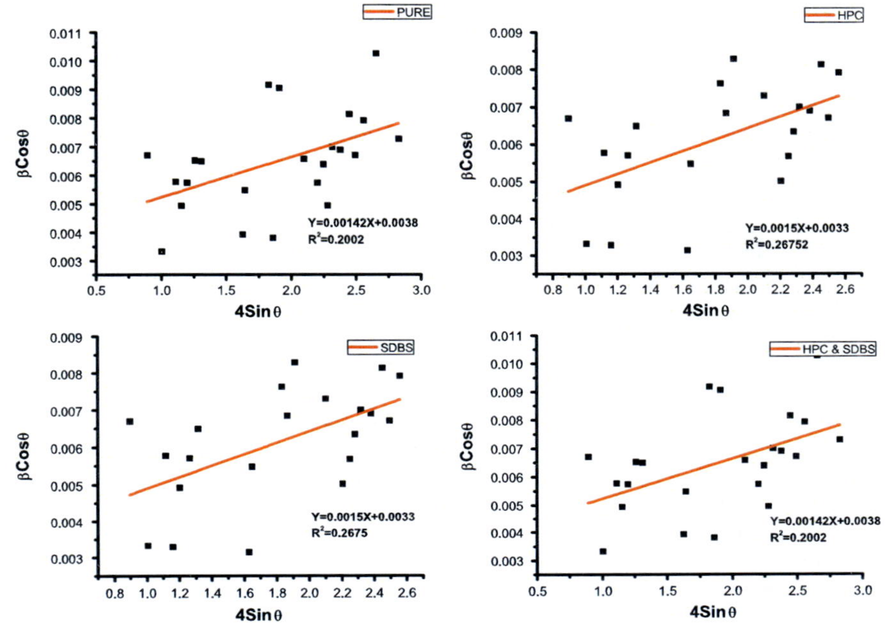

Fig. 4 Williamson–Hall plot of ZnO–TiO_2 doped with 0.1 M CeO_2 modified by HPC, SDBS, HPC, and SDBS

3.2 *Surface Morphology*

FESEM images of 0.1 M CeO_2 doped to ZnO–TiO_2 and modified with HPC, SDBS, HPC, and SDBS are as shown in Fig. 6. The FESEM results showed that the particles are spherical in shape and uniform in size. The average diameter of the particle is approximately 35 nm. Identification of ZnO, TiO_2, and CeO_2 is not possible in the image because of similar electron density of Zn and Ti. Hydroxy propyl cellulose modified composite appears as much bright and dispersed.

3.3 *UV Visible Spectroscopy*

Absorbance. UV–Visible absorption spectra are shown in Fig. 7. From the figure, all the bands exhibit red shift, composite shows maximum absorption in this UV region and shifts toward the visible region. Because of maximum absorption of TiO_2, CeO_2 is in UV region and absorption of ZnO is in visible region. The maximum absorption was observed at 351 nm is 88.76% due to hydroxy propyl cellulose. SDBS modified composite is at wavelength of 362 nm is 51.88%, SDBS and HPC mixed composites

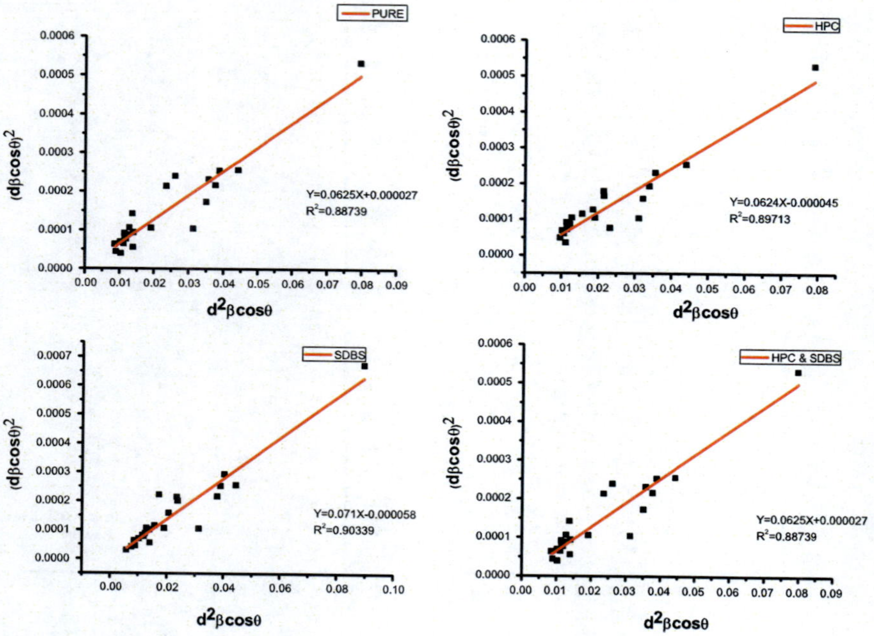

Fig. 5 Size–stain plot of $ZnO–TiO_2$ doped with 0.1 M CeO_2 modified by HPC, SDBS, HPC, and SDBS

maximum absorption is 368 nm and is only 33%, pure composite is at 363 nm and is 48.15%.

Bandgap calculation. The bandgaps are estimated from the Tauc-Plot Technique by plotting a line from the linear portion of high photon energy to zero absorption coefficients as shown in Fig. 8. The relation between the absorption edge and photon energy ($h\vartheta$) can be written as

$$\alpha h\vartheta = A\left(h\vartheta - E_g\right)^{n/2} \tag{5}$$

The list of bandgaps is tabulated in Table 3.

The optical bandgaps of ZnO, TiO_2, and CeO_2 are 3.3 eV, 3.2 eV, and 3.19 eV. But the estimated bandgap of the nanocomposite is 2.4–2.6 eV. The reduction in the bandgap is because of composite which generates new energy levels of TiO_2, ZnO incorporation with CeO_2. The maximum bandgap reported at 0.1 M CeO_2 is doped to $ZnO–TiO_2$ modified by HPC is 2.6 eV [14].

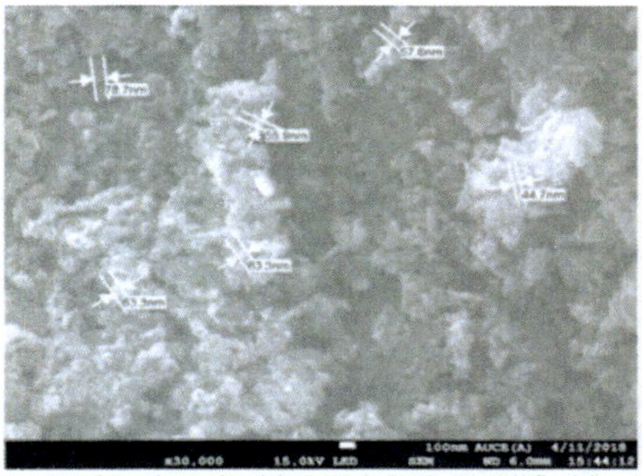

$ZnO\text{-}TiO_2$ doped with 0.1M CeO_2 modified by Pure Composite

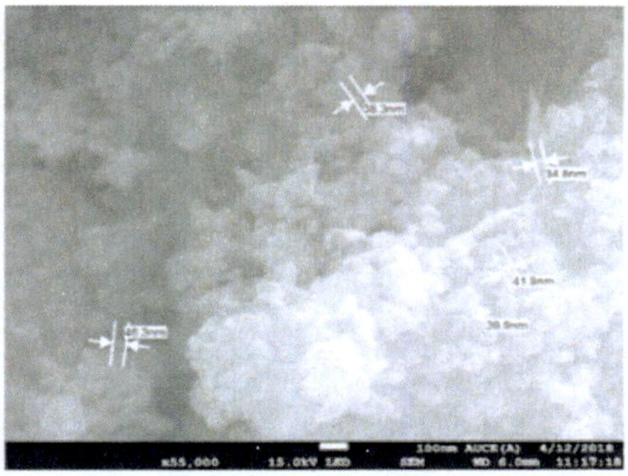

$ZnO\text{-}TiO_2$ doped with 0.1M CeO_2 modified by HPC

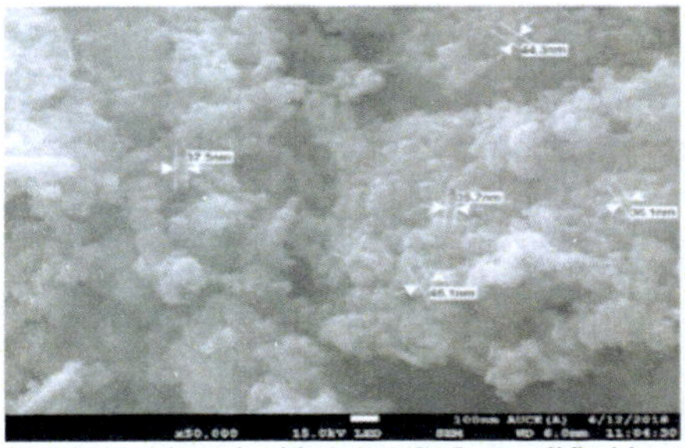

$ZnO\text{-}TiO_2$ doped with 0.1M CeO_2 modified by SDBS

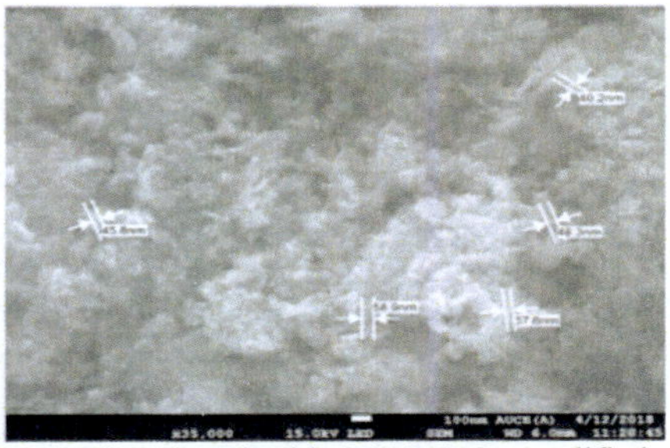

$ZnO\text{-}TiO_2$ doped with 0.1M CeO_2 modified by HPC&SDBS

Fig. 6 FE-SEM images of $ZnO–TiO_2$ doped with 0.1 M CeO_2 modified by HPC, SDBS, HPC, and SDBS

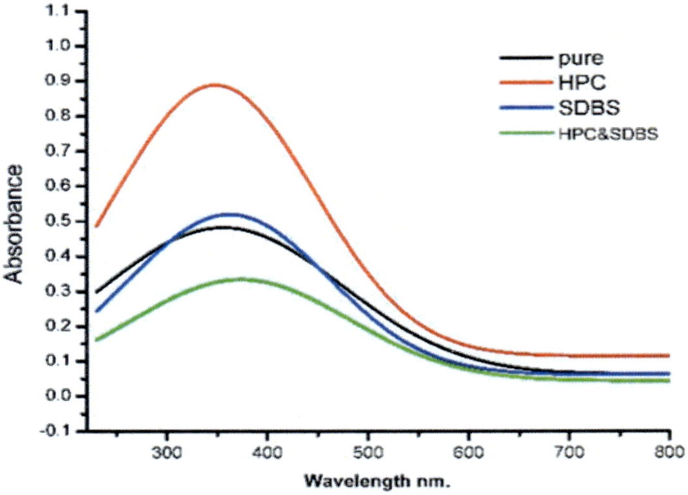

Fig. 7 UV–visible absorption spectra of $ZnO–TiO_2$ doped with 0.1 M CeO_2 modified by HPC, SDBS, HPC, and SDBS

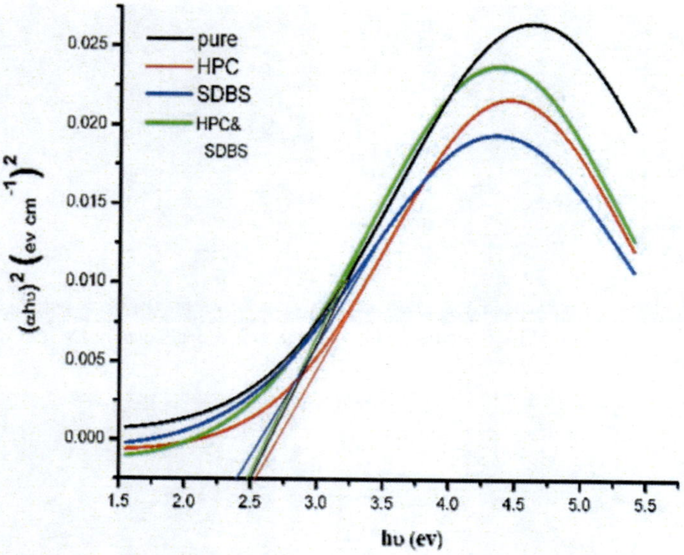

Fig. 8 Tauc-plot of ZnO–TiO_2 doped with 0.1 M CeO_2 modified by HPC, SDBS, HPC, and SDBS

Table 3 Tauc-Plot of ZnO–TiO_2 doped with 0.1 M CeO_2 modified by HPC, SDBS, HPC, and SDBS

S. No	Surfactant	The optical energy bandgap $E_g = 1240/\lambda$ in eV	Energy bandgap from Tauc-Plot in eV
1	Pure	3.36	2.5
2	HPC	3.53	2.6
3	SDBS	3.42	2.4
4	HPC and SDBS	3.36	2.5

4 Conclusion

A solid-state method was approached to prepare CeO_2–ZnO–TiO_2 semiconductor composites with different surfactants. TiO_2 holds anatase phase and no other impurity elements are observed in the composite. Incorporation of TiO_2 to ZnO with addition of CeO_2 generates a new energy level at 2.6 eV. Absorption of the composite modified by hydroxy propyl cellulose is greatly enhanced due to the weak binding and quick absorption at lower activation energy barrier of HPC. SDBS, mixed HPC & SDBS does not show much impact on composite, so surfactant HPC is more suitable to ZnO, TiO_2 semiconducting metal oxide nano composites and mixed HPC do not show much impact on composite, so HPC is more suitable surfactant to ZnO, TiO_2

semiconducting oxides. Prepared semiconducting materials with the obtained semiconducting properties may be useful to optoelectronic, microelectronic, and sensing applications.

References

1. Lin GM, Shang M, Zhang W (2014) Research on nanomaterials and its latest application. Adv Mater Res 912–914:305–308
2. Venkatanarayanan A, Spain E (2014) Review of recent developments in sensing materials. Compr Mater Process **13**:47–101
3. Prabhu S, Viswanathan T, Jothivenkatachalam K, Jeganathan K (2014) Visible light photocatalytic activity of CeO_2-ZnO-TiO_2 composites for the degradation of rhodamine B. Indian J Mater Sci 1–10
4. Yang G, Yan Z, Xiao T (2012) Preparation and characterization of SnO_2/ZnO/TiO_2 composite semiconductor with enhanced photocatalytic activity. Appl Surf Sci 258 (22):8704–8712
5. Wattanawikkam C, Pecharpa W (2015) Optical, dielectric and photocatalytic properties of perovskite $ZnTiO_3$ nanoparticle synthesized by sonochemical process. IEEE, pp 280–283
6. Yan X, Cui-lian Z, Yi-long Z, Zhen-jun W, Jian-min Y, Wen-sheng L (2015) Synthesis and characterization of $ZnTiO_3$ with high photocatalytic activity. Trans Nonferr Metals Soc China 25(7):2272–2278
7. Rajendar V, Raghu Y, Rajitha B, Chakra C, Rao K, Park S (2017) Synthesis, characterization, and photcatalytic behavior of nanocrystalline ZnO, TiO_2 and ZnO/TiO_2 nano composites. J Ovonic Res 13(3):101–111
8. Simon J, Protasenko V, Lian C, Xing H, Jena D (2010) Polarization-induced hole doping in wide-band-gap uniaxial semiconductor heterostructures. Science 327(5961):60–64
9. Walsh A, Buckeridge J, Catlow A, Jackson J, Keal T (2013) Limits to doping of wide band gap semiconductors. Chem Mater 25(15):2924–2926
10. Kendall K, Kendall M, Rehfeldt F (2010) Adhesion of nanoparticles. In: Adhesion of cells, viruses and nanoparticles. Springer, Dordrecht
11. Sujaridworakun P, Natrchalayuth K (2014) Influence of pH and HPC concentration on the synthesis of zinc oxide photocatalyst particle from zinc-dust waste by hydrothermal treatment. Adv Powder Technol
12. Ihsan KHHAA (2015) Restriction of particle size and lattice strain through X-ray diffraction peak broadening analysis of ZnO Nanoparticles 49(1):34–45
13. Zak A, Majid W, Abrishami M, Youse R (2011) X-ray analysis of ZnO nanoparticles by Williamson–Hall and size–strain plot methods. Solid State Sci 13
14. Prasannalakshmi P, Shanmugam N (2017) Fabrication of TiO_2/ZnO nanocomposites for solar energy driven photocatalysis. Mater Sci Semicond Process 13(61):114–124

TEC Variation Analysis During Near Peak Solar Year Activity Over Indian Subcontinent

Bharati Bidikar, Rajkumar Goswami, G. Sasibhushana Rao, and Ch. Babji Prasad

Abstract Earth's atmospheric layers are subjected to continuous changes in near-earth space. Hence, the space weather needs to be investigated, and the most reliable method is using satellite system. Widely used technique is Global Positioning System (GPS) which is primarily designed for tracking and navigation. Ionosphere, one of the earth's atmospheric layers which contains free electrons, is highly affected by space weather like solar flares, geomagnetic storms and seasons. Since the GPS satellite signal travels through the ionosphere, its propagation is affected, and signal gets delayed by plasma of free electron known as total electron content (TEC). TEC not only depends on space weather but also on geographic location, specially the low latitude region (23° above and below the equator). Indian subcontinent falls under this low latitude region, and hence, satellite system applications like precise positioning, navigation, tracking and satellite communication are affected. Hence to correct the GPS delay, accurate estimation of TEC is necessary. In this paper, TEC and rate of TEC (ROT) are calculated, and the variation of the TEC is analyzed. The ionospheric delay, TEC and ROT are calculated for GPS data received on March 10, 2013, from the dual-frequency GPS receiver of NovAtel make located in K L University, Guntur, India (Lat: 16.44° N/Long: 80.62° E). The analysis presented in this paper will help in precise estimation of composition of ionosphere which will improve navigational accuracy.

Keywords Ionospheric delay · TEC · ROT · Pseudorange · Elevation

B. Bidikar (✉) · R. Goswami · G. Sasibhushana Rao · Ch. Babji Prasad
Department of Electronics and Communication Engineering, Andhra University College of Engineering (A), Andhra University, Visakhapatnam, Andhra Pradesh 530003, India
e-mail: bharati.bidikar@gmail.com

G. Sasibhushana Rao
e-mail: sasigps@gmail.com

P. S. R. Chowdary et al. (eds.), *Microelectronics, Electromagnetics and Telecommunications*, Lecture Notes in Electrical Engineering 655,
https://doi.org/10.1007/978-981-15-3828-5_44

1 Introduction

The earth's atmosphere from the altitude of 30–600 miles constitutes to ionosphere. The plasma of free electrons is formed due to solar extreme ultraviolet (EUV) radiation which releases these charged particles and geo-electromagnetic waves [1]. Precise TEC estimation is essential to determine the delay caused in GPS signal propagation [2]. This also helps to investigate the space weather and the disturbances in the ionosphere due to sun storms and other cosmic activities [3]. The satellite signal-based study of ionosphere is done for specific geographical locations. The magnitude of the delay due to TEC is more toward near low latitude and equatorial region [4]. Hence, it is essential to analyze the TEC and rate of TEC (ROT) [5]. The following section gives the estimation of these parameters using the GPS observables.

2 Total Electron Content Calculation

The TEC irregularities vary with local time and the ray path between the satellite and the receiver. Hence, the pseudorange measured by each receiver is different from that of the other receiver, and also, it differs from satellite to satellite. Ionospheric TEC estimations can be done using the existing models like Klobucher model, IRI model, NeQuick model, etc. But the method explained below exploits the frequency-dependent scattering of the GPS signal while propagating in ionosphere. The GPS L_1 and L_2 signal refraction in ionosphere proportional to TEC and geomagnetic field [6]. For these received signals, the phase index of refraction (n_{p}) is derived from Appleton–Hartree formula [7].

$$n_{\mathrm{p}}^2 = 1 - \frac{X(1-X)}{(1-X) - \frac{1}{2}(Y\sin\theta)^2 \pm \left[\frac{1}{4}(Y\sin\theta) + (1-X)^2(1+\cos\theta)^2\right]} \quad (1)$$

where

$$X = \frac{\omega_p^2}{\omega^2};\quad Y = \frac{\omega_c}{\omega};\quad \omega_p^2 = \frac{Ne^2}{m\varepsilon_o};\quad \omega_c = \frac{Be}{m}; \quad (2)$$

Here,

'ω' is angular frequency of *L1* or *L2* signal,
'ε_o' is free space permittivity,
'B' is geomagnetic induction,
'ω_{p}' and 'ω_{c}' are electron plasma and cyclotron frequencies which describe electron oscillation,
'N' is electron density (el/m^2),

'e' is charge (c),
'm' is mass (kg),
'θ' is the angle between geomagnetic field and direction of signal propagation.

The range error of 1–2 m is contributed by the higher-order terms of Appleton–Hartree formula [8]. The simplified equation is shown below after approximating the higher-order terms,

$$n_{\mathrm{p}} \approx 1 - \frac{X}{2};$$
$$n_{\mathrm{p}} = 1 - \frac{1}{2}\frac{(2\pi)^2(Ne^2/m\varepsilon_o)^2}{(2\pi)^2 f^2} \tag{3}$$

For electron density (N) = 10^{12} el/m^2 and geomagnetic induction (B) = 32,000 nT, the expression is simplified to,

$$n_{\mathrm{p}} = 1 - 40.3\frac{N}{f^2} \tag{4}$$

The group index of refraction (n_{g}) is deduced from phase index of refraction as,

$$n_{\mathrm{g}} = n_{\mathrm{p}} + f\frac{\mathrm{d}}{\mathrm{d}f}(n_{\mathrm{p}})$$
$$\Rightarrow n_{\mathrm{g}} = 1 + 40.3\frac{N}{f^2} \tag{5}$$

The group index of refraction given by (5) indicates it to be greater than unity, which means that the group velocity is less than the velocity of light. The integration of group refractive index given by (5) along the signal propagation path gives ionospheric delay experienced by the signal (I_{L}) while propagating through ionosphere [9].

$$I_{\mathrm{L}} = \frac{40.3}{f_{\mathrm{L}}^2}\int_{\text{Receiver}}^{\text{Satellite}} n_{\mathrm{g}}$$
$$I_{\mathrm{L}} = \frac{40.3}{f_{\mathrm{L}}^2}\mathrm{TEC} \tag{6}$$

where

$$\mathrm{TEC} = \frac{1}{40.3}\left(\frac{1}{f_1^2} - \frac{1}{f_2^2}\right)(\mathrm{PR}_1 - \mathrm{PR}_2) \tag{7}$$

The speed of GPS signal while propagating through ionosphere is lesser than the speed of light, and hence, the distance between the satellite and the receiver, calculated using the signal propagation time, is not same as the geometric distance between the two. This can be inferred from Eqs. (5) and (6). Since the ionosphere affects differently for the signals of different frequencies as given by (7), the TEC can be calculated using two L band carriers received by the dual-frequency GPS receiver. The STEC along the propagation path of GPS satellite and the user and the VTEC are related through obliquity factor (OF) [10].

$$\mathrm{VTEC} = \frac{1}{\mathrm{OF}}\mathrm{STEC} \quad \text{where} \quad \mathrm{OF} = \left(1 - \left(\frac{\mathrm{RE} \times \cos(\mathrm{el})}{\mathrm{RE} + h}\right)^2\right)^{-1/12} \tag{8}$$

Here,

'*RE*' is radius of earth = 6371 km,
'*h*' is spherical shell height = 450 km,
'*el*' is elevation angle.

Here, slim layer mapping (SLM) [11] is used to model the obliquity factor. In this paper, the OF is calculated, and further slant TEC and vertical TEC are calculated using (8). The delay in transit time due to ionosphere (I_1) is calculated using (6).

- Rate of TEC (ROT) estimation

The ionosphere is variable in space and time. The earth's rotation around its axis causes diurnal variations. Ionization in the ionosphere changes with time in day, and it is less during dawn and dusk and relatively high during noon [12]. It characterizes the GPS phase fluctuations by measuring the time rate of differential phase of dual-frequency GPS signals. Because the differential phase can yield precisely the ionospheric TEC a long line-of-sight (LOS), the measure is also called the rate of TEC (ROT, in the unit of TECU/min, 1 TECU = 10^6 electrons/m^2). Equation (9) shows the algorithm to calculate ROT [13].

$$\mathrm{ROT} = \frac{\mathrm{TEC}_k^i - \mathrm{TEC}_{k-1}^i}{t_k - t_{k-1}} \tag{9}$$

where

'*i*' is the visible satellite ranging from 1 to 24,
'*k*' is the time of epoch with an epoch interval of 1 min.

3 Results and Discussion

The impact of the ionosphere on GPS satellite transmitted *L1* signal is estimated for the ephemeris data collected on near peak solar activity year. The data are collected from GPS receiver located in geographical location with geodetic latitude of 16.44° N

Table 1 TEC variation and ionospheric delay on SVPRN21

S. No.	Time (h)	Elevation angle (°)	TECU	ROT	Ionospheric delay (m)
1	05:00	06.41	9.1	−0.20	1.494
2	06:00	13.80	3.966	−0.1526	0.6346
3	07:00	24.62	1.520	0.0368	0.2432
4	08:00	38.91	4.927	0.0155	0.7655
5	09:00	59.93	10.62	0.0248	1.647
6	10:00	75.22	25.18	0.0817	4.029
7	11:00	54.84	48.30	0.0983	7.656
8	12:00	25.78	65.39	0.2066	10.57
9	13:00	04.98	130.7	0.5095	20.91

and longitude of 80.62° E. The elevation angle of the satellite SVPRN 21 for every hour from 05:00 h to 13:00 h is given in Table 1. The table also shows TEC and its corresponding ionospheric delay.

Figure 1a shows the total electron content on *L1* signal for the entire day on March 10, 2013. The highest TEC of 104 was observed at around 12:00 h. Since it is diurnal, the highest ionospheric delay is observed at noon which is about 16 m as shown in Fig. 1b.

The TEC and ionospheric delay are calculated for all the visible satellite, but the results pertaining to SVPRN are presented in this chapter. The highest TEC of 132 el/m^2 is observed at around 12:00 h. The change in elevation angle and the TEC is shown in Fig. 2 of (a) and (b), respectively. Figure 3a, b shows the change in TEC and the corresponding ionospheric for SVPRN 21.

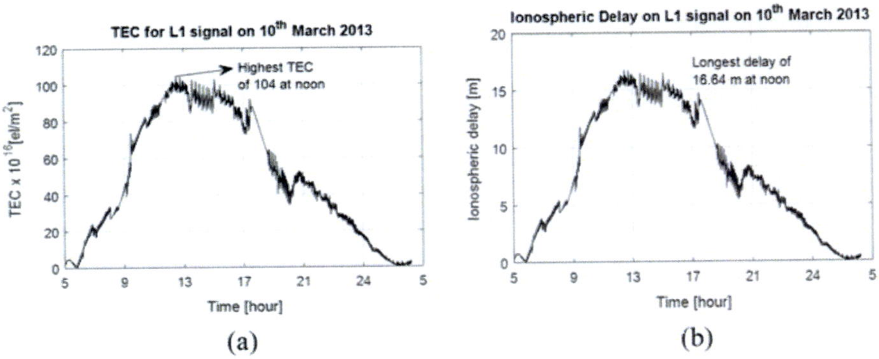

Fig. 1 Total electron content and ionospheric delay on GPS satellite transmitted *L1* signal on March 10, 2013, for the duration of 24 h starting from 5 AM Indian Standard Time

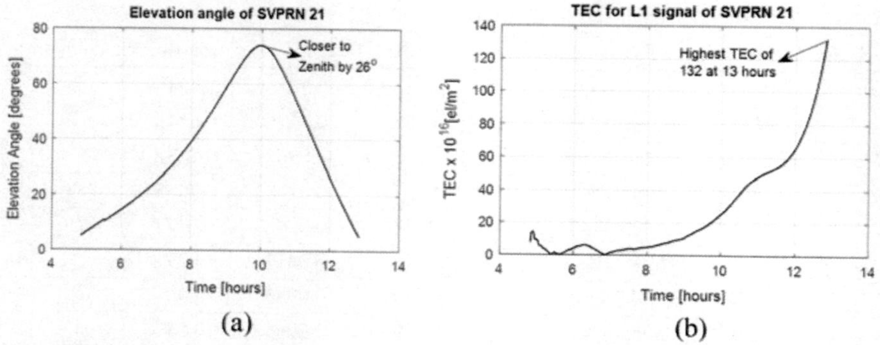

Fig. 2 Elevation angle and total electron content on *L1* signal transmitted by SVPRN 21

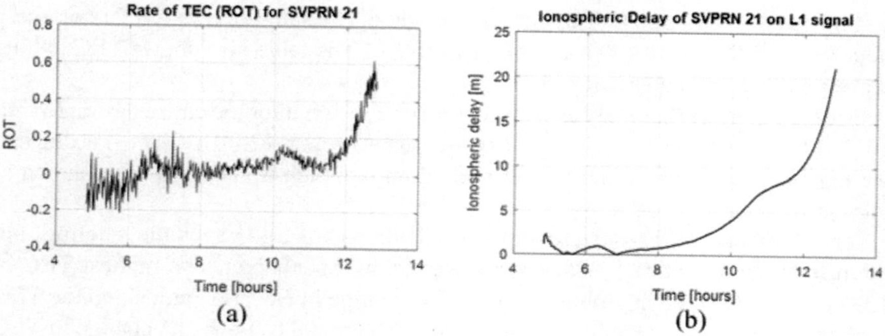

Fig. 3 Rate of TEC and ionospheric delay on *L1* signal of SVPRN 21

4 Conclusions

The impact of the ionosphere on the GPS satellite transmitted *L1* signal is analyzed in this paper. The variation in the TEC and ROT shows the ionospheric layer impact not only depends on the elevation angle but also on the solar activity. During noon, the TEC and ROT are the highest. It is observed that the longest delay of about 20 m is observed at noon on the *L1* signal of SVPRN 21. Similar delays on all the satellite signals will result in inaccuracy of the receiver position. Hence, it is important to estimate the delay and correct the pseudorange.

References

1. Davies K (1990) Ionospheric radio. IEE electromagnetic waves series. Peter Peregrinus Ltd
2. Aarons J, Mendillo M, Yantosca R (1996) GPS phase fluctuations in the equatorial region during the MISETA 1994 campaign. J Geophys Res 101(26):851–862

3. Kelley MC, Hysell D, Musman S (1996) Simultaneous global positioning system and radar observations of equatorial spread F at Kwajalein. J Geophys Res 101:2333–2342
4. Weber EJ, Basu S, Bullett TW et al (1996) Equatorial plasma depletion precursor signatures and onset observed at 110 south of the magnetic equator. J Geophys Res 101:26829–26838
5. Aarons J, Mendillo M, Yantosca R (1997) GPS phase fluctuations in the equatorial region during sunspot minimum. Radio Sci 32:1535–1550
6. Hunsucker RD, Clayton C, Jeffrey C, Lott G (1995) An investigation of the feasibility of utilizing GPS/TEC signatures for near real time forecasting of auroral-E propagation at high-HF and low-VHF frequencies. IEEE Trans Ant Prop 43:1313–1318
7. Coco DS, Gaussiran TL, Coker C (1995) Passive detection of sporadic E using GPS phase measurements. Radio Sci 30:1869–1874
8. Groves KM, Basu S, Weber EJ et al (1997) Equatorial scintillation and systems support. Radio Sci 32(5):2047–2064
9. Bidikar B, Sasibhushana Rao G et al (2016) Orthogonal decomposition algorithm for ionospheric delay estimation for precise GPS applications. Procedia Comput Sci 85:634–640
10. Miyazaki S (1997) Expansion of GSI's nationwide GPS array. Bull Geogr Surv Inst 43:23–34
11. Afraimovich EL, Kosogorov EA, Leonovich LA, Palamartchouk KS, Perevalova NP, Pirog OM (2000) Determining parameters of large-scale traveling ionospheric disturbances of auroral origin using GPS-arrays. J Atm Sol Terr Phys 62(7):553–565
12. Otsuka Y, Ogawal T, Saito A, Tsugawa T, Fukao S, Miyazaki S (2002) A new technique for mapping of total electron content using GPS network in Japan. Earth Planets Space 54:63–70
13. Pi X, Mannucci AJ, Lindqwister UJ, Ho CM (1997) Monitoring of global ionospheric irregularities using the worldwide GPS network. Geophys Res Lett 24:2283–2286

A Review on Schemes for Interconnecting Microgrids of Urban Buildings

S. N. V. Bramareswara Rao and Kottala Padma

Abstract Interconnected operation of microgrids is one of the prominent solutions to meet the increased demand of electricity by large consumer such as urban buildings. Besides, the present-day urbanization motivations further enhance the power crisis problem. So, this creates major burden on the utility grid and consequently leading to grid failures. So, these buildings can be integrated to share their generation and load appropriately. However, the microgrids deployed for all the individual buildings maybe equipped with heterogeneous energy sources and type of power generation with different capacities. So, clear understanding about the integration possibility has to be known to create above-mentioned integrating environment. Hence, this paper gives an overview of all the possible schemes for integrating microgrids of various urban community buildings.

Keywords Urban Community · Renewable energy sources (RES) · Microgrid

1 Introduction

In the present scenario, the demand for electricity is growing rapidly due to industrialization around the world. On the other hand, conventional grid is overloaded and faces lot of problems like global pollution by CO_2 emissions, investment costs, operating costs, etc. As the conventional fossil fuels are deploying, focus of research is moving toward searching of other alternative energy sources. So, microgrids are increasingly attractive to the customers, and also in the future, greater numbers of the microgrids are installed at consumer's premises [1–4]. This increases a burden to the operation and control of conventional grid, and also, some technical limitations will appear

S. N. V. Bramareswara Rao (✉)
Department of Electrical and Electronics Engineering, Sir C. R. Reddy College of Engineering, Eluru, Andhra Pradesh, India
e-mail: snvbrao.eee@sircrrengg.ac.in

K. Padma
Department of Electrical Engineering, Andhra University College of Engineering, Visakhapatnam, Andhra Pradesh, India

P. S. R. Chowdary et al. (eds.), *Microelectronics, Electromagnetics and Telecommunications*, Lecture Notes in Electrical Engineering 655,
https://doi.org/10.1007/978-981-15-3828-5_45

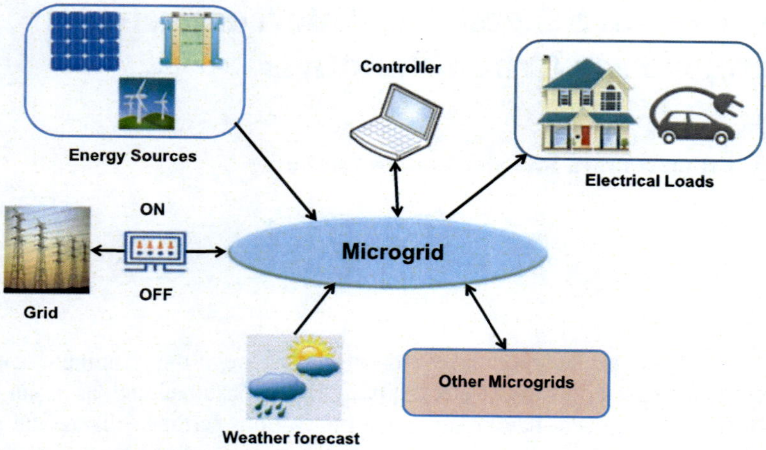

Fig. 1 Schematic view of microgrid architecture

when huge amount of distributed generations are installed. Microgrids interconnection is the unique opportunity in Indian power scenario due to key issues in rural electrification, power demands, peak loads, reliability and quality of power. Microgrid is either alternative current (AC), direct current (DC) or AC and DC (hybrid) facilitates the connection of any architecture and is operated in (1) standalone/off-grid/islanding mode (2) grid-connected mode. Interconnection of microgrids with utility at the level of distribution will enhance reliability of electrical power with some benefits to private electricity generators. The microgrid system thus designed feeds the power independently to remote/rural areas where the access of power is limited. Figure 1 shows a general microgrid architecture, which consists of several distributed energy sources such as wind, solar photovoltaic, fuel cell, diesel generator, controller unit, storage devices and AC and DC loads. Also, there is a possibility of interconnecting other microgrids to the existing microgrid/utility grid. Some control strategies are implemented for the interconnection of different microgrids.

This paper organized as follows: The literature details of interconnection of microgrids are presented in Sect. 2. The possible methods and techniques for interconnection of microgrids and their integral operation were discussed in Sect. 3. In Sect. 4, summary describes the importance of interconnection and integration of microgrids for urban community buildings.

2 Literature Review

Interconnection and integration of neighboring microgrids are the prominent solution for reducing burden on utility grid. So many research works were carried out on microgrids, but only limited researches have been focused on interconnection possibilities of microgrids. Various microgrid architectures were surveyed, and system configurations, energy management, control, generator unit sizing were discussed [5]. Two neighboring DC microgrids were connected by using DC cable and bidirectional converter. Robustness of the proposed system is also verified by transferring power from both sides [6]. Issues involved in the implementation and development, such as protections, power converters, economic analysis and availability, are discussed in [7]. Technical, environmental, economic benefits of integration of renewable sources were briefly discussed in [8]. Multimicrogrid architectures to form a grid were surveyed; thereafter, architectures were defined and compared in terms of layout, line technologies and interface technologies [9]. A brief survey on intelligent methods which are applied to microgrids to study stability aspects, uncertainty in parameters, tracking error has been illustrated in [10].

A multiobjective particle swarm optimization was introduced to minimize the operating cost and also to find the best configuration of system [11]. A parametric mixed-integer linear programming technique was introduced to reduce the burden, and uncertainty in parameters when the microgrids are interconnected was discussed in [12]. Operation cost, pollution rate are the two important factors to be discussed in the design of energy management of microgrid interconnections was described in [13]. Cost plays an important role in the formation of interconnecting microgrids, and different optimization techniques were used to solve economic dispatch problem [14]. A model predictive control coupled with mixed-integer linear programming technique used for load side of the microgrid [15]. A hybrid Nelder-Mead Cuckoo search algorithm was presented in [16] to minimize the power loss in hybrid AC/DC microgrid systems by optimizing the output power of renewable energy distributed generators. Microgrids are intertied together with the use of some droop control techniques and interlinking power converters [17]. An innovative load management tool was proposed for effective load management in intertied microgrids through artificial neural networks [18]. This paper mainly describes and discussed about various microgrid architectures available for interconnection. Figure 2a–c shows standard microgrid architectures [19].

3 Methods of Interconnection

The interconnected system may be operated in standalone/grid-connected mode, which will provide the reliable power to the rural areas where the people are perusing higher quality of life. The interconnected microgrid system developed will reduce the burden on utility grid and also providing stable and reliable power. This section summarizes possible methods and technologies such as switch gear mechanism, power electronic interface and static switches for interconnection of microgrids. There are so many standards which enumerate different aspects for interconnection of grid networks, out of which IEEE 1547 is most important. In this standard, main provision is of having network protections on the grid side.

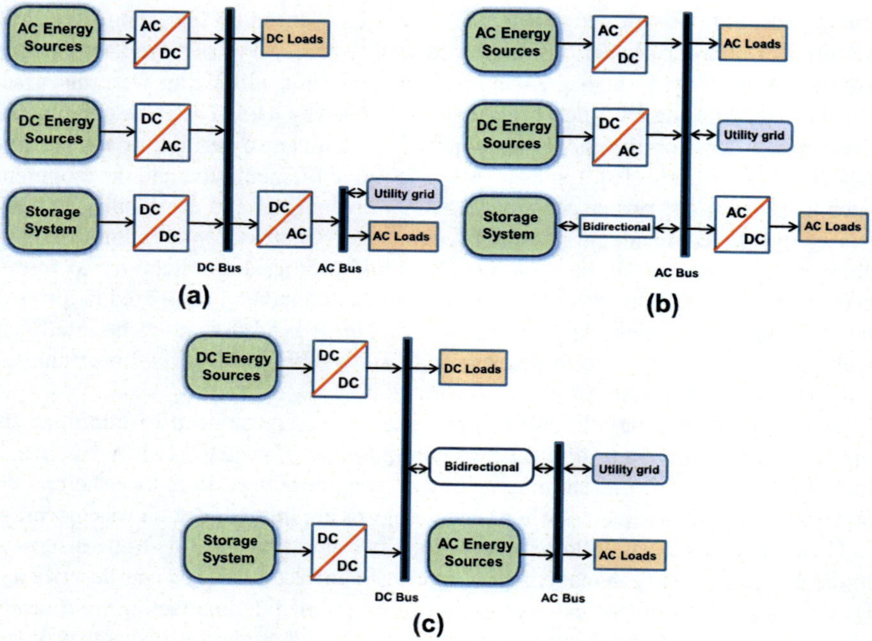

Fig. 2 Architectures of microgrid

3.1 *Interconnection of Two DC Microgrids*

Microgrid-1, Microgrid-2 are two neighboring dc microgrids (DCMGs), with different voltages and are tied together through a bidirectional DC/DC (BDC) converter and a DC cable shown in Fig. 3. Each microgrid consists of energy sources such as diesel, wind, photovoltaic, fuel cells and storage systems. The two microgrids are operating in islanded mode, which creates an opportunity to supply the power to rural areas.

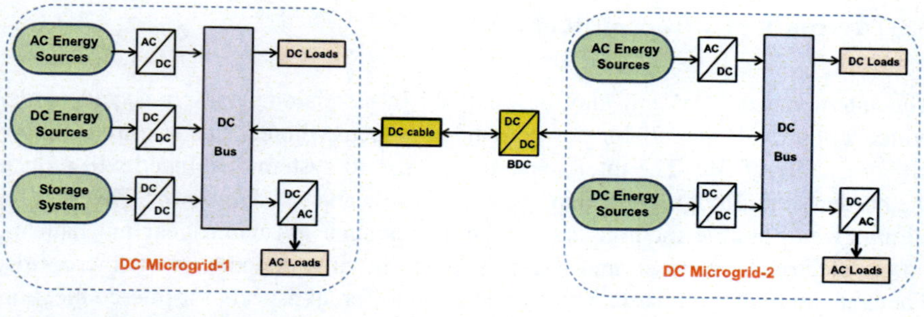

Fig. 3 Architecture of interconnecting DC microgrids

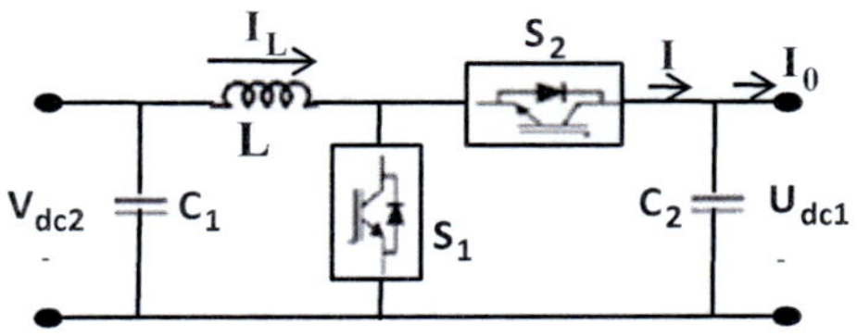

Fig. 4 Equivalent circuit of bidirectional DC/DC converter (BDC)

The distance between two DC microgrids is of any value. The DGs which are connected to each microgrid operate independently without any communication between microgrid-1 and microgrid-2. In the system shown in Fig. 4, a bidirectional DC/DC converter is modeled and functions as a boost converter when power flowing from low voltage microgrid to high voltage microgrid, and it works as buck converter when power flowing from high voltage microgrid to low voltage microgrid. The average value of voltage across an inductor and current in capacitor in boost converter mode of operation are given in Eqs. (1) and (2).

$$\frac{\mathrm{d}I_L}{\mathrm{d}t} = -\frac{rI_L}{L} - \frac{(1-D)U_{\mathrm{dc1}}}{L} + \frac{V_{\mathrm{dc2}}}{L} \tag{1}$$

$$\frac{\mathrm{d}U_{\mathrm{dc1}}}{\mathrm{d}t} = \frac{(1-D)I_L}{C_2} - \frac{I_0}{C_2} \tag{2}$$

where V_{dc2} is the average voltage across C_1, U_{dc1} is the average value of voltage across $C_{2,}$ and D is the duty cycle. Similarly, during the operation of buck converter, the average values of voltage across inductor and current through the capacitor are given by Eqs. (3) and (4).

$$\frac{\mathrm{d}I_L}{\mathrm{d}t} = -\frac{rI_L}{L} + \frac{DU_{\mathrm{dc1}}}{L} - \frac{V_{\mathrm{dc2}}}{L} \tag{3}$$

$$\frac{dV_{dc2}}{dt} = \frac{I_L}{C_1} - \frac{I_0}{C_1} \tag{4}$$

where 'I' is the average value of the current through C_2 and 'I_L' is the average value of current flowing through L.

3.2 *Interconnection of Two AC Microgrids*

Two AC microgrids are interconnected through the switchgear mechanism shown in Fig. 5. Each microgrid consists of AC and DC energy sources such as wind, solar, fuel cell and AC load. Each microgrid is designed to operate in standalone/grid-connected mode. The switchgear mechanism consists of circuit breaker, relay unit, communication system and monitoring system, all together acts as a digital signal

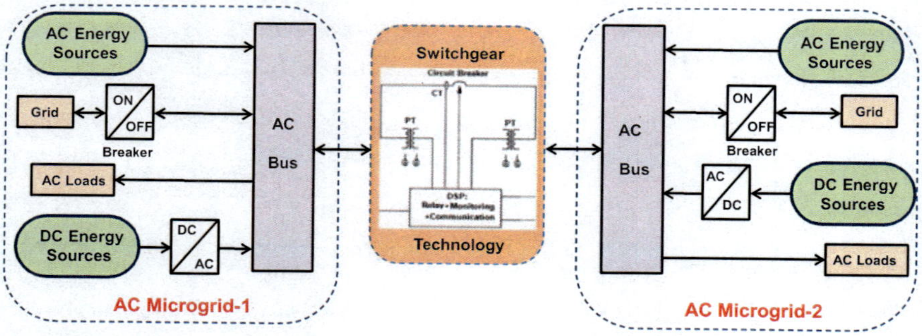

Fig. 5 Architecture of interconnecting AC microgrids

processing (DSP) unit. This unit will monitor the information transferring between the two AC microgrids. The total system is capable of supplying electricity to rural areas independently. This mechanism is relatively simple and inexpensive. Since the electrical characteristics on both the sides (microgrid side and utility grid side) of circuit breaker must be same, the microgrid characteristics are dependent on the utility grid characteristics.

3.3 *Interconnection of AC and DC (Hybrid) Microgrids*

AC microgrid and DC microgrid connected together to form a hybrid microgrid shown in Fig. 6. Both AC and DC microgrids were connected with the power electronic interface technology. This technology reduces multiple reverse connections in an individual grid has been reduced and is more flexible. To maintain the stability of the microgrid under variable supply and load conditions, advanced control strategies of power electronics are used to control the microgrid operation either in island mode or grid-connected mode. It is easy to control active, reactive powers with

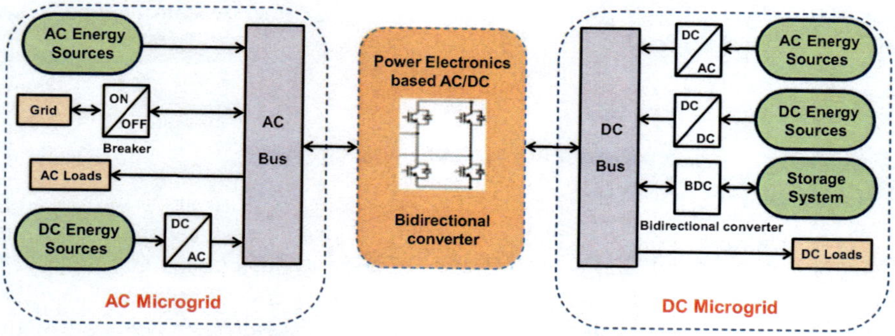

Fig. 6 Architecture of interconnecting AC and DC (hybrid) microgrids

this technology. The power transfers between AC microgrid and DC microgrid were continuously monitored and controlled by the switches in bidirectional converters.

4 Summary

Thus, in view of present-day energy requirements and types of energy resources and consumer loads, this paper summarizes the possibilities of interconnecting different microgrids located in an urban community. These presented architectures improve the reliability of the electrical power by facilitating power exchange among the interconnected microgrids. This further leads to overcome frequent utility grid outages. Besides, the effectiveness of this integrated operation can be further improved with the design of proper control mechanisms. So, the details in this paper will also be useful for the control engineers to develop the required control strategies for the building microgrids.

References

1. Keyhani A, Md Marwali N, Dai M (2009) Integration of green and renewable energy in electrical power systems. Wiley, New York
2. Ramakumar R, Hughes WL (1981) Renewable energy sources and rural development in developing countries. Special Issue on Energy Related Studies. IEEE Trans Educ 24(3):242–251
3. Nayar CV, Phillips SJ, James WL, Pryor TL, Remmer D (1993) Novel wind/diesel/battery hybrid energy system. Sol Energy 51(1):65–78
4. Bramareswara Rao SNV, Kumar YVP, Padma K (2018) Framework for forming minigrids by interfacing urban community buildings. Int J Pure Appl Math 118(24)
5. Nehrir MH, Wang C (2011) A review of hybrid renewable/alternative energy systems for electric power generation: configurations, control, and applications. IEEE Trans Sustain Energy 2(4)
6. Kumar M, Srivastava SC, Singh SN, Ramamoorty M (2015) Development of a control strategy for interconnection of islanded direct current microgrids. IET Renew Power Gener 9(3):284–296
7. Planas E, Andreu J (2014) AC and DC technology in microgrids: a review. Int J Renew Sustain Energy Rev 726–749
8. Adefarati T, Bansal RC (2016) Integration of renewable distributed generators into the distribution system: a review. IET Renew Power Gener 10(7):873–884
9. Bullich-Massague E, Francisco D-G, Monica A-P, Francesc G-L, Pol O-R, Andreas S (2018) Microgrid clustering architectures. Int J Appl Energy 212:340–361
10. Mahmoud MS, Alyazidi NM, Abouheaf MI (2017) Adaptive intelligent techniques for microgrid control systems: a survey. Int J Electr Power Energy Syst 90:292–305
11. Borhanazad H, Mekhilef S (2014) Optimization of micro-grid system using MOPSO. Int J Renew Energy 295–306
12. Umeozor EC, Trifkovic M (2016) Energy management of a microgrid via parametric programming. IFAC-Papers on Line 49-7, pp 272–277
13. Gholamreza A, Noradin G (2018) Multi-objective energy management in a micro-grid. Int J Energy Rep 4:218–225
14. Maulik Avirup, Das Debapriya (2017) Optimal operation of microgrid using four different optimization techniques. Int J Sustain Energy Technol Assess 21:100–120

15. Bartolucci L, Cordiner S (2017) Renewable sources integration through the optimization of the load for residential applications. In: 9th international conference on applied energy, ICAE2017, pp 21–24
16. Senthil Kumar J, Charles Raja S, Jeslin Drusila Nesamalar J, Venkatesh P (2018) Optimizing renewable based generations in AC/DC microgrid system using hybrid Nelder-mead—Cuckoo search algorithm. Int J Energy 158:204–215
17. Nutkani IU, Loh PC, Blaabjerg F (2013) Power flow control of intertied ac microgrids. IET Power Electron 6(7)
18. Barelli L, Bidini G, Bonucci F, Ottaviano A (2018) Residential micro-grid load management through artificial neural networks. Int J Energy Storage 17:287–298
19. https://link.springer.com/article/10.1007/s40565–015-0101-7

Performance Analysis of Multiwalled Carbon Nanotube, Composite Multiwalled Carbon Nanotube, and Copper-Based Antenna in X-Band Applications

V. Suryanarayana, M. Satya Anuradha, and S. Paul Douglas

Abstract In this paper, three different materials, such as copper, MWCNT, and MWCNT composite PU/MWCNT (polyurethane $C_{27}H_{36}N_2O_{10}$), are used as conducting structures for the design of a rectangular patch antenna, which resonate in X-Band (8–12 GHz). These antenna structures are implemented using FR-4 substrate, having a dielectric constant of 4.4 and with a loss tangent of 0.02. This work concentrates on the comparison of these three antennas in terms of their performance metrics like return loss, VSWR, impedance bandwidth, gain, and radiation efficiency. Initially, these antennas are simulated using HFSS and fabricated. Their performance is tested using ROHDE&SCHWARZ ZVB 20 vector network analyzer. From the results, it is clear that among the three antennas, MWCNT-based antenna design provides wider impedance bandwidth when compared to composite MWCNT and copper-based structures, whereas PU/MWCNT exhibits similar bandwidth as of MWCNT, but it offers more mechanical strength, electrical conductivity, and thermal conductivity. Thus, both MWCNT and composite MWCNT are equally applicable in the X-band range. Both simulated and fabricated results are presented in this paper.

Keywords MWCNT composite patch antenna · Impedance bandwidth · X-band · SAR

V. Suryanarayana (✉) · M. Satya Anuradha
Department of Electronics and Communication Engineering, Andhra University College of Engineering (A), Andhra University, Visakhapatnam, Andhra Pradesh, India
e-mail: surya.velamala@gmail.com

M. Satya Anuradha
e-mail: radhamsa@gmail.com

S. Paul Douglas
Department of Engineering Chemistry, Andhra University College of Engineering (A), Andhra University, Visakhapatnam, AP, India
e-mail: pauldouglas12@gmail.com

P. S. R. Chowdary et al. (eds.), *Microelectronics, Electromagnetics and Telecommunications*, Lecture Notes in Electrical Engineering 655,
https://doi.org/10.1007/978-981-15-3828-5_46

1 Introduction

Each nanotube present on the surface of CNT material to resonate electromagnetic waves individually influences the impedance bandwidth [1]. Nanomaterials exhibit excellent electrical conductivity, and its conductive properties can be enhanced with the addition of composite material like PANI, PU, CFC, and epoxy are used to overcome the drawbacks of metals such as weight, fabrication procedure, and corrosion resistance [2–4]. Carbon nanotube (CNT) and CNT with conducting polymers are used to replace the copper patch antennas [5, 6]. An antenna designed using CNT composite material is a lossy antenna in nature as ohmic loss of the composite material is greater than that of conventional metal, i.e., copper. Hence, the composite antenna bandwidth is wider than that of the standard copper antenna operating in the X-band frequency range. As the ohmic loss increases, the bandwidth increases, and the Q-factor decreases, which is the actual behavior of the CNT composite antenna [7].

The current study deals with the design and fabrication of Cu, MWCNT, and polymer-based MWCNT composite patch antenna. These antennas are fabricated with thin film technology by depositing a thin layer of MWCNT, PU/MWCNT materials using spin coating and drop casting techniques are used to fabricate the MWCNT antennas on the FR-4 substrate [8–10].

The rest of the paper is organized as follows: Sect. 2 material used for experimentation, Sect. 3 discussed antenna design considerations. In Sect. 4, results and discussions of above antennas are discussed, and finally, conclusions are made in Sect. 5.

2 Experimentation

2.1 *Material Used for Experimentation for MWCNT Antenna*

MWCNTs, sodium dodecyl sulfate (SDS) ($NaC_{12}H_{25}SO_4$), and ethanol (C_2H_5OH) were supplied by Sigma-Aldrich, which are used to fabricate the MWCNT antenna.

2.2 *Material Preparation for MWCNT Antenna*

The pure MWCNTs (98%) are dispersed in ethanol in the ratio of 1:1 by weight to acquire a stable and homogenous solution. The SDS surfactant is known to have the ability to disperse MWCNTs in aqueous media by establishing an excellent interface between the surface of MWCNT and ethanol. The weight concentration of MWCNT in ethanol was maintained at 0.1 mg ml^{-1}. A stable solution with a required viscosity is attained by sonicating the dispersion for 8 h, at room temperature of 25 °C, employing bath sonicators. After the sonication process, the centrifugation technique is applied to separate large MWCNT aggregates from small MWCNT aggregates.

2.3 Material Used for Experimentation for Composite MWCNT Antenna

MWCNTs, polyurethane (PU) $C_{27}H_{36}N_2O_{10}$, dimethylformamide (DMF) C_3H_7NO, were supplied by Sigma-Aldrich; materials are used to fabricate the MWCNT composite antenna.

2.4 Material Preparation for Composite MWCNT Antenna

For the preparation of composite material, initially Polyurethane (PU) $C_{27}H_{36}N_2O_{10}$ granules are dissolved in dimethylformamide (DMF) which is an organic compound in nature with the formula C_3H_7NO. The composition is stirred using magnetic stirring for approximately 6 h. Now, MWCNTs were dispersed in an alternative beaker in DMF for 6 h, using ultra-sonication. The dispersed MWCNTs and dissolved polyurethane in DMF were mixed thoroughly using magnetic stirring for another 6 h. The mixed solution is now poured into a petri dish and placed in a vacuum oven for 12 h at 80 °C temperature for the solvent to evaporate. The final product is PU/MWCNT composite material. Here, 97% of polyurethane (PU), 3% weight MWCNT, and 50 ml DMF are used.

3 Antenna Design

See Fig. 1; Tables 1 and 2.

4 Results and Discussion

In this section, simulated and fabricated results of MWCNT, polymer-based MWCNT composite rectangular patch antennas are presented for X-band (8–12 GHz) applications, and these antennas are compared with the standard copper patch antenna. Figure 1 depicts the proposed simulated and fabricated antenna models. Fig. 2 shows test setup using vector network analyzer to measure the return loss,

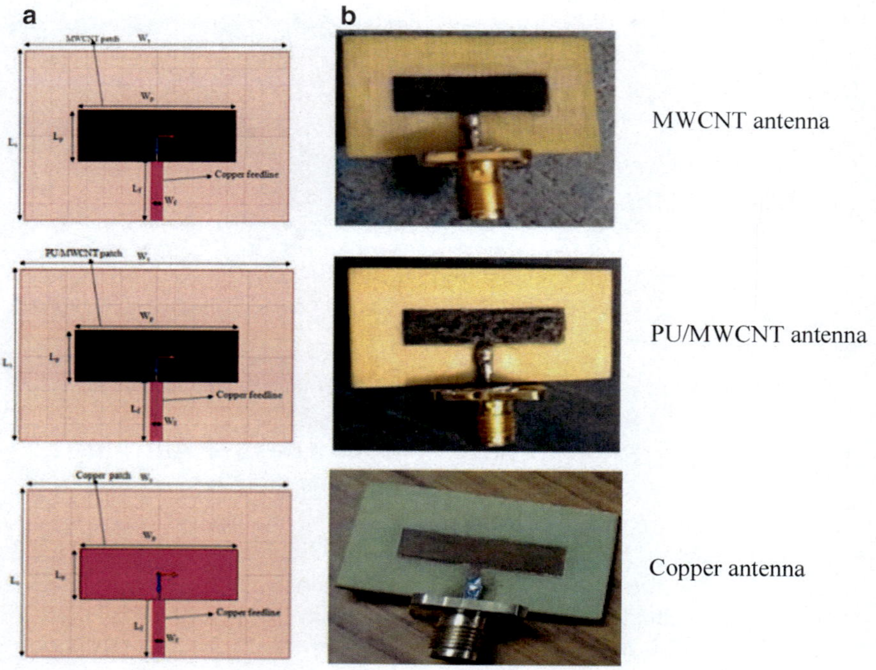

Fig. 1 Antenna simulation and fabrication models. **a** Simulated, **b** fabricated

Table 1 Material properties

Parameters	Copper antennas [8]	PU/MWCNT antenna [9]	MWCNT antenna [10]
Dielectric constant	1	18.1	8.5
Loss tangent	0	14.9	15
Conductivity	5.8×10^7 S/m	5 S/m	0.24×10^{-3} S/m
Tensile strength (GPa)	0.22	0.28	11–63
Thermal conductivity ($\times 10^{-3}$ W/m K)	0.385	300	3

VSWR, impedance of the feed line. Figure 3 illustrates return loss and VSWR comparison between simulated and fabricated antennas. The measured impedance bandwidth of the fabricated MWCNT antenna is 19% which exhibits 2% greater than MWCNT composite antenna and 4% wider than the copper patch antenna. Figure 4 shows the simulated and fabricated smith charts, it shows impedance of the feed line, and antenna impedance matching with microstrip feed line is done by tuning the width of the feed line. Among these antennas, copper antenna has better impedance matching than other two antennas, the simulated antennas impedance values are 49.02 Ω

Table 2 Dimensions of rectangular patch antenna resonating at 10 GHz

Parameters	Diameter (mm)		
	Copper antenna	PU/MWCNT antenna	MWCNT antenna
Substrate width and length ($Ws \times Ls$)	30 × 20	30 × 20	30 × 20
Rectangular patch width and length ($Wp \times Lp$)	6.2 × 10	6.2 × 10	6.2 × 10
Feed line width and length ($W_f \times L_f$)	1.8 × 7.6	1.8 × 7.6	1.8 × 7.6

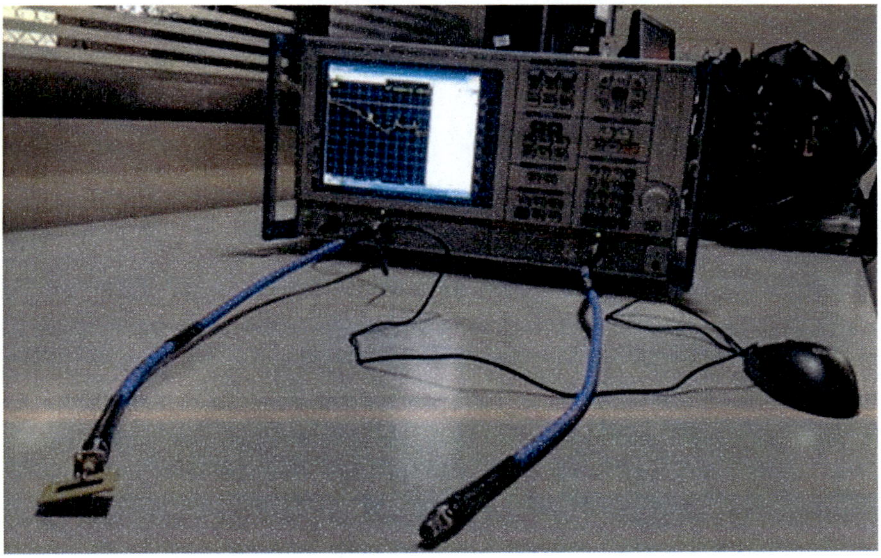

Fig. 2 Vector network analyzer

for copper, 43.4 Ω for MWCNT composite, and 47.4 Ω for MWCNT, and fabricated antennas impedance values are 47 Ω, 33 Ω, and 35 Ω, respectively.

The simulated gain and radiation efficiency versus frequency plots are shown in Fig. 5. It is observed that all the designed antennas give almost equal amount of gain and efficiency values in the operating band, and copper patch antenna provides better gain and efficiency values when compared to other antennas (6.24 dB and 83%). Table 1 shows the material properties, and Table 2 depicts the dimensions of the three antennas. Overall simulated and fabricated results are listed in Table 3.

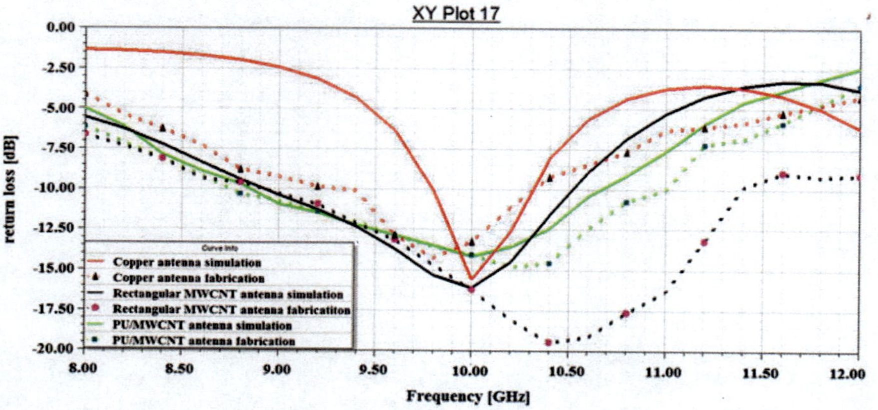

(a)

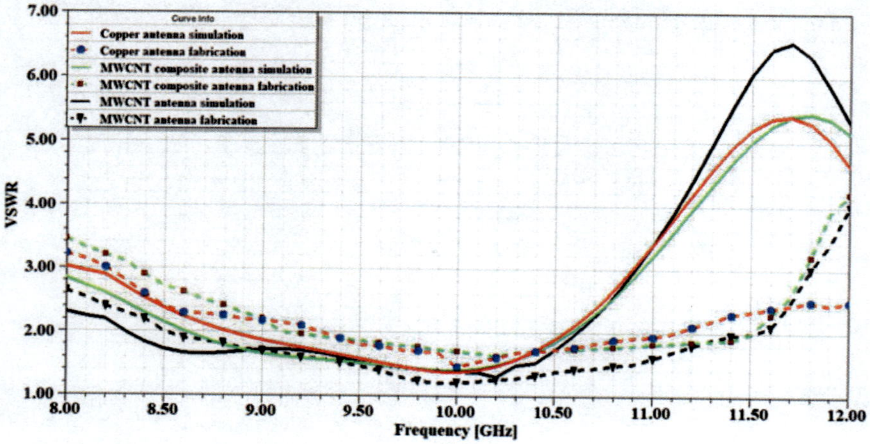

(b)

Fig. 3 **a** Return loss, **b** VSWR of simulated and fabricated antennas

5 Conclusion

In this paper, the performance of MWCNT and PU/MWCNT composite rectangular patch antennas in X-band (8–12 GHz) is simulated and fabricated. The performance of these antennas is compared with basic copper antenna. Various antenna parameters such as return loss, VSWR, and impedance of the feed line are compared. Among these antennas, MWCNT antenna provides wider impedance bandwidth. Improvement in the impedance bandwidth is due to resonant behavior of each nanotube present on the surface. From simulated and measured results, the MWCNT antenna shows wider impedance bandwidth of 19% in the range of 11.2–9.3 GHz. It can be suitable for synthetic aperture radar (SAR) applications. Material properties can be

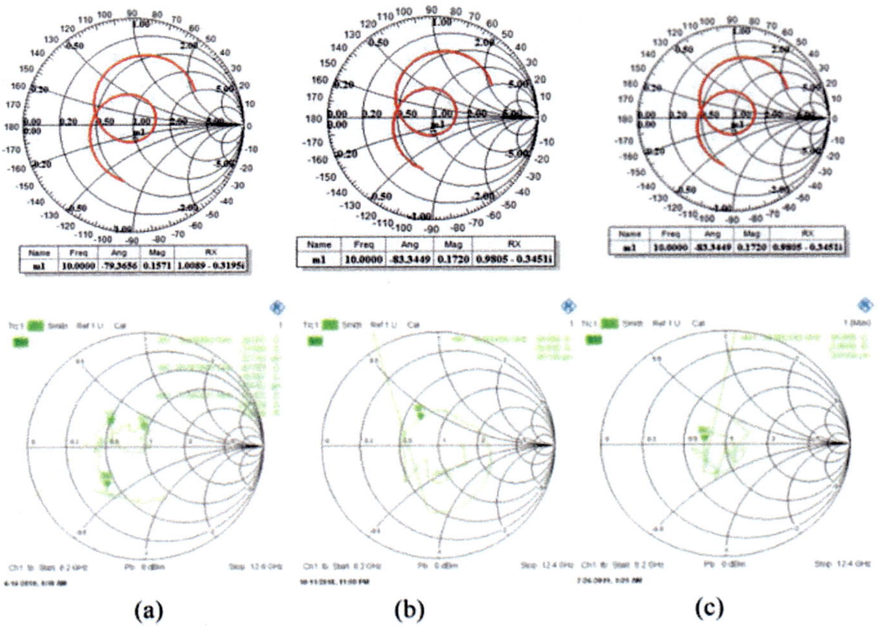

Fig. 4 Simulated and fabricated smith charts. a Copper antenna, b PU/MWCNT antenna, c MWCNT antenna

modified by controlling the material concentration in CNT materials, and this is not possible with the standard copper material.

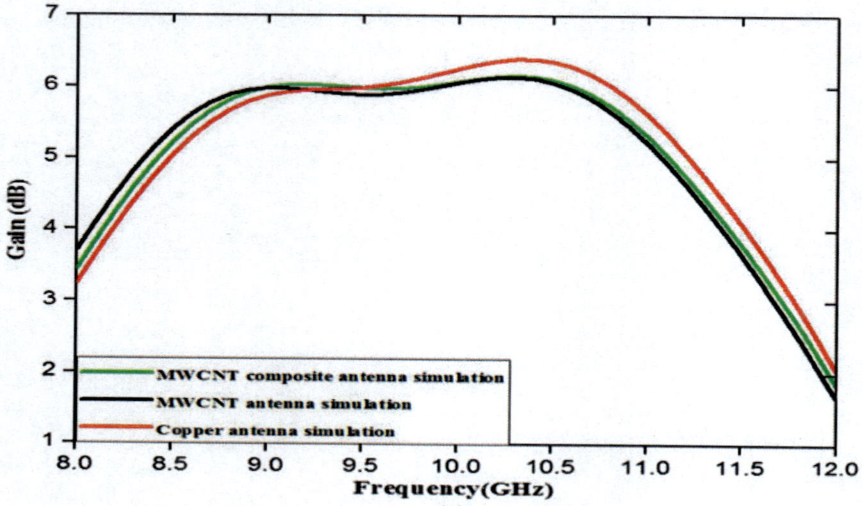

(a)

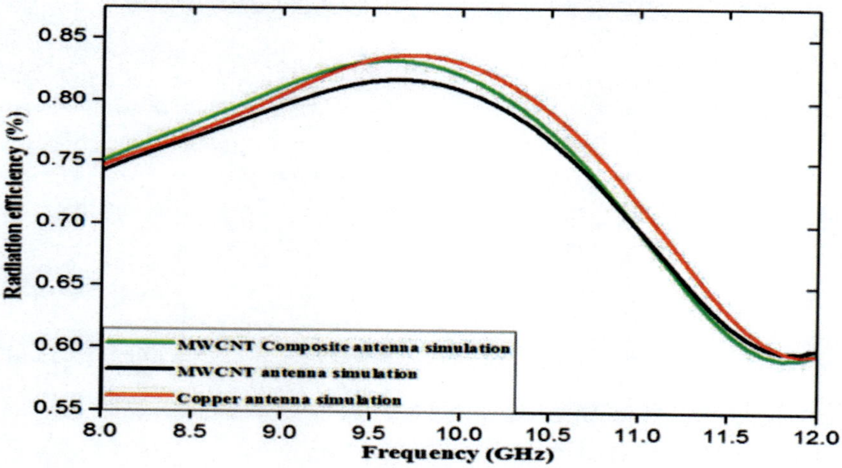

(b)

Fig. 5 Simulated. **a** Gain versus frequency, **b** radiation efficiency versus frequency of designed antenna

Table 3 Performance parameters of simulated and fabricated antenna

Parameter	Copper simulated antenna	Copper fabricated antenna	PU/MWCNT simulated antenna	PU/MWCNT fabricated antenna	MWCNT simulated antenna	MWCNT fabricated antenna
Resonant frequency (GHz)	10 GHz	9.8 GHz	10 GHz	10.2 GHz	10 GHz	10.4 GHz
Return loss ($S11$) (dB)	−16.07	−14.12	−14.84	−14.04	−14.27	−11.30
VSWR	1.37	1.54	1.41	1.58	1.38	1.21
Input impedance	49.02 Ω	47 Ω	43.4 Ω	33 Ω	47.4 Ω	35 Ω
Impedance bandwidth (%)	(10.3–9.8) GHz 5%	(10.2–9.4) GHz 8%	(10.8–8.9) GHz 19%	(11–8.8) GHz 18%	(10.5–8.9) GHz 16%	(11–9) GHz 20%
Gain (dB)	6.24	–	6.05	–	6.07	–
Directivity (dB)	7.11	–	7.09	–	7.15	–
Radiation efficiency (%)	83	–	82	–	81	–

References

1. Suryanarayana V, Aanuradha MS, Paul Douglas S, Uzwala N (2018) Performance analysis of multi-walled carbon nanotube based rectangular patch antenna for X-band applications. JARDCS 2147–2155
2. Nikitin PV, Lam S, Rao KVS (2005) Low cost silver ink RFID tag antennas. In: Proceedings of IEEE antennas propagation society international symposium, pp 353–356
3. Ludmerer S (2007) Conductive inks for RFID antenna: the low cost high speed route to RFID labels. Parelec, Inc., Rocky Hill
4. Anguera J, Daniel J-P, Borja C, Mumbru J, Puente C, Leduc T, Laeveren N, Roy PV (2006) Metallized foams for fractal-shaped microstrip antennas. IEEE Antennas Propag Mag 50(6):20–38
5. Holloway CL, Sarto MS, Johansson M (2005) Analyzing carbon fiber composite materials with equivalent-layer models. IEEE Trans Electromagn Compat 47(4):833–844
6. De Rosa IM, Sarasini F, Sarto MS, Tamburrano A (2008) EMC impact of advanced carbon fiber/carbon nanotube reinforced composites for next-generation aerospace applications. IEEE Trans Electromagnet Compat 50(3):556–5632(5):99–110
7. Mehdipour A, Sebak AR, Rosca l, Hoa SV (2014) Carbon nanotube composites for wide band millimeter-wave antenna applications. Research gate
8. Iijima S (1991) Helical microtubules of graphitic carbon. Nature 354:56–58
9. Shokry ASA, El Morsi AK, Sabaa MS, Mohamed RR, El Sorogy HE (2014) Synthesis and characterization of polyurethane based on hydroxyl terminated poly butadiene and rein forced by carbon nanotubes. Egypt J Pet
10. Li H, Xu C, Banerjee K (2010) Carbon nanomaterials: the ideal interconnect technology for next-generation ICs. IEEE. Des Test Comput 27(4):20–31

Wideband Printed Antenna Design Techniques: A Comprehensive Study

Gundapaneni Sri Latha and G. S. N. Raju

Abstract Wideband (WB) antennas are extensively used in biomedical communication systems, due to their cost-effectiveness and high data rate handling capacity. In these applications, printed antennas are preferred due to their low power requirement; these antennas are ready to cover the wide range of frequencies, starting from 3.1 to 10.6 GHz; it has been possible to generate required omnidirectional radiation patterns, and such antennas are simple and small in configuration. In this work, an attempt is made to bring out to design and analyze methods of WB printed antennas in the comprehensive background representation which enables designers to go for suitable structures depending on the application and user specifications.

Keywords HFSS-IE · Wideband antenna · Ultra-wideband antenna · Printed antennas

1 Introduction

Nowadays, due to the rapid development in the wireless communication system, several kinds of new antennas are being implemented every year to satisfy the required applications. Antennas are fundamental devices in making any wireless communication systems [1]. While designing modern wireless communication systems, the main concentration is designing antennas at a broad frequency range which has extended range of wireless communication services in a high data rate system rapidly [2]. Due to salient features of slot antennas such as low profile, wide bandwidth, low cost, low power consumptive, with good impedance matching, bandwidth efficiency and simple fabrication, these are more appropriate elements for wideband applications. Recently, due to low power consumptive and high data transmission speed requirements the wideband antennas have garnered much attention [3]. Considerable research effort has been put on wideband technology and finds its application

G. Sri Latha (✉) · G. S. N. Raju
Centurion University of Technology and Management, Visakhapatnam, India
e-mail: srilathacrttcs@gmail.com

P. S. R. Chowdary et al. (eds.), *Microelectronics, Electromagnetics and Telecommunications*, Lecture Notes in Electrical Engineering 655,
https://doi.org/10.1007/978-981-15-3828-5_47

in the 1950s for military communications and high-resolution radar's, satellite communication. A lot of work is reported on wide variety of WB antennas, and they have designed and simulated both analytically and experimentally. Particularly, the frequency band ranging from 3.3–10.5 GHz has allocated by the Federal Communication Commission (FCC). WB devices have been researched for commercial WB applications and found application in both industry and academia [4]. This technology is an ideal approach for short-distance and wide-range wireless communication [5]. WB systems transmit messages as very narrow pulses by generating radio energy at specified time intervals. WB systems additionally can also provide data rate up to thousands of K bits per second, and it establishes secure communication in military communications [6]. We have extended the antenna type by adopting printed antenna to wideband technology [7]. Printed antennas are being used for WB systems, because of numerous merits including mechanical strength, compactness, conformability and cheap manufacturing costs. The antenna engineer's main objective is always trying to provide high-performance antenna by providing good impedance bandwidth, compact size and gain [8]. Many of researches and engineers have evolved several simulation tools to make easier design steps, and constructively, antennas are designed to reach the given stipulations. To enhance or optimize designed antenna characteristics, various parameter investigations are conducted. From given radiation characteristics, designed antenna is synthesized and concerned antenna designs are improved by the proposed optimization techniques [9]. Consequently, a good effort on review work is done to achieve design and optimization process which gives an ideal solution efficiently and precisely. The rest of the paper is organized as follows: Sect. 2 consists of the literature survey; it consists of feed networks for WB antennas and different WB antenna designs for wireless communications.

2 Literature Survey

There are several methods to design printed WB antennas, and this section summarizes several investigations and developments of WB antennas during the last decade. In this survey, totally 42 papers are referred. Initially, discussion on design and analysis of CPW-fed WB antenna is presented and feeding techniques using micro-strip lines. Next, differentiate the UWB antennas with band-notched technique in comparison with the part 1 of UWB antennas design. The survey further discusses the reconfigurable WB antennas designed using state of the art techniques and MIMO-WB antennas. At last, the survey focus on some new designs of WB antennas that have a broadband, cognitive radio and multi purpose multiband applications; for which the system requirements are suitable in UWB applications.

2.1 Feed Network for WB Antennas

Compact coplanar waveguide (CPW) feeding and micro-strip line feeding for WB antennas are described.

WB antenna designs based on CPW-fed techniques: Compact antenna design using a coplanar waveguide (CPW) feeding technique offers better performance than line feed method. The CPW offers low dispersion and less radiation leakage the integration into the communication system is easy. Alignment error can be avoided using CPW feeding technique. In this method, center strip formed by the conductor and separated with narrow gap from two ground planes on both sides is demonstrated in the following literature. Li Singsong et al. demonstrated a fractal antenna with printed diversity Cantor set in UWB region. By employing Cantor set fractal technology, antenna is fabricated which includes antennas with two identical wide slots and the radiation patch. The designed antenna offers an isolation of 20 dB. In CPW design, very high isolation is provided by etching ground plane with multiple slots. The proposed design simulated by means of HFSS to examine the parameters is called S11 and S12. Generally, antenna diversity performance is determined by isolation and radiation patterns [10]. Likewise Zhong Shu-shi et al. designed a wideband printed monopole radiating elements, which was compounded of a circular or annular patch, and results are provided, exhibiting good correspondence and validating of patch which enhances the wideband characteristics [11]. By using CPW-fed line, designed antennas are easily integrated with monolithic microwave circuits (MMICs). Hence, CPW-fed antennas turn to more attractive part. The main advantage of CPW-fed antennas is that providing impedance matching to the fed line is easy.

WB antenna design based on micro-strip fed technique: Wideband antennas fed by micro-strip line were examined for WB and UWB applications. A study of a monopole antenna UWB applications was developed by Liang Jinxing et al. The proposed monopole element is printed on a dielectric substrate and considered a 50/spy omega/micro-strip line as feed line; monopole with a planar circular disk has been explained to produce 10 dB return loss with required radiation properties [10]. Printed wide-slot antenna with small in size was presented and experimentally studied by Razavi-Rad Majid et al. In this design, by embedding three rectangular notches at the edge of the symmetric polygon-like slot below the micro-strip fed line and by optimizing the feeding structure, resonant frequencies are generated and the measured impedance bandwidth was increased up to 134.6%, from 2.5 to 12.8 GHz, for a VSWR less than 2. This optimized design has dimensions of 35 $\times$ 44 mm which are compact in size and a peak antenna gain of approximately 4.7 dBi [12]. Printed quasi-self-complementary UWB antenna was demonstrated by Guolu et al. The presented antenna, with micro-strip feed line, designed to provide a 10 dB ultra-wide impedance bandwidth with good radiation properties. This design provides compact antenna in physically and electrically small dimension; the antenna key parameters which influence the performance are also calculated [13]. Yang Fan et al. have developed a single-patch E-shaped wideband micro-strip patch antenna. To expand its bandwidth, two parallel slots are included into the patch of a micro-strip antenna. Finally, E-shaped patch antenna with a 30.3%, resonating at 1.9 and 2.4 GHz was modeled, fabricated and measured [14].

2.2 *UWB Antennas for Wireless Communication*

A comprehensive review on several geometries, design methods, materials and analysis tools for broadband, multiband and UWB antennas has been presented.

Printed monopole wideband antennas: These antennas are designed and analyzed to provide improved antenna parameters with better performance. A competent method was defined for designing UWB antenna named as planar monopoles designed in the shape of elliptical and circular shape by Amin M. Abbosh and Marek E. Bialkowski; planar antennas are designed by using RT6010LM substrate; four different monopole antennas were fabricated, and their functionality was examined in view of return loss, radiation characteristics, gain and time-domain response. The manufactured antenna shows a high fidelity factor of greater than 90 [15]. Printed monopole antennas with regular shapes, with different feed positions, were obtained by Kamala Prasan [16]. Zhang described about printed elliptical monopole antennas in UWB systems, to extend the design flexibility; two slots are included in the edges of the UWB antennas [17]. In printed monopole antenna which was electrically small in size, empirically reported with a fractional bandwidth of 185% (115 MHz to 290 GHz) with S11 more than 10 dB, it resonates at 1.45 GHz with 2.35 dib as peak gain and 78.8% of radiation efficiency, represented by Mohammad et al. [18].

Reconfigurable WB antenna designs: From the past ten years, a significant amount of work was done on reconfigurable antennas; reconfigurability of antenna allows adopting their parameters dynamically as per the system requirements and nearby environment which effectively increases the spectrum utilization. Antennas are able to sense available white spaces in the spectrum by operating in the ultra-wideband mode; as per the observed outcome, it should be redesigned to operate in the predetermined band of frequencies. A compact microstrip monopole antenna designed for UWB, with reconfigurable polarization capability was presented for cognitive radio applications by Aboufoul Tamer et al. Proposed antenna able to changed its state from linear to right-hand circular polarization (RHCP) or left-hand circular polarization (LHCP). After the polarization reconfiguration, the measured 3dB fractional bandwidth is 59.5% at 5.9–10.9 GHz frequency band with RHCP or LHCP. This is suitable for cognitive radio systems [19]. Compact planar ultra-wideband monopole antenna with pattern reconfigurability was presented by Aboufoul Tamer et al. The pattern reconfigurability achieved by incorporating the 4-PIN diodes and 2-parasitic elements, which helped in directing the energy into particular desired direction while reducing the radiation in undesired areas. Example applications include smart radio (CR) as it adds the benefits of pattern diversity in multipath environments results in radio front-end flexibility; performance of antenna is improved [20]. A reconfigurable micro-strip monopole antenna in UWB region with multiband function was fabricated by Aboufoul Tamer et al. Reconfigurability was attained by attaching multiple stubs of different lengths to the main feed line of the monopole by using GaAs FET switches [21]. Denidni explained GaAs FET switches; it was demonstrated that an UWB region is formed by arranging some gap between two tapered slots. It could be utilized efficiently to merge a NB antenna also. The UWB and NB antennas accommodate a reasonable isolation between them with a VSWR of 2:1, and the band of frequencies ranges from 2.6 to 11 GHz and 4.9 to 6.2 GHz, respectively. The proposed design is suitable for cognitive radio applications [22].

MIMO-based wideband antennas: In recent years, due to the vast improvement in wireless communication systems, several kinds of new antennas are being implemented every year to satisfy the demand for ultra-high data rates which is an ongoing challenge. To develop 5G systems and other applications, massive MIMO technology is an ongoing trend. Log-periodic dipole antenna (LPDA) operated in wideband with compact size was designed by A. Moallemizadeh et al. Dipole arms mounted as second order fractals to resonate at more than one frequency. Consequently, size of LPDA antenna is greatly reduced. This antenna mainly consists of square-shaped substrate with dimensions of 90 * 60 mm and employed at 2–8 GHz band of frequencies with VSWR which is less than 2, and it offers constant radiation pattern over the entire frequency range [23]. Doose Niklas and Peter Adam Hoeher provided multi-mode antennas with massive MIMO system. The system was operated at 6–8.5 GHz frequency as an UWB system. This effort mainly focuses on design considerations of UWB systems and emphasizes the potential of employing multi-mode antennas in massive MIMO systems [24–27].

Survey on optimization techniques of wideband printed antenna: Aiting Wu et al. have presented a new design of optimized printed WB slot antenna modeled by a shape blending algorithm. Mainly, this design includes a tuning stub with wide rectangular slot, pie and a diamond-shaped profile and is modeled by the shape blending outcome. Modeled antenna geometry will modify the impedance bandwidth with different shapes of blending, and results have been examined and evaluated [28]. Printed monopoles with GA-optimized patch were examined by John Marthias and Max Ammann [15]. The design of printed planar monopole antennas in wideband range, with the help of GA optimization techniques was presented by Marthias and Ammann [18]. A patch antenna with elaborated optimization technique for multiband and wideband design was illustrated by Jin Nanbo and Yahya Rahmat-samii [29]. Optimum antenna was designed by combining particle swarm optimization (PSO), and the finite-difference time-domain (FDTD) design was proposed by Ankit et al. [25].

3 Conclusion

Therefore, in conclusion, some main investigations need to be considered while fabricating WB printed antennas with improved performance for biomedical applications; we have traced out some potential applications that could be taken into consideration while designing WB antennas. Through the survey it is observed that, the characteristics in time domain the measurements for practical applications could contribute further to WB antennas are testimony for future evolutions parallel to the rapid growth of wireless communications that we are observing today.

References

1. Raju GSNT (2005) Antennas and propagation. Pearson Education
2. Abdalla MA, Ibrahim AA, Boutejdar A (2015) Resonator switching techniques for notched ultra-wideband antenna in wireless applications. IET Microwav Antenna Propag 9(13): 1468–1477
3. Tsai L-C, Lai W-C (2018) Wideband CPW-fed spiral-shaped slot antenna for wireless applications. Radio Electron Commun Sys 61(4):180–183
4. Rushingabigwi G, Sun L, Zhu Q, Xia Y, Yu Y (2016) Design and realization of a dual wide band printed monopole antenna for WiFi and WiMAX systems. Int J Commun Netw Syst Sci 9(05):184
5. Chan D, Ahmed M, Hedayat A, Hart BD (2015) Dynamic bandwidth selection for wide bandwidth wireless local area networks. U.S. Patent 8,971,273. Issued: 3 Mar 2015
6. Singh A, Singh S (2015) A novel CPW-fed wideband printed monopole antenna with DGS. AEU-Int J Electron Commun 69(1):299–306
7. Xu K, Liu F, Peng L, Zhao W-S, Ran L, Wang G (2017) Multimode and wideband printed loop antenna based on degraded split-ring resonators. IEEE Access 5:15561–15570
8. Liang J, Chiau CC, Chen X, Parini CG (2004) Printed circular disc monopole antenna for ultra-wideband applications. Electron Lett 40(20):1246–1247
9. Abutarboush HF, Farooqui MF, Shamim A (2016) Inject-printed wideband antenna on resin-coated paper substrate for curved wireless devices. IEEE Antennas Wireless Propag Lett 15:20–23
10. Liang J, Chiau CC, Chen X, Parini CG (2005) Study of a printed circular disc monopole antenna for UWB systems. IEEE Trans Antennas Propag 53(11):3500–3504
11. Zhong S, Liang X, Yao F, Wang W, Bai X (2006) Ultra-wideband printed antennas for communication applications. J Shanghai Univ (English Edition) 10(1):48–52
12. Majid R-R, Ghobadi C, Nourinia J, Zaker R (2010) A small Printed ultra-wideband polygon-like wide-slot antenna with a fork-like stub. Microw J 53(3)
13. Guo L, Wang S, Chen X, Parini C (2009) A small printed quasi-self-complementary antenna for ultra wideband systems. IEEE Antennas Wirel Propag Lett 8:554–557
14. Fan Y, Zhang X-X, Ye X, Rahmat-Samii Y (2001) Wide-band E-shaped patch antennas for wireless communications. IEEE Trans Antennas Propag 49(7):1094–4400
15. Abbosh AM, Bialkowski ME (2008) Design of ultra wideband planar monopole antennas of circular and elliptical shape. IEEE Trans Antennas Propag 56(1):17–23
16. Kamala Prasan R (2008) Design aspects of printed monopole antennas for ultra-wideband applications. Int J Antennas Propag
17. Zhang H, Zhou R, Wu Z, Xin H, Ziolkowski RW (2010) Design of ultra wideband (UWB) printed elliptical monopole antennas with slots. Microw Opt Technol Lett 52(2):466–471

18. Mohammad A, Virdee BS, See CH, Abd-Alhameed R, Ali A, Falcone F, Limiti E (2018) Wideband printed monopole antennas for application in wireless communication systems. IET Microw Antennas Propag 12(7):1222–1230
19. Tamer A, Alomainy A, Parini C (2013) Polarization reconfigurable ultra wideband antenna for cognitive radio devices. In: IEEE antennas and propagation society international symposium (APSURSI), pp 636–1637. IEEE
20. Tamer A, Parini C, Chen X, Alomainy A (2013) Pattern-reconfigurable planar circular ultra-wideband monopole antenna. IEEE Trans Antennas Propag 61:10:4973–4980
21. Antenna for cognitive radio applications using GaAs FET switches. IEEE Antennas Wireless Propag Lett 11:392–394
22. Gijo A, Denidni TA (2012) An integrated ultra wideband/narrow band antenna in unipolar configuration for cognitive radio systems. IEEE Trans Antennas Propag 60(11):5479–5484
23. Moallemizadeh A, Hassani HR, Mohammad Ali Nezhad S (2012) Wide bandwidth and small size LPDA antenna. In: 2012 6th European conference on antennas and propagation (EUCAP), pp 1–3. IEEE
24. Niklas D, Hoeher PA (2015) Massive MIMO ultra-wideband communications using multi-mode antennas. In: SCC 2015; 10th international ITG conference on systems, communications and coding, pp 1–6. VDE
25. Ankit A, Srinivas K, Gupta M (2015) Optimal design of UWB antenna using differential evolution. ESSENCE Int J Environ Rehabil Conserv VI(2):155–160. ISSN 0975-6272]
26. Chaursia RK, Mathur V, Pareekh RL, Tamsir M, Srivastava VK (2018) A computational modeling of micro strip patches antenna and its solution by RDTM. Alexandia Eng J 57(3):1877–1881
27. Keskin Y, Oturanc G (2009) "Reduduced differential transform method for partial differential equations. Int J Nonlinear Sci Numer Simul 10(6):741–749
28. Mohammed HJ, Abdullah AS, Ali RS, Abd-Alhameed RA, Abdulraheem YI, Noras JM (2016) Design of a uniplanar printed triple band-rejected ultra-wideband antenna using particle swarm optimization and the firefly algorithm. IET Microw Antennas Propag 10(1):31–37
29. Marthias J, Ammann MJ (2007) Wideband printed monopole design using a genetic algorithm. IEEE Antennas Wireless Propag Lett 6:447–449

Comparison of Various Algorithms for Side Lobe Level Reduction in 5G Antenna Arrays

Y. Laxmi Lavanya and G. Sasibhushana Rao

Abstract 5G antenna arrays are being designed to generate dedicated stream of data to every single user. This results in more capacity and speed over the network along with the least possible latency rate. Depending upon the direction of the user demanding Internet access, the beam can be steered in that direction. Meanwhile, the occurrence of side lobes in such systems can be menacing. There are several algorithms that can be implemented to reduce side lobes. A comparison of three such algorithms—pattern search, simulated annealing and genetic algorithm—is made and the one which has the best side lobe level reduction is implemented. An 80-element uniform linear array is considered. MATLAB R2018b optimization toolbox has been used for simulation.

Keywords Latency rate · Pattern search · Simulated annealing · Genetic algorithm

1 Introduction

Ever since their evolution, cellular mobile communications have come a long way. Numerous applications have turned out to be reality. From the basic call application to text message services, from multimedia messages to data services, from high speed to very high speed data transmission, the capabilities of cellular mobile communications are continuously upgrading. There is still a lot more that these systems could do. For instance, automatic vehicular control which leads to unmanned vehicles, Internet of things (IoT) that could enable all our gadgets and appliances to communicate with each other and of course lightning fast data speeds with least response times till date [1]. All of these possibilities could be achieved with a new technology called 5G, i.e., 5th generation of cellular mobile communications.

The existing 4G LTE systems are at the verge of reaching their maximum potential in terms of data traffic. They can barely withstand the fast growing number of devices

Y. L. Lavanya (✉) · G. Sasibhushana Rao
Department of Electronics and Communication Engineering, AU College of Engineering (A), Andhra University, Visakhapatnam, Andhra Pradesh 530003, India
e-mail: lavanya9234@gmail.com

P. S. R. Chowdary et al. (eds.), *Microelectronics, Electromagnetics and Telecommunications*, Lecture Notes in Electrical Engineering 655,
https://doi.org/10.1007/978-981-15-3828-5_48

demanding Internet access. At this point of time, 5G comes to our rescue. The new technology can drastically enhance the capacity and speeds provided by current 4G LTE technology [2]. Further, the latency rate or the response time is expected to be far lower enabling 5G to be deployed in many real-time applications which are not possible till now.

Increased capacity and speed are a result of dedicated data stream for each user within the 5G coverage area. To achieve this, a smart antenna array is used, in which the beam can be steered in the desired direction. The primary concern in the generation and steering of such a dedicated beam is the occurrence of side lobes, which causes interference among users [3]. In a highly sophisticated technology like 5G, such interference is undesirable. Therefore, the reduction of side lobes has gained much significance. Several methods have been proposed for the reduction of side lobes in 5G antenna arrays [4]. This paper focuses on three algorithms: pattern search, simulated annealing and genetic algorithm, to reduce side lobe levels. A comparison is made to determine which algorithm performs best for the desired application. After selecting the best of the algorithms that can reduce side lobes to a nearly desired level, the further scope of the work is to generate multiple beams in different directions as per the users' demand.

2 Antenna Array Design

An 80-element linear antenna array with isotropic antenna elements has been considered with uniform element spacing $d = 0.5\ \lambda$, where λ is the wavelength. The radiation pattern of an antenna array is defined as the array factor multiplied by elemental radiation pattern, where array factor is a function of elemental excitations and spacing [5]. For an isotropic element, which radiates equally in all directions, the array radiation pattern is simply the array factor plot. The array factor at an angle θ is given by

$$\mathrm{AF}(\theta) = \left(\frac{\sin\left(\frac{N\phi}{2}\right)}{\sin\left(\frac{\phi}{2}\right)} \right) \tag{1}$$

Here, N is number of elements of array, $\phi = kd\cos\theta$, where $k = 2\pi/\lambda$ and d is a multiple of λ. Thus, $\mathrm{AF}(\theta)$ is frequency independent, but millimeter waves used for 5G implies shorter wavelengths. This means the antenna elements used are reduced in size as well as the distance between them (d) is very less. This enables more number of antennas to accommodate in less space, which makes 5G practically deployable. As the frequency is altered within the millimeter range, the spacing between antennas (d) changes.

Also, depending on the direction in which the user demands access, the beam is generated and steered. For a steering angle of α, the progressive phase shift to be given to each of the array elements is given as [6]

$$\Delta\emptyset = \frac{2\pi d}{\lambda}\sin\alpha \qquad (2)$$

Here, λ is wavelength and d is the spacing between elements of the array. MATLAB R2018b optimization toolbox 8.2 has been used for simulation of various algorithms.

2.1 Pattern Search Algorithm

The pattern search algorithm is a direct method that can be used to find the minimum or maximum value of a cost function. It does not involve any derivatives and is relatively simpler than algorithms like GA. The algorithm computes minimum of a function using the below steps [7]:

Step 1: Define the mesh size in which the search has to be done, maximum iterations and an initial point to begin the search.
Step 2: At the mesh points around the initial point, cost function values are calculated.
Step 3: If the value is less than the previous value, cost function value is updated and mesh size is increased, else cost function value is not updated and mesh size is decreased. The number of iterations is increased by 1.
Step 4: Continue until cost function value no longer changes significantly or maximum number of iterations is reached. The final value of cost function is the minima.

2.2 Simulated Annealing Algorithm

The algorithm is used to find global minima of a cost function with many variables. It avoids the solution getting trapped in local minima by slowly exploring the search space, which is analogous to slow cooling or gradual decrease in temperature. Its steps are as follows:

Step 1: Define a random initial point, set an initial temperature and maximum iterations.
Step 2: Find the cost function at a random neighborhood point and if it is less than that at the previous point, update the point and increase iteration by 1, else do not update and increase iteration by 1.
Step 3: Continue until iterations reach maximum value.
Step 4: Decrease the initial temperature (linear, logarithmic or exponential) and continue Steps 2–4 until temperature reaches minimum acceptable value. The cost function obtained at this point is the global minima.

2.3 Genetic Algorithm

Genetic algorithm is an evolutionary algorithm inspired by natural selection process. It can be used to find minimum or maximum of a function. The defined cost function is minimized using the following steps:

Step 1: Create a starting population of individuals and evaluate fitness function for each individual.
Step 2: Check for stopping criterion. If not met, create a new set of population using random selection.
Step 3: Current generation called parents are crossed over to generate offspring for the next generation. Also, changes in parents are made which is known as mutation. After selection, crossover and mutation, the new population formed is again evaluated for fitness function [8].
Step 4: Repeat Steps 2 and 3 until stopping criterion is met. The resulting value of cost function is the obtained global minima.

3 Results and Discussion

For minimizing side lobes of the designed antenna array, the cost function for all the three algorithms is defined as follows:

$$\text{Cost function} = \text{SLL}_{\text{desired}} - \max\left\{20\log\left|\frac{\text{AF}(\theta)}{\text{AF}(\theta)_{\max}}\right|\right\} \tag{3}$$

where $\text{AF}(\theta)_{\max}$ is the maximum value of array factor or main lobe level. Therefore, the second term in the above equation represents the first side lobe level in db (SLL1), $\text{SLL}_{\text{desired}}$ is the desired side lobe level in db. The difference is minimized, i.e., side lobe level close to desired SLL is obtained. The weights corresponding to minimum cost function value are the optimum weights to the antenna array.

The various minima and corresponding side lobe levels obtained by implementing pattern search, simulated annealing and genetic algorithm using the above objective function with $\text{SLL}_{\text{desired}} = -30$ db are shown in Table 1.

Table 1 Comparison of various algorithms for side lobe level reduction

Algorithm used	Cost function minima	Corresponding SLL1 from Eq. (3)
No optimization	−16.78	−13.22
Pattern search	−6.29	−23.71
Simulated annealing	−4.75	−25.25
Genetic algorithm	−3.24	−26.76

From Table 1, it is observed that the side lobe level when each subarray is given unity excitation (no optimization) is −13.22 db and it is reduced to −23.71 db, −25.25 db and −26.76 when pattern search, simulated annealing and genetic algorithms are applied, respectively (from Eq. (3) when $\mathrm{SLL}_{\mathrm{desired}} = -30$ db). It is evident that genetic algorithm has the best side lobe level reduction among the three.

The plots corresponding to algorithms in Table 1 are shown in Fig. 1.

The corresponding rectangular plot is shown in Fig. 2.

The side lobe level is significantly high. It can be calculated by substituting the Y values obtained in Fig. 2 into Eq. (3). For example, from Fig. 2, $\mathrm{AF}(\theta)_{\max} = 80$ and $\mathrm{AF}(\theta) = 17.45$ for first side lobe. Substituting in Eq. (3) with $\mathrm{SLL}_{\mathrm{desired}} = -30$ db, cost function obtained is −16.78 db implying side lobe level of −13.22 db. To minimize this problem, three different algorithms are implemented. Using pattern search algorithm, the cost function minima obtained is −6.29 db, which means the side lobe level has been reduced to −23.71 db. Taking the weights corresponding to this cost function value and applying them as excitations to the antenna array, the rectangular plot is shown in Fig. 3.

Similarly, by applying simulated annealing algorithm, the side lobe level has been reduced to −25.25 db. The rectangular plot of the antenna array with the simulated annealing optimized weights is shown in Fig. 4.

A further reduction in side lobe level has been observed by applying genetic algorithm. The cost function minima obtained is −3.24 db, which means that the side lobe level obtained is −26.76 db. The corresponding weights are called genetic algorithm optimized weights. When they are applied, the plot obtained is shown in Fig. 5.

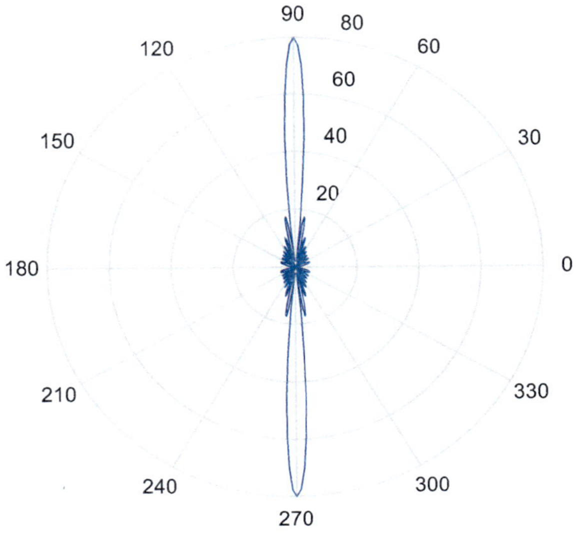

Fig. 1 Polar plot of 80-element array antenna with unity excitation (no optimization)

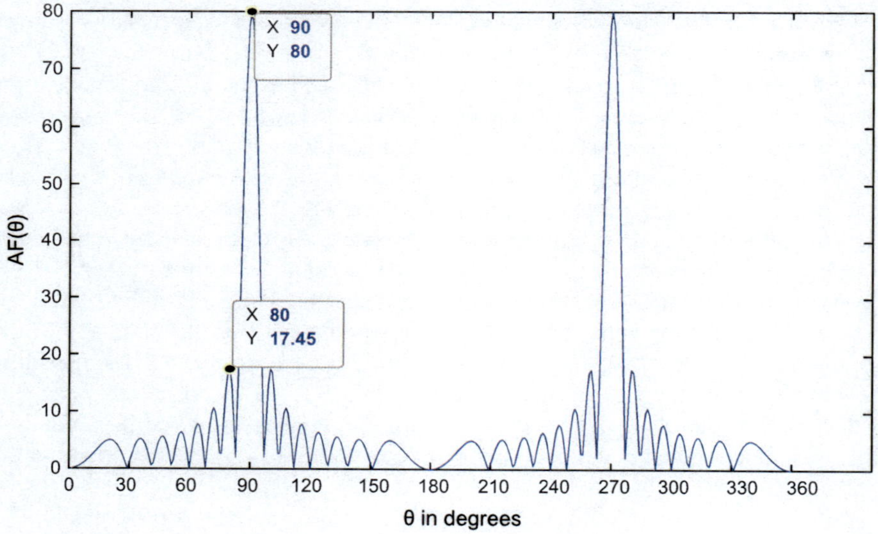

Fig. 2 Rectangular plot of 80-element array antenna with unity excitation (no optimization)

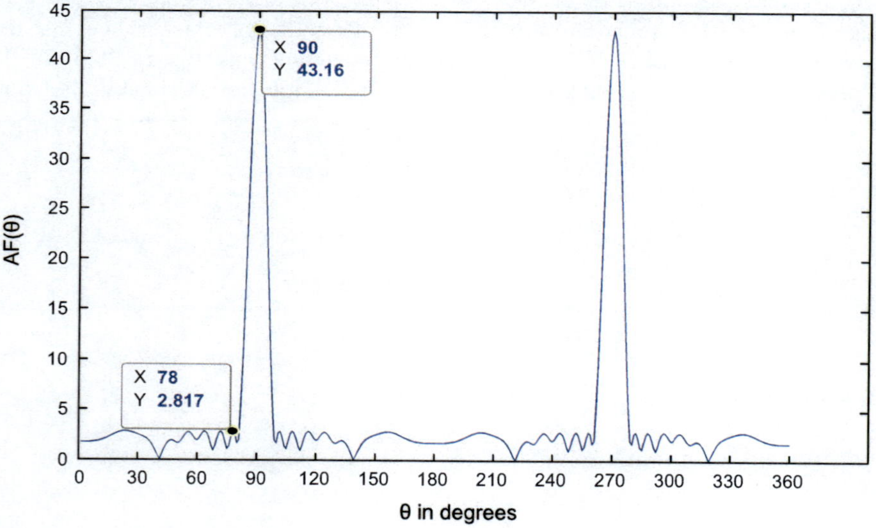

Fig. 3 Rectangular plot of 80-element array antenna with pattern search optimized weights

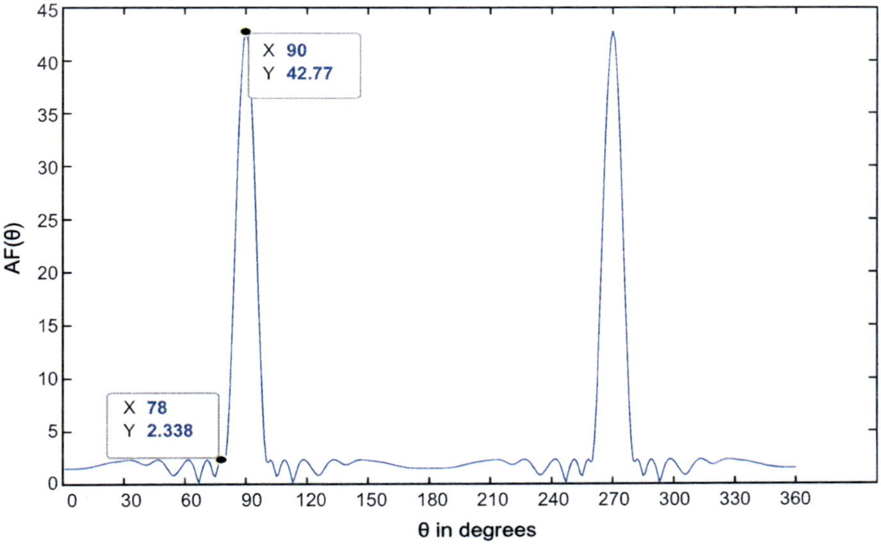

Fig. 4 Rectangular plot of 80-element array antenna with simulated annealing optimized weights

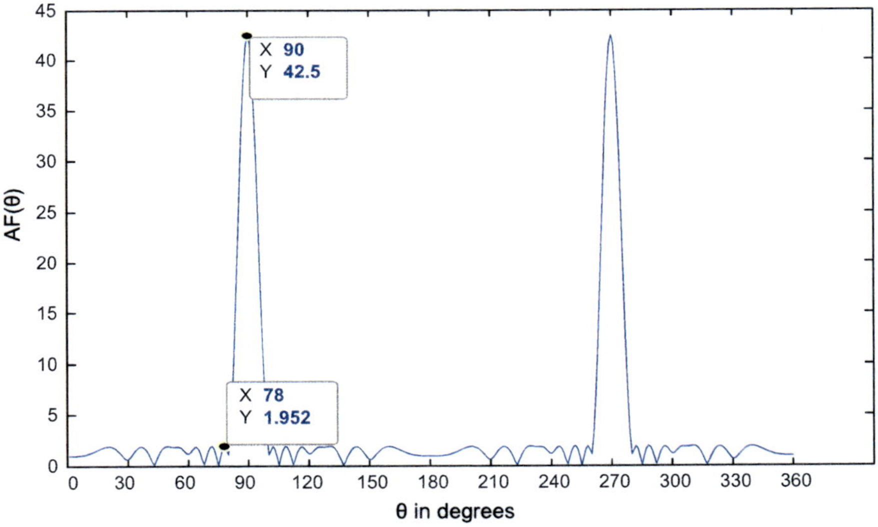

Fig. 5 Rectangular plot of 80-element array antenna with GA optimized weights

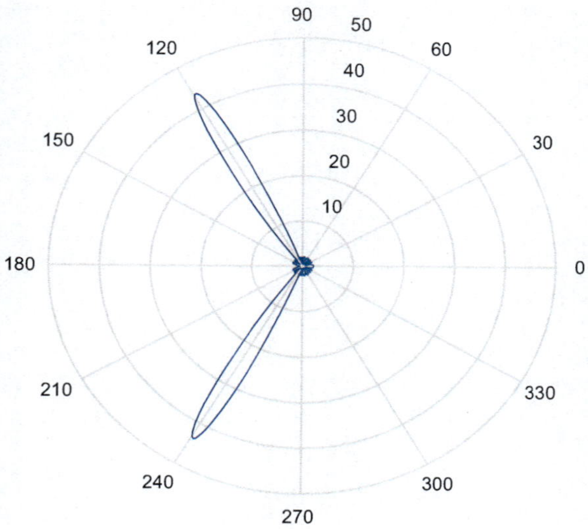

Fig. 6 Polar plot of 80-element array antenna with GA optimized weights and steering angle 30°

The level of main lobe is also reduced after optimization. This is the consequence of amplitude tapering in optimization techniques. Also, the beam can be steered in any direction by providing phase shifts to the antenna array elements. For example, to steer the beam by 30°, a phase shift of 90° is required between array elements. The polar plot of an antenna array with GA optimized excitations and a progressive phase shift of 90° between its elements are shown in Fig. 6.

From the above observations, it is evident that genetic algorithm has significantly reduced the side lobe levels and its performance is better than pattern search and simulated annealing algorithms in terms of SLL reduction.

4 Conclusions

In a technology like 5G, side lobes can cause severe interference, which is highly undesirable. To effectively reduce them, three different algorithms have been implemented and genetic algorithm is found to be the best one out of them. Yielding nearly desired side lobe levels, GA has also been simple yet efficient side lobe level reduction technique.

References

1. Gupta A, Jha A (2015) A survey of 5G network: architecture and emerging technologies. IEEE Access 3:1206–1232
2. Zhang J, Wu Z, Xie W, Yang F (2018) MEC architectures in 4G and 5G mobile networks. In: 2018 10th international conference on wireless communications and signal processing (WCSP). IEEE, pp 1–5

3. Zainal NA, Kamarudin MR, Yamada Y, Seman N, Khalily M, Jusoh M (2016) Sidelobe reduction of unequally spaced arrays for 5G applications. In: 2016 10th European conference on antennas and propagation (EuCAP). IEEE, pp 1–4
4. Sun G, Liu Y, Li H, Liang S, Wang A, Li B (2018) An antenna array sidelobe level reduction approach through invasive weed optimization. Int J Antennas Propag
5. Maharimi SF, Jamlos MF, Abd Malek MF, Neoh SC (2012) Impact of number elements on array factor in linear arrays antenna. In: 2012 IEEE 8th international colloquium on signal processing and its applications. IEEE, pp 296–299
6. Balanis CA (2005) Antenna theory: analysis and design. Wiley, Hoboken, NJ, USA
7. Zewail I, Saad W, Shokair M, El_dolil SA (2017) Maximization of total throughput using pattern search algorithm in underlay cognitive radio network
8. Yan K-K, Lu Y (1997) Sidelobe reduction in array-pattern synthesis using genetic algorithm. IEEE Trans Antennas Propag 45(7):1117–1122

Analysis of Electromagnetic Shielding Effectiveness Properties of Al6061 Metal Matrix Composites at X-Band for Aerospace Applications

Srinu Budumuru and M. Satya Anuradha

Abstract Over the last few years to save consumption of fuel by the aeroplanes, aerospace industries prefer composite materials (fiber-reinforced plastic) with less weight, good mechanical properties like high strength and stiffness instead of metals. But, the major drawback of the fiber-reinforced materials is that they cannot conduct high impulse electric currents and electric filed forces through them which precipitates the structural damage of airplane when lightning strikes. Lightning strike protection (LSP) solution should be considered by the designer to easily allow the flow of current and field line through the surface to provide the aircraft with EMI shielding. In the present work, electromagnetic characteristics of metal matrix composite AL6061 was considered based on reinforcement fly ash particulates at different levels of percentages at X-band. The shielding effectiveness of the composites is improved by 25 dB.

Keywords Shielding effectiveness · Reinforcement · Fly ash

1 Introduction

It is expected that an aircraft is struck by lightning frequently one each between 10,000 and 10,000 h of flight [1]. When lightning strikes the airplane, majority of destruction occurs at the incident or exit part where the high intensity of currents and electromagnetic field lines are unable to pass through it easily [2]. Immense current produced by the lightning flow through the surface of the structure has less resistance. Suppose, if the designer does not consider the lightning effect to protect the surface of airplane, there exists dreadful consequences like delamination, embrittlement and

S. Budumuru (✉)
Department of Electronics and Communication Engineering, GITAM University, Visakhapatnam, India
e-mail: srinubudumuru@yahoo.co.in

M. Satya Anuradha
Andhra University, Visakhapatnam, Andhra Pradesh, India
e-mail: radhamsa@gmail.com

P. S. R. Chowdary et al. (eds.), *Microelectronics, Electromagnetics and Telecommunications*, Lecture Notes in Electrical Engineering 655,
https://doi.org/10.1007/978-981-15-3828-5_49

structural failure. For industrial and defense applications, composite materials are the best replacements of metals because of their high strength-to-weight and modulus-to-weight ratios [3]. Daniel Lingvay et al. conducted an experiment for composites' add mixture of fly ash particle and concluded that reinforcement of fly ash improves the shielding effectiveness of the sample [4]. The porous carbon grain and iron oxides of fly ash cause for better absorption of electromagnetic when passing through it [5]. Yinsuo Dai et al. made attempt on reinforcement of fly ash particles with different materials and concluded that porous carbon grain and iron oxides of fly ash cause for good absorption of electromagnetic waves [6]. Zuoyong Dou, Gaohui Wu et al. measured the EMSE of aluminum composite reinforced with fly ash and obtained the shielding between 32 and 102.5 dB [7]. The total work is organized as investigation of lightning strike protection, composite material preparation, characterization of composite material chemical and electrical properties, finding shielding effectiveness from obtained electrical parameters of composite, finding tensile strength, hardness of material and finally results and conclusion.

2 Investigation on LSP Material

The durability and long-term maintenance of flight were limited by the damaged composite material which was affected by the lightning strike. In this study, to overcome the above drawback, investigations are made on composite material for good mechanical properties like tensile strength, hardness and improved SE. Further, fly ash is of 5, 10, 15% reinforced with the AL6061 metal matrix composite. Reinforcement of fly ash with AL6061 provides good alternating materials for regular FRP composites which are used at airplanes in terms of LSP.

3 Composite Preparation

Stir casting is the basic method for preparation of composite in a liquid state at a melting temperature of individual elements [8]. Here, pure AL6061 was heated at a temperature of 700 °C and mixed with fly ash melted at a temperature of 600 °C. This mixture is converted into specimens of required shape. Fly ash is a chemical combination of the following chemical compounds.

3.1 Fly Ash Chemical Composition

Compound	Al_2O_3	Na_2O	K_2O	SiO_2	CaO	Fe_2O_3	SO_3	MgO
wt%	26.23	0.21	0.95	53.1	4.59	3.17	0.62	1.10

4 Electrical Parameters Measurements

U. C. Hasar designed a two-port device to calculate the electrical parameters of specimen using vector network analyzer (VNA). This method concerns calibration of conductivity, permeability and permittivity of specimen which is kept inside the waveguide slot based on transmitted and reflected signals. Experiment is conducted on composite material based on vector network analyzer, and electrical parameters glean as in Table 1.

5 Shielding Effectiveness of Composite

The shielding effectiveness of the specimen is well-defined based on contribution of absorption loss and reflection loss; here, the internal reflections are not considered because of its low impact. Especially, for metallic materials, S. A. Schelkunoff proposed an intuitive method to relate material properties to shielding effectiveness as a sum of absorption loss, reflection loss and multiple reflections [9, 10]. Shielding effectiveness of conductive sheet explained in terms of transmitted and reflected waves is shown in Fig. 1.

Propagation constant of planar wave is defined as

$$\gamma = (\alpha + j\beta) = \sqrt{j\omega\mu(\sigma + j\omega\varepsilon)} \tag{1}$$

where 'α' represents attenuation constant and 'β' represents phase constant.

For a metallic sheet, parameters are given as follows

Attenuation constant, phase constant

$$\alpha = \beta = \sqrt{\pi\mu f\sigma} \tag{2}$$

Skin depth of material is defined as

$$\delta = \frac{1}{\alpha} = \sqrt{\frac{2}{\omega\mu\sigma}}\ \mathrm{m} \tag{3}$$

For a given wave, absorption loss with sheet thickness of 'l' is given as

$$A = 20\ \log_{10}\frac{\mathrm{E}_i}{\mathrm{E}_t} = 8.686\alpha l\ \mathrm{dB} \tag{4}$$

Table 1 Electrical parameters of the composite material with different percentages of fly ash

Frequency (GHz)	Pure AL6061			5% of fly ash reinforcement (experimental values)			10% of fly ash reinforcement (experimental values)			15% of fly ash reinforcement (experimental values)		
	μ_r	σ_r	ε_r	μ_r	σ_r	ε_r	μ_r	σ_r	ε_r	μ_r	σ_r	ε_r
8.2	−1.1	2.59	3.42	−8.6	0.23	4.94	5.91	0.26	2.69	5.91	0.26	2.69
9	0.47	0.42	2.49	0.37	−0.31	1.49	0.22	2.95	15.60	0.22	2.95	15.60
10	0.51	1.91	6.65	−0.5	0.79	5.65	0.81	1.57	13.20	0.81	−1.57	−13.2
11	2.12	2.82	5.75	0.31	−3.85	9.75	0.25	−8.72	15.10	0.25	−8.72	15.10
12	4.23	−1.8	4.31	0.03	−0.78	9.75	−0.04	−5.10	19.60	−0.04	−5.10	19.60

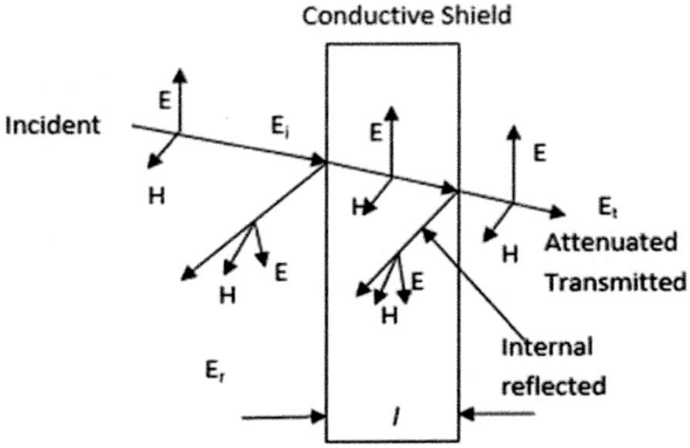

Fig. 1 Representation of reflected signal and transmitted signal

The normal input impedance of infinite planar sheet with thickness '*l*'

$$Z = \eta \frac{Z(l)\cosh \gamma l + \eta \sinh \gamma l}{\eta \cosh \gamma l + Z(l) \sinh \gamma l} \tag{5}$$

$Z(l)$ is the impedance right to the plane $x = l$. The intrinsic impedance of the planar sheet is given as 'η'. According to the above equations, reflection coefficients and transmission coefficients are given as

$$q_{\mathrm{E}} = \frac{E_r}{E_i} = \frac{Z(l) - \eta}{Z(l) + \eta} \qquad q_{\mathrm{H}} = \frac{H_r}{H_i} = \frac{\eta - Z(1)}{\eta + Z(1)} \tag{6}$$

$$p_{\mathrm{E}} = \frac{E_t}{E_i} = \frac{2Z(l)}{\eta + Z(l)} \qquad p_{\mathrm{H}} = \frac{H_t}{H_i} = \frac{2\eta}{\eta + Z(l)} \tag{7}$$

If two mismatched interfaces are considered, the net transmission coefficient is the product of transmission coefficients across two boundaries.

$$p = p_{\mathrm{E}} = p_{\mathrm{H}} = p_{\mathrm{E}}(0).p_{\mathrm{E}}(t) = p_{\mathrm{H}}(0).p_{\mathrm{H}}(t) = \frac{4Z(t)\eta}{(Z(t) + \eta)^2} \tag{8}$$

Total reflection loss is

$$R = -20 \log_{10}|p| \tag{9}$$

Shielding Effectiveness (SE) = Absorption loss (A) + Reflection loss (R).

Table 2 MATLAB simulation results of SE, reflection loss, absorption loss of AL6061 for different combinations of fly ash

S. No.	Composite material	Reflection loss range (dB)	Absorption loss range (dB)	Shielding effectiveness range (dB)	Tensile strength	Hardness
1	AL6061	34.33–82.41	0–29.99	26.36–99.15	110	33
2	AL6061 + 5% fly ash (Experimental result)	29.27–74.57	0–34.23	31.62–105.8	124	47
3	AL6061 + 10% fly ash (Experimental result)	24.22–75.31	0–45.26	22.16–119.7	126	55
4	AL6061 + 15% fly ash (Experimental result)	10.68–79.08	0–44.34	7.50–125.9	128	60

6 Tensile Strength, Hardness

Mechanical properties of the AL6061 composites were analyzed for different percentages of fly ash and compared with pure Al6061. Brinell hardness test worn for finding the hardness of the composite. Tensile strength of composite is measured employing universal testing machine, and the results are shown in Table 2.

7 Results and Discussion

SE is combination of both absorption loss and reflection loss; the following plots represent SE in terms of both '*A*' and '*R*'.

Shielding effectiveness is calculated in terms of both absorption and reflection loss at X-band. Figure 2 shows the SE of pure AL6061, maximum absorption loss of 29.99 dB, reflection loss in the range of 34.33–82.41 dB and the corresponding shielding effectiveness in the range of 26.36–99.15 dB is obtained in the required frequency range.

From Figs. 3, 4 and 5, maximum absorption loss of 34.23 dB, 45.26 dB and 44.34 is resulted for 5%, 10% and 15% of fly ash compositions, respectively. Reflection loss of 29–74 dB, 24–75 dB and 10–79 dB is observed for the compositions considered in this work, and shielding effectiveness in the range of 31–105 dB, 22–119 dB and 7–125 dB is achieved with 5%, 10% and 15% of fly ash compositions, respectively (Fig. 6).

It is evident from the results that for the composite material, reflection loss decreases and absorption loss increases with increase in frequency. Finally, the SE

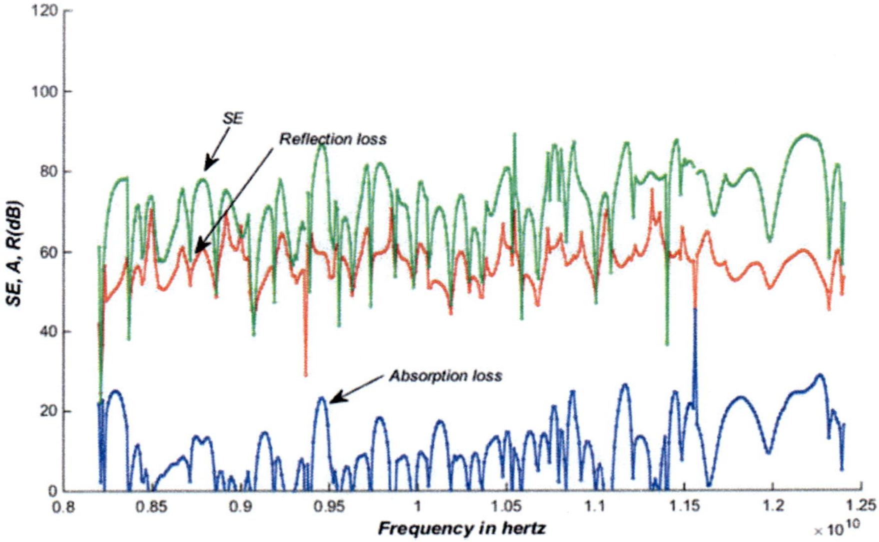

Fig. 2 Reflection loss, absorption loss and shielding effectiveness of pure AL6061 by varying frequency

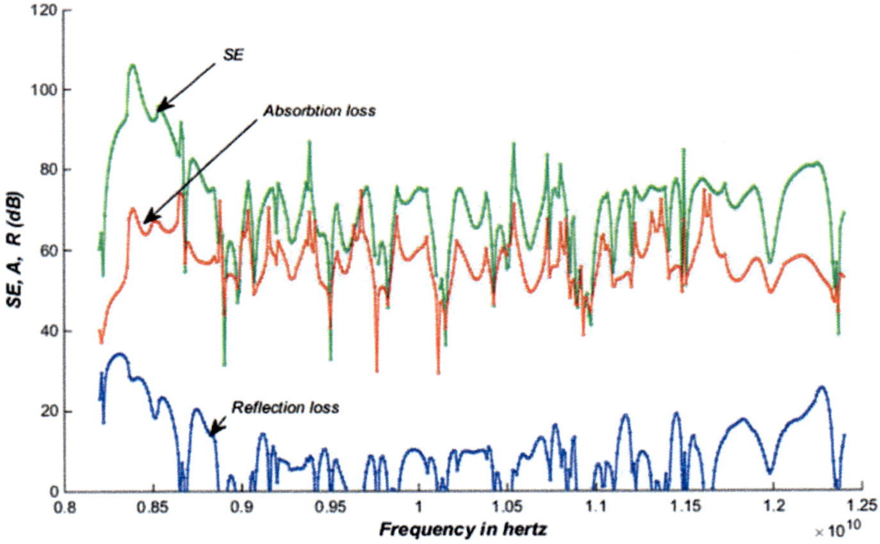

Fig. 3 Reflection loss, absorption loss and shielding effectiveness of AL6061 reinforced with 5% of fly ash by varying frequency

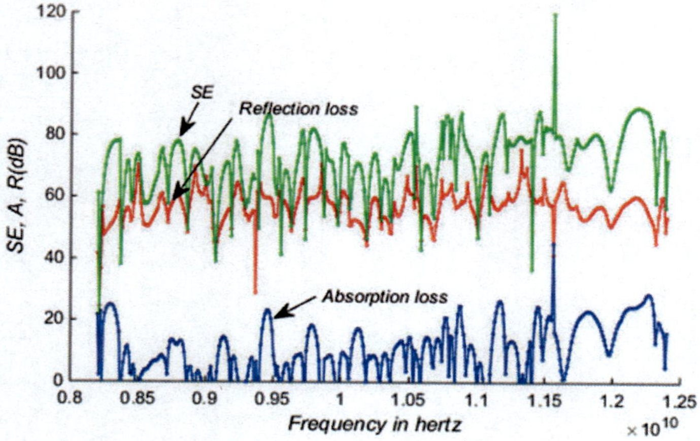

Fig. 4 Reflection loss, absorption loss and shielding effectiveness of AL6061 reinforced with 10% of fly ash by varying frequency

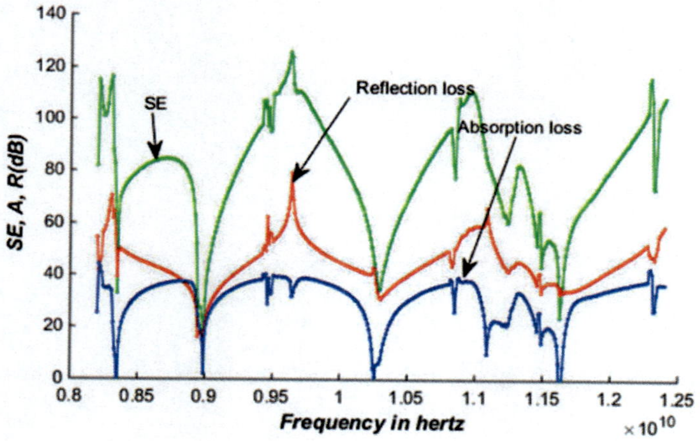

Fig. 5 Reflection loss, absorption loss and shielding effectiveness of AL6061 reinforced with 15% of fly ash by varying frequency

of a composite was improved as the percentage of the fly ash increases. The tensile strength and hardness of AL6061 composite also enhance as the add mixture percentage of fly ash increases. Zuoyong Dou and Gaohui Wu et al. obtained for aluminum composite EMSE which is about 102.5 dB, and Al6061 composite reinforced with fly ash results in better shielding effectiveness of 125.9 dB.

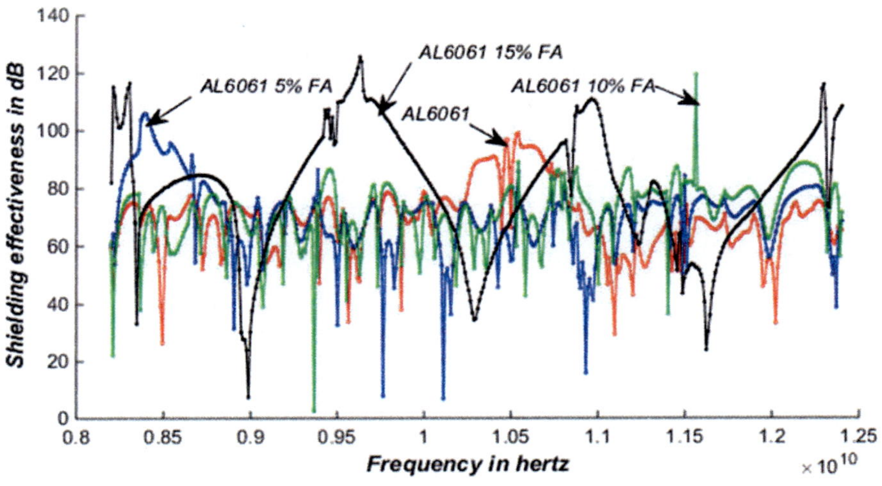

Fig. 6 Shielding effectiveness comparison plot by varying frequency

8 Conclusion

The present aerospace industry manufacturers are looking for the composite materials having lesser weight, good shielding and robust mechanical properties. In this work, AL6061 composite material has been taken and reinforced with flash particles at different percentages, and for each combination, reflection loss, absorption loss and shielding effectiveness were determined. As per the obtained results, add mixture of fly ash with composite materials shows improved SE properties with less weight when compared with the pure AL6061. Pure AL6061 has maximum shielding of 99.15, whereas composite with fly ash of 5%, 10%, 15% shows 105.5 dB, 119.7 dB, 125.9 dB, respectively.

References

1. Larsson A (2002) The interaction between a lightning flash and an aircraft in flight. C R Phys 3:1423–1444
2. Guo Y, Xu Y, Wang Q, Dong Q, Yi X, Jia Y (2019) Eliminating lightning strike damage to carbon fiber composite structures in Zone 2 of aircraft by Ni-coated carbon fiber nonwoven veils. Compos Sci Technol 169:95–102
3. von Klemperer CJ, Maharaj D (2009) Composite electromagnetic interference shielding materials for aerospace applications. Compos Struct 91:467–472
4. Lingvay1 D, Velciu G (2018) Fly ash for increased electromagnetic protection of constructions. Electrotehnica Electronica Automatica (EEA) 66
5. Selvam JDR, Smart DSR, Dinaharan I (2013) Microstructure and some mechanical properties of fly ash particulate reinforced AA6061 aluminum alloy composites prepared by compocasting. Mater Design 49:28–34

6. Dai Y, Wu J, Wang D (2018) Effect of mineralogical phase and chemical composition of fly ash on electromagnetic wave-absorbing properties. Mater Trans 59(6):876–882
7. Dou Z, Wu G, Huang X, Sun D, Jiang L (2007) Electromagnetic shielding effectiveness of aluminum alloy–fly ash composites. Compos: Part A 38:186–191
8. Manimaran R, Jayakumar I (2018) Mechanical properties of fly ash composites—A review. Energy Sour Part A: Recovery Util Environ Effects
9. Sai Srinivas NS, Chakravarthy V, Sudheer Kumar T (2016) Analysis of single-layered multiple aperture shield for better shield effectiveness. In: Satapathy S, Raju K, Mandal J, Bhateja V (eds) Proceedings of the second international conference on computer and communication technologies. advances in intelligent systems and computing, vol 380. Springer, New Delhi
10. Dharma Raj C, Sasibhushana Rao G, Jayasree PVY, Srinu, Lakshman P (2010) Analysis of reflectivity and shielding effectiveness of absorbing material–conductor laminate for electromagnetic compatibility. J Electromag Anal Appl

Design, Simulation and Experimental Validation of Patch Antenna in S-Band Satellite Communication

Karedla Chitambara Rao, P. Mallikarjuna Rao, B. Sadasiva Rao, and Pavada Santosh

Abstract In worldwide, almost geosynchronous satellites are using radio communication from broadband to narrowband which include broadcasting, positioning and telecommunication. At present, there are different types of satellite communications starting from UHF to Ka-band. Out of these satellite communications, S-band and C-band satellite communications are the suitable for high speed data, voice and video transmissions. For communicating with a satellite, a special antenna is required, and it should have some unique characteristics such as low VSWR, circular polarization, high gain, reasonable axial ratio and higher beam width. For satellite communication, micro-strip patch antenna is appropriate because it can be designed to use as a circular polarized antenna by modifying its shape and easily developing a 90° phase shift in the antenna. Square patch antenna has been taken and designed for S-band satellite communication. This antenna is designed for receiving purpose. After designing, the antenna is simulated in simulated software for the different values like gain, axial ratio, VSWR and 3 dB beam width. After simulation, the antenna is fabricated and measured for the same characteristics. Moreover, the simulated results are obtained using CST MW STUDIO 2015 software, and the antenna is fabricated using CNC machine. Apart from this, experimental results are obtained using network analyzer and anechoic chamber. VSWR measurement of antennas has been carried out

K. Chitambara Rao (✉)
Department of Electronics and Communication Engineering, Aditya Institute of Technology and Management, Tekkali, Andhra Pradesh, India
e-mail: rao.chiddubabu@gmail.com

P. Mallikarjuna Rao · P. Santosh
Department of Electronics and Communication Engineering, AU College of Engineering, Visakhapatnam, Andhra Pradesh, India
e-mail: pmraoauece@gmail.com

P. Santosh
e-mail: santosh.pavada14@gmail.com

B. Sadasiva Rao
Department of Electronics and Communication Engineering, GVP College of Engineering for UG & PG Rushikonda Campus, Visakhapatnam, Andhra Pradesh, India
e-mail: profbsrao7@gmail.com

P. S. R. Chowdary et al. (eds.), *Microelectronics, Electromagnetics and Telecommunications*, Lecture Notes in Electrical Engineering 655,
https://doi.org/10.1007/978-981-15-3828-5_50

using network analyzer. Gain and radiation patterns have been carried out using anechoic chamber. Validation of the antenna has been done based on the comparison of simulation and measurement results in S-band satellite communication.

Keywords Square patch antenna · Rectangular patch antenna · S-band communication

1 Introduction

At present, most of the communication has been done based on the satellites. These satellites are more useful for all network services like 3G, 4G and 5G. The satellites are not only used for network services but also used for positioning and navigation purpose. For these satellites, compact antennas are required for communicating with an allover world. On the satellite, the antenna should have circular polarized characteristics, and at the earth station, the antenna may or may not have circular polarization characteristics. If the satellite antenna and earth station antenna are same circular polarized antennas, then maximum power reception can be possible at satellite as well as at earth station. Patch antenna is the appropriate antenna for satellites as well as earth station because it has tremendous advantages like low cost, low weight, less volume. Moreover, patch antenna will be operated at multi-band operations by incorporating the 90° phase shift in the antenna. Patch antenna with rectangular shape was designed at 2.4 GHz for wireless communication with and without DGS technique. The antenna parameters like return loss, VSWR and beam width have been improved by using DGS technique when compared to the without DGS technique [1]. Rectangular patch antenna was designed at 1.88 GHz for WLL-Cor-DECT Technology. For designing this antenna, the silicon material was used with dimensions of height 31 mm and dielectric constant 11.9. Single feed technique is used for such an antenna. Return loss of −35 dB was obtained at an operating frequency [2]. Based on different values of substrate thickness, two types of antennas are designed such as patch with circular shape antenna and slotted patch antenna with circular shape. When the thickness of the substrate is increased from 1 to 2 mm for both two types of antennas, circular patch with slot antenna gave the better result in terms of gain when compared to the circular patch antenna without slot [3]. Based on the techniques like slotted partial ground and tuning stub, the ultra-wideband rectangular patch was designed, and the same antenna was simulated for different parameters like return loss, gain and radiation patterns in the frequency range of 3.2–15.7 GHz. Highest bandwidth 12 GHz and gain of 7.5 dB were obtained, but high value of return loss −40 dB is achieved [4]. Based on the three orthogonal sector slots, the elliptical antenna was designed, simulated, fabricated and experimented for the different parameters such as VSWR, gain, return loss and beam width. The measured and simulated results were compared. From the comparison results, it is known that antenna performance is slightly improved, and also, size is reduced [5]. For wireless communication applications, patch antenna was designed with the help of coaxial type of feeding technique. The same antenna is operated at 2.4 GHz frequency. The lowest values of gain 8.27 dB and VSWR of 1.18 were obtained for such type of feeding technique [6]. Based on the thickness of the dielectric constant, the rectangular and square patch antennas have been compared at 2.4 GHz frequency. With dielectric substrate, the parameters are improved like beam width, VSWR, gain and return loss when compared to the without dielectric substrate [7]. Symmetrical slotted U patch antenna and asymmetrical slotted U patch antenna are designed for 1.9–2.5 GHz. Symmetrical U-slot patch antenna and asymmetrical U-slot patch antennas have been compared in terms of size, bandwidth and dielectric constant. From the comparison of results, it is noted that patch antenna with symmetrical U-slot has advantages like reduction of size, more bandwidth and less dielectric constant [8]. Merits, demerits of patch antenna and various feeding techniques have been reviewed [9]. Patch antenna with DGS technique was designed for C-band satellite communication applications [10].

2 Antenna Design

Directly, square patch antenna can be designed in the simulation software by taking the equal lengths which shows that length and width of the square patch antenna. Theoretically, the square patch antenna cannot be designed directly, first design a rectangular patch antenna and then design a square patch antenna with the help of ratio of equating lengths. For satellite communication, circular polarized antennas are required, and the patch antenna can also be used as a circular polarized antenna by modifying its shape and providing a 90° phase shift in the antenna. For circular polarized antenna, 90° phase shift is required between the horizontal and vertical components of the radiated wave. There are so many techniques to develop a 90° phase shift in the antenna. One way of easy technique is to provide the feeding in the diagonal of the patch antenna. From the patch antenna, LHCP or RHCP can be developed by incorporating the feed point in the diagonal of the patch. Left-hand circular polarization (LHCP) may be obtained by incorporating a feed point along the right corner diagonal. Right-hand circular polarization (RHCP) may also be obtained by incorporating a feed point along the diagonal opposite to the left-hand circular polarization. The equations in [10] are used to calculate the design parameters of the rectangular patch antenna.

2.1 Patch Antenna

Figure 1 shows the simulated receiving patch antennas. This antenna is designed based on the design parameters which are presented in Table 1. The design parameters are calculated with the help of design equations in [10]. Probe method of feeding is used for this antenna.

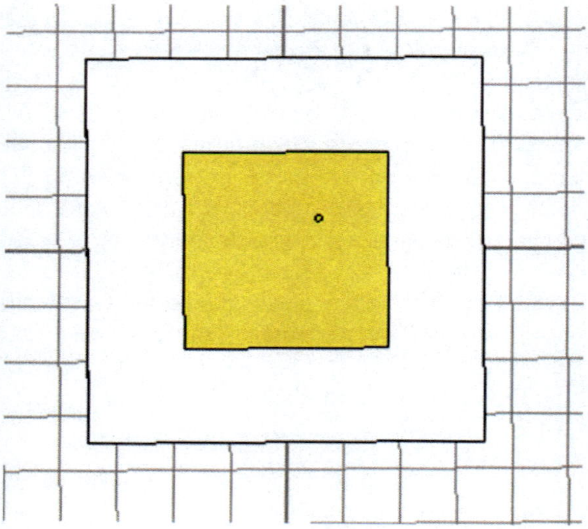

Fig. 1 Simulated patch antenna

Table 1 Antenna design parameters

S. No.	Design parameter	Antenna dimensions	Design specifications	Specification values
1	Resonant frequency	2.5 GHz	R_x antenna freq. range	2.5–2.52 GHz
2	Length of the ground plane	80 mm	VSWR	3:1(max)
3	Width of the ground plane	80 mm	Gain	−3 dBi (min) to 3 dBi (max)
4	Substrate (RT Duriod 5880)	1.5 mm	Polarization	LHCP
5	Substrate dielectric constant	2.2	Beam width	±45° from vertical
6	Patch (Copper)	0.1 mm	Axial ratio	<5 dB
7	Patch length	40 mm	Spatial coverage	Overtop hemisphere
8	Patch width	40 mm	Connector	SMA

Left-hand circular polarization technique is used for this antenna in S-band satellite communication. LHCP or RHCP may be used based on the user requirements. The specifications [10] in Table 1 are considered for receiving patch antenna. The specifications should be satisfied for proper working of an antenna in S-band communication.

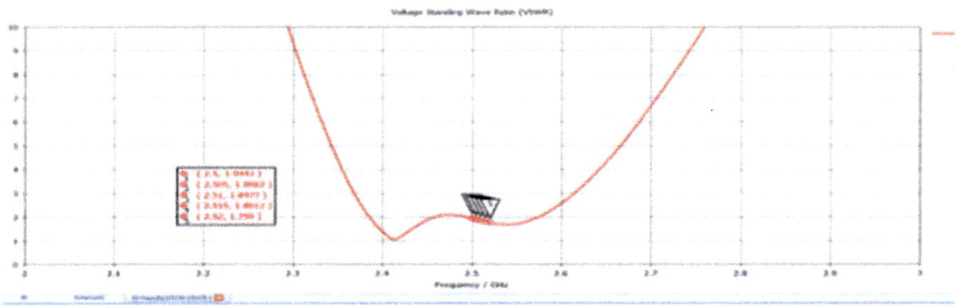

Fig. 2 VSWR curve

Table 2 VSWR at receiving frequencies

Freq. (GHz)	VSWR
2.5	1.9443
2.505	1.8962
2.51	1.8477
2.515	1.8012
2.52	1.759

3 Simulation Results

After designing a receiving antenna in simulation software, that is analyzed for the different parameters like gain, axial ratio, VSWR and 3-dB beam width at antenna operating frequencies. The operating frequencies of the S-band satellite communication are 2.5–2.52 GHz. The 2D radiation patterns are used to get the beam width values at receiving frequencies. 2D radiation patterns are obtained from 3D patterns by using different techniques.

3.1 Simulated VSWR

Figure 2 shows the VSWR against various receiving frequencies. The VSWR values are obtained from the VSWR curve and which are tabulated in Table 2. From Table 2, **it is seen that the VSWR values are low values according to given specifications.**

3.2 Simulated Gain

The receiving patch antenna is simulated for gain at 0°. Figure 3 shows the gain at 0° against frequency. From Table 3, it is known that highest gain values are obtained

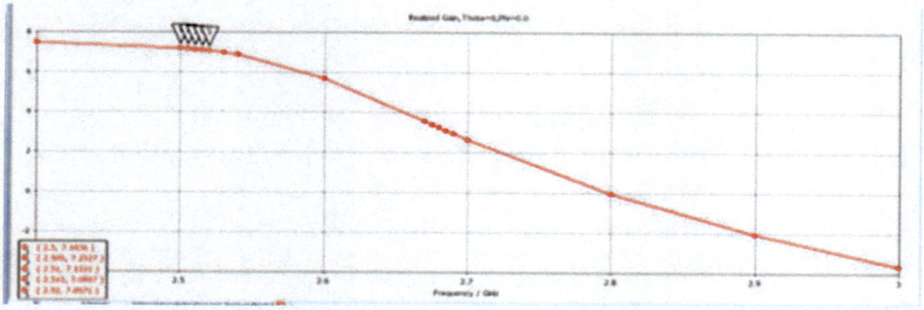

Fig. 3 Gain at $\theta = 0°$

Table 3 Gain at receiving frequencies

Freq. (GHz)	$\theta = 0°$
2.5	7.1836
2.505	7.1527
2.51	7.1221
2.515	7.0907
2.52	7.0571

at the operating frequencies.

3.3 *Simulated Axial Ratio (AR)*

Like gain, the axial ratio is obtained at 0° for different receiving frequencies. From Table 4, it is known that higher axial ratio values are obtained, and they are not a constraint for a receiving antenna (Fig. 4).

Table 4 AR at receiving frequencies

Freq. (GHz)	$\theta = 0°$
2.5	10.227
2.505	10.755
2.51	11.28
2.515	11.802
2.52	12.318

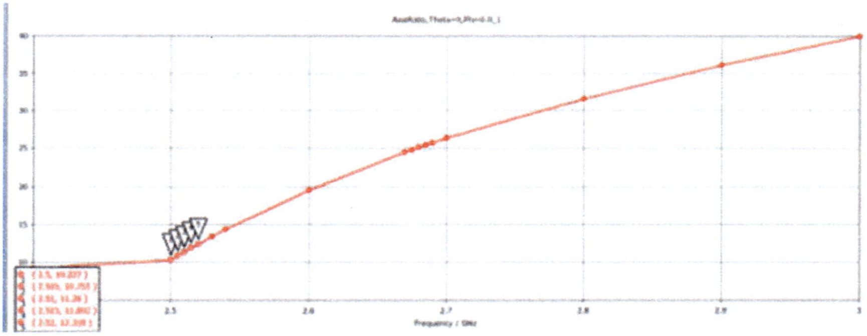

Fig. 4 Axial ratio at $\theta = 0°$

3.4 Simulated 2D Patterns at the Receiving Frequencies

The 3-dB beam width values in Table 5 are obtained from the 2D patterns. These are suitable for receiving antenna (Figs. 5 and 6).

Table 5 3-dB beam width at different frequencies

Freq. (GHz)	Beam width (°)
2.5	65.3
2.505	65.1
2.51	64.9
2.515	64.7
2.52	64.6

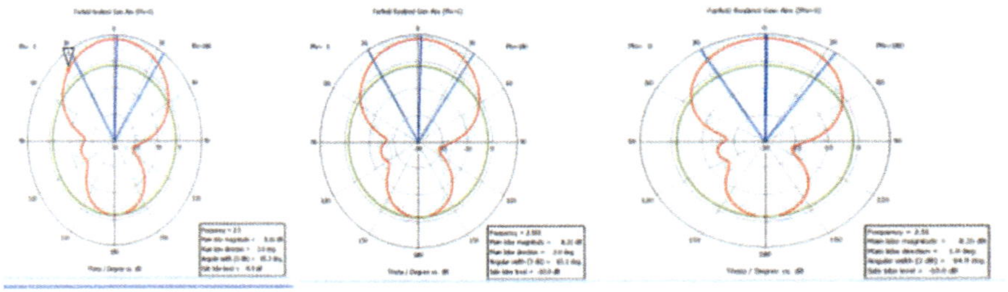

Fig. 5 2D radiation patterns at 2.5, 2.505 and 2.51 GHz frequencies

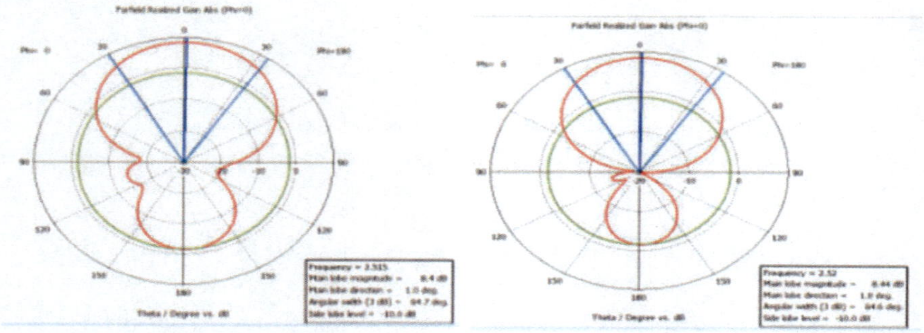

Fig. 6 2D radiation patterns at 2.515 and 2.52 GHz frequencies

4 Measurement Results

The receiving patch antenna is fabricated and measured for different parameters at its operating frequencies. Also, antenna is measured in two types of polarizations like horizontal polarization (HP) and vertical polarization (VP) (Figs. 7 and 8).

4.1 Measured VSWR

Figure 8 shows the VSWR against frequency. From this VSWR curve, the VSWR values are obtained and given in Table 6. The tabulated VSWR values are very low values and suitable for receiving antenna.

Fig. 7 Fabricated patch antenna

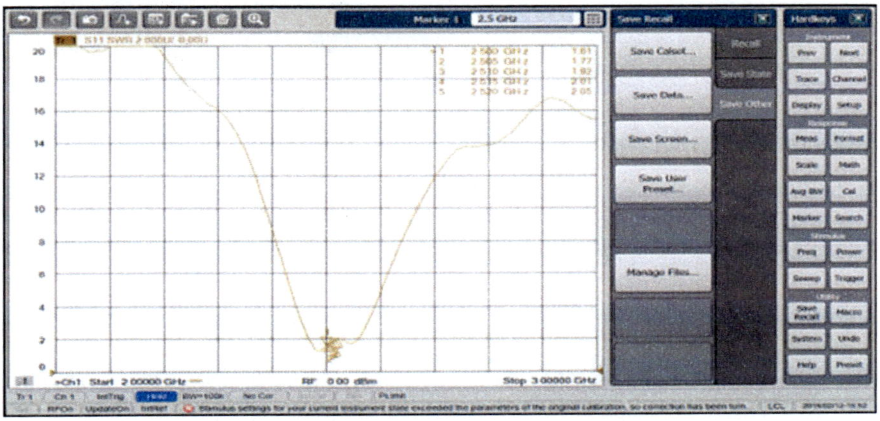

Fig. 8 VSWR curve

Table 6 VSWR at receiving frequencies

Frequency (GHz)	VSWR
2.5	1.81
2.505	1.77
2.51	1.92
2.515	2.01
2.52	2.05

4.2 *Measured Radiation Patterns of Receiving Antenna in Horizontal and Vertical Polarizations*

See Fig. 9.

4.3 *Measured Gain, Beam Width and Axial Ratio at the Receiving Frequencies*

Gain of the receiving patch antenna is measured in horizontal and vertical polarizations. In Table 7, the antenna under test (AUT) gain is calculated based on the equation in [10]. Moreover, the axial ratio in Table 7 is determined using the equation in [10]. The beam peak values in horizontal and vertical polarizations and 3-dB beam width values are obtained from the measured radiation patterns which are presented in the Sect. 4.2.

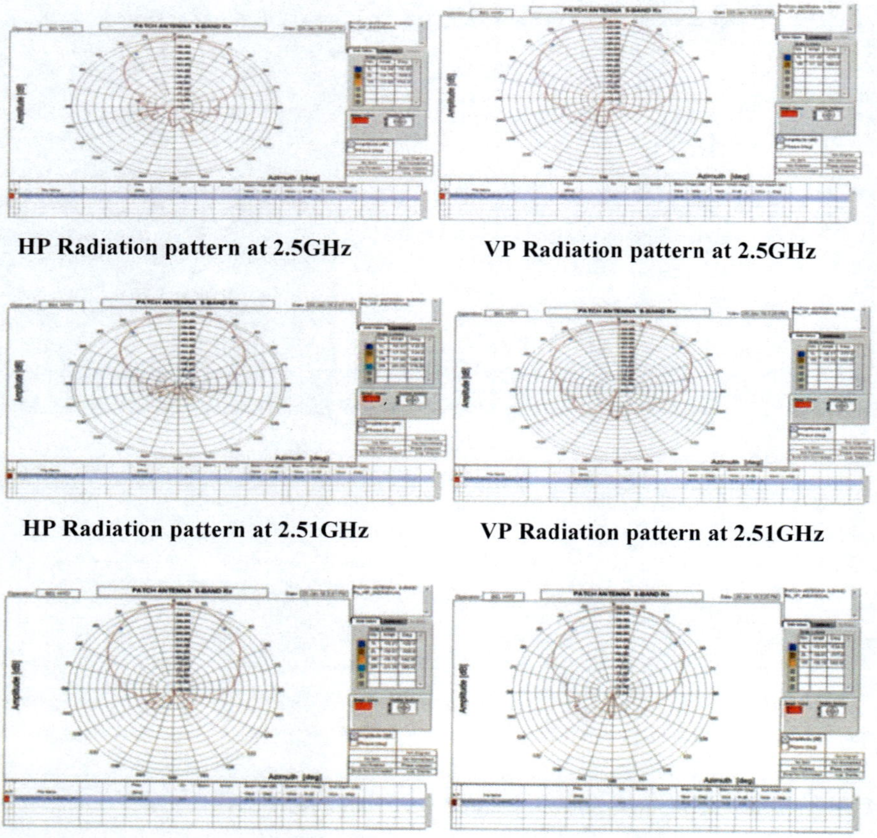

HP Radiation pattern at 2.5GHz **VP Radiation pattern at 2.5GHz**

HP Radiation pattern at 2.51GHz **VP Radiation pattern at 2.51GHz**

HP Radiation pattern at 2.52GHz **VP Radiation pattern at 2.52GHz**

Fig. 9 HP and VP patterns of receiving antenna at the receiving frequencies

4.4 Comparison of Simulated and Measured Results of Receiving Patch Antenna

The simulated and measured results are compared in terms of VSWR, gain at 0° in vertical polarization, 3-dB beam width and axial ratio at the receiving frequencies. From the comparison of results, it is known that all the simulated and measured values match very well (Table 8).

Table 7 Gain, beam width and axial ratio at the receiving frequencies

Freq. (MHz)	Beam Peak in HP (dB)	Beam peak in VP (dB)	Beam width in HP (°)	Beam width in VP (°)	STD peak (dB)	STD gain (dB)	AUT gain with polarization loss in HP at 0°	AUT gain with polarization loss in VP at 0°	Axial ratio (dB)
2500	−52.73	−49.99	65.84	78.84	−37.85	17.05	7.17	9.91	2.74
2510	−51.3	−50.7	68.34	76.18	−38.01	17.06	8.77	9.37	0.6
2515	−51.12	−51.78	68.14	75.34	−38.3	17.065	9.245	8.585	0.66
2520	−50.43	−52.19	69.53	74.85	−38.35	17.07	9.99	8.26	1.76

Table 8 Comparison of VSWR, gain, beam width and axial ratio

Freq. (MHz)	Simulated VSWR	Measured VSWR	Simulated gain at 0°	Measured gain at 0° in VP	Simulated 3-dB beam width	Measured 3-dB beam width in VP	Simulated axial ratio at 0°	Measured axial ratio at 0°
2500	1.9443	1.81	7.1836	9.91	65.3	78.84	10.227	2.74
2510	1.8477	1.92	7.1221	9.37	64.9	76.18	11.28	0.6
2515	1.8012	2.01	7.0907	8.585	64.7	75.34	11.802	0.66
2520	1.759	2.05	7.0571	8.26	64.6	74.85	12.318	1.76

5 Conclusion

Square patch antenna is designed, simulated, fabricated and experimentally validated for the reception of the signals in the S-band satellite communication. This antenna is used as a receiving antenna in the range of frequencies from 2.5 to 2.52 GHz. The receiving patch antenna has been simulated and experimentally tested for the required parameters like gain, VSWR, beam width, radiation patterns and axial ratio at the receiving frequencies. Apart from this, the simulated and measured results are compared to get the validation of the antenna in S-band satellite communication. From the comparison of results, it is known that patch antenna is suitable for the reception of signals in S-band satellite communication.

References

1. Nawale PA, Zope RG (2015) Rectangular micro strip patch antenna for 2.4 GHz. Communication using defected ground structure. Int J Adv Found Res Comput (IJAFRC) 2(1)
2. Mishra S, Awasthi S (2010) Design and optimization of patch antenna for WLL-Cor-DECT technology using IE3D. Inf Sci Technol 1(2):01–04. ISSN 0976-917X & ISSN:0976-9188
3. Madhav BTP, Phani Srinivas K, Divya A (2012) Design and analysis of circular patch & patch-slot antenna with dimensional characterization. Int J Recent Trends Electr Electron Eng. ISSN 2231–6612
4. Rafiqul IM, Zahirul AA, Khan MFAJ, Alkaraki S (2012) Design of micros trip patch antenna using slotted partial ground and addition of stairs and stubs For UWB application. Cyber J: Multidiscip J Sci Technol J Sel Areas Telecommun (JSAT)
5. Vijay Sharma V, Saxena K, Sharma KB (2011) Radiation performance of an elliptical patch antenna with three orthogonal sector slots. Roman J Inf Sci Technol 14(2):123–130
6. Muhammad A (2015) Micro strip patch antenna-designing at 2.4 GHz frequency "Science Stays True Here". Biol Chem Res 128–132 (Science Signpost Publishing)
7. Saidulu V, Srinivasa Rao K, Somasekhar Rao PVD (2013) The characteristics of rectangular and square patch antennas with superstrate. Int J Eng Sci Emerging Technol 6(3):298
8. Khodaei GF, Nourinia J, Ghobadi C (2008) A practical miniaturized U-slot patch antenna with enhanced bandwidth. Prog Electromagn Res B 3

9. Singh I, Tripathi V (2011) Micro strip patch antenna, and its applications: a survey. Int J Comp Tech Appl 2(5):1595–1599
10. ChitambaraRao K (2019) Design, simulation and experimental validation of wideband DGS circularly polarized MPA for C-band satellite communication applications. Int J Eng Adv Technol 8(6S2)

Performance Efficient Floating-Point Multiplication Using Unified Adder–Subtractor-Based Karatsuba Algorithm

K. V. Gowreesrinivas and P. Samundiswary

Abstract The execution of multiplication consumes more time, power and also requires more area than other arithmetic operations. Hence, in floating-point multiplication, performance-optimized mantissa multiplication is necessary to get efficient response. In this paper, unified adder–subtractor-based 24-bit mantissa multiplication is designed. First, single precision floating-point multiplication is designed with Karatsuba algorithm to improve the speed, and later for more better performance, an unified adder–subtractor-based carry-select adder is introduced in Karatsuba algorithm. In this, Karatsuba algorithm is developed by using Vedic multiplication along with unified adder–subtractor logic. Further, the performance metrics are analyzed for the existing techniques with the proposed techniques. All modules are developed with Xilinx ISE.

Keywords Unified adder–subtractor · Urdhva Tiryagbhyam · Karatsuba · Single precision

1 Introduction

In high-speed digital signal processors, performance efficient multipliers play a major role in affecting the processor metrics. In general, floating-point number system offers more dynamic range by representing number, with a mantissa and exponent. Most of the performance is influenced by multipliers in any digital processing system. So, the research on optimization of multipliers has formed a lot of space and attention for more research. The floating-point multiplication (FPM) realization requires mantissa multiplication and exponent addition for final result generation. In the past,

K. V. Gowreesrinivas (✉)
Anil Neerukonda Institute of Technology and Sciences, Visakhapatnam, Andhra Pradesh, India
e-mail: srinu43306@gmail.com

P. Samundiswary
Pondicherry University, Puducherry, India
e-mail: samundiswary_pdy@yahoo.com

P. S. R. Chowdary et al. (eds.), *Microelectronics, Electromagnetics and Telecommunications*, Lecture Notes in Electrical Engineering 655,
https://doi.org/10.1007/978-981-15-3828-5_51

many researchers have worked on improvement of floating-point multiplication by integrating several optimized logics [1–7].

Generally, to add partial products, various adders like as ripple and carry-save adders are used [8–15]. The prevailing carry-save adder is substituted by improved BEC-based CSLA to attain better power and area. Further, to achieve better performance of mantissa multiplication, Karatsuba technique is introduced in floating-point multiplier. Karatsuba algorithm offers better delay, but area is more. Floating-point multiplier is developed using three pipelining stages and eight clock cycles of latency to achieve high speed and accuracy. This method can provide high throughput and execute more instructions per unit time compared to that of the existing multipliers. Few researchers are also used the combination of KA and VM in floating-point multiplication to expand the delay and power efficiency of mantissa multiplication. In this, the proposed work is coded using HDL and targeted on Spartan-3E, Virtex-4 FPGA families. Subsequently, researchers are concentrated on DPFPM using Karatsuba algorithm to improve the delay of the system. Few authors made an attempt in developing DPFPM using combination of Vedic sutras and Karatsuba algorithm. Later on, unified adder–subtractor-based adders are introduced in multipliers to improve the multiplication performance [16]. The unification of adding and subtracting is done with unified adder–subtractor (UAS) for BEC-based MCM design to reduce the resource utilization, and power consumption is explained in by Ding et al. [8]. In general, carry-select addition is widely used block in most of the DSP processors to bring out fast arithmetic additions.

In this, SPFPM is analyzed using Karatsuba algorithm with normal adders and unified adder–subtractor module for partial products accumulation in mantissa multiplication. Performance metrics considered for the analysis are area and speed. From the verification outcomes, it is exhibited that the proposed unified adder–subtractor logic attains improved delay associated with that of the conventional adders. The remaining portion of the paper is planned as follows: In Sect. 2, IEEE 754 standard is conferred, Karatsuba algorithm is discussed in Sect. 3, unified adder–subtractor is explained in Sect. 4, simulation results are analyzed in Sect. 5 and conclusion is drawn in Sect. 6.

2 IEEE 754 Standard

The Institute of Electrical and Electronics Engineers (IEEE) established one standard for floating-point arithmetic is IEEE 754 [4]. The generalized floating-point multiplication is represented using the below equation for all floating-point numbers.

$$Y = (-1^{\text{sign}}) * 2^{E1+E2\text{-bias}} * (1.M)$$

Here, Bias = 127 for single precision numbers

The final multiplication result is produced by computing all the three results

Step1: Exponents addition and subtraction of the bias value from the result.
Step2: Multiplication of two mantissa operands.
Step3: Generation of sign bit by XORing the MSB bits.

3 Karatsuba Algorithm

Usually, more time is needed for the multiplication as the operand width increases. So, in order to design efficient multiplication in view of complexity and speed, Karatsuba algorithm (KA) with integrating Vedic multiplication is introduced [10]. In KA, higher bit length operands are decomposed into smaller bit length operands to make multiplication simple. In conventional procedure of KA, few multipliers task is carried out by taking extra adders in place of multipliers to enhance the multiplication performance, they took less time to execute the result. Let us consider an example that two operands of each taking '2' digits, it entails '4' multipliers, but Karatsuba algorithm can finish the same by employing '3' multipliers and '2' subtractors. This KA is most effective when the input operand width increases [1]. The efficacy of KA is optimum when the width of operands is more than 16 bits. The mantissa multiplier is designed by integrating KA with Vedic using high-speed adders which is discussed [3]. The multiplication is done for two 23-bit mantissa numbers. The Karatsuba algorithm is preferable to improve the performance when the number of bits is more. KA is explicated below: Let us take two significand portions of assumed '2' numbers T and X.

- Mantissa part is splitted into '3' parts, each taking '8' bits, i.e., T_0, T_1, T_2 and S_0, S_1, S_2 for the inputs T and S.
- Process of KA is described below. This idea of KA uses enough quantity of adders in places of few multipliers to advance hardware utilization.

$$\begin{aligned} T &= T_2\, 2^{2n} + T_1\, 2^n + T_0 \\ S &= S_2\, 2^{2n} + S_1\, 2^n + S_0 \\ TS &= T_2 S_2\, 2^{4n} + (T_2 S_1 + T_1 S_2)\, 2^{3n} + [(T_2 S_0 + T_0 S_2) + T_1 S_1] \\ &\quad 2^n + (T_0 S_1 + T_1 S_0)\, 2^n + T_0 S_0 \end{aligned}$$

This typical algorithm entails nine multipliers, eight adders to produce the final outcome. KA using Vedic multiplier using UAS is used to compute multiplication.

4 Unified Adder–Subtractor

From Fig. 1, it is depicted that $w1$ is a mutual signal for 'sum' and 'dif' output signals and turns as selection line for 2×1 MUXs to produce the 'car' and 'bor'. In this,

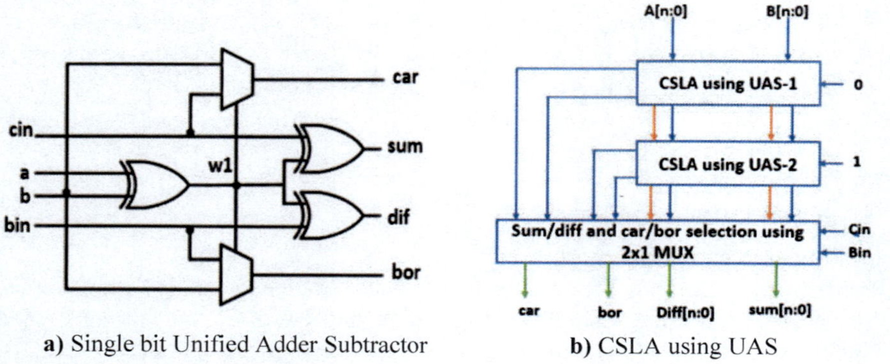

Fig. 1 **a** Single-bit unified adder–subtractor, **b** CSLA using UAS

UAS CSLA cell is cast off to develop 16-bit CSLA [16]. The UAS CSLA consists of UAS-1 and UAS-2 with 'cin', 'bin' equal to '0' and '1'. Lastly, the sum, dif, car, bor outputs are selected depending on the real 'car' and 'bor' input using the 2×1 multiplexer. The 24-bit KA is developed using UAS-based CSLA unit. In this, all the adder blocks are replaced with UAS-based CSLA in mantissa multiplication. The UAS-based KA is used in developing the mantissa multiplication (Fig. 2).

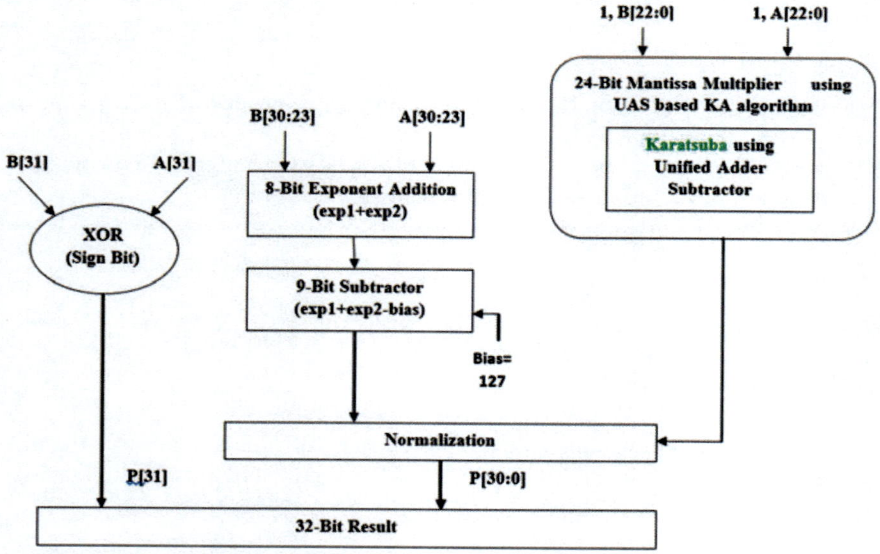

Fig. 2 SPFPM using UAS based KA

Fig. 3 Area of SPFPM using KA with UAS

Device Utilization Summary				
Logic Utilization	**Used**	**Available**	**Utilization**	**Note(s)**
Number of 4 input LUTs	553	9,312	5%	
Number of occupied Slices	283	4,656	6%	
Number of Slices containing only related logic	283	283	100%	
Number of Slices containing unrelated logic	0	283	0%	
Total Number of 4 input LUTs	553	9,312	5%	
Number of bonded IOBs	96	232	41%	
Average Fanout of Non-Clock Nets	3.27			

5 Simulation Results

The SPFPM using Karatsuba algorithm using conventional adders and unified adder–subtractor is designed and analyzed. In this, performance is restrained in terms of resource utilization and delay. From the resulting simulation report, it is proved that SPFPM with Karatsuba algorithm using Vedic with unified adder–subtractor gives better performance compared to KA using conventional adder-based mantissa multiplication with respect to resource utilization which is measurable in terms of slices and slice LUTs and speed. All the blocks involved in multiplication are verified with HDL language and simulated using Xilinx ISE.

5.1 Area Utilization of SPFPM Using KA with UAS

See Fig. 3.

5.2 Area Utilization of SPFPM Using KA with RCA

It is observed from Table 1 that SPFPM using UAS-based KA is more efficient in terms of delay and area compared to that of RCA-based KA algorithm (Fig. 4).

Table 1 Assessment of SPFM with KA using UAS and RCA

SPFPM	Slice LUTs	Slices	Delay (ns)
KA with UAS	553	283	17.89
KA with RCA	763	395	43.41

Fig. 4 Area of SPFPM using KA with RCA

Device Utilization Summary				
Logic Utilization	Used	Available	Utilization	Note(s)
Number of 4 input LUTs	763	9,312	8%	
Number of occupied Slices	395	4,656	8%	
Number of Slices containing only related logic	395	395	100%	
Number of Slices containing unrelated logic	0	395	0%	
Total Number of 4 input LUTs	763	9,312	8%	
Number of bonded IOBs	96	232	41%	
Average Fanout of Non-Clock Nets	3.54			

6 Conclusion

Multiplication for IEEE 754 standard single precision numbers is developed and analyzed with the help of unified adder–subtractor to improve the performance with respect to speed and area. In this, SPFPM is analyzed by integrating Karatsuba algorithm with Vedic multiplier using unified adder–subtractor in mantissa multiplication, and performance is measured in respect of area and delay metrics which are verified by Xilinx ISE. From the verification results, it is depicted that SPFPM using UAS-based KA is more efficient in delay and area compared to that of other SPFPM algorithms.

References

1. Manish Kumar J, So KWH (2017) DSP48E efficient floating point multiplier architectures on FPGA. In: Proceedings of 30th international conference on VLSI design and 16th international conference on embedded systems, Hyderabad, India, pp 1–6
2. Ramesh AP, Tilak AVN, Prasad AM (2013) An FPGA based high speed IEEE-754 double precision floating point multiplier using Verilog. In Proceedings of international conference on emerging trends in VLSI, embedded system, nano electronics and telecommunication system (ICEVENT), Tamil Nadu, India, pp 1–5
3. Srinivasa Rao Y, Kamaraju M, Ramanjaneyulu DVS (2015) An FPGA implementation of high speed and area efficient double precision floating point multiplier using Urdhva Tiryagbhyam technique. In: Proceedings of conference on power, control, communication and computational technologies for sustainable growth (PCCCTSG), Kurnool, India, pp 271–276
4. Pranal DK, Prof. Thakre MN, Prof. Mrs. Mandavgane RN (2014) A survey on 64 bit floating point multiplier based on vedical multiplication techniques. Int J Sci Eng Res 5(2):234–238
5. Sharma B, Bakshi A (2015) Comparison of 24 × 24 bit multipliers for various performance parameters. In: Proceedings of international conference on advent trends in engineering, science and technology, Maharashtra, pp 146–149
6. Neelima K, Rohit S, Paidi S (2012) Performance comparison of fast multipliers implemented on variable precision floating point multiplication algorithm. Int J Comput Appl Eng Sci 2:55–59

7. Kanhe A, Das S, Singh A (2012) Design and implementation of floating point multiplier based on Vedic multiplication technique. In: Proceedings of international conference on communication, information computing technology, Mumbai, pp 1–4
8. Ding J, Chen J, Chang C-H (2016) A new paradigm of common sub expression elimination by unification of addition and subtraction. IEEE Trans Comput Aided Des Integr Circ Syst 35(10):1605–1617
9. Hormigo J, Villalba J, Zapata EL (2013) Multi operand redundant adders on FPGAs. IEEE Trans Comput 62:10, 2013–2025
10. Han L, Zhang H, Ko S-B (2015) Area and power efficient decimal carry-free adder. IET J Mag Electron Lett 51(23):1852–1854
11. Ramkumar B, Kittur HM (2012) Low-power and area efficient carry select adder. IEEE Trans Very Large Scale Integr Syst 20(2):371–375
12. Konstantinos V, Al-Khalili AJ (2007) Performance of parallel prefix adders implemented with FPGA Technology. In: Proceedings of international conference on circuits and systems, Nothesat
13. Naga Sravanthi V, Terlapu SK (2020) Design and performance analysis of rounding approximate multiplier for signal processing applications. In: Satapathy S, Bhateja V, Mohanty J, Udgata S (eds) Smart intelligent computing and applications. smart innovation, systems and technologies, vol 160. Springer, Singapore
14. Dake JL, Terlapu SK (2016) Low complexity digit serial multiplier for finite field using redundant basis. Ind J Sci Technol 9(S1):1–5
15. Ashok Chaitanya Varma R, Venkata Subbarao M, Srinivasa Raju GRLVN (2020) High throughput VLSI Architectures for CRC-12 computation. In: Satapathy S, Raju K, Shyamala K, Krishna D, Favorskaya M (eds) Advances in decision sciences, image processing, security and computer vision. Learning and analytics in intelligent systems, vol 3. Springer, Cham
16. Solomon R, Nirmal Singh N, Arun Samuel TS (2019) Resource minimization and power reduction of ESPFFIR filter using unified adder/subtractor. Analog Integr Circ Sig Process 98(1):169–179

Implementation of Program Page, Read Page and Block Erase Operations in NAND Flash Memory Controller

Ch. Harini, B. Keerthi Priya, and D. V. Rama Koti Reddy

Abstract In underwater, vehicles like AUV'S, submarines, etc., will travel underwater in hours or days depending on the requirement. The sensors and many subsystems are interfaced in these vehicles. The information from the sensors like direction, voltage and pressure parameter, etc., are collected and stored in the memory. These parameters from the memory and saved into the controlling system. The data is changed from min to min. Earlier NVRAMs used as a memory unit. It is heavy in size and suitable for larger ships and not for smaller in size vehicles. So the unit "NAND flash memory" is being used to store data and as chip. So it is suitable for all types of vehicles. Storage can be done in the form of bytes, pages and block types. "NAND flash memory controller" performs operations like program page, read page and block erase. A VHDL program is developed using a state machine for program page, read page and block erase operations. The simulation of the program and waveforms of program page, read page and block erase operations are using Xilinx software. The complete VHDL program testing is done in FPGA. A Xilinx Spartan-6 FPGA is used to interface the "NAND flash memory controller."

Keywords NAND flash memory controller · Flash memory

C. Harini (✉) · B. Keerthi Priya
Gayatri Vidya Parishad College of Engineering (A), Visakhapatnam, India
e-mail: Chithakala.harini@gmail.com

B. Keerthi Priya
e-mail: keerthipriya@gvpce.ac.in

D. V. Rama Koti Reddy
Andhra University College of Engineering (A), Andhra University, Visakhapatnam, Andhra Pradesh 530008, India
e-mail: rkreddy_67@yahoo.co.in

P. S. R. Chowdary et al. (eds.), *Microelectronics, Electromagnetics and Telecommunications*, Lecture Notes in Electrical Engineering 655,
https://doi.org/10.1007/978-981-15-3828-5_52

1 Introduction

1.1 NAND Flash Memory Organization

Flash memory is a non-volatile memory which can replace EEPROM because of its structure of interconnection in memory cells [1]. It is of two types: NOR and NAND flash type memories. NOR flash type memory is a memory for random access and NAND flash type is a memory for sequential access. It has less size compared to NOR flash memory [2]. So, high-density memory can be obtained by NAND flash memory which comprises an array organization internally. It comprises of 2 GB, of 2 planes. The 2 GB of 2 planes are even plane and odd plane means even number of blocks are in even plane and odd number of blocks in odd plane. The total array organization of 2 Gb memory divided into 2048 blocks of 2 planes. Each plane consists of 1024 blocks. Each block consists of 64 pages and each page consists of 1056 words of data. The NAND flash controller needs to perform program page, read page and block erase operations successfully. The data in the memory array works on 16-bits of data at a time internally, where it is loaded into 1024 bytes and programmed into the NAND flash array. The array addressing formatted and set back into the registers when a read operation is initiated. The data at registers in NAND flash is clocked out 16-bits of data at a time by user.

The flash memory consists of "bad blocks" within the memory which can affect the operations because of it. To store the bad block in the spare area of the NAND flash module, ECC method is implemented to separate from the other blocks. Architecture of NAND flash memory organization is discussed in Sect. 2, Methodology and implementation of program page, read page and block erase are discussed in Sect. 3. Results of program page, read page and block erase implemented in Xilinx are discussed in Sect. 4 (Fig. 1).

2 Architectures

The architectural design of the NAND flash comprises of transmitting data, multiplexing instructions and addresses on the pins and receiving data via the I/O control unit.

Instructions also received in input/output units and are latched with the command register and sent to control logic to produce control signals for control operations [3].

The address has been latched with the address register, which was sent to row/column decoder. Row decoder should be used to choose page and block address. Column decoder should be used to choose the column address (1024 words of a page).

The data transmitted as byte by byte from memory array via data and cache register. Data and cache registers, however, in normal operations, are used as a single register [4].

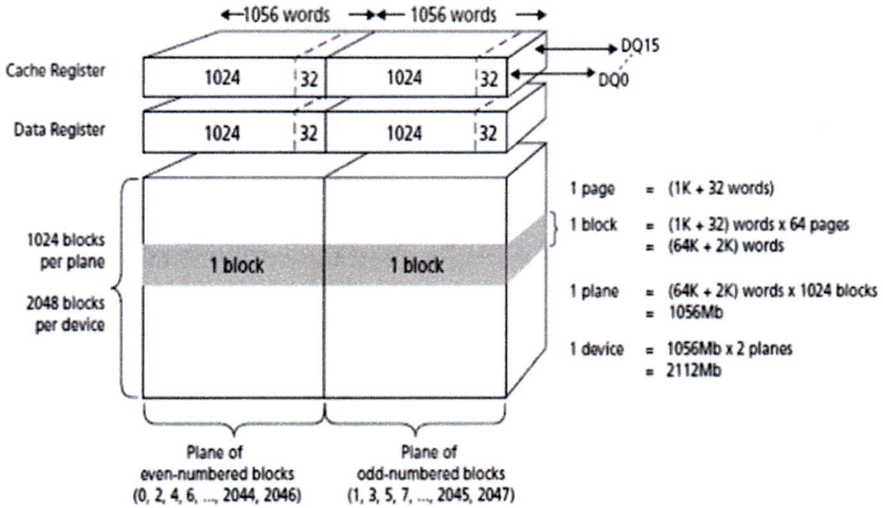

Fig. 1 Array organization of NAND flash

After programming NAND flash memory array, read can be performed using page based operations, and block-based operations are used to erase [5–7].

3 Methodology

In this paper implementing page program, read and block erase operations [8]. To perform any operation in NAND flash memory, the command code needs to be loaded into the command register. In addition to the command code, command resister also holds either the data or address to execute the command. Table 1 consists of commands with representing command cycle and number of address cycle required. The main controlling signals are CE, CLE, ALE, RE, WE signals. Here, CE is chip enable signal to enable the memory. CLE is command latch enable is enable when the command is given. ALE is address latch enable is enabled when the address is

Table 1 Commands and address cycles

Name of the command	Command of cycle 1 (h)	Number of address cycles required	Command of cycle 2 (h)	Valid during busy state
Block erase	60	3	D0	No
Program page	80	5	10	No
Read page	00	5	30	No

given. WE is write enable, is enabled when the data is written into the memory. RE is read enable, is enable when the data is read from the memory [9, 10].

3.1 Program Page (80–10 h)

This allows input data to the cache register from the buffer and transfers the data from cache to page and block address indicated in the array. The program page command is required command cycle of 80–10 h.

First, we have enable the memory using CE = 1, and the next CLE = 1 and WE = 1 are for writing the command into the register. The command cycle start to write 80 h to the command register. After giving the command, CLE and WE become disabled and write the address of the memory. So that WE and ALE are enabled and then write 5 address cycles which contain page and block address. After giving the address, disables the ALE and WE signals. End of the program page command is 10 h. For giving end command, we have to enable the CLE and WE signals. Whenever the 10 h command is given to the register, it notifies that the program page operation is terminated [11]. All the signals will be disabled. If status command 70 h is given within the program page operation, it shows the address where the particular operation is performing in the array (Fig. 2).

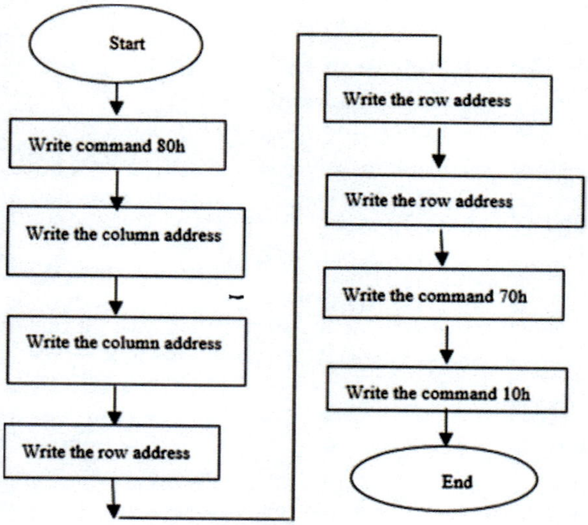

Fig. 2 Flowchart for program page operation

Fig. 3 Flowchart for read operation

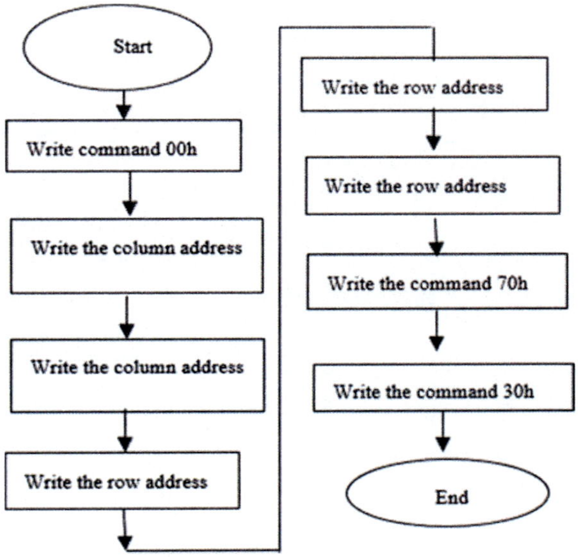

3.2 Read Page (00–30 h)

This command enables the input operands toward the cache memory from buffer and transfers the operands from cache to specified page and block address in the array program page command is required command cycle of 00–10 h.

First, we have enable the memory using CE = 1, and the next CLE = 1 and WE = 1 are for writing the command into the register. The command cycle start to write 00 h to the command register. After giving the command, CLE and WE become disable and write the address of the memory. So that WE and ALE are enable and then write 5 address cycles which contain page and block address. After giving the address, disables the ALE and WE signals. End of the program page command is 10 h. For giving end command, we have to enable the CLE and WE signals. Whenever the 30 h command is given to the register, it notifies that the program page operation is terminated. All the control signals will be disabled. If status command 70 h is given within the read page operation, it shows the address where the particular operation is performing in the array (Fig. 3).

3.3 Block Erase (60–D0 h)

The erase operation is done through the block wise. This command erases the data in the particular block. The command cycle for block erase is 60–D0 h.

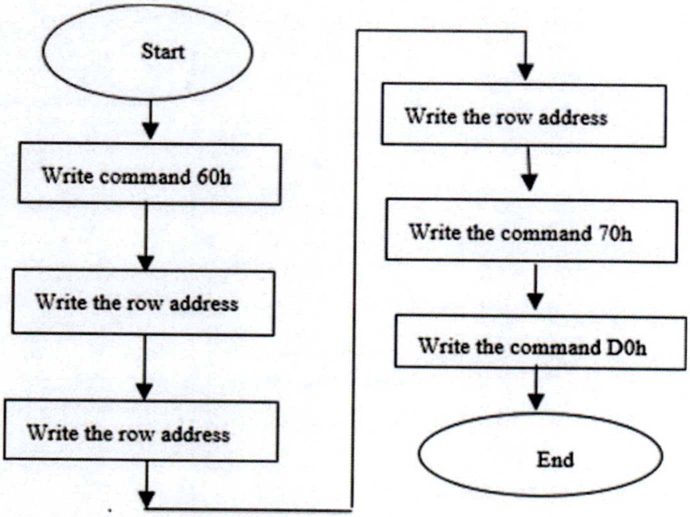

Fig. 4 Flowchart for block erase operation

First, we have enable the memory using CE = 1, and the next CLE = 1 and WE = 1 are for writing the command into the register. The command cycle start to write 60 h to the command register. After giving the command, CLE and WE become disabled and write the address of the memory. So that WE and ALE are enabled and then write 3 address cycles which contain block address. After giving the address, disables the ALE and WE signals. End of the program page command is 10 h. For giving end command, we have to enable the CLE and WE signals. Whenever the D0 h command is given to the register, it notifies that the program page operation is terminated. All the control signals will become disabled. If status command 70 h is given within the operation, it shows the address where the particular operation is performing in the array (Fig. 4).

4 Results

A VHDL program is developed using state machine and simulation of the program is doing using Xilinx software. Complete testing is done in FPGA. In this work, A Xilinx Spartan-6 FPGA is used to interface the "NAND flash memory controller." Program page operation waveform is shown in Fig. 5, read page operation waveform is shown in Fig. 6 and block erase operation waveform is shown in Fig. 7.

Below waveform shows the operation of program page. This waveform shows the command cycle and address are given in data input/output (DIO). It contains the command cycle 80 h followed by address cycle of 5 cycles which contain page

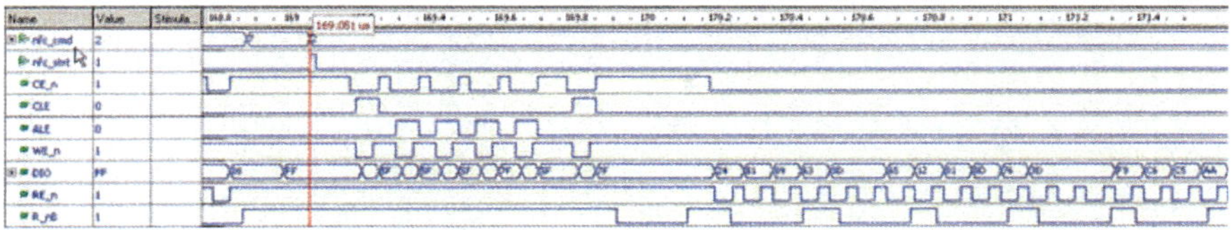

Fig. 5 Program page operation waveform

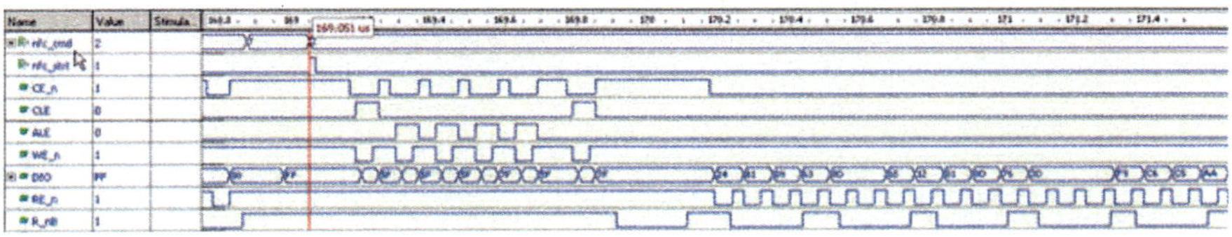

Fig. 6 Read page operation waveform

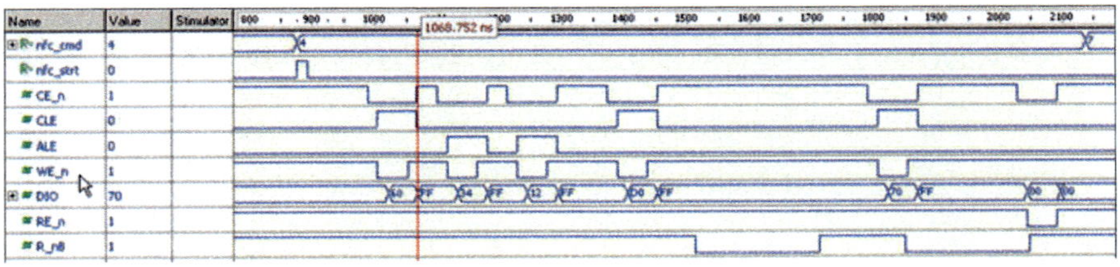

Fig. 7 Block erase operation waveform

and block address. This waveform also shows that CE, CLE, ALE, WE, RE signals, which vary according to the operation.

The below waveform shows the operation of read page. This waveform shows the command cycle and address are given in DIO. Its contains the command cycle 00 h followed by address cycle of 5 cycles which contain page and block address and the end of the command cycle 30 h. This waveform also shows that CE, CLE, ALE, WE, RE signals, which vary according to the operation.

The below waveform shows the operation of block erase. This waveform shows the command cycle and address are given in DIO. It contains the command cycle 60 h followed by address cycle of 3 cycles which contain block address and end of command cycle DOH. This waveform also shows that CE, CLE, ALE, WE, RE signals, which vary according to the operation.

5 Conclusion

The sensors interfaced in under water vehicles. The information from the sensors like direction, voltage and pressure parameter, etc., are collected and stored in the memory. The NAND flash memory controller designed to meet the characteristics of flash memory. The NAND flash controller is used to control the operations of page program, read page and block erase. These operations are implemented in VHDL. A Xilinx Spartan-6 FPGA is used to interface the "NAND flash memory controller." Future scope is bad block management process in VHDL language and simulation using Xilinx software.

References

1. Yu X, Wang X, Yan W (2014) A high performance solid-state storage system for high speed data acquisition applications. In: IEEE international on electron devices and solid-state circuits (EDSSC), China, pp 1–2
2. Park C, Talawar P, Won D, Jung MJ, Im JB, Kim S, Choi Y (2006) A high performance controller for NAND flash-based solid state disk (NSSD). In: Proceedings of the 21st IEEE non-volatile semiconductor memory workshop. NVSMW, pp 17–20
3. Mitiukhina N, Overview of NAND flash high speed interfacing and controller architecture. IEE 5008 memory systems final report 0060805
4. Micron Technology Inc (2005) The datasheet of Nand flash memory by Micron Technologies. Memory management in NAND flash arrays overview. Technological note TN-29-28
5. Micron Technology Inc (2009) 64 Gb, 128 Gb, 256 Gb, 512 Gb asynchronous/synchronous NAND features
6. Kleinfelder SA, The ARIANNA Collaboration (2013) Design and performance of the autonomous data acquisition system for the ARIANNA high energy neutrino detector. IEEE Trans Nucl Sci 60(2):612–618
7. Kaynak MN, Khayat PR, Parthasarathy S (2017) On die bit error rate estimator for NAND flash memory. IEEE Trans Circuits Syst II 64(7):772–776
8. Venkata Swathi A, Raju IBK (2017) ASIC implementation of NAND flash memory. Int J Electr Electron Data Commun. ISSN: 2320-2084
9. Lin CS, Dung LR (2007) A NAND flash memory controller for SD/MMC flash memory card. IEEE Trans Magn 43(2):933–935
10. Uchigaito H, Miura S, Nito T (2018) Efficient data-allocation scheme for eliminating garbage collection during analysis of big graphs stored in NAND flash memory. IEEE Trans Comput 67(5):646–657
11. Rodríguez-Olivares NA, Gómez-Hernández A, Nava-Balanzar L, Jiménez-Hernández H, Soto-Cajiga JA (2018) FPGA-based data storage system on NAND flash memory in RAID 6 architecture for in-line pipeline inspection gauges. IEEE Trans Comput 67(7):1046–1053

Design of Low Standby Power 10T SRAM Cell with Improved Write Margin

R. Manoj Kumar and P. V. Sridevi

Abstract There is a growing concern regarding the increased standby power and reduced stability of SRAM due to continued scaling in technology node. So, there is a necessity to design a new SRAM cell which addresses the concerns related to SRAM cell. So, 10T SRAM cell is proposed with reduced standby power and enhanced stability in read, write and hold modes of operation. There is a reduction in standby power because of the usage of stacked transistors. P10T SRAM cell has decreased the standby power while holding 1 by 4.9%, 15.99% and 1.68% compared to 8T, 8TG and 9T respectively at the worst process corner FF at 0.9 V V_{DD}. There is an increase of 262.89, 47.566, 261.75% write margin compared to 6T, 8TG, 9T SRAM cells at 0.9 V supply voltage for TT corner. The influence of process and voltage variations on write margin was studied on available and proposed SRAM cells. All designs are designed using in cadence virtuoso in 45 nm CMOS technology node.

Keywords Static noise margin · Write margin · Process corners · Standby power

1 Introduction

In computing systems, static random access memory (SRAM) forms the basic building block which occupies considerable area and consumes considerable amount of power. This necessitates stable and low standby power SRAM for portable devices [1]. Applications like implantable biomedical devices, wireless sensor networks (WSN), and battery-based electronic devices often require low power consumption in the present trend. Devices using FPGA are increasingly preferred day by day, because of its reconfigurability and prototyping capability, to develop advanced embedded applications [2, 3]. FPGA based on SRAM forms a significant portion of overall FPGA market. Reduction in power supply voltage quadratically decreases active power and linearly decreases standby power but restricted by PVT variations

R. Manoj Kumar (✉) · P. V. Sridevi
Department of Electronics and Communication Engineering, Andhra University College of Engineering (A), Andhra University, Visakhapatnam, Andhra Pradesh 530008, India
e-mail: manojkumar.rongali@gmail.com

P. S. R. Chowdary et al. (eds.), *Microelectronics, Electromagnetics and Telecommunications*, Lecture Notes in Electrical Engineering 655,
https://doi.org/10.1007/978-981-15-3828-5_53

[4]. As the supply voltage is decreased, standby power dominates the active power [5]. This significant increase in standby power is going to have an increased impact in SRAM cells as most of the cells tend to operate in standby mode [6].

6T SRAM cell suffers from read–write sizing conflict and also the RSNM degradation. In [7], asymmetric sizing was used to improve the stability of SRAM cell but with increased leakage currents. In [8], 4T read port is used to reduce the leakage but at the expense of increased area and less write margin. Chang et al. [9] proposed an 8T SRAM cell which isolates the read path from the storage nodes to increase RSNM. But it suffers from low write margin. A multi-threshold 9T SRAM was proposed in [10] for the energy efficiency, but it requires extra boosted supply. SRAM cell in [11] has P-P-N devices. It has low leakage power, but the cell has degraded write ability. In [12, 13], reduction in leakage power was reported with slight increase in area. In [14], data-dependent power supply SRAM cell for low leakage has been reported.

Static noise margin is used to determine the SRAM cell stability in hold and read modes of operation, respectively. In write operation, stability is determined by write margin (WM).

The paper is comprised of the following. Section 2 explains the existing SRAM designs. Section 3 describes the proposed SRAM design with operation. Section 4 compares the existing SRAM cells with the proposed SRAM cell w.r.t. standby power, HSNM, RSNM, and write margin. Section 5 concludes.

2 Existing SRAM Designs

6T SRAM cell shown in Fig. 1 is the conventional design which is formed using two cross-coupled inverters to latch the data while the access transistors are separately used to perform either read or write operation. Cell ratio and pull up ratio decide the stability in read and write operation, respectively. In 8TG SRAM cell [15] is similar to 6T SRAM cell, but the access transistors are replaced with transmission gates. In [9], read port has been isolated which gives an option to optimize the read and write SNMs individually. In [16], an addition NMOS has been added to the read port for further reduction of leakage power.

Fig. 1 6T SRAM cell

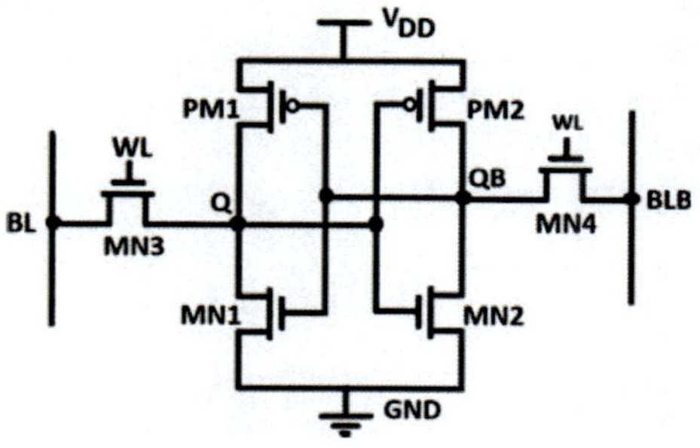

3 Proposed 10T SRAM Cell (P10T)

Existing SRAM designs suffer from one of the aspects like high leakage power, less write stability, and less read stability. So, in order to maintain the stability with less leakage power, a new SRAM cell is proposed with 10 transistors (P10T) as shown in Fig. 2. Table 1 lists different control signals status in different modes of operation. Stacked transistors PM3 and MN2 were used which reduces leakage. PM3, PM4, MN2, and MN4 assist in high write ability.

3.1 Write Operation

While performing write operation PM3 and MN2 are disconnected from the power supply rails by keeping the control signals WLPU and WLPD at Logic 1 and 0, respectively. This brings down either *Q* or QB from its stored value to its complement at a faster pace. When Logic 0 is stored, WL is enabled, BL is set to Logic 1 and its complement to BLB, RWL is disabled, WLB is enabled which performs write 1 into

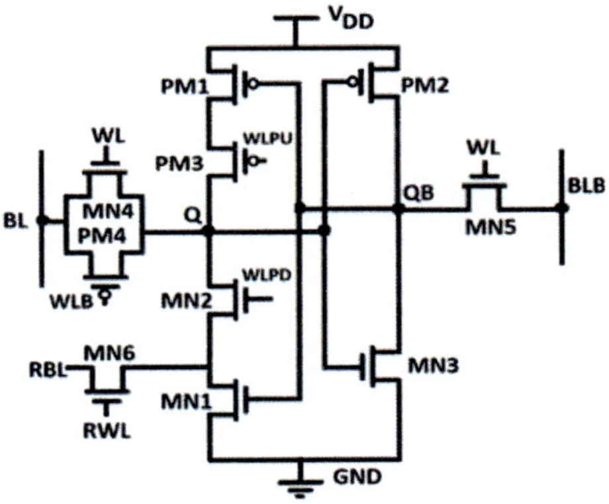

Fig. 2 Proposed 10T SRAM cell

Table 1 Control signals for P10T SRAM cell

	Hold	Read	Write 1/0
BL	1	1	1/0
BLB	1	1	0/1
WL	0	0	1
WLB	1	1	0
WLPU	0	0	1
WLPD	1	1	0
RWL	0	1	0
RBL	1	1	0

the storage node Q. When Logic 1 is stored, WL is enabled, BL is set to Logic 0 and its complement to BLB, RWL is disabled, WLB is enabled which performs write 0 into the storage node Q.

3.2 Read Operation

In read mode, WL and WLB are disabled, WLPU and WLPD are kept at Logic 0 and Logic 1, respectively, RWL is enabled which turns ON MN6, RBL is precharged to V_{DD}. When $Q = 0$, MN1 is ON creating a path for RBL to be discharged to ground. When $Q = 1$ and QB $= 0$, MN1 is OFF creating no path for RBL to discharge to ground. Based on the value of RBL, sense amplifier senses and gives the output.

3.3 Hold Operation

WL, WLB, and RWL are disabled, WLPU and WLPD are kept at Logic 0 and Logic 1, respectively, which forms the cross-coupled inverters to latch the data.

4 Results and Discussion

4.1 Standby Power

As the majority of SRAM cells operate in standby mode, standby power or leakage power reduction is essential. It is estimated as the multiplication of supply voltage and the current flowing through it [17]. Standby power is analyzed in the proposed SRAM (P10T) cell along with the other SRAM cells available in the literature. Figure 3 shows standby power at different process corners at 0.9 V V_{DD}. There is

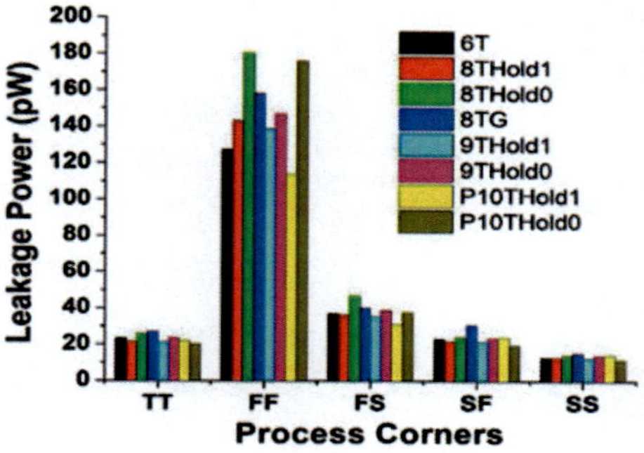

Fig. 3 Standby power comparison at 0.9 V supply voltage for all process corners

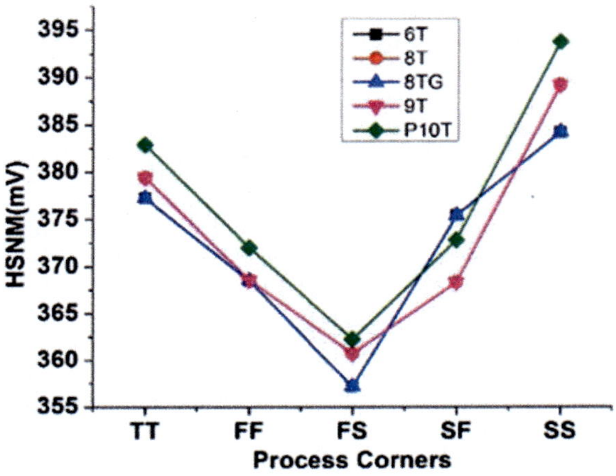

Fig. 4 HSNM comparison at 0.9 V supply voltage for all process corners

a significant reduction in standby power due to the usage of stacked transistors in pull down and pull up network. P10T SRAM cell decreased the standby power while holding 1 by 4.91%, 15.99% and 1.68% compared to 8T, 8TG and 9T respectively at the worst process corner FF. Hold 1 power of P10T has also decreased by 7.5%, 4.62%, 19.03% compared to 8TG at TT, FS and SF corners respectively.

4.2 *Hold Stability*

While SRAM is operating in hold mode, stability can be determined by HSNM which can be estimated by the butterfly curve. As the control signals of transistors PM3 and MN2 are kept at Logic 0 and 1, respectively, both the transistors are turned ON which replicates the two back to back inverters similar to 6T SRAM cell. From Fig. 4, HSNM slightly increased at all process corners except at SF corner.

4.3 *Read Stability*

Read stability is analyzed using the metric RSNM. Side length of largest square that can be fitted in the butterfly curve gives RSNM [18]. Read current through the storage nodes degrades RSNM. Due to this 6T, 8TG, P10T SRAM cell has lower RSNM. 8T, 9T SRAM cells have isolated the flow of read current through the storages nodes which leads to higher RSNM as shown in Fig. 5.

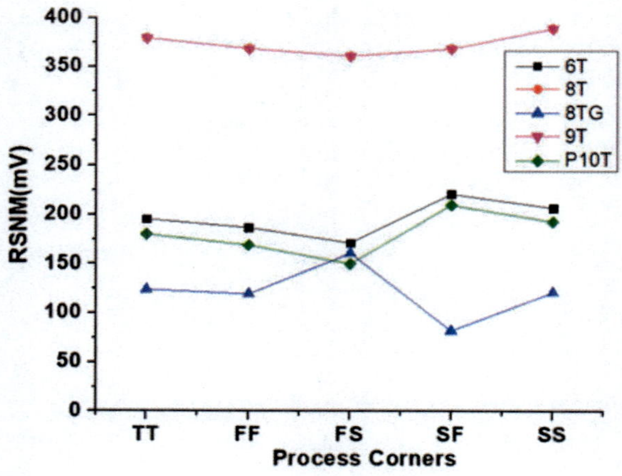

Fig. 5 RSNM comparison at 0.9 V supply voltage for all process corners

4.4 *Write Stability*

In 6T, 8TG SRAM cells, enhancement of write margin is restricted by the sizing of access transistor, pull down and pull up transistors. So read port is isolated in 8T, 9T, and P10T SRAM cells which gives freedom to separately size the devices which further enhances read and write stability individually. In P10T SRAM cell, PM3 and MN2 are disconnected to the power rails by the control signals WLPU and WLPD, respectively, and the usage of transmission gate enhances the write stability. Write margin is calculated using World Line sweep method [19]. Figure 6 shows the estimated write margin values at all process corners.

In Fig. 7, write margin is also observed for wide supply voltage range at TT corner. It clearly indicates P10T has enhanced write stability than other SRAM cells.

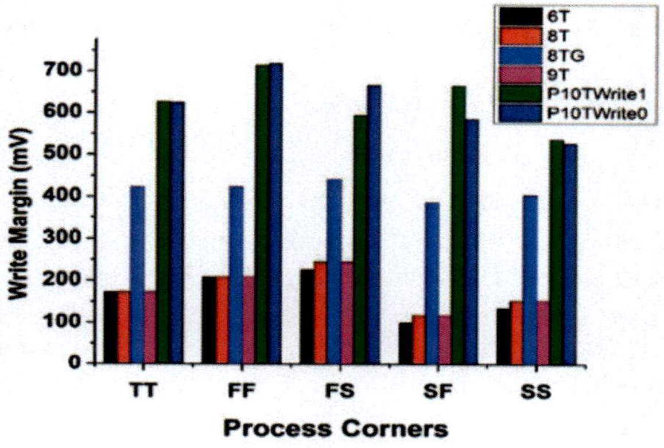

Fig. 6 Write margin comparison at 0.9 V supply voltage for all process corners

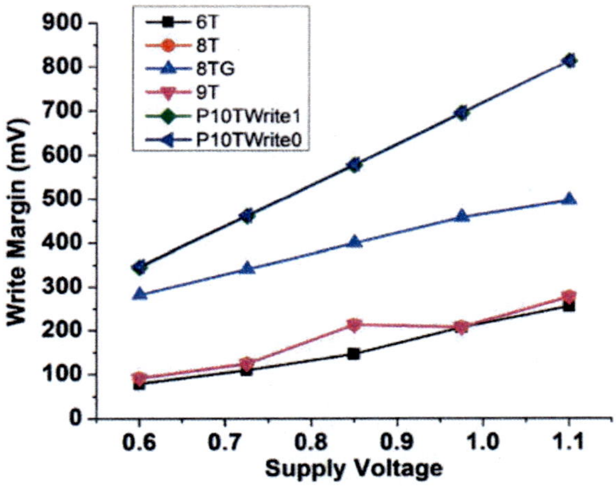

Fig. 7 Write margin comparison at various supply voltages for TT corner

5 Conclusion

Design metrics like standby power and stability are the major parameters that need to be considered while designing the SRAM cell. Though available SRAM cells excel in one of the design metrics, they suffer degradation in other design metrics. The proposed 10T (P10T) cell provides less standby power and enhanced write margin while having decent HSNM.

References

1. Rad JS, Guthaus M, Hughey R (2014) Confronting the variability issues affecting the performance of next generation SRAM design to optimize and predict the speed and yield. IEEE Access 2:577–601
2. Safarulla IM, Manilal K (2014) Design of soft error tolerance technique for FPGA based soft core processors. In: 2014 international conference on advanced communication control and computing technologies (ICACCCT). IEEE, pp 1036–1040
3. Ullah Z, Jaiswal MK, Cheung RC (2015) Z-TCAM: an SRAM-based architecture for TCAM. IEEE Trans Very Large Scale (VLSI) Integr Syst 23(2):402–406
4. Toh SO, Guo Z, Liu T-JK, Nikolic B (2011) Characterization of dynamic SRAM stability in 45 nm CMOS. IEEE J Solid-State Circuits 46(11):2702–2712
5. Aly RE, Faisal MI, Bayoumi MA (2005) Novel 7T SRAM cell for low power cache design. In: Proceedings of the IEEE international SOC conference, Sept 2005, pp 171–174
6. Pasandi G, Fakhraie SM (2015) A 256-kb 9T near-threshold SRAM with 1 k cells per bitline and enhanced write and read operations. IEEE Trans VLSI Syst 23(11):2438–2446
7. Nalam S, Calhoun BH (2011) 5T SRAM with asymmetric sizing for improved read stability. IEEE J Solid-State Circuits 46(10):2431–2442
8. Calhoun BH, Chandrakasan AP (2007) A 256-kb 65-nm sub-threshold SRAM design for ultra-low-voltage operation. IEEE J Solid-State Circuits 42(3):680–688

9. Chang L, Montoye RK, Nakamura Y, Batson KA, Eickemeyer RJ, Dennard RH, Haensch W, Jamsek D (2008) An 8T-SRAM for variability tolerance and low-voltage operation in high-performance caches. IEEE J Solid-State Circuits 43(4):956–963
10. Wang B, Nguyen TQ, Do AT, Zhou J, Je M, Kim TT-H (2015) Design of an ultra-low voltage 9T SRAM with equalized bitline leakage and cam-assisted energy efficiency improvement. IEEE Trans Circ Syst I Reg Pap 62(2):441–448
11. Lo C-H, Huang S-Y (2011) PPN based 10T SRAM cell for low-leakage and resilient subthreshold operation. IEEE J Solid-State Circuits 46(3):695–704
12. Gavaskar K, Ragupathy US, Malini V (2019) Design of novel SRAM cell using hybrid VLSI techniques for low leakage and high speed in embedded memories. Wirel Pers Commun 108(4):2311–2339
13. Mishra JK, Srivastava H, Misra PK, Goswami M (2019) Analytical modelling and design of 9T SRAM cell with leakage control technique. Anal Integr Circ Sig Process 101(1):31–43
14. Sharma V, Gopal M, Singh P, Vishvakarma SK, Chouhan SS (2019) A robust, ultra low-power, data-dependent-power-supplied 11T SRAM cell with expanded read/write stabilities for internet-of-things applications. Analog Integr Circ Sig Process 98(2):331–346
15. Islam A, Hasan M (2012) A technique to mitigate impact of process, voltage and temperature variations on design metrics of SRAM cell. Microelectron Rel 52(2):405–411
16. Lin Sheng, Kim Yong-Bin, Lombardi Fabrizio (2010) Design and analysis of a 32 nm PVT tolerant CMOS SRAM cell for low leakage and high stability. Integration 43:176–187
17. Singh P, Vishvakarma SK (2017) Ultra-low power high stability 8T SRAM for application in object tracking system. IEEE Access 6(11):2279–2290
18. Do AT et al (2011) An 8T differential SRAM with improved noise margin for bit interleaving in 65 nm CMOS. IEEE Trans Circ Syst I Reg Pap 58(6):1252–1263
19. Gierczynski N, Borot B, Planes N, Brut H (2007) A new combined methodology for write margin extraction of advanced SRAM. In: Proceedings of the IEEE international conference on microelectronic test structures (ICMTS), pp 97–100

Design and Analysis of Symmetric and Asymmetric Staircase Patch Antenna

Bammidi Deepa, P. Chaya Devi, and M. Syamala

Abstract The symmetric staircase patch antenna for multiband resonant frequencies is presented. The work is extended to an asymmetric staircase patch antenna to obtain a wideband frequency response and better impedance match. A clear improvement in the performance characteristics is observed. The staircase antennas are designed using FR4 epoxy material and Rogers-RT Duroid material using the simulator software and their performance characteristics are compared and analyzed. It is observed that the asymmetric staircase patch antenna gives an impedance bandwidth of about 5.1 GHz, which makes the design very attractive in terms of wider bandwidth.

Keywords Symmetric staircase · Asymmetric · Bandwidth

1 Introduction

The microstrip patch antennas suffer the low-bandwidth drawback and a change in the shape of the patch or structural design modifications is required to achieve wider bandwidth. In all practical applications, microstrip patch antenna design has challenges in terms of their gain and bandwidth specifications [1, 2]. The reduction of antenna size is also an important requirement of the antenna design. Much research work was done to satisfy the industrial and real-time requirements of antenna design. A staircase patch with L-shaped defected ground structure was designed [3]. Staircase patch design with T-shaped slots was discussed for the enhancement of bandwidth [4]. C-shaped symmetrical slots design of patch was discussed [5]. Slotted partial

B. Deepa (✉) · P. Chaya Devi
Anil Neerukonda Institute of Technology and Sciences, Sanghivalasa, Visakhapatnam, Andhra Pradesh, India
e-mail: deepa.ece@anits.edu.in

P. Chaya Devi
e-mail: chayadevi.ece@anits.edu.in

M. Syamala
Andhra University, Visakhapatnam, Andhra Pradesh 530003, India
e-mail: msyamala869@gmail.com

P. S. R. Chowdary et al. (eds.), *Microelectronics, Electromagnetics and Telecommunications*, Lecture Notes in Electrical Engineering 655,
https://doi.org/10.1007/978-981-15-3828-5_54

ground and addition of stairs and stubs were designed to achieve UWB characteristics [6]. Stepped feed, partial slotted ground, two-level stairs are described to achieve bandstop characteristics [7]. Notch band characteristics are achieved using fractal structures [8, 9].

In the paper, the basic design equations of the antenna are given in Sect. 2. Also, the designed dimensions of both symmetric and asymmetric antennas are given. Section 3 shows the performance characteristics of both symmetric and asymmetric antennas using FR-4 and Rogers-RT Duroid. Conclusions derived from the symmetric and asymmetric antenna designs are shown in Sect. 4.

2 Antenna Design

2.1 Design Equations of the Rectangular Patch Antenna

$$\text{Width}(w) = \frac{C_0}{2f_r}\sqrt{\frac{2}{\varepsilon_r + 1}} \tag{1}$$

w Width of the patch
C_0 Speed of light
ε_r Dielectric constant of the substrate
f_r Resonant frequency

$$L_{\text{eff}} = \frac{C_0}{2f_r\sqrt{\varepsilon_{\text{eff}}}} \tag{2}$$

$$\varepsilon_{\text{eff}} = \frac{(\varepsilon_r + 1)}{2} + \frac{(\varepsilon_r - 1)}{2}\left\{\frac{1}{\sqrt{1 + \frac{12h}{w}}}\right\} \tag{3}$$

$$\Delta L = 0.412h\frac{(\varepsilon_{\text{eff}} + 0.3)\left(\frac{w}{h} + 0.264\right)}{(\varepsilon_{\text{eff}} - 0.258)\left(\frac{w}{h} + 0.8\right)} \tag{4}$$

$$\text{Actual length of patch} = L = L_{\text{eff}} - 2\Delta L \tag{5}$$

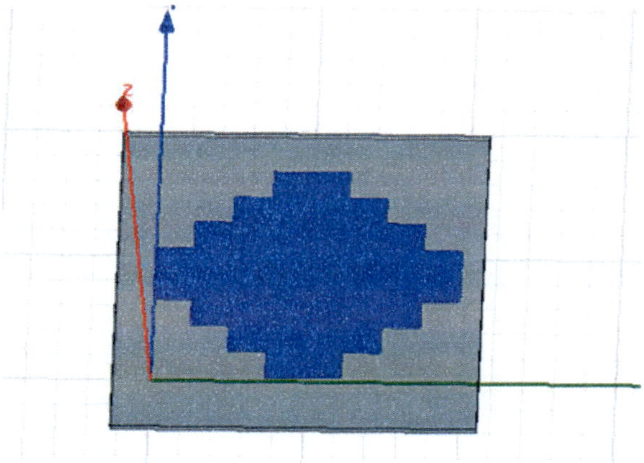

Fig. 1 Design of symmetrical staircase patch using FR4 epoxy

2.2 Symmetrical Staircase Patch Antenna Using FR4 Epoxy Material

Coaxial probe feed is used for the antenna design. Figure 1 shows the design of the symmetric staircase patch with the following dimensions using Eqs. (1)–(5) [1].

Ground Dimensions: *X*-size = 47.64 mm, *Y*-size = 39.753 mm, Position = (−19.876, −23.82, 0) 40; Substrate Dimensions: *X*-size = 47.64 mm, *Y*-size = 39.753 mm, *Z*-size = 1.6 mm, Position = (−19.876, −23.82, 0). Patch Dimensions: *X*-size = 8.04 mm, *Y*-size = 30.153 mm, Position = (−15.076, −19.02, 1.6). Inner conductor one dimension: Height = 1.6 mm, Radius = 2 mm, Position = (0, 0, 0); Inner conductor 2 dimensions: Height = −5 Radius = 2 Position = (0, 0, 0). Outer conductor dimensions: Height = 4 mm, Radius = −5 mm, Position = (0, 0, 0), and Air Box Dimensions: *X*-size = 47.64 mm, *Y*-size = 39.753 mm, *Z*-size = 100 mm, Position = (−19.876, −23.82, 0).

2.3 Symmetrical Staircase Patch Antenna Using Rogers-RT Duroid Material

To improve the frequency response, the same symmetrical staircase antenna is designed using Rogers-RT Duroid material with the following dimensions. Ground Dimensions: *X*-size = −59.01 mm, *Y*-size = −59.05 mm, Position = (−6, 10, 0); Substrate Dimensions: *X*-size = −59.01 mm, *Y*-size = −59.05 mm, *Z*-size = 1.6 mm, Position = (−6, 10, 0); Patch Dimensions: X-size = −49.41 mm, *Y*-size = −41.34 mm, Position = (0, 0, 1.6); Inner conductor one dimension: Height = −7 mm, Radius = 1.6 mm, Position = (18.705, 20.67, 0); Inner conductor two dimensions: Height = 2 mm, Radius = 1.6 mm, Position = (18.705, 20.67, 0); Outer conductor dimensions: Height = −7 mm, Radius = 5 44 mm, Position = (18.705,

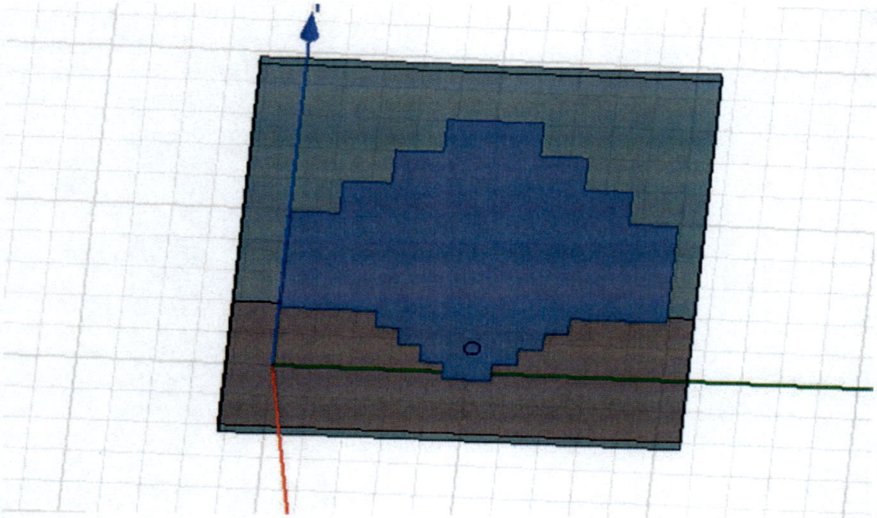

Fig. 2 Design of asymmetrical staircase patch using Rogers RT Duroid

20.67, 0); Air Box Dimensions: X-size = 59.01 mm, Y-size = −59.05 mm, Z-size = 100 mm, Position = (−6, 10, 0).

2.4 Asymmetrical Staircase Patch Antenna

To improve the bandwidth in the resonant frequency range, an asymmetric staircase design is adopted as shown in Fig. 2, with following dimensions: Ground Dimensions: X-size = 59.01 mm, Y-size = −20 mm, Position = (−6, 10, 0); Substrate Dimensions: X-size = 59.01 mm, Y-size = −20 mm, Z-size = 1.6 mm, Position = (−6, 10, 0); Patch Dimensions: X-size = 2.495 mm, Y-size = −10 mm, Position = (23.505, 0, 1.6); Air Box Dimensions: X-size = 59.01 mm, Y-size = −20 mm, Z-size = 10 mm, Position = (−6, 10, 0).

3 Results and Discussions

3.1 Symmetric Staircase FR4-Epoxy Patch

Figure 3 shows the return loss characteristics of the FR-4 Epoxy symmetric staircase patch antenna. The antenna exhibits resonance at different frequencies : fr_1 = 5.5 GHz, fr_2 = 6.2 GHz, fr_3 = 6.9 GHz, fr_4 = 7.9 GHz, fr_5 = 8.9 GHz, fr_6 = 9.6 GHz. The VSWR values shown in Fig. 4 are not much satisfactory at those frequencies.

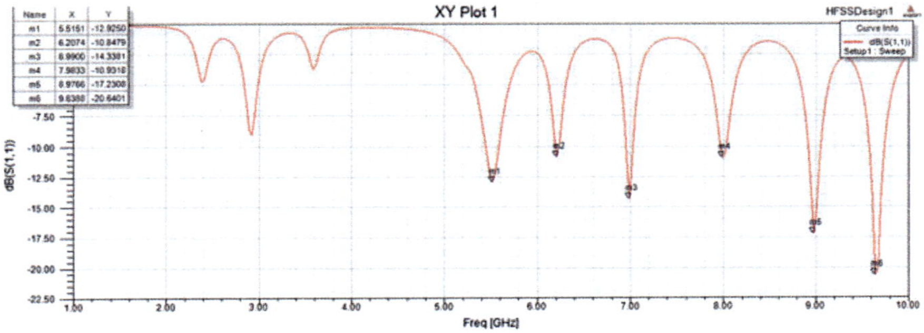

Fig. 3 Return loss of FR4-Epoxy symmetrical staircase patch

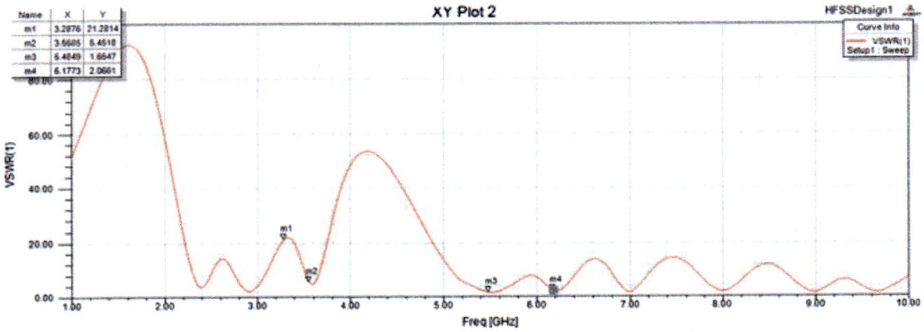

Fig. 4 VSWR of FR4-Epoxy symmetrical staircase patch

3.2 *Symmetric Staircase Rogers-RT Duroid Patch*

To improve the impedance matching, the antenna designed using RT Duroid material, whose return loss characteristics exhibit resonance at 2.4, 5.57, 7.86, and 9.45 GHz, as shown in Fig. 5. The VSWR values of the respective frequencies are below 2, which is a satisfactory result of impedance matching as shown in Fig. 6.

The gain characteristics are also satisfactory at 6 dB at $f_r = 2.4$ GHz, 4.3 dB for $f_r = 5.53$ GHz, 8.4 dB at $f_r = 7.75$ GHz, 4.95 dB at $f_r = 9.93$ GHz as shown in Figs. 7, 8, 9, and 10, respectively.

3.3 *Asymmetric Staircase Patch Using Rogers-RT Duroid*

To improve the bandwidth, an asymmetric staircase antenna design is adopted whose return loss shown in Fig. 11 exhibits an impedance bandwidth of about 5.1 GHz, with −318 dB return loss, resulting in a VSWR about 1.05. The antenna provides a satisfactory gain of about 2.72 dB at $f_r = 7.48$ GHz, as shown in Fig. 12. The radiation pattern of the antenna exhibits near omnidirectional characteristics.

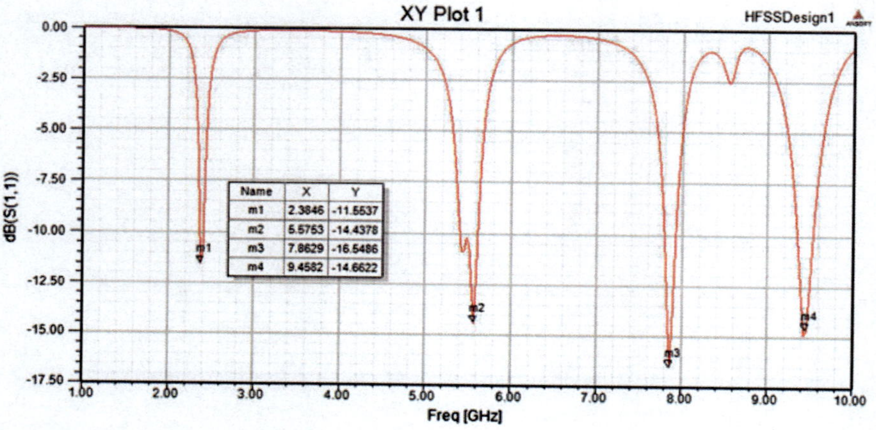

Fig. 5 Return loss of Rogers-RT Duroid symmetrical staircase patch

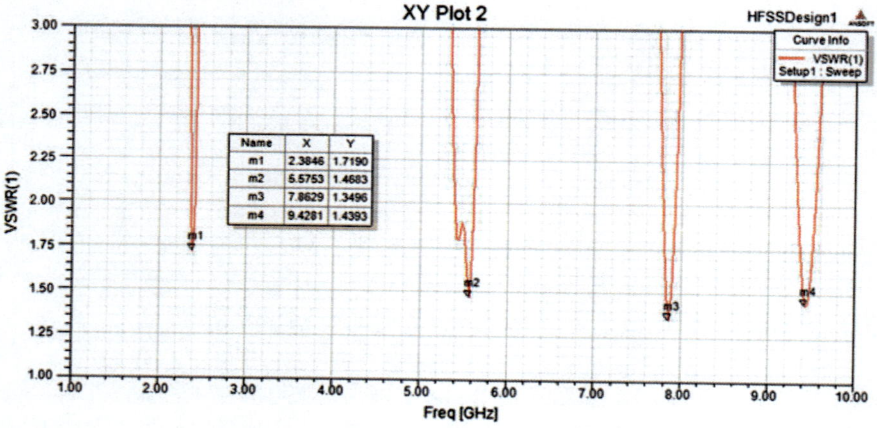

Fig. 6 VSWR of Rogers-RT Duroid symmetrical staircase patch

4 Conclusions

From the design of symmetric patch antennas, it is observed that Rogers-RT Duroid substrate gives better impedance matching results, but with narrow bandwidth. As there is a growing demand for the antennas which have wide bandwidth, an asymmetrical staircase patch has been designed which is fed by the coaxial line feed and a partial ground using the Rogers-RT Duroid. The results show that the feed position is accurate because there is good impedance matching between the feed and the antenna. It has a wider bandwidth than the symmetrical antenna. This antenna has achieved good impedance matching, stable radiation patterns, VSWR, and return

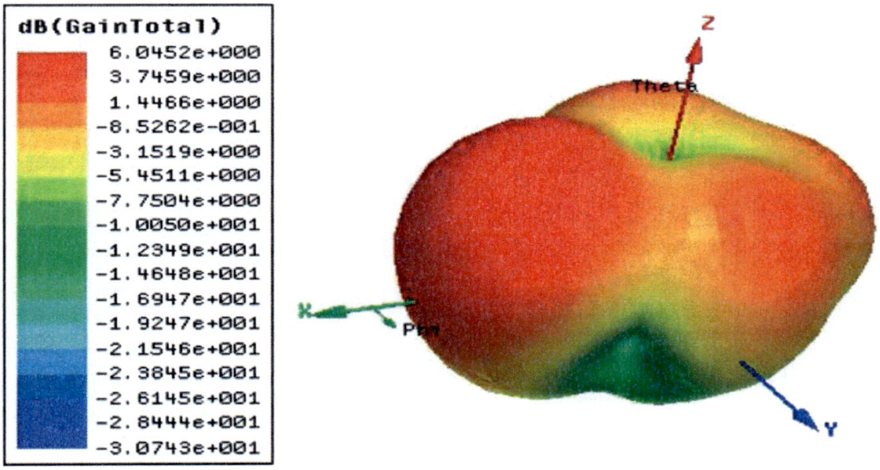

Fig. 7 Gain of Rogers-RT Duroid symmetrical staircase patch at $f_r = 2.4$ GHz

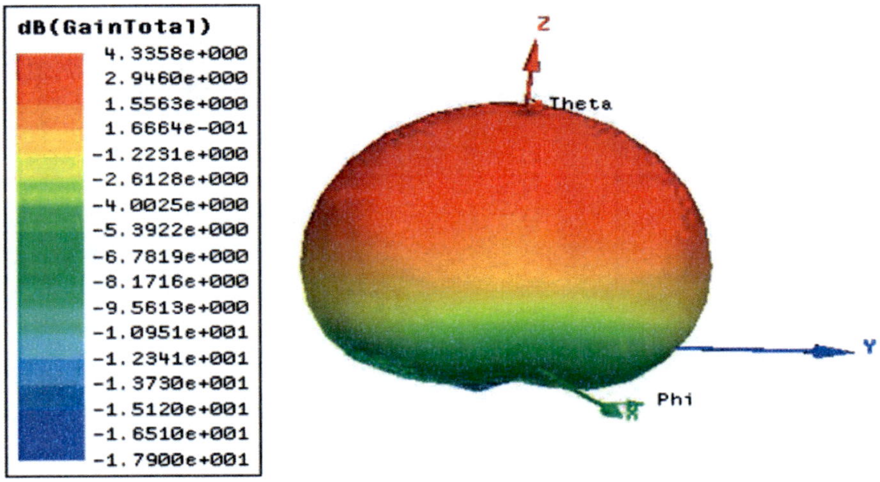

Fig. 8 Gain of Rogers-RT Duroid symmetrical staircase patch at $f_r = 5.53$ GHz

loss. The designed asymmetrical patch antenna exhibits an impedance bandwidth of 5.1 GHz with a return loss of −31.803 dB and it is resonating at 2.9 GHz.

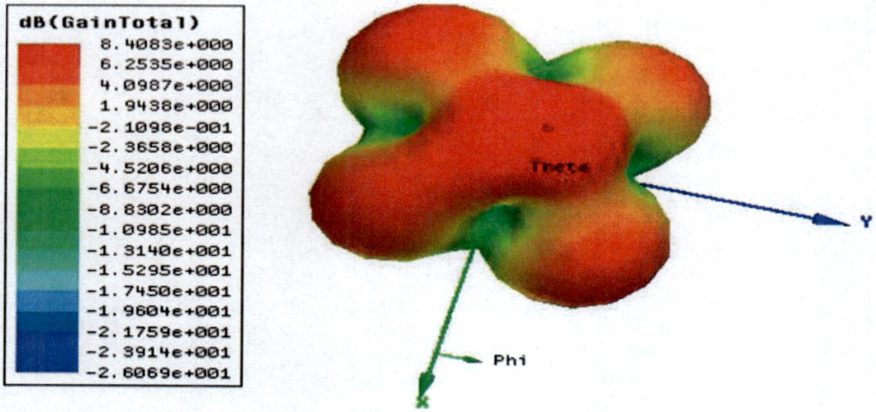

Fig. 9 Gain of Rogers-RT Duroid symmetrical staircase patch at $f_r = 7.79$ GHz

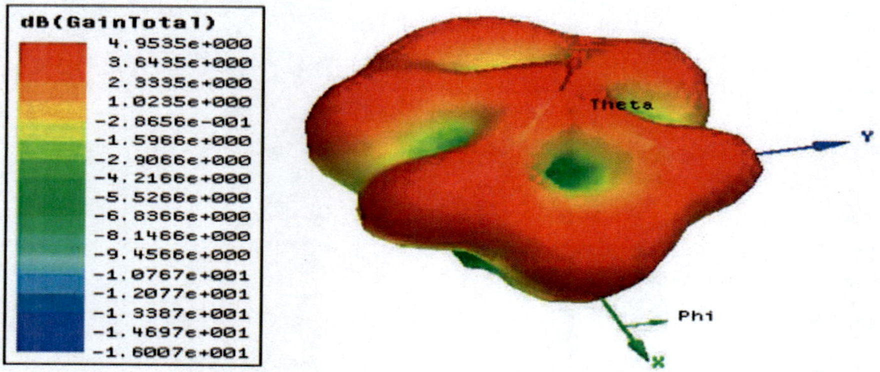

Fig. 10 Gain of Rogers-RT Duroid symmetrical staircase patch at $f_r = 9.93$ GHz

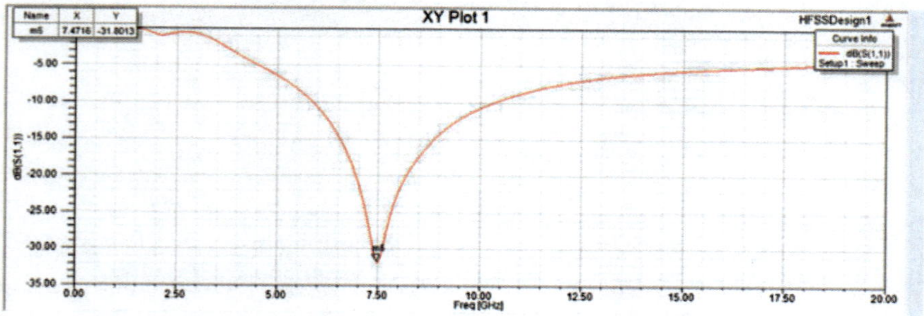

Fig. 11 Return loss plot of Rogers-RT Duroid asymmetrical staircase patch

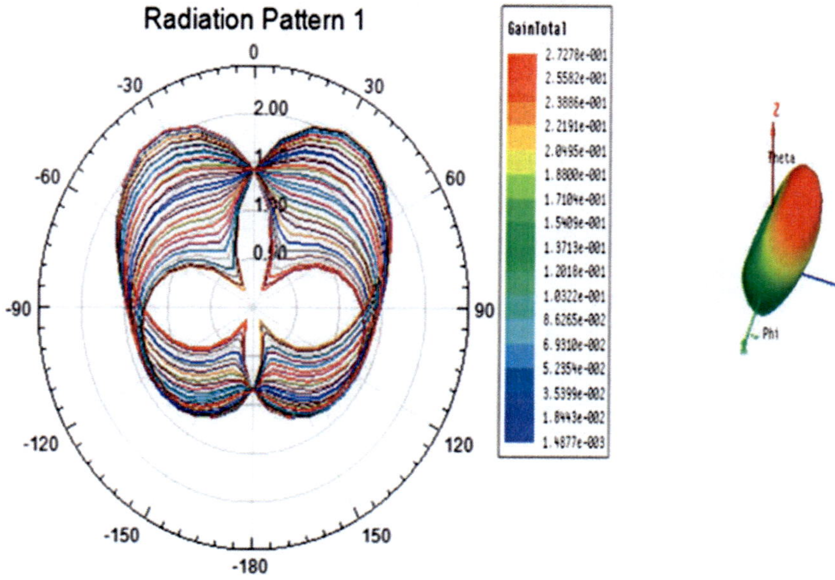

Fig. 12 Gain and radiation pattern characteristics of Rogers-RT Duroid asymmetrical patch

References

1. Balanis CAT (1997) Antenna theory analysis and design. Wiley, New York
2. Stutsman WL, Thiele GAT (1981) Antenna theory and design. Wiley, New York
3. Singh J, Kaur J, Kaur N (2018) Novel staircase microstrip antenna with L-shape slotted DGS for WLAN and WiMAX application. In: 2018 2nd international conference on micro-electronics and telecommunication engineering (ICMETE). IEEE Xplore
4. Dargar SK, Bharti S, Nyati A (2013) Compact UWB antenna with T-shaped slots and staircase ground plane for enhanced bandwidth. IOSR J Electron Commun Eng (IOSR-JECE) 4(5):12–17. ISSN: 2278-2834, ISBN: 2278-8735
5. Shukla S, Shivam (2017) Design of a symmetrical C shaped microstrip patch antenna with rectangular ground slot for WLAN and WIMAX applications. J Commun Eng Syst 7(1):12–15
6. Rafiqul IM, Zahirul AA, Feroze Akbar M, Khan J, Alkaraki S (2012) Design of microstrip patch antenna using slotted partial ground and addition of stairs and stubs for UWB application. Cyber J Multi J Sci Technol (JSAT), May edn
7. Raj RK, Gurjar S, Sharma M (2014) Design of stair and slotted UWB antenna using stepped-feed with modified slotted ground plane. Int J Comput Appl (IJCA) (0975-8887). National Seminar on Recent Advances in Wireless Networks and Communications, NWNC
8. Terlapu SK, Jaya C, Srinivasa Raju GRLVN (2017) On the Notch Band characteristics of Koch fractal antenna for UWB applications. Int J Control Theory Appl 10(06):701–707
9. Terlapu SK, Chowdary PSR, Jaya C, Sameer Chakravarthy VVSS, Satpathy SC (2018) On the design of fractal UWB wide-slot antenna with Notch Band characteristics. In: Anguera J, Satapathy S, Bhateja V, Sunitha K (eds) Microelectronics, electromagnetics and telecommunications. Lecture notes in electrical engineering, vol 471. Springer, Singapore

Performance Analysis of Multilevel Converter with Reduced Number of Active Switches

Sudheer Vinnakoti and Venkata Lakshmi Vasamsetti

Abstract In a multi-level inverter, as the number of levels increases there exists the problem of an increase in the number of active power switches particularly in high-power and low-voltage applications such as wind energy conversion system, UPS and photovoltaic inverters. Hence, this paper presents a novel technology which is capable of providing the same or more number of levels as that of the fundamental with reduced number of active switches. The proposed topology utilizes a DC link which is the combination of number of DC cells. The DC interface provides a variable DC connection voltage along with the regulated path through floating capacitors and provides the required degree of control for all inverter phases. The numbers of levels of the converter are increased by connecting the DC cells in a multi-cell structure. The five-level reduced multi-level converter (RMC) is compared with various other topologies such as single-phase five-level and seven-level cascaded H-bridge inverter in terms of number of active switches, and the results are validated through the percentage total harmonic distortions.

Keywords Multi-level converters · DC link

1 Introduction

With the increase in demand for electricity day by day, countries around the world are focusing mainly on green energy in order to reduce the effects caused by the conventional energy resources like thermal, nuclear, etc., which are emitting harmful gases and adversely affecting the environment. In view of this number of distributed

S. Vinnakoti (✉)
Department of Electrical and Electronics Engineering, Raghu Engineering College (A), Dakamarri, Visakhapatnam, Andhra Pradesh 531162, India
e-mail: sudheeer.218@gmail.com

V. L. Vasamsetti
Department of Electrical and Electronics Engineering, Vignan's Institute of Information Technology (A), Visakhapatnam, Andhra Pradesh, India
e-mail: lakshmi.vasamsetti270@gmail.com

P. S. R. Chowdary et al. (eds.), *Microelectronics, Electromagnetics and Telecommunications*, Lecture Notes in Electrical Engineering 655,
https://doi.org/10.1007/978-981-15-3828-5_55

generation (DG) has come into the picture to provide green energy such as solar, wind, hydro and tidal power. This provides clean and efficient energy for the community.

The present-day electricity infrastructure operates on alternating current (AC). However, the power generated from the green energy resources is direct current (DC). Hence, there is a need of converting this direct current into alternating current or DC/AC power converters called as inverter to meet the required load demand, for making electricity coming from green energy resources useable. A conventional two-level inverter can be used for this purpose; however, the pulsating current and voltage waveforms at the output involve the requirement of filters to get the required fundamental sinusoidal waveform. Moreover, a large amount of energy gets wasted in the higher-order harmonics; ultimately, the efficiency and the effectiveness of the system get reduced. Hence, there is a need to develop a new methodology capable of converting DC to AC which is efficient, cheap and compact which enables the green energy utilization to a wider area. In view of this initially two-level voltage source inverters (VSIs) emerged as a solution, however, the blocking voltage limit of 2-L VSI is overcome by multi-level inverters [1–6] and enabled VSIs applicable for medium voltages, with an added advantage of reduced dv/dt and higher efficiency.

The benefits of increasing the switching levels include (a) reduction in harmonics. (b) Output voltage and power levels can be increased with the increase in the number of switching levels, (c) filter requirement is greatly reduced, and (d) voltage stress on the power switches can be greatly reduced.

The basic multi-level topologies include diode clamped or neutral clamped multi-level inverter [7], capacitor clamped or flying capacitor multi-level inverter [8] and cascaded H-Bridge multi-level inverters. Although all the topologies serve the purpose of increasing the levels of the inverters to meet the required demand, there are some limitations. Among the three types of multi-level inverters, the cascaded inverters have the least number of components for a given level. It utilizes a series of H-bridge cells to synthesize the desired voltage level from several separate DC sources.

All these converters were primarily devised to reach medium-voltage and high-power operations. Hence, the device blocking voltage levels, switching frequencies, were the main drivers behind the multi-level converter topology research and development efforts. More recently, multi-level converters have entered into the low-voltage system, i.e., below 690 V. However, when these applications are reaching high-power levels, they require number of converters to be connected in parallel, interleaved or multi-channel modes which leads to the increase in the requirement of active switches, and in turn the increase in capacitor leading to increase in the probability of failure.

In view of the above discussion, this paper tries to solve the problem by presenting an improved multi-level converter methodology with reduced number of active switches by utilizing the ideology of the author presented in [9]. This improved methodology is capable of generating the same or more number of output voltage levels similar to the traditional multi-level converters. The operating principle is based on a multi-level DC link voltage, which is being generated from the multiple DC cells. Combination of several DC links can improve the number of voltage levels proportional to the three phases without any increase in the number of active switches and, hence, gives the name reduced multi-level converter (RMC) which in turn focused on basic demands like device switching frequency, filter size and power quality that are posed for grid connection applications. In this paper, a configuration of five levels is considered using three DC cells with two voltage source inverters. The rest of the paper is organized as follows. Part II presents the reduced switch multi-level converter, and the control scheme and the generation of the gating signal for the switches are discussed in part III. Part IV presents the simulation results of the novel topology and compared with the other multi-level inverter topologies in terms of number of active switches required, such as single-phase five-level cascaded H-bridge inverter and seven-level cascaded H-bridge inverter, and the results are validated through the percentage total harmonic distortions; finally, the conclusion is presented in part V.

2 Reduced Switch Multi-level Converter

The basic block diagram of the method is shown in Fig. 1 with three DC links and V_{dc} is the input voltage to the system. To generate the positive and negative half-cycles of the required output voltage, the output from the dc link is fed to the two-level voltage source inverter. The basic DC cell is shown in Fig. 1 where each DC cell has three power switches and one capacitor. The possible switching states obtained from each DC cell will connect the voltage of the capacitor in such a way to add, remove or even reroute the capacitor voltage. The *P* and *N* terminals act like the coupling grid and join the DC cell with the 2L inverter. The operating principle heavily relies on the DC link. The DC cell includes three possible switching states in the DC–DC converter unit.

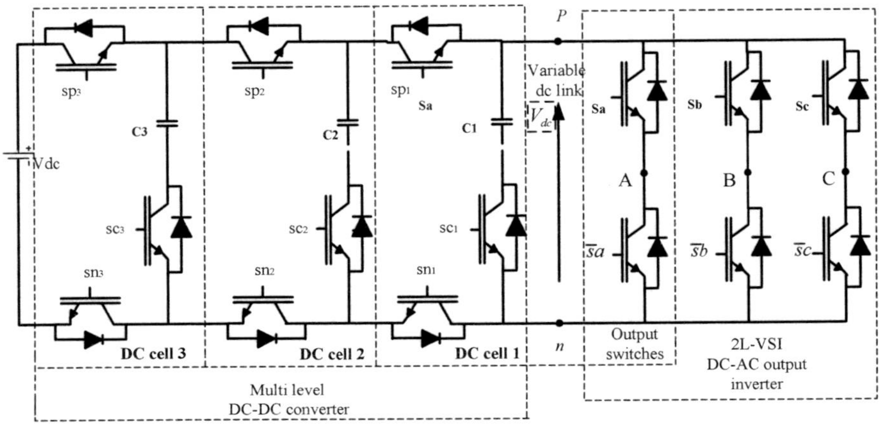

Fig. 1 Five-level RMC with DC cells and a 2-L VSI output inverter stage

Therefore, the complete permissible switching states can be represented as the function of DC cell, i.e., for N_{dc} states, we have $3^{N_{dc}}_{cells}$. The half-bridge section for the DC–AC 2-L VSI output stage consists of two switches. In each leg, there are two power switches which are switched ON and OFF in a complementary manner, and there are three output phases, and hence, the total number of possible switching states is $2^3 = 8$ on a whole and the total number of switching can be possible are given by $N_{states} = (N_{dc\text{–}dc}) * (N_{dc\text{–}ac}) = 3^3 * 2^3 = 27 * 8 = 216$ switching states. It is necessary that the voltage across each capacitor is to be followed a ratio of V_{dc1}:V_{dc2}:V_{dc3}:V_{dc4} = 4:3:2:1. From the ratio, let total voltage is equal to V_{dc} such that $V_{dc1} = V_{dc}$, $V_{dc2} = \frac{3V_{dc}}{4}$, $V_{dc3} = \frac{V_{dc}}{2}$ and $V_{dc4} = \frac{V_{dc}}{4}$.

The following constraints should be followed while operating the proposed multi-level inverter topology. It is always mandatory that six numbers of switches are needed to be kept ON at any instant of time. Since the entire row of switches S_{n1}, S_{n2}, S_{n3} is connected to neutral point N when all switches are kept ON state, we find no voltage that flows in reverse order, i.e., these switches are kept ON state to produce higher levels voltages, i.e., $\frac{V_{dc}}{4}$, $\frac{V_{dc}}{2}$, $\frac{3V_{dc}}{4}$ and finally V_{dc}.

From Fig. 1a [9], it is observed that when switches,s_{c1}, s_{c2}, s_{c3}, s_{n1}, s_{n2}, s_{n3}, are at ON state at terminal $V_{pn,}$ we have $V_{c1} = \frac{V_{dc}}{4}$ and $V_{nN} = 0$, respectively. Hence, the output voltage is given as follows, $V_0 = \left(V_{pn} - V_{nN}\right) = (V_{c1} - V_{nN}) = \frac{V_{dc}}{4} - 0 = \frac{V_{dc}}{4}$ when s_{p1}, s_{c2}, s_{c3}, s_{n1}, s_{n2}, s_{n3} switches are in active state as shown in Fig. 1b [9], the terminal V_{pn} gets $V_{c2} = \frac{V_{dc}}{2}$ and since, the capacitors are previously charged, at terminal $V_{nN} = 0$.

From Fig. 1c [9], when s_{p1}, s_{p2}, s_{c3}, s_{n1}, s_{n2}, s_{n3} switches are in active state, the terminal V_{pn} gets $V_{c2} = \frac{3V_{dc}}{4}$ and $V_{nN} = 0$. Since capacitors are previously charged. Hence, the output voltage is given as $V_0 = \left(V_{pn} - V_{nN}\right) = (V_{c3} - V_{nN}) = \frac{3V_{dc}}{4} - 0 = \frac{3V_{dc}}{4}$. From Fig. 1d [9], when s_{p1}, s_{p2}, s_{p3} and s_{n1}, s_{n2}, s_{n3} switches are in active state, terminal V_{pn} gets $V_{C4} = V_{dc}$ which is the input voltage and $V_{nN} = 0$ due to previously charged capacitors. Finally, the output voltage is given by the equation $V_0 = \left(V_{pn} - V_{nN}\right) = (V_{c4} - V_{nN}) = V_{dc} - 0 = V_{dc}$. So far, with different topologies peak level voltage, i.e.,$V_0 = V_{dc}$ is achieved as shown in Fig. 2, and the following topologies are concerned with the decrement voltage levels and can be achieved with the following switching state conduction topologies.

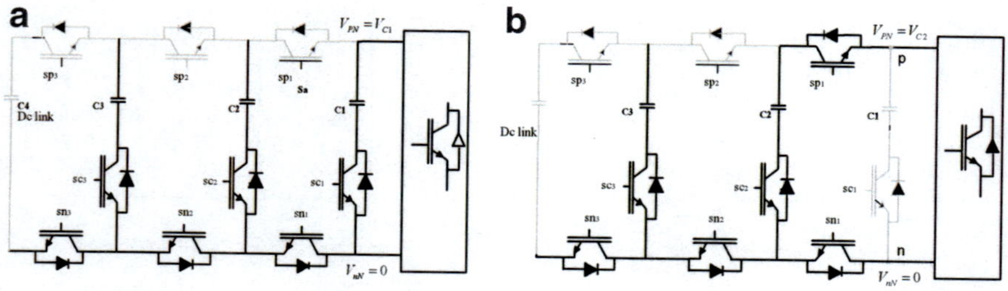

Fig. 1.1 **a** Switching state for an output voltage level of $V_{dc}/4$. **b** Switching state for an output voltage level of $V_{dc}/2$

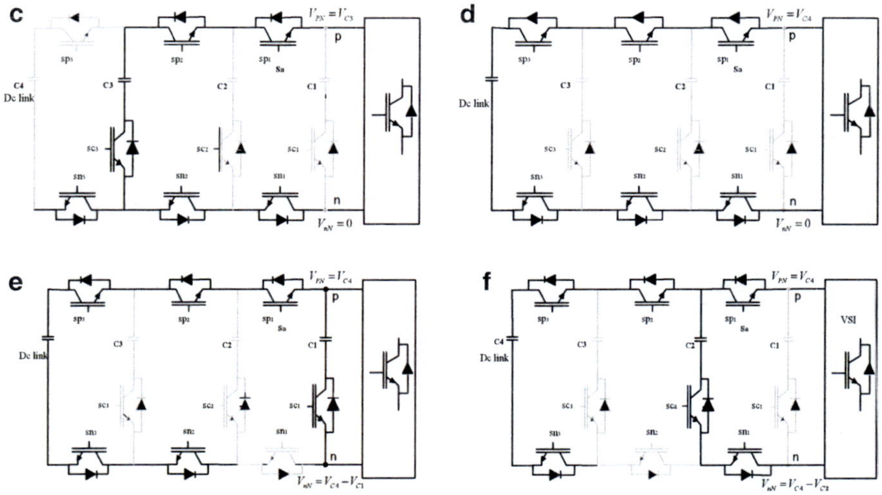

Fig. 1.1 **c** Switching state for an output voltage level of $3V_{\text{dc}}/4$. **d** Switching state for an output voltage level of V_{dc}. **e** Switching state for an output voltage level of $V_{\text{dc}/4}$. **f** Switching state for an output voltage level of $V_{\text{dc}}/2$

From Fig. 1e, when the switches s_{p1}, s_{p2}, s_{p3} and s_{c1}, s_{n2}, s_{n3} are kept active state, at V$_{pn}$ there exists a potential of $V_{pn} = V_{c4}$ and $V_{nN} = V_{c4} - V_{c1}$ and the total output is given as $V_0 = \left(V_{pn} - V_{nN}\right) = V_{c4} - (V_{c4} - V_{c1}) = V_{c1} = \frac{V_{\text{dc}}}{4}$.

From Fig. 1f [9], switches s_{p1}, s_{p2}, s_{p3} and s_{c2}, s_{n1}, s_{n3} are kept in active state which gives, $V_{pn} = V_{c4}$ and $V_{nN} = V_{c4} - V_{c2}$, and the total output voltage is given as $V_0 = V_{c4} - V_{c4} + V_{c2} = V_{c2} = \frac{V_{\text{dc}}}{2}$

From Fig. 1g [9], we have switches s_{p1}, s_{p2}, s_{p3} and s_{c3}, s_{n1}, s_{n2} are kept in active state; hence at terminal $V_{pn,}$ the potential is $V_{pn} = V_{c4}$ and $V_{nN} = V_{c4} - V_{c3}$. The total output voltage is given by the equation $V_0 = V_{c4} - V_{c4} + V_{c3} = V_{c3} = \frac{3V_{dc}}{4}$. In Fig. 1h, j, the alternative switching methods to produce voltages are discussed below. From Fig. 1h [9], $V_{pn} = V_{c3} = \frac{3V_{\text{dc}}}{4}$ and $V_{nN} = V_{c3} - V_{c2} = \frac{V_{\text{dc}}}{4}$, then final output voltage $V_0 = \frac{V_{\text{dc}}}{2}$. From Fig. 1i, $V_{pn} = V_{c3} = \frac{3V_{\text{dc}}}{4}$ and $V_{nN} = V_{c4} - V_{c1} = \frac{V_{\text{dc}}}{2}$, and the final output voltage is given as $V_0 = \frac{V_{\text{dc}}}{4}$. From Fig. 1j, $V_{pn} = V_{c2} = \frac{V_{\text{dc}}}{2}$ and $V_{nN} = V_{c2} - V_{c1} = \frac{V_{\text{dc}}}{2}$. The final output voltage is given as $V_0 = \frac{V_{\text{dc}}}{4}$.

3 Control Scheme

There are various pulse width modulation techniques developed to generate pulses to the switches such as pulse width modulation [8], pulse amplitude modulation [10] and pulse position modulation [10] in all the above-mentioned control techniques, the pulses are generated by comparing the reference signal with a carrier signal which

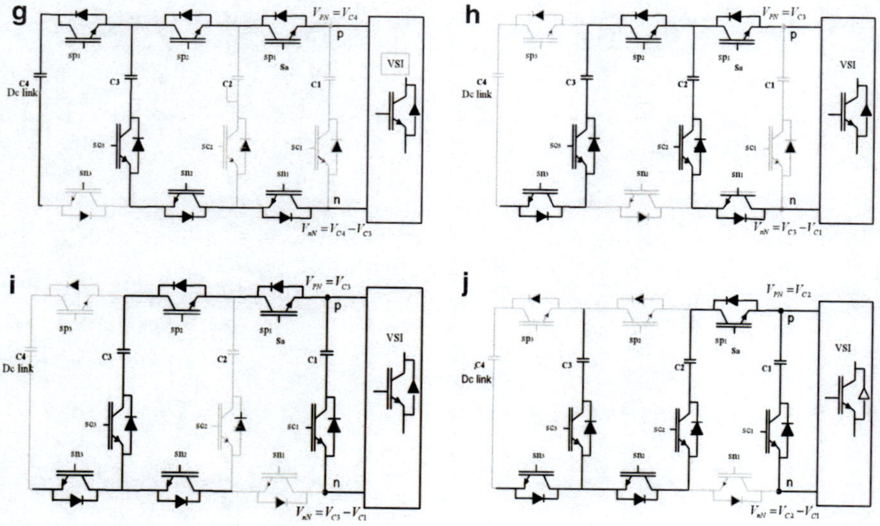

Fig. 1.1 **g** Switching state for an output voltage level of $3V_{dc}/4$. **h** Switching state for an output voltage level of $V_{dc}/2$. **i** Switching state for an voltage level of $V_{dc}/4$. **j** Switching state for an output voltage level of $V_{dc}/4$

leads to high switching frequency and it is undesirable to the proposed method. Gating signals for the switches are to be generated with low switching frequency (Fig. 2). The firing pulses to the switches s_{c1}, s_{c2}, s_{c3}, s_{n1}, s_{n2}, s_{n3}, s_{p1}, s_{p2}, s_{p3} are generated, by taking the total time period for a complete cycle as 0.02 s as shown in Fig. 3.

The output obtained from the following DC–DC link is sent through the 2-L inverter. The output obtained from the DC link might be of n level but as observed, this is actually a part of the full output which means it is just a positive half of the total output, which can be converted into full form of output with the help of an

Fig. 2 Change in voltage levels with different switching states

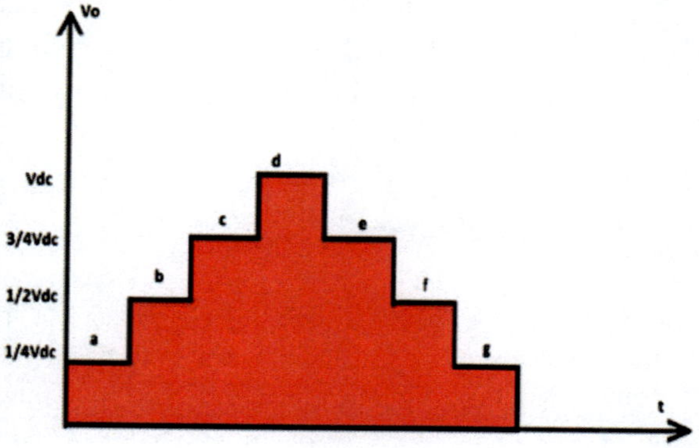

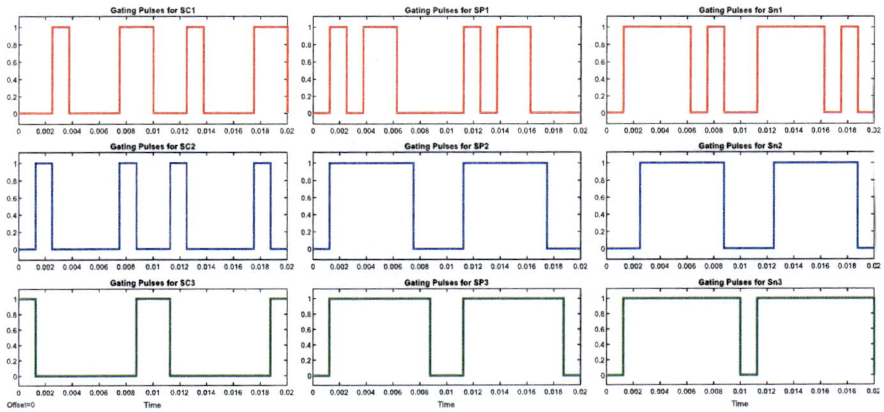

Fig. 3 Gating signals to the switches

inverter, whereas the total output results to be a $(2N - 1)$ level because of n levels in a positive and n negative half-cycle with a common zero level.

4 Simulation Results and Analysis

4.1 Single-Phase Five-Level Topology

The simulation of single-phase five-level reduced switched topology with DC cells connected to the 2-L VSI inverter is shown in Fig. 4. A five-level output voltage requires nine IGBT switches out of which only six switches are utilized to generate the desired voltage levels, with single DC source as input, and each leg has a DC source having a specified voltage. The gate pulse for different switches is generated at variable switching states. The switching operations are followed by different DC

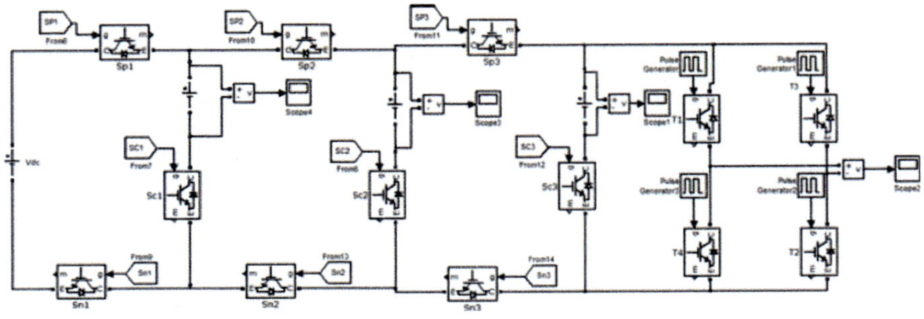

Fig. 4 Simulation diagram of 5-L RMC

cell switching states. For each level generation, one of the simple switching states is applied as shown in Table 1. The different output voltage levels are obtained by different combinations of the nine switches S_{p1}, S_{p2}, S_{p3}, S_{c1}, S_{c2}, S_{c3}, S_{n1}, S_{n2} and S_{n3}. The constructional block diagram of modified single-phase five-level topology is as shown in Fig. 4. The five-level output voltage can be observed in scope, and by using FFT tool, the analysis of output voltage for required number of cycles for different frequencies with THD% can be determined.

The nominal conventional cascaded H-bridge inverter method utilizes $2(m-1)$ switches for a m-level inverter, and hence for a five-level inverter, it utilizes eight IGBT switches, out of which all the switches are utilized to generate the desired output voltage. Hence, the stress on the switches is increased due to continuous turn ON and OFF of the switches to produce the desired voltage, and the number of switches needs to be increased if the switching level increases. The advantage of the novel method is that although nine switches are taking place, only six switches will be utilized and the remaining switches act as redundancy, which is the advantage of the proposed method. Moreover, the number of levels can be increased with the same number of active switches, i.e., six switches are active at any instant of time and the remaining switches serve as redundancy. The same topology can be applied to the three-phase inverter without any change in the number of switches, which cannot be achieved with the conventional method. The staircase output by different combinations of the nine switches, S_{p1}, S_{p2}, S_{p3}, S_{c1}, S_{c2}, S_{c3}, S_{n1}, S_{n2} and S_{n3}, is obtained by applying the switching table shown in Table 1. The output voltage waveform of single-phase five-level reduced multi-level inverter is shown in Fig. 6.

The constructional block diagram of the single-phase five-level reduced multi-level converter is shown in Fig. 4, and the modified 5-level reduced multi-level inverter is shown in Fig. 5 which utilizes only six switches along with the 2-level voltage source inverter to produce the desired output voltage as shown in Fig. 6. The harmonic analysis is done with FFT analysis, and it is observed to be 54.3% for a single-phase reduced multi-level inverter (Fig. 7).

4.2 Single-Phase Seven-Level Topology

The constructional block diagram of single-phase seven-level new topology with DC cells connected with 2-L VSI inverter is shown in Fig. 8. For a conventional method, the generation of seven-level output voltage requires 12 number of active switches, and however, the proposed method makes use of nine IGBT switches, out of which three switches were utilized, i.e., the same circuit constructed for a five-level inverter can be utilized to generate seven levels with different combinations of switching topologies as shown in Table 1. For seven-level output voltage requires 9-IGBT switches, out of which only seven switches are in use along with the 2-level voltage source inverter to generate the desired voltage levels and the remaining three IGBT switches serve the purpose of redundancy. But here a DC source is reduced for generating zero level. The gate pulse for different switches is generated by using

Table 1 Switching states for generating desired voltage levels

Switching sequence	S_{p3}	S_{p2}	S_{p1}	S_{n3}	S_{n2}	S_{n1}	S_{c3}	S_{c2}	S_{c1}	Output voltage
A	0	0	0	1	1	1	0	0	1	$V_{dc}/4 = V_{c1}$
B	0	0	1	1	1	1	0	1	0	$V_{dc}/2 = V_{c2}$
C	0	1	1	1	1	1	1	0	0	$3V_{dc}/4 = V_{c3}$
D	1	1	1	1	1	1	0	0	0	$V_{dc} = V_{c4}$
E	1	1	1	0	0	0	0	0	1	$V_{c4} - (V_{c4} - V_{c1}) = V_{c1}$
F	1	1	1	0	0	1	0	1	0	$V_{c4} - (V_{c4} - V_{c2}) = V_{c2}$
G	1	1	1	0	1	1	1	0	0	$V_{c4} - (V_{c4} - V_{c3}) = V_{c3}$
H	0	1	1	1	0	1	1	1	0	$V_{c3} - (V_{c3} - V_{c2}) = V_{c2}$
I	0	1	1	1	1	0	1	0	1	$V_{c3} - (V_{c3} - V_{c1}) = V_{c1}$
J	0	0	1	1	1	0	1	1	1	$V_{c2} - (V_{c2} - V_{c1}) = V_{c1}$

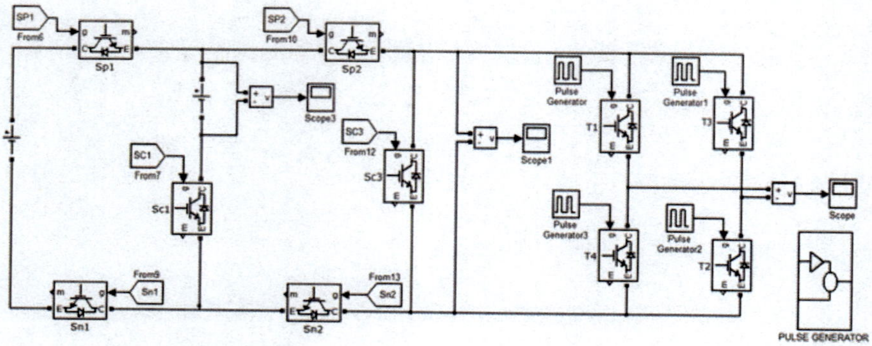

Fig. 5 Simulation diagram of modified 5-L RMC

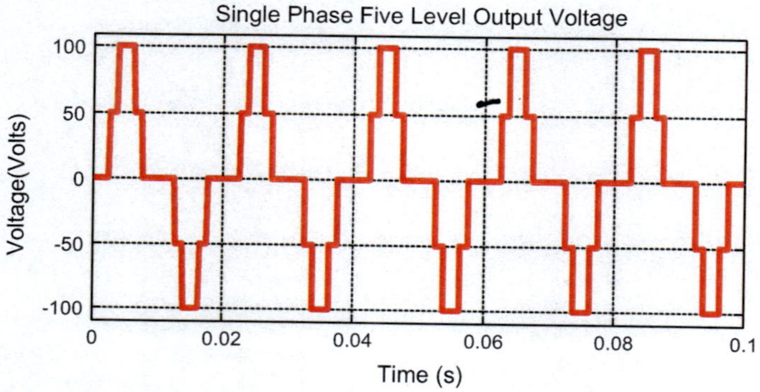

Fig. 6 Output voltage of a modified 5-L RMC

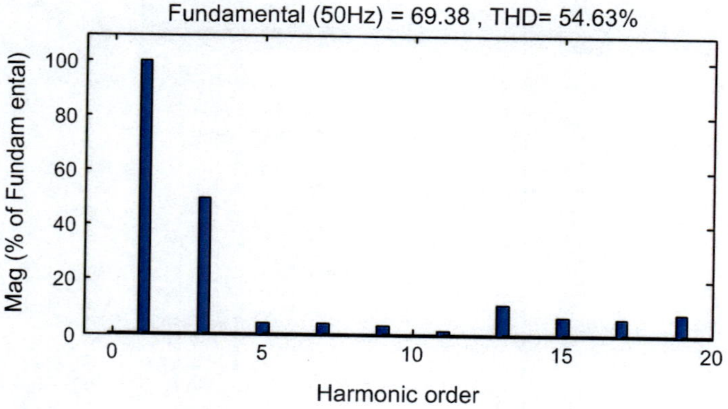

Fig. 7 %THD of the modified 5-L RMC

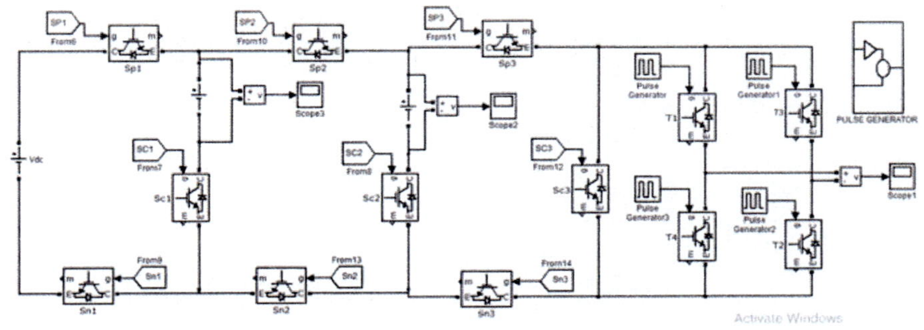

Fig. 8 Simulation model of seven-level RMC

control scheme which is stored in subsystem named pulse generator as triggering the switches at variable switching states. The switching operations are followed by different DC cell switching states. For each level generation, one of the simple switching states is applied as shown in Table 1. The output voltage levels are obtained by connecting the DC source by different combinations of the six switches, S_{p1}, S_{p2}, S_{p3}, S_{c1}, S_{c2}, S_{n1}, S_{n2} and $S_{n3,}$ as shown in Fig. 8.

The simulation model of single-phase seven-level reduced multi-level inverter is shown in Fig. 8, and its corresponding output voltage is illustrated in Fig. 9. By using FFT tool, the analysis of output voltage can be determined using %THD. The THD% for single-phase seven-level reduced multi-level converter is 25.47% as shown in Fig. 10.

As observed from the simulation results, %THDs of seven level inverter topology are reduced when compared to five level RMC which proves that as the voltage level increases, the total harmonic distortions decrease. The simulated value of THD is 25.47% observed in Fig. 10, whereas for five-level inverter, the simulated value of THD is 54.63% as observed in Fig. 7.

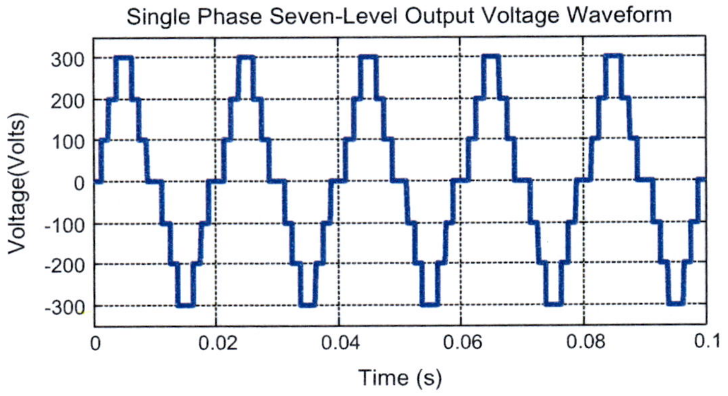

Fig. 9 Seven-level output voltage waveform of new topology

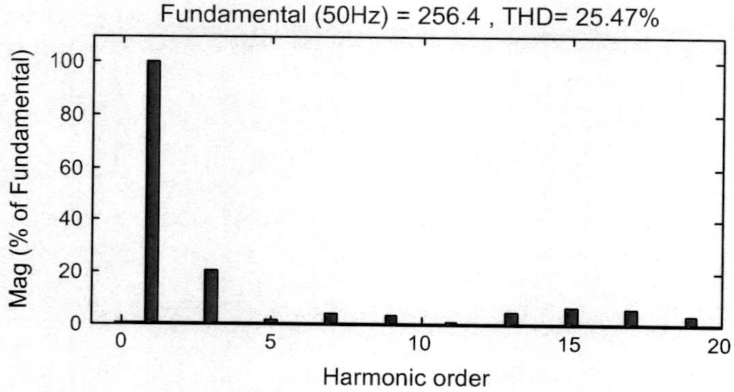

Fig. 10 %THD of seven-level voltage

5 Conclusion

For high-power and high-voltage applications, multi-level converters play an important role. Although with these converters, the percentage of THD is reduced, and the price of the converters has increased tremendously. The work on the multi-level converter is increased with a reduced number of switches to reduce costs. Based on switched DC cell and variable DC link operation principle, a new multi-level converter topology is built in this research, which can achieve the same number of output voltage rates as current multi-level topologies with a fraction of active switching devices and DC sources, as the DC reference variable is shared by all inverter output phases.

From the proposed topology, it is observed that any number of rates can be achieved at the output stage by combining DC–DC circuit and two-level VSI. Five-level, seven-level output voltages are generated, and their percentage of THDs is evaluated. It is found that due to its decreased switches, their percentage of THDs is increased slightly.

References

1. Sandhiya M, Meena S, Mahesh Manivanna Kumar G (2018) Multilevel inverter topology with reduced number of switches. Int Res J Eng Technol (IRJET) 05(01):72
2. Moses Stephen P, Ponraj M, Sathesh N, Balamurugan K, Karthikeyan K (2018) Modified cascaded H-bridge multilevel inverter for household appliances. Int Res J Eng Technol (IRJET) 05(03)
3. Vijay Kumara E, Guptab SK, Singh B, Malad D (2018) New multilevel inverter topology with reduced number of switches using advanced modulation strategies for improved power quality. World J Technol Eng Res 3(1):397–406
4. Gupta SR, Anitha GS (2015) PD-PWM based cascaded H-bridge multilevel inverter for photovoltaic systems. Int J Adv Res Electr Electron Instrum Eng 4(7):2278–8875

5. Thombre N, Rawat R, Rana P, Umashankar S (2014) A novel topology of multilevel inverter with reduced number of switches and DC sources. Int J Power Electron Drive Syst (IJPEDS) 5(1):56–62
6. Maheshri S, Khampariya P (2014) Simulation of single phase SPWM (unipolar) inverter. Int J Innov Res Adv Eng (IJIRAE) 1(9)
7. Jariwala AJ, Shah NV (2015) A new multilevel inverter topology with reduced number of switches. Int J Curr Eng Sci Res (IJCESR) 2(7)
8. Xu J, Nie Z, Zhu J (2015) Random switching frequency pulse width modulation: analysis, design and simulation. In: 2015 Chinese automation congress (CAC). Wuhan, pp 1975–1978
9. Norambuena M, Kouro S, Dieckerhoff S, Rodriguez J (2018) Reduced multilevel converter: a novel multilevel converter with a reduced number of active switches. IEEE Trans Industr Electron 65(5):3636–3645
10. Madany W, Rashdan M, Hasaneen E (2016) Pulse-position and pulse-amplitude modulation interface for serial data link applications. In: 2016 fourth international Japan-Egypt conference on electronics, communications and computers (JEC-ECC). Cairo, pp 95–98

Reliability Assessment of a Hybrid PV/Battery Converter

Shaik Daryabi, Allamsetty Hema Chander, B. G. Madhuri, and V. Pramadha Rani

Abstract Renewable energy systems with individual converters have drawn the attention of researchers in the recent past due to several reasons. However, individual converter for sources increases the size, control complexity and cost of the system. In this regard, this paper proposes a multiple input converter that integrates PV and battery. The proposed converter possesses the advantages like simple structure, high control flexibility and efficient integration of sources. Most of the emerging MIC topologies fail to estimate the lifetime of the converter which is an essential parameter for real-time implementation. Therefore, this paper aims to evaluate the reliability and effect of various parameters on reliability of the proposed MIC. Further, the reliability of the system depends on various factors such as on the on-state resistance of the switch, junction temperature and values of passive elements. In this regard, the effect of major factors on reliability has been discussed and the mean time to failure has been evaluated. This evaluation aids in estimating the failure rate of the components and thereby improving the reliability of the proposed MIC.

Keywords Multi-input DC–DC converter · Hybrid PV · Reliability assessment

Nomenclature

λ_s System failure rate
π_A Application factor
λ_b Base failure rate
π_C Capacitance factor

S. Daryabi (✉) · B. G. Madhuri · V. Pramadha Rani
Department of Electrical and Electronics Engineering, Raghu Engineering College (A), Dakamarri, Visakhapatnam, Andhra Pradesh 531162, India
e-mail: munnidarya@gmail.com

A. Hema Chander
Department of Electrical and Electronics Engineering, Gayatri Vidya Parishad College of Engineering for Women, Visakhapatnam, Andhra Pradesh, India

P. S. R. Chowdary et al. (eds.), *Microelectronics, Electromagnetics and Telecommunications*, Lecture Notes in Electrical Engineering 655,
https://doi.org/10.1007/978-981-15-3828-5_56

π_Q Quality factor
π_E Environmental factor
π_T Temperature factor
θ_{JA} Junction to thermal resistance
T_a Ambient temperature
π_S Electrical stress factor
π_{SR} Series resistance factor
T_J Junction temperature in °K
π_c Contact construction factor
π_V Voltage stress factor
T_{HS} Hot spot temperature
P_S Power dissipated across the switch
P_D Power dissipated across the diode
T_A The ambient temperature

1 Introduction

The renewable energy systems especially with energy storage systems have been the recent trend to meet the load requirement in a reliable and efficient manner. To achieve effective integration of sources, individual DC–DC converters are used. However, increased size, cost and control complexity are the major limitations of such an approach. To overcome these limitations, the multi-input converters (MICs) have drawn more attention of researchers in the recent years, especially to integrate sources with different voltage–current characteristics. The key features of MICs include high control flexibility among the sources and loads besides modular and simple structure [1–4]. Initially, MICs focused on integrating the sources comprising different voltage–current characteristics for grid-connected applications. Later, the development has focused on the electric vehicle applications and photovoltaic applications employing energy storage systems.

A numerous MIC topologies have been discussed in the state of the art for a variety of applications. However, the traditional single input converters employing individual DC–DC converters are extensively used in most of the applications. This might be due to the limitation of literature for review of the effect of various parameters on MIC's performance. The key attribute of the performance of any converter is its reliability assessment.

In general, the reliability of a converter can be defined as the maximum possible duration of a device to perform the definite operation under a set of predefined operational and environmental conditions without failure [5, 6]. Reliability factor highly depends on the time and usually decreases with time. The other factors that likely affect the reliability include the power modules and devices employed in the converter. Further, it also quantifies the involved degree of risk and determines the most prone to failure components/parameters.

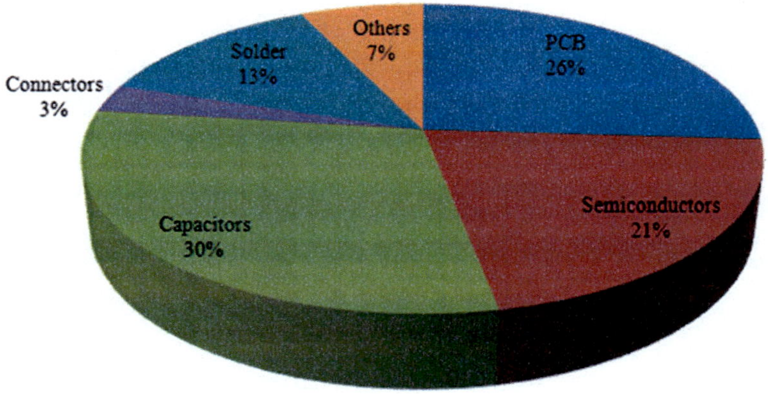

Fig. 1 Components in a system that are most fault prone

A comprehensive review on reliability of converters has been presented in [7]. In [8], a study on converters reliability has been discussed and the quantitative analysis on the critical components summarizes that capacitors and semiconductor switching devices are more prone to faults [9] as depicted in Fig. 1. Further, the present state of health of the system can be estimated through the reliability analysis [8].

Thanks to the power semiconductor sector and its technological advancements that aided in increased efficiency of power converters to approximately 98%. However, in renewable energy applications, one of the major concerns is reliability [10–12]. This mainly results due to high lifetime of the renewable systems, especially photovoltaic systems which have lifetime of about 20 years and hence the low failure rate is preferred for converters. This necessitates the development of a reliable converter for lowering the probability of failure of the device resulting in reduced cost of maintenance during operation. This motivated the researchers and design engineers to develop a reliable converter topology and/or assess the reliability of the existing topologies for real-time implementation.

Thus, it is important to predict the chances of failure, defined as (RoF) of the components, and estimate their lifetime with respect to the employed atmospheric conditions to assess the reliability. In this scenario, the major expansion and breach of MICs for different applications have been investigated. It is noteworthy that the assessment of MIC's reliability is a serious issue and has not yet been performed.

Due to the discussed key features, an MIC topology is proposed in this paper with a single inductor and employs individual switch for each source providing high control flexibility. Further, RoF has been estimated for assessment of the reliability of the converter. This paper first briefs the operation of the proposed MIC topology in Sect. 2. Section 3 discusses the expressions and conditions for assessment of reliability. The reliability evaluation of the proposed MIC has been detailed in Sect. 4 followed by Sect. 5 which concludes the paper.

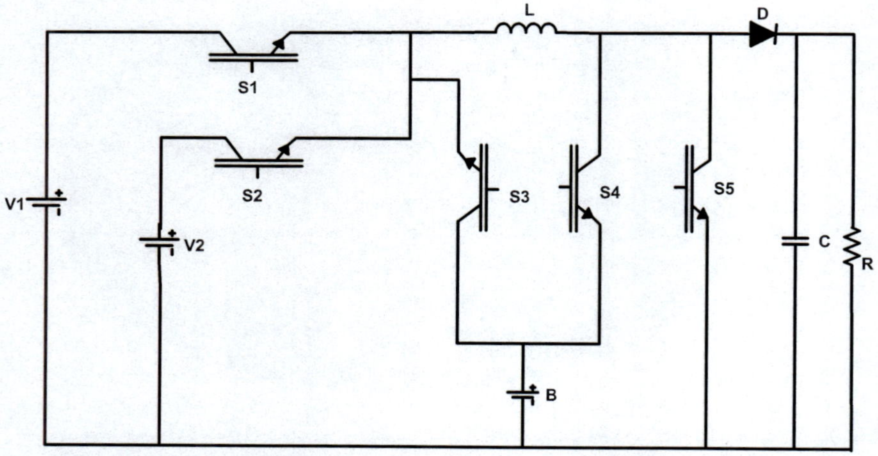

Fig. 2 Proposed MIC topology

2 Proposed MIC Topology

The schematic of the proposed MIC topology is depicted in Fig. 2. Switches S1 and S2 that are in series with the PV sources aid to realize maximum power point of individual panel and also control the amount of power delivered by each PV panel. The switches S3 and S4 determine the charging and discharging of the battery. Further, buck mode is employed for battery charging and discharged in boost mode.

The battery is charged when the generated PV power is more than the load power, and the battery is not fully charged. Similarly, battery aids the PV source when the load demands more power than the generated PV power, provided the battery is not deeply discharged. The state of charge of the battery determines the charging and discharging and is controlled by the algorithm depicted in Fig. 3.

3 Estimation of Reliability

The estimation of reliability discussed in this section is as per the guidelines discussed in MIL-HDBK-217F. The assessment is [13]. The possibility of the converter to operate without fault for 't' time period is represented by ($R_{\text{sys}}(t)$) as

$$R_{\text{sys}}(t) = \mathrm{e}^{-(\lambda_s \times t)} \tag{1}$$

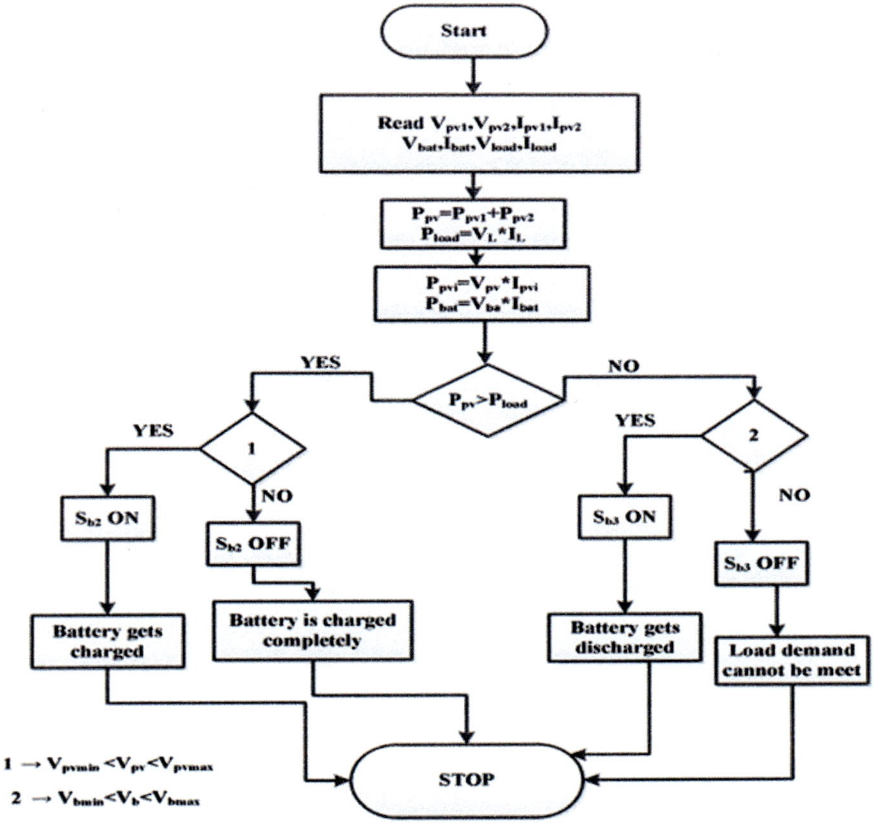

Fig. 3 Control architecture

The mean time to failure (MTTF) is evaluated through (1) as

$$\text{MTTF} = \frac{1}{\lambda_s} = \int_0^{\infty} R_{\text{sys}}(t)\text{d}t \tag{2}$$

All λ'*s* are expressed in failures per million hours (F/MH).
The diode RoF (λ_D) is estimated by

$$\lambda_D = \pi_T \lambda_b \pi_S \pi_Q \pi_C \pi_E \tag{3}$$

$$\pi_S = V_s^{2.43} \tag{4}$$

$$V_s = \frac{\text{Applied Voltage}}{\text{Rated Voltage}} \tag{5}$$

$$\pi_{\mathrm{T}} = \exp\left[-2100\left(\frac{1}{T_{\mathrm{J}}} - \frac{1}{298}\right)\right] \tag{6}$$

With

$$T_{\mathrm{J}} = T_{\mathrm{a}} + (\theta_{\mathrm{JA}})P_{\mathrm{D}} \tag{7}$$

The switch (MOSFET) RoF is obtained using

$$\lambda_{\mathrm{S}} = \pi_{\mathrm{A}}\lambda_{\mathrm{b}}\pi_{\mathrm{T}}\pi_{\mathrm{E}}\pi_{\mathrm{Q}} \tag{8}$$

With

$$\pi_{\mathrm{T}} = \exp\left[-1925\left(\frac{1}{T_{\mathrm{J}}} - \frac{1}{298}\right)\right] \tag{9}$$

$$T_{\mathrm{J}} = T_{\mathrm{a}} + (\theta_{\mathrm{JA}})P_{\mathrm{S}} \tag{10}$$

The inductor RoF (λ_{ind}) is

$$\lambda_{\mathrm{ind}} = \pi_{\mathrm{Q}}\lambda_{\mathrm{b}}\pi_{\mathrm{E}}\pi_{\mathrm{T}} \tag{11}$$

$$\pi_{\mathrm{T}} = \exp\left[\frac{-0.11}{8.617 \times 10^{-5}}\left(\frac{1}{T_{\mathrm{HS}} + 273} - \frac{1}{298}\right)\right] \tag{12}$$

with

$$T_{\mathrm{HS}} = T_{\mathrm{A}} + 1.1(\Delta T) \tag{13}$$

with ΔT being the average increase in temperature over T_{A} depending on the inductor's power loss (W_{L}) and A, area of case (in^2)

$$\Delta T = \frac{125 W_{\mathrm{L}}}{A} \tag{14}$$

The capacitor RoF (λ_{cap}) is

$$\lambda_{\mathrm{cap}} = \pi_{\mathrm{C}}\pi_{\mathrm{T}}\pi_{V}\lambda_{\mathrm{b}}\pi_{\mathrm{E}}\pi_{\mathrm{SR}}\pi_{\mathrm{Q}} \tag{15}$$

With

$$\pi_{\mathrm{C}} = (C \text{ in } \mu\mathrm{F})^{0.23} \tag{16}$$

$$\pi_{V} = \left(\frac{S}{0.6}\right)^{17} + 1 \tag{17}$$

With S being the ratio of operating voltage to the rated and π_T is given by

$$\pi_T = \exp\left[\frac{-E_a}{8.617 \times 10^{-5}}\left(\frac{1}{T} - \frac{1}{298}\right)\right] \quad (18)$$

The RoF of the complete system is given by

$$\lambda_s = \sum_{i=1}^{5} \lambda_{S_i} + \lambda_D + \lambda_{cap} + \lambda_{ind} \quad (19)$$

4 Reliability Assessment for the Proposed MIC

The reliability of the proposed MIC using the above expressions has been assessed in this section. Further, the effects of perturbation of various parameters and the corresponding impact on reliability have been discussed. Tables 1 and 2 list the parameters considered for the design of MIC and assessment of reliability, respectively. With the values in Table 2 and estimating the RoF of the converter for two inputs, λ_{sys} resulted in 15.26 F/MH or 0.0314 failures/year. This results in the mean time between the failures (MTTF) to be 0.06 million hours. The reliability for 15 years has been evaluated and illustrated in Fig. 4.

It is noteworthy that the reliability depends on different factors mostly the $R_{DS(ON)}$ of MOSFET and the ambient temperature. The effect of the discussed perturbations on reliability is detailed:

Table 1 Parameters considered

Parameter	Value	Parameter	Value	Parameter	Value
V_b (V)	48	d_3 and d_4	0.65	d_5	0.4
d_1 and d_2	0.5	R_{Load} (Ω)	10	L_{dc} (mH)	2.5
V_1 (V)	50	V_2 (V)	50	C_{dc} (μF)	1200

Table 2 Parameters considered for reliability assessment

Parameter	Value	Parameter	Value	Parameter	Value
θ_{JA}	5 °C/W	π_{SR}	1	T	50 °C
λ_b of C	0.0004	E_a	0.35	λ_b of L	0.000030
A	7	π_E	9	T_a	25 °C
π_Q	8	λ_b of S	0.012	λ_b of D	0.0038

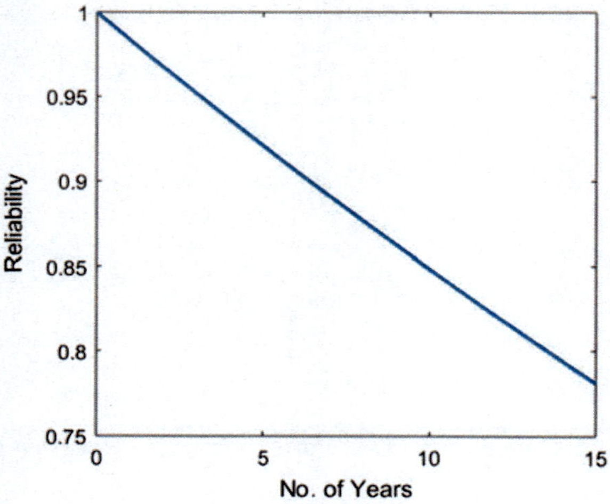

Fig. 4 Estimated reliability for 15 years

i. Perturbation in $R_{DS(ON)}$

   (a) Effect of $R_{DS(ON)}$ on power loss of semiconductor
   A perturbation in $R_{DS(ON)}$ has severe impact on the switch power loss and has been assessed. For variation of $R_{DS(ON)}$ from 0.03 to 0.1 Ω, the power loss has been calculated and plotted in Fig. 5a. It is observed that a variation in $R_{DS(ON)}$ by 70% results in 29% increase in power loss.

   (b) Effect of $R_{DS(ON)}$ on temperature
   The effect of variation in $R_{DS(ON)}$ on temperature has been evaluated. For variation of $R_{DS(ON)}$ from 0.03 to 0.1 Ω, the temperature has been calculated and plotted in Fig. 5b.
   It is observed that a variation in $R_{DS(ON)}$ by 70% results in 16% increase in temperature. Further, the perturbation in $R_{DS(ON)}$ on reliability has been evaluated and plotted. With the increase in $R_{DS(ON)}$, it was observed that the power loss across increased by approximately 29% and the temperature increased by approximately 16%. Thus, decreasing the reliability by 2.12% is illustrated in Fig. 5c.

ii. Perturbation in temperature
   The perturbation in temperature corresponds to the proportional perturbation in power loss across the semiconductor, thus resulting in perturbation of the reliability of the component and hence the overall converter. Three different temperatures are considered to assess the reliability (25, 30 and 40 °C) and is depicted in Fig. 6. An approximately 8% reduction in reliability has been observed for variation in temperature from 25 to 40 °C.

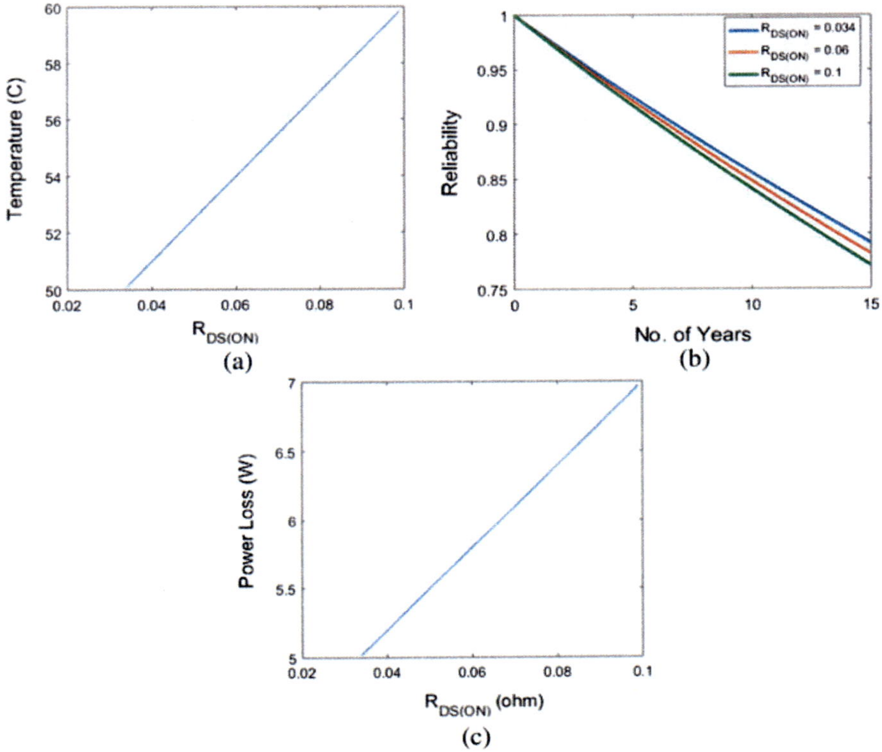

Fig. 5 Perturbation in $R_{DS(ON)}$ and its effect on **a** temperature, **b** power loss, **c** reliability

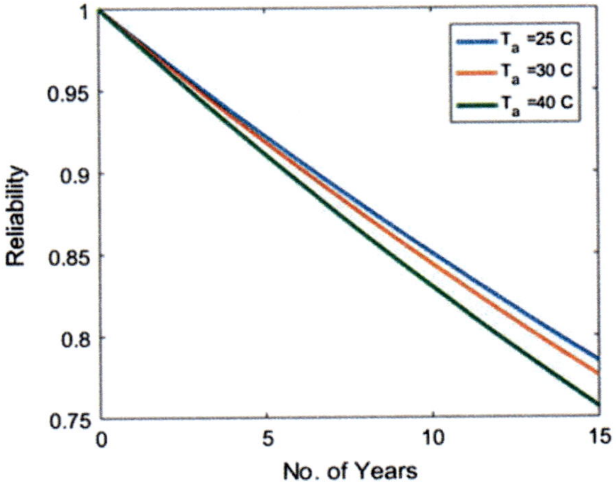

Fig. 6 Effect of temperature on reliability

5 Conclusion

A PV/battery hybrid MIC has been proposed in this paper. Further, the detailed reliability analysis has been performed with various parametric variations considered. The key factor for decrease in reliability is the aging of components. Further, the perturbation in $R_{DS(ON)}$ has severe impact on power loss and temperature which has been evaluated individually. A proportionate increase in power loss and temperature has been observed with increase in $R_{\mathrm{DS(ON)}}$, thereby reducing the reliability of the converter. The temperature has also severe impact on the reliability and was studied. It was observed that the component power loss increases with the rise in temperature, thereby reducing the converter's reliability.

References

1. Di Napoli A, Crescimbini F, Solero L, Caricchi F, Capponi FG (2002) Multiple-input DC–DC power converter for power-flow management in hybrid vehicles. In: Conference record of the 2002 IEEE industry applications conference. 37th IAS annual meeting (Cat. No. 02CH37344), pp 1578–1585
2. Hintz A, Prasanna UR, Rajashekara K (2015) Novel modular multiple-input bidirectional DC–DC power converter (MIPC) for HEV/FCV application. IEEE Trans Ind Electron 62(5):3163–3172
3. Filsoof K, Lehn PW (2016) A bidirectional multiple-input multiple-output modular multilevel DC–DC converter and its control design. IEEE Trans Power Electron 31(4):2767–2779
4. Kumar L, Jain S (2013) A multiple source DC/DC converter topology. Int J Electr Power Energy Syst 51:278–291
5. Tu P, Yang S, Wang P (2019) Reliability and cost-based redundancy design for modular multilevel converter. IEEE Trans Ind Electron 66(3):2333–2342
6. Ristow A, Begović M, Pregelj A, Rohatgi A (2008) Development of a methodology for improving photovoltaic inverter reliability. IEEE Trans Ind Electron 55(7):2581–2592
7. Sang Y, Wang B (2013) Survey on reliability of power electronic systems. IEEE Trans Power Electron 28(1):591–604
8. Alam MK, Khan FH (2014) Reliability analysis and performance degradation of a boost converter. IEEE Trans Ind Appl 50(6):3986–3994
9. Falck J, Felgemacher C, Rojko A, Liserre M, Zacharias P (2018) Reliability of power electronic systems: an industry perspective. IEEE Ind Electron Mag 12(June):24–35
10. De Leon-Aldaco SE, Calleja H, Aguayo Alquicira J (2015) Reliability and mission profiles of photovoltaic systems: a FIDES approach. IEEE Trans Power Electron 30(5):2578–2586
11. Petrone G et al (2008) Reliability issues in photovoltaic power processing systems. IEEE Trans Ind Electron 55(7):2569–2580
12. Kumar L, Jain S (2013) Multiple-input DC/DC converter topology for hybrid energy system. IET Power Electron 6(8):1483–1501
13. Military handbook: reliability prediction of electronic equipment-MIL-HDBK-217F

Speech Enhancement Using Beamforming and Kalman Filter for In-Car Noisy Environment

G. Ramesh Babu and G. V. Sridhar

Abstract The effectiveness of the communication system gets seriously degraded in car by noises like engine sounds and ambient noise, thus decreasing the quality of speech. In modern cars, a lot of effort is put on reducing the background noise. In this paper, speech enhancement cascaded scheme named BEAM-KAL is developed to get the better intelligibility and quality of speech. For this, multichannel beamforming techniques are combined with single channel Kalman filter to get better quality of speech signals which suffer in-car noises. In beamforming, microphone arrays are used to extract the speech signal of interest from a specific desired direction, whereas signals contaminated with noises from various directions are attenuated. However, this technique does not appear to provide enough improvement by itself. Hence, the Kalman filter has been used for its further enhancement. Experiments are performed with real recordings taken while driving in a noisy automobile environment. The performance is investigated with SNR, PESQ and spectrograms and has been shown to produce a better quality of speech.

Keywords Speech enhancement · Kalman filter · Beamforming · BEAM-KAL · Car noise

1 Introduction

Nowadays, speech-based applications like automatic speech recognition systems (ASR) are becoming very important. But the performance of the ASR system in-cars is deteriorated very much by background noises and other various disturbances [1]. Hence, the essential step is to develop effective speech enhancement methods to suppress the background disturbing noises and improve the intelligibility of capturing speech under in-car noisy situations. In recent years, the devices which are activated

G. Ramesh Babu (✉) · G. V. Sridhar
Raghu Engineering College (A), Dakamarri, Visakhapatnam 531162, India
e-mail: drgramesh24@gmail.com

G. V. Sridhar
e-mail: sridhar1209@gmail.com

P. S. R. Chowdary et al. (eds.), *Microelectronics, Electromagnetics and Telecommunications*, Lecture Notes in Electrical Engineering 655,
https://doi.org/10.1007/978-981-15-3828-5_57

by human voice commands have been given much attention by car manufacturers. However, the performance of currently available commercial products is degrading substantially under real-world conditions. The noises originated from pumps, engines, audio equipment, wind, road and air-conditioning, radio and communication are usually non-stationary in nature and time-varying [1, 2]. So, some speech enhancement techniques dealing with in-car noises were developed in recent years to extract the original speech from the distorted noisy speech. The present scheme involves two main algorithms in which beamforming exploits the time correlation of speech signals captured by microphones and Kalman filter uses the different statistics in speech and noise signals to separate them. The purpose of combining the beamforming and Kalman filter is to exploit the strength of each individual method and to improve the quality of speech. When these two methods are combined, a strongly interfering source signal is first separated by beamforming and remaining noise signals are denoised by an adaptive Kalman filter [3]. Thus, they produce the best performance when worked jointly rather independently.

2 Microphone Array-Based Speech Enhancement System

The in-car acoustic ambient contains various sources of disturbances besides the speaker. In order to separate the speech signal of the speaker from these disturbances, multichannel speech enhancement system involving microphone arrays is being used. By processing the microphone array signal appropriately, we can achieve the direction-dependent sensitivity of the source. This technique is named as beamforming.

2.1 *Beamforming*

In beamforming, it isolates a source from a specific direction, while still maintaining some semblance of directionality of the receiving signal. The beamforming be able to represented by linearly arranged of the sensor outputs

$$R(l) = \sum_{i=1}^{M} w_i(l) Y_i(l) \tag{1}$$

where $w_i(l)$ are the weights associated to each ith sensors, and M represents the l number of sensors. The data is described by the vector,

$$\begin{aligned} R(l) &= w^T(l) Y(l) \\ w(l) &= [w_1(l) \ldots w_M(l)]^T \end{aligned} \tag{2}$$

Beamforming can be classified as conventional and adaptive where in conventional the weights across time are fixed, and in adaptive the weights are varied according to the acoustic surroundings of speech.

2.1.1 Fixed Beamforming

In conventional beamforming, [4] the weights $w_i(k)$ are fixed and determined by minimizing the signal power at the beamformer output and subjected to a constraint ensures that the desired signal is unvarnished [4], i.e., the optimal weights are the solution to

$$\begin{aligned} &\text{mini } w^*(l)\boldsymbol{\Psi}_{yy}(l)w(l) \text{ subject to } w^*(l)1 = 1 \\ &w(l) \end{aligned} \tag{3}$$

where * represents the complex conjugate transpose and whereas $\Psi_{yy}(l)$ is the power spectral density matrix of size $M \times M$ of the noisy speech signal with (i, j)th value which is $E\left[X_i(l)X_i^*(l)\right]$. The constraint of zero distortion in the desired direction is given by means of a vector of one's while we consider the array is pre-steered [5] towards the preferred signal direction. The solution is the minimum variance distortion less response beamformer with the constrained optimization [6]

$$w(l) = \frac{\Psi_{ww}^{-1}(l)1}{1^T\Psi_{ww}(l)1} \tag{4}$$

where $\Psi_{ww}(l)$ is the noise PSD matrix of size $M \times M$ whose $(i; j)$th entry is $E\left[W_i(l)W_i^*(l)\right]$. By considering the noise field as homogeneous, the solution is in terms of coherence matrix

$$w(l) = \frac{\Gamma_{ww}^{-1}(l)1}{1^T\Gamma_{ww}(l)1} \tag{5}$$

The $(i; j)$th value of the coherence matrix $M \times M$ is

$$\Gamma_{ij}(l) = \frac{\Psi_{w_iw_j}(l)}{\sqrt{\Psi_{w_iw_i}(l)\Psi_{w_jw_j}(l)}} \tag{6}$$

$$= \frac{\Psi_{w_iw_j}(l)}{\Psi_{ww}(l)} \tag{7}$$

In the above equation, $\Psi_{w_iw_j}(l)$ is the cross spectral density between e ith and jth sensors and the noise signals. By the assumption of a homogeneous noise field, $\Psi_{w_iw_i}(l) = \Psi_{ww}(l)$ for i. The incoherence noise fields, $\Gamma_{ww} = \mathrm{I}$, $w = \frac{1}{M}1$ and the minimum variance distortion less response beamformer reduced to a delay-and-sum beamformer, in which first the sensor output speech signals are delayed and then followed by average. The pre-steering corresponds to the delay and the speech signal components at various sensors added beneficially, and at the same time the noise components are get cancelled. In a delay-and-sum beamformer, the weights of the amplitude are fixed and the weights of phase introduce the delay. But both the amplitude and phase weights vary in filter-and-sum beamformer (FSB). These FSBs are used in designing beamformers with a specific pattern of direction for microphone arrays. Most of the noises fall into the category of noise fields, and the coherence function is in the form of

$$\Gamma_{ij}(l) = \sin c\left(\frac{2\Pi k}{l}\frac{D_{ij}}{c}\right) \tag{8}$$

where k is the frame length and c is the velocity of sound in air, $c = 339$ m/s. Where D_{ij} is the distance between the ith and jth sensors in the array and $\sin c(z) = \sin(z)/z$.

In the coherence matrix, if we use the equivalent expression for the resultant beamformer, it is known as a super directive beamformer (SDB). Although SDB is used in diffuse noise fields, it has a disadvantage of amplifying uncorrelated noises at low frequencies. It can be overcome by incorporating white noise gain restriction in the design.

2.1.2 Adaptive Beamforming

In this, the weights are changed according to the acoustic surroundings. The optimized weights are taken by means of minimizing the variance of the output signal. To make sure that the desired speech signal is not cancelled out or distorted, a distortion less constraint is forced on the desired signal. The generalized side lobe canceller (GSC) [6] is an efficient implementation of the linearly constrained minimum variance (LCMV) procedure, which converts the constrained optimized problem to an unconstrained one. This will give a better performance for the updated weights. The block diagram of GSC is shown in Fig. 1. The structure of GSC contains three modules a beamformer (BF), blocking matrix (BM) and noise canceller (NC). The BF is having a pre-steering module which aligns the desired speech components Y_{BF} by designing its coefficients. Blocking matrix is orthogonal to the beamformer and resulting outputs, called the noise reference signals by blocks the desired speech signal. By taking the differences between adjacent sensor signals, noise references will be formed. The noise cancellation in Fig. 1 removes any remaining noise residual in the speech reference that is correlated with the noise references.

3 Kalman Filter

When speech signal is corrupted with noise, the output y(l) is given as

$$y(l) = x(l) + v(l) \tag{9}$$

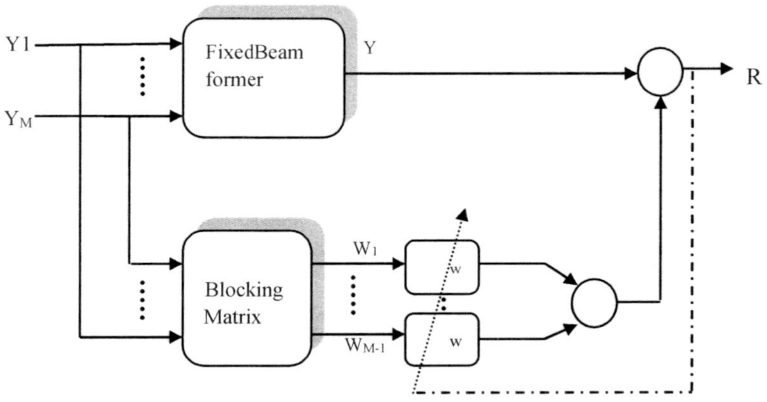

Fig. 1 Implementation of the generalized side lobe canceller in the frequency domain

$x(l)$ is the clean speech.

A qth order autoregressive AR predictor is used to model the speech signal.

Where $x(l)$, the present sample, depends on the linear combination of previous q samples added with a noise.

$$x(l) = \sum_{i=1}^{q} a_i x(l-i) + u(l) \tag{10}$$

where $x(l)$ is the lth sample of the clean signal, and $y(l)$ is the lth sample of the noisy speech, and $a_i(l)$ is ith autoregressive process parameter. This can be modelled by the following state-space expression. Where, the sequences $u(l)$ and $v(l)$ are uncorrelated Gaussian white noise sequences with the mean $\bar{\mathrm{u}}$ and $\bar{\mathrm{v}}$ and the variances σ_u^2 and $\sigma_v^2.x(l)$ is the $Q \times 1$ state vector.

$$X(l) = [s(l-q+1), \ldots, s(l), v(l-q+1), \ldots, v(l)]^T \tag{11}$$

The Kalman filter gives the updating state vector estimator equations

$$e(l) = y(l) - H\hat{X}(l/(l-1) \tag{12}$$

$$K(l) = q(l/(l-1))H \times \left[Hq\left(l/(l-1)H^T\right]^{-1} \tag{13}$$

$$\hat{X}(l/l) = \hat{X}(l/(l-1)) + k(l)e(l) \tag{14}$$

$$q(l/l) = [I - k(l)H]q(l/l-1) \tag{15}$$

$$\hat{\mathrm{x}}(l+1/l) = \mathrm{F}(l)\hat{\mathrm{x}}(l/l) + \mathrm{G}_{\bar{\mathrm{u}}} \tag{16}$$

$$Q(l+1/l) = F(l)P(l/l)F^T(l) + GG^T\sigma_u^2 \tag{17}$$

where $\hat{x}(l+1/l)$ is the minimum mean square estimation of the state vector X(l) given the past l − 1 observations y(l), …, y(l − 1).

The predicted state error vector is $\hat{x}(l/l-1) = x(l) - \hat{x}(l/l-1)$.

$Q(l/l-1) = E[\tilde{x}(l/l-1)\tilde{x}^T(l/l-1)]$ is *predicted* state error correlation matrix where

$\hat{x}(l/l)$ is the filtered estimation of the state vector.

$\hat{x}(l/l) = x(l) - \hat{x}(l/l)$ is the filtered state error vector.

$Q(l/l) = E[\tilde{x}(l/l-1)\tilde{x}^{-T}(l/l)]$ is the filtered state error correlation vector.

$K(l)$ is the Kalman gain and $e(l)$ is the innovation sequence.

The estimated signal can be retrieved from the state vector estimator

$$\hat{s}(l) = H\hat{x}\left(\frac{l}{l}\right) \tag{18}$$

4 Experimental Results

The efficiency of the proposed system BEAM-KAL is tested under the speech signal corrupted with car interior noise at various inputs SNRs −6, 3 dB [7] which was taken from NOIZEUS database [7]. Perceptual evaluation of speech quality scores (PESQ) [8] for the proposed BEAM-KAL [9] are found to be consistently good at lower input SNR, i.e., −6 dB than individual filters shown in Fig. 2. The time domain graphs

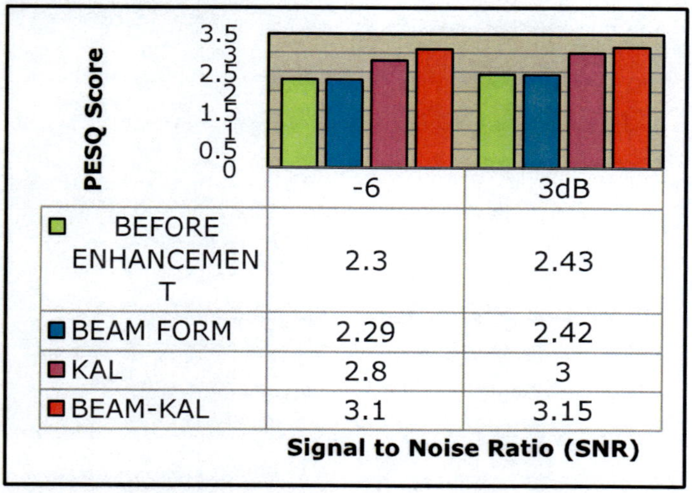

Fig. 2 PESQ scores for the proposed filter and with individual filters

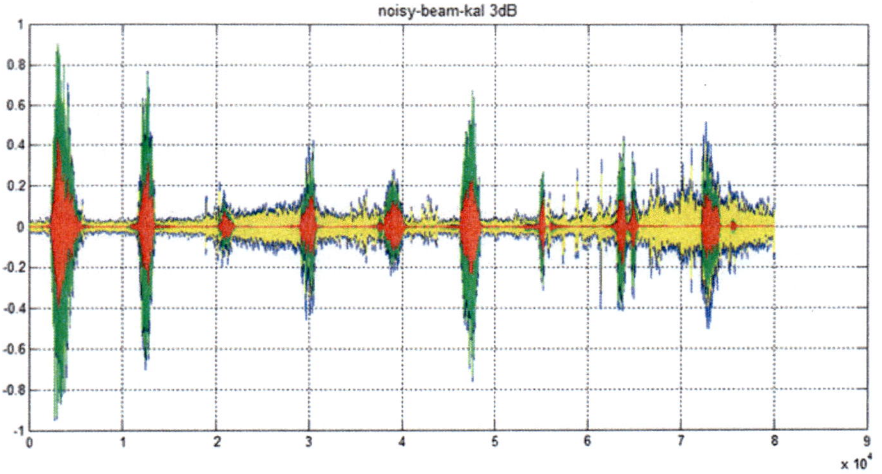

Fig. 3 Comparison of time domain plots at 3 dB. **a** Noisy signal (blue colour). **b** The enhanced beamform (yellow colour) **c**. The enhanced Kalman (green in colour). **d** Enhanced signal of the proposed BEAM-KAL (red colour) and the black circle shows the noise and it was removed

and spectrogram of noisy, beamform, Kalman and BEAM-KAL enhanced speech signals are done at 3 dB where the circles show the noise removal as in Figs. 3 and 4. From this, it is clear that BEAM-KAL combination is superior when compared to others in case of noise reduction. Speech enhancement must satisfy the listener by improving speech quality and hence subjective tests, like informal *A–B* testing, are used to evaluate the performance. In *A–B* testing, a group of people listening to a number of pairs of speech files (labelled *A* and *B*) are involved, and they decide which is better in each case. The obtained results show that the proposed system is extremely good at removing in-car noise, with excellent speech quality and high intelligibility even at low noise exhibits better performance than those obtained with the beamforming and Kalman filter (Table 1).

5 Conclusion

In this paper, a cascaded scheme BEAM-KAL based upon a combination of generalized side lobe canceller (GSC) beamformer and Kalman filter was proposed for the enhancement of speech signals corrupted with car noise. Simulation results were conducted, and the results of the proposed scheme are compared to the beamformer and Kalman filter individually at various input SNRs. The overall performance of the proposed method is shown to outperform beamforming and Kalman filter. Using an objective speech quality measure, spectrogram analysis, as well as formal subjective

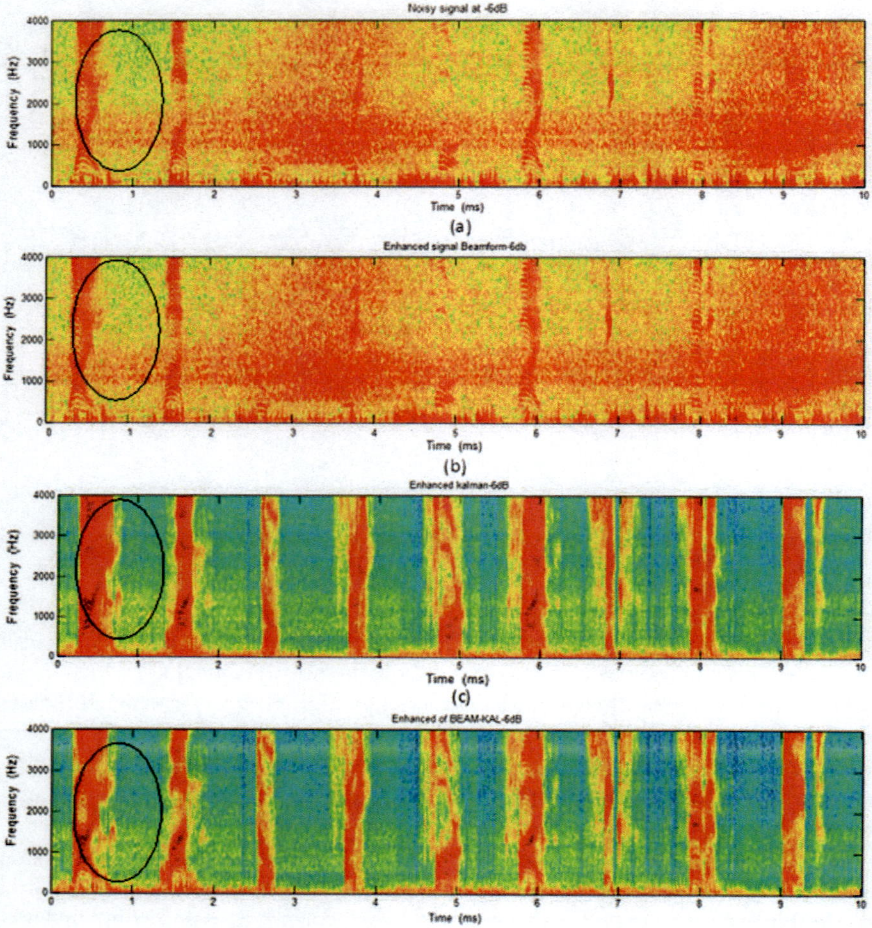

Fig. 4 Spectrogram of the speech sample sp01 from NOIZEUS data corrupted with car noise. **a** Corrupted speech with car noisy signal at −6 dB SNR. **b** Enhanced beamform the speech obtained signal at −6 dB. **c** The output of Kalman filter (enhanced signal at −6 dB). **d** Output signal of the proposed BEAM-KAL at input −6 dB SNR

Table 1 Subjective *A–B* test

Test	Listener preference		
	BEAM-KAL (%)	Other (%)	Undecided (%)
BEAM-KAL/noisy signal	100	0	0
BEAM-KAL/Beamform	96	4	0
BEAM-KAL/KAL	92	1	7

listening tests, showed that the proposed method is capable of reducing noise resulting in improved speech quality. Hence, this scheme provides a promising solution for real-time speech enhancement in noisy car environments.

References

1. Abut H, Hansen JHL, Takeda K (2005) DSP for in-vehicle and mobile systems. Springer
2. Poulat LD (2004) Robust speech recognition techniques evaluation for telephony server based in-car applications. In: ICASSP 2004, I-65–I-68
3. Paliwal K, Basu A (1987) A speech enhancement method based on Kalman filtering. In: Proceedings of IEEE international conference on acoustics speech and signal processing (ICASSP)
4. Van Veen Barry, Buckley Kevin M (1988) Beamforming: a versatile approach to spatial filtering. IEEE Sig Process Mag 5:4–24
5. Bitzer J, Simmer KU (2001) Superdirective microphone arrays. In: Brandstein MS, Ward DB (eds) Microphone arrays: signal processing techniques and applications. Springer, Berlin, pp 19–38 (Chapter 2)
6. Breed BR, Strauss J (2002) A short proof of the equivalence of LCMV and GSC beamforming. IEEE Sig Process Lett 9(6):168–169
7. Noizeus: a noisy speech corpus for evaluation of Speech enhancement algorithms, http://www.utdallas.edu/~loizou/speech/noizeus
8. ITU-T P.862 (2000) Perceptual evaluation of speech quality (PESQ) and objective method for end-to-end speech quality assessment of narrow band telephone networks and speech codecs. ITU-T Recommendation, p 862
9. Ramesh Babu G, Rao R (2012) Combination of beamforming and Kalman filter techniques for speech enhancement. Int J Comput Sci Commun Netw 3(1):338–343

Performance Evaluation of UWB Waveforms in High-Resolution Radar

Ch. Srinivasu, D. Monica Satyavathi, and N. Markandeya Gupta

Abstract The radar waveform design evaluates the target detection, range, and velocity. The performance depends on its structure and resolution. In this paper, ultra-wideband waveform (UWB) design techniques are presented in two-dimensional and three-dimensional scenarios. The ambiguity function is a three-dimensional and two-variable mathematical tool, which demonstrates both range and Doppler resolution capabilities of the waveform. The ambiguity function and autocorrelation functions are used to evaluate the target parameters by studying their surfaces. This paper demonstrates the important results of two-dimensional and three-dimensional plots of non-sinusoidal chirp signals and Barker code of length five. The three-dimensional plots are generated for the above UWB coded signals, UWB chirp signal, and Barker code of length five, and the results are presented. In this paper, range resolution plots of the UWB waveforms are presented and discussed their results in the applications of high-resolution radar for testing detection capabilities. The ambiguity surfaces of the above signals are plotted to determine the range–velocity resolution. The ultra-wideband nonlinear frequency modulated waveform can simultaneously achieve high peak energies and shows high a range and Doppler resolution performance. The delay-Doppler ambiguity function of Barker code of length five using UWB pulses is derived, and their surfaces are examined.

Keywords UWB signals · Pulse compression · Ambiguity · Barker code

1 Introduction

The radar transmits [1] the waveforms toward the target, and echoes are reflected back. The transmitted waveform may use different types of waveforms which include narrowband and wideband. The radar which uses UWB waveforms is called as UWB radar [2]. The delay happens between two waveforms [3] which derive the important

C. Srinivasu (✉) · D. Monica Satyavathi · N. Markandeya Gupta
Department of Electronics and Communication Engineering, Raghu Engineering College (A), Dakamarri, Visakhapatnam, Andhra Pradesh 531162, India
e-mail: vasuch145@gmail.com

P. S. R. Chowdary et al. (eds.), *Microelectronics, Electromagnetics and Telecommunications*, Lecture Notes in Electrical Engineering 655,
https://doi.org/10.1007/978-981-15-3828-5_58

results like a range of target. Similarly, the changes in the frequency show velocity of target by calculating its Doppler shift. And also the UWB signal processing techniques explain the resolution of the target in a more effective way. There are two important types of resolutions: One is range, and the other one is Doppler. Range resolution may give out discrimination between two closely placed targets. And Doppler resolution may result two closely moving targets. These important results may be derived using UWB waveforms. More interestingly, these waveforms show very high resolution [4] at a short range. There are many advantages of using UWB radar instead of radars with continuous waves described. It requires less power, accuracy of resolution, high data speed at short range, etc. Hence, it has wide applications in defense, medical, and advanced communications.

The minimum bandwidth must occupy more than 25% of the center frequency [5]. The minimum bandwidth must exceed 500 MHz. The total bandwidth that could be occupied as defined by the FCC is from 3.1 to 10.6 GHz.

2 Power Spectral Density (PSD) of UWB Pulses

The power spectral density (PSD) of the Gaussian pulse can be calculated as $u(t)$ is a UWB signal. $u(f)$ is Fourier transform of $u(t)$.

$$u(f) = \int_{-\infty}^{\infty} u(t)\exp(-j2\pi ft)\mathrm{d}t \tag{1}$$

$u(f)^2$ power spectral density of $u(t)$

$$|u(f)|^2 = \left|\int_{-\infty}^{\infty} u(t)\exp(-j2\pi ft)\right|^2 = \exp\left(-(2\pi\sigma f)^2\right) \tag{2}$$

The below plots show the performance of the derivations of the basic pulse shape can control the PSD of the monocycle, which in this case is modeled as a Gaussian pulse as shown in Fig. 1. These results show how monocycle shifts for higher-order differentiations.

Another issue is the pulse width, as a Gaussian pulse has infinite duration. This is not, of course, practical and limit to the effective pulse width. Pulse width has always the limitation of such type of waveforms. An effective way to define the pulse width is to choose the pulse width is so that a given percentage of the power in the infinite duration pulse is captured.

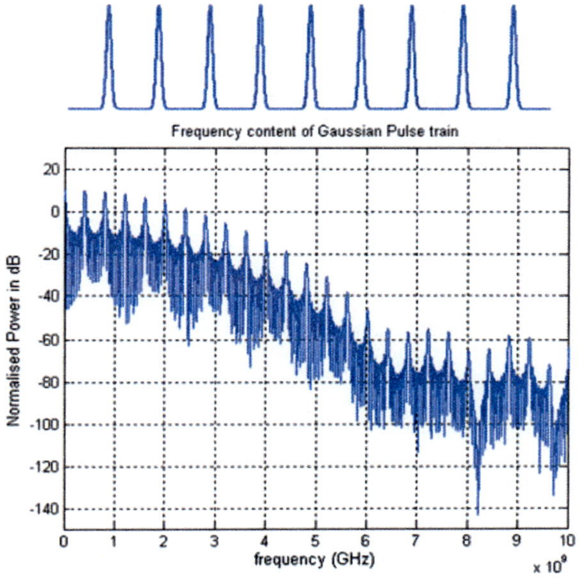

Fig. 1 Power spectral density of different pulse shapes of UWB signal

3 Waveform Design of UWB Ambiguity Diagrams

The matched filter performs as a correlate receiver, which results both compression and noise suppression. Autocorrelation [6] and ambiguity function show a vital role in the design [4] of radar waveforms.

In this paper, ambiguity surfaces [2] of different UWB waveforms are presented and their performance is discussed in the conclusions. Performance criterion of ambiguity diagrams: Ambiguity diagrams are three-dimensional surfaces [7] and are plotted by taking time delays on the x-axis, Doppler shifts on the y-axis, and resultant amplitudes on the z-axis. Time delays measure range, and Doppler shifts measure the velocity of the targets. The surfaces [8] of ambiguity diagrams demonstrate the range and Doppler resolution of the radar for the transmitted waveforms [9]. The diagram should have a large peak main lobe and minimum side lobe values.

These are also called as "thumbtack" diagrams. These plots avoid the ambiguities between the range and velocity of targets. And minimum side lobes [10] are advantageous to avoid the wrong detection of targets and reducing the constant alarm rate (CFAR).

4 Methodology

The following methodologies are implemented in this paper.

i. UWB waveforms are modeled using ambiguity functions in the field of radar.

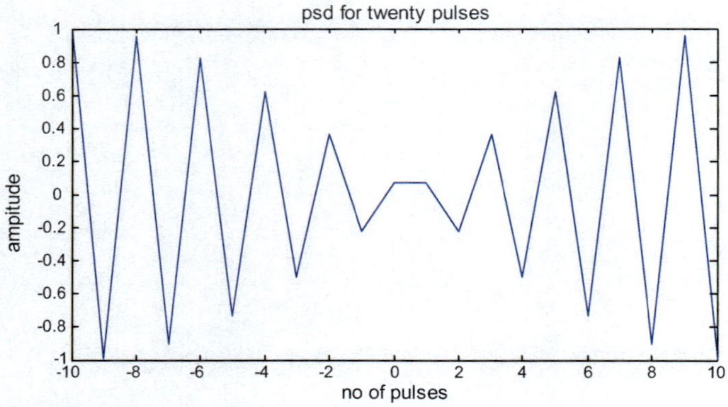

Fig. 2 Plot of energy levels of twenty UWB pulses

ii. Different types of waveforms like UWB Chirp, Barker code, and multiple UWB are simulated using ambiguity functions.
iii. UWB 3D ambiguity surfaces are generated and presented.
iv. The important implications are derived and presented.
v. Suitability of these surfaces in radar design is discussed.

4.1 Range–Velocity Resolution of Multiple UWB Signals

This paper presents the power spectral density curves of multiple ultra-wideband pulses. These plots present how the autocorrelation energy increases as a number of pulses are increased, which is the main criterion of the implementation of these plots to improve the detection capabilities of UWB pulses. The range resolution functions are observed analytically and are presented as two-dimensional surfaces as a function of time delay and autocorrelation energy as shown in Figs. 2 and 3. It is evident from the plots that they show high-energy spectral characteristics and sharp main lobe ambiguity surfaces.

4.2 Range Resolution of UWB Chirp Signal

The We refers to $s_t(t)$ as the UWB chirp signal in similar to the sinusoidal chirp signal $c(t)$ with the pulse duration in the order of nanoseconds.

A chirp signal having a huge train of UWB pulses derives large energy levels needed for good target detection and the large bandwidth required for greater and greater resolution performance (Figs. 4 and 5).

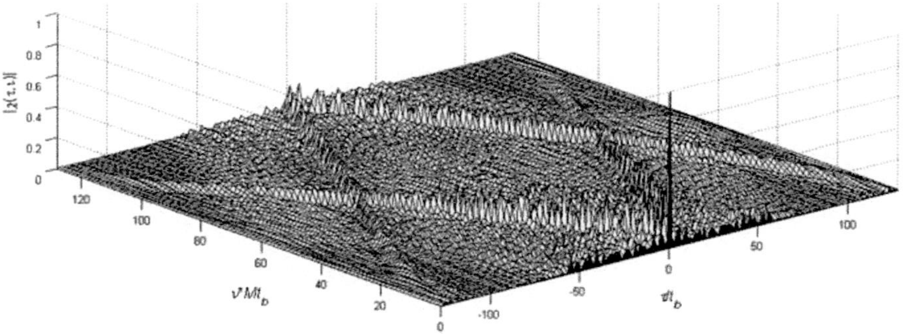

Fig. 3 Ambiguity surface of chirp signal

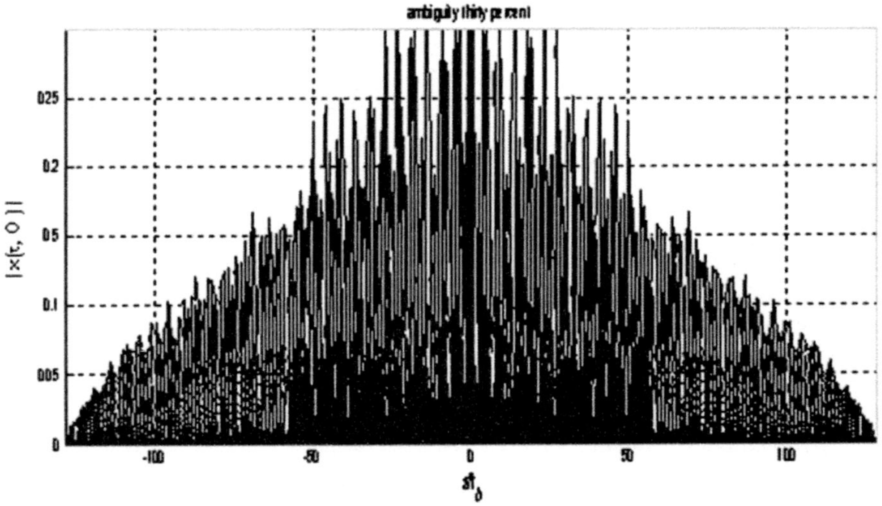

Fig. 4 Plot of ambiguity surface of UWB chirp signal cut along z-axis

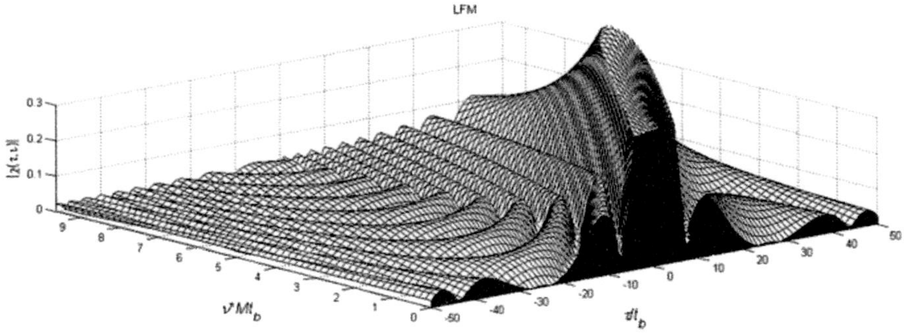

Fig. 5 Plot of ambiguity surface of sinusoidal chirp signal

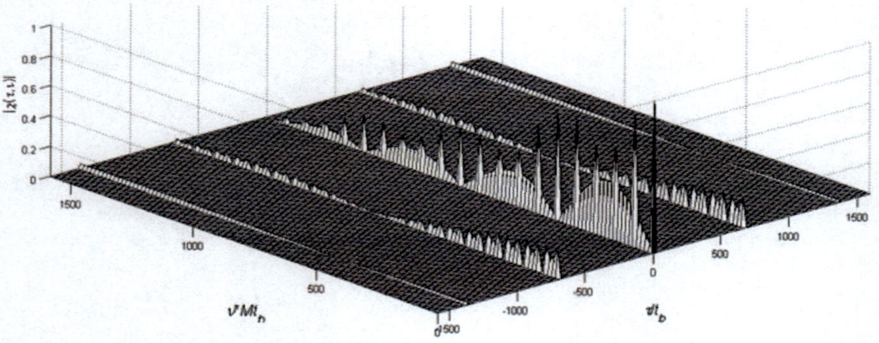

Fig. 6 Plot of ambiguity surface of UWB Barker code of length five for pulse width of 0.02 ns

Barker codes are the popular bi-phase codes which are used in radar applications. The following results as shown in Fig. 6 represent the range and velocity plots of Barker code of length five using UWB pulses with varying pulse widths of 2 and 0.02 ns. Figure 6 shows better resolution characteristics. But side lobes show some significant energy levels, which are undesirable.

5 Conclusions

The ambiguity surfaces of UWB waveforms show narrower main lobe width, and lower side lobes are advantageous in high-resolution radar to resolve the closely spaced targets. The linear frequency modulated waveform shows energy loss in its spectrum in the form of ridges. Analogous to the conventional chirp signal, a UWB chirp signal that is composed of a large number of impulses, provide simultaneously the radiation energy needed for improved target detection and the ultra-wide bandwidth required for achieving high resolution. The features of UWB chirp signal depict the characteristics of the "thumbtack" ambiguity function which is the ideal ambiguity function for high-resolution radar. The UWB chirp waveforms avoid the energy losses because the ridges disappear in its spectrum. The range resolution can be improved because of the high peak autocorrelation values. Because of high autocorrelation peak values and power spectral densities, the detection probability increases. The PSD curves are also observed, and more number of pulses results in higher spectral energy, which is useful for high threshold detection of targets in the presence of noise.

References

1. Win MZ, Scholtz RA (1997) Comparisons of analog and digital impulse radio for multiple-access communications. In: Proceedings of IEEE international conference on communications. Montreal, PQ, Canada, 91–95 June 1997
2. Srinivasu C, RajaRajeswari K (2011) Ambiguity surfaces of bi-phase codes using ultra wide band pulses (UWB) in high-resolution radar. Int J Digital Signal Process 3(4):159–164. Coimbatore Institute of Information Technology (CIIT)
3. Hussain MGM (2007) Waveform design and generalized ambiguity function for ultra wideband non-sinusoidal signals. Electromagn Phenom J 7(1):45–75
4. Hussain MGM (2007) Waveform design and modulation schemes for impulse communications and radar. In: Proceedings of IEEE third international waveform diversity and design (WDD) conference, Pisa, Italy, pp 260–264
5. Srinivasu C, RajaRajeswari K (2011) Autocorrelation properties using multiple ultra wideband (UWB) pulses. Int J Eng Sci Technol (IJEST) 3(4):2939–2943
6. Gill GS (1997) Ultra wideband radar using Fourier synthesized waveform. IEEE Trans Electromagnet Compat 39(2):124–131
7. Woodward PM (2014) Probability and information theory with applications radar. McGraw-Hill, New York
8. Fontana RJ, Foster LA, Fair B, Wu D (2012) Recent advances in ultra wide band radar and ranging systems. IEEE Trans Veh Technol 61(8):3630–3642
9. Scientific Reports (2018) Ultra wideband impulse radar through wall detection of vital signals. Open Online 06 Sep 2018
10. Van Thi Phuoc N, Tang L (2019) Review-microwave radar sensing systems for search and rescue purposes. Sensors 19(13):2879

Single-Stage AC-DC Integrated Double Buck-Boost LED Driver

Thanikonda Yedukondalu, Mopidevi SubbaRao, S. Satyanarayana, and M. V. Sudarsan

Abstract The proposed converter is equivalent to the cascade connection of buck-boost converters, by using a single controlled switch between two stages. It is able to supply a lamp with good efficiency, high power factor and less %THD. In this paper, linear peak current mode control (LPCMC) technique is presented which is a simplified power factor correction technique for single-stage AC to DC converters, the advantages of this circuit include no multiplies are required and voltage sensing circuits. The implementation of this controller is made easy by the use of standard PWM control IC's. From the AC mains, the IDBB converter will power the LED lamps which provides low ripple current through LED, less total harmonic distortion and high power factor.

Keywords IDBB converter · LED lamp · Linear peak current mode control

T. Yedukondalu (✉)
Department of Electrical and Electronics Engineering, SSN Engineering College, Ongole, India
e-mail: tkondalu@gmail.com

M. SubbaRao (✉)
Department of Electrical and Electronics Engineering, VFSTR, Vadlamudi, Andhra Pradesh 531001, India
e-mail: subbarao.mopidevi@gmail.com

S. Satyanarayana
Department of Electrical and Electronics Engineering, Raghu Institute of Technology, Dakamarri, Visakhapatnam, Andhra Pradesh 531162, India
e-mail: ssn_professoreee@yahoo.com

M. V. Sudarsan
Department of Electrical and Electronics Engineering, Vignan's LARA Institute of Technology and Science, Vadlamudi, India
e-mail: mvsudarsan.eee@gmail.com

P. S. R. Chowdary et al. (eds.), *Microelectronics, Electromagnetics and Telecommunications*, Lecture Notes in Electrical Engineering 655,
https://doi.org/10.1007/978-981-15-3828-5_59

1 Introduction

LED is a semiconductor device which emits the visible light when current passes through it and it converts the electric current into light. The lamps are having good efficiency, mercury free and long lasting. LED gives control over light distribution with lenses or small reflectors and provides flexibility in lighting fixtures designing [1].

LED's cannot feed from direct AC or DC supply, because of their constant-voltage behavior. In order to limit the current through lamp, a current-limiting circuit must be used. Under strict operating conditions only, the efficiency of power LEDs can be maintained high, which include low junction temperature and low direct current [2].

Integrated double buck-boost (IDBB) LED driver is proposed in this paper to supply LED lamps from AC mains by providing less ripple current, low %THD and high power factor.

The proposed converter is equivalent to cascade connection of buck-boost converters, at which single switch is connected between the two stages. The converter thus proposed includes two capacitors and two inductors which features good reliability and less cost for LED applications.

In order to provide well- regulated drive current and to control the current through LED for different input voltage, long lifetime LED drivers are required. These may also have specialized dimming control protocols to interface with building control systems. For this, peak current mode control techniques are proposed.

The block diagram representation of an IDBB converter is shown in Fig. 1. A diode rectifier is used which converts AC to DC, while according to its reference the controller will operate the switch properly to shape the input current.

2 Converter Design

The converter behaves as the combination of the two buck-boost converters. By elements L_1, S, D_1 and C_1 the input side buck-boost converter are made and the output side buck-boost converter is made up of L_2, D_2, S, C_2, and D_3. Reversing polarity

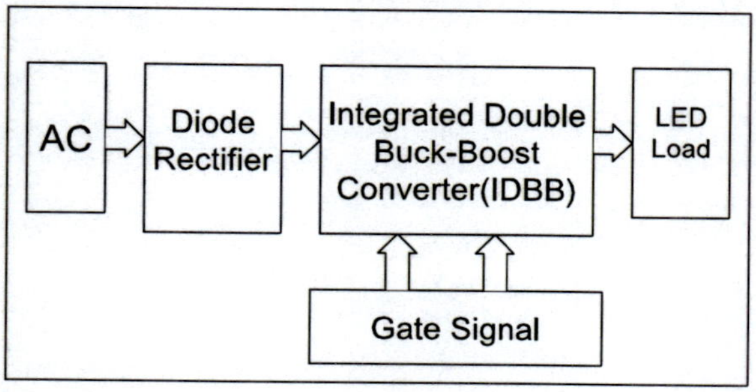

Fig. 1 IDBB converter block diagram

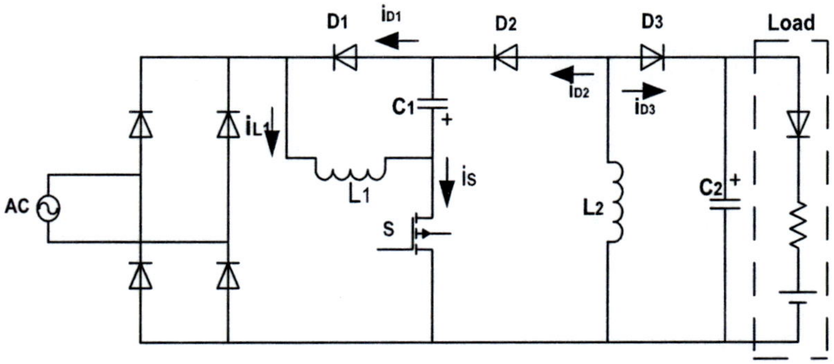

Fig. 2 IDBB converter circuit diagram

in capacitor C_1 in first converter can be corrected by means of second converter, which gives output voltage as positive with reference to ground which simplifies load current measurement for operation of closed loop, thus cost can be reduced (Fig. 2).

Input inductor L_1 is operated in DCM, at which the line voltage will be proportional to average current of the line, thus nearly a unity PF is providing. Output side inductor L_2 can be operated in CCM for reduced capacitance value.

3 Analysis of Driver

3.1 Interval I [$0 < t < DT_s$]

In this interval, $S \rightarrow$ ON, L_1 is charged to V_{L1}, such that $V_{L1} = Vs$. At that time, the capacitor C_1 is discharged through L_2 and D_2, such that $V_{L2} = V_{C1}$ and V_o is supplied by the C_2 (Fig. 3a).

3.2 Interval II [$DT_s < t < (DT_s + t_1)$]

In this interval $S \rightarrow$ OFF and through D_1, input inductor L_1 discharges to C_1. Through diode D_3, the charge saved in the output inductor L_2 is given to load and output capacitor C_2 (Fig. 3b).

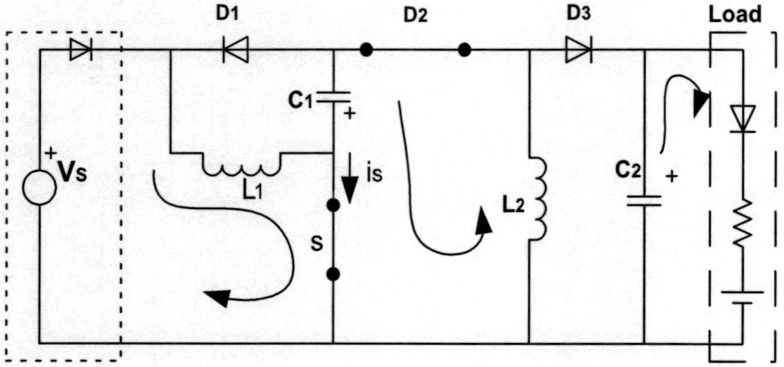

Fig. 3 **a** Equivalent circuit of driver in interval I

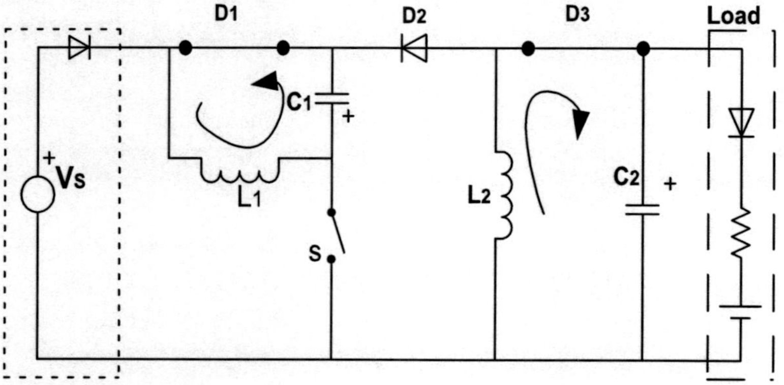

Fig. 3 **b** Equivalent circuit of driver in interval II

3.3 *Interval III [($DT_s + t_1$) < t < T_s]*

In third interval switch $S \rightarrow$ OFF. L_1 completely discharges and current becomes zero. The diode is turned OFF. Through the diode, the charge saved in L_2 is given to load and C_2 (Fig. 3c).

4 Design Example

Input inductance L_1 can be calculated for a given output power as

$$L_1 = \frac{D^2 V_g^2}{4 P_o f_s} \tag{1}$$

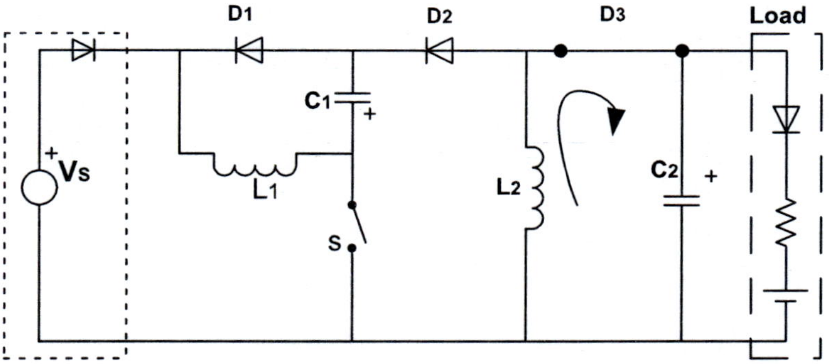

Fig. 3 c Equivalent circuit of driver in interval III

In order to limit ripple of low-frequency, the C_1 can be calculated as

$$C_1 = \frac{D^2}{8V_B\pi L_1 f_s f_L \Delta V_{B\_LF}} V_g^2 \tag{2}$$

L_2 and C_2 are calculated by using expressions of buck-boost converter

$$L_2 = \frac{DV_B}{0.5\Delta I_{\mathrm{LO\_HF}} f_s} \tag{3}$$

$$C_2 = \frac{DV_O}{\Delta V_{O\_\mathrm{HF}} f_s} \tag{4}$$

where $\Delta I_{Lo\_\mathrm{HF}}$ represents peak-to-peak ripple current of high-frequency, I_O is load current and $\Delta V_{\mathrm{O\_HF}}$ is peak-to-peak output ripple voltage of high-frequency.

5 Control Strategy for the Integrated Double Buck-Boost Converter

In order to eliminate sub-harmonic oscillations, the usual control technique in CCM is by using a current control loop with a compensating ramp and to gain stability consequently [3]. In this control to ensure the stability of feedback loop and to obtain PF correction, a compensating ramp issued.

In this control technique, a current loop is designed whose gain is dependent upon the off duty cycle of switch linearly. Thus, the input voltage is proportional to the gain of the current loop and current loop inherently controls the current through inductor. Reference current is obtained by multiplying a gain is set by the controller output of the voltage error amplifier which subtracts from the compensated ramp. Figure 4 shows LPCM implementation [4].

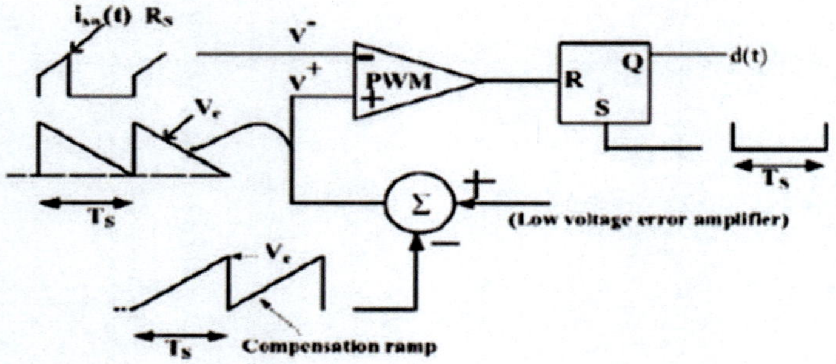

Fig. 4 Implementation of LPCM

Table 1 Design parameters

Parameter	Value
Switching frequency (f_s)	50 kHz
Input inductance (L_i)	2.5 mH
Input capacitance (C_1)	80 μF
Output inductance (L_o)	9 mH
Output capacitance (C_o)	40 μF
Load resistance (R)	570 Ω
LED voltage (V_γ)	170 V

The advantages of LPCM control technique is it eliminates the controller multiplier, implementation is easy by using less cost PWM control current loop unconditional stability and input voltage sensing circuits.

6 Design Parameters

At 230 V_{rms} line voltage and with a frequency of 50 Hz. The load current is 350 mA, for 70 W output power. The design parameters are shown in Table 1.

7 Simulation Results

For the proposed converter simulation, parameters of different circuit components are taken as:

Input power factor is as shown in Fig. 5.

The %THD of LED driver for peak current control is shown in Figs. 6 and 7.

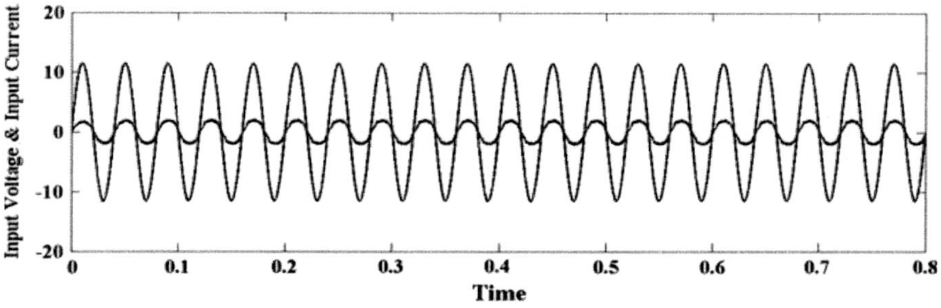

Fig. 5 Input voltage and current

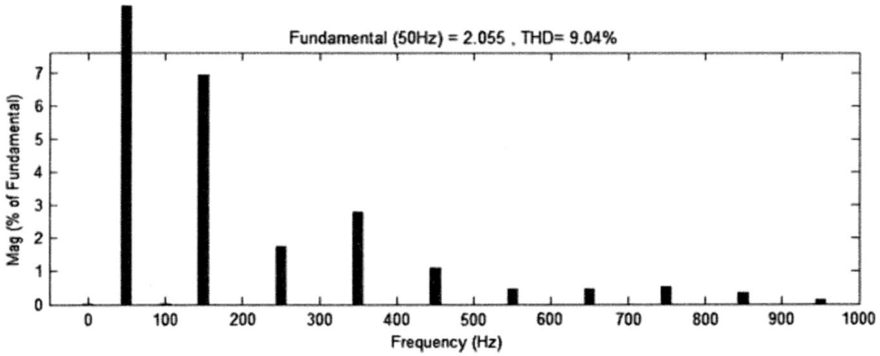

Fig. 6 THD representation for 230 V supply voltage

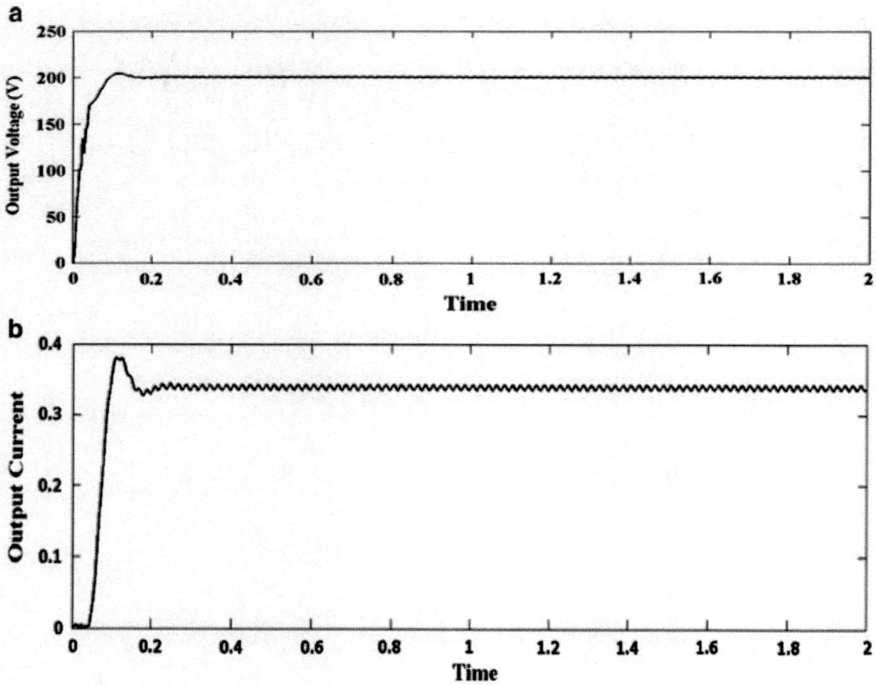

Fig. 7 **a** Output voltage at 230 V supply voltage. **b** Output current at 230 V supply voltage

Fig. 8 Hard ware model of LPCM controlled IDBB converter

The output current and voltage waveforms when peak current mode applied are shown in Figs. 7 and 8. For 230 V supply voltage output voltage, output current waveforms are shown in Fig. 7a, b.

8 Experimental Results

Experimental model was developed to verify the simulation results. Figure 8 is the hardware model of the proposed converter. Figures 9 and 10 show the source voltage and current waveforms at supply voltage of 230 V, 50 Hz at rated load. Figure 11 is the pulse signal generated from the linear peak current controller to the control switch used in IDBB converter.

The variation of output voltage V_o, output current I_o, THD with respect to different supply voltages is shown in Table 2.

9 Conclusion

The proposed LED driver for lighting applications consists of two buck-boost converters are connected by means of only one switch. A high PF at input can be obtained by operating converter at input stage in DCM. Low LED ripple current

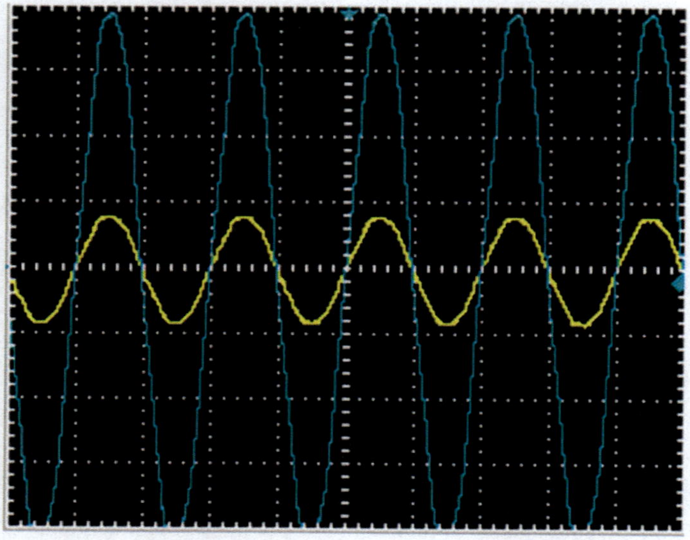

Fig. 9 Input voltage and current

Fig. 10 Load voltage

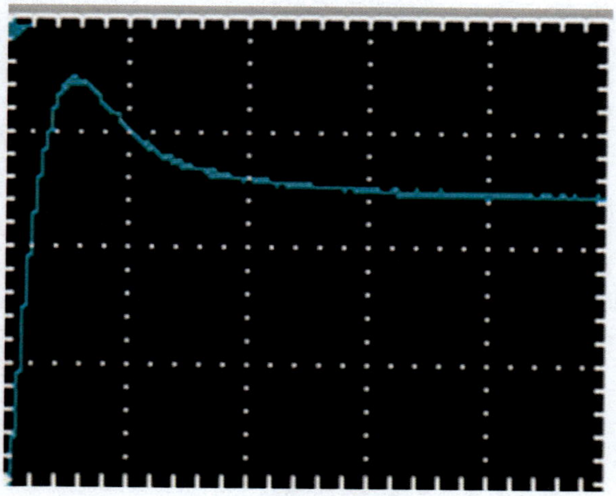

can be obtained by operating the second stage in CCM without using a high output capacitance value. By using only film capacitors, the converter can be implemented and avoided the use of electrolytic capacitors. It increases meantime of converter between failures. Thus, the proposed converter with the linear peak control technique can provide low %THD and high power factor. The proposed scheme is implemented in MATLAB/Simulink tool and experimental results validate the satisfactory with simulation results.

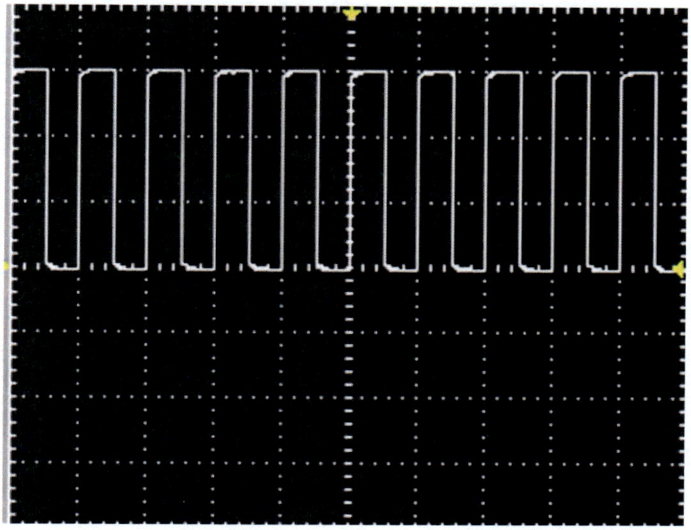

Fig. 11 Pulse signal

Table 2 Variation of output voltage V_o, output current I_o, THD with respect to different supply voltages

V	190	200	210	220	230
V_O	186.6	190	193.5	197	200.6
I_O	0.18	0.22	0.26	0.301	0.341
THD	9.25	9.20	9.15	9.09	9.04

References

1. SubbaRao M, SaiBabu C, Satyanarayana S (2018) Design and analysis of variable switching frequency controlled integrated switched mode power converter for Class C & Class D appliances. Ain Shams Eng J 9:2849–2858
2. SubbaRao M, SaiBabu C, Satyanarayana S (2018) Digital fuzzy current mode controlled integrated PFC converter with external ramp compensation. J Circuits Syst Comput (JCSC) 27(9):1850147-1-23. doi: https://doi.org/10.1142/S0218126618501475. ISSN: 1793-6454
3. SubbaRao M, SaiBabu C, Satyanarayana S (2017) Charge current controlled single phase integrated switched mode power factor correction converter. Energy Proc 117:71–78
4. SubbaRao M., SaiBabu, C, Satyanarayana, S (2017) DSP based current controlled single stage single phase integrated converter with external compensating signal for Class-C & Class-D appliances. Cogent Eng J 4:1–13. ISSN: 2331-1916

Comparison of Incremental Conductance with Fuzzy Controller for a PLL-Less Scheme for Grid-Interfaced PV System

Alluvada Bala Raja Ram, T. Srinivas Sirish, M. V. Suresh Kumar, and A. Sai Sita Ram Murthy

Abstract In modern days, necessity of electrical power has increased due to rapid industrial growth. With this, dependence on conventional energy increased, the depletion of its sources started, and here, renewable energy sources like solar, wind and tidal energy sources started to act as alternate energy source and are playing a crucial role among which solar is quiet important. The proposed scheme serves two purposes such as extraction of maximum power from a PV cell along with mitigation of harmonics. Secondly, a notch filter scheme is designed to eliminate phase-locked loop, and this filter extracts the real component of load current which is previously done by PLL. As phase-locked loop is eliminated, system dependence on PI controller tuning decreases and thereby dynamic response and robust nature of system increases. Along with this, a fuzzy logic-based maximum power point tracker (MPPT) is proposed and compared with incremental conductance-type power point tracker.

Keywords MPPT · PV system · PLL-less scheme · Incremental conductance · Fuzzy logic-based MPPT

A. Bala Raja Ram (✉) · M. V. Suresh Kumar · A. Sai Sita Ram Murthy
Department of Electrical and Electronics Engineering, Raghu Engineering College (A), Dakamarri, Visakhapatnam, Andhra Pradesh 531162, India
e-mail: balaram993@gmail.com

M. V. Suresh Kumar
e-mail: mvsk258@gmail.com

A. Sai Sita Ram Murthy
e-mail: 4rams33635@gmail.com

T. Srinivas Sirish
Department of Electrical and Electronics Engineering, Gayatri Vidya Parishad College of Engineering (A), Vishakhapatnam, Andhra Pradesh, India
e-mail: srinivas.sirish@gmail.com

P. S. R. Chowdary et al. (eds.), *Microelectronics, Electromagnetics and Telecommunications*, Lecture Notes in Electrical Engineering 655,
https://doi.org/10.1007/978-981-15-3828-5_60

1 Introduction

With a continuous raise in industrial era, more demand for power and energy has increased which affected the conventional energy sources, which is becoming more costly day-by-day and also ejecting more amount of carbon quantity into atmosphere. Renewable energy sources like solar energy, wind and tidal energy have been an answer for above issues Out of these, solar energy proved to be both efficient and economic which has become a main reason for a large number of models. Among these pros, the only drawback the solar model carries is its efficiency, i.e., to extract the maximum power from the solar cell irrespective of nature calamities.

As extraction of maximum power has become essential and for this to happen, different types of MPPTs are used, and this is discussed by many researches mainly in [1]. Algorithms which are based on artificial intelligence like genetic algorithms and neural networks are used. Along with the above-mentioned MPPTs like incremental conductance (INC), hill climbing (HC), and perturb and observe (P&O), conventional-type methods have also been proposed to extract maximum power. These MPPTs are discussed in [1]. Of these, hill-climbing and P&O-type MPPTs are regularly used because of its easy working. When it comes to operation of MPPT with P&O, it works on the voltage of PV by perturbing in the desired direction and perturbs in the opposite direction when point moves away from the maximum power point. Because of this easy operation, many papers have focused on this algorithm along with hill-climbing MPPT. But, in case of incremental conductance MPPT, its working is based on the PV curve slope v/s voltage at zero at its maximum power point. This algorithm uses fixed step size iteration mechanism, and this is one of the main reasons for its accurate and fast response which decreases the fluctuation time along the MPP when it is tracked. But, the only drawback of this algorithm is the fixing of the step size which will decrease the quick response of the system. Here, the working of incremental conductance MPPT along with fuzzy-based MPPT and its operation are discussed. The output of this block is given as input to the DC–DC converter for duty cycle. Normally, buck or boost or buck–boost converters are used according to the requirement. In this proposed paper, boost converter is used to step-up voltage by varying the input (duty cycle) given by MPPT block. This converter is then integrated to grid. Different types of system have been developed to integrate the solar PV with different type of networks.

A reference frame theory for three phases has been discussed in [2], and by using this, zero voltage regulation can be achieved. Also a three-phase single-stage OCC is designed using CSI-based topology [3]. Apart from these, a six-switch inverter based on phase-locked loop is designed and discussed [4]. Bi-directional converters for both single and three phases using modified OCC are discussed in [5] the increase in the use of nonlinear load, the tricky nature of harmonics increase. Non-sinusoidal currents have no cooperation to active power components but will upsurge the losses in the distribution system and distort the voltage at PCC. Some researchers have proposed a filter for estimation of frequency and also which can also estimate the frequency component in the distorted voltage signal two- and three-wire single-phase single-stage PV system with harmonic elimination are discussed in [6] where control algorithm is grounded on load power approximation. The assessed load power involves second order harmonics, if not filtered or removed, will cause misrepresentation in reference grid current. As related to the above, the two-stage single-phase grid-interfaced system uses a notch filter strategy as discussed in [7]. This filter serves two purposes, one to estimate active load power component and estimation of unit vectors. The peak PCC voltage is assessed from identified PCC voltage, and unit vectors are derived and give inherent frequency locking capacity to the system. Modelling and simulation of solar photo voltaic models is discussed briefly in [8]. In this paper, INC-based MPPT is used instead of P&O-type MPPT and compared with FLC algorithm. The total harmonic distortion (THD) is under IEEE-519 standard (less than 5%) for nonlinear load also.

2 System Configuration

The integration of two- stage single-phase solar photovoltaic generation system with grid is shown in Fig. 1. This configuration has two phases, at first stage, power electronic converters are used to track power from cells and transferred it to grid. At the second stage, the photovoltaic modules of series–parallel combination of desired rating are interfaced with boost converter, whose duty cycle is determined by using INC MPPT. Then, this is interfaced to a voltage-source converter through a DC link capacitor, and this in-turn will be interfaced to a grid, load and a ripple factor through and inductor at point of common coupling. Variables i_{PV} and v_{PV} are PV current and voltage, where i_{VSC} is current inflowing point of common coupling from VSC and i_S is the grid current. Developing a PLL-less scheme with a filter along with extraction of maximum power by using INC MPPT and fuzzy logic-based MPPT will be discussed in control algorithm. The control algorithm is explained in two parts. (1) VSC block and (2) MPPT control block.

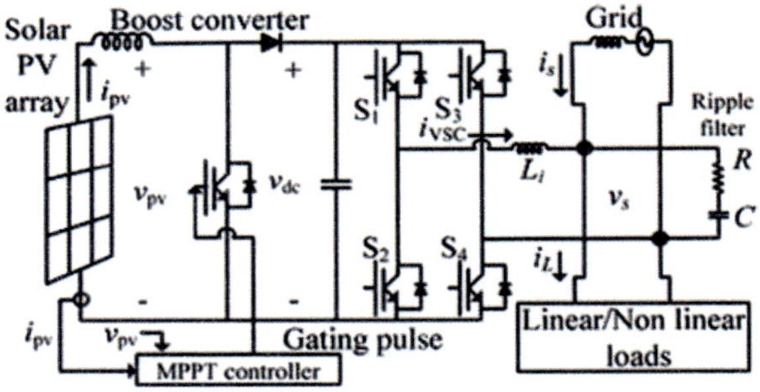

Fig. 1 Integration of two-stage single-phase PV system to grid [7]

3 Control Algorithm

The VSC block generates controlling pulses via an indirect current regulator scheme for controlling the VSC. The VSC controller discussed in [7] is shown in Fig. 2. The real element from load current is derived using a zero crossing detector, a filtering scheme, S&H logic circuit. The loss element of voltage-source controller is determined using a proportional integral controller. By using a U_V of PCC voltage, predictable reference current is matched with grid voltage.

3.1 VSC Control Block

VSC control block is designed by determining (i) load current i_l, (ii) real component i_{lp}; (iii) fundamental and out-of-phase component by 90° of PCC voltage.i_{lp} is derived using 90° out of phase component of PCC voltage, ZCD, sample and hold logic. This phase component of load current will not contain any harmonics.

$$i_{l\alpha} = i_m \sin(\text{wt} - \theta) \tag{1}$$

$$i_{l\alpha} = i_m\left[\sin(\text{wt})^* \cos\theta - \cos(\text{wt}) - \sin\theta\right] \tag{2}$$

$$[i_m \sin(\text{wt} - \theta)]_{\cos(\text{wt}=0)} = i_{\text{LP}} = i_m \cos\theta \tag{3}$$

The difference in phase between load current and PCC voltage is termed as θ, and wt is measured from positive zero crossing PCC voltage.

$v_{s\alpha}$ and $v_{s\beta}$ values are derived by a filtering scheme, and to calculate unit template u_v, amplitude of PCC voltage is required which is used for synchronizing the reference current with grid voltage. Mathematically, it is represented as

$$V_{sp} = \sqrt{v_{s\alpha}^2(k) + v_{s\beta}^2(k)} \tag{4}$$

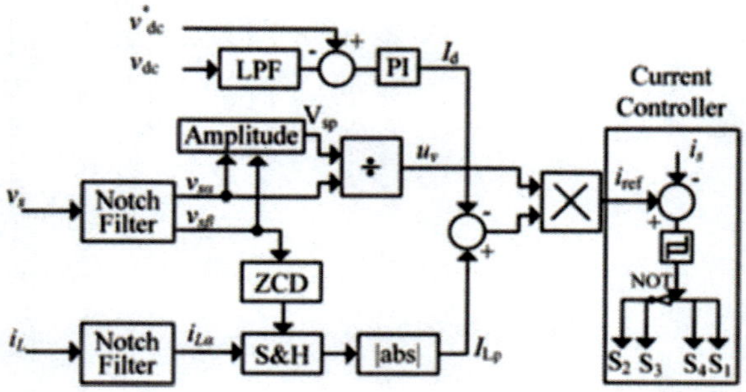

Fig. 2 Block diagram of VSC controller [7]

$$U_V = v_{s\alpha}(K)/v_{s\beta}(K) \tag{5}$$

A compensation current signal is produced from PI controller, and it is represented as

$$i_d(k) = i_d(k-1) + K_P[e_{\mathrm{dc}}(k) - e_{\mathrm{dc}}(k-1)] + K_I[e_{\mathrm{dc}}(k)] \tag{6}$$

$$e_{\mathrm{dc}}(k) = V_{\mathrm{dc}}^*(K) - V_{\mathrm{dc}}(k-1) \tag{7}$$

From Eqs. (2), (3) and (4), real component of load current's magnitude is obtained and is subtracted from PI controller's output, and this is the compensation current signal after removal of losses. The output of this is synchronized with grid voltage by multiplying with u_v of PCC voltage to produce reference grid current. This signal is then deducted from identified grid current to produce error signal. This error between sensed and grid reference signal is equated with upper and lower threshold for every sample moment, and this is known as indirect current control technique. S1 and S4 are the switches used if the error is more than higher threshold, and the complimentary is decided for S2 and S3. Mathematically, it is represented as

$$i_{\mathrm{ref}}(K) = \left[i_{\mathrm{Lp}}(K) - i_d(K)\right]u_V(K) \tag{8}$$

The interfacing inductor is set at 6 mH and ripple filters R and C at 5 Ω and 5 μF. PI controller is tuned, and K_P and K_I values are set at 0.5 and 2.5.

3.2 MPPT Control Block

In the incremental conductance technique, speed of operation is defined by the step size used. As mentioned before, this is a disadvantage in case of incremental conductance as this affects the quick response of the system. If the step size is large, operation is fast with high-speed less-numbered oscillations, and if step size is small, it operates slowly with more number of oscillations. Incremental conductance (INC) is one kind of an algorithm where it uses both voltage and photovoltaic current to produce the duty cycle. In this case, maximum power point (MPP) is obtained only with a condition that the instantaneous conductance $\frac{I_{\mathrm{PV}}}{V_{\mathrm{PV}}}$ must be equal to the negative value of $\frac{\mathrm{d}I_{\mathrm{PV}}}{\mathrm{d}V_{\mathrm{PV}}}$. The sum of both instantaneous and incremental conductance is equal to $\frac{\mathrm{d}P}{\mathrm{d}V}$. The principle is grounded on the slope of PV system if it is zero at MPP $\left(\frac{\mathrm{d}P}{\mathrm{d}V} = 0\right)$. It defers in the case where the summation is positive, i.e., the power point tracker will be located on left side $\left(\frac{\mathrm{d}P}{\mathrm{d}V} > 0\right)$. On the other side, if the case changes out to summation that is negative, the power point tracker will be on the right side $\left(\frac{\mathrm{d}P}{\mathrm{d}V} < 0\right)$.

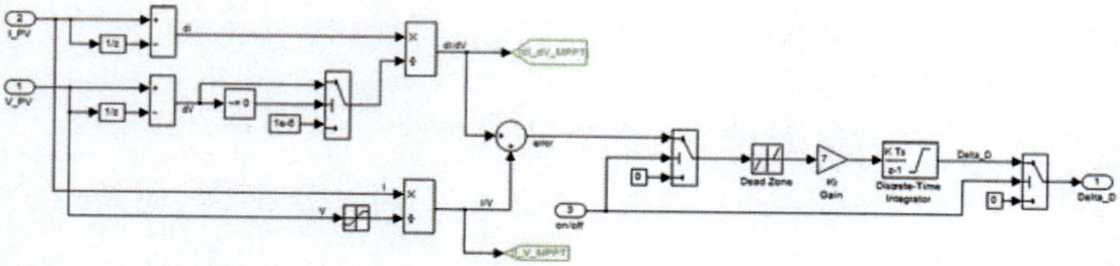

Fig. 3 MATLAB-implemented INC-based MPPT

$$\frac{\mathrm{d}I_{\mathrm{PV}}}{\mathrm{d}V_{\mathrm{PV}}}+\frac{I_{\mathrm{PV}}}{V_{\mathrm{PV}}}=0 \tag{9}$$

$$\frac{I_{\mathrm{PV}}}{V_{\mathrm{PV}}}=-\frac{\mathrm{d}I_{\mathrm{PV}}}{\mathrm{d}V_{\mathrm{PV}}} \tag{10}$$

For attaining the maximum power point condition, incremental conductance uses a searching method to track the point by considering a reverence voltage as a point (V_{REF}). If the maximum power point is obtained, then $\frac{\mathrm{d}I_{\mathrm{PV}}}{\mathrm{d}V_{\mathrm{PV}}}$ is equal to zero. If $\frac{\mathrm{d}I_{\mathrm{PV}}}{\mathrm{d}V_{\mathrm{PV}}}$ is greater than zero, then V_{REF} is increased; else, if $\frac{\mathrm{d}I_{\mathrm{PV}}}{\mathrm{d}V_{\mathrm{PV}}}$ is less than zero, then V_{REF} will be decreased. $\frac{\mathrm{d}I_{\mathrm{PV}}}{\mathrm{d}V_{\mathrm{PV}}}$ specifies that the sum of these is higher than zero. In contrast, $\frac{I_{\mathrm{PV}}}{V_{\mathrm{PV}}}$ (right side of the above equation) that the summation of incremental conductance is lesser than zero (Figs. 3 and 4).

4 Proposed System

FLC is one type of intelligent controller in which this is obtained by achieving the adoption of human-type behavior or type-taking decision. One of the main advantages is that it does not require any mathematical calculations or functions like other type MPPTs and works using imprecise input. In simple, it comprises of three stages, namely fuzzification process, inference system, rule and defuzzification, and the arithmetical variable input is transformed into a fuzzy input through the membership function during the process of fuzzification. FL is the fuzzy logic input variable, E is the error, and C is the change in the error.

Below are the rules through which the variables are processed

$$E(n)=\frac{P(n)-P(n-1)}{V(n)-V(n-1)} \tag{11}$$

$$CE(n)=E(n)-E(n-1) \tag{12}$$

Now, after generation of FL, defuzzification is done. Output of this is the change in duty cycle (dD). In the process of defuzzification, the output will be transformed

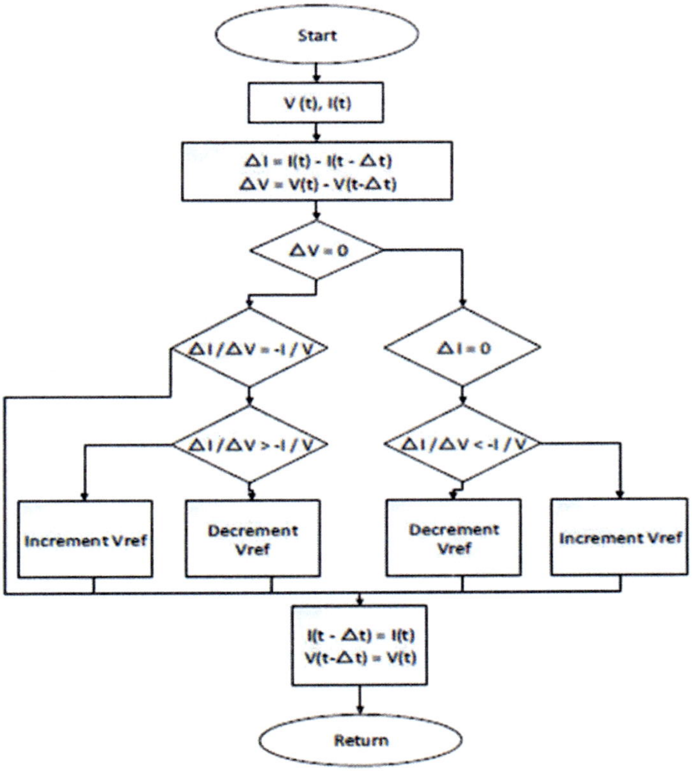

Fig. 4 Flowchart of INC-based MPPT

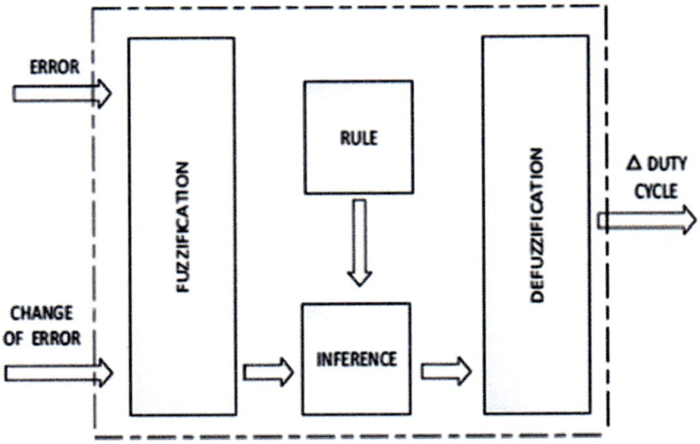

Fig. 5 Block diagram of fuzzification process

into an arithmetical output. The inputs are *E* and *CE* as shown in Fig. 6a, b. All membership functions are classified into eight parts: NB is negative big, NM is negative medium, NS is negative small, ZO is zero, PS is positive small, PM is positive medium and PB is positive big. The *MAMDANI* inference system (FIS) is employed to generate the FLC (Fig. 5; Table 1).

Execution of a fuzzy logic-based MPPT in MATLAB is shown in Fig. 7.

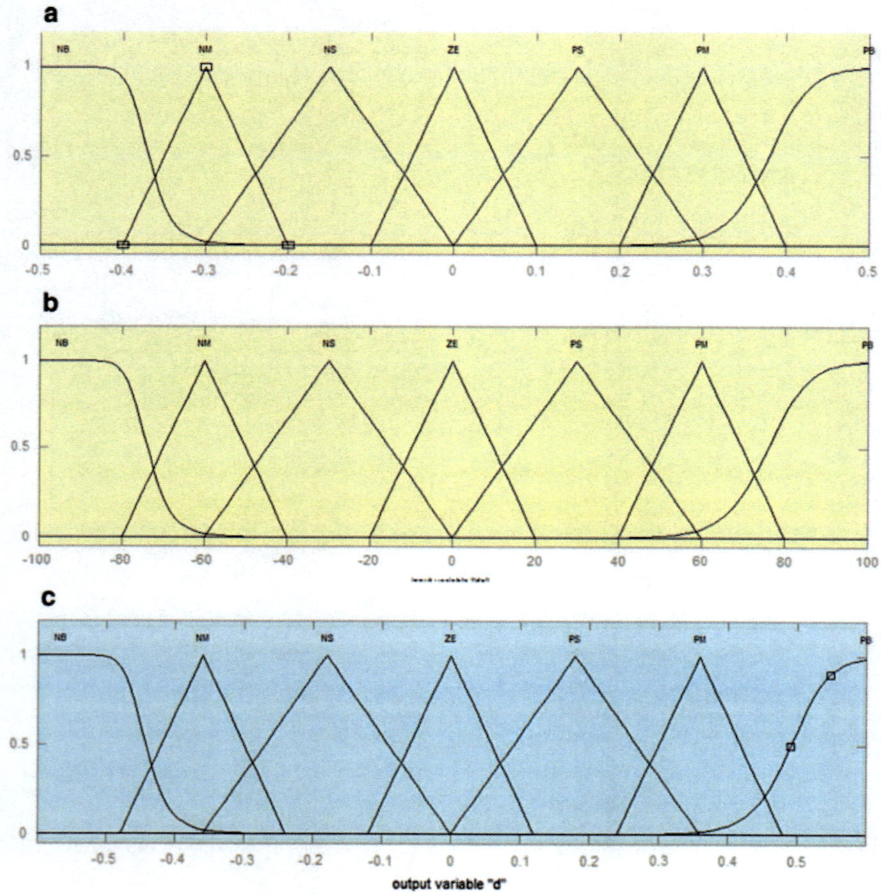

Fig. 6 **a** Membership functions of error (*E*). **b** Membership functions of change in error (*CE*). **c** Membership functions of change in duty cycle (dD)

Table 1 Membership functions

E, DE	NB	NM	NS	ZE	PS	PM	PB
NB	ZE	NM	ZE	NM	NS	PS	PB
NM	NM	ZE	NM	NM	ZE	PS	PM
NS	ZE	NS	ZE	PS	PS	PS	PS
ZE	NM	NM	NS	ZE	PS	PM	PB
PS	NM	NS	NS	NS	ZE	PS	PM
PM	NM	NS	ZE	PS	PS	ZE	PM
PB	NB	NM	NS	NS	PS	PS	ZE

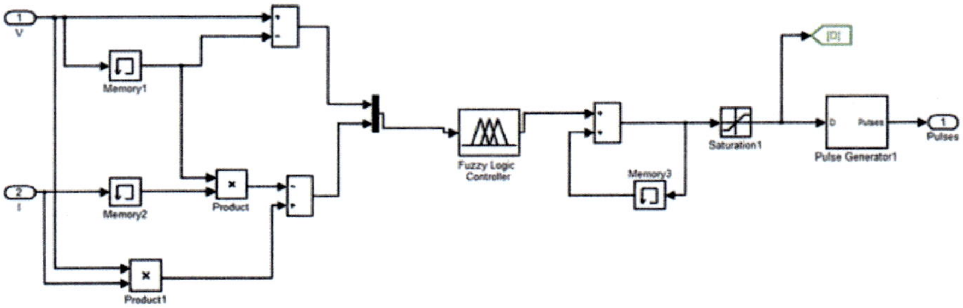

Fig. 7 Design of MPPT based on fuzzy logic controller

5 Simulation Results and Analysis

It is never easy to pick an algorithm for MPPT. At times, it depends on the need of the system. If the user is with the need of moderate complexity and oscillations, then incremental conductance technique can be considered. If fast convergence with very less oscillations is the need of the hour, then fuzzy logic is much suited. The above- discussed algorithms incremental conductance and fuzzy logic-based power point tracker are implemented in a simulation model in Simulink/MATLAB for the validation, and the results are compared.

Figure 8 shows the THD content of current which is being injected into the grid along with the power when incremental conductance MPPT is used. From figure, we can see that the injected current has harmonic content (4.23) which is under specified IEEE-519 standard.

Fig. 9 shows the power that is injected into the grid and that is the reason which is in the negative phase.

The above done work is compared with the proposed fuzzy logic controller, and its current and power graphs are shown below. From Figure 10, it is clear that the injected current is having harmonic content (2.90) (THD) which is under IEEE-519 specified limit. Figure 11 shows the power that is injected into the grid and that is the reason which is in the negative phase.

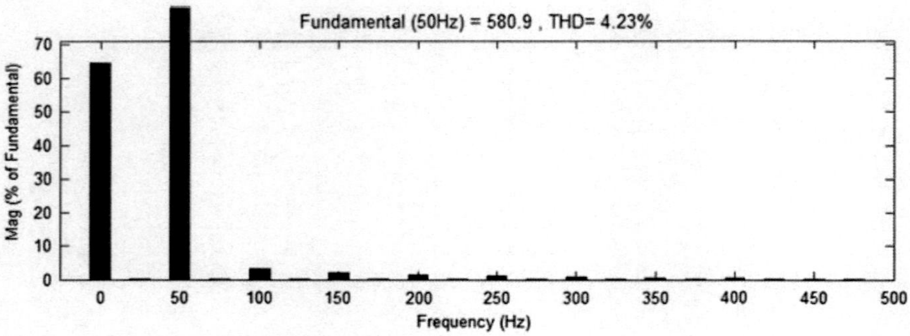

Fig. 8 THD content in injected current

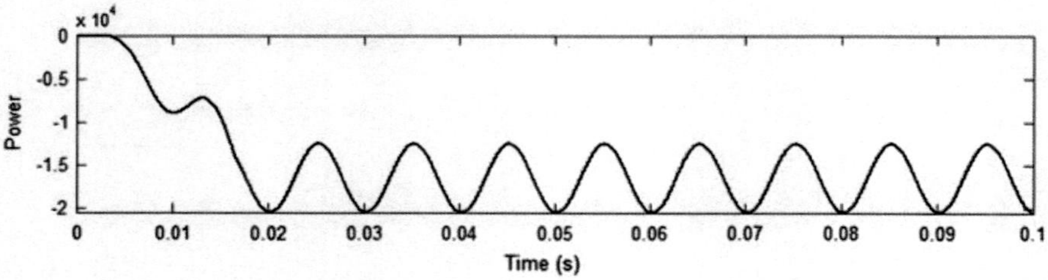

Fig. 9 Waveform of power injected into the grid

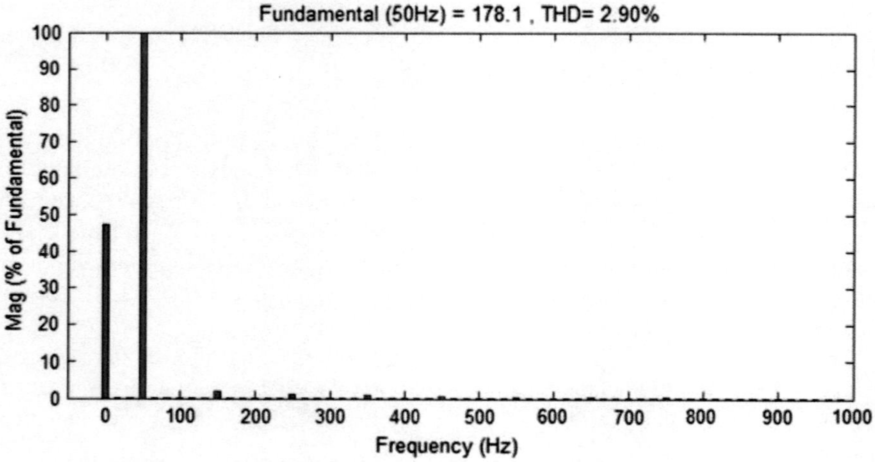

Fig. 10 THD content in injected current

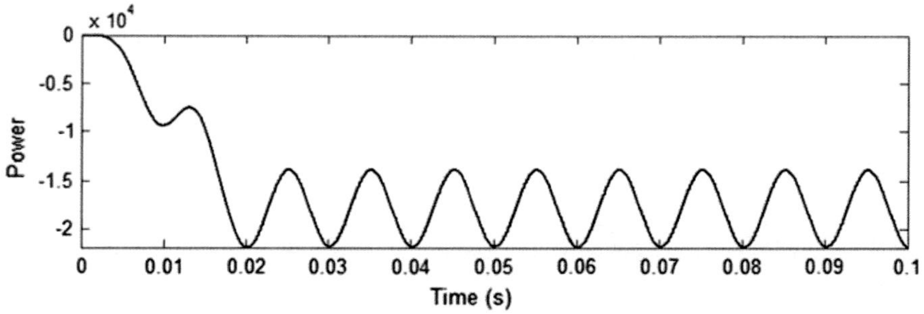

Fig. 11 Waveform of power injected into the grid

6 Conclusion

In this paper, incremental conductance MPPT method is compared with proposed fuzzy logic controller, and in the comparison, THD content (2.9%) is generated by using fuzzy-based MPPT which is less to that in the harmonic content (4.23%) generated by using incremental conductance. In both the cases, THD is under the specified limit of IEEE-519, but fuzzy logic controller-based MPPT has less THD content and is much better than that of incremental conductance.

References

1. Raj MP, Joshua AM (2017) Modeling and performance analysis of perturb & observe, incremental conductance and fuzzy logic MPPT controllers. In: IEEE international conference on advances in electrical technology for green energy (ICAETGT'2017)
2. Verma AK, Singh B, Shahani DT (2012) Grid interfaced solar photovoltaic power generating system with power quality improvement at AC mains. In: Proceedings of IEEE third international conference on sustainable energy technologies (ICSET), Sept 2012, pp 177–182
3. Fortunato M, Giustiniani A, Petrone G, Spagnuolo G, Vitelli M (2008) Maximum power point tracking in a one-cycle-controlled single-stage photovoltaic inverter. IEEE Trans Ind Electron 55(7):2684–2693
4. Wang Z, Fan S, Zheng Y, Cheng M (2012) Control of a six-switch inverter based single-phase grid-connected PV generation system with inverse park transform PLL. In: Proceedings of IEEE international symposium on industrial electronics (ISIE), pp 258–263
5. Ghodke DV, Chatterjee K, Fernandes BG (2008) Modified one-cycle controlled bidirectional high-power-factor AC-to-DC converter. IEEE Trans Ind Electron 55(6):2459–2472
6. Wu T-F, Shen C-L, Chang C-H, Chiu J (2003) 1φ 3 W grid-connection PV power inverter with partial active power filter. IEEE Trans Aerosp Electron Syst 39(2):635–646
7. Deo S, Jain C, Singh B (2015) A PLL-less scheme for single-phase grid interfaced load compensating solar PV generation system. IEEE Trans Ind Inf 11(3):692–699
8. Priyanka M, Mittarpal A (2016) Modeling, simulation and analysis of solar photovoltaic system with maximum power point tracker. Int J Adv Res Comput Sci Softw Eng 6(9)

A Real-Time Audio Transmission and Reception Over Wireless Channel Using PXI System

M. Padmaja, K. Prasuna, and K. Murali

Abstract Audio signal transmission is a significant part of wireless communication. There has been a great interest in transmission and reception of audio/video signals over wireless links for future mobile radio communication. In this paper, a simple VI has been developed for both transmission and reception of audio signal over the wireless channel by using LabVIEW and PXI systems (PXI-1065). The proposed scheme is based on real-time modulation and demodulation of the audio signal with LabVIEW and configuration of PXI module in real time.

Keywords Audio transmission and reception · LabVIEW · PXI system

1 Introduction

Nowadays, the advance of communication technology has given growth to a massive demand on high-performance communication systems. Audio transmission is considered to be one of the important applications in recent communication trends. In the recent years, the quality of audio has gained importance in the increasing technologies. Transmission of audio over wireless channels has achieved wide challenges. Therefore, to meet the demands, it is necessary to develop and implement high-quality transmissions with the latest devices as required, to transmit audio complexity should be reduced.

The concept of power plays a vital role in decreasing the noise and enhancing the quality of audio transmission over wireless channels. The system model classically involves a device of measuring the excellence of the channel seen by the receiver and provided that such in sequence to the transmitter to adjust the amount of transmitted

M. Padmaja
Department of Electronics Communications Engineering, VR Siddhartha Engineering College, Vijayawada, Andhra Pradesh, India

K. Prasuna · K. Murali (✉)
Department of Electronics Communications Engineering, Vijaya Institute of Technology for Women, Vijayawada, Andhra Pradesh, India
e-mail: kalipindimurali@gmail.com

P. S. R. Chowdary et al. (eds.), *Microelectronics, Electromagnetics and Telecommunications*, Lecture Notes in Electrical Engineering 655,
https://doi.org/10.1007/978-981-15-3828-5_61

power. Most challenging tasks involve super signal acquisition, scaling and multiple visualizations.

Virtual instrumentation gives new potential for traditional acoustic capacity applications. Using the influence of an industry-standard computer and the flexibility of LabVIEW, custom audio measurement can be performed.

LabVIEW, PXI system [1] and instrument driver toolset are used to obtain, and analyze nearby audio data. The most common measurements are obtainable as well as the LabVIEW code to perform numerous responsibilities of measuring audio.

Recent software and hardware trends make influential engineers to analyze many aspects of a sound signal. Software programming like LabVIEW provides ease of flexibility, presentation and powerful functionality to make complex measures in very low time duration.

This paper provides the steps to expand an audio transmission and reception system based on LabVIEW industry software that delivers better efficiency while on the condition of separation of errors [2]. Recent audio transmission is among the most challenging operations for a cellular phone system. To carry out unbeaten audio transmission in real time, the software must be talented to execute a number of responsibilities. National instruments hardware and software are put together effortlessly to replace many devices and offer much more awareness to users.

In this paper, a simple audio transmission and reception VIs by using modern programming software Laboratory Virtual Instrument Engineering Workshop (LabVIEW) is proposed, with which PXI systems can be used in order to transmit an audio signal. The proposed system model and the utilization of VIs in LabVIEW and configuration of PXI modules, simulation results and conclusions are presented.

2 System Description

The block diagram considered is a FM modulation and demodulation communication system for transmission and reception in real time. Initially, the audio signal is taken by considering only one channel for audio data and then going for modulation, next to that, initializing the vector signal generator and going for configuration of an arbitrary waveform and finally, transmitting data in real-time using PXI system as shown in Fig. 1.

When we are talking about the real-time environment, noise is one of the constraints which can disturb both power and bandwidth. In order to avoid the noise,

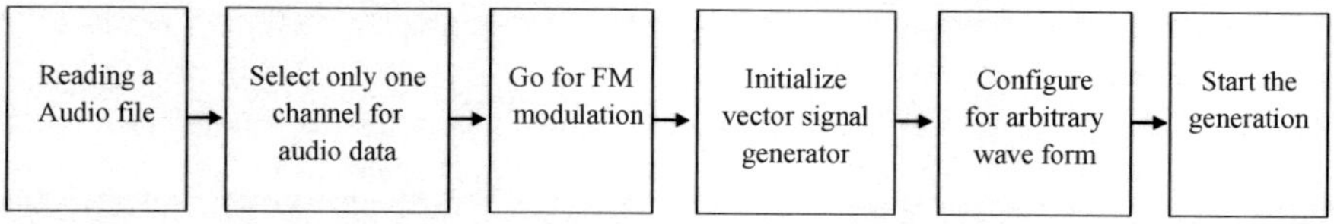

Fig. 1 Block diagram of the proposed audio transmission scheme

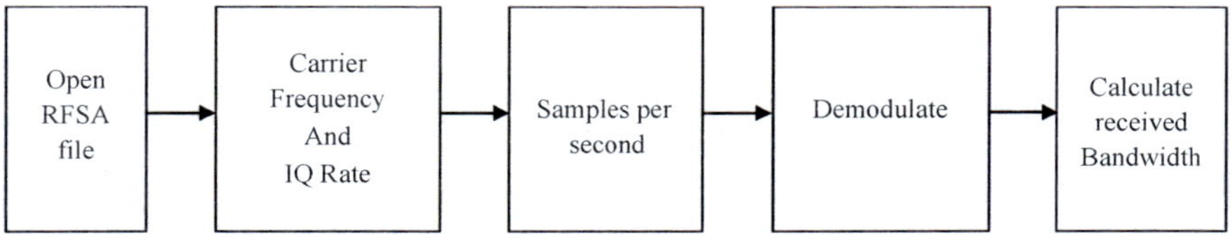

Fig. 2 Block diagram of the proposed audio reception scheme

both RFSG and RFSA are configured on single system by which we are going to transmit the complex waveform through the wireless channel using PXI systems. The received signal is processed using a RF signal analyzer (RFSA) session. In a real-time environment, configuration of reference clock, carrier frequency, IQ rate and samples per second are considered as important aspects. Demodulation and bandwidth calculation are performed. The sequence of steps is illustrated in Fig. 2.

Peripheral Component Interconnect (PCI) eXtension Instrumentation (PXI) is a strong PC-based device. In this paper, RF communication modules (PXIe-1065) are used in order to transmit the audio signal. PXIe-1065 is capable of transmitting the signals over a wide range of frequencies, i.e., 500 kHz to 6.6 GHz. Even though PXI costs are expensive, there is no need of designing a new hardware again and again for a specific application, and even it is easy to build the hardware by using different modules.

3 VIs and Hardware Modules

LabVIEW is a text-based programming mostly based on code. This is a user-friendly software which uses icons as visuals to highlight subroutines which are used to represent the flow of the program. PXI is closely associated with the integration of architectures. These architectures are to the requirement of industry needs. In order to reduce time for building codes for any system, one should prefer for special tools like LabVIEW rather than using regular tools [3, 4]. For example, if we want to build a code for image to array, there is a need to spend some time for writing codes. Whereas, LabVIEW tools are made up of inbuilt VIs and modules [5]. These tools help the user to build the code as fast as possible. Instead of writing the code, we can place the VI block as shown in Fig. 3.

PXI is having two types of modules: software and hardware.

Blocks used for transmission of audio signal using PXI: In order to build the block diagram, several VIs have been used [6]. Some of the important VIs like MT modulation FM, Configure Generation Mode, Resample and write are mentioned below.

MT modulate FM VI: This module performs frequency modulation on an RF carrier wave according to a baseband information signal passed to the message signal

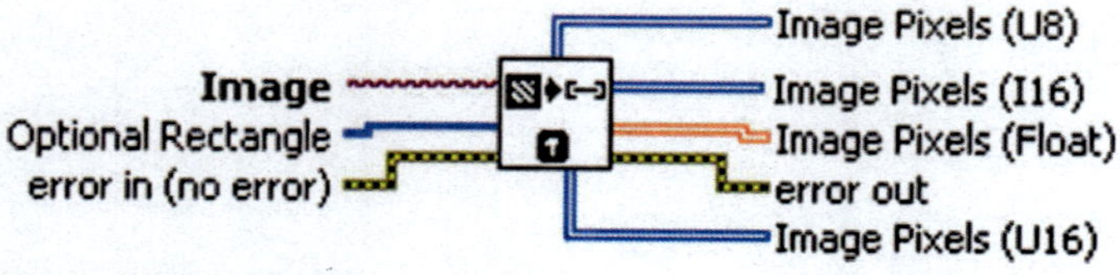

Fig. 3 VI for the image to array conversion

parameter. The complex envelope of the FM modulated signal in the FM modulated waveform parameter can be observed as shown in Figs. 4 and 5.

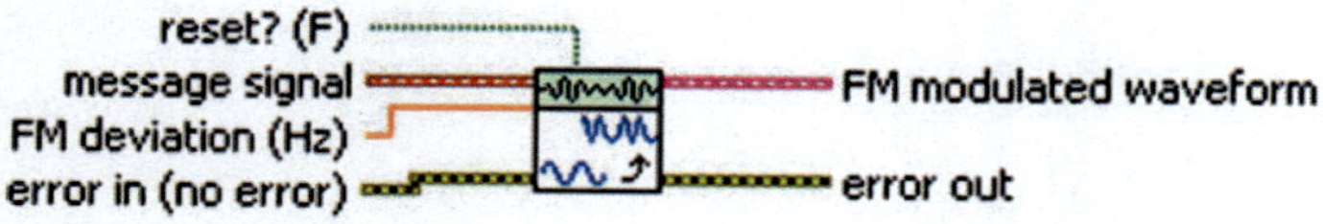

Fig. 4 VI for the modulation of FM

Fig. 5 VI to generate a sine tone

Resample and write: Given a phase continuous complex waveform, this VI preserves the phase continuity while resampling. It optimizes memory usage by doing the resampling and writing in blocks of 100 k samples and by de-allocating unused memory after the VI is executed as shown in Fig. 6.

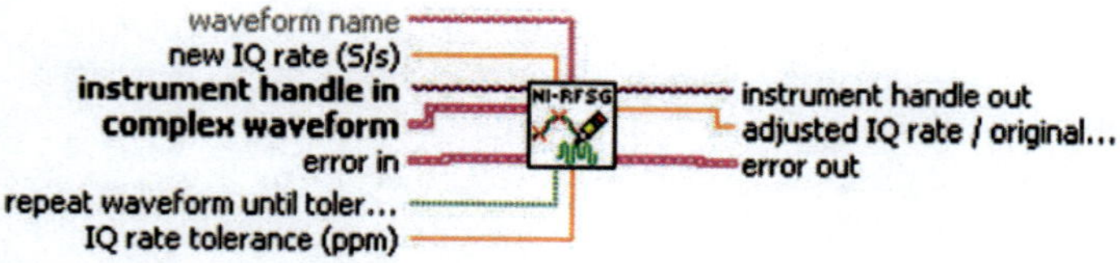

Fig. 6 VI to resample and write the complex waveform

4 Blocks Used for Reception of Audio Signal Using PXI

Several blocks for the reception which are available in the PXI are listed in this Section as follows:

A. **Demodulate FM**: Frequency demodulation on the received I/Q signal is performed. The improved sequence signal is returned in the FM demodulated waveform parameter. The illustration is mentioned in Fig. 7.

B. **Read IQ**: Initiates an acquisition and fetches a single I/Q data record. Do not use this VI if you have configured the device to continuously acquire data samples or to acquire multiple records. The representation of the block is given in Fig. 8.

C. **Modules used for PXIe-1065**: Even though there are many other tools where we can replace our code with simple VI blocks or in build commands, the reason for using LabVIEW is the hardware configuration of PXI systems can be implemented only through this software tool. PXIe-1065 is a RF communication specified hardware ranging from 500 kHz to 6.6 GHz. The important modules used for configuring the hardware are mentioned below.

NI PXIe-5641R IF Transceiver: It consists of two IF inputs and two IF outputs. It has an extra future of built-in digital up converter and down converter as shown in Figs. 9 and 10 [7–10].

A small audio signal has been transmitted using radio frequency signal generator (RFSG) which has been configured using PXI systems [8]. Figure 11 shows the transmission of a small audio signal over the wireless channel at 1G with a power level of −30 dBm. The audio data has been shown at the bottom of figure on a graph.

Figure 12 shows reception mode through radio frequency signal analyzer (RFSA) which has been configured using chassis of the PXI system. The spectrum can be seen on screen with both sidebands at 1G along with a spam of 20M. Figure 13 follows the same result but without any sidebands. Figure 14 shows the result of the received signal in a data form (message form) and even with power spectrum calculation. Figure 15 follows the same but this time with RMS calculation.

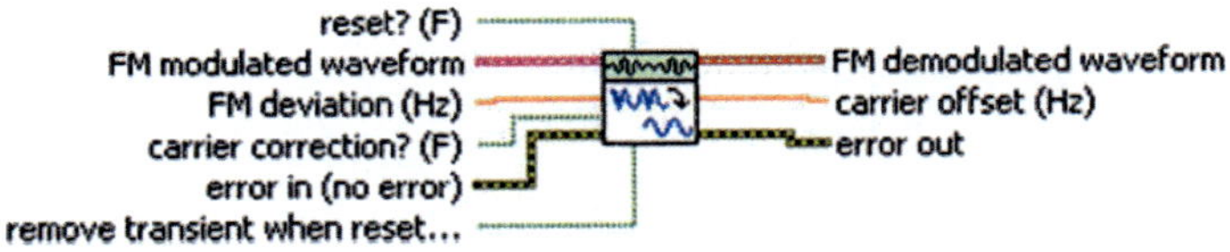

Fig. 7 VI for demodulation of FM signal

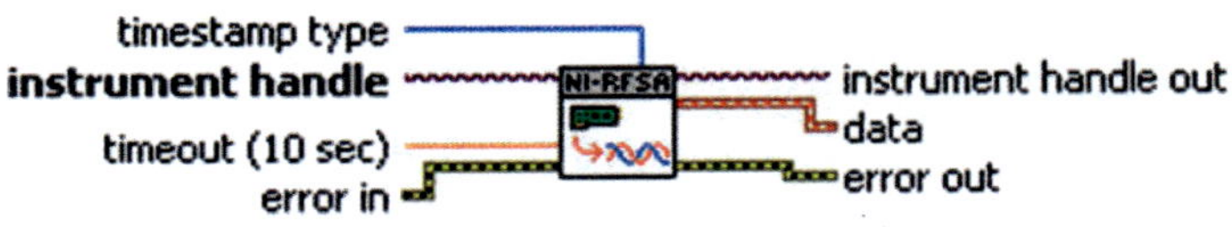

Fig. 8 VI to fetch a single I/Q data

Fig. 9 Hardware module of the IF transceiver

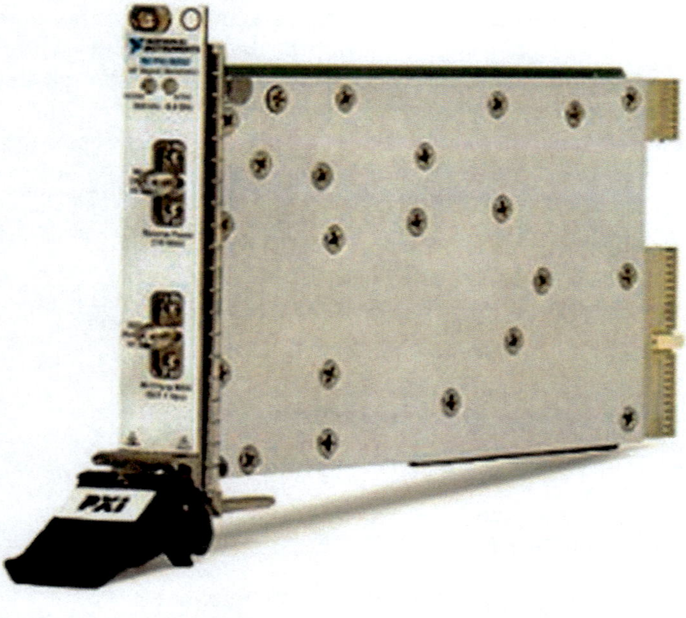

Fig. 10 Hardware module of the signal generator

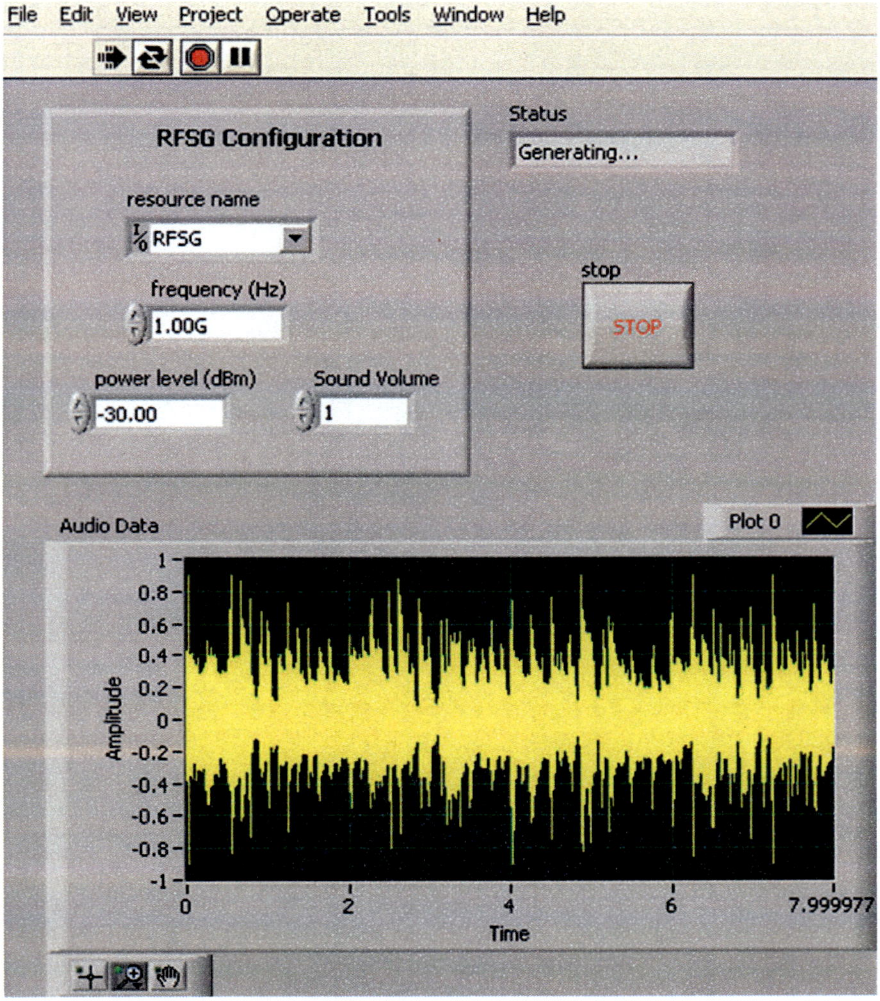

Fig. 11 Transmission of the audio signal at 1G

5 Conclusion

In real-time modulation and demodulation of audio signal with LabVIEW and configuration of PXI module, the received signal shows a better quality signal with less distortion. The original transmitted audio signal and the signal in RFSA with and without RMS are illustrated with block diagrams. The reception of sound signal other than 1G and reception of sound signal at 1G with RMS are also demonstrated. Optimization of data and transmission over wireless channels in real-time environment are advent and widely developing area. It also has a high scope in future for image,

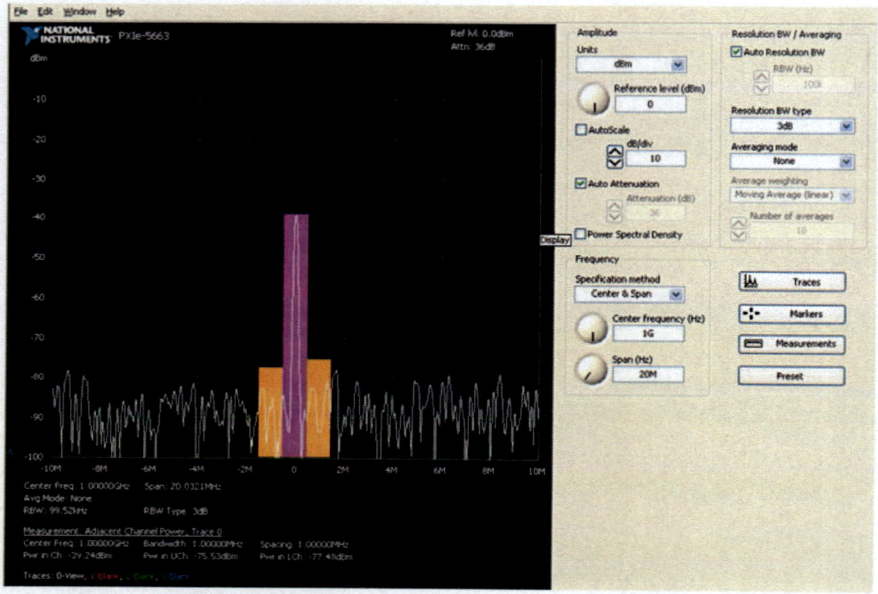

Fig. 12 Received signal in RFSA with sidebands

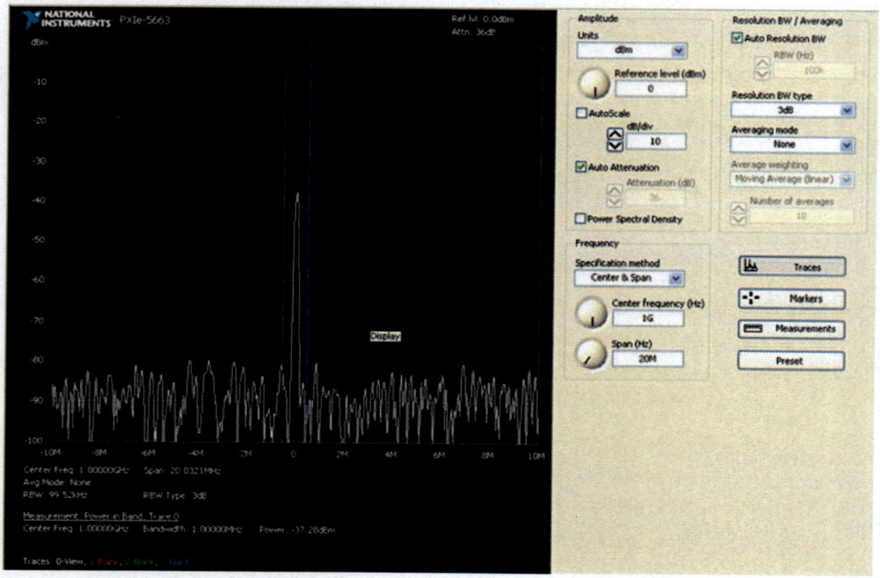

Fig. 13 Received signal in RFSA

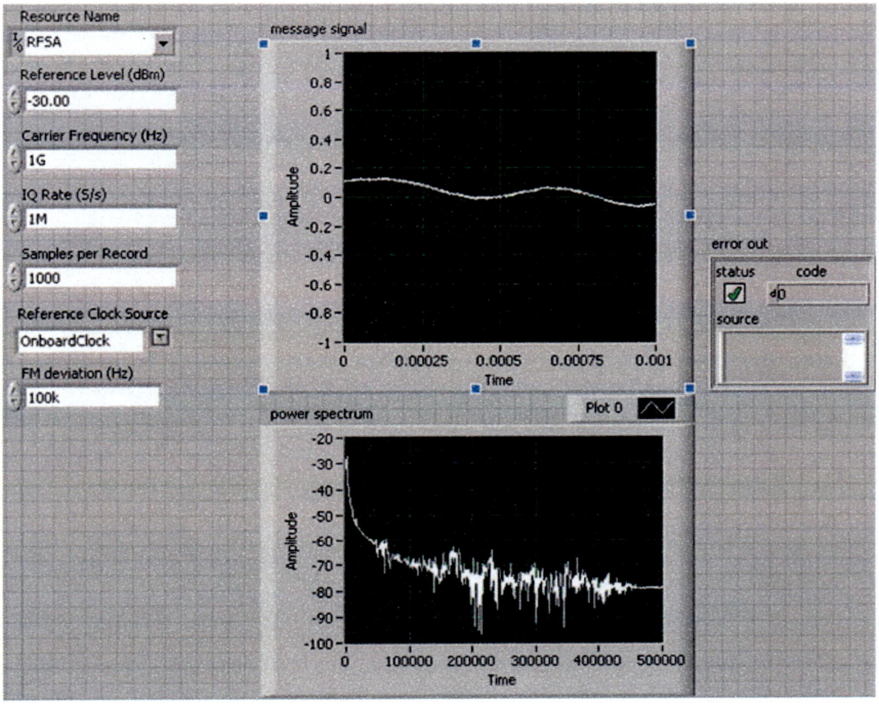

Fig. 14 Reception of the sound signal other than 1G (90M)

video transmission over wireless channels, especially for high profile systems. Power allocation and coding techniques can play a major role for transmission of a better quality of data and quality with high gain.

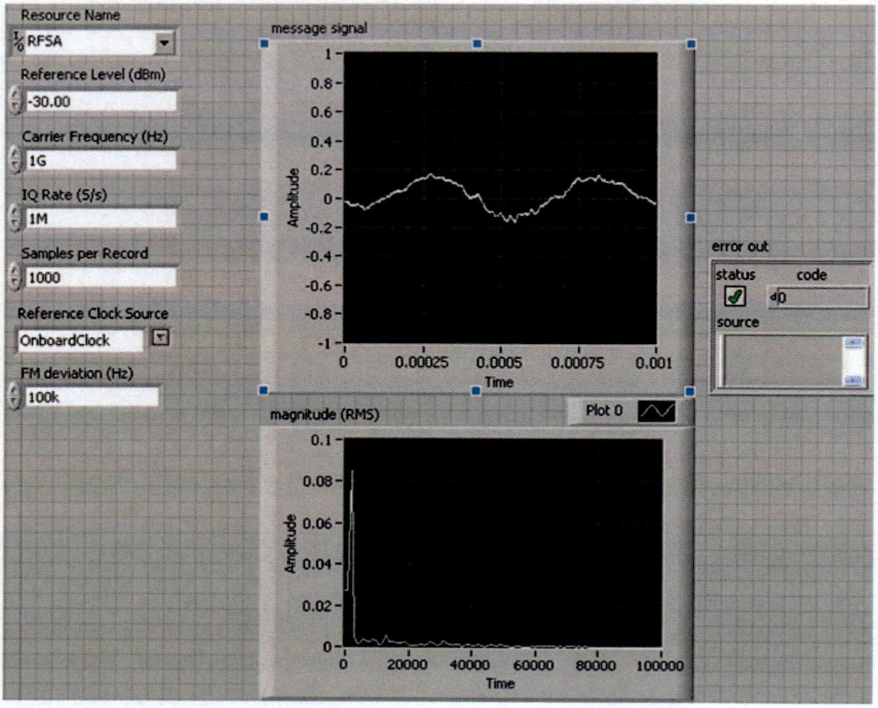

Fig. 15 Reception of the sound signal at 1G with RMS

References

1. Mori T, Kaneko T, Moriya T, Ikeda K (2001) A real-time IMT-2000 audio transmission system. IEEE Trans Consum Electron 47(4):860–866
2. Khalifeh A, Yousefi'zadeh H (2010) Optimal audio transmission over error-prone wireless links. IEEE Trans Multimedia 12(3):204–214
3. Mohan Y (1995) Test system for mobile field-strength recording. Electromagnet Interference Compat 242–245
4. Tikander M (2005) Sound quality of an augmented reality audio headset. In: Proceedings of the 8th international conference on digital audio effects (DAFx-05), Madrid, Spain
5. Boxer S (2005) Art that puts you in the picture, like it or not. New York Times, 27 Apr 2005
6. Tikander M, Karjalainen M, Riikonen V (2008) An augmented reality audio headset. In: Proceedings of the 11th international conference on digital audio effects (DAFx-08), Espoo, Finland
7. Härmä A et al (2004) Augmented reality audio for mobile and wearable appliances. J Audio Eng. Soc. 52(6):618–639
8. Open Sound Control home page. http://opensoundcontrol.org
9. Tan Y, Xu X, Liang C, Zhang X, Zhang Q, Li Y (2018) An end-to-end covert channel via packet dropout for mobile networks. Int J Distrib Sens Netw 14(5)
10. Paladuga CS, Vedula CV, Anguera J, Mishra RK, Andújar A (2018) Performance of beamwidth constrained linear array synthesis techniques using novel evolutionary computing tools. Appl Comput Electromag Soc J 33(3):273–278

A Safe and Cost-Effective Algorithm for Automation of LPG Cylinder Booking Using ESP8266

Srinivasa Naidu Nalla and K. V. Gowreesrinivas

Abstract Toward a safe kitchen and stress-free LPG cylinder booking system especially in the case of illiterate households, we are proposing a cost-effective method for automation of LPG cylinder for both LPG consumer and LPG dealer along with gas leakage alert. In the current scenario of the LPG booking system, human intervention is required at the consumer end to book a cylinder. In this paper, a ESP8266 (System on Chip with onboard Wi-Fi from Espressif Systems) microcontroller with its inbuilt Wi-Fi through Internet updates the percentage of gas cylinder weight in a designated pre-designed Web site maintained by concerned LPG dealer, which makes LPG cylinder booking completely automatic. The ESP8266 microcontroller employs a load cell, whose objective is to calculate the force exerted by the LPG cylinder and generate an equivalent analog signal. This system also looks out for any gas leakage and warns the household the same.

Keywords ESP8266 microcontroller · Load cell · Booking of LPG cylinder · MQ6 sensor · Networking of things · Spillage of gas

1 Introduction

Networking of things brought a lot of comfort to the households in terms of home automation, etc. Active LPG connections in India are around twenty-two crores. In the current scenario, human intervention is required to refill a LPG cylinder. In this paper, we are proposing an algorithm to make LPG cylinder refilling completely automatic with help of Internet of Things and other technologies along with gas leakage detection.

S. N. Nalla (✉) · K. V. Gowreesrinivas
Department of Electronics and Communication Engineering, ANITS, Visakhapatnam, Andhra Pradesh, India
e-mail: srinivasanaidun@gmail.com

K. V. Gowreesrinivas
e-mail: srinu43306@gmail.com

P. S. R. Chowdary et al. (eds.), *Microelectronics, Electromagnetics and Telecommunications*, Lecture Notes in Electrical Engineering 655,
https://doi.org/10.1007/978-981-15-3828-5_62

Earlier methods have been proposed to detect gas leakage [1–5]. In-addition to natural gas detection, MQ6 gas sensor can also responds rapidly to butane and propane where as MQ4 [1] and MQ5 [2] responds to natural gas only. Alerting with buzzer [1] and Global System for Mobile Communications (GSM)-based Short Message Service (SMS) is implemented in case if any gas leakage is detected [2, 7]. Status of gas content is sent to user thru SMS to LPG dealer by using GSM under the control of Arduino Uno development board [7, 8]. The detail of gas content is updated to a Web site using the Internet [9] with the help of a wired data connection shoots up the overall system cost.

2 Proposed System

The hardware requirements of the LPG cylinder booking system along with safety are illustrated in Fig. 1. The proposed work is discussed in two fronts. One is at LPG consumer end and another one is LPG dealer end.

2.1 LPG Consumer End

Weighing the cylinder: The gas cylinder is placed on the load cell, which will convert the force exerted by cylinder into an analog signal, which further feeds to an analog to digital converter (ADC) like HX711. The digital output produced by ADC is compared with a preset value by ESP8266 (System on Chip with onboard Wi-Fi from Espressif Systems). If the compared value is less than the preset value,

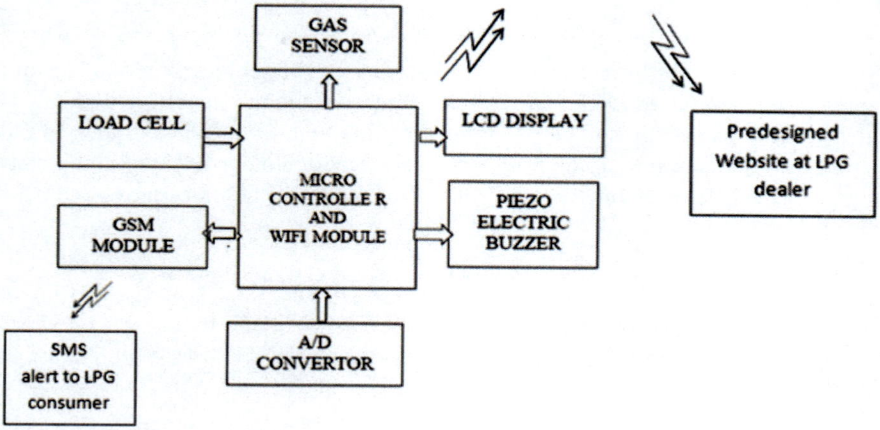

Fig. 1 Architecture of automated LPG booking system

the ESP8266 immediately triggers two actions, viz. sending SMS to consumer and activating Web site access as shown in Fig. 2.

Gas spillage detection: The LPG gas spillage from the cylinder is detected by using a MQ6 gas sensor, which is placed in the cylinder vicinity. The algorithm of leakage detection is shown in Fig. 2. MQ6 sensor generates an analog voltage whenever it

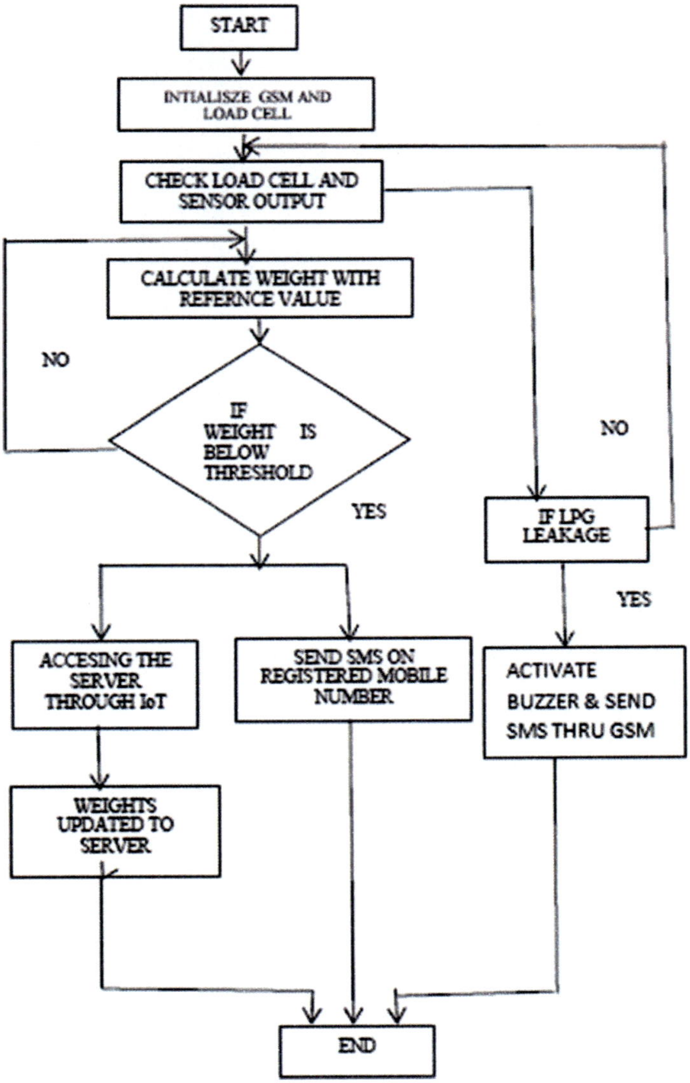

Fig. 2 Flowchart of proposed system at consumer

detects a gas spillage, and this is further fed to an analog input of ESP8266. Now ESP8266 will not only activate the buzzer but also send an alert SMS thru GSM.

2.2 LPG Agency End

At the LPG agency (LPG cylinder delivery office), a Web site has been designed to update the weights of each consumer. As indicated in Fig. 3, each consumer module will access the Web site and update their weights when the weight is below

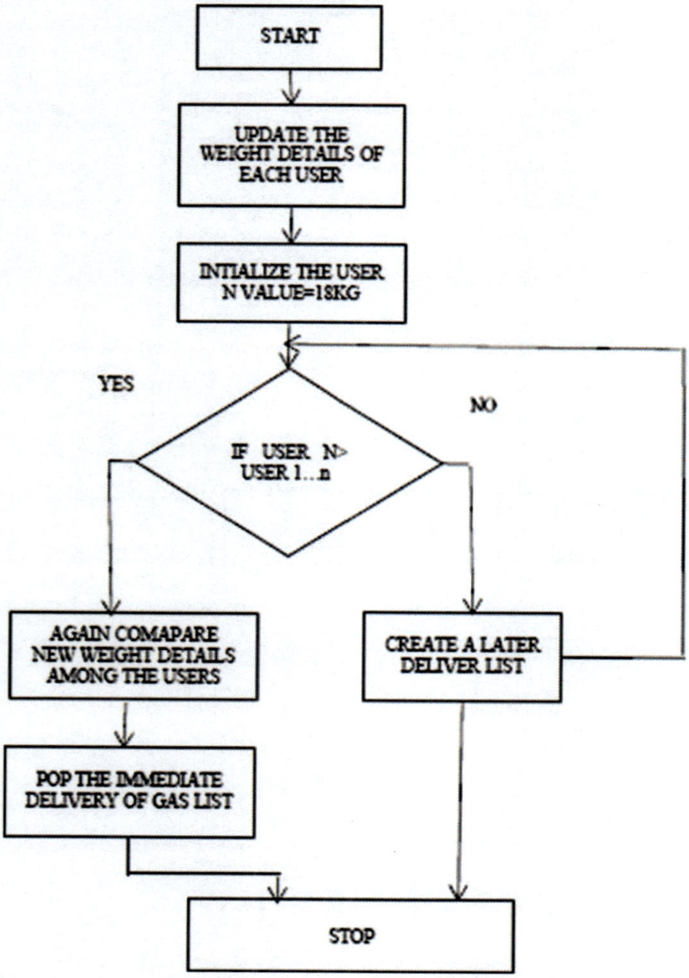

Fig. 3 Flowchart of proposed system at agency side

Table 1 Information of weights of different user and accessing of Web based on weights

Users (1, 2, …, *N*)	If weight is above the threshold value (threshold < 20%)	If the weight is below the threshold value (threshold > 20%)	Web accessing by the client
User1	No	Yes	Yes
User2	Yes	No	No

the threshold value. As indicated in Table 1, initially the weight of each user is compared and a list is maintained for each user separately. When the weight reaches below to a predefined critical value, for example 10% of total weight of cylinder, those user ID's are moved to a new window named as 'Immediate delivery'.

Employing ESP8266 will make the proposed system cost-effective as it costs only $7 when compared to Arduino Uno [8, 9] which is $23 around. Arduino Uno requires an external data connectivity module such as an "Ethernet shield" costing 24 $ or a Wi-Fi module (costs around $3). The proposed system is simplified the framework required at agency side along with effective cost, which is shown in Fig. 3.

3 Results

The results are organized into two subsections, viz. weight measuring—booking cylinder and gas spillage detection.

3.1 *Weight Measuring*

The system initially with no cylinder is shown in Fig. 4. Since no cylinder is placed on the load cell, the load cell shows 0 kg, and the same is updated to pre-designed Web site as described in Sect. 2.1. This is illustrated in Fig. 5, as −0.01 kg which is a null value. This is a screenshot of the Web site indicating the weight detail of chosen consumer.

Now LPG cylinder is placed on the load cell carefully. In place of the domestic cylinder for practical purpose, some weighing stones weighing 20.3 kg have been placed as shown in Fig. 6a. As the weight is above the threshold value, it does not send any message to the user. The same weight detail is communicated to Web site, hence Web site is displaying the weight as 20.30 kg as shown Fig. 6b.

Now, as shown in Fig. 7a, weights have been removed (3 kg) to resemble the gas usage in the LPG cylinder. The same is updated to Web site, hence Web site is displaying 17.3 kg. As per the proposed system, this is the "threshold value-20% of total weight." Hence, an SMS alert is sent to user's predefined mobile number as shown in Fig. 7c.

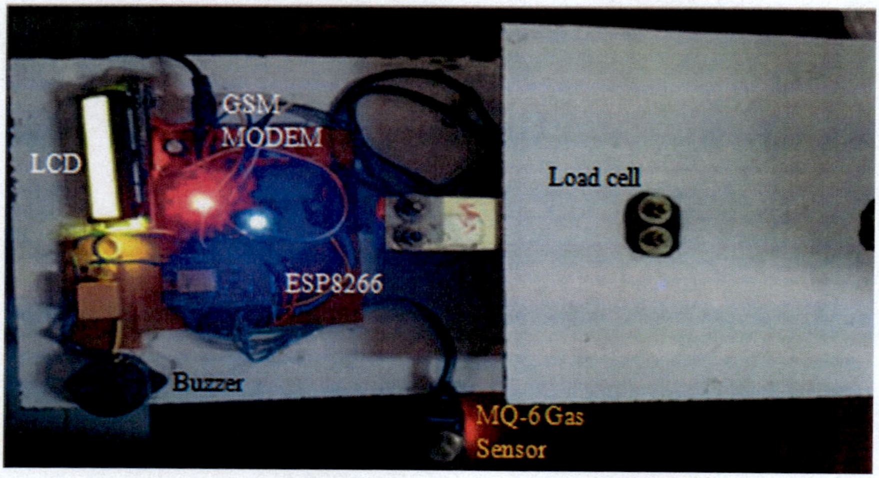

Fig. 4 Proposed architecture's hardware setup with "No Cylinder" being placed on load cell

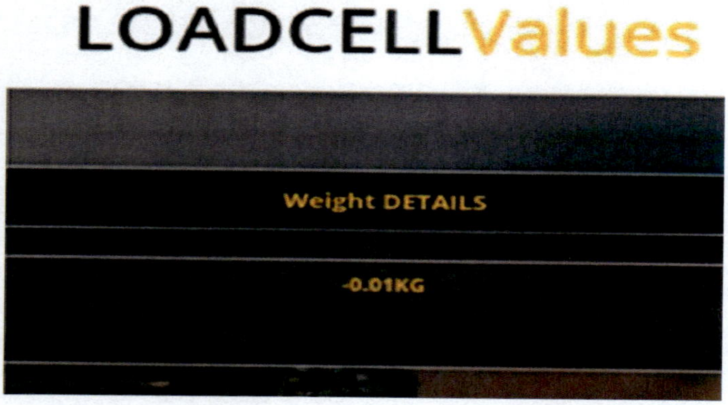

Fig. 5 Web site indicating a null value

Plot between LPG gas usage—gas cylinder weight against time as recorded by Web site in the chosen period is shown in Fig. 8.

Gas leakage: Here, MQ-6 gas sensor detects the gas leakage. Here, we created a gas leakage by using a domestic lighter. After detecting gas spillage by gas sensor, it automatically enables buzzer and sends a "GAS LEAKAGE" alert SMS to the registered number using GSM as depicted in Fig. 9.

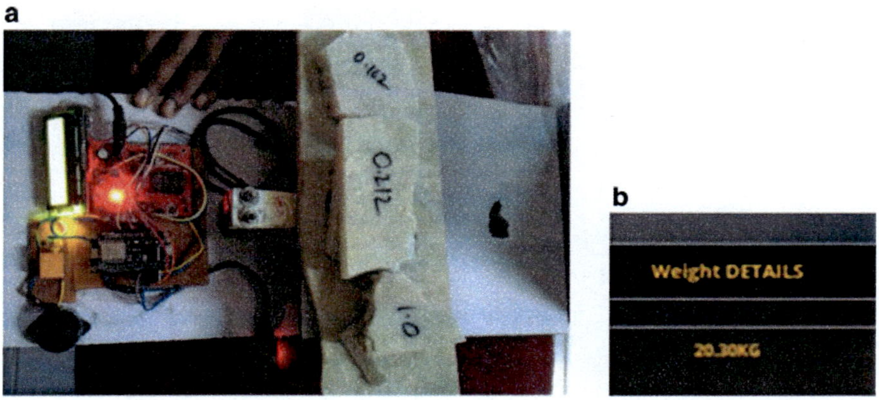

Fig. 6 **a** Placing of 20.3 kg weight on the load cell. **b** Web site showing "20.30 kg"

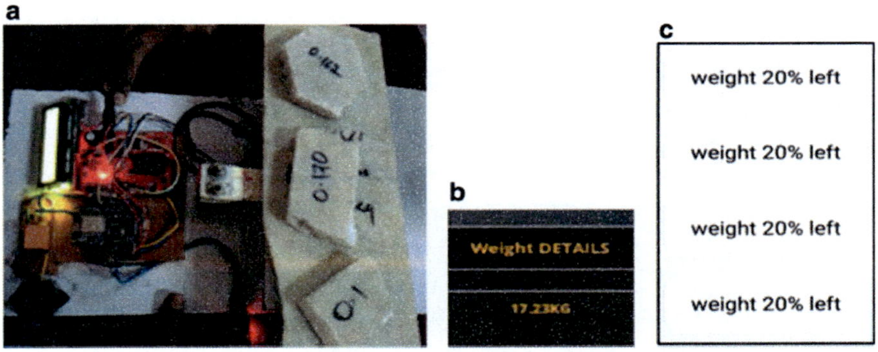

Fig. 7 **a** Some weights are removed, **b** Weight details from the Web site "17.3 kg", **c** SMS alert received by the user

4 Conclusion

The proposed system integrates the household cylinders with the gas agency through a Web site. Automation is done in this paper for gas cylinder booking. The ESP8266 is interfaced with weight sensing equipment to measure the weight of gas cylinder thru an ADC. The measured weights are continuously updated to the Web site with the help onboard Wi-Fi using Internet. Thereby at the agency Web site, a pop-up window is created based on the threshold value. Now, gas cylinders are delivered to the list of users in this window. This system not only automates the gas booking system but also detects the gas spillages through the gas sensor and alerts the user with a buzzer and SMS.

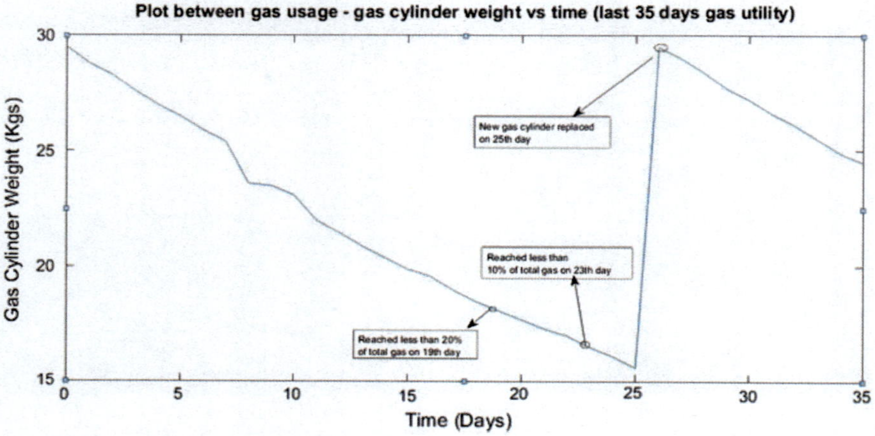

Fig. 8 Plot between LPG gas usage—gas cylinder weight versus time in the chosen period

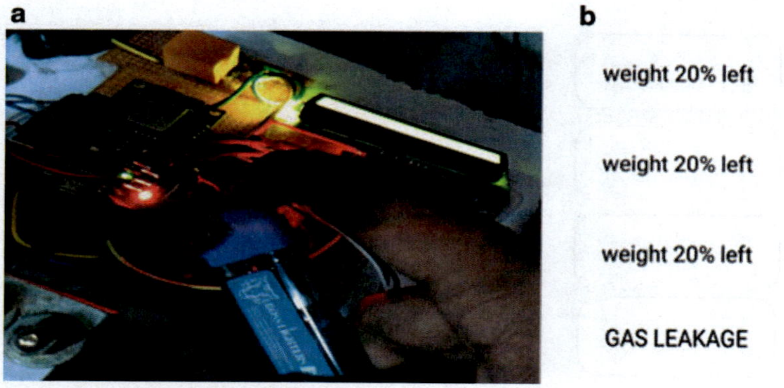

Fig. 9 **a** Gas leakage detection using the proposed system, **b** SMS alert received by user

References

1. Mahalingam A, Naayagi RT, Mastorakis NE (2012) Design and implementation of an economic gas leakage detector. Recent Res Appl Electr Comput Eng 20–24
2. Macker A, Shukla AK, Dey S, Agarwal J (2018) Arduino based LPG gas monitoring and automatic cylinder booking with alert system. In: Proceedings of IEEE international conference on trends in electronics and informatics, pp 1209–1212
3. Chengjun D, Liu X, Ping D (2011) Development on gas leak detection and location system based on wireless sensor networks. In: Proceedings of 3rd international conference on measuring technology and mechatronics automation, pp 1067–1070
4. Fraiwan L, Lweesy K, Bani-Salma A, Mani N (2011) A wireless home safety gas leakage detection system. In: Proceedings of 1st Middle East conference on biomedical engineering, pp 11–14

5. Lopes dos Santos P, Azevedo-Perdicoúlis T-P, Ramos JA, Jank G, Martins de Carvalho JL, Milhinhos J (2010) Gas pipelines LPV modelling and identification for leakage detection. In: Proceedings of American control conference, pp 1211–1216
6. Technical data MQ6 gas sensor. www.sparkfun.com/datasheets/Sensors/Biometric/MQ-6.pdf
7. Padma Priya K, Surekha M, Preethi R, Devika T, Dhivya N (2014) Smart gas cylinder using embedded system. Int J Innov Res Electr Electron Instrum Control Eng 2(2):958–961
8. Naresh Naik R, Siva Nagendra Reddy P, Nanda Kishore S, Tharun Kumar Reddy K (2016) Arduino based LPG gas monitoring and automatic cylinder booking with alert system. IOSR J Electron Commun Eng (IOSR-JECE) 11(4):6–12
9. Tamizharasan V, Ravichandran T, Sowndariya M, Sandeep R, Saravanavel K (2019) Gas level detection and automatic booking using IoT. In: Proceedings of IEEE 5th international conference on advanced computing and communication systems, pp 922–925

Intelligent Traffic Signal Control System Using Machine Learning Techniques

Mohammad Ali, G. Lavanya Devi, and Ramesh Neelapu

Abstract Traffic congestion is a huge problem in almost every developing country as the people using private vehicles are increasing each day and the capacity of the road networks is still not up to the mark. Vehicular traffic problem is very common in urban areas as both private vehicles and other public transportation services are huge in number due to the dense population. This problem affects the functioning of the city. Every individual has to schedule his/her day within the 24 hours time limit. However, traffic volumes in urban areas kill potential time of the individuals. Also, huge amounts of fuel is wasted due to the increasing waiting time, particularly at signal points. Additionally, many urban areas are facing severe air pollution issues. This has very high impact on the health and well-being of the society. To address this issue, we need better and efficient infrastructure of the city and proper management of road traffic. Nowadays, the artificial intelligence (AI) and machine learning (ML) are playing an important role in solving many of the real-world problems. We may use these ML techniques to address road traffic management problem. As the manual maintenance is difficult and not sufficient with the increasing number of vehicles on roads, automation of traffic signal management with ML may result in better traffic conditions in urban areas. The idea is to divide the system into two phases. In the first phase, we classify the traffic signal junctions into one of the three different zones. High-level, medium-level, and low-level traffic zones. Support Vector Machine (SVM) algorithm is used for classification. In second phase, we optimize the signal configuration of high-level traffic zones to bring them to either medium-level or low-level traffic zones.

M. Ali (✉) · G. Lavanya Devi · R. Neelapu
Department of Computer Science and Systems Engineering, AU College of Engineering (A), Andhra University, Visakhapatnam, India
e-mail: md.ali401@gmail.com

G. Lavanya Devi
e-mail: lavanyadevig@yahoo.co.in

R. Neelapu
e-mail: rameshauvsp@gmail.com

P. S. R. Chowdary et al. (eds.), *Microelectronics, Electromagnetics and Telecommunications*, Lecture Notes in Electrical Engineering 655,
https://doi.org/10.1007/978-981-15-3828-5_63

Keywords Simulation of urban mobility (SUMO) · Traffic signal junction · Traffic control interface (TraCI) · Traffic zone

1 Introduction

Road transportation is one of the important modes of transportation in India which connects almost the entire nation. It has an important role in the economical growth of India. Though there are many advantages with road transportation, it causes a major problem in urban areas because of the insufficient capacity of road networks and incompatible manual management of traffic. Traffic congestion is more and more frequent in urban areas and it delays long destination transports and disturbs the daily lives of urban people.

Problems in Urban Road Transport System

Traffic Movement and Congestion: Traffic congestion becomes frequent when urban road transport networks are no longer capable of accommodating the volume of vehicles that use them.

Difficulties for Pedestrians: The majority number of victims of road accidents are pedestrians. The pathways are mostly blocked by parked cars and the increasing pollution of the urban environment with traffic noise, and exhaust fumes affecting most directly those on feet.

Traffic Noise: The major environmental problem caused by urban traffic is traffic noise. It is very annoying and disturbing. Walking and other activities in urban areas can be difficult and more importantly, traffic noise penetrates through the interior of buildings. Therefore, working hours becomes a frustration for workers since noise disturbs concentration. Domestic life as sleeping and relaxation also become affected.

Atmospheric Pollution: Traffic fumes emitted from poorly maintained vehicle engines can be very offensive and contribute to the unpleasantness of life in urban areas. Urban streets are polluted with the effects of engine fumes and these almost certainly contribute toward the formation of smog [1].

These problems can be solved with the help of AI and ML. ML is a part of AI, in which we can teach a system to learn from data. These learning will help in taking appropriate decisions to different situations. ML has already proven itself worth in many real-world applications. We will use these ML techniques to address our road traffic issues.

The idea is to classify all traffic signal junctions into different classes. So that, each traffic signal junction would belong to one of the high traffic zone, medium traffic zone, or low traffic zones. And once we have divided them, we will try to bring down the high traffic at high traffic zones to medium or low traffic which results in decreased number of high traffic zones.

Classification is supervised ML in which a model is trained with some training data with known labels, and then the model will be able to label new observations with unknown labels. These labels are classes that each observation belongs to.

As the AI and ML became more powerful with the latest development of memory capacity and powerful processors, many researchers started proposing solutions to many kinds of real-world problems and successfully applied them to a variety of fields. Use of the road transport system has limitations with inefficient maintenance techniques. AI and ML can be used to remove these limitations and to provide much better and more sophisticated maintenance techniques.

SUMO tool was created to observe and solve the road transport problems. A group of researchers explained about the traffic simulation in SUMO in a paper "Microscopic Traffic Simulation using SUMO" at International Conference of Intelligent Transportation Systems in the USA. They have discussed about different types of traffic simulation tools and explained the working flow in microscopic traffic simulation tool SUMO with an example [2].

Codeca and Harri [3] from France used SUMO to create a realistic simulation model for road transport system in which vehicles and other users of road network are connected and can communicate with each other. This information provided by SUMO is a similar copy of real-world vehicular information. Cooperative intelligent transportation systems (C-ITS) uses this information about mobility of vehicles and vulnerable road users to improve road traffic and safety of users in urban environment. They also discussed advanced parking management solutions, and different transport modes applications using the Monaco SUMO Traffic (MoST) Scenario [3].

Major advances in reinforcement learning techniques have shown highly potential results in solving complex traffic control problems. Researchers suggested reinforcement learning techniques for the maintenance of road traffic control. Reinforcement learning is a part of machine learning and comes under semi-supervised learning. Deep neural network (DNN) and convolutional neural network (CNN) algorithms are deep learning algorithms that are used to apply reinforcement learning in traffic controls system [2, 4–8].

2 Methodology

City Visakhapatnam is taken for implementing the proposed intelligent traffic signal controlling system (ITCS). To classify each traffic zone into one of the three classes that include high traffic zone, medium traffic zone, and low traffic zones, video surveillance of each junction is required. This video surveillance can be used to detect the traffic level in each traffic signal junction. As acquiring streaming video data and applying machine learning techniques on each traffic signal junction and observing results in real-time is not allowed to individuals, a road traffic simulation tool SUMO is used. Using SUMO, a simulation of the road network of city Visakhapatnam is created.

2.1 Sumo

SUMO is designed to handle large road networks. It is available freely as an open-source microscopic and multi-model road traffic simulation package. It is highly portable and allows to simulate vehicle movement in given road network. It allows users to address a large number of traffic management issues [9].
OpenStreetMap (OSM): OSM is the map of the whole world which is free of cost and anyone can download and edit the map for their use. This free and editable map is built by the volunteers largely from scratch and it is under an open-content license.

The OSM License allows users to download map images from Internet and it gives access to all underlying map data. The need for creation and growth of OSM came as there are many restrictions on use of map data across the world.

It provides locations that can be downloaded along with road networks. Using OSM, any area that is available can be imported. City Visakhapatnam (see Fig. 1) is downloaded and imported into SUMO and road traffic is simulated. The simulation provides almost all the road networks that are present in the city. Certain traffic signal

Fig. 1 Simulation of Visakhapatnam

junctions where some noticeable traffic is visible are marked for testing our system. These junctions are loaded with default signal configurations.
TraCI: Using Python programming language, we can communicate with this simulation. A Python library TraCI is used to access the running road traffic simulation in SUMO. Important information about every traffic signal junction is retrievable through TraCI. This information is used to observe all the traffic signal junctions and to classify them.

For every traffic signal junction, the number of vehicles that are in halt is stored for every second till the given time period. This series of data is then calculated to get aggregate values like mean, max, mode, and variance. These aggregate values of each traffic signal junction are used for classification by applying classification algorithms.

2.2 Algorithm

INPUT: A running simulation of road traffic from SUMO and training data with labels.
OUTPUT: Optimization of road traffic in SUMO.
PROBLEM DESCRIPTION: Optimization of traffic conditions.
Step 1: Begin.
Step 2: Train the SVM model with training data.
Step 3: Begin simulation by running SUMO through TraCI.
Step 4: Get the traffic information of every signal junction from SUMO.
Step 5: Classify traffic signal junctions using the data collected from SUMO.
Step 6: Change traffic signal configurations of high-level traffic zones.
Step 7: Repeat from Step 4 to Step 7 as much as needed.
Step 7: End simulation.

2.3 Classification

SVM is used for classification of traffic signal junctions. In this algorithm, each data object is considered as a point in an n-dimensional space, where n is the number of features. Then, we will find a hyperplane that distinctly classifies the data points into different classes.

2.4 Training Data

This SVM model is trained with data that is created by SUMO. Training data is stored by creating different scenarios in simulation for all three high traffic zone, medium traffic zone, and low traffic zones. This data with their respective classes is used to train the model which is being used for classification of other unknown data from simulation of city traffic.

A certain amount of time period is chosen to classify the junctions continuously. Every traffic signal junction is periodically checked and classified into one of the three classes at each time period.

Active traffic signal junctions that were present in the simulation are tl_bst, tl_chlm, tl_jd, tl_jr, tl_nad, tl_rlcn, tl_srp, tl_stjn, tl_pdwtr, tl_bchrd (see Fig. 2).

To lower the traffic level at high traffic zones, after each interval of classifying the junctions, the signal configuration of each high traffic zone is optimized by replacing the current configuration with a better one, which results into either medium traffic zone or low traffic zone in most cases. A back-up signal configuration is stored as default which will be applied if all the other signal optimizations resulted into more problematic situations.

Table 1 includes ten traffic signal junctions that were considered for experiment. Each junction is classified and resultant class is mentioned in third column. The

Fig. 2 Selected traffic signal junctions in Visakhapatnam

Table 1 Traffic signal junctions after classifying

S. No.	Traffic signal junction name	Class
1	tl_bst	Medium
2	tl_chlm	Medium
3	tl_jd	High
4	tl_jr	Low
5	tl_nad	Medium
6	tl_rlcn	Medium
7	tl_srp	High
8	tl_stjn	High
9	tl_pdwtr	Low
10	tl_bchrd	Medium

traffic signal junctions tl_jd, tl_srp, tl_stjn are high traffic zones. A better signal configuration is applied to these junctions.

Signal configurations of tl_jd, tl_srp and tl_stjn were replaced with optimized signal configuration that was designed by the user for high traffic zones. As a result, we observed that on the next periodic classification step, traffic level at tl_jd, tl_srp, tl_stjn was successfully reduced to medium from high (see Table 2).

At present, the traffic signals in most of the cities work on fixed time (FT) mode. In FT system, irrespective of the volume of vehicles in a lane, the traffic lights change after a fixed time. Recently, vehicle-actuated controlled (VAC) system is introduced at major traffic signal junctions in city Visakhapatnam. This system will maintain traffic signal based on the flow of vehicles. Basically, in VAC systems, sensors placed at selected places give inputs on the length of the line of vehicles. This input is used to detect the volume of traffic. Certain situations can fool the system if sensors do not work properly. Even though VAC systems work well, they can't handle situations like passing of ambulances, VIP visits, and other emergencies.

Table 2 Traffic signal junctions after signal optimization

S. No.	Traffic signal junction name	Class
1	tl_bst	Medium
2	tl_chlm	Medium
3	tl_jd	Medium
4	tl_jr	Low
5	tl_nad	Medium
6	tl_rlcn	Medium
7	tl_srp	Medium
8	tl_stjn	Medium
9	tl_pdwtr	Low
10	tl_bchrd	Medium

Our proposed system can accurately detect the volume of vehicles and as the traffic signal junctions have video surveillance (in real-time) which is being monitored by powerful machine learning models, additional importance can be given to vehicles of emergency services like ambulance and fire engine by quickly clearing the routes that they are taking. Heavy vehicles that travel through the city can also be diverted to better routes to avoid traffic in the city.

3 Conclusions and Future Work

In this paper, an intelligent system that manages traffic signal control system with the help of machine learning is implemented. Even though this system solves the problems in most cases, the application of a different signal configuration at a high-level traffic junction will not always result in lowering the traffic at that junction. A better approach would be to use some techniques to select the appropriate signal configuration. This system can be further developed by including reinforcement learning to select signal configuration efficiently for traffic signal junctions after classifying them. Other components can be included to connect the drivers of vehicles to this system so that real-time route suggestions can be given to them to avoid traffic congestion. A more sophisticated system can be developed which will be able to help the city administration and city civil engineers, by giving them better infrastructure plans for roads for better road traffic management.

References

1. http://www.geographynotes.com
2. Lin Y, Dai X, Li L, Wang F-Y (2018) An efficient deep reinforcement learning model for urban traffic control, 24 Aug 2018
3. Codeca L, Harri J (2017) Towards multimodal mobility simulation of C-ITS: the Monaco SUMO traffic scenario. In: IEEE vehicular networking conference (VNC), Torino, Italy
4. Zhang R, Ishikawa A, Wang W, Striner B, Tonguz O (2019) Intelligent traffic signal control: using reinforcement learning with partial detection. 4 Feb 2019, USA
5. Mousavi SS, Schukat M, Howley E (2017) Traffic light control using deep policy-gradient and value-function based reinforcement learning. 27 May 2017, Galway
6. Yang K, Tan I, Menendez M (2017) A reinforcement learning based traffic signal control algorithm in a connected vehicle environment. In: 17th swiss transport research conference
7. Guo M, Wang P, Chan C-Y, Askary S (2019) A reinforcement learning approach for intelligent traffic signal control at urban intersections
8. Wei H, Yao H, Zheng G, Li Z (2018) IntelliLight: a reinforcement learning approach for intelligent traffic light control. In: 24th ACM SIGKDD international conference on knowledge discovery and data mining, KDD 2018, London, United Kingdom
9. https://sumo.dlr.de

10. Lopez PA, Behrisch M, Bieker-Walz L, Erdmann J, Flotterod Y-P, Hilbrich R, Lucken L, Rummel J, Wagner P, WieBner E (2018) Microscopic traffic simulation using SUMO. In: 21st international conference on intelligent transportation systems (ITSC), Maui, HI, USA
11. Koh SS, Zhou B, Yang P, Yang Z, Fang H, Feng J (2018) Reinforcement learning for vehicle route optimization in SUMO. In: 20th IEEE international conference on high performance computing and communications, 28–30 June 2018, Exeter, United Kingdom

An Analytical Review on Log Periodic Dipole Antennas with Different Shapes of Dipole Elements

Swetha Velicheti and P. Mallikarjuna Rao

Abstract Nowadays, for the current and future communication applications like radio wave direction-finding systems, 5G mobile and wireless communication, airborne applications, UWB radar, and mobile imaging applications, the log periodic dipole array (LPDA) antenna is the most suitable, and it can be used as in either direction-finding applications or UWB applications based on its properties. In this paper, various miniaturization techniques of log periodic dipole array antenna are studied and analyzed. Summary of different types of LPDA structures such as dual-band elements, folded planar helix dipole, sinusoidal dipoles, T-shaped and H-shaped top loading dipoles with respect to operating frequency, bandwidth, type of substrate, gain, and dimensions of the antenna is presented at the end. Miniaturization of the antenna is achieved with these different configurations. A high gain and VSWR value which are less than 2 can be obtained by applying these miniaturization techniques. These antennas can be designed and simulated using electromagnetic tools like HFSS and CST MW Studio.

Keywords Log periodic dipole array (LPDA) antenna · Antenna miniaturization · Dielectric loading · Top loading

1 Introduction

In recent times, some of the frequency-independent antennas such as LPD arrays [1–5] are used for various applications. To form a co-planar array dipoles are placed alongside by using design parameters such as scaling factor τ and spacing factor σ.

Figure 1 shows the basic schematic of LPDA with some design parameters which are shown below:

S. Velicheti (✉) · P. Mallikarjuna Rao
Department of Electronics and Communication Engineering, AU College of Engineering (A), Visakhapatnam, Andhra Pradesh, India
e-mail: swethavelicheti1@gmail.com

P. Mallikarjuna Rao
e-mail: pmraoauece@yahoo.com

P. S. R. Chowdary et al. (eds.), *Microelectronics, Electromagnetics and Telecommunications*, Lecture Notes in Electrical Engineering 655,
https://doi.org/10.1007/978-981-15-3828-5_64

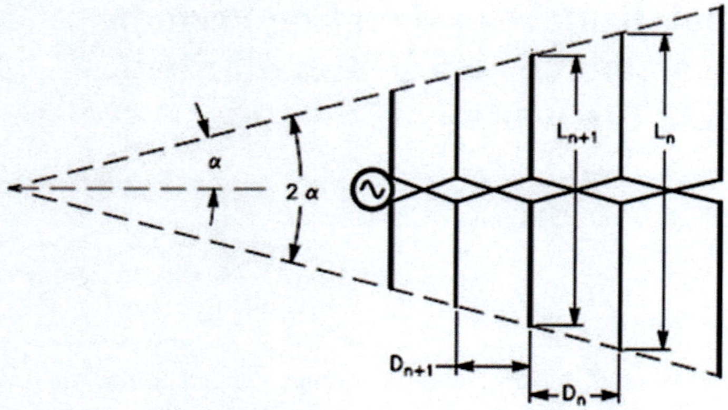

Fig. 1 Basic schematic diagram of LPDA

$$\text{Scaling factor } \tau = \frac{L_n}{L_{n+1}} = \frac{d_{n,n-1}}{d_{n-2,n-1}} \quad (1)$$

$$\text{Spacing constant } \sigma = \frac{d_{n,n-1}}{2L_{n-1}} \quad (2)$$

$$\text{Half aperture angle } \alpha = \tan^{-1}\left[\frac{1-\tau}{4\sigma}\right] \quad (3)$$

$$\text{Active region bandwidth Ba} = 1.1 + 7.7(1-\tau)2\cot\alpha \quad (4)$$

$$\text{Desired bandwidth } B = f_{\max}/f_{\min} \quad (5)$$

$$\text{Designed bandwidth } B_s = B * B_a = B[1.1 + 7.7(1-\tau)2\cot\alpha] \quad (6)$$

$$\text{Number of dipole elements } N = 1 + \frac{\ln(B_s)}{\ln(1/\tau)} \quad (7)$$

$$\text{Length of the longest dipole } L_{\max} = \frac{\lambda_{\max}}{2} \quad (8)$$

$$\text{Wavelength related to } f_{\min}\lambda_{\max} = \frac{\nu}{f_{\min}} \quad (9)$$

The relation between directivity, σ and τ, is shown in Fig. 2 [2]. First we choose σ and τ values according to the directivity using Fig. 2, and then, we find the design parameters using above relations.

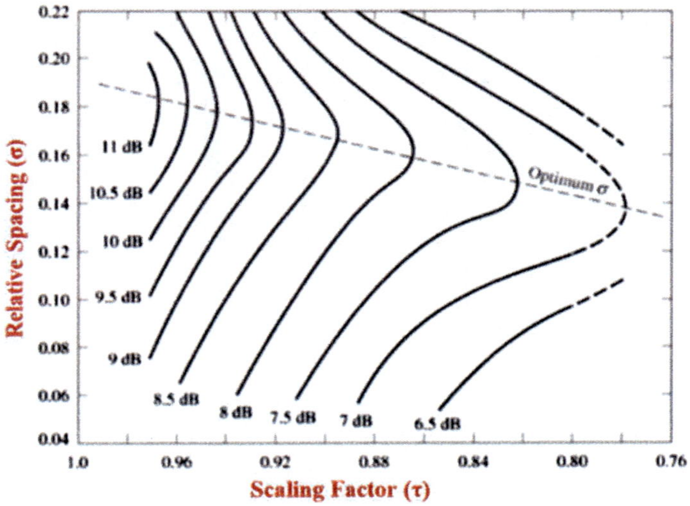

Fig. 2 Contours of constant directivity versus σ and τ of LPDA

Consider a conventional λ/2 element LPDA operating from $f_{\min} = 0.95$ GHz to $f_{\max} = 2.2$ GHz. The scaling factor and spacing factor are chosen from Fig. 2.

Let $\tau = 0.8$ and $\sigma = 0.148$, then from Eq. 3 $\alpha = \tan^{-1}\left[\frac{1-0.8}{4\times 0.148}\right] = 18.66°$

From Eqs. 4 and 5 Ba $= (2.2/0.95)[1.1 + 7.7(1 - 0.8)2 \cot 18.66] = 10.22$

From Eq. 6 Bs $= 23.7$

From Eq. 7 $N = 1 + \frac{\ln(23.7)}{\ln(1/0.8)} = 12, \quad N/2 = 6$

From Eqs. 8 and 9, length of the largest element $l_6 = \frac{3\times 10^8}{0.95\times 10^9} = 315.8/2 = 158$ mm

From Eq. 1 $l_5 = \tau\, l_6 = 0.8 \times 158 = 126$ mm.

In the similar way, lengths of all the elements are calculated and the values are shown in Table 1.

Now spacing between the elements is calculated using Eq. 2.

That is, $d_5 = 2\sigma\, l_5 = 2 \times 0.148 \times 126 = 38$ mm

Similarly d_4, d_3, d_2, and d_1 are calculated and shown in Table 1.

Table 1 Similar way lengths of all the elements are calculated and the values

S. No.	Length of the element (mm)	Spacing between the elements (mm)
1	$l_1 = 52$	$d_1 = 15$
2	$l_2 = 65$	$d_2 = 20$
3	$l_3 = 81$	$d_3 = 24$
4	$l_4 = 101$	$d_4 = 30$
5	$l_5 = 126$	$d_5 = 38$
6	$l_6 = 158$	

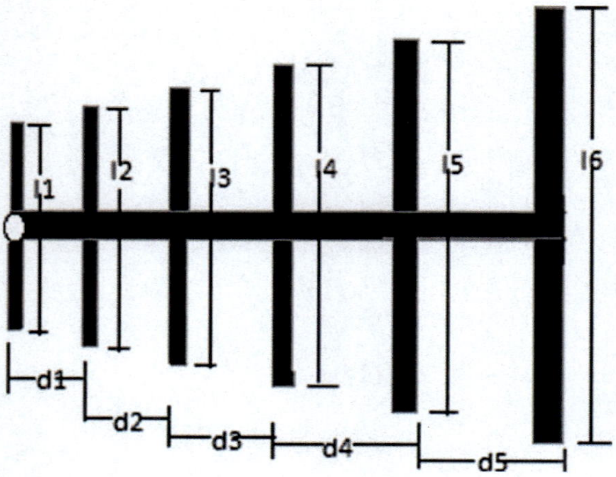

Fig. 3 Conventional λ/2-element log periodic dipole array

By using all these dimensions, the conventional log periodic array antenna is designed, which is shown in Fig. 3. But due to the size constraint, these conventional LPD arrays are unable to utilize in mobile wireless systems. So in order to overcome this problem, various miniaturization methods are available such as fractalization, top loading and also by changing the shape of dipoles.

2 Shortened LP Dipole Antenna Using Printed Dual-Band Dipole Elements

The configuration of the shortened log periodic dipole antenna with dual-band elements [6, 7] is shown in Fig. 4. This antenna contains a double-sided printed principal half-wave dipole and an auxiliary dipole.

This antenna is designed without using ground to obtain minimized structure. In this design, due to feeding mechanism of the dipole tuning of the antenna is easy for matching purpose. This antenna has average gain of 3–6.01 dBi and VSWR < 2 with a limited number of dual-band dipole elements. Its axial length is 218 mm, and dimensions of this antenna are 218 mm × 260 mm × 0.508 mm.

3 Gain Characteristics of LPD Antenna Using Partially Applied FPH Dipoles

This antenna can minimize the gain degradation phenomenon that is generally appeared in antenna miniaturization. This can be accomplished by replacing conventional λ/2 dipoles with FPH dipole. The radiation resistance and efficiency of

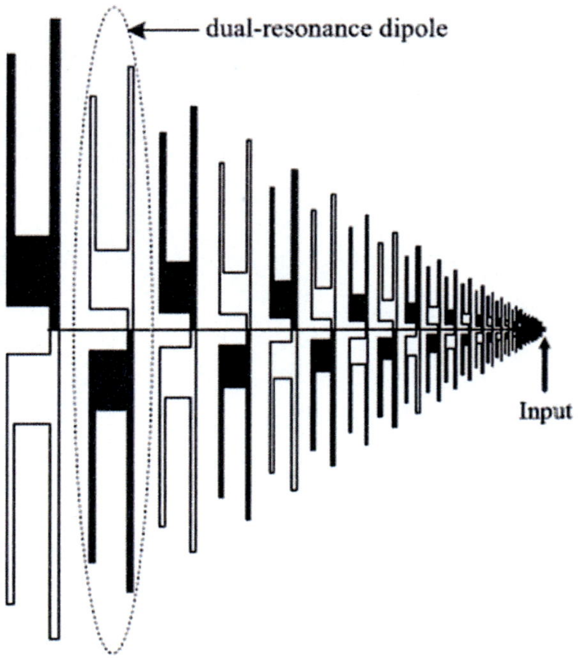

Fig. 4 LPDA antenna with dual-band dipole elements

this folded planar dipole are almost similar to those of conventional half-wave dipole with a size reduction of 39%.

In this configuration, longest and third longest half-wave dipoles are replaced by the FPH element and second longest λ/2 dipole with a meander line [8–10] as shown in Fig. 5.

This arrangement gives better impedance matching and an improved FBR. The total dimensions of this antenna are 445 mm × 275 mm × 3.2 mm. The operating band of this antenna is 400–865 MHz with the average gain of 5.7 dBi.

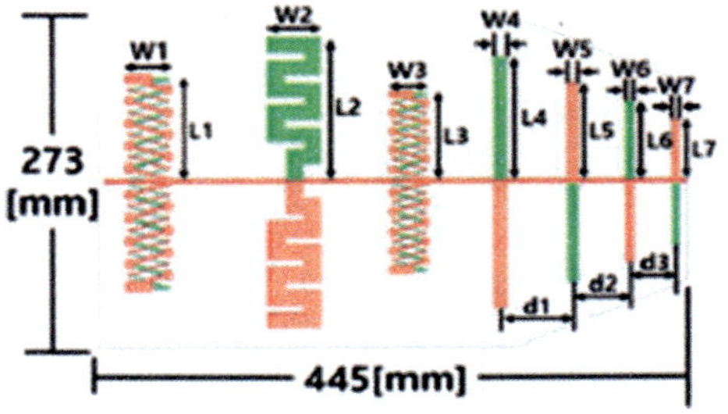

Fig. 5 LPDA antenna using partially applied folded planar helix dipoles

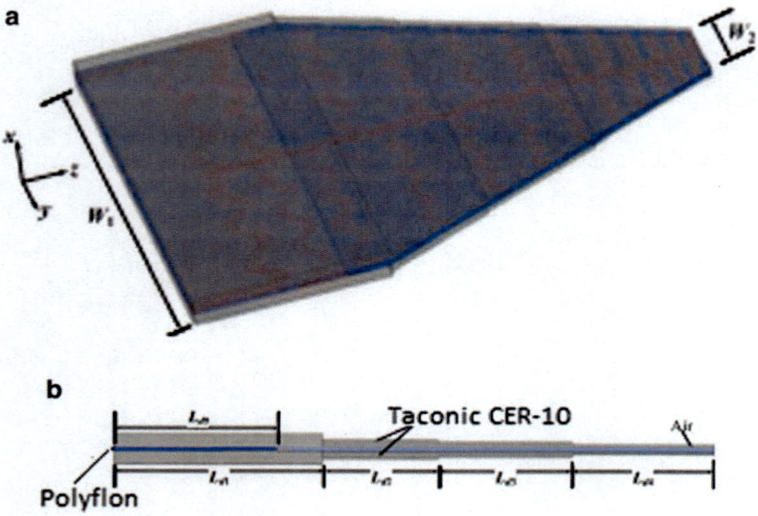

Fig. 6 **a** 3-D view of compact LPDA and **b** side view

4 A Compact Dielectric Loaded LPDA Antenna

In this antenna, sinusoidal dipoles [11–13] are used for miniaturization. Figure 6 shows that this antenna contains two-stepped dielectric material (Taconic CER-10).

By using these stepped dielectric materials, low frequency performance is obtained, partially loaded air layer improves the impedance matching performance, and sinusoidal shaped dipoles can extend the current path. These sinusoidal-shaped dipoles have a very low resonant frequency when compared to conventional dipole of the same length. This antenna is operating from 200 to 803 MHz for VSWR < 2. By using sinusoidal-shaped dipole elements, the antenna is miniaturized with a width of 420 mm and has high power capacity. So, this antenna can be used for high-power radiation applications in air-borne platforms.

5 Compact LPDA Antenna with BW Enhancement Techniques for the LF Band

There are different top loading methods which are used to minimize the size of the LPDA antenna with a very small spacing factor between the elements.

But miniaturization of antenna limits the antennas wide band characteristics over the desired range of frequencies. In addition to top loading technique, a feed line meander and a trapezoidal stub [14, 15] have been integrated onto this design to

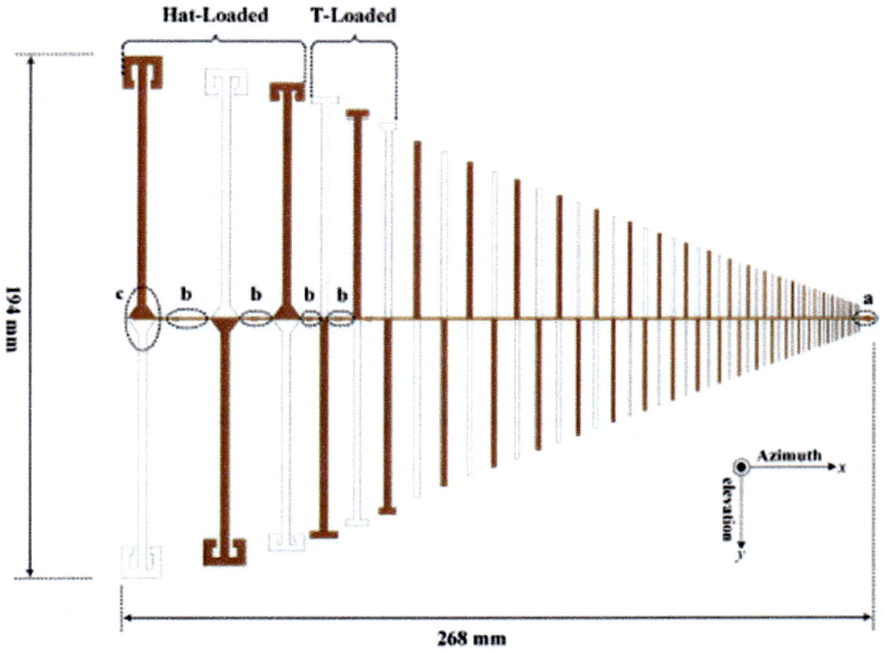

Fig. 7 Structure of the proposed planar LPDA antenna

increase the bandwidth for the operating frequency range (0.55–9 GHz). The antenna arrangement is shown in Fig. 7.

6 Design of LPDA with T-Shaped Top Loadings for Miniaturization

In this antenna configuration, conventional λ/2 dipoles are replaced with top loading of T-shaped dipoles [15–17]. So the lengths of the individual elements are reduced by 45% compared to the conventional λ/2 dipoles. The arrangement of the antenna is shown in Fig. 8.

This technique increases the radiation resistance of the antenna with a gain of 5.1 dBi and impedance bandwidth of 87.3%.

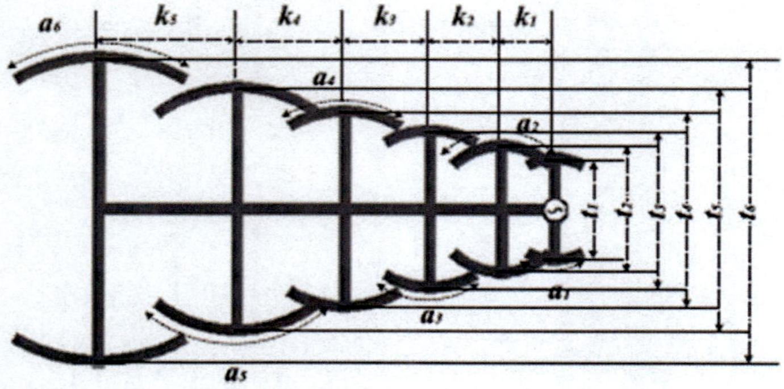

Fig. 8 LPD antenna array with top loadings

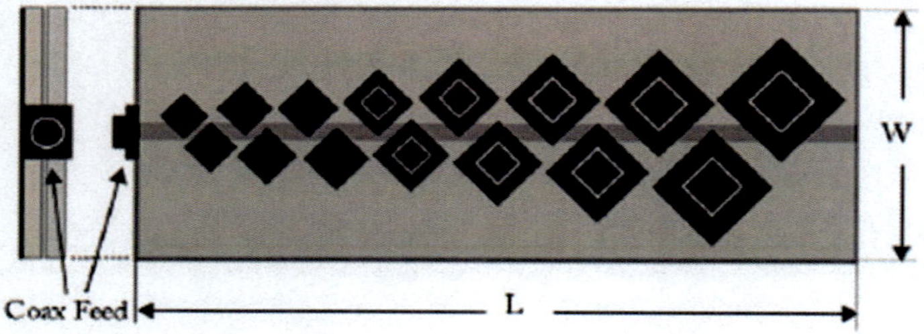

Fig. 9 Log periodic square fractal antenna

7 Miniaturized UWB Log Periodic Square Fractal Antenna

This is a miniaturized log periodic antenna for the UWB services operating in the frequency range of 3.6–10.6 GHz. The antenna with square fractal elements [18–20] is shown in Fig. 9.

By using this antenna, a broadside radiation in ultra wideband services is accomplished, where this antenna contains square fractals to sustain good radiation characteristics at higher frequencies. A miniaturization factor of 23% is achieved by using square fractals with a stable gain in the desired band of frequencies.

8 Summary of Different LPDA Structures

Table 2 shows the various parameters such as operating frequency, bandwidth, gain, VSWR, miniaturization percentage, substrate material used, and dimensions of various miniaturized printed LPDA antennas. By using above-mentioned miniaturized elements (such as dual-band elements, FPH dipoles, meander line dipoles, sinusoidal dipoles, square fractal elements, T-loaded and hat-loaded dipoles), the size of

Table 2 Various parameters in LPDA antennas

Antenna type	Operating frequency	Bandwidth	Substrate	Gain (dB)	VSWR	Length × Width × Height	Miniaturization	Applications
Shortened LP dipole antenna using printed dual-band elements [6]	0.5–10 GHz	9.5 GHz	Rogers RO4003 (ε_r = 3.55)	3–6.01	≤2	218 mm × 260 mm × 0.508 mm	19% shorter dipole elements	Source antenna for (EMC) measurement
Gain characteristics of LPD antenna using partially applied FPH dipole [8]	400–800 MHz	400 MHz	FR4 (ε_r = 4.5)	5.7 dBi	≤2	445 mm × 273 mm × 3.2 mm	39% of total size is reduced	Mobile radio wave direction-finding applications
A compact dielectric loaded LPDA antenna [11]	200–803 MHz	≈600 MHz	Polyflon (ε_r = 2.55)	3.2–5.5	≤2	576.6 mm × 420 mm × 29.6 mm	0.28 λ_L of width is reduced	Air-borne applications
Compact LPDA antenna with BW enhancement techniques for the LF Band [14]	0.55–9 GHz	8.45 GHz	Rogers RO4003 (ε_r = 3.55)	2.4–7.8	<2	268 mm × 194 mm × 0.508 mm	27%, 20% of width and length is reduced	EMC applications
Design of LPDA with T-shaped top loadings for miniaturization [16]	0.85–2.2 GHz	1.35 GHz	Styrofoam (ε_r = 1.06)	2.3–6.1	<2	124 mm × 85 mm	45% of dipole element is reduced	Direction-finding applications
Miniaturized UWB log periodic square fractal antenna [17]	3.1–10.6 GHz	7.5 GHz	Rogers RO4003 (ε_r = 3.55)	8 dBi	<2	173 mm × 70 mm	23% of total size is reduced	UWB radar and medical imaging

the antenna is reduced as compared to the conventional half-wave dipole log periodic array with VSWR < 2 and maintaining reasonable gain.

9 Conclusion and Future Scope

It is concluded that log periodic dipole array (LPDA) antenna is the good wide band antenna with a modest level of gain. The bandwidth of the LPDA antenna may be increased based on the concept of frequency independent. Moreover, log periodic dipole array (LPDA) antenna may also provide large bandwidth at low and high frequencies. Apart from this, various miniaturization techniques of log periodic dipole array antenna are studied and analyzed for various applications based on the frequency ranges starting from 0.5 to 10 GHz, 400–800 MHz, 200–803 MHz, 0.55–9 GHz, 0.85–2.2 GHz and 3.1–10.6 GHz. In the future work, these antennas will be modified by using different feeding techniques such as co-planar waveguide (CPW), substrate integrated waveguide (SIW), and half-mode substrate integrated waveguide (HMSIW) and also different kinds of dipole shapes, and the same antenna will be simulated for the characteristics like VSWR, gain, beam width, and radiation patterns. After simulation, the antenna will be fabricated and the experimental validation is also considered for future scope of the work.

References

1. Rumsey V (1967) Frequency independent antennas. In: Proceedings of IRE international convention record, New York, NY, USA, pp 114–118
2. Isbell D (1960) Log periodic dipole arrays. IRE Trans Antennas Propag 8(3):260–267
3. Carrel R (1961) The design of log-periodic dipole antennas. In: IRE international convention record, vol 9, pp 61–75
4. Balanis CA (2005) Antenna theory: analysis and design. Hoboken, NJ, USA: Wiley, pp 621–633
5. Thomas A (2005) Milligan: modern antenna design. Wiley Inc, Publication
6. Anim K, Jung Y-B (2018) Shortened log-periodic dipole antenna using printed dual-band dipole elements. IEEE Trans Antennas Propag 66(12):6762–6771
7. Yu J, Le Y, Lim S (2014) Design of a dual-band, electrically small, parasitic array antenna. IEEE Antennas Wireless Propag Lett 13:1453–1456
8. Shin G, Kong M, Lee S-H, Kim S-T, Yoon I-J (2018) Gain characteristic maintained, miniaturized LPDA antenna using partially applied folded planar helix dipoles. IEEE Access, vol 6, pp 25874–25880
9. Lee JM, Ham HJ, Ryu HK, Woo JM, Park BJ, Lee KS (2012) Miniaturization of log-periodic dipole array antenna using triangular meander structure. In: Proceedings IEEE international conference wireless information technology System, pp 1–4
10. Gheethan AA, Anagnostou DE (2008) Reduced size planar log-periodic dipole arrays (LPDAs) using rectangular meander line elements. In: Antennas and propagation society international symposium, AP-S 2008. IEEE, pp 1, 4, 5–11
11. Chang L, He S, Zhang JQ, Li D (2017) A compact dielectric-loaded log-periodic dipole array (LPDA) antenna. IEEE Antennas Wireless Propag Lett 16:2759–2762

12. Anagnostou DE, Papapolymerou J, Tentzeris MM, Christodoulou CG (2008) A printed log-periodic Koch-dipole array (LPKDA). IEEE Antennas Wireless Propag Lett 7:456–460
13. Hsu HT, Huang TJ (2013) A Koch-shaped log-periodic dipole array (LPDA) antenna for universal ultra-high-frequency(UHF) radio frequency identification (RFID) handheld reader. IEEE Trans. Antennas Propag 61(9):4852–4856
14. Kyei A, Sim D-U, Jung Y-B (2017) Compact log-periodic dipole array antenna with and width-enhancement techniques for the low frequency band. IET Microw Antennas Propag 11(5):711–717
15. Jardon-Aguilar H, Tirado-Mendez JA, Flores-Leal R, Linares-Miranda R (2011) Reduced log-periodic dipole antenna using a cylindrical-hat cover. IET Microw Antennas Propag 5(14):1697–1702
16. Chen J, Ludwig J, Lim S (2017) Design of a compact log-periodic dipole array using T-shaped top loadings. IEEE Antennas Wireless Propag Lett 16:1585–1588
17. Kuo S (1970) Size-reduced log-periodic dipole array. Proc IEEE Antennas Propag Soc Int Symp 8:151–158
18. Amini A, Oraizi H, Chaychi zadeh MA (2015) Miniaturized UWB log-periodic square fractal antenna. IEEE Trans Antennas Propag 14(12):1322–1325
19. James JR, Hall PS (1989) Handbook of microstrip antennas. Peter Peregrinus Ltd.
20. Hall P (1980) New wideband microstrip antenna using log periodic technique. Electron Lett 16(4):127–128

Adaptive Level Cross Sampling for Next-Generation Data-Driven Applications

Viswanadham Ravuri, Sudheer Kumar Terlapu, and S. S. Nayak

Abstract Processing of sensor signals in modern wireless networks becomes a challenging task because of power limitation at sensor nodes. The data collected by sensors such as pressure, humidity, temperature, speech and electrocardiograms contains only low frequencies. These signals have very negligible variations for such a long time. Therefore, traditional uniform sampling analog-to-digital (A/D) converters make more samples even if no variations in the signal, which may result in the data processing system, becomes computationally complex and needs higher power. In order to overcome this, an adaptive level cross sampling scheme (ALCSS) is presented in this paper. An iterative algorithm is used in this paper in order to reconstruct the original signal. The performance of the proposed ALCCS is compared with the traditional level cross scheme (LCS) in terms of computational complexity.

Keywords Adaptive sampling · LCS · ALCSS · Activity selection

1 Introduction

Wireless sensor networks (WSN) and Internet of Things (IoT) will play a major role in the next few years. In the near feature, everything will be controlled with information extracted from the sensor signals [1, 2]. In order to exchange the data, these applications require more bandwidth because of conventional A/D converters. Uniform sampled A/D converters always follow the Nyquist criterion and they do not consider the time-varying nature of the sensor signal [3–5]. The major drawbacks of conventional A/D converters are more redundant samples which are present in the sampled signal, it requires higher bandwidth, and it consumes more power to process the signal, low-performance speed and independent to signal variations. Because of this reason, uniform sampling is not always the best sampling method for

V. Ravuri (✉) · S. S. Nayak
Centurion University of Technology and Management, Odisha, India
e-mail: viswamravuri@gmail.com

S. K. Terlapu
Department of ECE, Shri Vishnu Engineering College for Women, Bhimavaram, India

P. S. R. Chowdary et al. (eds.), *Microelectronics, Electromagnetics and Telecommunications*, Lecture Notes in Electrical Engineering 655,
https://doi.org/10.1007/978-981-15-3828-5_65

data acquisition [6]. In the past, several techniques are developed which are adapted to time-varying frequency properties of the signal [7–9]. Current level crossing sampling schemes require a large amount of samples to faithfully reconstruct the analog representation [10–13]. This technique not only accommodates burst signals, but also drastically reduces the number of samples needed. Moreover, the technology outperforms current ones by saving dynamic power. By doing so, this technology also saves bandwidth resources during inactive signals to allocate toward other parts of a system. Therefore, this more efficient method could potentially free up resources in electronic devices to be used elsewhere in the system. So, in mobile communications also, we can provide bandwidth for more number of users within the available channel [14–17]. To overcome the limitations of LCS, an ALCSS is developed in this paper with the motivations of the rest of the paper which is organized as follows. Section 2 gives the information about non-uniform sampling using LCS. ALCSS and signal reconstruction are discussed in Sect. 3. In Sect. 4, the performance of the proposed ALCSS is evaluated for different inputs. Section 5 finally concludes the article.

2 Level Crossing Sampling (LCS)

The LCS is an asynchronous sampling scheme where the local variations of the signal can be tracked. In this scheme, whenever the signal $x(t)$ crosses one of the predefined levels then a sample is captured. As a result, the samples which are captured are not uniformly spaced. The amplitude of the sample at time t_n is x_n. However, x_n is obviously equivalent to any one of the predefined level. The present sample time t_n can be expressed as [18]

$$t_n = t_{n-1} + \mathrm{d}t_n \tag{1}$$

here,

t_n is instant of the present sample
t_{n-1} is instant of the past sample and
$\mathrm{d}t_n$ is time over the current sample with the previous sample.

The graphical representation of the LCS is shown in Fig. 1. Here, q is the voltage separation between two uniformly arranged predefined levels.

3 Adaptive Level Crossing Sampling Scheme (ALCSS)

In ALCSS, when the signal is crossed by any one of the predefined levels, then a sample is considered. In this scheme, the samples will be updated by considering the position of the samples. Samples will be updated by using distance between the two

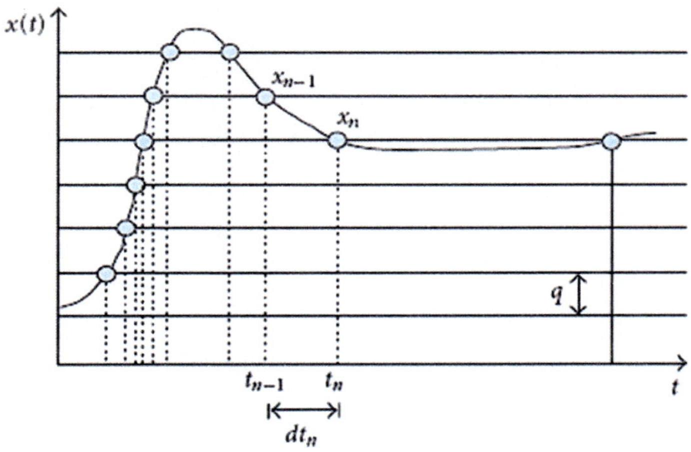

Fig. 1 Concept of LCS

levels and an offset parameter δ. The updating procedures of two successive samples are as shown in Fig. 2.

Case-1: The successive samples, q_1 and q_2, are in the signal positive slopes as shown in Fig. 2. In this case, the updated values of q_1 and q_2 are given in Eq. 2.

$$\begin{aligned} q_1^1 &= q_1 - \delta + d \\ q_2^1 &= q_2 - \delta + d \\ q_1 &= q_1^1, q_2 = q_2^1 \end{aligned} \tag{2}$$

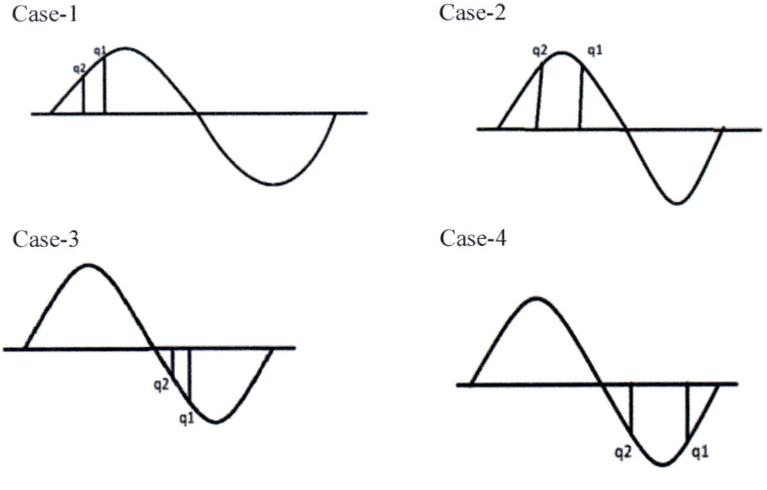

Fig. 2 Updating of q_1 and q_2 in ALCSS

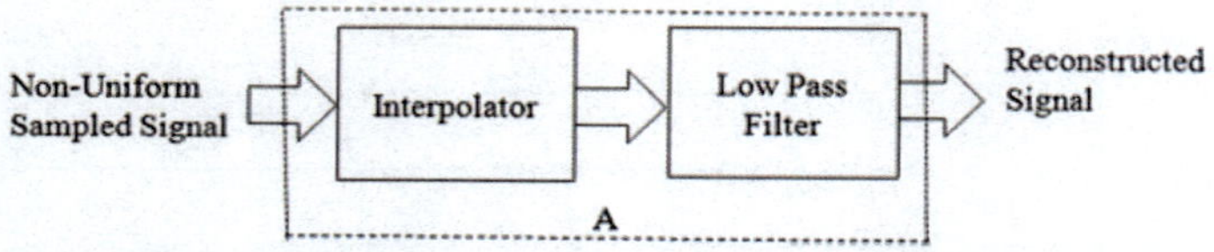

Fig. 3 Signal reconstruction

Case-2: The sample q_1 intersects the signal in the negative slope region, and the sample q_2 intersects the signal in the positive slope region as shown in Fig. 2. In this case, the updated values of q_1 and q_2 are given in Eq. 3.

$$\begin{aligned} q_1^1 &= q_1 - \delta - d \\ q_2^1 &= q_2 - \delta - d \\ q_1 &= q_1^1, q_2 = q_2^1 \end{aligned} \tag{3}$$

Case-3: The samples, q_1 and q_2, are present in the negative signal slopes as shown in Fig. 2. In this case, the updated values of q_1 and q_2 are given in Eq. 4.

$$\begin{aligned} q_1^1 &= q_1 + \delta - d \\ q_2^1 &= q_2 - \delta + d \\ q_1 &= q_1^1, q_2 = q_2^1 \end{aligned} \tag{4}$$

Case-4: The sample q_1 intersects the signal in the positive slope region, and the sample q_2 intersects the signal in the negative slope region as shown in Fig. 2. In this case, the updated values of q_1 and q_2 are given in Eq. 5.

$$\begin{aligned} q_1^1 &= q_1 + \delta + d \\ q_2^1 &= q_2 + \delta + d \\ q_1 &= q_1^1, q_2 = q_2^1 \end{aligned} \tag{5}$$

Signal Reconstruction

For non-uniform sampling, the reconstruction of the signal is not possible with simple low pass filter. Hence, the band-limited operator (A) is used to reconstruct the original signal by eliminating inverse problem. The reconstruction of the original signal from non-uniform sampled signal is shown in Fig. 3.

4 Simulations and Discussions

For simulation, we have considered two different speech signals. The signal shown in Fig. 4 is taken as input for the case-1 which has less number of samples than case-2 signal.

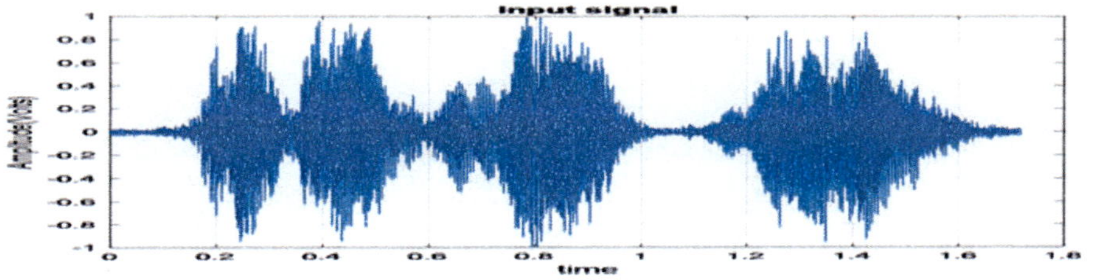

Fig. 4 Input speech signal 1

The uniform sampled signal and the level crossing output are shown in Figs. 5 and 6, respectively. From these, it can be observed that the number of samples obtained from uniform sampling is more than the LCS.

Figure 7 shows the output of the ALCSS in which the number of samples will be greater than LCS signal but lesser than uniform sampled signal. The extra samples can be useful for the better reconstruction of the original signal.

Figure 8 shows the reconstructed signal of the speech-1 through an iterative algorithm. **Interpolator and FIR low pass filter with the Kaiser window are used for** reconstruction of the original signal.

The second speech signal, ALCSS signal, and its reconstructed signal are shown in Figs. 9, 10 and 11, respectively.

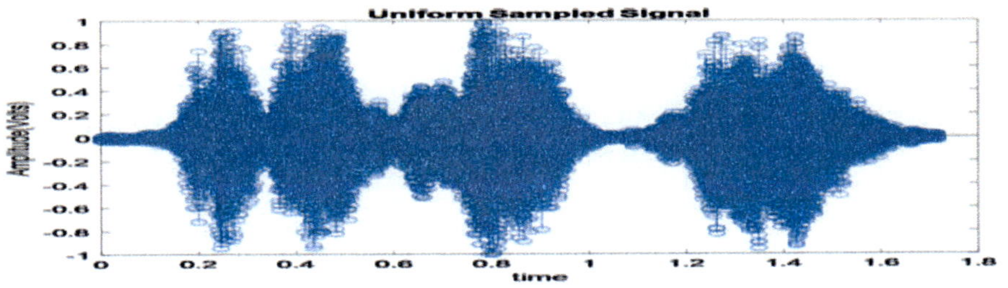

Fig. 5 Uniform sampled signal

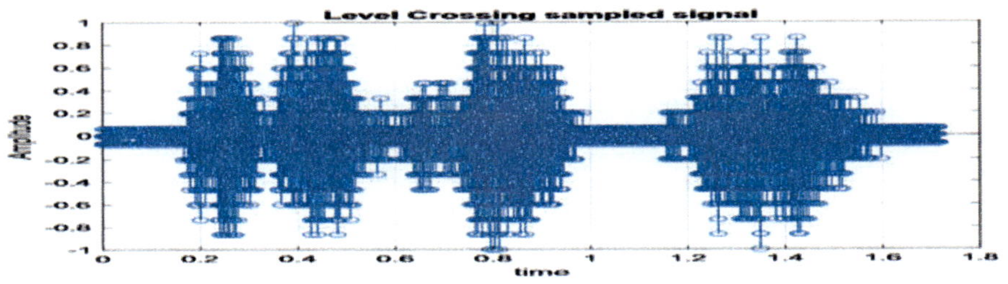

Fig. 6 Signal obtained through LCS

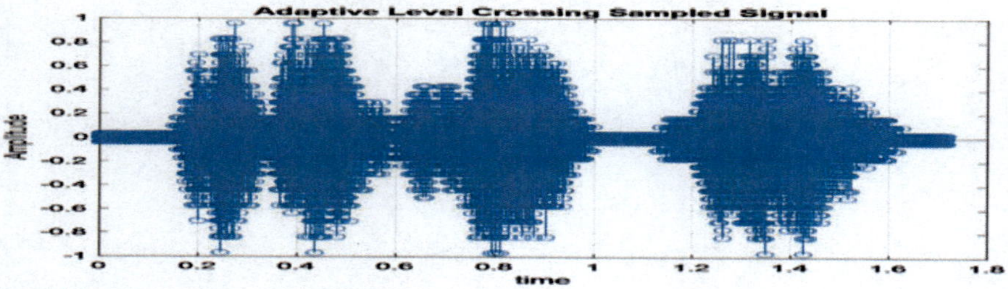

Fig. 7 Signal obtained through ALCSS

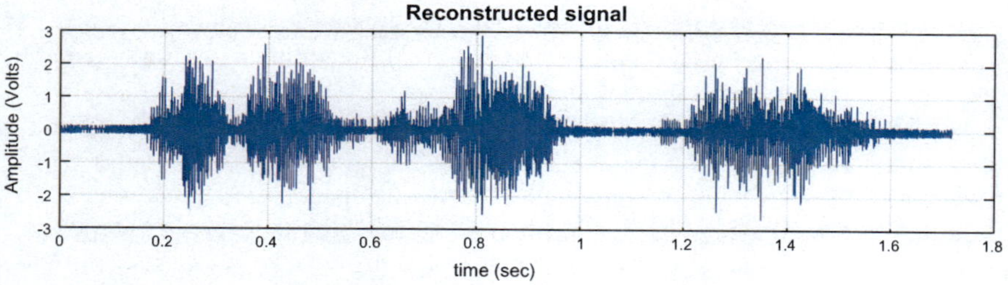

Fig. 8 Reconstructed signal

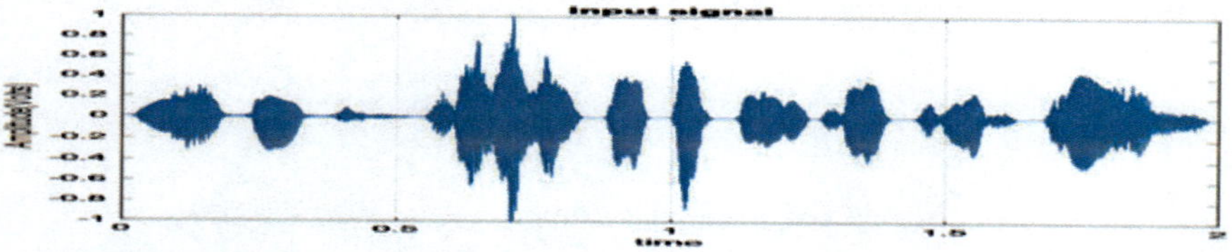

Fig. 9 Input speech signal 2

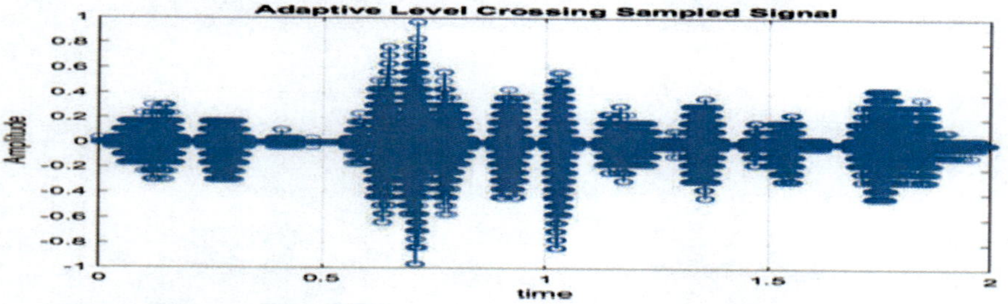

Fig. 10 Signal obtained through ALCSS

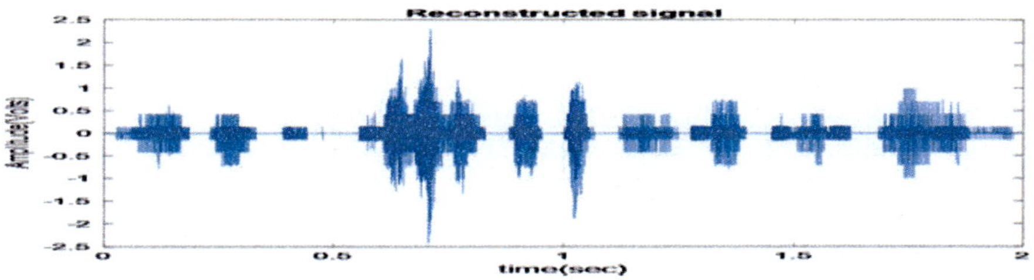

Fig. 11 Reconstructed signal

Table 1 Performance analysis of LCS and ALCSS in terms of number of samples

Type of signal	Uniform sampling	Level crossing sampling	Adaptive level crossing sampling
Signal 1	18,944	7,779	10,815
Signal 2	96,000	7,992	13,096

Table 2 Performance analysis of LCS and ALCSS in terms of bandwidth

Type of signal	Uniform sampling (kHz)	Level crossing sampling (kHz)	Adaptive level crossing sampling (kHz)
Signal 1	75.7	31.1	43.2
Signal 2	384	31.9	52.3

From the results, it is clear that ALCSS consists of less number of samples and also suitable for reconstruction.

From Tables 1 and 2, it is clear that ALCSS is superior in performance when compared to uniform sampling. Even though LCS is superior in performance but the reconstructed signal is not exact as ALCSS approach.

5 Conclusion

In order to reduce bandwidth and computational complexity, non-uniform sampling techniques are implemented in this work. The proposed techniques consider more number of samples if any signal activity is present, and it considers very few samples when no activity is present in the signal. An adaptive reconstruction scheme is developed in order to reconstruct the original signal from non-uniform sampled signal. The performance of the proposed sampling technique is superior to the traditional uniform sampling techniques.

References

1. Guan KM, Singer AC (2006) A level-crossing sampling scheme for both deterministic and stochastic non bandlimited signals. In: Proceedings of Sarn off symposium
2. Duan Z, Zhang J, Zhang C, Mosca E (2006) A simple design method of reduced-order filters and its application to multi-rate filter bank design. Elsevier J Sig Process 86(5):1061–1075
3. Kofman E, Braslavsky JH (2006) Level crossing sampling in feedback stabilization under data rate constraints. In: Proceedings of IEEE conference decision and control, pp 9423–4428
4. Guan KM, Singer AC (2007) Opportunistic sampling by level-crossing. In: ICASSP'07, pp 1513–1516
5. Malmir Chegini M, Marvasti F (2007) Performance improvement of level-crossing A/D converters. In: IEEE international conference on telecommunications and Malaysia international conference on communications ICT-MICC, pp 438–441
6. Nairn D (2008) Time-interleaved analog-to-digital converters. In: IEEE 2008 custom integrated circuits conference (CICC), pp 289–296
7. Wang T, Wang D, Hurst PJ, Levy BC, Lewis SH (2009) A level-crossing analog-to-digital converter with triangular dither. IEEE Trans Circ Syst I Reg Pap 56(9):2089–2099
8. Kurchuk M, Tsividis Y (2010) Signal-dependent variable-resolution clock less A/D conversion with application to continuous-time digital signal processing. IEEE Trans Circ Syst 57:982–991
9. Sun M, Senay S, Chaparro LF, Sclabassi RJ (2010) Adaptive level-crossing sampling and reconstruction. In: Proceedings of EUSIPCO 2010, pp 1296–1300
10. Trakimas M, Sonkusale (2011) An adaptive resolution asynchronous ADC architecture for data compression in energy constrained sensing applications. IEEE Trans Circ Syst I Reg Pap 58(5):921–934
11. Greitans M, Shavelis R, Fesquet L, Beyrouthy T (2011) Combined peak and level-crossing sampling scheme. IEEE Trans Commun
12. Venkata Subbarao M, Samundiswary P (2016) An intelligent cognitive radio receiver for future trend wireless applications. Int J Comput Sci Inf Secur (IJCSIS) 14:7–12
13. Kafashan M, Ghorbani M, Marvasti F (2012) A sigma-delta analog to digital converter based on iterative algorithm. EURASIP J Adv Signal Process 2012(1):149
14. Weltin-Wu C, Tsividis Y (2013) An event-driven clock less level-crossing ADC with signal-dependent adaptive resolution. IEEE J Solid-State Circ 48(9):2180–2190
15. Venkata Subbarao M, Samundiswary P (2018) Spectrum sensing in cognitive radio networks using time–frequency analysis and modulation recognition. Springer lecture notes in electrical engineering, vol 471, pp 827–837
16. Qaisar SM, Fesquet L, Renaudin M (2014) Adaptive rate filtering a computationally efficient signal processing approach. EURASIP J Adv Signal Process 94:620–630
17. Martinez-Nuevo P, Patil S, Tsividis Y (2015) Derivative level-crossing sampling. IEE Trans Circ Syst II 62(11):11–15
18. Viswanadham R, Sudheer Kumar T, Venkata Subbarao M (2018) A level cross-based nonuniform sampling for mobile applications. Springer lecture notes in electrical engineering, vol 471, pp 767–777

Image Fusion of X-ray Mammography Using Weighted Averaging GA-Based SWT Technique

M. Prema Kumar, V. Veer Raju, and P. Rajesh Kumar

Abstract Nowadays, image fusion of medical images plays a crucial role and enables the radiologists to diagnosis the diseases. X-ray mammogram has become a universal practice for the identification of cancer cells in the breast. Sometimes, a single X-ray mammography is not alone sufficient to express full details about breast cancer to the radiologist. To overcome this problem, equalization of the histogram is chosen to improve the superiority of the X-ray mammography and then image fusion using weighted averaging genetic algorithm-based stationary wavelet transform is planned to enhance the quality of the fused image. An experimental result shows this proposed technique of medical image fusion which gives improved results w.r.t. existing methods.

Keywords Image fusion · X-ray mammography · GA and SWT

1 Introduction

Nowadays, image fusion of medical images plays a vital role and enables the radiologists to diagnosis the diseases. X-ray mammography has become a universal practice for the identification of the cancer cells in the breast. Image fusion combines harmonizing the significant information available from various image sensor data such that the obtained images are pertinent to human visual perception and auxiliary processing. The booming image fusion performed from available imaging types is of giant worth in several fields in image processing. The exploit of a variety of medical imaging systems are fast-growing, so image fusion of multi-modality images plays

M. P. Kumar (✉)
Department of ECE, SVECW (A), Bhimavaram, Andhra Pradesh, India
e-mail: medapatiprem@gmail.com

V. Veer Raju
Department of ECE, Sir CRRCoE, Eluru, Andhra Pradesh, India

P. Rajesh Kumar
Department of ECE, AU College of Engineering (A), Andhra University, Visakhapatnam, Andhra Pradesh, India

P. S. R. Chowdary et al. (eds.), *Microelectronics, Electromagnetics and Telecommunications*, Lecture Notes in Electrical Engineering 655,
https://doi.org/10.1007/978-981-15-3828-5_66

a significant role in the area of medical imaging. The fusion of the multimodality images habitually leads to extra significant medical details that are not manifested in the separate images [1, 2].

The process of image fusion allows the integration of different images from different available sources. The fused image is proficient to balance both the domain's characteristics of resolution.

The image fusion of X-ray mammograms has been used rarely. Normally, one X-ray mammogram has quite a lot of significant information implanted for the doctor to analyze the condition. If more than one input has taken and fused, then it would be further enhanced the inherent data in the final image. Then it would make the radiologist suitable and present with the improved quality output for trouble-free diagnosis of cancer cells in the breast [3–5].

This paper focuses on a new image fusion by using a weighted averaging GA-based SWT technique which is proposed. This technique is enhanced with intensity and exhibits improved experimental results in identifying cancer tissues in the breast when compared with weighted averaging SWT technique of image fusion.

2 Stationary Wavelet Transform-Based Image Fusion

The normal discrete wavelet transform is of time-variant transform in nature. This approach is to restore the shift invariance back to a little special DWT, named undecimated DWT and to define the stationary wavelet transform (SWT) which is shift invariant in nature; it does so by holding back the decimated algorithm in the down-sampling step and by padding zeros between the filter coefficients as an alternative to the up-sampling the filters. SWT with holes is the same as DWT but with an only follow up of down-sampling is concealed which interns the SWT is shift invariant [6–9].

2.1 2-D Stationary Wavelet Transforms

In level 1 SWT, DWT with the decimated form of a signal would be calculated by exploiting the operation of convolution between the signal and impulse responses, i.e., the filter coefficients as in the case of DWT but without using the down-sampling block [10–12] (Fig. 1).

2.2 Design of Basic Genetic Algorithm (GA)

The process of implementation of GA for the selection of weights is explained in Fig. 2.

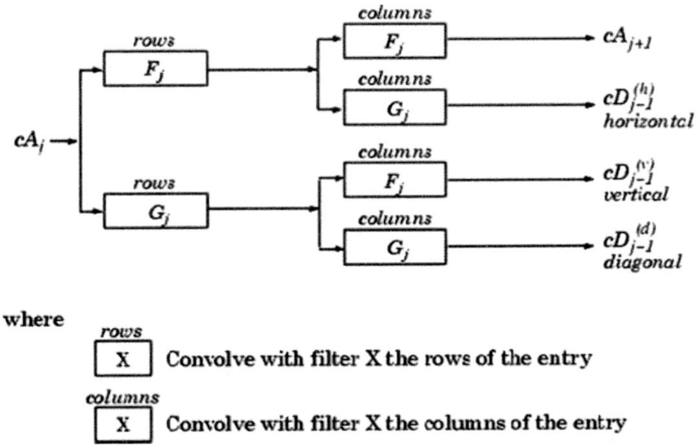

Fig. 1 2-D SWT

Fig. 2 Flow of basic GA

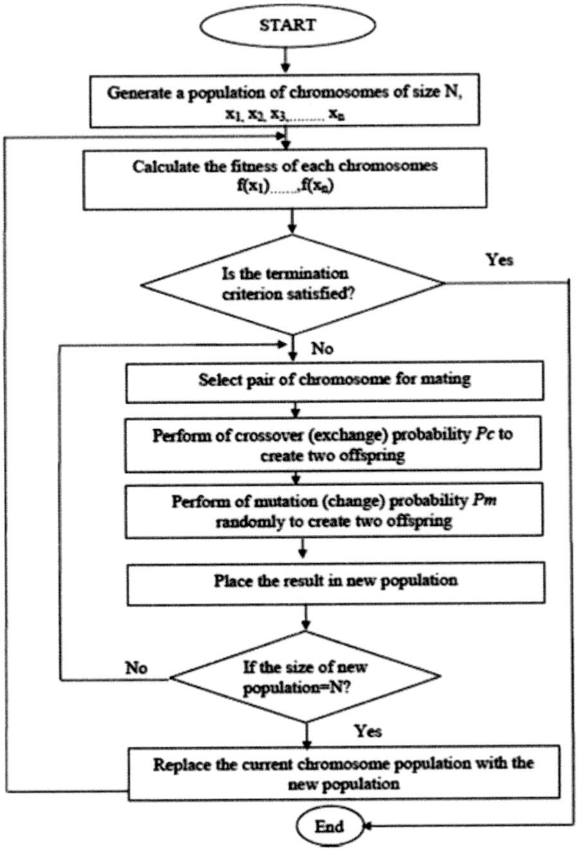

2.3 Justifications for Employing GA

The weights were selected randomly in the case of weighted averaging SWT technique. On trial and error basis, the results were subjectively and objectively observed and the final weights were fixed. But in GA-based weighted averaging SWT technique, the weights are dynamically fixed by the genetic algorithm [13, 14].

3 Implementation

It is proposed in this technique of image fusion that X-ray mammogram from one subject is taken and performs fusion operation by using the following procedure:

- Read input X-ray mammogram.
- Perform histogram equalization on the input image.
- Perform SWT on the input image and histogram equalized image for one level of decomposition.
- Reconstruct the decomposed images.
- Fuse the two output images using weighted averaging method.
- Repeat the above procedure for all the available X-ray mammograms.

The image fusion of X-ray mammogram is performed using weighted averaging GA-based SWT technique and weights are optimized using GA.

SWT-based Simple Averaging Image Fusion:

$$I(a,b) = \frac{\mathrm{SWT}\{I_1(a,b)\} + \mathrm{SWT}\{I_2(a,b)\}}{2} \quad (1)$$

SWT-based Weighted Averaging Image Fusion:

$$I(a,b) = \frac{w_1 * \mathrm{SWT}\{I_1(a,b)\} + w_2 * \mathrm{SWT}\{I_2(a,b)\}}{w_1 + w_2} \quad (2)$$

Weighted Averaging GA-based SWT Technique:

$$I(a,b) = \frac{\mathrm{GA}(w_1) * \mathrm{SWT}\{I_1(a,b)\} + \mathrm{GA}(w_2) * \mathrm{SWT}\{I_2(a,b)\}}{\mathrm{GA}(w_1) + \mathrm{GA}(w_2)} \quad (3)$$

where

$I(a, b) \rightarrow$ Fused image
$I_1(a, b) \rightarrow$ Image 1
$I_2(a, b) \rightarrow$ Image 2
$w_1, w_2 \rightarrow$ Weights.

Finally, the image 1 and image 2 were fused and the image will be generated for quality assessment with the above-mentioned image quality index parameters.

4 Quality Assessment of Fused Images

The quality assessment of fused images is shown qualitatively from Figs. 3, 4, 5, 6, 7, and 8 and quantitatively in Table 1. The Quantitative analysis is carried out by using these image quality assessment parameters like peak signal-to-noise ratio, mean square error, root-mean-square error, mutual information, image quality index, normalized cross-correlation, fusion factor, and fusion symmetry. The mathematical representations of the above-mentioned image quality parameters are given below.

$$\text{PSNR} = 10\log_{10}\left\{\frac{255^2}{\text{MSE}}\right\} \tag{4}$$

$$\text{MSE} = \sum_{m=1}^{M}\sum_{n=1}^{N}[I_1(m,n) - I_2(m,n)]^2 \tag{5}$$

$$\text{NCC} = \frac{\sum_{m=1}^{M}\sum_{n-1}^{N} I_1(m,n)I_2(m+x,n+y)}{\sqrt{\sum_{m=1}^{M} I_1^2(m,n)\sum_{n=1}^{N} I_2^2(m+x,n+y)}} \tag{6}$$

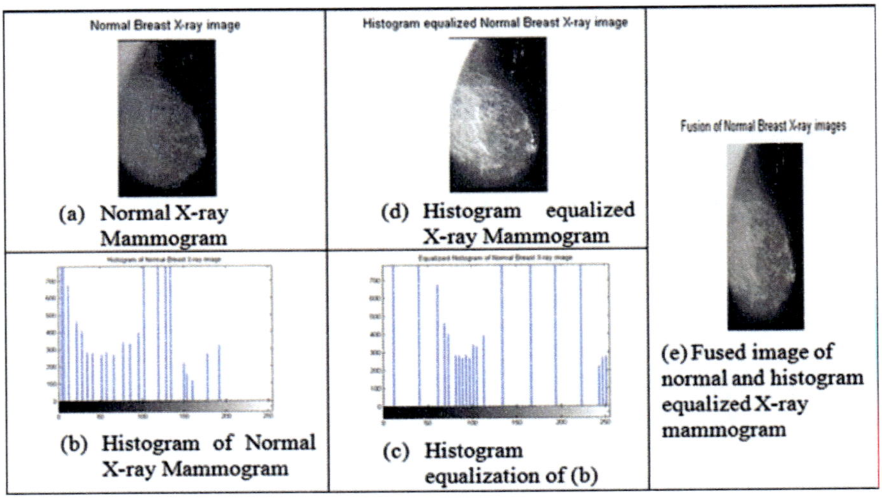

Fig. 3 Image fusion using simple averaging SWT method for normal X-ray mammogram

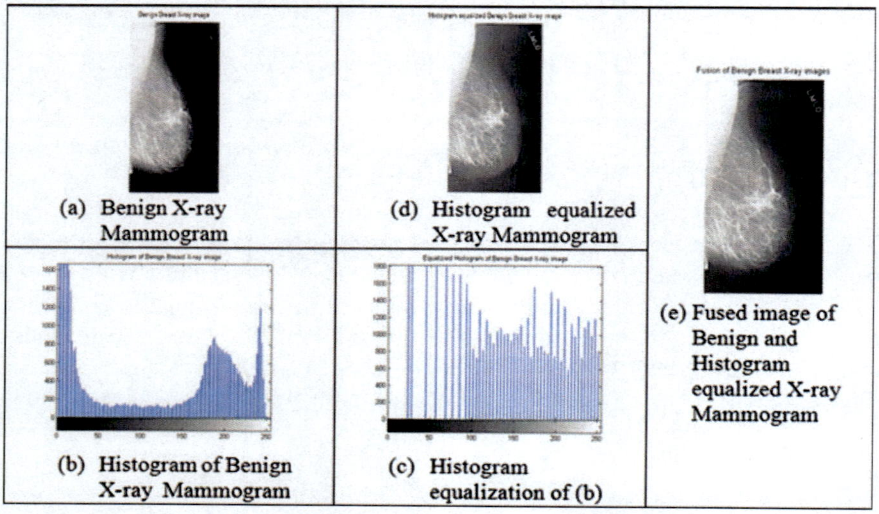

Fig. 4 Image fusion using simple averaging SWT method for benign X-ray mammogram

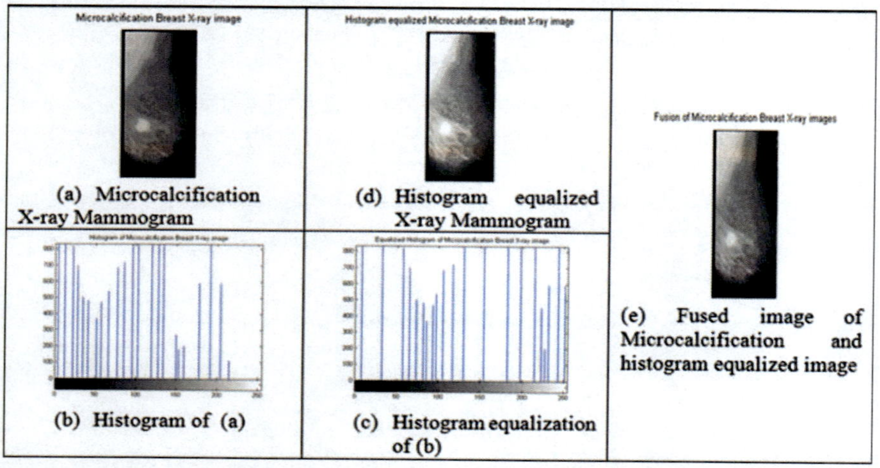

Fig. 5 Image fusion using simple averaging SWT method for microcalcification X-ray mammogram

$$\text{IQI} = \frac{m_{ab}}{m_a m_b} \frac{2xy}{x^2 + y^2} \frac{2m_a m_b}{m_a^2 + m_b^2} \tag{7}$$

$$\text{MI} = \sum_{(f,a)} P_{\text{FA}}(f, a) \log_2 \frac{P_{\text{FA}}(f, a)}{P_F(f) P_A(a)} + \sum_{(f,b)} P_{\text{FB}}(f, b) \log_2 \frac{P_{\text{FA}}(f, b)}{P_F(f) P_b(b)} \tag{8}$$

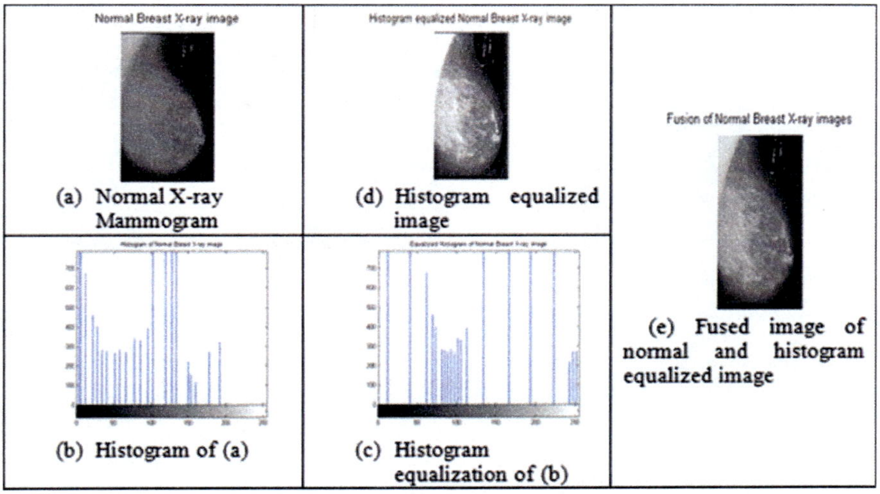

Fig. 6 Image fusion using GA-based SWT technique for normal X-ray mammogram

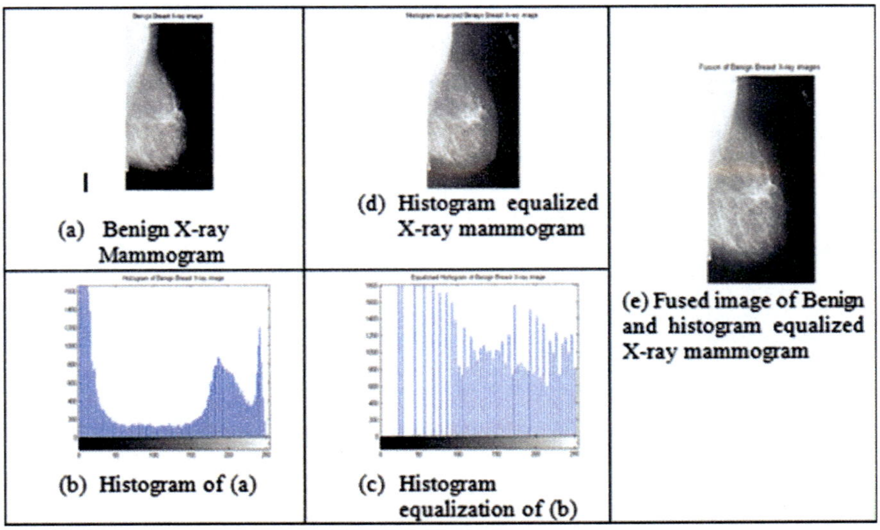

Fig. 7 Image fusion using GA-based SWT technique for benign X-ray mammogram

$$\mathrm{FS} = \mathrm{abs}\left(\frac{\mathrm{MI}_{\mathrm{FA}}(f, a)}{\mathrm{MI}_{\mathrm{FA}}(f, a) + \mathrm{MI}_{\mathrm{FA}}(f, b)} - 0.5\right) \tag{9}$$

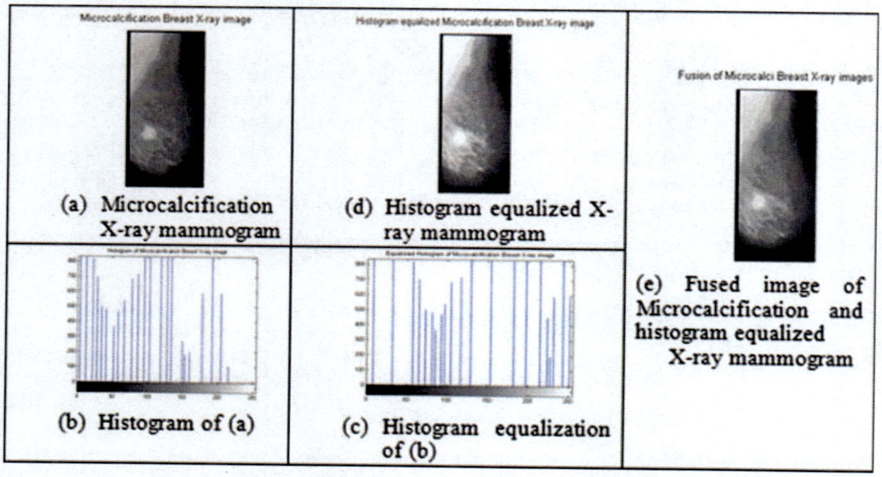

Fig. 8 Image fusion using GA-based SWT technique for microcalcification X-ray mammogram

Table 1 Parameter analysis of image fusion using weighted averaging GA-based SWT technique

S. No.	Parameter	X-ray mammography		
		Normal	Benign	Microcalcification
1	PSNR	21.358	24.259	25.691
2	RMSE	21.810	15.616	13.243
3	IQI	0.845	0.899	0.929
4	FS	0.001	0.081	0.028
5	NCC	0.913	0.851	0.923
6	MI	1.652	1.994	1.614

5 Results

Weighted averaging GA-based SWT technique is proposed to fuse two X-ray mammograms. One X-ray mammogram is taken as input 1 and another X-ray mammogram as input 2 is generated using histogram equalization process of input image 1. An experimental result shows that this method of fusion gives better results than existing methods.

This weighted averaging GA-based SWT technique is verified, and image quality assessment parameters were calculated with the three available types of X-ray mammograms. Then, finally fused output images for available three types of X-ray mammograms using weighted averaging SWT technique and weighted averaging GA-based SWT techniques were shown in Figs. 3, 4, 5, 6, 7, and 8. For better experimental results, the weights were optimized using GA.

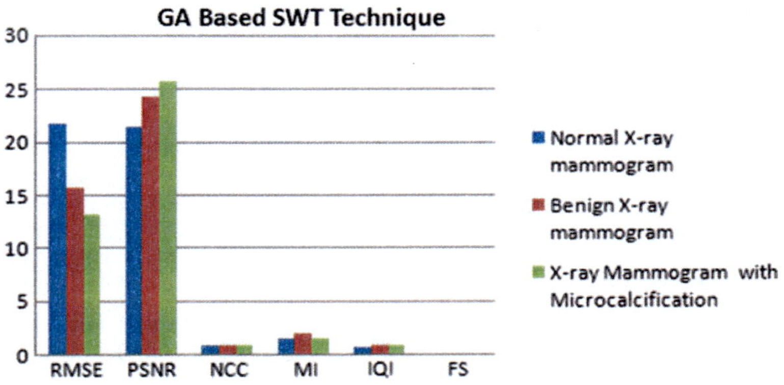

Fig. 9 Parameter analysis of weighted averaging GA-based SWT technique

6 Conclusion

It is accomplished from Table 1 that the weighted averaging GA-based SWT technique gives improved experimental results compared with the weighted averaging SWT technique of image fusion [1].

It is observed from Table 1 and Fig. 9 that is RMSE value significantly reduced which interns PSNR value constantly increasing for X-ray mammogram with microcalcification and benign X-ray mammogram. Figure 9 shows the parameter comparison. Image quality index is also having better value for microcalcification X-ray mammogram when compared with normal and benign X-ray mammograms. As in the case of mutual information benign X-ray mammogram is having better value when compared with normal and microcalcification X-ray mammogram. Coming to fusion symmetry, it is having slightly high value for microcalcification X-ray mammogram when compared with normal and benign X-ray mammograms. In normalized cross-correlation, microcalcification X-ray mammogram has slightly higher values compared with normal and benign X-ray mammograms.

References

1. Prema Kumar M, Sowjanya N, Rajesh Kumar P (2017) Weighted averaging SWT technique for enhanced image fusion in X-ray mammography. In: International conference on microelectronics, electromagnetics and telecommunications ICMEET 2017. LNEE Springer book series, vol 471, pp 739–747
2. Prema Kumar M, Rajesh Kumar P (2014) Weighted averaging DWT technique for enhanced image fusion in mammography. Int J Image Process Vis Commun 3(2):10–15
3. Naidu VPS, Rao JR (2008) Pixel-level image fusion using wavelets and principal component analysis. Defence Sci J 58(3):338–352
4. James AP, Dasarathy BV (2014) Medical image fusion: a survey of the state of the art. In: Information fusion (2014)

5. Wang Z, Clavijo CA, Roessl E, van Stevendaal U, Koehler T, Hausergy N, Stampanoni M (2012) Image fusion scheme for differential phase contrast mammography. In: 7th medical applications of synchrotron radiation workshop (MASR 2012) Shanghai synchrotron radiation facility (SSRF), 17–20, Published By IOP Publishing For Sissa Media lab
6. Venkata Subbarao M, Sayedu Khasim N, Jagadeesh T, Sastry MHH (2013) Hybrid image compression using DWT and neural networks. Int J Adv Sci Technol 53:31–41
7. Vijay Kumar Raju V, Prema Kumar M (2014) Denoising of MRI and X-ray images using dual tree complex wavelet and curvelet transforms. In: International conference on communication and signal processing (ICCSP 2014) on 3–5 Apr 2014, at Adhiparasakthi Engineering College, Melmaruvathur, India, pp 1135–1139
8. Vijay Kumar Raju V, Prema Kumar M (2014) Performance evaluation of image fusion algorithms based on Wavelet Transforms. Int J Electron Commun Eng Adv Res (IJECEAR) 2(2). ISSN: 2347-9329 (Online)
9. Aymaz S, Kose C (2017) Multi-focus image fusion using Stationary Wavelet Transform (SWT) with Principal Component Analysis (PCA). In: 10th international conference on electrical and electronics engineering (ELECO), pp 1176–1180
10. Li M, Dong Y, Wang X (2013) Pixel level image fusion based the wavelet transform. In: 6th international congress on image and signal processing
11. Borwonwatanadelok P, Rattanapitak W, Udomhunsakul S (2009) Multi-focus image fusion based on stationary wavelet transform. In: 2009 international conference on electronic computer technology. 978-0-7695-3559-3/09 $25.00 © (2009) IEEE
12. Naidu VPS, Rao JR (2008) Pixel-level image fusion using wavelets and principal component analysis a comparative analysis. Defence Sci J 58(3):338–352
13. Prema Kumar M, Rajesh Kumar P (2015) Image fusion of mammography images using genetic algorithm (GA). Aust J Basic Appl Sci 9(33):45–50
14. Jyothi V, Rajesh Kumar B, Rao PK, Reddy DVRK (2012) Image fusion using evolutionary algorithms (GA). Int J Comput Technol Appl 2(2)

Comparison on Radar Echo Cancellation Techniques for SAR Jamming

Ch. Anoosha and B. T. Krishna

Abstract To protect a target from detection by SAR, radar jamming is used. Echo cancellation of target is an inventive method which protects the target being detected by radar. There are different types of jamming methods which can prevent the target from detection. In this paper studied different types of echo cancellation methods, their advantages and disadvantages and also the difference between LFM and NLFM radar signals used in echo cancellation methods. Proposed jamming technique uses a nonlinear frequency modulation (NLFM) waveform to increase the output SNR and a nonperiodic interrupted sampling for generation of echo cancellation signal to produce high-amplitude side lobes to perform continuous jamming at the target area.

Keywords NLFM · SAR · Jamming · ISRJ · Echo cancellation

1 Introduction

The environment of jamming is complicated and diverse in real processing of actual combat, thus to bring great challenge to modem radar system [1]. Synthetic aperture radar (SAR) plays a important role in civil and military applications [2]. To protect the target of interest (TOI) from getting detected by the enemy radar, different approaches were developed in past decades called SAR jamming. Echo cancellation is an inventive method that protects the target by cancelling target echo at receiver. The main idea for echo cancellation lies in generating an echo signal except 180° phase shift by an active source [2].

Ch. Anoosha (✉)
ANITS, Visakhapatnam, India
e-mail: anushachukka@gmail.com

B. T. Krishna
JNTU, Kakinada, India

P. S. R. Chowdary et al. (eds.), *Microelectronics, Electromagnetics and Telecommunications*, Lecture Notes in Electrical Engineering 655,
https://doi.org/10.1007/978-981-15-3828-5_67

2 Jamming Signal Methods

Three different types of jamming methods were discussed with model figures.

2.1 *Method 1: To Produce False Target Echo Signals Using Interrupted Sampling Repeater Jamming*

Interrupted Sampling Repeater Jamming (ISRJ) technique produces a sequence of false targets after matched filter. ISRJ generates periodic sampling pulse at some constant rate which samples the radar signal at repeating segments to produce many false echo signals with real target echo signals which forms a deception jamming at SAR receiver. The ISRJ produces multiple false targets with different amplitudes [3]. Model results [3] are shown in Fig. 1.

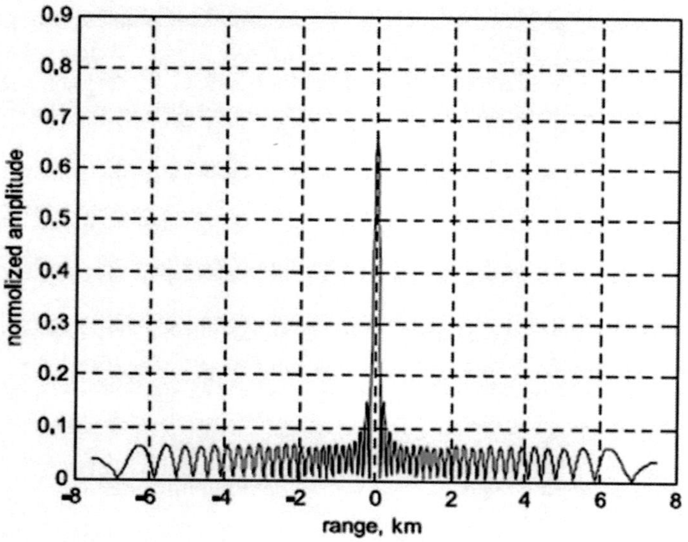

Fig. 1 Amplitude of target echo after match filtering

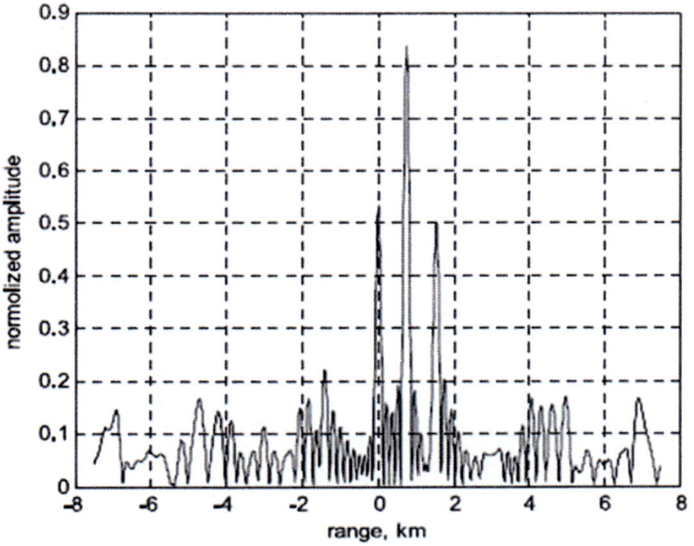

Fig. 2 Amplitude of jamming signal after match filtering

2.2 *Method 2: To Cancel the Radar Target Echo Signal by Generating Synthesized Replica of Target Echo with 180° Phase Using ISRJ Method*

In this, the idea is to produce a cancellation echo same as target echo but with 180° phase shift due to which both target echo and cancellation echo cancel each other when received simultaneously by the radar receiver and also to produce a high-amplitude false echoes using ISRJ to deceive the SAR [4]. With this idea, echo cancellation method using ISRJ with a phase shifter is proposed. So, the ISRJ signal ideally cancels radar target echo and also generates multiple false targets. Therefore, a better jamming is performed compared with the Method 1. Model results are shown in Figs. 2 and 3 [4].

2.3 *Method 3: Cancellation of Radar Target Echo Signal Using a Frequency Shift Modulation*

In the above method, echo cancellation signal generated is same as target echo with 180° phase shift. After receiving at the same instant by the radar antenna, the original target echo and generated cancellation echo cancel each other as they are of 180° out of phase and synchronized in time [2]. But there is a possibility that the cancellation echo may differ with the desired echo in time domain. This is due to the processing

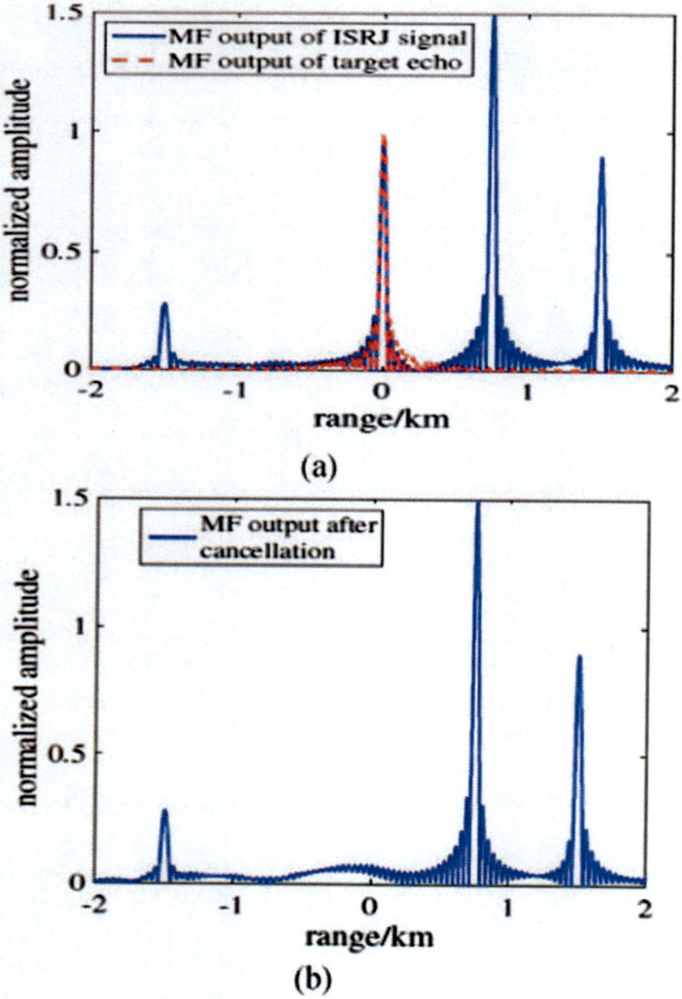

Fig. 3 **a** Output of the matched filter before cancellation and **b** matched filter output after echo cancellation

time, required to generate cancellation echo when SAR transmits and receives which results in cancellation failure. The proposed method can generate accurately the cancellation echo [4]. By using the property called range-Doppler coupling for linear FM signal [5], which shifts the pulse in time proportional to change in Doppler frequency. By using this property, we can make LFM radar as frequency shifting deception jamming [6]. In a positive-slope LFM, the peak occurs earlier in time when compared to stationary target. By using this concept, if we add a frequency shift on the target echo assuming existence of delay, then the problem discussed above

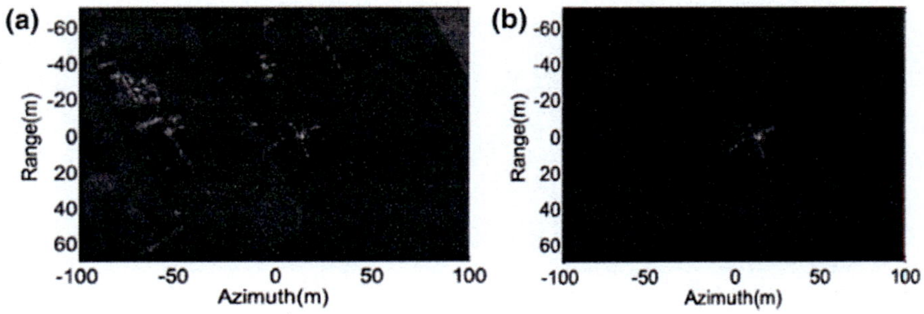

Fig. 4 **a** SAR image of real scene and **b** SAR image of the target

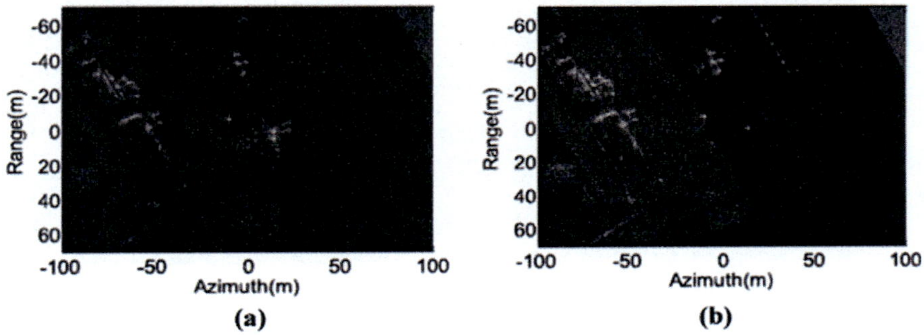

Fig. 5 **a** SAR image after cancellation of echo signal when delay is present in cancellation signal and **b** SAR image after cancellation of echo signal when frequency shift f_d is added in cancellation signal to avoid the delay present

can be solved. So, the cancellation echo can be generated by modulating intercepted target echo with frequency shift *fd*. Model results are shown in Figs. 4 and 5 [4].

3 Signal Model

The sampling function is a rectangular function with n number of pulses denoted as $g(t)$, then

$$g(t) = \text{rect}\left(\frac{t}{\tau_p}\right) * \sum_{-\infty}^{\infty} \delta(t - (n * T_s)) \tag{1}$$

where t denotes the time, τ_p is width of the pulse, T_s is the time period between pulses called pulse repetition interval (PRI), δ(.) is the impulse or delta function, n is number of pulses count and * is convolution operation in time domain [3, 4].

The Fourier transform of $g(t)$ is given as

$$G(f) = T_p f_s \sum_{-\infty}^{\infty} \sin c\big(nT_p f_s\big)\delta(f - nf_s) \tag{2}$$

where $\sin c(x) = \sin(px)/(px)$, and $f_s = 1/T_s$.

The LFM signal is denoted as

$$x(t) = \frac{1}{\sqrt{T}} \mathrm{rect}\left(\frac{t}{T}\right) \exp\big(j\pi\mu t^2\big) \tag{3}$$

T = pulse width, μ = chirp rate.

The interrupted sampling LFM signal

$$d(t) = g(t) * x(t) \tag{4}$$

The matched filter output of ISRJ signal is a 'n' no. of false targets at radar receiver which is given by

$$y(t) = \sum_{-\infty}^{\infty} a_n y_{sn}(t) \tag{5}$$

where

$$a_n = T_p f_s \sin c\big(nT_p f_s\big) \tag{6}$$

$$y_{sn}(t) = \sin c[(\mu t + nf_s)(T - |t|)]\left(1 - \frac{|t|}{T}\right) \exp(j\pi n f_s t) \tag{7}$$

represents the final LFM output of nth-order false target.

After adding frequency delay, the ISRJ signal becomes

$$d'(t) = d(t)\exp(j2\pi f_d t) = g(t)x(t)\exp(j2\pi f_d t) \tag{8}$$

Let $x'(t) = x(t)\exp(j2\pi f_d t)$ is the linear frequency-modulated signal after adding delay in terms of frequency f_d. The output at the receiver of the matched filter is [4]

$$d_{mf}(t) = \sin c[(\mu t + nf_s)(T - |t|)]\left(1 - \frac{|t|}{T}\right) \exp(j\pi f_d t) \tag{9}$$

The matched filter output of sampling repeater jamming signal with frequency delay modulation will be

$$y(t) = \sum_{-\infty}^{\infty} a_n y'_{\mathrm{sn}}(t) \tag{10}$$

$$y'_{\mathrm{sn}}(t) = \sin c[(\mu t + nf_{\mathrm{s}} + f_{\mathrm{d}})(T - |t|)]\left(1 - \frac{|t|}{\mathrm{T}}\right)\exp(j\pi(f_{\mathrm{d}} + nf_{\mathrm{s}})t) \tag{11}$$

represents the final LFM output of the *nth*-order false target.

4 Proposed Method

Improved radar target echo cancellation using nonperiodic ISRJ for a nonlinear FM pulse waveform:

The first proposed method is if there is an amplitude mismatch between target echo and generated cancellation echo, then the original echo cannot be cancelled completely and remains visible to the SAR image. Even with the mismatch in amplitudes, the TOI can be jammed by applying an improved echo cancellation method. An improved cancellation method is based on nonperiodic interrupted sampling repeater jamming (NP-ISRJ), which generates very high-amplitude side lobes which form a constant jamming in the area where target is present [6]. If this proposed modified method is applied, the TOI area is totally masked by jamming; thus, a stronger echo cancellation is obtained even there is an amplitude mismatch. Secondly, in all previous methods the linear frequency modulation (LFM) waveform is commonly used radar signal. The pulse-compressed LFM signal at the receiver which uses a weighing function produces the first side lobe at a level of -13 dB with respect to main lobe [7] which is high and reduces the output SNR by 1–2 dB. In an attempt to attain low side lobe level without applying weighting function, nonlinear frequency modulation (NLFM) signal can be used [8]. So, to increase the echo cancellation signal amplitude at the output of matched filter, a NLFM is used. Model results are shown in Fig. 6 [6].

5 Conclusion

In this paper studied the detailed echo cancellation techniques to protect the target of interest to get detected by SAR. In Method 1, a train of false targets is generated using ISRJ technique. But any how the SAR receives the echoes of target with the false targets which cannot be ignored. Therefore, SAR image can target by removing the jamming signal and retains desired echo signal. In Method 2 produced a cancellation echo same as target echo in amplitude but with 180° phase shift by ISRJ. As the cancellation echo and target echo are same in amplitude and in the time domain, both get cancel each other perfectly which is shown in Fig. 2. The techniques developed are on assumption that cancellation echo and target echo synchronize in the time

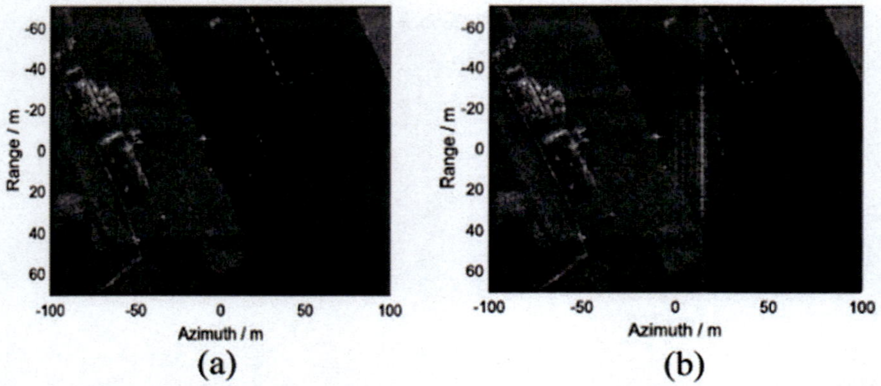

Fig. 6 **a** Cancellation using Method 2 and Method 3 with amplitude match and **b** cancellation of target echo using proposed method

domain, but they both can lag behind due to processing time. Thus, they cannot be cancelled perfectly. In Method 3, frequency shift modulation is used to synchronize the cancellation echo and target echo in the time domain to cancel the original echo totally. In the proposed method, NLFM radar signal is used to increase the output SNR and to reduce side lobes, and when there is a mismatch in amplitude, the real echo cannot be cancelled completely. Therefore, a better echo cancellation method using nonperiodic interrupted sampling repeater jamming (NP-ISRJ) is used, which has very high-amplitude side lobes and gives a continuous jamming. For understanding model results are shown from already published papers.

References

1. Anoosha Ch, Kusmakumari Ch, Nirmala M (2013) Detection and cancellation of jamming signal noise using digital filters for radar applications. Int J Electron Commun Comput Eng 4(5). ISSN (Online): 2249–071X, ISSN (Print): 2278–4209
2. Xu L, Feng D, Liu Y, Pan X, Wang X (2015) A three-stage active cancellation method against synthetic aperture radar. IEEE Sens J 15:6173–6178
3. Feng D, Xu L, Wang W, Yang H (2014) Radar target echo cancellation using interrupted—sampling repeater. IEICE Electron Exp 11(8):1–6
4. Wu Q, Zhao F, Wang J, Liu X, Xiao S (2019) Improved ISRJ-based radar target echo cancellation using frequency shifting modulation. Electronics 46. https://doi.org/10.3390/electronics8010046
5. Skolnik M (2001) Introduction to radar systems. McGraw-Hill, Boston, MA, USA
6. Wu Q, Liu J, Wang J, Zhao F, Xiao S (2018) Improved active echo cancellation against synthetic aperture radar based on non periodic interrupted sampling modulation. IEEE Sens J 18:4453–4461
7. Parwana S, Kumar S (2015) Analysis of LFM and NLFM radar waveforms and their performance analysis, IRJET 2(2)
8. Lesnik C (2009) Nonlinear frequency modulated signal design, optical and acoustical methods in science and technology. Acta Physica Pol A 116

Design of Nanoscale Square Ring Resonator Band-Pass Filter Using Metal–Insulator–Metal

Surendra Kumar Bitra and M. Sridhar

Abstract In this paper, metal–insulator–metal (MIM)-based single-band plasmonic band-pass filters (BPF) design and analysis is presented. The basic design comprises a square ring resonator (SRR) which is coupled using straight waveguide with gap. The straight waveguide SRR is responsible for a single-band-operating wavelength (1355 nm, i.e., 221.24 THz). Design and simulations are carried out using finite-difference time-domain (FDTD)-based solver embedded in the Computer Simulation Technology (CST) Microwave Studio suite. The proposed filters are used for plasmonic single-band band-pass filter (BPF) applications in photonic integrated circuits (PICs).

Keywords MIM · Photonic integrated circuits · SRR · BPF

1 Introduction

The surface plasmon (SP) refers to the sea of electrons, which are freely commuting along with the interface of two different materials with the positive and negative reflection coefficients, respectively. Typically, such an interface is emphasized by the metal–insulator–metal (MIM) arrangement. The recent research on surface plasmon yielded several outcomes is referred to as SP-based devices. These devices are meant for optical and electromagnetic applications [1, 2]. The charge carriers moving in the field at the interface of two different materials are called surface plasmon polaritons (SPP), thereby the whole structure serves as a guiding means of the electromagnetic signal. In due course, with reference to the available literature, it is evident that the structure took several geometrical shapes preserving the guiding mechanism. Recently, the nanoparticle array and nanowires are developed using this technology [3–5]. Further, the SPP-based MIM waveguides were developed along with other geometry like grooves and wedges [6–9].

S. K. Bitra (✉) · M. Sridhar
Department of Electronics and Communication Engineering, Koneru Lakshmaiah Educational Foundation, Vaddeswaram, Guntur, Andhra Pradesh, India
e-mail: bitrasurendrakumar@gmail.com

P. S. R. Chowdary et al. (eds.), *Microelectronics, Electromagnetics and Telecommunications*, Lecture Notes in Electrical Engineering 655,
https://doi.org/10.1007/978-981-15-3828-5_68

The MIM-based waveguides and other structures make use of their inherent characteristic nature of supporting propagation through the SPP mode [2]. Hence, they have emerged as devices which can replace or perform like several other optical and microwave devices [4, 5]. This phenomenon is intelligently utilized to design and develop different-shaped waveguides and other waveguide components. These components include couplers, splitters and combiners of power [5–10]. Recent developments in the surface plasmon devices attribute to the above statement where the MIM-based devices have slowly started replacing all microwave devices with optical systems. SP is an excellent choice with the good slope in being an active part of modern devices [1–5].

The MIM structures form a strong candidate which facilitates surface plasmon polaritons (SPPs). The SPP mode can be easily excited and propagated through the interface. Very high frequencies of the range of terahertz have a very low wavelength. These shorter wavelengths are interestingly necessary while using nanoscale devices [10]. MIM-based rectangular ring resonator band-pass filter is designed and analysed using coupling gaps between input and output ports which are implemented in [11, 12]. Recently, several ring-shaped resonators are proposed for filter applications [13, 14]. These rings are typically square and rectangular. Similarly, some literature suggested ring which is tooth-shaped [15] and Bragg grating [10] filters. The merit of high transmission at 90° bends rectangular ring resonators is advantageous than circular ring resonator due to ease of fabrication.

In this paper, MIM-based waveguide structure is developed for single-band application. The analysis is carried out in terms of reflection and transmission coefficients primarily. However, the impedance characteristics curve and the effective index plots are useful for further analysis of the designs. The simulation is carried out in the efficient EM modelling tool known as CST Microwave Studio suite. Further, the paper is organized into four sections. A brief insight of the MIM modelling is reviewed in Sect. 2. The design of single-band filter is demonstrated in Sect. 3 along with simulation results. The overall conclusion is mentioned in Sect. 4.

2 Modelling of MIM Waveguide

In MIM waveguide, the insulator is sandwiched between two metal waveguides as shown in Fig. 1. The silica ($\varepsilon_i = 2.50$) is used as the insulator, and silver is used as metal. The corresponding electrical properties like the dielectric constant of the material/metal used are generalized with the following expression [16].

$$\varepsilon(\omega) = \varepsilon_\infty - \frac{\omega_p^2}{\omega(\omega + j\gamma)} \tag{1}$$

Here, ε_∞ (=3.7) is permittivity of the metal, ω_p (=1.38×10^{16}) is the plasma frequency, and γ (=2.73×10^{13} rad/s) is the damping frequency of the metal. The MIM

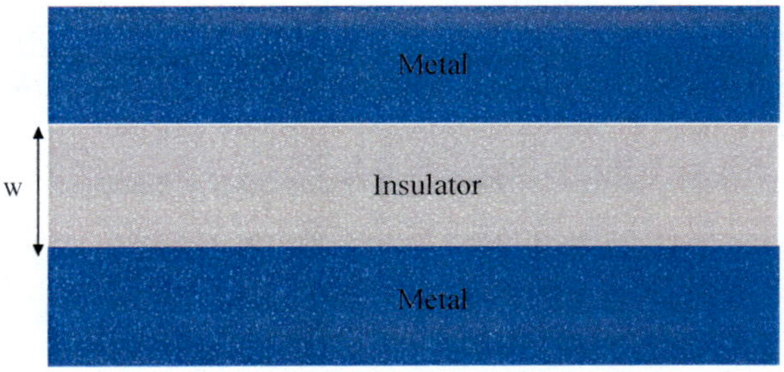

Fig. 1 Basic MIM waveguide

waveguide is designed, simulated and analysed using FDTD-based CST Microwave Studio suite. The perfect-matched layer (PML) boundaries are used to investigate the waveguide, and the mesh sizes are taken as 5 nm × 5 nm. Blue colour indicates Metal and grey colour indicates insulator throughout the paper.

The propagation constant of the transmission line is $\gamma = \alpha + j\beta$. Here, the attenuation constant is referred by α. Similarly, the phase constant is represented by β. Fig. 2 represents the propagation length of the SPP waveguide which is $L_{\text{spp}} = 1/2|\alpha|$. The real part of the effective index (EI) is given by the expression $\text{Re}(N_{\text{eff}}) = \beta/k_0$. Here, k_0 is the wavenumber. The variation of the EI with respect to the MIM waveguide width (W) is shown in Fig. 2. It can be inferred from the plot that the **real part of the EI significantly diminishes with the increase in waveguide width and** decreases relatively slowly as wavelength (λ) increases. It shows that the waveguide dispersion properties are dependent on width variation.

As the width of the insulator region increases, the characteristic impedance (Z_0) is also decreasing as shown in Fig. 3. For $W = 54$ nm width, the maximum characteristic impedance is 85 Ω at 1000 nm and minimum Z_0 is 71 Ω at 2000 nm. According to the observations, as the height of the MIM waveguide increases, the characteristic impedance decreases. Figure 4 **shows the field distribution of MIM waveguide.**

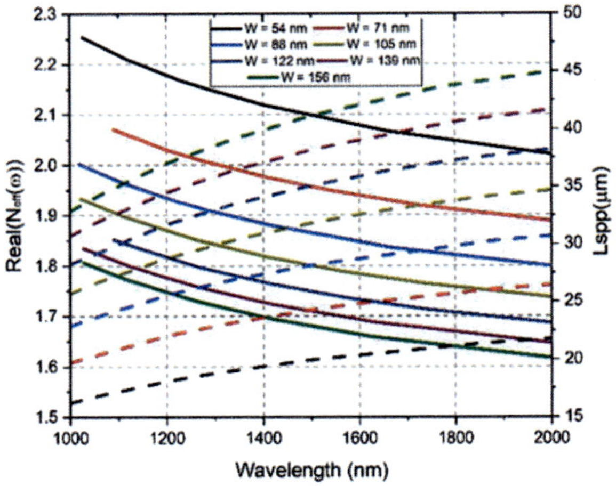

Fig. 2 Real part of EI versus propagation length of SPP (L_{spp}) for several widths of insulator in MIM waveguide

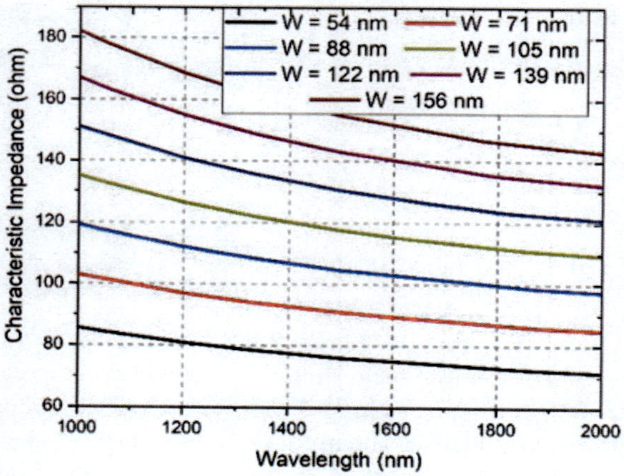

Fig. 3 Characteristic impedance versus wavelength for several widths of MIM waveguide

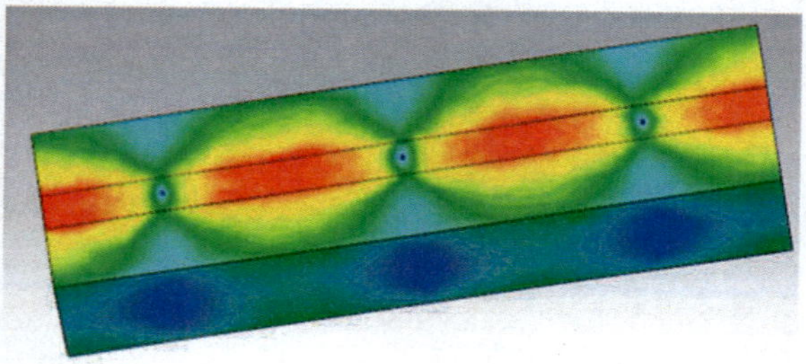

Fig. 4 Field distribution of MIM waveguide

3 Single-Band Filter

The design of the single-band filter is presented in this section. The typical geometry of the SRR has a re-entrant waveguide. The waveguide is arranged in the close proximity of a linear and straight waveguide. The length of square loop is generally given as $L = N\lambda_g$. Here, the term N refers to an integer. When the frequency of the input is proportional to the cavity's resonance length, the energy is coupled to the cavity. As a result, the output typically falls down. The waveguiding length in the ring is the average of the inner and outer perimeter, $L = 4 \times L1$ (=810 nm), where $L1$ is the edge length. d is the thickness of the ring and is 50 nm. The proposed single-band SRR is shown in Fig. 5. The gap between the ring and the waveguide has an effect on the coupling characteristics of the cavity and waveguide. The single-band SRR reflection and transmission coefficients are shown in Fig. 6. Fig. 6 shows the single-band SRR which is operating in 1345 nm with the reflection coefficient of −47.3 dB, and the corresponding transmission coefficient is ~1 dB. The field distribution of the SRR is shown in Fig. 7.

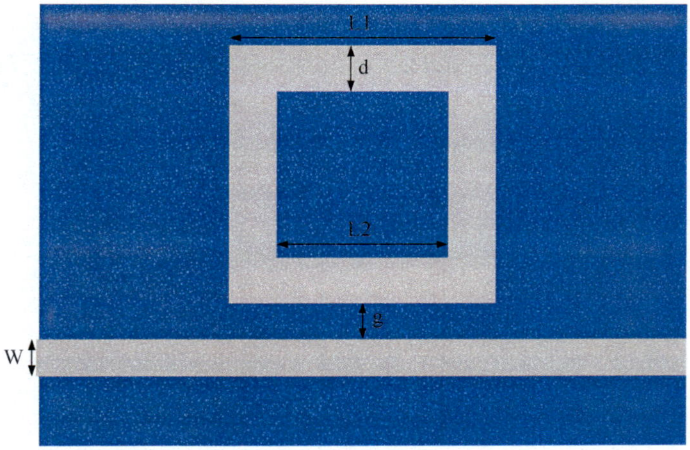

Fig. 5 Single-band SRR

Fig. 6 Reflection and transmission coefficients of single-band SRR

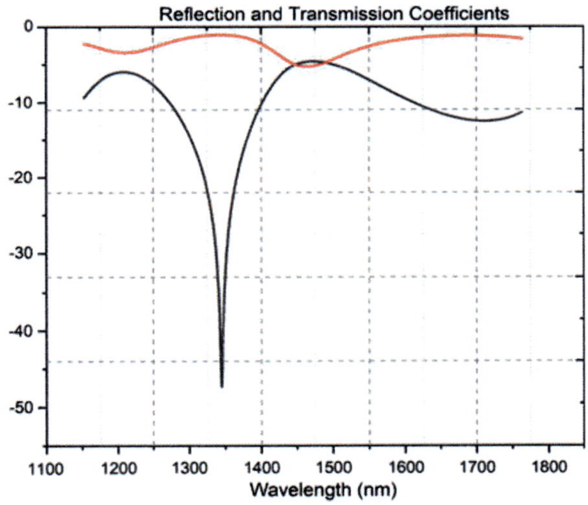

4 Conclusions

The single-band SRR is simulated and numerically analysed in this paper. The length of the square ring is 810 nm for single filter. The square ring resonators are chosen due to ease of fabrication. The single-band filter is designed to operate in 1345 nm (221.24 THz). The proposed resonate filter transmission characteristics are in a good agreement with theoretical calculations. The single-band BPF is used mainly in photonic integrated optical circuits.

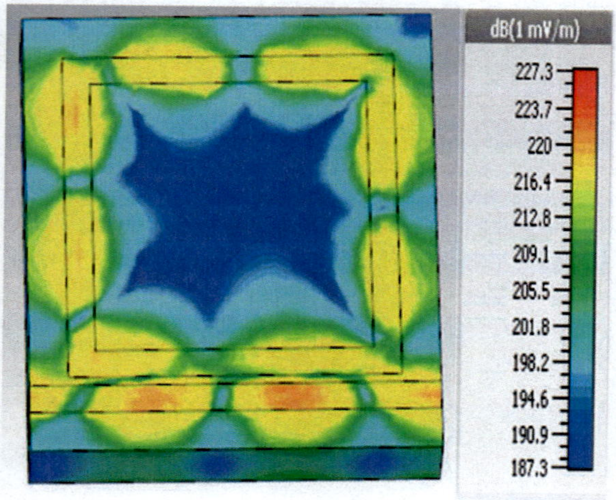

Fig. 7 Field distribution of single-band SRR at 1355 nm (221.24 THz)

References

1. Ozbay E (2006) Plasmonics: merging photonics and electronics at nanoscale dimensions. Science (80-.).311:189–193
2. Atwater HA (2007) TiN Harsha 13 the promise of plasmonics 56–63
3. Barnes WL, Dereux A, Ebbesen TW (2003) Surface plasmon subwavelength optics. Nature. 424:824–830
4. Wassel HMG et al (2012) Opportunities and challenges of using plasmonic components in nanophotonic architectures. IEEE J Emerg Sel Top Circ Syst 2:154–168
5. He XY, Wang QJ, Yu SF (2012) Numerical study of gain-assisted terahertz hybrid plasmonic waveguide. Plasmonics 7:571–577
6. Okamoto K et al (2004) Surface-plasmon-enhanced light emitters based on InGaN quantum wells. Nat Mater 3:601–605
7. Reineck P et al (2012) A solid-state plasmonic solar cell via metal nanoparticle self-assembly. Adv Mater 24:4750–4755
8. Pile DFP, Gramotnev DK (2005) Plasmonic subwavelength waveguides: next to zero losses at sharp bends. Opt Lett 30:1186
9. Dionne JA, Sweatlock LA, Atwater HA, Polman A (2005) Planar metal plasmon waveguides: frequency-dependent dispersion, propagation, localization, and loss beyond the free electron model. Phys Rev B Condens Matter Mater Phys 72:1–11
10. Pile DFP et al (2005) Two-dimensionally localized modes of a nanoscale gap plasmon waveguide. Appl Phys Lett 87:1–3
11. Hosseini A, Massoud Y (2007) Nanoscale surface plasmon based resonator using rectangular geometry. Appl Phys Lett 90:2–5
12. **Yun B, Hu G, Cui Y (2010) Theoretical analysis of a nanoscale plasmonic filter based on a** rectangular metal-insulator-metal waveguide. J Phys D Appl Phys 43
13. **Xiao S, Liu L, Qiu M (2006) Resonator channel drop filters in a plasmon-polaritons metal** 14:847–848
14. Wang T-B et al (2003) Surface plasmons; (140.4780) optical resonators; (130.7408) wavelength **filter. Nature 424:824–830**
15. **Lin X-S, Huang X-G (2008) Tooth-shaped plasmonic waveguide filters with nanometeric sizes.** Opt Lett 33:2874
16. Maier SA (2007) TiN Modulator 7 Plasmonics: fundamentals and applications

Comparative Analysis of Sentiment Analysis Between All Bigrams and Selective Adverb/Adjective Bigrams

Mounicasri Valavala and Hemalatha Indukuri

Abstract Increase in the amount of unstructured data across different platforms serves as a valuable resource for predicting market trends, analyzing product features, and considering the customer sentiment in designing new features/products. The sentiment of unstructured data such as tweets, Facebook comments, and web reviews is calculated by using the polarity and intensity of the words, whereas polarity indicates positive or negative sentiment, and intensity indicates the strength of polarity. In this paper, a comparative study of sentiment analysis performance and accuracy between all bigrams and selective adverb/adjective bigrams is done. The outcome of this research will serve as a metric for both academia and industry to implement sentiment analysis projects.

Keywords Sentiment analysis · Bigrams · POS tagging

1 Introduction

The topics discussed in social media range from politics to market trends, technology to entertainment, and finance to sports, and so on. The analysis of opinions from social media content drives the decision-making process in business today [1] and can be done at different levels such as document level and aspect level [2]. The data needs to be preprocessed prior to analysis [3], as the raw data contains considerable amount of outliers [4]. Sentiment analysis can be performed at different levels such as document level and aspect level [2]. Sentiment analysis can be done in two different ways, such as lexicon-based and machine learning [5]. Lexicon-based design is based on rule-based algorithm that classifies the data based on "if–then" rules [6].

M. Valavala (✉)
University of the Cumberlands, Williamsburg, KY 40769, USA
e-mail: mvalavala1456@ucumberlands.edu

H. Indukuri
S.R.K.R. Engineering College, Bhimavaram, Andhra Pradesh 534204, India
e-mail: indukurihemalatha@gmail.com

P. S. R. Chowdary et al. (eds.), *Microelectronics, Electromagnetics and Telecommunications*, Lecture Notes in Electrical Engineering 655,
https://doi.org/10.1007/978-981-15-3828-5_69

In this approach, the sentence is represented as a bag of words [7] and important aspects that contribute to sentiment are polarity and intensity [8]. The advantage of using a lexicon-based approach over machine learning is that there is no need to train the dataset [7, 9]. Bigram analysis is an important aspect of lexicon-based model, as the sentiment of preceding word impacts its succeeding [3].

Adverbs and adjectives are of high intensity, whereas verbs are of marginal strength and nouns are of slight intensity [10]. The negation words can reverse the polarity, whereas modifiers contribute to intensity [11]. The sentiment is calculated by applying rule-based algorithm using decision tree induction (DTI) architecture. DTI is a non-parametric algorithm, which can deal with complex and large datasets [12].

2 Methodology

Sentiment analysis is performed with two different approaches, and both the ideas are discussed below in detail. Bigrams are chosen, as there is a high chance of predecessor word impacting the next word. We used Python to implement the algorithms. Intensity and polarity of the sentiment are considered as the key features for the classification. The algorithms are not subjected to overfitting as there are not too many features, and the values of features are all in similar scaling, i.e., from 0 to 1. The POS tagger available in TextBlob module of Python is used in order to identify parts of speech (POS). The final result for each sentence runs from 0 to 1.

The sentences are divided into bigrams, and the analysis is done in two ways. In the first approach, i.e., the all bigrams algorithm, polarities of both the words are taken into consideration, whereas in the second approach, i.e., selective adverb/adjective bigrams algorithm, the polarity of the first word is considered if and only if it is an adverb or an adjective. In both the approaches, the polarity of the second word is always taken into account in the sentiment calculation process. All the sentences in the document will be classified as positive or negative or neutral. The sentiment strength is assigned from SentiWordNet [13] (Fig. 1).

2.1 Dataset

The labeled datasets with tweets about iPhone, Apple, Google, Android, and iPad are retrieved [14]. The dataset contains columns such as text, product, and sentiment. The text column of the dataset contains the tweets content, product column contains the subject of the analysis, and sentiment column contains polarity of the tweet. The results from the algorithms are compared against these sentiment columns of the dataset, to calculate precision and recall, which are further used for F-score calculation. However, the dataset does not have intensity of the tweets, which is calculated by both all bigrams and selective adverb/adjective bigrams algorithms described in this paper (Fig. 2).

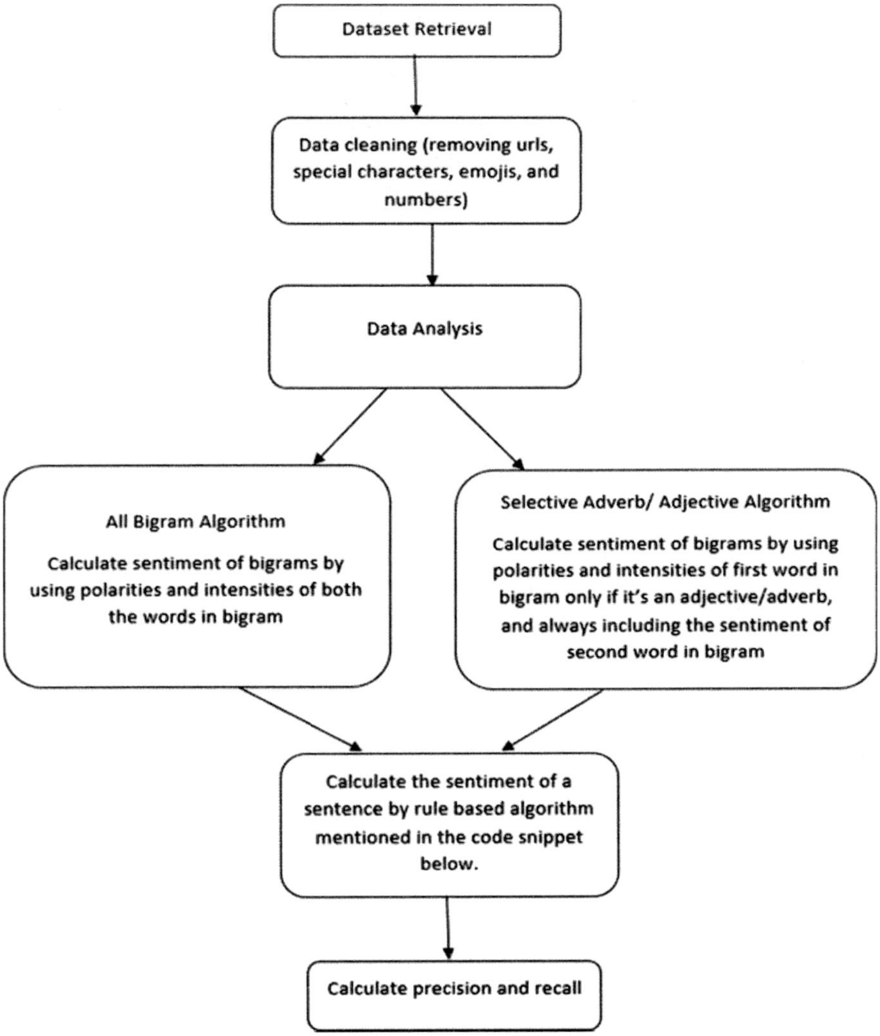

Fig. 1 Flowchart of sentiment analysis

2.2 Data Cleaning

The tweets mentioned in this dataset have special characters, numbers, URLs, and hashtags, which does not contribute to sentiment. Hence, all these are cleaned up from the text, in order to reduce the processing time and memory usage. Regex patterns are used in Python in order to perform the cleanup. Emojis are not considered as part of the analysis; hence, the emojis are also eliminated.

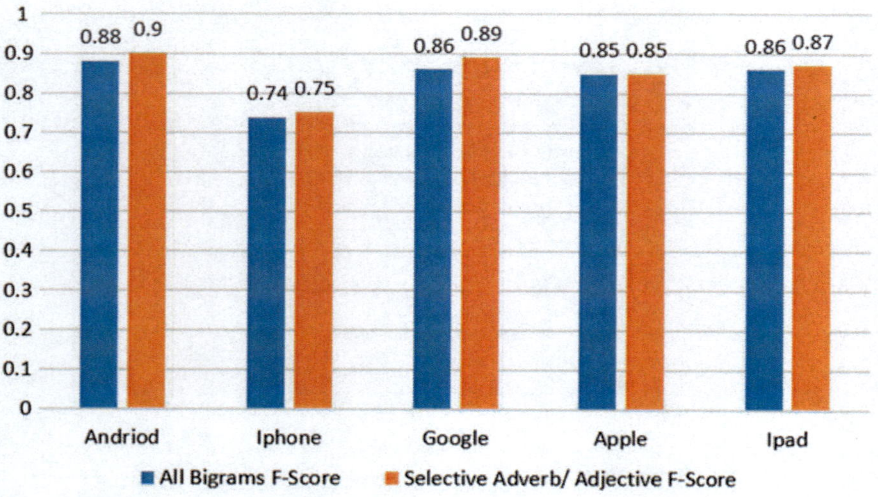

Fig. 2 Comparison of *F*-score between the bigram algorithms

2.3 Algorithms

All Bigrams Algorithm: The polarity and intensity of both the words in the bigram are considered, and the sentiment of the bigram is calculated by calculating mean of sentiments of both the words. This algorithm does not consider parts of speech of a sentence. The bigram polarities are fed into the rule-based algorithm, which calculates the final sentiment of the sentence. Below is an example that divides the tweet into bigrams and assigns sentiment value.

Tweet: The iPhone is a transient device used in short bursts, and the iPad is an afterpm on the couch device mention.

All Bigram Sentiment: {['the', 'iPhone']: [0, 0], ['iPhone', 'is']:[0, 0], ['is', 'a']:[0, 0], ['a', 'transient']:[0, 0.0], ['transient', 'device']:[0.0, 0.21143], ['device', 'used']:[0.21143, 0], ['used', 'in']:[0, 0], ['in', 'short']:[0, 0.15983], ['short', 'bursts']:[0.15983, 0], ['bursts', 'the']:[0, 0], ['the', 'iPad']:[0, 0], ['iPad', 'is']:[0, 0], ['is', 'an']:[0, 0], ['an', 'afterpm']:[0, 0], ['afterpm', 'on']:[0, 0], ['on', 'the']:[0, 0], ['the', 'couch']:[0, 0.0], ['couch', 'device']:[0.0, 0.21143], ['device', 'mention']:[0.21143, 0.12903]}

Selective Adverb/Adjective Algorithm: The polarity and intensity of the first word in bigram are considered only when it is an adverb or adjective. However, the sentiment of the second word is always taken into consideration, irrespective of its parts of speech tagging. The sentiment of the bigram is calculated as the mean of sentiments of both the words and is fed into the analysis process. Below is an example where the tweets are divided into bigrams, and sentiment is assigned to predecessor based on its POS tagging.

Tweet: The iPhone is a transient device used in short bursts, and the iPad is an afterpm on the couch device mention.

Bigram Sentiment: {['the', 'iPhone']:[0, 0], ['iPhone', 'is']:[0, 0], ['is', 'a']:[0, 0], ['a', 'transient']:[0, 0.0], ['transient', 'device']:[0, 0.21143], ['device', 'used']:[0, 0], ['used', 'in']:[0, 0], ['in', 'short']:[0, 0.15983], ['short', 'bursts']:[0.15983, 0], ['bursts', 'the']:[0, 0], ['the', 'iPad']:[0, 0], ['iPad', 'is']:[0, 0], ['is', 'an']:[0, 0], ['an', 'afterpm']:[0, 0], ['afterpm', 'on']:[0,0], ['on', 'the']:[0, 0], ['the', 'couch']:[0, 0.0], ['couch', 'device']:[0, 0.21143], ['device', 'mention']:[0, 0.12903]}

Observation: The sentiment of the word "device" which is a noun is 0.21143, when it is in the second position of bigram, whereas it is 0 when it is in the first position of bigram. The reason for being 0 is that it is not an adverb or adjective; hence, its sentiment contribution on the following word is ignored according to selective adverb/adjective algorithm.

2.4 Data Analysis

Rule-Based Analysis: The rule-based algorithm calculates the sentiment and intensity based on the polarity scores of the words. If the polarity of the first word is negative, then it adds up the negative impact to the second word, which makes the bigram polarity negative; similarly, if the polarity of the first word is positive and polarity of the second word is negative, then both add up to negative polarity for bigram. On the other hand, if the polarities of both the words are positive, it will increment the positive sentiment. The total intensity and polarity of the sentence are calculated by adding up all the bigram polarities and divide by the number of words, which keeps the sentiment to fall on the same scale.

2.5 Program Code

The below program code shows the logic of rule-based algorithms used to perform lexicon-based sentiment analysis.

```
polarity = 0;
sentiment = 0;
for i in range(len(bigrams)):
      word1 = bigrams[i][0]
      word2 = bigrams[i][1]
if sentiment(word2) != 0:
      if sentiment(word1) == 0:
            polarity += sentiment(word2)
      else if sentiment(word1) < 0:
            if sentiment(word2) > 0:
                  polarity += (sentiment(word1) - sentiment(word2))/2
            else
                  polarity += (sentiment(word1) + sentiment(word2))/2

      else if sentiment(word1) > 0:
           if sentiment(word2)> 0:
                  polarity += (sentiment(word1) + sentiment(word2))/2
           else
                  polarity += (sentiment(word1) - sentiment(word2))/2
sentiment = polarity/len(bigrams)
```

3 Results

The precision and recall are used to identify the deviation of the actual results from expected results. The number of true positives, false positives, true negatives, and false negatives are calculated for both the algorithms. Precision, recall, and *F*-Score are calculated using the below formulae.

$$\text{Precision} = \frac{\text{True Positives}}{(\text{TruePositives} + \text{FalsePositives})}$$

$$\text{Recall} = \frac{\text{True Positives}}{(\text{True Positives} + \text{False Negatives})}$$

$$F\text{ - Score} = \frac{2(\text{Precision} * \text{Recall})}{(\text{Precision} + \text{Recall})}$$

The *F*-score shows that selective adverb/adjective algorithms have more or equivalent *F*-score when compared to all bigrams algorithm. In addition, the computational efficiency is improved as a result of excluding sentiment of the first word of bigrams based on POS tag, when we used selective adverb/adjective algorithm. The graph below shows the comparative analysis of *F*-scores between all bigrams and selective adverb/adjective bigrams algorithm (Table 1).

The precision and recall give more insight into the behavior of these algorithms in term of false positives and false negatives. The selective adverb/adjective algorithm is efficient when compared to the first one, when the focus of the problem is mainly to minimize the false positives, rather than on false negatives.

Table 1 Comparison of precision, recall, and *F*-score between all bigrams and selective adverb/adjective algorithms

Dataset	All bigrams	Selective adverb/adjective
Android	Precision: 87.5% Recall: 90.3% *F*-Score: 0.88	Precision: 87.6% Recall: 91.9% *F*-Score: 0.90
iPhone	Precision: 64.5% Recall: 88.7% *F*-Score: 0.74	Precision: 65.1% Recall: 90% *F*-Score: 0.75
Google	Precision: 80.2% Recall: 94.4% *F*-Score: 0.86	Precision: 84.1% Recall: 94.9% *F*-Score: 0.89
Apple	Precision: 83.7% Recall: 86.5% *F*-Score: 0.85	Precision: 83.9% Recall: 86.0% *F*-Score: 0.85
iPad	Precision: 85.1% Recall: 88.4% *F*-Score: 0.86	Precision: 85.2% Recall: 88.01% *F*-Score: 0.87

4 Conclusion

The results show that assigning scores to adjective/adverbs in bigrams if they appear in preceding position yields more accuracy and performance, than assigning sentiment to both words in bigrams. The limitation of this research is that the results may slightly vary with the dataset. The future work of this paper is to extend this research to include emoticons and implement selective adverb/adjective bigrams algorithm using machine learning approaches and compare the accuracy with rule-based algorithm introduced in this paper.

References

1. Singh J, Singh G, Singh R (2017) Optimization of sentiment analysis using machine learning classifiers. In: Human-centric computing and information sciences. https://doi.org/10.1186/s13673-017-0116-3
2. Wogenstein F, Drescher J, Reinel D, Rill S, Scheidt J (2013) Evaluation of an algorithm for aspect-based opinion mining using a lexicon-based approach. ACM. https://doi.org/10.1145/2502069.2502074
3. Hemalatha I, Saradhi Varma GP, Govardhan A (2012) Preprocessing the informal text for efficient sentiment analysis. Int J Emerg Trends Technol Comput Sci
4. Kharde VA, Sonawane SS (2016) Sentiment analysis of twitter data: a survey of techniques. Int J Comput Appl 139:5–15. https://doi.org/10.5120/ijca2016908625
5. Alaoui IE, Gahi Y, Messoussi R et al (2018) A novel adaptable approach for sentiment analysis on big social data. J Big Data. https://doi.org/10.1186/s40537-018-0120-0
6. Asghar MZ, Khan A, Ahmad S, Qasim M, Khan IA (2017) Lexicon-enhanced sentiment analysis framework using rule-based classification scheme. PLoS ONE 12(2):e0171649. https://doi.org/10.1371/journal.pone.0171649

7. Jurek A, Mulvenna MD, Bi Y (2015) Improved lexicon-based sentiment analysis for social media analytics. Secur Inf. https://doi.org/10.1186/s13388-015-0024-x
8. Tian L, Lai C, Moore JD (2018) Polarity and intensity: the two aspects of sentiment analysis. Assoc Comput Linguist 40–47. http://dx.doi.org/10.18653/v1/w18-3306
9. Palanisamy P, Yadav V, Elchuri SH (2013) Simple and practical lexicon based approach to sentiment analysis. Assoc Comput Linguist 543–548
10. Nicholls C, Song F (2009) Improving sentiment analysis with Part-of-Speech weighting. IEEE. https://doi.org/10.1109/ICMLC.2009.5212278
11. John A, John A, Sheik R (2019) context deployed sentiment analysis using hybrid lexicon. IEEE. https://doi.org/10.1109/ICIICT1.2019.8741413
12. Song YY, Lu Y (2015) Decision tree methods: applications for classification and prediction. Shanghai Arch Psychiatry 27:130–135
13. Gatti L, Guerini M, Turchi M (2016) SentiWords: deriving a high precision and high coverage lexicon for sentiment analysis. IEEE Trans Affect Comput 7(4):409–421
14. Data for Everyone. In: Figure eight. https://www.figure-eight.com/data-for-everyone. Accessed (2019/09/1)

Compact Quad Band Radiator for Wireless Applications

V. Saritha, C. Chandrasekhar, and K. Murali

Abstract Currently available wireless technologies require antenna of compact size, less cost and weight for multiple band operation. In this paper, a planar ring antenna with dimensions of 34 × 18 × 1.6 mm^3 is proposed that can operate in quad band centred at 2.5 GHz (Bluetooth), 3.5 GHz (WiMAX), 5.5 GHz (WLAN) and 8.0 GHz (SAT com). The proposed antenna comprises fork-shaped strip carved inside a rectangular ring, fed by a microstrip line, having a ground plane that is defected in structure with three small rectangular slots and a parasitic rectangular patch placed behind the fork-shaped strip. It is observed that simulated and experimental results are in very good agreement.

Keywords Quad band radiator · Fork-shaped strip · Parasitic rectangular patch and slots

1 Introduction

Of late, tremendous development has taken place due to continuous research in the field of wireless communication systems. Technologies like Wi-Fi, WiMAX and SAT com are deployed in portable wireless devices. These technologies demand multiband, multifunction antenna of low profile, less weight, low cost and easy in fabricating. To meet these requirements, various designs of multiband antenna have been reported in the literature [1–12] for WiMAX and WLAN applications.

V. Saritha (✉)
Department of Electronics and Communication Engineering, Jawaharlal Nehru Technological University, Anantapur, Anantapuramu, Andhra Pradesh, India
e-mail: sarithagreen@gmail.com

C. Chandrasekhar
Department of Electronics and Communication Engineering, SV Engineering College, Tirupati, India

K. Murali
Department of Electronics and Communication Engineering, Vijaya Engineering College, Vijayawada, India

P. S. R. Chowdary et al. (eds.), *Microelectronics, Electromagnetics and Telecommunications*, Lecture Notes in Electrical Engineering 655,
https://doi.org/10.1007/978-981-15-3828-5_70

An attempt to enhance the number of operating bands reported and the structure of antenna is modified. Here a compact microstrip ring antenna with fork-shaped strip carved inside and having a ground plane with defected structure along with parasitic slots and patch is proposed and designed for WLAN, WiMAX and SAT com applications. It is a quad band antenna being compact compared to those reported in the references given at the end. The antenna is optimized using HFSS software, fabricated and tested for its performance. The measured results for S11 (return loss) and radiation patterns are tallying closely with those of simulated ones. Section 2 covers designed antenna configuration with dimensions, evolution steps followed in optimized antenna, Sect. 3 presents a comparison of simulated and measured results with discussions, and Sect. 4 concludes the findings followed by acknowledgements and references.

2 Antenna Design

2.1 Structure of Radiator

Figure 1 details the physical structure of the proposed compact quad band microstrip radiator fed by microstrip line.

When a rectangular patch with the dimensions of 3 GHz, with full ground is considered, there is no resonating frequency obtained at 3 GHz. When a defected ground plane with optimized dimensions is considered, there is a resonating frequency obtained at 3 GHz. When a rectangular slot is etched inside the patch forming a rectangular ring, it is found to operate at almost three frequencies rather than a single frequency. When a monopole of length λ/4 at one of the desired frequencies 8 GHz is inserted inside the ring, the antenna resonated at 2.5, 4, 5.5, 8.5 GHz, where 4 and 8.5 GHz are not the desired frequencies. When a 'c'-shaped strip is loaded in the monopole which forms a fork shape at an appropriate position, the antenna has resonance at 2.5, 3.5, 5.5 and 8.5 GHz. Now three out of the four desired bands are

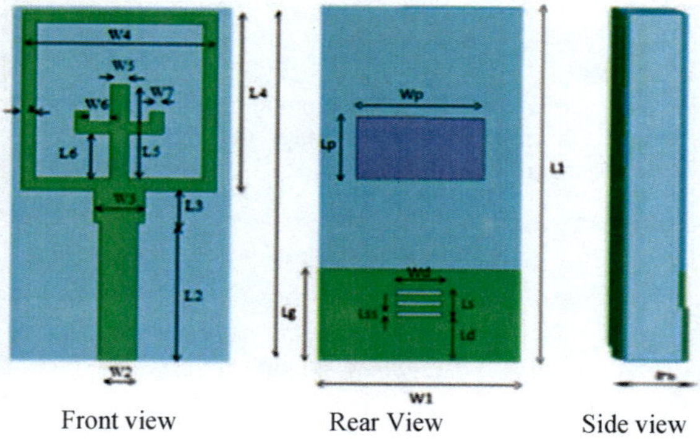

Fig. 1 Structure of the proposed monopole antenna

obtained, and there is a need to obtain the resonance at 8 GHz. When a parasitic patch in the defected ground plane is placed with the dimensions at 8 GHz with optimizing its length, width and position beneath the fork, then the antenna has resonances at 2.5, 3.5, 5.5 and 8 GHz.

The rectangular ring in the front view with dimensions $W4$ and $L4$ is designed by using the equation

$$f_r = \frac{c}{4(l_r - W_r)\sqrt{\varepsilon_{\text{reff}}}} \tag{1}$$

and its dimensions are $W4 = 15.5$ mm, $L4 = 17$ mm and the thickness of the ring = 1 mm.

A rectangular monopole is placed inside the rectangular ring to resonate at 8 GHz, and its size is obtained by using the following equations

$$L = \frac{c}{4f} \quad \text{for } W < 0.02\,\lambda \tag{2}$$

The length of the monopole is about one-quarter wavelength at 8 GHz. The optimized dimensions are $L5 = 9$ mm and $W5 = 1.4$ mm.

The dimensions of the defected ground plane are found by using the following equations

$$L(g) = 6h + L \tag{3}$$

$$W(g) = 6h + W \tag{4}$$

The dimensions obtained by using the above equations are optimized. The width and length at 5.5 GHz are found to be 18 and 8.7 mm.

The dimensions of the parasitic patch in the ground plane (to resonate at 8 GHz) are found by using the equations.

$$W = \frac{c}{2f}\sqrt{\frac{2}{\varepsilon r + 1}} \tag{5}$$

$$\varepsilon_{\text{reff}} = \frac{(\varepsilon r + 1)}{2} + \frac{(\varepsilon r - 1)}{2}\left[1 + 12\left(\frac{h}{W}\right)\right]^{\frac{-1}{2}} \tag{6}$$

$$\frac{\Delta L}{h} = 0.412\frac{(\varepsilon_{\text{reff}} + 0.3)\left(\frac{W}{h} + 0.264\right)}{(\varepsilon_{\text{reff}} - 0.258)\left(\frac{W}{h} + 0.8\right)} \tag{7}$$

$$L = \frac{1}{2f_r\sqrt{\varepsilon_{\text{reff}}}} - 2\Delta L \tag{8}$$

The width and length at 8 GHz are found to be 11.4 and 5.9 mm.

Table 1 Dimensions of the optimized proposed antenna

Parameters	Values (mm)	Parameters	Values (mm)
*L*1	34	*w*1	18
*L*2	13.4	*w*2	3
*L*3	3	*w*3	4
*L*4	17	*w*4	15.5
*L*5	9	*w*5	1.4
*L*6	4.4	*w*6	1.8
Lg	8.7	*w*7	1
Lp	5.85	*w*8	8.2
Ld	4.3	*wp*	11.4
Lss	0.5	*wd*	4
Ls	5.5	*h*	1.6

With this configuration, there is a dip in the gain at 5.5 GHz. Observing the distribution of current at 5.5 GHz, there is more concentration in the ground plane. As a result, to improve the gain in a step-by-step manner, three rectangular slots are etched in the ground plane.

When the number of slots and dimensions of these slots and their positions is optimized, there is good improvement in the gain.

The designed antenna with overall dimensions of 34 mm × 18 mm is printed on FR4 substrate whose thickness is of 1.6 mm and relative dielectric constant with a value of 4.4. It has a strip with a fork shape, carved inside a rectangular ring patch radiator with the defected ground plane, parasitic patch and slots fed by a 50 Ω microstrip line. The length of the microstrip line is found by using the equation

$$Z_o = \frac{60}{\varepsilon_{\text{reff}}} \ln\left[\frac{8h}{W} + \frac{W}{4h}\right] \tag{9}$$

The strip line width is found to be 3 mm, and a quarter-wave impedance matching with width 4 mm is used.

Table 1 presents the dimensions of the optimized proposed antenna.

2.2 Evolution Steps

Figure 2 shows various stages that are involved in evolution of the proposed antenna structure to arrive at the optimized geometry. Rectangular ring microstrip antenna and the microstrip feed structures (Fig. 2; Type A) play a major role in the generation of the lower and higher radiating frequency bands. The concentration of current distribution at mid-band is found to be on the fork-shaped strip carved inside the ring (Fig. 2; Type B) and radiates at 3.5 GHz. Also, it is found that the presence of the slot

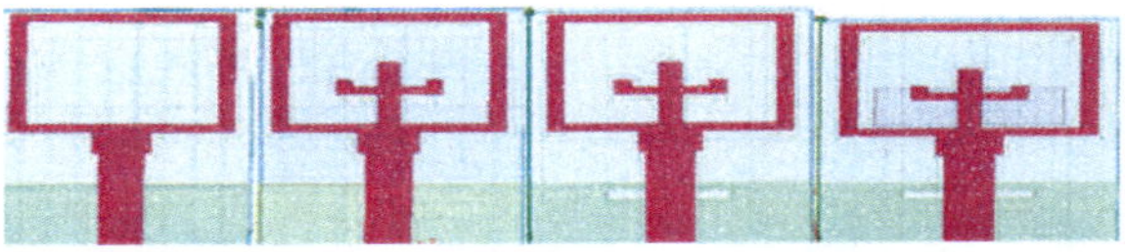

Fig. 2 Stages of the iteration process

Fig. 3 Fabricated antenna

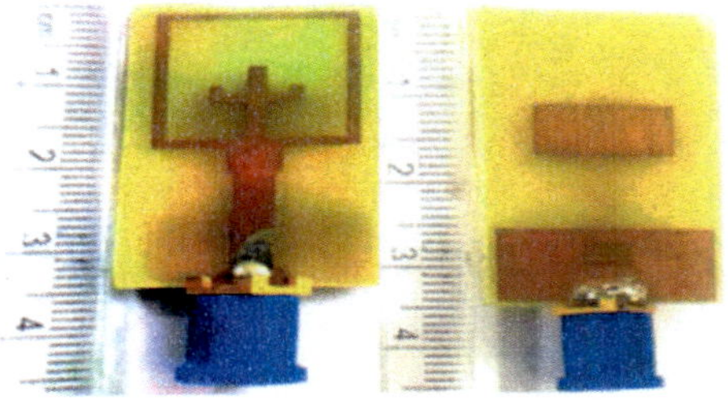

(Fig. 2; Type C) beneath the feed line in the ground plane effects the distribution of current on the feed line and defected ground to radiate at 5.5 GHz. The rectangular parasitic patch on the ground plane below the fork-shaped strip is found to radiate at 8 GHz (Fig. 2; Type D).

The proposed antenna is found to radiate at frequencies 2.5 GHz, i.e. operating from 2.1 to 2.8 GHz with a bandwidth of 700 MHz, 3.5 GHz operating from 3.2 to 3.9 GHz with a bandwidth of 700 MHz, 5.5 GHz operating from 5.2 to 5.7 GHz with a bandwidth of 500 MHz and 8 GHz operating from 7.7 to 8.4 GHz with a bandwidth of 700 MHz.

2.3 Fabrication of Proposed Antenna

From the optimized dimensions arrived and given in Table 1 is fabricated using FR4 substrate and is shown in Fig. 3

3 Results and Discussions

The computer simulation is carried out using HFSS tool. The parameters of the fork-shaped strip, parasitic patch and with slots in the defected ground plane are optimized till the required values of return loss are obtained at the desired radiating bands.

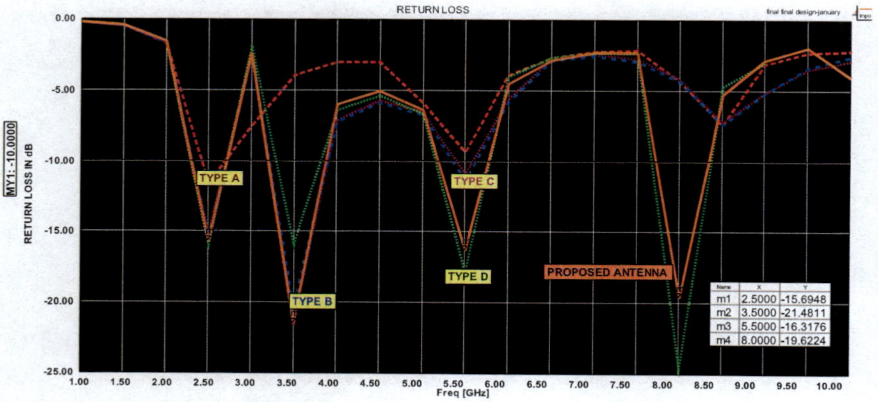

Fig. 4 Comparison of S_{11} for Type A (red colour), B (blue colour), C (pink colour), D (green colour) and proposed antenna (orange colour)

Figure 4 compares the return loss of the proposed antenna evolution—Type A, B, C and D that are involved in the process of evolution and the optimized/proposed antenna for quad band operation.

Figure 5 illustrates the plot of return loss for the fabricated prototype antenna which is measured using a vector network analyser.

Table 2 gives the return loss comparison for the antenna simulated and fabricated one at the four operating frequencies.

As an attempt to improve the gain at operating frequencies, a number of slots in the defected ground plane are varied and their dimensions are optimized to achieve reasonable gain whose comparison is provided in Table 3.

The elevation and azimuthal plane patterns are plotted for simulated results, whereas only elevation plane patterns are plotted for measured ones. Figures 6, 7, 8,

Fig. 5 Return loss plot for the fabricated prototype antenna

Table 2 Comparison of return loss

Frequency (GHz)	Simulated value (dB)	Measured value (dB)
2.5	−15.7	−14
3.5	−21.5	−24
5.5	−16.3	−14
8	−19.6	−16

Table 3 Gain of proposed antenna at operating frequencies

Frequency (GHz)	Antenna with 3 slots (dB)	Antenna with 2 slots (dB)
2.5	2.2	2.3
3.5	1.9	1.9
5.5	1.2	−1.5
8	3.0	1.5

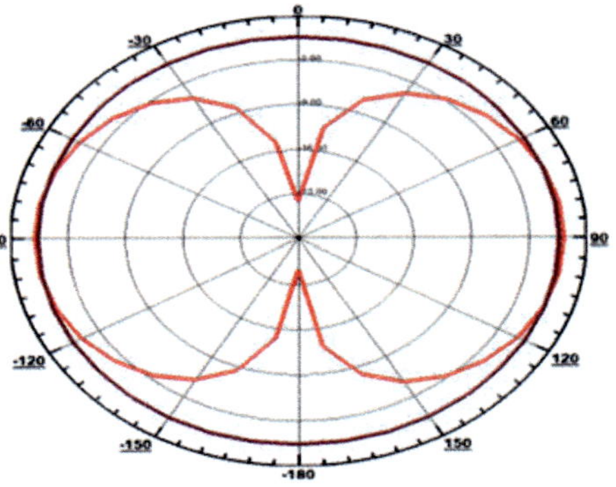

Fig. 6 Simulated pattern at 2.5 GHz

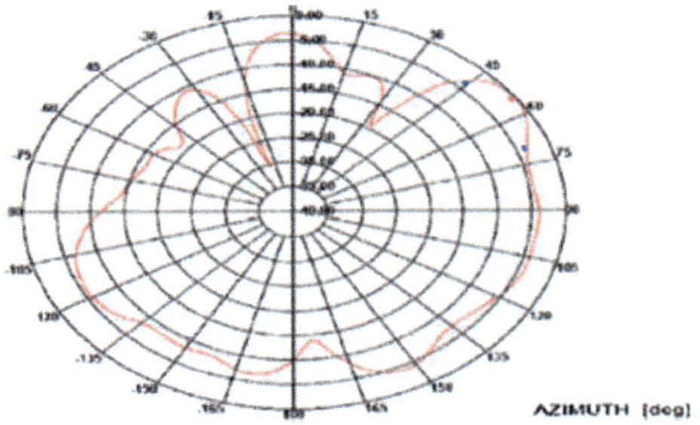

Fig. 7 Measured pattern at 2.5 GHz

9, 10, 11, 12 and 13 show the radiation patterns at the four radiating frequencies for simulated and measured ones in the principle planes.

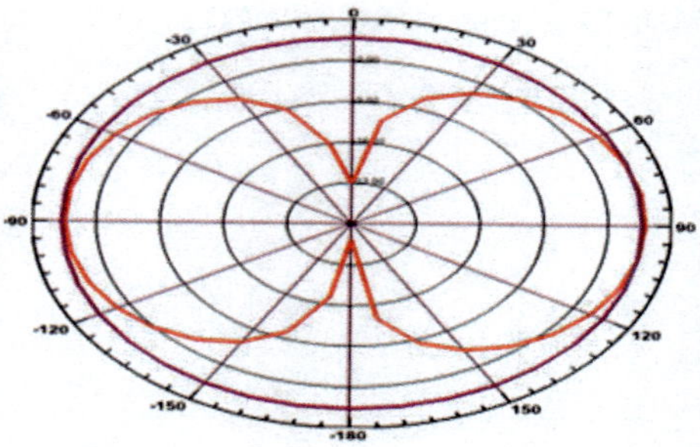

Fig. 8 Simulated pattern at 3.5 GHz

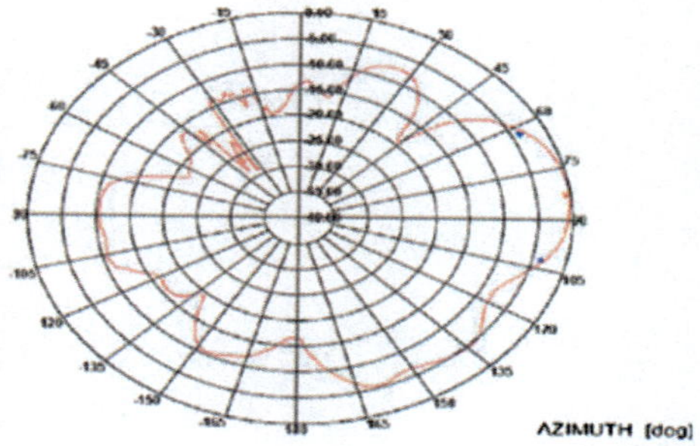

Fig. 9 Measured pattern at 3.5 GHz

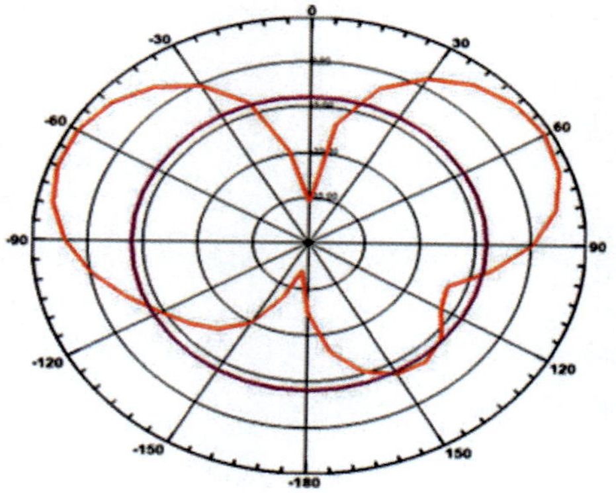

Fig. 10 Simulated pattern at 5.5 GHz

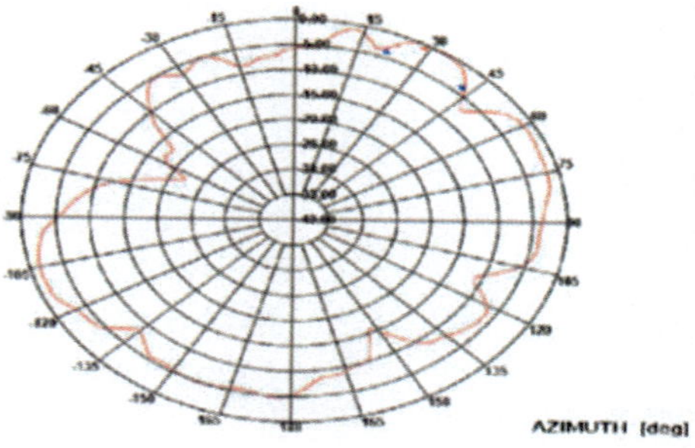

Fig. 11 Measured pattern at 5.5 GHz

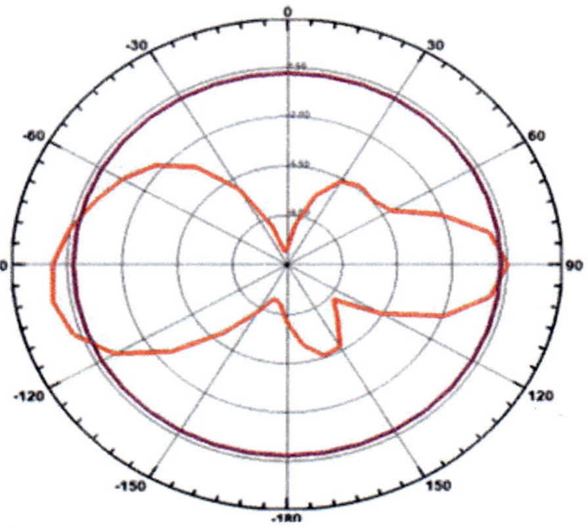

Fig. 12 Simulated pattern at 8 GHz

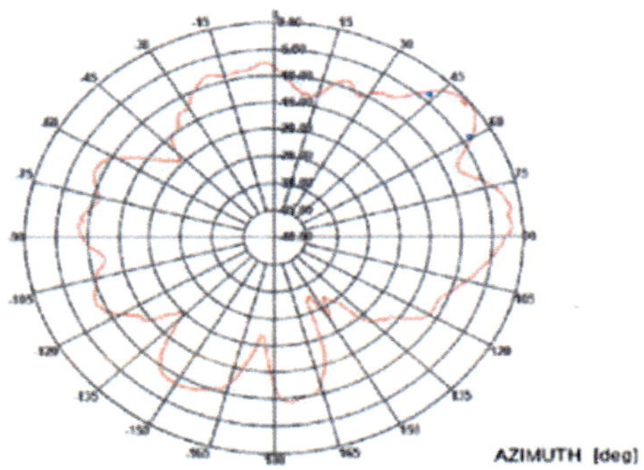

Fig. 13 Measured pattern at 8 GHz

Distribution of surface current densities for the radiating element on the top plane and on the ground plane (bottom) is shown in Fig. 14.

From the distributions of surface current at the radiating frequencies 2.5, 3.5, 5.5 and 8 GHz, it is observed that the current distributes on the rectangular ring at

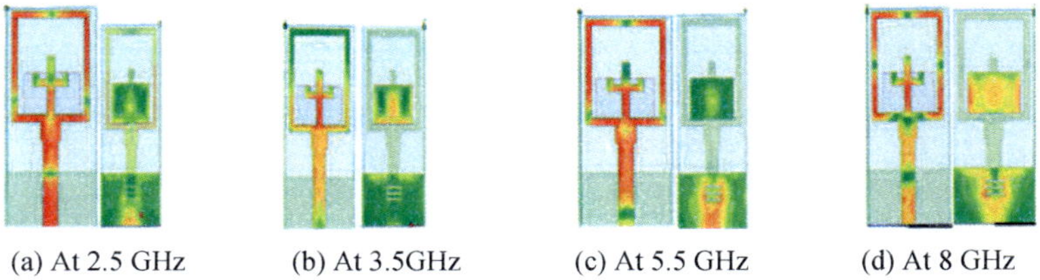

(a) At 2.5 GHz (b) At 3.5GHz (c) At 5.5 GHz (d) At 8 GHz

Fig. 14 Current density distribution of radiating element and the ground plane

Table 4 Comparison of frequencies and sizes of quad band radiators operating 2–10 GHz

S. no	No. of bands	Frequency (GHz)	Size (mm^3)
[12]-2012	4	2.07–2.77, 3.3–3.8, 5.15–5.35, 5.7–5.89	25 × 20 × 1.6
Proposed	4	2.4–2.6,3.4–3.69, 5.125–5.825, 7.9–8.4	34 × 14 × 1.6

low frequency (2.5 GHz) and on the parasitic patch on the ground plane at higher frequency (8 GHz). The surface current distribution concentrates on the fork-shaped strip inside the rectangular ring at 3.5 GHz and the parasitic slot beneath the feed line modify the surface current distribution on the ground and the feed line to radiate at 5.5 GHz.

For all the measurements and simulations, the antenna is mounted vertically with feed structure also being vertical.

It is noticed from the plots of return loss and radiation patterns that there lies a closed match between the simulated results and measured results for the quad band rectangular ring patch antenna proposed.

Table 4 elucidates the comparison of frequencies and also sizes of various quad band radiators in the literature.

This proposed antenna structure is compact compared to the quad band antennas existing in the literature as quoted above and useful for applications in four bands covering Wi-Fi, WiMAX and SATcom wireless communication systems. Omnidirectional patterns in azimuthal planes and elevation planes, low profile and ease in fabricating make this proposed antenna a good candidate for the above applications.

4 Conclusions and Future Scope

This proposed antenna operates in four bands centred at 2.5, 3.5, 5.5 and 8.0 GHz with an impedance bandwidth of minimum 500 MHz. This method of inserting parasitic patches beneath the radiator and slots in the defected ground plane can be extended to achieve other band of frequencies for any planar antenna.

References

1. Ali T, Khaleeq MM, Biradar RC (2018) A multiband reconfigurable slot antenna for wireless applications. AEU Int J Electr Commun 84:273–280
2. Singh A, Singh S (2014) A novel CPW-fed wideband printed monopole antenna with DGS. AEU Int J Electron Commun 69:299–306
3. Saladi A, Panda J, Kshetrimayum R (2010) A compact printed 9-shaped dual-band monopole antenna for WLAN and RFID applications. In: 2010 international conference on computing communication and networking technologies (ICCCNT). IEEE, pp 1–4

4. Manouare A, El Bakouchi R, Idrissi A, Ghammaz A, Ibnyaich S (2014) A compact triple-band patch antenna for WLAN/WiMAX applications. In: 2014 international conference on multimedia computing and systems (ICMCS). IEEE, pp 1322–1327
5. Tu DTT, Ngoc NT, Zhu F, Nguyen DN, Dutkiewic E (2017) Quad-band antenna for GSM/WSN/WLAN/LTE-A application in IoT devices. In: International symposium on communications and information technologies (ISCIT)
6. Bian F, Zhang F, Yang Y, Zhang Q, Huang JX (2010) A compact dual band printed monopole antenna for WLAN applications. In: International symposium on signals systems and electronics, vol 1, pp 1–4
7. Chu HB, Shirai H (2017) Design of multiband antennas using asymmetric E-CRLH TL unit cells. In: IEEE conference
8. Lu J, Zeng B, Li Y (2014) Planar multi-band monopole antenna for WLAN/WiMAX applications. In: 2014 international symposium on antennas and propagation (ISAP). IEEE, pp 475–476
9. L Li, X Zhang, X Yin, Zhou L (2014) A compact triple-band printed monopole antenna for WLAN/WiMAX applications. IEEE Antennas Wirel Propag Lett 13
10. Lin P, Ruan C (2011) A microstrip fed patch antenna with two parasitic invert L stubs for dual-band WLAN applications. Wireless Pers Commun 57(4):727–734
11. Swaraj P, Kumar M (2014) Quad-band U-slot microstrip patch antenna. Int J Sci Res Eng Technol 3:145–148
12. Liu W, Yin Y, Xu W (2012) Compact self-similar triple-band antenna for WLAN/WiMAX applications. Microw Opt Technol Lett 54(4):1084–1087

Synthesis of Non-uniformly Spaced Linear Antenna Array Using Firefly Algorithm

Nagavalli Vegesna, G. Yamuna, and T. Sudheer Kumar

Abstract In this paper, the radiation characteristics of a non-uniformly spaced linear antenna array are synthesized to reduce the sidelobe level (SLL) and increase the directivity of the main lobe. A firefly algorithm (FFA) has been used to determine the optimum spacing between the antenna elements. The computations are carried out for a different number of elements ranging from $N = 40$ to 80. Finally, it has been evident that for a fixed number of iterations, the sidelobe level has reduced.

Keywords Linear array · Firefly algorithm · Optimization · Sidelobe level · Convergence

1 Introduction

The high-end applications of antennas are efficient to be achieved by using antenna arrays. It is proved to have high directivity and gain which is usually achieved by the narrow beams of the radio waves. The shape of the radiation pattern plays an important role in forming the patterns of main lobe and controlling their angular placement. Beamforming techniques and null positioning can be achieved by the genetic algorithm (GA) with optimized amplitude and phase. Basic GA uses binary coding and binary crossover. Unlike that, here array excitation weighting vectors are used as complex number chromosomes. The advantage of GA is that it provides simple, clear and fast results by avoiding encoding and decoding processes. So, computational speed will be increased [1–4]. The reduction of sidelobes by the narrow beams produced due to the synthesis of sum patterns is a process used in the flower

N. Vegesna (✉) · G. Yamuna
Annamalai University, Chidambaram, Tamil Nadu, India
e-mail: valli.vegesna@gmail.com

T. Sudheer Kumar
Shri Vishnu Engineering College for Women, Bhimavaram, Andhra Pradesh, India

P. S. R. Chowdary et al. (eds.), *Microelectronics, Electromagnetics and Telecommunications*, Lecture Notes in Electrical Engineering 655,
https://doi.org/10.1007/978-981-15-3828-5_71

pollination algorithm. It is achieved by the amplitude excitation coefficients. The circular array can be synthesized, and positioning nulls can be achieved effectively using FPA than GA [5–8]. The method of particle swarm optimization proved to be more efficient than the conventional genetic algorithm as it has provided a maximum reduction in sidelobe level with a beamwidth of major lobe being constant. Also, there is a comparative improvement in the array pattern of the particle swarm method [9, 10]. Another approach to reducing the sidelobe level and the null control of isotropic antenna arrays is a bio-geography-based optimization (BBO) technique. Array parameters are varied to achieve the optimization of sidelobes. BBO is an effective technique than genetic, self-adaptive differential evolution and sequential quadratic programming techniques [11]. The one-dimensional linear antenna array radiation pattern can be synthesized using an artificial bee colony optimization algorithm. The optimization process happens to be in two stages to achieve efficient results in less computational time. The algorithm proved to effectively reduce the number of iterations to 1500 when compared to that of PSO and GA which requires over 10,000 iterations [12]. Unevenly spaced linear antenna arrays can be effectively synthesized using the firefly algorithm (FFA) when compared to PSO and conventional methods [13–15]. But from the literature, it is observed that only for small arrays the synthesis has been performed. In this paper, computations are carried out for a different number of elements ranging from $N = 40$ to 80.

The paper is organized into five sections. In Sect. 2, the non-uniform linear array formulation is described. The optimization using the firefly algorithm is elaborated in Sect. 3. The synthesized results related to the radiation pattern and spacing values are presented in Sect. 4. To the end, in Sect. 5, a detailed outcome of the firefly algorithm is stated.

2 Antenna Array Geometry and Mathematical Formulation

The radiation pattern of a uniform linear array is shown in Fig. 1. The maximum sidelobe level for this pattern is −13.50 dB. In order to reduce electromagnetic interference effects, the sidelobes are to be reduced. This can be achieved by controlling parameters like amplitude, spacing and phase. In this paper, non-uniform spacing is considered.

Let N isotropic elements be placed non-uniformly in a linear plane as shown in Fig. 2. Where $N = 2$ M, d is the distance between elements. Instead of N elements, only M elements position is optimized.

The individual radiation pattern of each element and spacing between the elements affect the radiating field of the array.

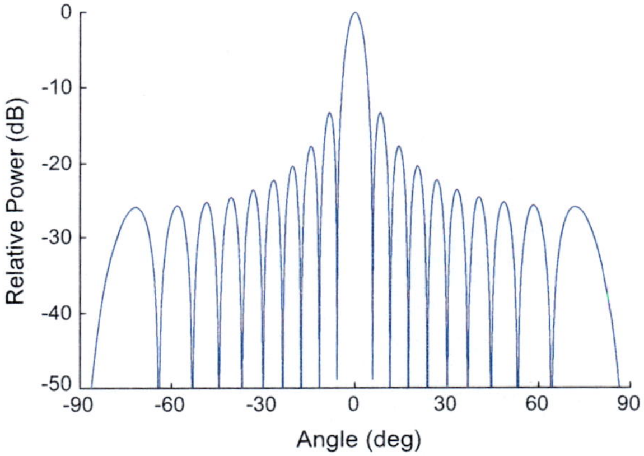

Fig. 1 Normalized radiation pattern of the uniform linear array

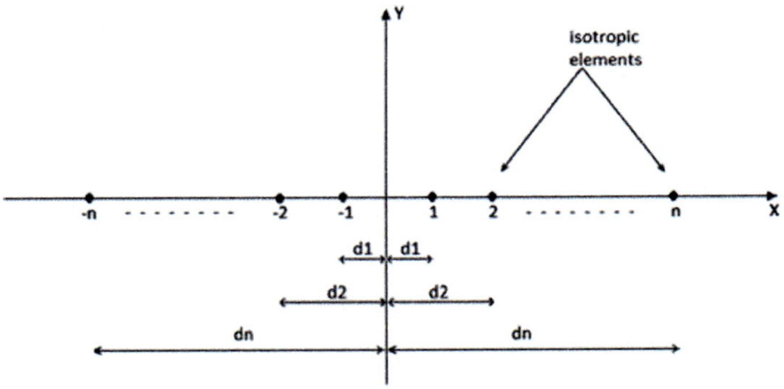

Fig. 2 Antenna array geometry

The array factor (AF) is given as follows:

$$\mathrm{AF}(\theta) = 2\sum_{n=1}^{N} A_n \cos[\pi (K S_n)\sin\theta] \tag{1}$$

where

n	is an nth element,
N	is no. of elements in an array,
K	is the wavenumber,
$K = 2\pi/\lambda$	where $\lambda = 1$,
θ	is the angle of observation,
A_n	is the nth element's excitation amplitude,
S_n	is the space between the reference point and the nth element

Normalized far field is expressed in dB as

$$E(\theta) = 20\log_{10}\frac{|\mathrm{AF}(\theta)|}{|\mathrm{AF}(\theta)|_{\max}} \tag{2}$$

∴ The fitness function can be written as [15]

$$F = \mathrm{Max}(E(\theta))\forall\theta \neq 0 \tag{3}$$

3 Firefly Algorithm

3.1 Nature of Fireflies

Fireflies are familiar insects living in tropical settings as shown in Fig. 3, producing cold light. This happens on specific light-emitting organs, generally on the lower abdomen of a firefly, owing to a sort of chemical reaction known as "bioluminescence". The light is generated by the luciferase enzyme when it is with oxygen or magnesium ions. The larva of a firefly is known as a glow-worm.

The fireflies blinking light are an incredible sight in the tropical and temperate regions of the summer sky. There are approximately 2000 species of fireflies, and most fireflies generate brief and rhythmic flashes. For a specific species, the pattern of flashes is often distinctive. However, attracting mating partners (communication) and attracting prospective prey are two basic functions of such flashes. Flashing can also serve as a mechanism for protection by producing a warning rhythmic flash. Blinking rates and time amount are important for two sexes to combine. In some species, females react to the flashing pattern of males, while in species like photuris, the female fireflies attract and eat the male species.

It is evident that the intensity of light decreases as the distance increases, i.e., if a firefly is acting as a source by flashing light with luminous intensity "I_v" and light traveling distance "d", then according to the inverse square law, they are related by

Fig. 3 Firefly insect

$$I_V \alpha \frac{1}{d^2} \quad (4)$$

Distance "d" and the light intensity "I_v" are inversely proportional. So, only several hundreds of meters are covered at night which is enough for fair communication. Within that limit, the firefly with a lower intensity of light is attracted to the brighter firefly. If there is no firefly with brighter intensity in that range, then it moves randomly.

3.2 Algorithm

Xin-She Yang has developed this algorithm at Cambridge University. Heuristic implies finding or discovering by path and mistake. This solution can be discovered in a fair quantity of time, and there is no assurance that optical solutions will be achieved. Whereas in metaheuristic, meta implies beyond or above the stage. Generally, this works better than simple heuristics. There are many swarm generated optimizations that are inspired by nature. Among them, the firefly is an algorithm which is generated by the inspiration of the firefly nature of flashing behavior and their moving motion. The flashing pattern of each firefly differs for the firefly algorithm which has many advantages as compared to other algorithms like PSO and DEA. The algorithm is generated by the consideration of three main rules. Figure 4 shows the firefly attraction pattern.

- Irrespective of the sex, one firefly is attracted to the other due to its unisexual nature.
- The attractive nature of the firefly depends on the flashing brightness produced by each firefly. The attractive nature increases with the flashlight brightness, and they

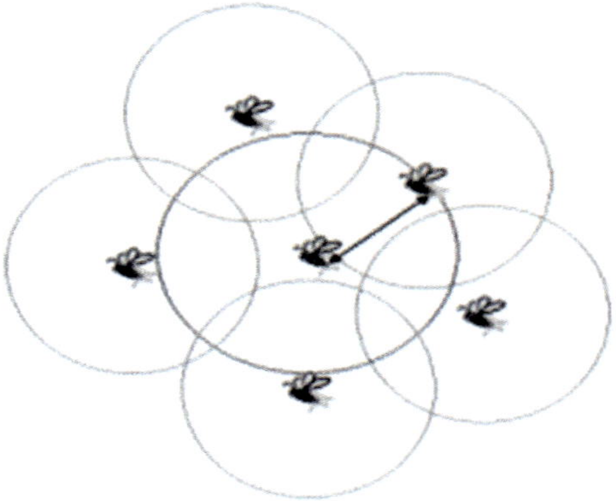

Fig. 4 Attraction pattern of fireflies

decrease as distance increases. The firefly with more brightness gets attracted to the one with the least ones.

- The firefly brightness nature is obtained by the function landscape.

At the beginning of the firefly algorithm all the required constraints, fitness functions are initialized and then random variables are generated. The loop starts iterating each agent or firefly. The loop terminates only if several iterations are greater than the maximum generation. The light intensities of fireflies are evaluated based on the fitness value function. The light intensity of fireflies is the fitness value in this algorithm. The firefly with low fitness value moves toward the one with high fitness value. The light intensity (fitness value) is updated for each firefly. The fitness value is the characteristic which ranks the firefly and then is used to update its position. Later on, it checks the stopping condition and then gives the best solution. This completes the firefly algorithm.

4 Results

The firefly algorithm is successfully applied to the synthesis of array antennas to yield radiation patterns with reduced sidelobe levels. The firefly algorithm is used to obtain optimum values of spacing between the elements. The optimized spacing values are then used to evaluate their respective radiation patterns. The results are obtained from $N = 40$ to 80 in steps of 20. Figures 5, 6 and 7 show the optimum spacing and their respective radiation patterns.

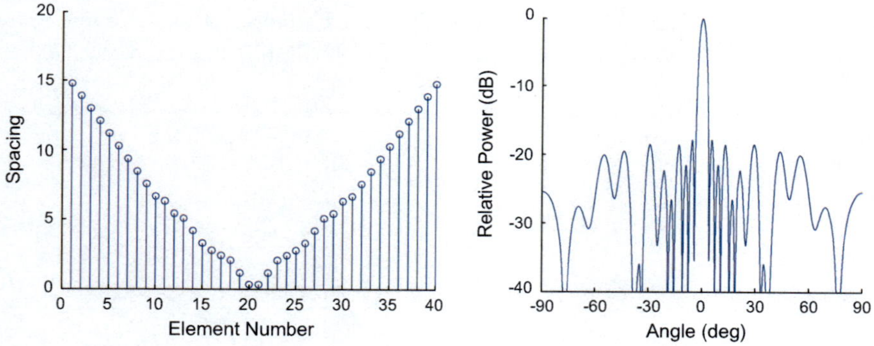

Fig. 5 Optimized spacing between the elements and radiation pattern for $N = 40$ elements using FA

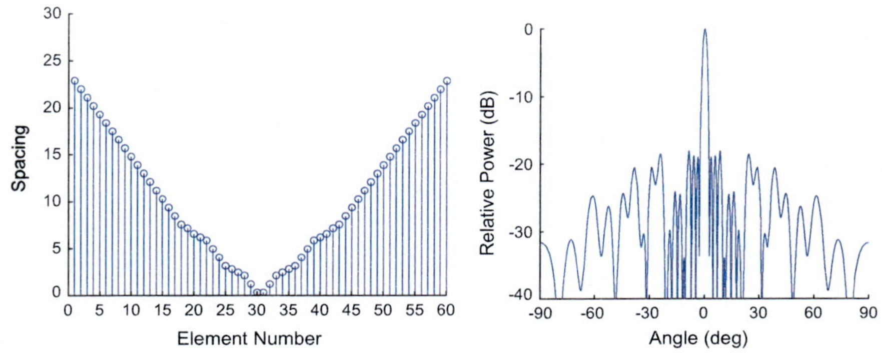

Fig. 6 Optimized spacing between the elements and radiation pattern for $N = 60$ elements using FA

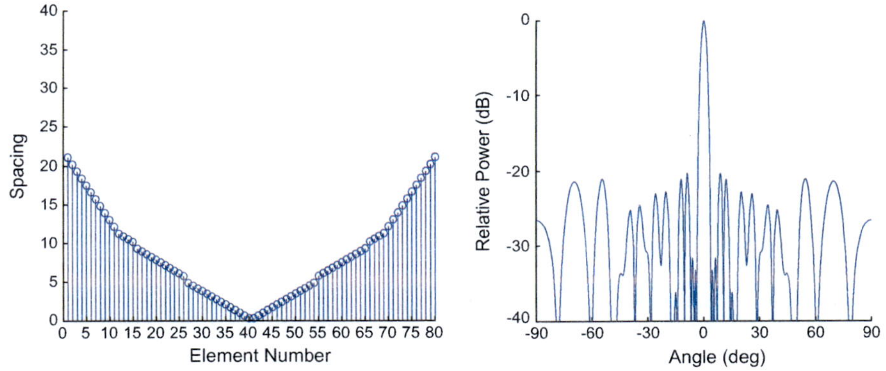

Fig. 7 Optimized spacing between the elements and radiation pattern for $N = 80$ elements using FA

5 Conclusion

Radiation patterns with reduced sidelobes are produced by applying the firefly algorithm. Firefly algorithm effectively searches for the optimum spacing between the elements to produce low sidelobes. The methodology is extended for both small and large array. Further, the work can be extended for beam scanning for smart antenna applications and to produce shaped beams like sector beams.

References

1. Yan KK, Lu Y (1997) Sidelobe reduction in the array-pattern synthesis using a genetic algorithm. IEEE Trans Antennas Propag 45(7):1117–1122
2. Chakravarthy VS, Rao PM (2015) Amplitude-only null positioning in circular arrays using a genetic algorithm. In: 2015 IEEE international conference on electrical, computer and communication technologies (ICECCT), IEEE, pp 1–5, March 2015
3. Terlapu SK, Raju GSN (2018) Synthesis of sector beams from array antennas using real coded genetic algorithm. In: Data engineering and intelligent computing, Springer, Singapore
4. Lommi A, Massa A, Storti E, Trucco A (2002) Sidelobe reduction in sparse linear arrays by genetic algorithms. Microw Opt Technol Lett 32(3):194–196
5. Terlapu SK, Raju GRLVNS, Raju GSN (2016) Array pattern synthesis using flower pollination algorithm. In: 2016 International conference on electromagnetic interference & compatibility (INCEMIC), IEEE, pp 1–4, December 2016
6. Vedula VSSS, Paladuga SR, Prithvi MR (2015) Synthesis of circular array antenna for sidelobe level and aperture size control using flower pollination algorithm. Int J Antennas Propag
7. Chakravarthy VS, Terlapu SK, Chowdary PSR, Rao TV, Satapathy SC (2018) On the convergence of synthesis of desired nulls from circular arrays using flower pollination algorithm. In: Proceedings of 2nd international conference on micro-electronics, electromagnetics and telecommunications, Springer, Singapore, pp 475–489
8. Singh U, Salgotra R (2018) Synthesis of linear antenna array using flower pollination algorithm. Neural Comput Appl 29(2):435–445
9. Najjar MSY, Khodier NDM (2008) Design of non-uniform circular antenna arrays using particle swarm optimization. J Electr Eng 59(4):216–220
10. Recioui A (2012) Sidelobe level reduction in linear array pattern synthesis using particle swarm optimization. J Optim Theory Appl 153(2):497–512
11. Sharaqa A, Dib N (2014) Design of linear and elliptical antenna arrays using biogeography based optimization. Arab J Sci Eng 39(4):2929–2939
12. Zaman MA, Gaffar M, Alam MM, Mamun SA, Matin MA (2011) Synthesis of antenna arrays using artificial bee colony optimization algorithm. Int J Microw Opt Technol 6(4):234–241
13. Zaman MA, Matin A (2012) Nonuniformly spaced linear antenna array design using firefly algorithm. Int J Microw Sci Technol
14. Yang XS (2009) Firefly algorithms for multimodal optimization. In: International symposium on stochastic algorithms. Springer, Berlin, Heidelberg, pp 169–178, October, 2009
15. Terlapu SK, Raju GSN (2016) Synthesis of linear array for sidelobe reduction using particle swarm optimization. a A, 2(1): 1

Timeout-Aware Inter-Queuing for QoS Provisioning of Real-Time Secondary Users in Cognitive Radio Networks

K. Annapurna and B. Seetha Ramanjaneyulu

Abstract Assurance of quality of service (QoS) to secondary users in cognitive radio networks is a challenge as the secondary users have to use the spectrum on an opportunistic basis, which is random. To address this challenge, the secondary users are classified into four priority classes by maintaining a separate queue for each priority class. The assignment of channels to secondary users starts from the highest priority queue towards the least priority queue. Further, the QoS to real-time secondary users is improved by introducing timeout-aware inter-queuing (TAIQ), in which shifting of the secondary user is allowed from low-priority queue to high-priority queue when its time deadline is about to reach the time deadlines of its next high level priority queue SUs. The results showed that the proposed four priorities with timeout-aware inter-queuing mechanism give less blocking to real-time secondary users with respect to conventional real-time and non-real-time secondary users.

Keywords Cognitive radio networks · Real-time secondary users · Queues · Blocking probability · QoS · Timeout-aware inter-queuing

1 Introduction

Day to day the need of spectrum is increasing drastically to address the new emerging wireless applications and high demand of data transfers. But the available spectrum is a valuable and limited resource. On the other hand, most of the licensed users are not even utilizing the spectrum up to 30% [1]. To address these spectrum scarcity and underutilization problems, cognitive radio networks are one of the best solutions [2–4].

K. Annapurna (✉) · B. Seetha Ramanjaneyulu
VFSTR (Deemed to be University), Guntur, India
e-mail: arya.anu85@gmail.com

B. Seetha Ramanjaneyulu
e-mail: ramanbs@gmail.com

P. S. R. Chowdary et al. (eds.), *Microelectronics, Electromagnetics and Telecommunications*, Lecture Notes in Electrical Engineering 655,
https://doi.org/10.1007/978-981-15-3828-5_72

In cognitive radio networks, the unlicensed users can access the licensed spectrum such that they do not interfere with the transmissions of licensed users, primary users (PUs). These unlicensed users which are exploiting the channels opportunistically are termed as secondary Users (SUs).

Adequate care is taken in cognitive radio networks, to protect the transmissions of primary users (PUs). But there are no such measures available to protect the transmissions of secondary users (SUs). In this backdrop of fluctuating nature of PUs' arrivals and departures, offering QoS to SUs is a challenging task.

When delay-sensitive data like real-time audio/video need to be transmitted by cognitive radio devices, satisfying their QoS in terms of delivering the data within the stipulated time deadline is very essential. In such cases, prioritizing the SUs based on their urgency and allocating the vacant channels depending on the priorities become more relevant.

When SUs are dealing with delay-sensitive real-time data, then they may not complete their communication within their time deadline, if all the SUs are treated equally. It is observed more when the PUs' activity is high or demand from SUs is more or both. In such cases, to support the real-time data, the SUs are classified into four priorities, namely SU1s, SU2s, SU3s and SU4s, based on their time deadlines. Among these, SU1s, which are hard real-time users, should be treated as the highest priority among all SUs. SU2s and SU3s are soft real-time users, which would take next levels of priority, out of which SU2s have more priority than SU3s. Finally, SU4s are non-real-time users with the least priority.

In addition to four levels of priorities, timeout-aware inter-queuing (TAIQ) is proposed here. The objective of TAIQ is if SU2s and SU3s do not get the transmission opportunity within their priority queue and the time deadline is about to reach, then they are moved to next higher priority queue, otherwise they continue with their own priority class. The advantage of this method is that all SUs do not come for competition with SUs having time-critical data from the beginning itself.

The remaining part of the paper is organized as follows: Literature review and related work are covered in Sect. 2, proposed work is described in Sect. 3, system model is explained in Sect. 4, results are discussed in Sect. 5, and Sect. 6 covers the conclusion and future scope.

2 Related Work

In [5], priority is given to real-time traffic than the best effort traffic as real-time systems like health care, where abnormal condition of patient should reach the doctor immediately, are of higher priority. Two methods of channel allocation are proposed, namely reservation-based channel allocation method and absolute priority channel allocation method. In reservation-based channel allocation method, SUs will be given channel only if channel is existed during pre-allocated time intervals, whereas in absolute priority channel allocation method, whenever free channel is available, real-time traffic is served. Best effort traffic is served when free channels are available and no further real-time users are waiting for service.

The authors of [6] considered heterogeneous real-time (RT) and non-real-time (NRT) SUs. In view of offering better QoS to RT users, the available spectrum holes are first assigned to RT users and leftover spectrum holes to NRT users. They considered the QoS in terms of blocking and dropping probabilities.

In the above two works, SUs are classified into two types although in real world, in many applications, two types of SUs cannot cover all priority levels.

The authors of [7] proposed different inter-frame spaces in view of giving different priorities based on applications. Less inter-frame space is used for high-priority applications and high inter-frame space for low-priority applications, for example, 10 μs for voice, 12 μs for video, 14 μs for best effort and 16 μs for background type data. They classified SUs into four types in view of various applications like voice, video, best effort and background packet types. But giving priority to different types of SUs would be better dealt with the help of queues rather than inter-frame spaces.

In [8], the authors treated all SUs as single group only. But they used two buffers, one for handoff SUs and the other for new SUs. Some channels are reserved for handoff SUs as call dropping is more hurting than call blocking. They also considered the scenario, where SUs are leaving the queues due to impatience because of long waiting times. Under these circumstances, they derived blocking and dropping probabilities using Markov chain models. Here, they have given importance to handoff SUs than new requesting SUs.

In [9], the best channel with high channel holding time is assigned to users with the highest priority to overcome the disturbances to be experienced by high-priority SUs. For this, it needs prediction to anticipate the holding times of the channels.

In [10], to support the QoS to SUs, SUs are categorized into four priority groups based on their time-critical data and plan the usage of vacant channels based on the importance of data to be transmitted by them. Now, it is extending to add the TAIQ.

In [11], the authors considered two types of priorities for two types of traffic of SUs: Streaming and Non-Streaming. Their idea is to find successive channels such that it will satisfy the bandwidth requirement of secondary users. They have derived the optimum number of channels to be reserved for streaming type SUs. Here, they support the real-time SUs by reserving some channels for them. But reservation of channels may increase unnecessary blocking. When no channels are free except the reserved channels and if a non-streaming SU requests for service and no streaming SU need the service, it will be blocked even when reserved channels are free. Another challenge here is that assigning contiguous channels may not be possible all the times.

In [12], QoS to SUs is offered through an auction mechanism in which bid values are predefined, and priority of SUs is considered in terms of their willingness to pay.

The authors of [13] have considered 'M' channels, and each channel contains 'N' subchannels. They have considered that one channel is required for PUs, and one subchannel is sufficient for SUs. They have divided the SU traffic into two groups: SU1s and SU2s, out of which SU1s are given high priority compared to SU2s. Channel assignment to PUs is done from left most and towards right. Channel allocation of SU1s is done from rightmost and towards left. SU2s' channel allocation is done from middle and towards left. From this, it can be observed that SU1s will experience less blocking and dropping compared to SU2s. But this type of channel allocation is suitable to systems which do not have fixed frequency bands for their PUs like the cellular but not suitable for other cases like television white spaces. A survey is carried out on research efforts in the area of cognitive radio networks [14].

3 Proposed Timeout-Aware Inter-Queuing Technique

To support real-time SUs, SUs are classified into four types SU1s, SU2s, SU3s and SU4s, out of which first three belong to real-time traffic. The most delay-sensitive SUs come under SU1s, treated as the highest priority SUs. From SU1s to SU3s, delay sensitivity is considered to be decreasing in that order, and so, priorities will be decreased. The last type is non-real time, which does not have any time deadline.

By categorizing the real-time SUs into three groups, the probability of giving service to real-time SUs within their time deadline can be improved compared to the case where there are no priority categories or only two types of SU priority types exist. In addition to the distribution of real-time SUs into three groups, TAIQ is proposed here, to further improve the chances of offering channels to real-time SUs within their time deadlines. The results of proposed technique are compared with the two-level priority SUs counterpart.

4 System Model

Centralized decision making, cooperative type cognitive radio network is considered. That is, all sensing information and channel requests from individual SUs are sent to a central coordinator. Based on the available spectrum holes and number of type-wise SU requests, the central coordinator assigns the channels. Initially, it assigns the available vacant channels to SU1s. After fulfilling all SU1 requests at that time, if still vacant channels are available, then it proceeds to allot them to SU2s, then to SU3s and finally to SU4s.

The arrival times and conference times of PUs and the four types of SUs are taken as uniform distribution type. 'N' non-overlapping channels are considered. Each PU is considered to have its own channel. The PUs have freedom to use the channels at any time of their interest as they are licensed users. Each SU of the network has its own transceiver. SUs should use the channels opportunistically after knowing the status of spectrum holes through sensing. Whenever SU knows that a free channel is available, then SU can use it and thereafter should sense the spectrum periodically to notice the reappearance of PU. Whenever it observes the reappearance of the PU, the SU should vacate and occupy another free channel if it is available or drop the transmission otherwise. Four queues are considered for the four types of SUs.

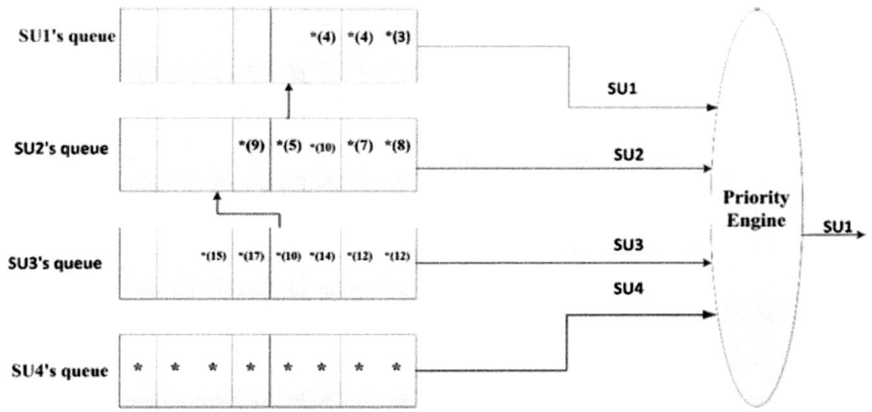

Fig. 1 Concept of timeout-aware inter-queuing

Further refinement of this scheme can be obtained by incorporating the mechanism of TAIQ. Deadline times are considered here to implement this mechanism. For example, within a priority class, if there are some users whose transmissions are having less time deadlines compared to some other users whose transmission deadlines are little higher, then the later users can be put into next priority class. Then, they continue with that less priority and try to get the transmission opportunities. As the time elapses, their deadline times would be approaching the values of former users. At those instances, at which the deadline values fall below the threshold value that decides the criticality, these later users are also brought to high-priority users' category. This is illustrated in Fig. 1.

Advantage of this approach is that these later devices which have higher time deadlines do not compete immediately with those later devices which have smaller time deadlines. They enter into the competition for high-priority, only when their deadlines also reach that criticality. Due to this mechanism, critical transmissions are handled effectively, with little impact on the performance of non-critical transmissions. It is proposed as TAIQ, in which the user of a low-priority queue will be shifted to high-priority queue when it is about to reach its time deadline.

In the simulated environment, all the devices are able to read all the available data (variables) from MATLAB environment. Hence, the central entity knows about the availability of free channels and priority requirements of all the devices. However, in the real environment, this information will be transmitted on the forward and reverse control channels, in the specified frame formats.

The system supports pre-emption, in which a low-priority SU needs to vacate the channel when a high-priority SU needs it. In this case, when the winner SU1 is unable to detect a free channel to its use as all channels are completely occupied by PUs and SUs, then the least priority SU which is using the channel needs to vacate the channel for the sake of honouring the request from SU1.

5 Results and Discussion

Blocking probability of SUs where the secondary users are categorized only into two types is shown in Fig. 2. The two types of SUs are real time (RT) and non-real time (NRT). The spectrum hole usage opportunities are first offered to RT SUs as they belong to high priority, then to NRT SUs if still spectrum holes are available. So, the blocking probability of RT SUs is less compared to NRT SUs.

Generally, RT SUs will have a time deadline and if service to the RT SUs is offered after their time deadline, then there is no use. So, if channel allocation to RT SUs only within their time deadline conditions is considered, then their corresponding blocking is increased with respect to the case where channel allocations are done even after their time deadlines. The drawback of this technique is that some RT SUs are missing the opportunity of getting the channel as all RT SUs are in single queue, which will not allow the SUs whose time deadline is about to reach. They are offering services to the SUs based on first come first serve principle. The blocking of NRT SUs is very large as its priority is least. It can also be observed that the blocking probability of users is decreasing with increasing number of channels as the opportunities are increasing.

The blocking probability of secondary users with TAIQ is shown in Fig. 3. It can be noticed that the blocking probability of the highest priority SUs, that is SU1s, is less compared to the remaining SUs. It is also observed that its value is 0.71 when only one channel is available and is gradually decreasing and reaches '0' when number of channels become five.

The blocking probability of next highest priority SUs, that is SU2s, is '1' when only one channel is available and gradually decreasing and reaches zero when number of channels become eight. Blocking of next level RT SUs is '1' up to five channels and then starts decreasing and becomes 0.12 when number of channels reaches ten. The last priority secondary users, called NRT SUs or SU4s, are experiencing a highest

Fig. 2 Blocking probability of RT and NRT secondary users for high PU occupancy and high SU demand

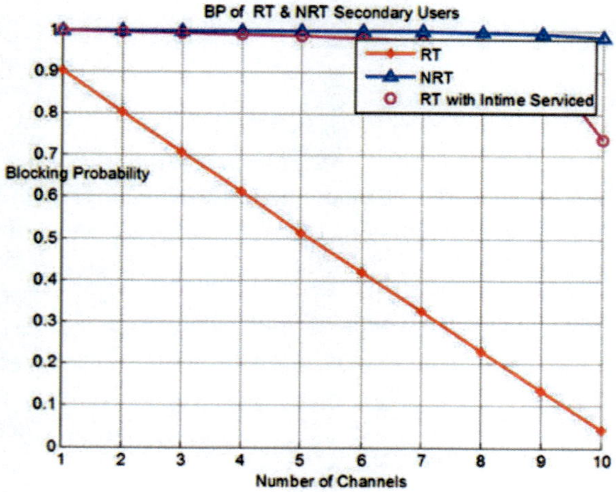

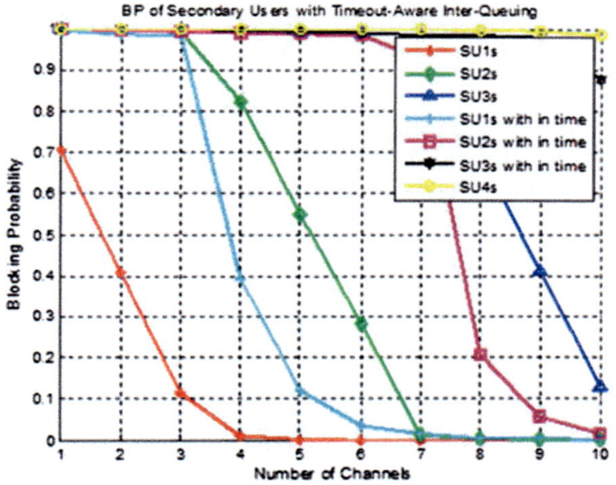

Fig. 3 Blocking probability of secondary users with timeout-aware inter-queuing for high PU occupancy and high SU demand

blocking probability of '1' up to six channels case and then start decreasing and become 0.75 when number of channels reach ten. It also describes the blocking scenario of SU1s, SU2s and SU3s when channel access is given to them within their time deadline.

The comparative blocking of real-time SUs with and without TAIQ is shown in Fig. 4. It can be observed that the blocking probability of RT SUs is very less in the TAIQ concept. It is due to its opportunity of getting channel prior to its actual scenario.

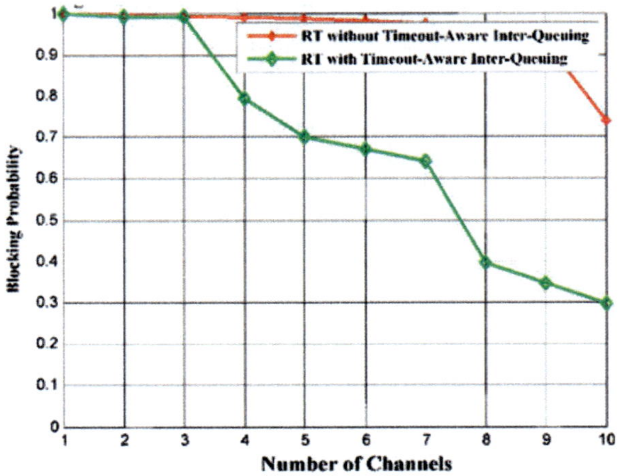

Fig. 4 Blocking probability of RT SUs with and without timeout-aware inter-queuing for high PU occupancy and high SU demand

6 Conclusion and Future Scope

Quality of service to secondary users based on multiple priorities is explored in this paper. Users of four priority levels, SU1s to SU4s, are considered, with SU1s being the highest priority and SU4s the least priority. In addition to priority levels, TAIQ, i.e shifting of the users from one priority level to its next higher priority level queue when their time deadlines are approached, is also introduced. The proposed 'timeout-aware inter-queuing' results are compared with the existing method of RT and NRT user classification. Performance improvement in the form of reduced blocking probabilities in the range of 18–31% is observed for the high-priority users.

References

1. Spectrum policy task force report (2002) Technical report 02–135, Federal Communications Commission, Nov. (2002)
2. Mitola J, Maguire GQ (1999) Cognitive radio: making software radios more personal. IEEE Pers Commun 6(4):13–18
3. Haykin S (2005) Cognitive radio: brain-empowered wireless communications. IEEE J Sel Areas Commun 23(2):201–220
4. Akyildiz IF, Lee W-Y, Vuran MC, Mohanty S (2006) Next generation/dynamic spectrum access/cognitive radio wireless networks: a survey. Elsevier J Comput Netw 50:2127–2159
5. Liang Z, Feng S, Zhao D, Shen XS (2011) Delay performance analysis for supporting real-time traffic in a cognitive radio sensor network. IEEE Trans Wireless Commun 10(1):325–335
6. Alshamrani A, Shen XS, Xie LL (2011) QoS provisioning for heterogeneous services in cooperative cognitive radio networks. IEEE J Sel Areas Commun 29(4):819–830
7. Kunert K, Jonsson M, Bilstrup U (2012) Deterministic real-time medium access for cognitive industrial radio networks. In: Proceedings of 9th IEEE international workshop on factory communication systems, Lemgo, pp 91–94
8. Wang J, Huang A, Wang W, Quek TQ (2013) Admission control in cognitive radio networks with finite queue and user impatience. IEEE Wirel Commun Lett 2(2):175–178
9. Motamedi N, Kumar S, Hu F, Rowe N (2013) A priority-aware channel selection scheme for real-time data transmission in cognitive radio networks. In: 2013 international conference on computing, networking and communications (ICNC), San Diego, CA, 734–739
10. Kumar KS, Annapurna K, SeethaRamanjaneyulu B (2015) Supporting real-time traffic in cognitive radio network"s. In: 2015 International conference on signal processing and communication engineering systems. https://doi.org/10.1109/spaces.2015.7058201
11. Doost-Mohammady R, Naderi MY, Chowdhury KR (2014) Spectrum allocation and QoS provisioning framework for cognitive radio with heterogeneous service classes. IEEE Trans Wireless Commun 13(7):3938–3950
12. Annapurna K, SeethaRamanjaneyulu B (2019) QoS maintenance in cognitive radio networks with priority-supporting novel channel allocation method. Int J Bus Data Commun Netw
13. Park J-H, Chung JM (2016) Prioritized channel allocation-based dynamic spectrum access in cognitive radio sensor networks without spectrum handoff. EURASIP J Wirel Commun Netw 2016(266):1–8
14. Meghalatha CK, SeethaRamanjaneyulu B, Jagadeesh RS, Krishna Chaitanya K (2017) Survey of research issues and standardization efforts for the utilization of TV white spaces. *in proc* JARDCS, pp 233–244

Inter-user Interference Mitigation Scheme for IEEE 802.15.4

C. K. Meghalatha, K. S. Sravan, K. Krishna Chaitanya, and B. Seetha Ramanjaneyulu

Abstract Wireless sensor networks have a wide range of applications. However, these networks suffer interference from each other when many nodes belonging to different WSNs are deployed closely at the given geographical location. To mitigate this problem, a mechanism of prioritized access is proposed in this work. The proposed mechanism is carried out in two phases. In the first phase, requests received from the devices are analyzed. The analysis obtained in the first phase is applied to a dynamic slot assignment in the second phase. Simulation results show a considerable improvement in throughput performance, for high priority data. The effect of varying the active period in IEEE 802.15.4 superframe is also analyzed to know its impact on throughput.

Keywords Slotted CSMA/CA · Superframe structure · Interference mitigation · WSN

1 Introduction

Wireless sensor networks are being used in many application areas such as healthcare, habitat monitoring, disaster prediction and security alarming. Low power consumption, less cost and numerous methods of data collection and transmission are the main reasons for the wide use of sensor networks. WSNs are usually deployed densely to cover the given region properly. In this context, when multiple WSNs are deployed

C. K. Meghalatha (✉) · K. S. Sravan · K. Krishna Chaitanya · B. Seetha Ramanjaneyulu
Vignan's Foundation for Science, Technology and Research (Deemed to be University), Vadlamudi, Andhra Pradesh 531001, India
e-mail: meghalatha11@gmail.com

K. S. Sravan
e-mail: sraonekumar92@gmail.com

K. Krishna Chaitanya
e-mail: krishna.kandregula444@gmail.com

B. Seetha Ramanjaneyulu
e-mail: ramanbs@gmail.com

P. S. R. Chowdary et al. (eds.), *Microelectronics, Electromagnetics and Telecommunications*, Lecture Notes in Electrical Engineering 655,
https://doi.org/10.1007/978-981-15-3828-5_73

in the same geographical location, it may lead to interference among them if enough care is not taken. In applications like body area networks (BAN), when multiple patients fitted with numerous sensors are stationed at one location, it can lead to inter-user interference among their wireless-enabled devices because large number of sensors that belong to multiple WSNs are deployed within short range.

Prioritized access of transmission slots to important data is proposed and analyzed. The throughput improvement of this prioritized access when compared with that of 'best effort' traffic is analyzed. Also, the impact of Superframe Order (SO) on the throughput of the devices has been analyzed for both slotted and unslotted CSMA/CA modes. Through this, the effect of changing the active periods of the frame is analyzed.

Section 2 of the paper describes the related work and IEEE 802.15.4 Superframe structure. Section 3 describes the system model, different traffic priorities and the proposed prioritized access mechanism. Section 4 describes the simulation results, and Sect. 5 concludes the paper.

2 IEEE 802.15.4 Frame

Superframe structure given by IEEE 802.15.4 [1] standard is as shown in Fig. 1. It is a MAC frame for personal area networks (PANs), which is an optional structure where manufacturers can modify it as per their requirements. In wireless sensor networks (WSNs), each PAN has a coordinator or mediator who decides the optional Superframe structure. In general, Superframe structure is bounded by network beacons (a signal which appears at the start and end of a frame) in which active portion is divided into 16 equally sized time slots, and the rest is left over as inactive portion.

In the first time slot, Superframe transmits a beacon, which synchronizes the devices attached to coordinator and identifies the PAN in the first place. All the remaining time slots are used by its competitors during contention access period

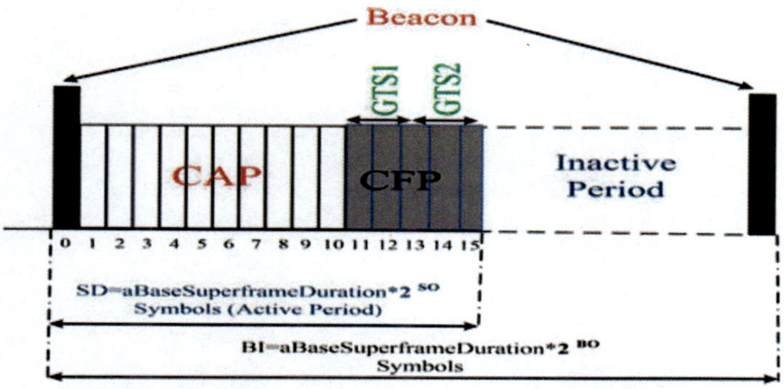

Fig. 1 Superframe structure of IEEE 802.15.4

(CAP). All devices use slotted carrier-sense multiple access with collision avoidance (CSMA/CA) mechanism. If the channel is found to be idle, a random back off value is selected, and after completion of the back off period, data will be transmitted [2]. If all the devices finish their communication before the end of CAP, then no collisions occur. Guaranteed time slots (GTS) [3] will be provided by PAN coordinator to meet the latency and bandwidth requirements of different applications, which could not complete their communication within the CAP. The number of GTSs may extend up to the end of contention free period (CFP).

The MAC Beacon Order (BO) [4] determines the length of the Superframe, which may find values in between 0 and 14, and if BO exceeds the maximum Superframe length, PAN coordinator shall not transmit beacon frames in future, unless there arises a request.

The MAC Superframe Order (SO) determines the CAP of the current structure. If SO $=$ BO, then there will not be any inactive period, and the entire Superframe can be used for packet transmissions.

In [5], authors carried out an experimental analysis on two kinds of hopping methods namely lightweight hopping and heavyweight hopping, to mitigate the heterogeneous interference. In these methods, slotted CSMA/CA mode is used to solve the inter-interference problem.

In [3], various interference scenarios are analyzed based on the QoS parameters. They have used fixed energy detection mechanism to compare with their proposed adaptive energy detection mechanism. The results show that it has better performance when compared to fixed energy detection mechanism.

In [6], authors proposed a hybrid channel scheme to access the devices in a densely populated sensor network for their inter-device communication. The proposed scheme was carried out in two phases. In the first phase, interference was analyzed for pre-defined device requirements. If its throughput does not meet the requirement, it will move to the second phase. In second phase, a hybrid channel access scheme has been proposed which meets the user's requirement.

A heterogeneous body area network [7–9] and [2] scenario is analyzed under CSMA/CA-based mechanism, in which channel switching is done by using a Markov chain model. An analysis has been performed to choose the contention window size randomly, according to the load on the channel.

In [4], to support the QoS to SUs, SUs are categorized into four priority groups based on their time-critical data and plan the usage of vacant channels based on the importance of data to be transmitted by them.

3 System Model

A network model with multiple devices that belong to different WSNs is considered. These devices are configured to carry data traffic of different priorities. NetSim is used in this work to simulate the proposed model. It is a network simulator software that is used to create, simulate, analyze and emulate various communication networks.

In this paper, a network consisting of 20 sensor nodes has been deployed. In order to cover a geographical region of 500 meters, these are spaced uniformly with 20 meters distance among each other. Devices are given access as per their priority levels. Three levels of priority namely best effort, low priority and high priority are considered to the data, whose priority levels increase in that order. Before applying priorities, impact of varying the duty cycle is identified. Duty cycle in IEEE 802.15.4 frame is defined as the ratio of active duration to the total duration of frame. This is described in terms of Beacon Order (BO) and Superframe Order (SO). Based on the Beacon Order value, interval between the successive beacons is determined. This interval is nothing but the total frame duration. Then, depending on Superframe Order, active duration of the frame (which is also known as Superframe duration), is determined. These are indicated in Superframe structure shown in Fig. 1.

4 Results and Discussion

Figure 2 shows a comparison of varying Superframe Order with respect to the throughput for different priorities like best effort (BE), low priority (LPR) and high priority (HPR). As we can see, at different priorities, the throughputs of the devices increase proportionally with respect to the Superframe Order. We can observe that the plot of traffic of HPR reaches its maximum throughput when compared to LPR and BE traffic.

Figure 3 shows a comparison of varying Superframe Order with respect to the throughput for different BO values (BO = 10 and BO = 12). Here, the SO order is increased gradually by keeping the BO value as constant. The result shows that the throughput is maximum when Beacon Order is minimum.

Figure 4 shows the variation of throughput with respect to the number of devices. Here, a normalized value of throughput is considered by running the simulation many times. There is a gradual decrease in the normalized throughput value as the number of devices increase.

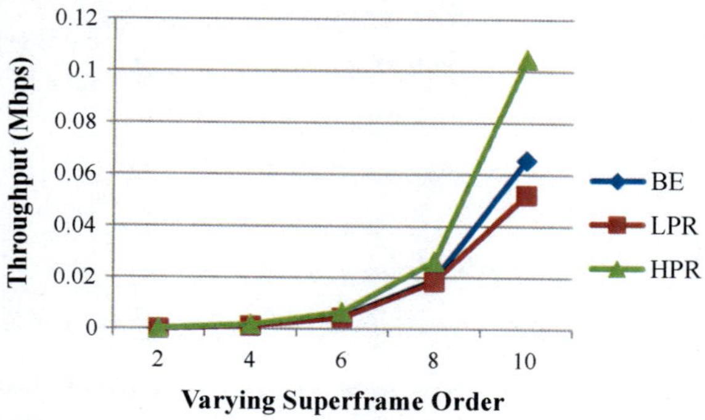

Fig. 2 Varying SO for different traffic priorities

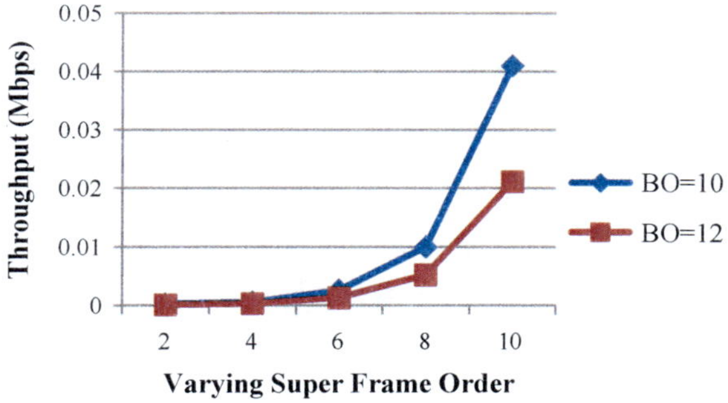

Fig. 3 Throughput analysis for varying SO at BO = 10 and 12

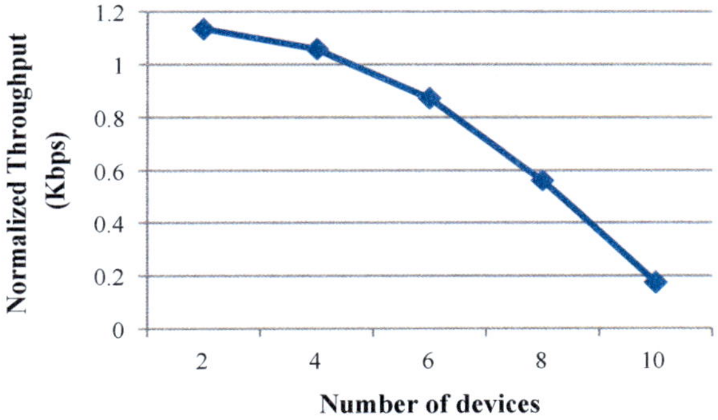

Fig. 4 Throughput analysis for varying number of devices

Figure 5 shows the variation of throughput with respect to different priorities for slotted and unslotted CSMA/CA. Here, a normalized value of throughput is considered by running the simulation many times. The results show that slotted CSMA/CA illustrates better performance than unslotted CSMA/CA in all kinds of traffic conditions.

Figure 6 shows the variation of throughput with respect to the number of devices

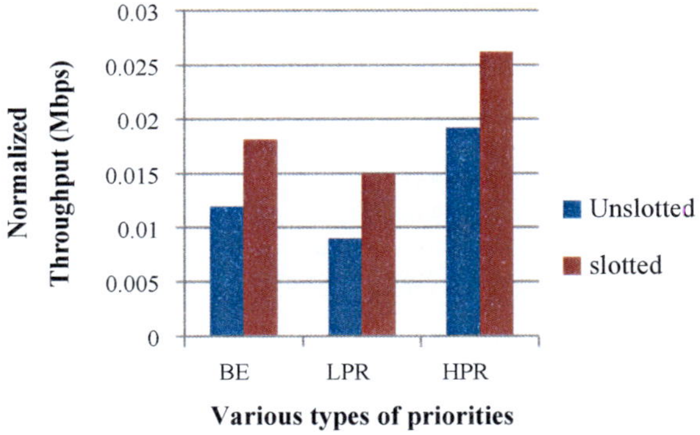

Fig. 5 Comparison of slotted and unslotted CSMA/CA at different priorities

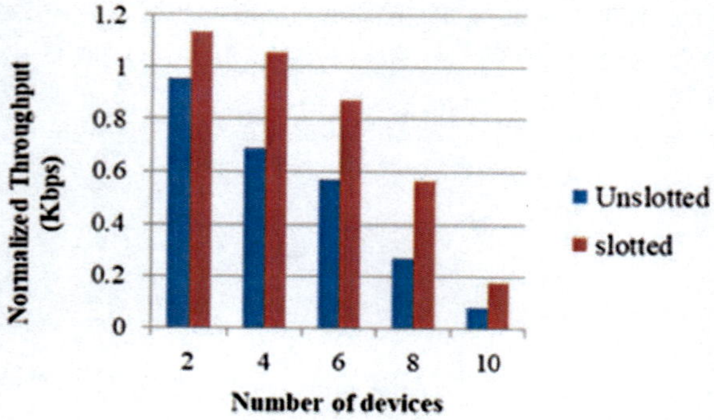

Fig. 6 Comparison of slotted and unslotted CSMA/CA for varying number of devices

for slotted and unslotted CSMA/CA. Here also, a normalized value of throughput is considered by running the simulation many times. The result shows that slotted CSMA/CA illustrates better performance than unslotted CSMA/CA for varying number of devices.

5 Conclusion

In this paper, Superframe structure of IEEE 802.15.4 based on WSN is analyzed for various traffic types and scenarios. The results show that there is considerable increase in throughput for slotted CSMA/CA mechanism compared to unslotted CSMA/CA. It also indicates that prioritized access of the channel by devices offers higher throughputs when compared to non-prioritized access. Impact of Beacon Order and Superframe Order is analyzed to identify the throughput impact of duty cycle variations in 802.15.4 frame.

References

1. IEEE standard for local and metropolitan area networks—Part 15.4: low-rate wireless personal area networks (LR-WPANs) Amendment1: MAC Sublayer, IEEE Standard 802.15.4e-2012 (Amendment to IEEE Standard 802.15.4-2011), pp. 1–225 (2012)
2. Hsiao C, Lin C, Kuo J, Chen W (2018) LAMP: load adaptive MAC protocol for inter-BAN interference mitigation. In: 2018 IEEE wireless communications and networking conference (WCNC), Barcelona, pp 1–6
3. Marbukh V, Sayrafian K (2017) Regret minimization based adaptation of the energy detection threshold in body area networks. In: 2017 global internet of things summit (GIoTS), Geneva, pp 1–6
4. Kumar KS, Annapurna K, SeethaRamanjaneyulu B (2015) Supporting real-time traffic in cognitive radio network"s. In: 2015 international conference on signal processing and communication engineering systems. https://doi.org/10.1109/spaces.2015.7058201
5. Cao B, Ge Y, Kim CW, Feng G, Tan HP, Li Y (2013) An experimental study for inter-user interference mitigation in wireless body sensor networks. IEEE Sens J 13(10):3585–3595

6. Kim B, Choi J, Cho J (2017) A hybrid channel access scheme for coexistence mitigation in IEEE 802.15.4-based WBAN. IEEE Sens J 17(21):7189–7195
7. Li N, Li C, Zhang B, Zhang, Y (2017) An interference mitigation MAC protocol for wireless body area network. In: 2017 IEEE/CIC International Conference on Communications in China (ICCC), Qingdao, pp 1–6
8. IEEE 802.15 WPAN Task Group 6 (TG6) Body area networks. Accessed on 2017/5/1. [Online] Available: http://www.ieee802.org/15/pub/TG6.html
9. IEEE standard for local and metropolitan area networks—Part 15.6: wireless body area networks, IEEE Standard 802.15.6-2012, pp 1–271 (2012)

Analysis of CPW-Fed Modified Z-Shaped Reconfigurable Antenna for Automotive Communications

T. Anilkumar, B. T. P. Madhav, R. Venkata Abhiram, K. Nikhil Sai Radhesh, J. Harish, and M. Venkateswara Rao

Abstract In this article, a coplanar waveguide-fed planar antenna is proposed for vehicular communication applications. The proposed antenna consists of a modified Z-shaped radiator and designed on a FR4 substrate with a footprint of $40 \times 30 \times 1.6\ mm^3$. The design is further made reconfigurable structure for providing switching among the operating bands of the antenna by incorporating the BAR64-03 W PIN diodes. The proposed antenna operates the switching among the complimentary operating bands exist under 2.39–2.8 GHz, 3.32–4.11 GHz, 7.34–9.03 GHz has been observed under different switching conditions. The proposed design also exhibits circularly polarized radiation characteristics under operating bands makes this antenna as a hybrid reconfigurable design. The measured results of fabricated antenna are in good correlation with simulated results. The radiation performance is analysed for stand-alone reconfigurable antenna and the virtual environment on a vehicle body.

Keywords Coplanar waveguide · Hybrid reconfigurable · Vehicular communication · Z-shaped radiator

1 Introduction

Nowadays, the wireless systems have been emerging in every domain that our lives revolve around. These systems adopt multi-standard wireless technologies such as cellular communication (GSM/UMTS/WCDMA/LTE, etc.), Wi-fi and WiMAX. [1–11], and antenna is a very important non-replaceable wireless communication entity since several decades which are serving till now [12–14] in radio transmission. Vehicular communications are emerging field in current scenario for enabling intelligent transportation systems. Effective communication systems with efficient antenna

T. Anilkumar · B. T. P. Madhav (✉) · R. Venkata Abhiram · K. Nikhil Sai Radhesh · J. Harish · M. Venkateswara Rao
Department of ECE, Koneru Lakshmaiah Education Foundation, Antennas and Liquid Crystals Research Center, Vaddeswaram, Guntur, India
e-mail: btpmadhav@kluniversity.in

P. S. R. Chowdary et al. (eds.), *Microelectronics, Electromagnetics and Telecommunications*, Lecture Notes in Electrical Engineering 655,
https://doi.org/10.1007/978-981-15-3828-5_74

modules are required to establish such networks [15]. People opt to access multiple services with very feasible and very small antennas, which cover most of the amenities in a vehicle like forecasting the weather information, efficient traffic status, tracking of vehicles [16] and quality entertainment purposes [17]. Most of the current day automotive antennas are of monopole/whip type operates for cellular, GPS and other communication interfaces. In such scenario, the antennas with wide impedance bandwidth are required to cover the multiple wireless interfaces [17–33]. The optimized antenna solutions for different constraints can be obtained using several techniques as explained in [34] using computational resources. In this scenario, reconfigurable antennas are much suitable for communication systems which can dynamically change according to the spectrum switchable circumstances.

In this article, a reconfigurable Z-shaped antenna with fractal geometries is designed with coplanar waveguide feed which resulted in three radio spectrums like L-band, Bluetooth, Wi-fi, C-band and X-band vehicular applications. This article is organized as follows: Sect. 2 contains the antenna design aspects and hierarchical geometrical evolution and their result of Z-shaped reconfigurable fractal antenna. Section 3 describes the results and discussions of the Z-shaped reconfigurable fractal antenna and concluded in Sect. 4.

2 Antenna Design

The antenna design is performed on FR4 substrate having relative permittivity of $\varepsilon_r = 4.4$, dielectric loss tangent of 0.02. The substrate is having the dimensions of $40 \times 30 \times 1.6\ \text{mm}^3$. The antenna design is concentrated on reconfiguring the antenna such that same antenna can be used for different frequencies. The development of the final model has gone through four iterations in which a tilted Z-shaped structure is considered as radiating element. The upper and lower arms of the Z-shaped structure are made of arc shape with thickness '*t*', and the same thickness is maintained long the rectangular strip of Z-shaped structure as shown in iteration-1 in Fig. 1. The remaining iterations-2, 3, 4 represent the designs which are introduced with gap

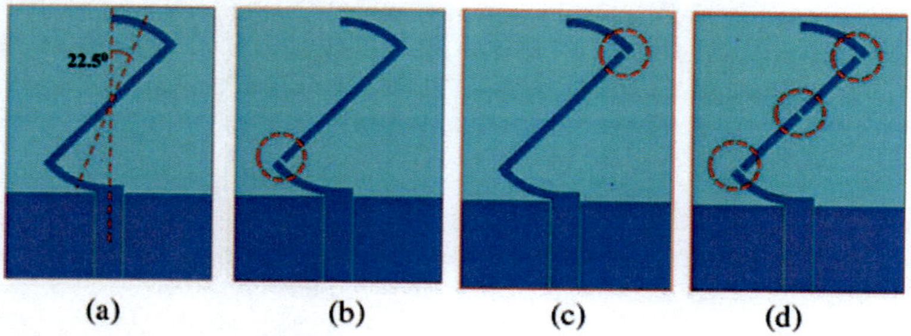

Fig. 1 Geometrical iterations of the antenna geometry with proposed locations of switching elements

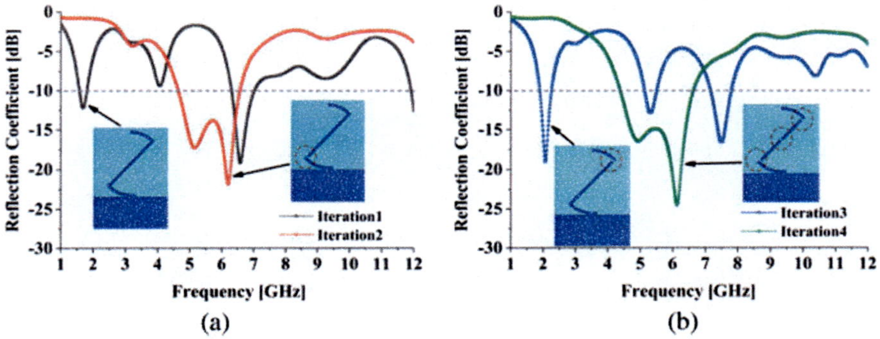

Fig. 2 Reflection coefficient versus frequency characteristics of antenna iterations-1 to 4

discontinuity at the lower bend, upper bend, along with mid-region of the rectangular strip, respectively. The entire structure is connected to the coplanar waveguide feed structure having feed width of 'W_f' and gap between the ground and feed line being 'g'.

The iterations from 1 to 4 (shown in Fig. 1) are simulated for their operating band characteristics through reflection coefficient parameter and presented in Fig. 2. The structure in iteration-1 produces two operating bands 1.58–1.81 GHz and 6.31–7.02 GHz, whereas the iteration-2 yields a single operating band that spans 4.65–6.57 GHz. The iteration-1 yielded two narrow bands, and the gap inserted in the lower bend of the Z-shaped structure has shifted the second band towards the lower side. The bandwidth enhancement is also observed when compared to iteration-1. The similar case has been observed for iteration-3 which is almost similar to iteration-1 except for the gap introduced at the upper bend of Z-shaped structure. This created one more resonant frequency at 5 GHz and thus has three operating bands 1.90–2.25, 5.13–5.49, 7.17–7.79 GHz.

The iteration-4 has two gap discontinuities at lower and upper bends of the Z-shaped structure and one more in the middle of the rectangular connecting strip between those arcs. This resulted again a single band performance from 4.38 to 6.63 GHz which is due to the effect of effective electrical length caused by the gaps. The gap discontinuities which are introduced in the design resemble the ON or OFF conditions of an ideal switch. The operating bands attained under these conditions can be further extended towards controlled switching which is further discussed in Sect. 2.1.

2.1 Reconfigurable Antenna Design

The Z-structure is further made reconfigurable with some geometrical changes as shown in Fig. 3. In this design, the rectangular connecting strip is replaced with fractal geometry which consists of collinear connected scaled circles. The dimensions of the

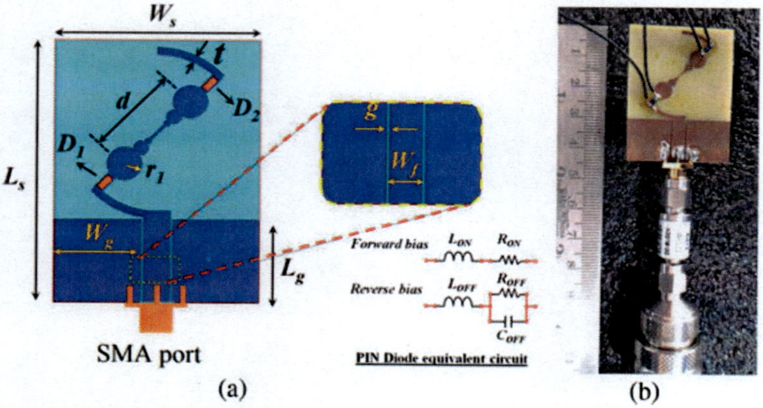

Fig. 3 Geometry of proposed reconfigurable antenna **a** structural representation, **b** fabricated prototype

Table 1 Geometrical parameters of proposed reconfigurable antenna

Parameter	L_s	W_s	t	L_g	W_g	g	W_f	h	s_f	r_1	d
Size/mm	40	30	1	12.73	12.5	0.35	3.84	1.6	0.42	2.5	13.4

circle are downscaled (with scaling factor $s_f = 0.42$) to produce such structure. Two PIN diodes (D1 and D2) BAR64-03 W are incorporated in the gaps of the design to establish controlled switching between the operating bands.

The modelling of PIN diode switches in the ANSYS HFSS simulation tool has been realized with the help of lumped RLC boundary condition. This diode consists of inductance $L_{ON} = 1.8$ nH, forward resistance $R_{ON} = 2.1\ \Omega$ in ON condition, whereas in OFF condition, the inductance $L_{OFF} = 1.8$ nH, reverse resistance $R_{OFF} = 3\ \text{K}\Omega$ and capacitance C_{OFF} is 0.17 pF as shown in Fig. 3 (Table 1).

3 Results and Discussion

The performance of the antenna is analysed through the simulations conducted in ANSYS HFSS and explained with different characteristics as follows.

3.1 Reflection Coefficient and Axial Ratio Versus Frequency Characteristics

The characteristics are plotted in Fig. 4a for the possible switching conditions of the incorporated PIN diodes. The binary numbers '0 and 1' refer to the OFF (reverse

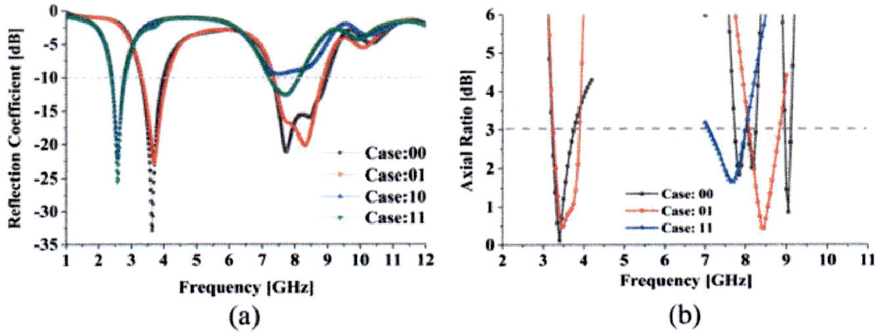

Fig. 4 **a** Reflection coefficient versus frequency characteristics. **b** Axial ratio versus frequency characteristics of the proposed reconfigurable antenna

bias) and ON (forward bias) conditions of the PIN diode, respectively. The impedance bandwidth is computed by considering −10 dB margin as reference. When the diode D1 (lower diode) is reverse biased and D2 (upper diode) is either forward or reverse biased, then the antenna operates in two bands that exist between 3–5 GHz and 7–9 GHz. When the lower diode is forward biased and upper diode is either forward or reverse biased, then the fundamental resonance shifted to 2–3 GHz region which is due to the establishment of conductive path of lower arc structure and the fractal structures.

The polarization characteristics are also studied for the proposed reconfigurable antenna and presented in terms of axial ratio versus frequency plots in Fig. 4b. It is observed that the circular polarized characteristics are observed for three out of four switching cases namely Case: 00, 01 and 11. Here, the axial ratio bandwidth is calculated according to the 3 dB margin. When both the diodes are in reverse bias, then the circular polarized radiation is observed across three bands 3.22–3.76 GHz, 7.73–8.25 GHz and 8.96–9.11 GHz. The axial ratio bandwidth is lying across two bands 3.25–3.89 GHz, 8.08–8.83 GHz. When both the diodes are turned ON, the circular polarization can be seen at single band 7.08–8.01 GHz (Table 2).

3.2 *Radiation Patterns*

The far-field radiation patterns of the stand-alone antenna are presented in Fig. 5. The patterns are visualized in 3D distribution at various operating frequencies. These operating frequencies are considered according to the obtained antenna characteristics at different switching cases in which the cases D1-OFF, D2-ON and D1-ON, D2-ON are chosen.

At lower frequencies like 3.7 and 2.58 GHz, the pattern is having omnidirectional radiation in H-plane and overall pattern in tilted doughnut-shaped. This tilting of pattern is due to the angular orientation of tilted Z-shaped structure. The direction of

Table 2 Characteristics of the proposed reconfigurable antenna

Switching case	Operating bands [GHz]	f_r [GHz]	S_{11} [dB]	Peak gain [dB]	Radiation efficiency	LHCP/RHCP	
						Phi = 0°	Phi = 90°
D1-OFF D2-OFF	3.32–4.05	3.67	−32.87	1.654	0.8396	RHCP	RHCP
	7.37–9.03	7.78	−20.80	4.763	0.8364	LHCP	LHCP
D1-OFF D2-ON	3.32–4.11	3.70	−22.77	2.517	0.8385	RHCP	RHCP
	7.34–8.87	8.31	−20.17	5.390	0.8353	RHCP	LHCP
D1-ON D2-OFF	2.40–2.79	2.59	−22.06	0.28	0.8549	LHCP	LHCP
D1-ON D2-ON	2.39–2.80	2.58	−25.46	0.32	0.8504	LHCP	LHCP
	7.2–8.15	7.70	−12.51	6.09	0.8240	LHCP	RHCP

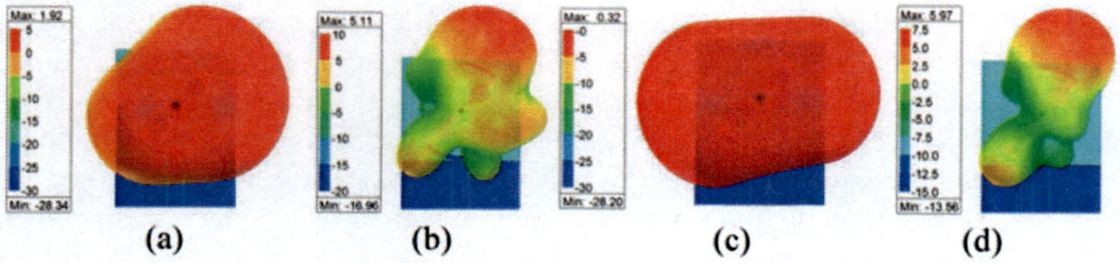

Fig. 5 3D far-field patterns of the proposed antenna **a** at 3.7 GHz, **b** at 8.31 GHz, **c** at 2.58 GHz and **d** at 7.70 GHz

maximum radiation is oriented towards the region at which the diode is turned ON. At higher frequencies like 7.70 and 8.31 GHz, the pattern exhibits directional nature showing its orientation towards 30° in azimuth plane of the stand-alone antenna.

3.3 Platform Mounted Antenna Radiation Characteristics

The proposed antenna is intended to utilize as a radiator for rooftop vehicular applications. The far-field studies are conducted for the proposed reconfigurable antenna by placing on rooftop of the vehicle and presented in Fig. 6. It is evident that the far-field radiation for the antenna installed on a vehicle is such that it can cover omnidirectionally around the vicinity of the vehicle which can be observed at lower frequencies (LTE/WiMAX, etc.). At higher frequencies (X-band frequencies), the case is similar, but the maximum is oriented upward direction to communicate with satellites. After installation of the antenna on the vehicle, the far-field radiation supports for communicating with other vehicles as well as the base stations.

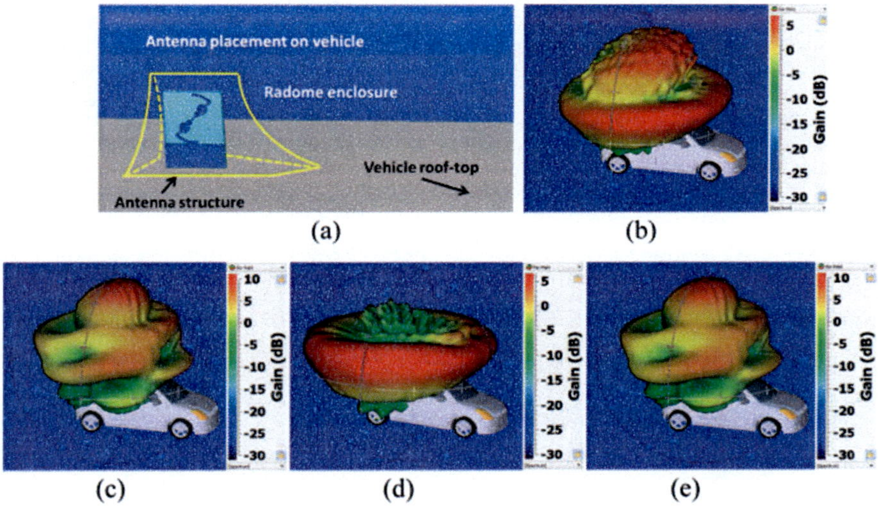

Fig. 6 **a** Placement location of the antenna on vehicle, 3D far-field patterns of the proposed antenna with virtual vehicular environment **b** at 3.7 GHz, **c** at 8.31 GHz, **d** at 2.58 GHz and **e** at 7.70 GHz

4 Conclusion

A modified Z-shaped antenna with CPW feed is presented in this article for hybrid reconfigurable applications suitable for vehicular communications. The proposed structure is a compact one and produces reconfigurable bands 2.39–2.8 GHz, 3.32–4.11 GHz, 7.34–9.03 GHz at different switching conditions. The proposed antenna also exhibits circular polarization at corresponding operating bands, and polarization switching between left-handed and right-handed circular polarization is attained. The radiation characteristics for the virtual placement of the antenna on the vehicular structure is also studied, and the enhancement in the gain is noticed as 5.8 dB, 6.2 dB and 6.5 dB at 2.6 GHz, 2.7 GHz and 7.8 GHz, respectively, which shows the proposed antenna is a good candidate for Bluetooth, WLAN, WiMAX and X-band applications.

Acknowledgements Thanking to the department of ECE, KLEF and DST-India through grants ECR/2016/000569 and EEQ/2016/000604.

References

1. Pokkunuri P, Sai GK, Venkateswararao M, Ganesh B, Tarakaram N, Teja DP (2019) Metamaterial inspired reconfigurable fractal monopole antenna for multiband applications. Int J Intelli Eng Sys 12(2):53–61

2. Bhavani KVL, Khan H (2015) Multiband slotted aperture antenna with defected ground structure for C and X-band communication applications. J Theor Appl Info Tech 82(3):454–461
3. Babu MA, Vaishnavi DN, Radhakrishna P, Bharath N, Madhuri K, Prasad KB, Harish K (2015) Flared V-shape slotted monopole multiband antenna with metamaterial loading. Int J Comm Ant Prop 5(2):93–97
4. Sundar PS, Harsha DS, Manasa P, Manikanta G, Brahmaiah K (2014) Fabric substrate material based multiband spike antenna for wearable applications. Res J Appl Sci Eng Tech 8(3):429–434
5. Taraka B, Tirunagari A (2018) Design and study of multiband planar wheel-like fractal antenna for vehicular communication applications. Microw Opt Tech Lett 60(8):1985–1993
6. Kranthi DL, Devi CK, Navyasanthi N, Reddy BT (2015) A multiband mimo antenna for S and C-Band communication applications. ARPN J Eng Appl Sci 10(14):6014–6022
7. Reddy MVKC, Rao MV, Krishna CM, Raj PC, Jaya G (2017) Quad band filtenna using split ring resonators to notch unwanted frequencies in medical application bands. J Theor Appl Info Tech 95(9):2070–2077
8. Manjeera M, Navya MS, Devi DS, Sumanth V (2016) Novel metamaterial loaded multiband patch antenna. Indian J Sci Tech 9(38):93377
9. Krishna TVR, Lekha KDS, Bhavya, D, Teja VSD, Reddy TM (2018) Multiband semicircular planar monopole antenna with spiral artificial magnetic conductor. In: Anguera J, Satapathy S, Bhateja V, Sunitha K (eds) Microelectronics, electromagnetics and telecommunications. Lecture notes in electrical engineering, vol 471, Springer, Singapore, pp 591–607
10. Kishore MP, Rao MV (2019) A CPW-fed elliptically curved antenna design for multiband operation with metamaterial loading. Int J Innov Tech Expl Eng 8(7):120–125
11. Ugendra V, Khan H, Joshna C (2018) Multiband fractal slot antenna with closed ground structure. In: smart innovation, systems and technologies, vol 78. 75–83
12. Rajiya S, Nadh BP, Kumar MS (2018) Frequency reconfigurable monopole antenna with DGS for ISM band applications. J Elect Eng 69(4):293–299
13. Rao DS, Supraja K, Tejaswini A, Phanikumar A, Prakash RS, Kumari ANM (2017) K15 nematic phase liquid crystal material based double-dipole reconfigurable antenna. Rasayan J Chem 10(3):866–872
14. Babu MA, Reddy BM, Chaitanya RDC, Satish T (2017) A dual-polarization reconfigurable antenna with beam switching characteristics for S-band applications. ARPN J Eng Appl Sci 12(16):4841–4847
15. Rao DS, Lalitha G, Parvez SM, Naveen J, Deepak DM, Kumari ANM (2018) A frequency reconfigurable spiral F-shaped antenna for multiple mobile applications. In: Anguera J, Satapathy S, Bhateja V, Sunitha K (eds) Microelectronics, electromagnetics and telecommunications. Lecture notes in electrical engineering, vol 471. Springer, Singapore, pp 571–580
16. Babu MA, Banu PF, Teja GHS, Prashanth P, Yamini KL (2018) Octagonal shaped frequency reconfigurable antenna for Wi-Fi and Wi-MAX applications. In: Anguera J, Satapathy S, Bhateja V, Sunitha K (eds) Microelectronics, electromagnetics and telecommunications. Lecture notes in electrical engineering, vol 471. Springer, Singapore, pp 581–588
17. Kumar CVN, Sruthi PS, Sahithi M, Manikanta KV (2019) Design of a frequency reconfigurable fractal antenna for internet of things (IoT) in vehicular communication. Int J Recent Tech Eng 7(6):1605–1611
18. Prasad BS, Rao PM (2017) CPW fed T-stub and U-slot reconfigurable antenna for Wi-Fi and WLAN communication applications. J Adv Res Dyn Cont Sys 9(14):2104–2116
19. Murthy KSR, Umakantham K, Murthy KSN (2017) U-shaped slotted reconfigurable monopole with Wimax band notching. J Adv Res Dyn Cont Sys 9(14): 1911–1919
20. Dhanade YB, Sreelakshmi K, Bora P, Mudliar M (2017) Frequency reconfigurable dual band antenna for wireless communications. J Adv Res Dyn Cont Sys 9(14):2328–2345
21. Rajiya SK, Monika M (2018) Circular slotted reconfigurable antenna for wireless medical band and X-band satellite communication applications. Indian J Public Health Res Dev 9(6):296–300
22. Monika M, Rajiya SK (2018) Fractal shaped concentric ring structured reconfigurable monopole antenna with DGS for GPS, GSM, WLAN and ISM band medical applications. Indian J Public Health Res Dev 9(6):285–289

23. Murthy KSR, Umakantham K, Murthy KSN (2019) Design and analysis of a reconfigurable antenna for dual band ISM medical and Wi-Fi applications. Int J Innov Tech Expl Eng 8(10):3630–3634
24. Varma SST, Pavani V, Kumar GM (2019) Defected ground structure based reconfigurable UWB antenna for IoT applications. Int J Recent Tech Eng 8(1):3030–3036
25. Saravanan RA, Venkateswararao M, Sahiti VS, Swetha V, Priyanka VK, Anand PS (2019) Frequency and pattern reconfigured multi band CPW antenna for Wimax and X-band applications. Int J Innov Tech Expl Eng 8(6):1202–1208
26. Allam VK, Maloji S (2019) A novel reconfigurable bandpass filtering antenna for IoT communication applications. Prog In Electromagn Res C 96:13–26
27. Kolusu SSR, Modala AB, Nadh BP (2019) Concentric circular ring loaded triple band antenna for wireless applications. Int J Intelli Eng Sys 12(5):241–248
28. Lakshmi MLSNS, Jayasree G, Vijay G, Nagarjuna P, Kumar CA (2015) Analytical study on Folded-Slot Koch fractal antenna. Indian J Sci Tech 8(17):66573
29. Gupta GS, Rahul M, Lahari OK, Sameera M (2016) Linearly polarized microstrip planar filtenna for X and Ku band communication systems. Indian J Sci Tech 9(38):97115
30. Sundar PS, Kotamraju SK, Sairam SSV, Srinivas A, Neelima K, Sravanthi G (2016) Radial stub loaded antenna with tapered defected ground structure. ARPN J Eng Appl Sci 11(19):11293–11299
31. Rao MV, Tirunagari A (2018) Conformal band notched circular monopole antenna loaded with split ring resonator. Wirel Pers Comm 103(3):1965–1976
32. Sowmithri P, Renuka M (2019) A triangular slot band notched MIMO antenna for UWB applications. J Adv Res Dyn Cont Sys 11(2):1002–1009
33. Monika M, Kumar MS, Prudhvinadh B (2019) Dual band reconfigurable compact circular slot antenna for WiMAX and X-band applications. Radioelec Comm Sys 62(9):474–485
34. Paladuga CSR, Vedula CS, Anguera J, Mishra RK, Andújar A (2018) Performance of beamwidth constrained linear array synthesis techniques using novel evolutionary computing tools. Appl Comp Electromag Soc J 33(3):273–278

Performance of Error Correction Codes for 5G Communications

B. Surendra babu, Idrish shaik, and N. Venkateswar Rao

Abstract The coded modulation scheme emerged as an efficient method to enhance the bandwidth utilization quality of the system in 5G communications. It typically makes use of both the channel coding as well as the simple modulation method. This process effectively adds to the performance of the modulation phenomenon comparatively with respect to un-coded. Among the coding schemes, the low-density parity check coding (LDPC) scheme is highly performing. In this paper, the performance analysis of the codes is carried for the BPSK system in AWGN environment.

Keywords BPSK · LDPC · 5G communications · Error correction codes

1 Introduction

The communication system (CS) typically serves as a medium for transferring mutual information through free space in the case of wireless communications or a pair of wires, coaxial cable, or optical fibers in wired communications. The input signal is processed in the transmitter considering the characteristic of the transmission channel, while the process almost always involves modulation and in some cases coding [1–4]. Then, transmission is conducted via the channel that can cause attenuation or loss of signal. The goal of the receiver is to retrieve the input signal that might have changed as a result of distortion, interference, or noise. Demodulation is performed on this received signal. The detected signal usually contains errors which can be minimized through the use of coding. Both modulation and coding are required for reliable long-distance transmission. To improve digital communication, different

B. Surendra babu (✉) · I. shaik · N. Venkateswar Rao
Bapatla Engineering College, Bapatla, Guntur, Andhra Pradesh, India
e-mail: surendrabachina@gmail.com

I. shaik
e-mail: idrishshaik@gmail.com

N. Venkateswar Rao
e-mail: nvrao68@gmail.com

P. S. R. Chowdary et al. (eds.), *Microelectronics, Electromagnetics and Telecommunications*, Lecture Notes in Electrical Engineering 655,
https://doi.org/10.1007/978-981-15-3828-5_75

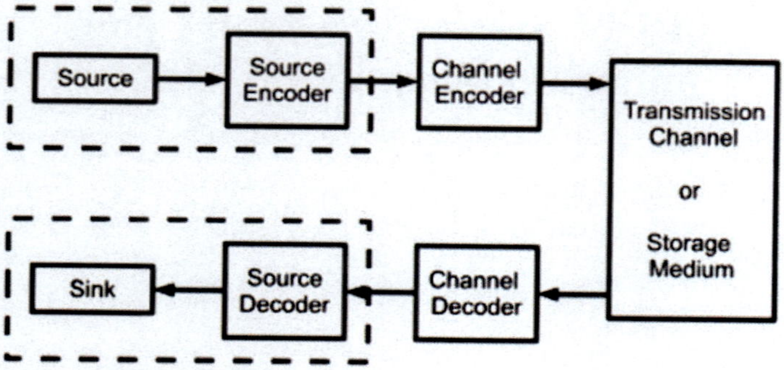

Fig. 1 CS block description

modulation and coding methods were introduced. The typical structure of a CS is given in Fig. 1.

There are many coding techniques, and still, many improvisations are suggested by many researchers to improve the coding block. The LDPC is most popular technique. In this paper, a brief description of the LDPC in the 5G modulation techniques is discussed.

2 Error Control Codes

The input signal may be either continuous or discrete as the case of analog or digital data. Consequently, transmission will be carried on in analog or digital domain, respectively [3–8]. The emphasis of this thesis is digital communication systems. Hereafter for the sake of simplicity, we will assume that the information alphabet is binary; i.e., A $\in$ {0, 1}, and a memoryless binary-input continuous-output channel is used. For binary data, the simplest modulation scheme is binary phase-shift keying (BPSK), the choice of the modulation in this thesis. Before modulation, binary input signals with alphabet A are mapped to $\{+1, -1\}$ accordingly. The converted sequence is then mapped into a set of antipodal signals that will be converted to the desired modulation frequency for transmission through additive white Gaussian noise (AWGN) channel, the most commonly used model for communication channels.

The process of encoding takes place by generating a codeword (CW) from the input sequence (IS). In this, the redundant bits are identified initially from the IS and are further added to itself. While the CW is generated, it is necessary to ensure the performance. This can be enhanced by simply optimizing the distances between them. While generating the CW, the size of the information bits is to be k. Accordingly, the corresponding binary code takes a length of 2^*k represented by M.

The k length sequence is further transformed as blocks. Here, each block size is referred as n. However, it is ensured that the value of n is always greater than that of k. Finally, the respective code is presented as (n, k). The magnitude of redundancy is essential to be computed in such cases. For this particular case of (n, k), it is referred as $1 - R$ while R represents the code rate. The LDPC is implemented and proposed during early 1960s [1]. The proposed technique was soon considered as the best way to get rid of the errors. Quickly, they attracted the interest of the engineers working in the field of information their and data encoding for high data rates. Moreover, the LDPC has the capability much close to that of Shannon's limit. Similarly, often it is considered that these codes have the ease of implementation.

They are well equipped with the excellent error correcting potential. The name is coined out of its characteristics from the parity check matrix. It typically consists of majorly zeroes than ones. Moreover, these codes have potential to deal with multiple channels as well as complicated algorithm during the decoding process. There are several merits to list with the LDPC over conventional turbo codes. The first being the very important decoding process. This process significantly adopts parallelism. As a result, it can perform at high speed. The other advantage is associated with the less complexity in the decoding process. In addition to the above, the other advantage is trouble-shootable decoding form. For an enhanced efficiency, the CW can be verified on the run.

It is possible with LDPC to generate optimal as well as short length binary codes. Citing these advantages, these codes can be used in various applications like WiMax (IEEE 802.16). Similarly, it is also applicable to several high-speed devices. Interestingly, the feature of parallel implementations made the LDPC a potential candidate which operates on iterative message-passing algorithms. The reduction in the complexity associated with algorithms and implementation of LDPC can lead to reduction in the chip size along with power consumption. However, the respective throughput can be maintained with ease effectively.

Typically, the LDPC codes are generated using sparse bipartite graphs which resemble as linear codes. The process accommodates several types of nodes which are named as check nodes and message nodes. Finally, the CW appears as the vector. It is constituted by sum of the neighbouring points for all check nodes among the message nodes which are equaled to zero. The demonstration of the codes with different nodes is shown in Fig. 2 [2].

This is the very simple and easy form of phase-shift keying. In PSK, the phase of carrier is switched between two or more value based on input modulating digital signal. In this case, 1800 phase shift is applied at the bit transitions. In this technique, only two phases are taken into consideration separated by 180°. Due to this factor, it can be termed as 2-PSK. It is also known as phase reversal keying. The simple formulation of the modulation technique is given as follows

$$S_n(t) = \sqrt{\frac{2E_b}{T_b}} \cos(2\pi f_c(t) + \pi(1 - n)), \quad n = 0,\ 1. \tag{1}$$

Only two phases are taken into consideration, i.e., $(0, \pi)$.

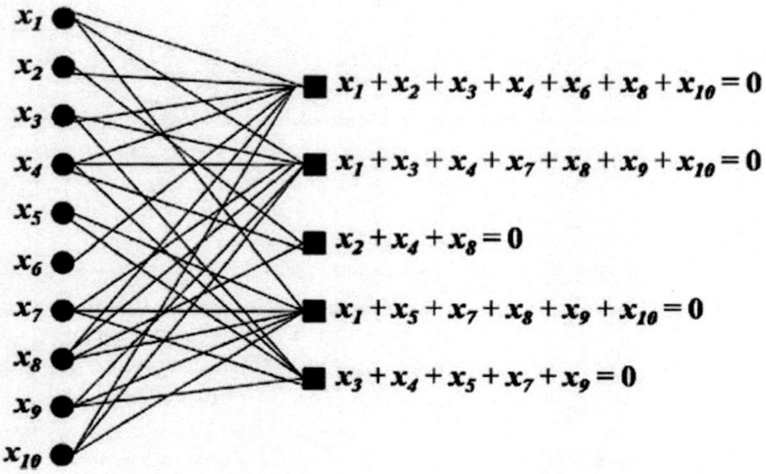

Fig. 2 Demonstration of LDPC

$$S_0(t) = \sqrt{\frac{2E_b}{T_b}} \cos(2\pi f_c t + \pi) = -\sqrt{\frac{2E_b}{T_b}} \cos(2\pi f_c t) \tag{2}$$

For symbol "0"

$$S_1(t) = \sqrt{\frac{2E_b}{T_b}} \cos(2\pi f_c t) \tag{3}$$

Here, the CW frequency is fc. Similarly, if the following function is used to represent the signal in the spatial domain,

$$\Phi(t) = \sqrt{\frac{2}{T_b}} \cos(2\pi f_c t) \tag{4}$$

The corresponding binary transformation is given as

$$\begin{aligned} 1 &= \sqrt{E_b}\phi(t) \\ 0 &= -\sqrt{E_b}\phi(t) \end{aligned} \tag{5}$$

3 LDPC Coding and Decoding

In spite of several advantages associated with the implementation and other features of the LDPC, it suffers from some disadvantages and demerits. Mostly, these demerits are associated with the corresponding encoding process [3–9]. The matrix which is employed is not practically sparse in nature. It directly impacts several processes which produce more number of complexity processes. The first encoding scheme is to deal with lower triangular shape parity check matrices. The other encoding schemes are mainly to deal with cyclic parity check matrices.

The most popular codes that can be easily encoded are the cyclic or pseudo-cyclic ones. Okamura (2003) used a Gallager [1] cyclic-based encoder. Finite geometry LDPC codes are also cyclic or pseudo-cyclic.

The LDPC decoders, namely the bit flipping (BF) algorithms and the message-passing algorithms, are widely used. Both of these methods have been introduced along with the LDPC codes by Gallager [1] and have been studied by various researchers. Considering the three aforementioned tradeoff properties of an error correction scheme, the efficiency depends on the code (the code rate to be specific) and is independent of the decoder selection. However, the complexity and the reliability depend both on the code and the decoder. The performance metrics are shown in Table 1. Here, **Eb/No** is classically defined as the ratio of energy per bit (**Eb**) to the spectral noise density (**No**).

Table 1 Performance metrics of the BPSK-LDPC

Number of iterations	Bit error rate	Number of error bits	Number of error blocks	Number of blocks
5	0.13857	116	5853	120
	0.12491	97	4397	100
	0.12354	79	3479	80
10	0.097017	92	4098	120
	0.12085	76	4254	100
	0.11602	57	3267	80
15	0.090507	76	3823	120
	0.10091	63	3552	100
	0.097869	52	2756	80
20	0.090672	65	3830	120
	0.091165	54	3209	100
	0.094212	47	2653	80

4 Conclusion

The performance of the LDPC is analyzed through simulation and the corresponding parameters like Eb/No, bit error rate, and the number of errors bits. It can be concluded that the LDPC is an excellent error correcting code. It is found that it is well suitable for different modulation schemes like BPSK. Further, the study can be extended to various other modulation schemes for the analysis.

References

1. Gallager RG (1963) Low-density parity-check codes. MIT Press, Cambridge, MA
2. Lentmaier M Error control coding. Lecture notes. EIT, Lund University
3. Rappaport TS (1996) Wireless communications: principles and practice. Prentice-Hall, Upper Saddle River, NJ
4. Tan W, Cruz JR (2004) Performance evaluation of low-density parity-check codes on partial-response channels using density evolution. IEEE Trans Comm 52(8):1253–1256
5. Shaghaghi M, Cai K, Guan YL, Qin ZL (2011) Markov chain Monte Carlo based detection for two-dimensional intersymbol interference channels. IEEE Trans Magn 47(2):471–478
6. LDPC Code using MATLAB, http://sites.google.com/site/bsnugroho/ldpc
7. Yao J, Teh KC, Li K. H (2012) Iterative detection scheme with LDPC codes for two-dimensional interference channels. In: 2012 IEEE international conference on communication systems (ICCS), IEEE, pp 398–402
8. Yao J, Teh KC, Li KH (2012) Joint iterative detection/decoding scheme for two-dimensional interference channels. IEEE Trans Commun 60(12):3548–3555
9. Paladuga CS, Vedula CV, Anguera J, Mishra RK, Andújar A (2018) Performance of beamwidth constrained linear array synthesis techniques using novel evolutionary computing tools. Appl Comput Electromag Soc J 33(3):273–278

Pattern Recognition of Time-Varying Signals Using Ensemble Classifiers

M. Venkata Subbarao, Sudheer Kumar Terlapu, V. V. S. S. S. Chakravarthy, and Suresh Chandra Satapaty

Abstract A new classification approach for time-varying power quality (PQ) signals using ensemble classifiers (EC) is proposed in this paper. To achieve high performance, existing expert systems require several signal features so that these systems have more computational complexity. In order to reduce the computational cost and to improve the accuracy further, a new set of features called moments and cumulants are introduced in this paper to classify PQ events. Further, the performance of various ensemble classifiers is analyzed with the proposed feature set. Moreover, the analysis is carried out with different training and testing rates. Finally, the performance comparison is made with that of the existing techniques to prove the superiority of the proposed features and classifiers.

Keywords PQ signals · Ensemble classifiers · Moments and cumulants · Boosting · Bagging

1 Introduction

The exploitation of sensitive electronic components in applications of smart cities, smart buildings, and homes is growing exponentially [1]. These sensitive devices are easily affected by PQ disturbances, such as sags, surges, interruptions. [2, 3]. In real-time applications, failure of these sensitive devices may cause serious damage, especially in smart applications [4]. An automatic or blind recognition system is required to detect and identify the occurrence of PQ problem, so that devices can be

M. V. Subbarao (✉) · S. K. Terlapu
Department of Electronics and Communication Engineering, Shri Vishnu Engineering College for Women, Bhimavaram, India
e-mail: mandava.decs@gmail.com

V. V. S. S. S. Chakravarthy
Department of Electronics and Communication Engineering, Raghu Institute of Technology, Visakhapatnam, India

S. C. Satapaty
Department of Computer Science and Engineering, KIIT, Bhubaneswar, India

P. S. R. Chowdary et al. (eds.), *Microelectronics, Electromagnetics and Telecommunications*, Lecture Notes in Electrical Engineering 655,
https://doi.org/10.1007/978-981-15-3828-5_76

prevented from damage [5]. In the last two decades, the researchers have developed different classification methods such as rough set-based [6], discrete wavelet-based [7, 8], neural network-based S-transform and modified ST-based [9–12], etc., for recognition of PQ events. Recently, artificial neural networks, rule-based expert systems, data mining-based classifiers, and fuzzy logic classification systems [13] are developed for classification of PQ events. All these methods extracted some specific features such as skewness, energy, mean, kurtosis, standard deviation, variance, and an average of the squared absolute values, etc., from the single and multiple disturbances for classification. Proper selection of features from the feature vector set is required to get more accuracy in classification and to reduce the classification time of the classifier. Some other approaches for classification of PQ events are discussed in Table 1.

To overcome the drawbacks of existing PQ classification approaches, a new attempt has been made in this present work to develop efficient algorithms for PQ

Table 1 PQ classification approaches

Ref. no.	Feature extraction method	Classifier	Feature selection	Number of disturbances	Accuracy		Year
					30 dB	20 dB	
[16]	WT	DT	x	5	96	x	2018
[17]	VMD	DT	x	9	93.8	x	2015
[18]	VMD	DT	x	9	96.7	x	2018
[19]	FDST	DT	x	13	x	98.3	2018
[20]	FDST	DT	x	11	95.3	x	2014
[21]	ST	DT	x	11	97.9	x	2013
[22]	TT	DT	ACO	9	91.2	x	2011
[22]	TT	DT	PSO	9	95.6	x	2010
[23]	DWT	SVM	x	5	98	95.6	2009
[24]	DWT	SVM	x	9	97	96.3	2011
[25]	WPT	SVM	x	8	93.4	x	2002
[26]	WPT	SVM	GA-FKNN	8	96.2	x	2012
[27]	DWT + HST	SVM	Gram-Schmidt orthogonal transform	10	99.3	98.7	2014
[28]	DWT + HST	SVM	Mutual forward selection	8	99.2	98.4	2014
[19]	FDST	Quadratic SVM	x	13	x	94.1	2018
[29]	VMD	SVM	x	6	x	x	2015
[30]	TQWT	SVM	x	14	96.4	96.4	2018

event classification. In order to develop the new algorithms, the following assumptions are considered. First, the classifier needs to be accurate. The accuracy is measured by the true-positive rate and false-negative rate. The lower false-negative rate gives better classifier for PQ classification. Second, the PQ event classifier needs to be robust under various noisy conditions. Third, the classifier is required to meet the essential requirements such as better classification accuracy even with more number of PQ classes in the pool.

The rest of the paper is organized as follows. Section 2 gives about the system model and feature extraction. Proposed EC is discussed in Sect. 3. In Sect. 4, the performance of the proposed EC is evaluated for different PQ signals. Section 5 concludes the article.

2 System Model

The proposed framework for PQ signal recognition is shown in Fig. 1. It involves feature extraction and training followed by testing.

Training of the classifier is performed with the extracted features. Finally, recognition is done with the trained EC.

The moments of a time-varying signal $y(k)$ are given by

$$M_{pq} = E\left[y(k)^{p-q} y^{*}(k)^{q}\right] \tag{1}$$

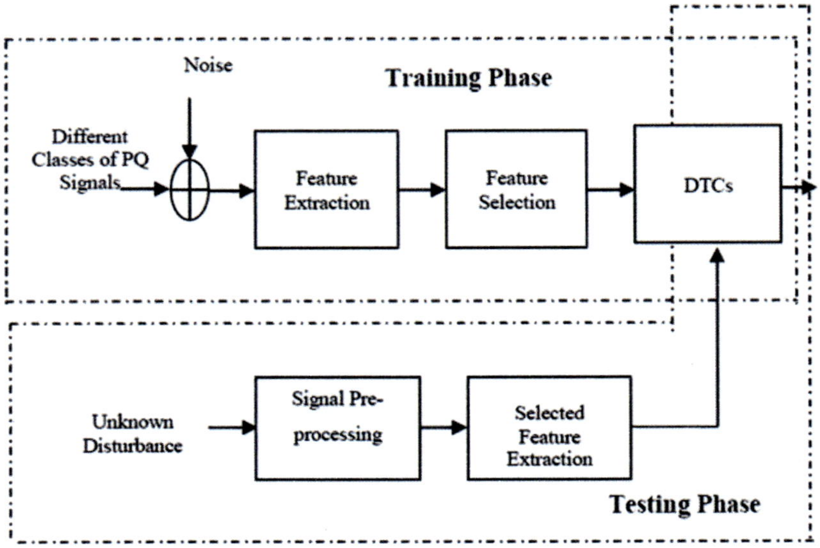

Fig. 1 PQ recognition framework

where p is the order of moment.

These moments are further used to derive the multi-order cumulants as follows:

$$C_{20} = E\left[y^2(k)\right] \tag{2}$$

$$C_{40} = M_{40} - 3M_{20}^2 \tag{3}$$

$$C_{60} = M_{60} + 30M_{20}^3 - 15M_{20}M_{40} \tag{4}$$

Here, C_{20}, C_{40} and C_{60} are second-, fourth-, and sixth-order cumulants, respectively.

3 Ensemble Classifier

The prediction accuracy of various PR classifiers is differed across data sets due to the algorithmic variability of the classifiers. It is not possible to predict the best working classification algorithm for different data sets. To overcome this problem, proposed ensemble classifiers are constructed with a set of different classifiers in order to allow them to vote for decision making. Through majority vote, these classifiers provide the best prediction accuracy. Ensemble classification is a nonlinear ML approach. An EC contains a set of autonomously trained classifiers, and their predictions are collected when recognizing new object. Finally, based on their majority vote, prediction will be done.

The most widely used ensemble technique is called boosting [14], and it works with the weighted training set. Once boosting is completed, then ensemble classifier model is ready for testing. In the testing phase, the unknown testing data is applied to ensemble model for PQ prediction. The detailed process using ensemble classifier is shown in Fig. 2 [15].

In this section, five types of ensemble classifiers such as bagged tress (BaT), boosted trees (BoT), subspace discriminant KNN (SDKNN), subspace KNN (SKNN), and RusBoosted trees (RBT) are developed for PQ recognition in combination with features listed in the Sect. 2. Bagged and boosted trees are constructed from deep trees, and these are slower in speed, whereas, subspace KNN and subspace discriminant KNN uses discriminant analysis and KNNs. In RusBoosted trees, weak learners will be boosted based on random under-sampling.

4 Results and Discussions

The recognition accuracy of proposed ECs is analyzed with seven types of PQ signals which are listed in Table 3. The detailed simulation parameters are discussed in

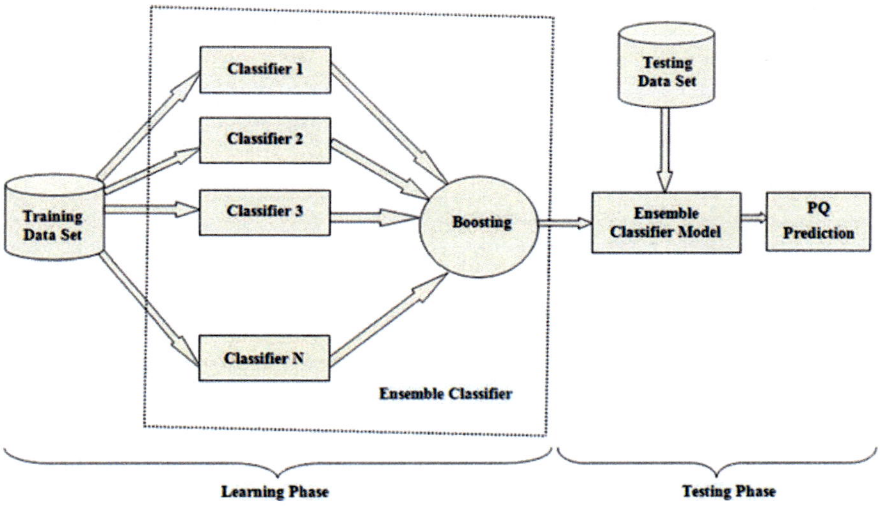

Fig. 2 PQ recognition with EC

Table 2. For the experiments, each PQ disturbance is considered for 1000 times at different SNR values. A set of features C_{20}, C_{40} and C_{60} are extracted from each copy signal and are used for training and testing. The accuracy of proposed ECs is tested at 50–90% training and 50–40% testing.

Table 3 shows the numerical values of the three selected features for all classes of PQ disturbances which are considered for simulation. These selected features are used for training the proposed ECs where LB and UB are lower bound and upper bound, respectively.

The performance of the proposed ECs with various percentages of training set is shown in Table 4. Among all the proposed classifiers, BaT and BoT classifiers have superior performance.

The recognition accuracy comparison of proposed ECs with various existing methods is shown in Table 5. From Table 5, it is proved that, the proposed EC classifiers

Table 2 Simulation parameters

Parameter	Description
PQ signals	Momentary interruption, sag, flicker, spikes, harmonics, swell, transient
Size of data set	7000*3 (1000 copies of each PQ class under varying noise conditions, 3 statistical features)
Size of training set	50–90%
Size of testing set	10–50%
Performance indices	Accuracy
Classifiers	BaT, BoT, SDKNN, SKNN, and RBT

Table 3 Cumulants of PQ events

Label	PQ signal	Features					
		C_{60}		C_{40}		C_{20}	
		LB	UB	LB	UB	LB	UB
S1	Sag	8.30	9.20	−1.42	−1.32	0.00	0.39
S2	Swell	9.37	9.51	−1.45	−1.43	0.03	0.10
S3	Outage	2.60	2.81	−0.83	−0.81	0.04	0.14
S4	Transient	65.7	71.0	5.30	5.50	0.02	0.07
S5	Flicker	6.90	7.18	−1.24	−1.21	0.03	0.14
S6	Harmonics	7.60	7.86	−1.30	−1.28	0.03	0.12
S7	Spikes	9.50	9.67	−1.46	−1.44	0.02	0.10

achieved the optimal classification accuracy even with less training. The recognition capability of proposed ECs is superior to that of existing approaches. The training time of proposed ECs is also very less.

5 Conclusion

A new PQ recognition approach using ensemble classifiers is proposed in this paper. Three new PQ signal features C_{20}, C_{40} and C_{60} are used along with the ensemble classifiers for PQ signal classification. Further, the recognition performance of proposed ECs is analyzed at various training and testing rates. From the experiments, it is clear that the performance of proposed ECs is superior to that of the existing techniques. Further, the research can be extended by incorporating optimization algorithms along with the proposed pattern recognition algorithms to improve classification accuracy and to reduce the training time.

Table 4 Performance of proposed ECs

Classifier/ % of training	% of recognition accuracy								
	90	85	80	75	70	65	60	55	50
BaT	100	100	100	100	100	99.9	99.9	99.9	99.9
BoT	100	100	100	100	100	99.9	99.9	99.8	99.8
RBT	100	100	100	99.9	99.9	99.8	99.8	99.8	99.8
SDKNN	100	100	99.8	99.8	99.6	99.5	99.1	98.9	98.9
SKNN	99.8	99.8	99.7	99.6	99.6	99.5	98.9	98.8	98.8

Table 5 Performance comparison of different methods

Method	Number of features	Training time (sec)	Accuracy (%)
ST, fuzzy C-means and APSO [13]	8	–	96.33
ST, fuzzy C-means and GA [13]	8	–	96.45
ST and PNN [11]	7	–	96.10
WPT and ANN [12]	7	31.26	95.25
ST and PNN [10]	3 4	0.9 (CPU)	95.91 97.40
DWT [2]	8	–	97.81
WPT and SVM [12]	7	27.68	97.25
Proposed BaT Training with $\geq$ 70% Training with 50–65%	3	2.21–2.39	100 99.9
Proposed BoT Training with $\geq$ 70% Training with 60–65% Training with 50–55%	3	2.19–2.41	100 99.9 99.8

References

1. He H, Starzyk JA (2006) A self-organizing learning array system for power quality classification based on wavelet transform. IEEE Trans Power Deliv 21(1):286–295
2. Masoum MAS, Jamali S, Ghaffarzadeh N (2010) Detection and classification of power quality disturbances using discrete wavelet transform and wavelet networks. IET Sci Meas Technol 4(4):193–205
3. Gaing Z-L (2004) Wavelet-based neural network for power disturbance recognition and classification. IEEE Trans Power Delivery 19(4):1560–1568
4. Santoso S, Power EJ, Grady WM, Parsons AC (2000) Power quality disturbance waveform recognition using wavelet-based neural classifier. I. Theoretical foundation. IEEE Trans Power Deliv 15(1): 222–228
5. Liao CC (2010) Enhanced RBF network for recognizing noise-riding power quality events. IEEE Trans Instrum Meas 59(6):1550–1561
6. Dalai S, Chatterjee B, Dey D, Chakravorti S, Bhattacharya K (2013) Rough-set-based feature selection and classification for power quality sensing device employing correlation techniques. IEEE Sens J 13(2):563–573
7. Zhao F, Yang R (2007) Power-quality disturbance recognition using S-transform. IEEE Trans Power Deliv 22(2):944–950
8. Gargoom AM, Ertugrul N, Soong WL (2008) Automatic classification and characterization of power quality events. IEEE Trans Power Deliv 23(4):2417–2425
9. Dash PK, Chilukuri MV (2004) Hybrid S-transform and Kalman filtering approach for detection and measurement of short duration disturbances in power networks. IEEE Trans Instrum Meas 53(2):588–596
10. Mishra S, Bhende CN, Panigrahi BK (2008) Detection and classification of power quality disturbances using S-transform and probabilistic neural network. IEEE Trans. Power Del 23(1):280–287
11. Lee CY, Shen YX (2011) Optimal feature selection for power-quality disturbances classification. IEEE Trans Power Deliv 26(4):2342–2351

12. Tong W, Song X, Lin J, Zhao Z (2006) Detection and classification of power quality disturbances based on wavelet packet decomposition and support vector machines. In: Proceedings of 8th International Conference on Signal Processing, IEEE Xplore, pp 1–4
13. Biswal B, Dash PK, Panigrahi BK (2009) Power quality disturbance classification using fuzzy C-means algorithm and adaptive particle swarm optimization. IEEE Trans Ind Electron 56(1):212–220
14. Bhattacharyya S, Konar A, Tibarewala DN, Khasnobish A, Janarthanan R (2014) Performance analysis of ensemble methods for multi-class classification of motor imagery EEG signal. In: Proceedings of international conference on control, instrumentation, energy and communication (CIEC), Calcutta, pp 712–716
15. Tuysuzoglu G, Moarref N, Yaslan Y (2016) Ensemble based classifiers using dictionary learning. In: proceedings of international conference on systems, signals and image processing (IWSSIP), Bratislava, pp 1–4
16. do Amaral Monteiro D, Zvietcovich WG, Braga MF (2018) Detection and classification of power quality disturbances with wavelet transform, decision tree algorithm and support vector machines. In: 2018 Simposio Brasileiro de Sistemas Eletricos (SBSE), May 2018, pp 1–6
17. Soman KP, Poornachandran P, Athira S, Harikumar K (2015) Recursive variational mode decomposition algorithm for real time power signal decomposition. Procedia Technol 21:540–546
18. Achlerkar PD, Samantaray SR, Manikandan MS (2018) Variational mode decomposition and decision tree based detection and classification of power quality disturbances in grid-connected distributed generation system. IEEE Trans Smart Grid 9(4):3122–3132
19. Reddy MV, Sodhi R (2018) A modified S-transform and random forests-based power quality assessment framework. IEEE Trans Instrum Meas 67(1):78–89
20. Erişti H, Yıldırım Ö, Erişti B, Demir Y (2014) Automatic recognition system of underlying causes of power quality disturbances based on Stransform and extreme learning machine. Int J Electr Power Energy Syst 61:553–562
21. He S, Li K, Zhang M (2013) A real-time power quality disturbances classification using hybrid method based on S-transform and dynamics. IEEE Trans Instrum Meas 62(9):2465–2475
22. Biswal B, Dash PK, Mishra S (2011) A hybrid ant colony optimization technique for power signal pattern classification. Expert Syst Appl 38(5):6368–6375
23. Erişti H, Uçar A, Demir Y (2010) Wavelet:based feature extraction and selection for classification of power system disturbances using support vector machines. Electr Pow Syst Res 80(7):743–752
24. Moravej Z, Abdoos AA, Pazoki M (2009) Detection and classification of power quality disturbances using wavelet transform and support vector machines. Electr Power Compon Syst 38(2):182–196
25. Zhang M, Li K, Hu Y (2012) Classification of power quality disturbances using wavelet packet energy and multiclass support vector machine. COMPEL Int J Comput Math Electr Electron Eng 31(2):424–442
26. Ray PK, Mohanty SR, Kishor N, Catalão JP (2014) Optimal feature and decision tree-based classification of power quality disturbances in distributed generation systems. IEEE Trans Sustainable Energy 5(1):200–208
27. Hajian M, Foroud AA (2014) A new hybrid pattern recognition scheme for automatic discrimination of power quality disturbances. Measurement 51:265–280
28. Hajian M, Foroud AA, Abdoos AA (2014) New automated power quality recognition system for online/offline monitoring. Neurocomputing 128:389–406
29. Aneesh C, Kumar S, Hisham PM, Soman KP (2015) Performance comparison of variational mode decomposition over empirical wavelet transform for the classification of power quality disturbances using support vector machine. Procedia Comput Sci 46:372–380
30. Thirumala K, Prasad MS, Jain T, Umarikar AC (2018) Tunable: Q wavelet transform and dual multiclass SVM for online automatic detection of power quality disturbances. IEEE Trans Smart Grid 9(4):3018–3028

Design and Analysis of Koch Fractal Slots for Ultra-Wideband Applications

Sudheer Kumar Terlapu, M. Venkata Subba Rao, P. Satish Rama Chowdary, and Suresh Chandra Satapaty

Abstract In this paper, a design of a dual-band Koch fractal antenna is proposed. Patch antenna has dimensions of 30 mm × 18 mm × 0.76 mm which operates in ultra-wideband (UWB) with a notch band from 5 to 6 GHz. Further, the design is modified using different Koch fractal geometries to observe the multi-band resonant behavior of the antenna. The modified antenna has dimensions of 18 mm × 18 mm × 0.8 mm. By modifying the shape of radiating patch and ground with fractal shapes and by introducing slots, impedance bandwidth and radiation characteristics were improved. The proposed antenna works well for WiMAX and WLAN applications with approximately 95% radiation efficiency. The parametric analysis is carried out for various parameters in order to obtain better return loss characteristics. The design is simulated using HFSS software, and the performance is measured by obtaining the antenna parameters.

Keywords Koch fractal · WiMAX · WLAN · UWB · HFSS

1 Introduction

The growing demand for wireless communications in recent years requires antennas which can provide large impedance bandwidths, high data rate, and higher gain. Ultra-wideband (UWB) is one such technology that provides very large bandwidth and higher data rates [1]. But the main disadvantage of using UWB technology is electromagnetic interference (EMI). To eliminate the interference problems, UWB

S. K. Terlapu (✉) · M. V. Subba Rao
Department of Electronics and Communication Engineering, Shri Vishnu Engineering College for Women, Bhimavaram, India
e-mail: profsudheer@ieee.org

P. S. R. Chowdary
Department of Electronics and Communication Engineering, Raghu Institute of Technology, Visakhapatnam, India

S. C. Satapaty
Department of Computer Science and Engineering, KIIT, Bhubaneswar, India

P. S. R. Chowdary et al. (eds.), *Microelectronics, Electromagnetics and Telecommunications*, Lecture Notes in Electrical Engineering 655,
https://doi.org/10.1007/978-981-15-3828-5_77

antennas are designed to achieve band notch characteristics [2–6]. Designing an antenna for wireless applications requires compactness and cost-effectiveness. Using the microstrip antennas for these applications is the right choice because of the numerous advantages provided by microstrip antennas [7–10]. Several designs have been proposed to achieve UWB response using microstrip antennas [11–15]. Single-, double-, and triple-band rejections are achieved by rearranging the radiating element or by cutting slots on patch, ground plane [16] or by introducing additional stubs on any of the radiating patch or ground plane [17]. Achieving wide bandwidth offers a low gain in the resonant band. The compromise between bandwidth and gain should be made before designing the antenna. Several techniques have been proposed to improve the gain while maintaining the required bandwidth [18].

The antenna miniaturization is getting much attention in the recent years since the fractal geometries are introduced into antenna design [19]. Fractal structures are used widely to reduce the size of the antenna and to achieve multi-band resonant behavior. Several fractal designs have been proposed to get multiple band resonances and at the same time more compact and high-impedance bandwidth structures [20–23]. The idea of fractals can also be used to achieve UWB range. By using the different fractal geometries, UWB with band rejection can also be designed [24].

In this paper, a rectangular printed patch antenna is designed to achieve the ultra-wideband behavior. By introducing additional stub on ground plane, band notch characteristics are achieved. Frequency band from 5 to 6 GHz is rejected from the ultra-wideband to eliminate the interference with WLAN. Implementing Koch fractal in antenna design provided the multiple band resonances. Different Koch fractal geometries were designed and compared.

2 Design Methodology

The designed antenna was fabricated using a low-cost FR4 substrate with 0.8 mm thickness and 4.4 relative permittivity. The dimensions of patch, ground plane, and slots were given in Table 1. Making use of defected ground structures and slots suppresses the surface waves, which in turn improves the bandwidth significantly.

Figure 1a shows the slotted ultra-wideband patch antenna. Frequency notch band characteristics are obtained with the addition of radiating stub as shown in Fig. 1b.

2.1 Design of Fractal Antenna

By using the fractal geometries in antenna design, multi-band response can be achieved. To achieve multi-band characteristics, this design Koch fractal geometry as shown in Fig. 2a–c is considered, and the results were compared for different fractal designs. The dimensions of slots were presented in Table 2.

Table 1 Dimensions of UWB antenna

Description	Symbol	Dimension in mm
Patch length	L	18
Patch width	W	14.4
Ground plane length	l_1	14.4
Substrate thickness	H	0.76
Dielectric constant	€r	4.4
Slot length	h_1	3
Slot width	w_1	3.5
Stub length	A	5.3
Stub width	B	7
Feed line width	w_f	1.4
Feed line length	l_f	15.6

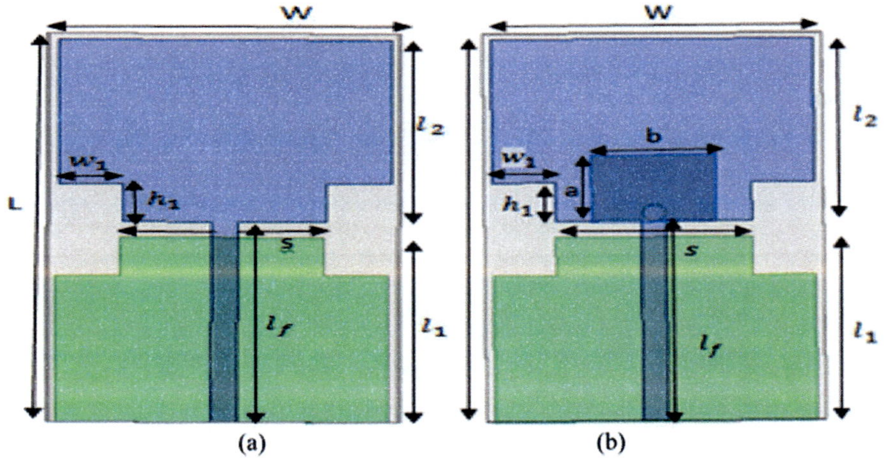

Fig. 1 **a** UWB antenna without notch element. **b** UWB antenna with notch element

3 Results and Discussions

3.1 Parametric Study

Ultra-wideband response for the microstrip antenna is obtained by introducing bevel slots on both radiating patch and ground as shown in Fig. 1a. The return loss plot for design 1 (UWB antenna without notch element) is given in Fig. 3a.

Notch band characteristics for the rectangular microstrip printed patch antenna are obtained by optimizing the width and length of the stub placed on ground plane. The parametric analysis results for UWB antenna with additional stub in ground

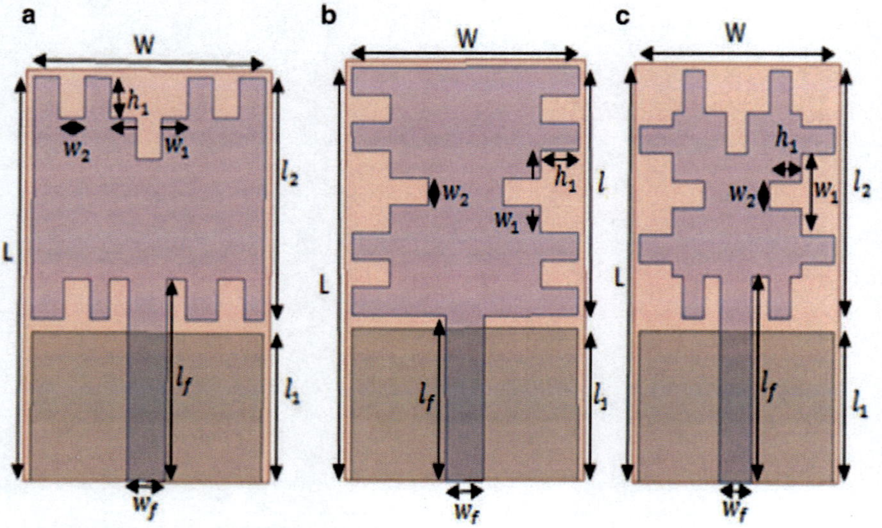

Fig. 2 **a** Koch fractal antenna with upper and bottom slots. **b** Koch fractal antenna with right and left slots. **c** Koch fractal antenna with slots in all sides

Table 2 Dimensions of fractal antenna

Description	Symbol	Dimension in mm
Substrate length	L_s	30.5
Substrate width	W_s	19
Patch length	L	18
Patch width	W	18
Ground plane length	l_1	14.4
Patch length	l_2	18
Substrate thickness	H	0.76
Dielectric constant	€r	4.4
Slot length	h_1	3
First slot width	w_1	6
Second slot width	w_2	2

plane were presented in Fig. 3b, c. The optimized values for stub length and width are given in Table 1. For stub length and width A and B, designed antenna operates in UWB range, rejecting the band from 5 to 5.9 GHz. The optimized return loss plot was given in Fig. 3d.

To achieve multi-band resonance, a Koch fractal is designed by two iterations. The return loss characteristics were given in Fig. 3e.

Fractal geometries act as combination of inductors and capacitors. By properly designing the fractal structure, required multi-band resonance can be obtained. By

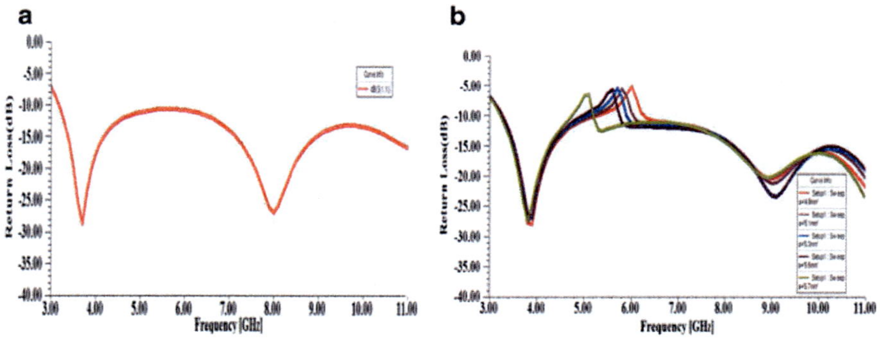

Fig. 3 **a** Return loss of UWB antenna without stub. **b** Effect of stub length on return loss

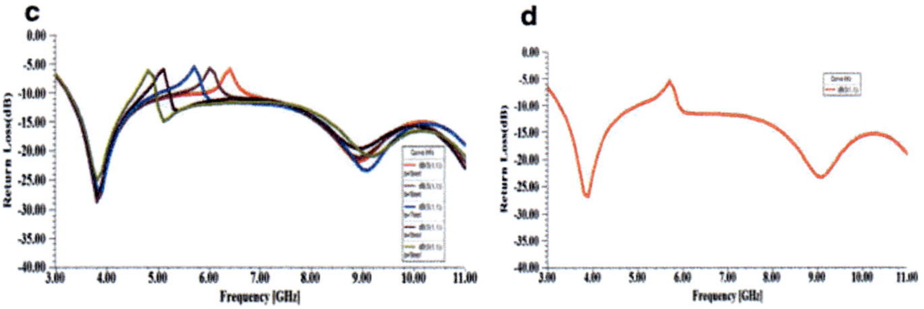

Fig. 3 **c** Effect of stub width on return loss. **d** Optimized return loss plot with single notch

using Koch fractal with all side slots, dual-band resonance at frequencies 3.3 and 5.5 GHz is obtained. The fabricated antenna and return loss plot for Koch fractal antenna are shown in Fig. 3f, g. Table 3 gives the comparison between three Koch fractal designs and notch band antenna design.

3.2 Return Loss

Return loss plots for simulated and fabricated antenna are shown in Fig. 3g. Results were same except for few deviations. From Fig. 3g, it should be noted that designed antenna works for both WiMAX (3.3 GHz) and WLAN (5.5 GHz) bands.

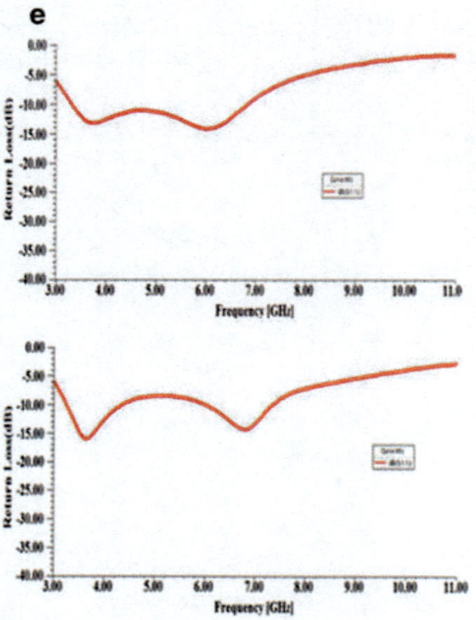

Fig. 3 **e** Koch fractal antenna return loss for designs 1, 2, and 3, respectively

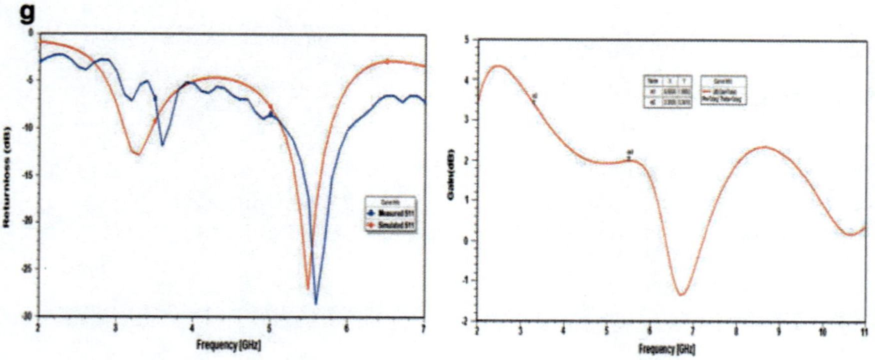

Fig. 3 **f** Fabricated and simulated antenna. **g** Return loss plot for Koch fractal antenna with all side slots (*red curve*-simulated, *blue curve*-measured) and gain versus frequency plot

Table 3 Comparison of resonant frequencies

Description	Resonant band (GHz)	Gain (dB)
UWB antenna without stub	3.2–11	2.064
UWB antenna with stub	3.2–11 with notch at 5–5.9	2.39
Koch fractal with bottom and upper slots	3.4–6.8	2.25
Koch fractal with right and left slots	3.7, 8.8	2.05
Koch fractal with all side slots	3.3, 5.5	3.3, 1.9

h

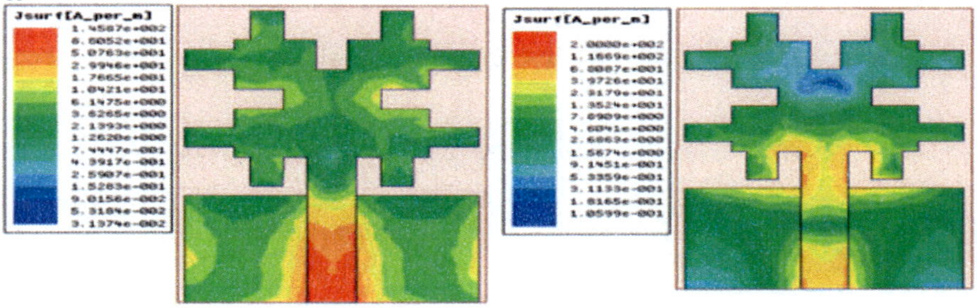

Fig. 3 **h** Electric field distributions at frequencies 3.3 and 5.5 GHz

3.3 Gain Versus Frequency

As shown in Fig. 3h, gain values obtained at frequencies 3.3 GHz and 5.5 GHz are 3 dB and 2 dB, respectively.

3.4 Electric Field Distribution

Field distribution at two resonant frequencies is shown in Fig. 3h. Maximum current is distributed at the edges of the patch which is responsible for maximum radiation at the edges.

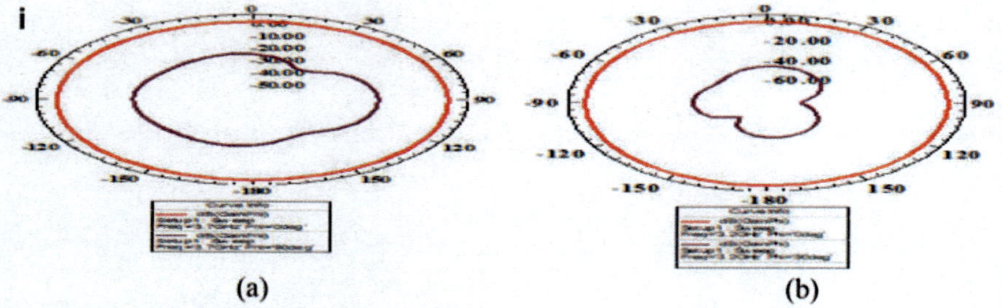

Fig. 3 **i** 2-D radiation pattern for UWB antenna. **a** Without stub and **b** with stub

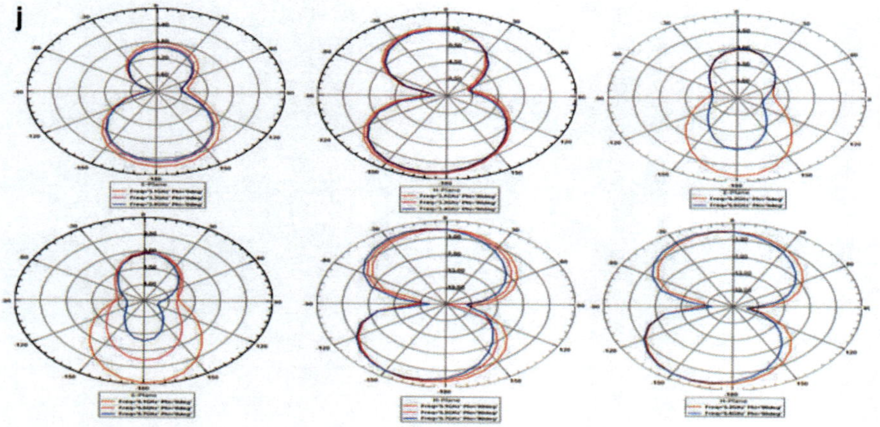

Fig. 3 **j** Radiation patterns in both E-plane and H-plane for frequencies 3.14, 3.3, 3.4, 5.16, 5.5, and 5.76 GHz

3.5 Radiation Pattern

The two-dimensional far-field pattern for UWB antenna without and with stub is presented in Fig. 3i, j. Figure 3k shows the radiation patterns for resonant frequencies and lower and higher cutoff frequencies. It shows that in both cases, the antenna radiates in nearly omni-directional manner.

The radiation plots for Koch fractal antennas were given in Fig. 3a–c.

4 Conclusion

The performance of the proposed antenna is evaluated for different fractal configurations. The antenna is simple and compact. The results show that the antenna can

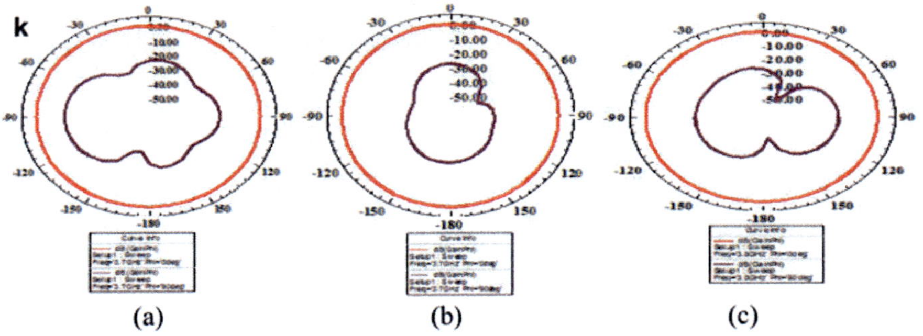

Fig. 3 **k** 2-D radiation pattern for Koch fractal antenna. **a** Upper and bottom slots, **b** right and left slots, and **c** all side slots

be used for multiple bands. As a future work, printed fractal antenna can be further extended to achieve the ultra-wideband characteristics.

References

1. Schantz HG (2004) A brief history of UWB antennas. IEEE Aerosp Electron Syst Mag 19(4):22–26
2. Peng L, Ruan CL (2011) UWB band-notched monopole antenna de-sign using electromagnetic-bandgap structures. IEEE Trans Microwave Theo and Tech 59(4): 1074–1081
3. Guo L, Wang S, Chen X, Parini CG (2010) Study of compact antenna for UWB applications. Electron Lett 46(2):115–116
4. Amini F, Azarmanesh MN, Ojaroudi M (2010) Small semi-circle-like slot antenna for ultra-wideband applications. Prog Electromagn Res C 13:149–158
5. Mishra SK, Gupta RK, Vaidya A, Mukherjee J (2011) A compact dual-band fork-shaped monopole antenna for bluetooth and UWB applications. IEEE Ant Wireless Propag Letts 10:627–630
6. Lin S, Cai R-N, Huang G-L, Wang J-X (2011) A miniature UWB semi-circle monopole printed antenna. Prog Electromagn Res Lett 23:157–163
7. Bahl IJ, Bhartia P (1980) Micro-strip Antennas. Artech House, Dedham, MA
8. James JR (1989) Handbook of microstrip antenna. Peter Peregrinus Ltd., London
9. Bugaj M, Przesmycki R, Nowosielski L, Piwowarczyk K (2012) Analysis different methods of microstrip antennas feeding for their electrical parameters. In: PIERS proceedings, Kuala Lumpur, Malaysia, 27–30 March 2012
10. Sidhu SK, Sivia JS (2015) Comparison of different types of microstrip patch antennas. Int J Comput Appl (ICAET) 0975(8887)
11. Barbarino S, Consoli F (2010) UWB circular slot antenna provided with an inverted-L notch filter for the 5 GHz WLAN band. Prog Electromagn Res 104:1–13
12. Li W-M, Ni T, Quan T, Jiao Y-C (2011) A compact CPW-fed UWB antenna with WiMAX-band notched characteristics. Prog Electromagn Res Lett 26:79–85
13. Mehranpour M, Nourinia J, Ghobadi Ch, Ojaroudi M (2012) Dual band-notched square monopole antenna for ultra-wideband applications. IEEE Antennas Wirel Propag Letters 11
14. Jiang W, Che W (2012) A novel UWB antenna with dual notched bands for Wi-MAX and WLAN applications. IEEE Antennas Wirel Propag Letters 11

15. Terlapu SK, Chowdary PSR, Jaya C, Sameer Chakravarthy VVSS, Satpathy SC (2018) On the design of fractal UWB wide-slot antenna with notch band characteristics. In: Anguera J, Satapathy S, Bhateja V, Sunitha K (eds) Microelectronics, electromagnetics and telecommunications. Lecture notes in electrical engineering, vol 471. Springer, Singapore
16. Sim CYD, Chung WT, Lee CH (2010) Compact slot antenna for UWB application. IEEE Trans Antennas Propag Letter 9:63–66
17. Weng YF, Cheung SW, Yuk TI (2011) Compact ultra-wideband antennas with single band-notched characteristic using simple ground stubs. Microw Opt Technol Lett 53(3):523–529
18. Naik GMM, Audre AA, NaveenKumar SK (2016) Slotted UWB antenna for bandwidth and gain enhancement. IRJET 3, July 2016
19. Mandelbrot BB (1982) The fractal geometry of nature. W. H. Freeman and Company
20. Yu YH, Ji CP (2011) Research of fractal technology in the design of multi-frequency antenna. In: CJMW proceedings
21. Jeemon BK, Shambavi K, Alex ZC (2013) A multi-fractal antenna for WLAN and WiMAX applications. In: Proceedings of IEEE conference on information and communication technologies (ICT)
22. Puente C, Romeu J, Cardama A (2001) The Koch monopole: a small fractal antenna. IEEE Trans Antennas Propag 48:1173–1781
23. Jeemon BK, Shambavi K, Alex ZC (2013) A multifractal planar antenna for wireless applications. In: International conference on communication and signal processing, India
24. Terlapu SK, Cheruku J, Raju G (2017) On the notch band characteristics of koch fractal antenna for UWB applications. Int J Control Theory Appl 10(6):701–707

Pattern Recovery in Linear Arrays Using Grasshopper Optimization Algorithm

V. V. S. S. S. Chakravarthy, P. Satish Rama Chowdary, Jaume Anguera, Divya Mokara, and Suresh Chandra Satapathy

Abstract In this paper, the technique of restoring the pattern even after the element failure is demonstrated in linear arrays (LA). The process involves in determining the amplitude excitation coefficients of each element in the linear array using grasshopper algorithm (GHA). Linear array with 20 elements is considered for the implementation, while the analysis is carried out using the radiation pattern in terms of side lobe level. Two cases of element failure are considered. In the first case, second element failure is inflicted, while in the second case, the same is repeated with 30-element linear array. The simulation is carried out using MATLAB, and the results are analyzed using the corresponding radiation pattern plots and convergence plots.

Keywords Grasshopper optimization · Linear array antenna · Element failure · Pattern recovery

V. V. S. S. S. Chakravarthy (✉) · P. S. R. Chowdary · D. Mokara
Department of Electronics and Communications Engineering, RIT, Visakhapatnam, India
e-mail: sameervedula@ieee.org

P. S. R. Chowdary
e-mail: satishchowdary@ieee.org

D. Mokara
e-mail: mdivya539@gmail.com

J. Anguera
Department of Electronics and Telecommunications, Universitat Ramon Llull, Barcelona, Spain
e-mail: jaume.anguera@salle.url.edu

Technology Department, Fractus Antennas, Barcelona, Spain

S. C. Satapathy
School of Computer Engineering, KIIT Deemed to be University, Bhubaneswar, India
e-mail: sureshsatapathy@gmail.com

P. S. R. Chowdary et al. (eds.), *Microelectronics, Electromagnetics and Telecommunications*, Lecture Notes in Electrical Engineering 655,
https://doi.org/10.1007/978-981-15-3828-5_78

1 Introduction

Every element of the array has contribution to the corresponding radiation pattern [1, 2]. If any element malfunctions or loses its characteristics, it leads to element failure. In few cases, the antenna array might hold hundreds of radiating elements which increase the possibility of element failures. It causes edge variations in the field intensity, abrupt changes in the side lobe level (SLL), beam width might be widened, changes the shape of the pattern, disruption in the null positions, and many other parameters [2–12]. All these are considered as some serious and considerable effects caused due to element failure. It is necessary to restore or recover the pattern in order to attain the original desired radiation pattern.

The performance degradation in element failure can be mitigated by using various conventional and soft computing techniques. Also, by recalculating the excitation of remaining working elements in the array, we can re-synthesize the original radiation pattern. This is another cost-effective alternative instead of replacing the hardware in the array set. This degradation can be mitigated by pattern recovery methods which use design parameters like amplitude, phase, and inter-element spacing [3–5]. Determining the current distribution of each radiating element of an antenna array results in producing desired radiation pattern. One or more design parameter can be used in the process [6–8].

Further, the paper is organized into five sections. The formulation of the problems is discussed in Sect. 2. A brief introduction to the GHA is given in Sect. 3. Results and discussions are presented in Sect. 4. Overall conclusions and future scope are given in Sect. 5.

2 Problem Formulation

For synthesis of LA using the evolutionary computing tools, it is important to formulate the problem. The mathematical model includes array factor and the corresponding objective function. Both are discussed in the subsequent subsections.

2.1 *Array Factor (AF)*

The spatial distribution and orientation of the LA is as shown in Fig. 1. The geometry produces a symmetric radiation pattern with the first element as reference [1]. There is another configuration of linear array in which the center element is the reference, and the amplitude distribution is also symmetric along with the pattern symmetry. However, in this work the first configuration is used. The AF formulation of the LA is given as [1].

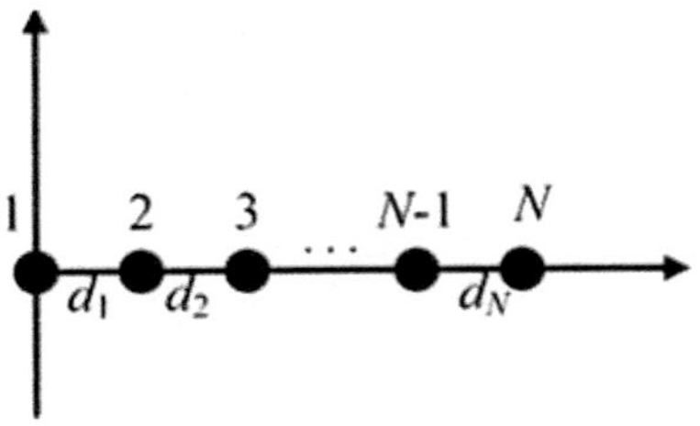

Fig. 1 Distribution of elements in LA

$$\mathrm{AF}(\theta) = \sum_{m=1}^{M} I_n e^{j(m-1)(\mathrm{kd}_n \cos\theta + \alpha_n)} \tag{1}$$

2.2 *Fitness Formulation*

As mentioned earlier and according to problem statement, except the failure element the amplitudes of current excitation have to be re-determined, such that the original pattern is recovered or reconstructed. For this, in the simulation process, the corresponding failure element is awarded with amplitude of '0'. Accordingly, the original patterns are those in which all the elements are functioning. Here for study, the Chebyshev patterns are considered as original patterns with SLL given as SLL (cheb). Accordingly, the fitness function is formulated as follows:

$$\mathrm{SLL_{diff}} = |\mathrm{SLL_{cheb}}| - \left|\max[\mathrm{AF}(\theta)]_{\mathrm{FN}}^{90^\circ}\right| \tag{2}$$

$$\mathrm{Fitness}\ (f) = \mathrm{SLL_{diff}} \quad \text{if} \quad \mathrm{SLL_{diff}} > 0 \tag{3}$$

$$\mathrm{Fitness}\ (f) = 0 \quad \text{if} \quad \mathrm{SLL_{diff}} \le 0 \tag{4}$$

Here, the difference between the obtained SLL and the uniform distribution SLL is given as $\mathrm{SLL_{diff}}$.

3 Grasshopper Algorithm

The grasshopper algorithm (GHA) mimics the behavior of the grasshopper (GH) swarm. The GH is an insect and is often considered as the enemy to the farmer. The insects invade the crop and degrade the production by directly destroying the plants. The GH lifecycle typically takes three stages starting from egg to nymph and finally adult. Most of the harm to the farmers through crop destruction is during the stage of nymph. Further, in their adult stage, these GH swarms form and spread to long distances. In addition to the general process of exploring and exploiting, the algorithm has efficient target-seeking strategy. Unlike other algorithms, this strategy exists inherently in the GHA. These strategies for survival are mathematically formulated and are given as

$$X_t = S_t + G_t + A_t \tag{5}$$

Here, for the tth individual the corresponding position, social conviction, gravity parameter, and wind advection are represented by X, S, G, and A, respectively. For every iteration, every individual is updated using the following equation.

$$X_t = \sum_{\substack{k=1 \\ k \neq t}}^{M} s(|x_k - x_t|)\frac{x_k - x_t}{d_{t,k}} - g\widehat{e_g} + u\widehat{e_w} \tag{6}$$

Here, $\widehat{e_g}$ and $\widehat{e_w}$ are unit vectors, and their coefficients are gravitational and drift constants.

4 Results and Discussion

The simulation-based experimentation is broadly divided into two cases with the objective of recovering the pattern in terms of the degraded SLL with BW as constraint. In the first case, a 20-element linear array is considered, while a 30-element array is synthesized as a second case. The experimentation framework involves in the following steps.

(a) Generate Chebyshev pattern: Synthesize N-element linear array with −30 dB of desired SLL, and note the corresponding BW.
(b) Defective element pattern: Analyze the radiation pattern by considering second element as defective or failure. The corresponding change in the SLL and BW is noted.
(c) Pattern recovery: Using GHA, the corresponding amplitude and inter-element spacing of each element in the array are determined which produce radiation pattern with desired SLL and BW as mentioned in Step 1.

Here N refers to number of elements in the linear array which is 20 and 30 for Case 1 and Case 2, respectively. In Case 1, the Chebyshev radiation pattern with −30 dB SLL is as shown in Fig. 2a. The corresponding BW can be read as 17.2°. As in Step 2, the second element failure pattern as shown in Fig. 2b has a degraded SLL when compared with pattern in Step 1 (Fig. 2a). The recovered pattern using GHA as shown in Fig. 2c reported the magnitudes of SLL and BW as similar as that of Fig. 2a. The corresponding convergence plot is given in Fig. 2d.

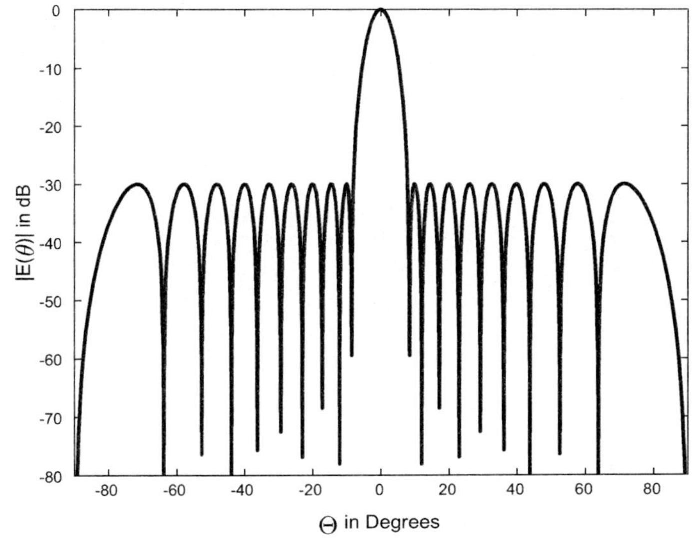

(a)

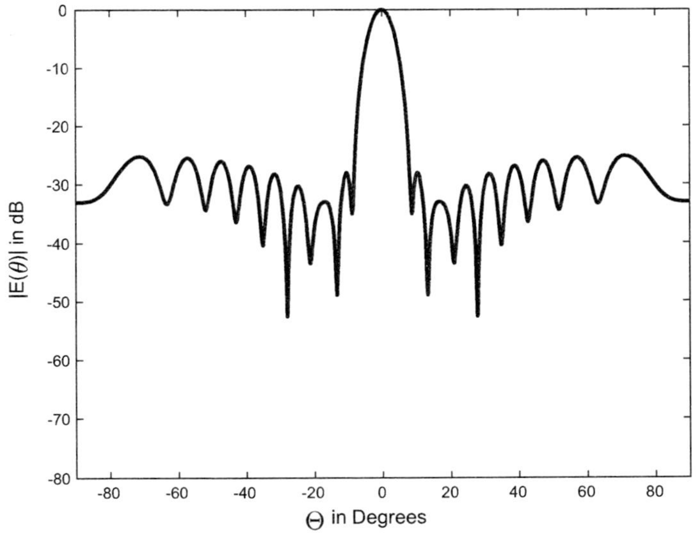

(b)

Fig. 2 20-element linear array radiation pattern for **a** Chebyshev optimization, **b** second element failure, **c** recovered and corresponding, and **d** convergence plot

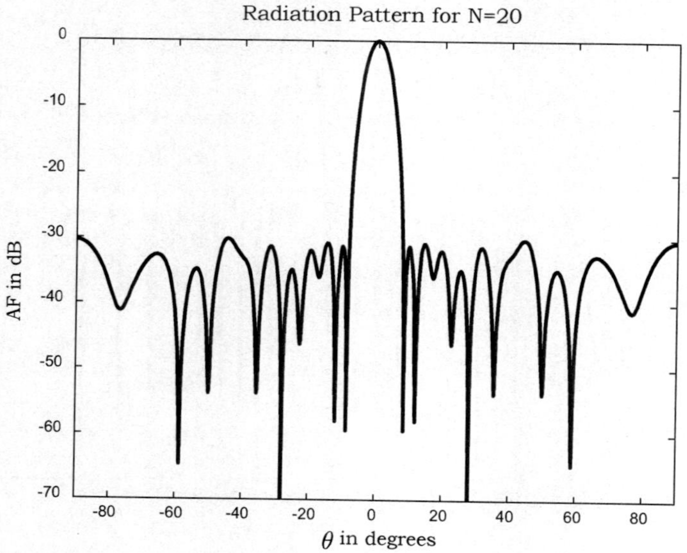

(c)

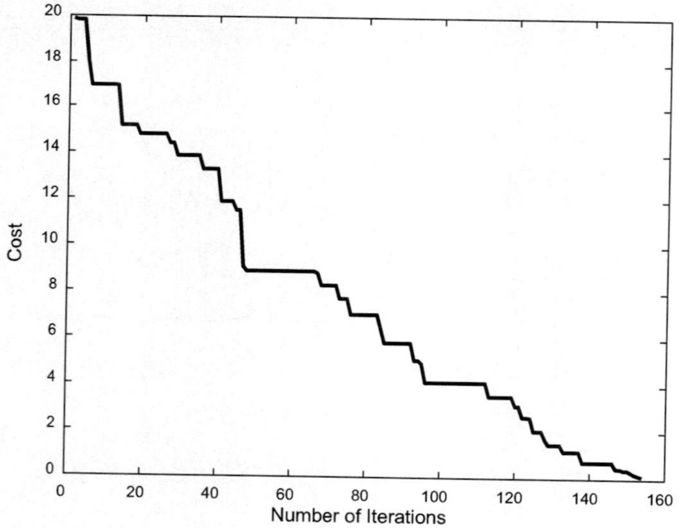

(d)

Fig. 2 (continued)

As a part of Case 2, initially the Chebyshev radiation pattern with −30 dB SLL is synthesized and is as shown in Fig. 3a. Similar to Case 1, now according to Step 2, the second element failure pattern condition is inducted and the respective radiation

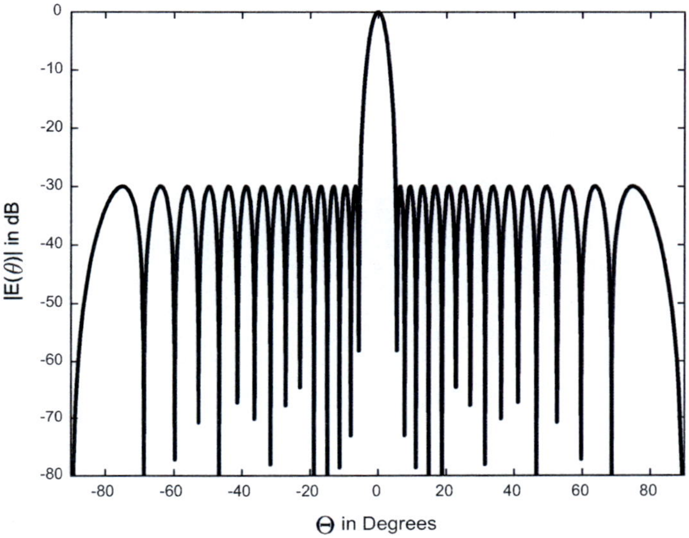

(a)

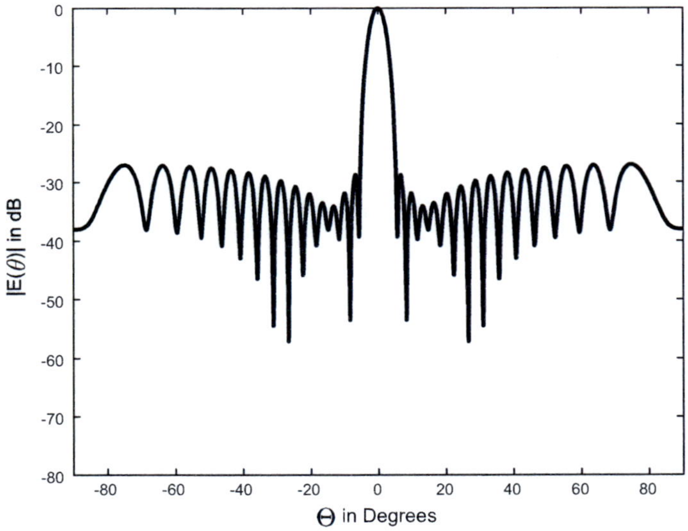

(b)

Fig. 3 30-element linear array radiation pattern for **a** Chebyshev optimization, **b** second element failure, **c** recovered and corresponding **d** convergence plot

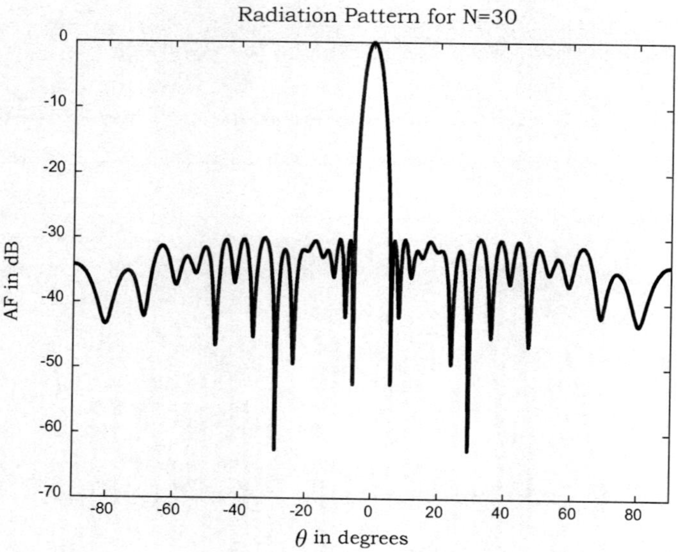

(c)

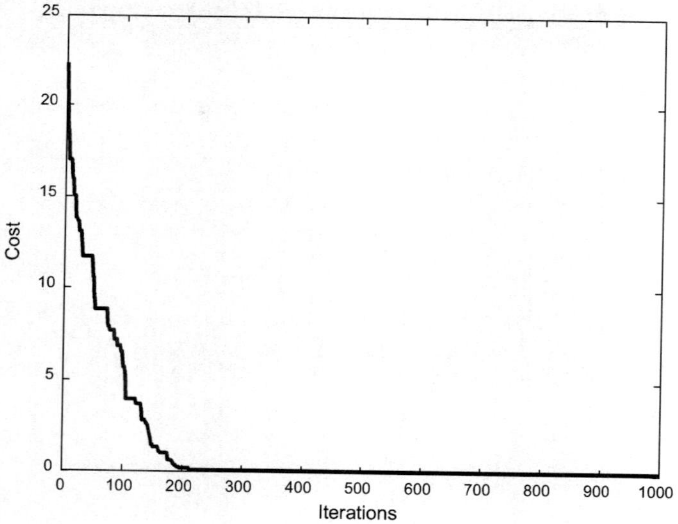

(d)

Fig. 3 (continued)

pattern is as shown in Fig. 3b. It can be observed that there is degradation in SLL when compared with pattern in Step 1 (Fig. 3a). Now, GHA used to determine the optimal coefficients of current excitation amplitudes and inter-element spacings to produce recovered radiation pattern as shown in Fig. 3c reported the magnitudes of SLL and BW as similar as that of Fig. 3a. The corresponding convergence plot is given in Fig. 3d.

The non-uniform amplitude and spatial distribution of the elements are determined by the GHA and are given in Tables 1 and 2 for Case 1 and Case 2 linear arrays. Similarly, the corresponding SLL and BW for comparison are presented in Table 3.

Table 1 Non-uniform amplitude and spacing distribution of 20-element linear array

Element number	Amplitude distribution	Inter-element spacings (λ)	Element number	Amplitude distribution	Inter-element spacings (λ)
1	0.208	0	11	0.946	0.294
2	0	0.452	12	0.063	0.548
3	0.352	0.309	13	0.95	0.116
4	0.485	0.708	14	0.778	0.508
5	0.733	0.702	15	0.949	0.573
6	0.889	0.717	16	0.778	0.625
7	0.055	0.401	17	0.674	0.621
8	0.997	0.363	18	0.629	0.716
9	0.297	0.438	19	0.415	0.780
10	0.208	0.188	20	0.287	0.720

Table 2 Non-uniform amplitude and spacing distribution of 20-element linear array

Element number	Amplitude distribution	Inter-element spacings (λ)	Element number	Amplitude distribution	Inter-element spacing (λ)
1	0.104	0	16	0.823	0.535
2	0	0.324	17	0.925	0.568
3	0.134	0.731	18	0.496	0.457
4	0.219	0.318	19	0.771	0.410
5	0.302	0.511	20	0.097	0.386
6	0.381	0.629	21	0.712	0.291
7	0.568	0.610	22	0.339	0.377
8	0.567	0.661	23	0.291	0.322
9	0.408	0.428	24	0.666	0.306
10	0.738	0.569	25	0.727	0.670
11	0.848	0.693	26	0.676	0.691
12	0.814	0.621	27	0.418	0.746
13	0.806	0.577	28	0.375	0.849
14	0.627	0.482	29	0.301	0.673
15	0.837	0.452	30	0.199	0.715

Table 3 Comparison of SLL and BW

Case	Number of elements	Case	SLL (dB)	BW (degrees)
Case-1	20	Chebyshev	−30	17.1
		Defective second element	−25	17.6
		Recovered	−30.12	17.1
Case-2	30	Chebyshev	−30	11.4
		Defective second element	−24	35.5
		Recovered	−30	11.4

5 Conclusion

The GHA is successfully applied to the antenna array synthesis problem. The linear array of different sizes is designed using amplitude-spacing technique and analyzed in terms of SLL and BW. The BW-constrained synthesis of linear array for pattern recovery is achieved using GHA. Further, the single-objective synthesis problem can be extended to multi-objective synthesis problem using multi-objective version of the GHA.

References

1. Raju GSN (2004) Antennas and wave propagation, Pearson Education (Singapore) Pvt. Ltd., International Edition
2. Hansen RC (2009) Phased array antennas, 2nd edn. Wiley, New York
3. Chakravarthy VVSSS, Rao PM (2015) Amplitude-only null positioning in circular arrays using genetic algorithm. In: 2015 IEEE international conference on electrical, computer and communication technologies (ICECCT), Coimbatore, pp 1–5, Mar 2015
4. Swathi AVS, Chakravarthy VVSSS (2020) Synthesis of constrained patterns of circular arrays using social group optimization algorithm. In: Satapathy S, Bhateja V, Mohanty J, Udgata S (eds) Smart intelligent computing and applications. Smart innovation, systems and technologies, vol 160. Springer, Singapore
5. Chakravarthy VVSSS, Rao PM (2015) On the convergence characteristics of flower pollination algorithm for circular array synthesis. In: 2nd international conference on electronics and communication systems ICECS 2015, no. Icecs, pp 485–489
6. Chakravarthy VVSSS, Chowdary PSR, Panda G, Anguera J, Andújar A, Majhi B (2018) On the linear antenna array synthesis techniques for sum and difference patterns using flower pollination algorithm. Arab J Sci Eng 43(8):3965–3977
7. Chakravarthy VS, Chowdary PSR, Satpathy SC, Terlapu SK, Anguera J Antenna array synthesis using social group optimization. In: Microelectronics, Electromagnetics and Telecommunications, pp 895–905
8. Paladuga CS, Vedula CV, Anguera J, Mishra RK, Andújar A Performance of beamwidth constrained linear array synthesis techniques using novel evolutionary computing tools. Appl Comput Electromag Soc J 33(3): 273–278
9. Mailloux RJ (1996) Array failure correction with digitally beamformed array. IEEE Trans Antennas Propag 44(12):1543–1550

10. Acharya OP, Patnaik A (2017) Antenna array failure correction [antenna applications corner]. IEEE Antennas Propag Mag 59(6):106–115. https://doi.org/10.1109/MAP.2017.2752683
11. Patnaik A, Chowdhury B, Pradhan P, Mishra RK, Christodolou C (2007) An ANN application for fault finding in antenna arrays. IEEE Trans Antennas Propag AP 55(3):775–777
12. Saremi S, Mirjalili S, Lewisa A (2017) Grasshopper optimisation algorithm: theory and application. Adv Eng Softw 105:30–47

Exponential Fourier Moment-Based CBIR System: A Comparative Study

J. Surendranadh and Ch. Srinivasa Rao

Abstract Content-based image retrieval (CBIR) is a technique for browsing, searching, and retrieving images from a large database of image collections. The availability of large amount of image collections necessitates powerful algorithms for image retrieval. CBIR system extracts image information known as features that are used to retrieve relevant images from image database that best match with query image. A moment-based content-based image retrieval system is explored in this paper. Exponential Fourier moment-based CBIR system, improved exponential Fourier moment-based CBIR system, and accurate and fast exponential Fourier-based CBIR system are developed. Comparative analysis is performed on three moment-based CBIR systems in terms of average precision and retrieval time for the two benchmark databases GT face and COIL-100. Among the three moment-based CBIR systems, it is observed that accurate and fast exponential Fourier-based CBIR system delivers good results.

Keywords Exponential Fourier moments · Improved exponential Fourier moments · Accurate and fast exponential Fourier moments

1 Introduction

Nowadays, digital applications increasing anywhere due to that image collections also increase because of powerful tools required for browsing and retrieval of the same type of images from extensive image collections. To overcome this, CBIR system is used to retrieve the relevant image from the database. Generally, image

J. Surendranadh (✉)
Jawaharlal Nehru Technological University Kakinada, University College of Engineering Vizianagaram, Vizianagaram, Andhra Pradesh, India
e-mail: surendranadh100@gmail.com

Ch. Srinivasa Rao
Electronics and Communication Engineering, Jawaharlal Nehru Technological University Kakinada, University College of Engineering Vizianagaram, Vizianagaram, Andhra Pradesh, India
e-mail: chsrao.ece@jntukucev.ac.in

P. S. R. Chowdary et al. (eds.), *Microelectronics, Electromagnetics and Telecommunications*, Lecture Notes in Electrical Engineering 655,
https://doi.org/10.1007/978-981-15-3828-5_79

retrieval is performed based on the feature extraction. Different types of features are colure, texture, shape, and combination of those. In CBIR system image retrieval performance based on the quire image. Initially, the quire image features are extracted and matched with already stored database features.

The shape is one of the essential factors in any objective description. Because of that reason, shape features of image plays an important role. One of the shape feature extraction techniques is moment feature extraction. In 1962 Hu introduced invariant moments [1] which are extended to geometric moments [2], and complex moments [3] which are extended to orthogonal moments in the due course of time they are named as Zernike moments (ZMs) [4], Pseudo Zernike moments (PZMs), Legendre moments [5] which are based on the discrete orthogonal polynomials. So many shape-based CBIR systems based on the moments initially Zernike moment based CBIR system developed by Whoi-Yul Kim ins 2000 [6] in this system Zernike moments are used to extract shape features. In this paper, different Exponential Fourier moments are used for feature extraction. Such Fourier exponential moments are Exponential Fourier moments [7], improved Exponential Fourier moment [8], and Accurate and Fast Exponent Fourier Moment [9]. Exponential Fourier moments are orthogonal and rotational invariant moments.

2 Proposed Method

CBIR system implemented by using generalized block diagram as shown in Fig. (1), feature extraction technique, and feature matching technique play a vital role. Exponential Fourier moments, improved exponential Fourier moments, and accurate and fast exponential Fourier moments are used in the feature extraction process.

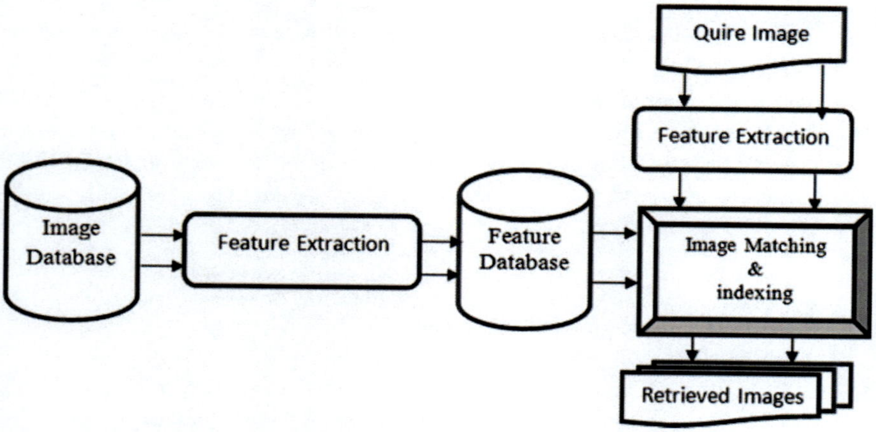

Fig. 1 Generalized block diagram of CBIR system

The query can be any image from the database. When a query image is applied to the CBIR system for processing, it first computes the features of the query image, and then similarity matching is carried out with the already stored feature database. In this work, GT face and the COIL-100 standard databases are used. The feature extraction process plays a significant role in the CBIR system. Moments are used to extract shape features those are exponential Fourier moments [7], improved exponential Fourier moments [8], and accurate and fast exponential Fourier moments [9]. These orthogonal moments are explained farther.

Image matching is used to determine the similarity between the features of query image and database image. Image matching is performed by using distance metrics; different distance metrics are Euclidian distance measure, Manhattan distance measure, Canberra distance measure, squared chord distance measure, Braycurtis distance measure, and squared chi-square distance measure. In the retrieval process, the system selects the number of images having higher overall similarity with the given query image and is present to the user as retrieved image. These images have the highest similarity with those images considered as similar images.

2.1 *Exponential Fourier Moments*

In the process of calculation of moment, two approaches are taken into consideration those are inner unit disk mapping and outer unit disk mapping use, respectively. Use the inner unit disk mapping approach; the pixels which fall outside the border of the unit disk are ignored during the moment computation. All the information carried by these pixels is lost. Hence, the approach of outer unit disk mapping is used during the computation of all the moments in this method (Fig. 2).

In the above approach, R1 is inside the region of image and R2 is outside of the image. By using these approaches, equations are written like this.

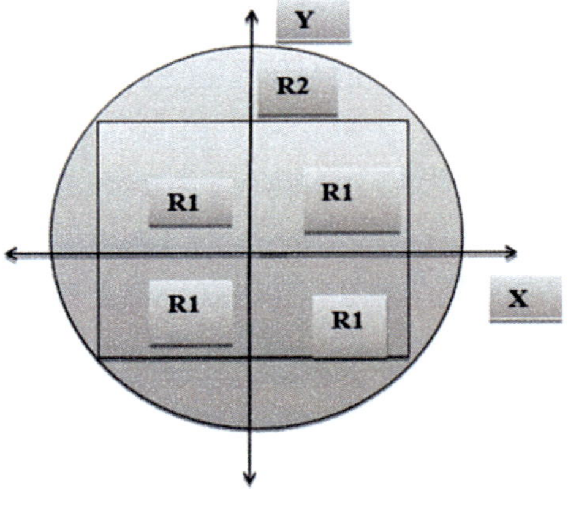

Fig. 2 Outer disk mapping

$$M_{\mathrm{kl}} = \frac{1}{2\pi} \int_{0}^{2\pi} \int_{0}^{1} f(r, \theta) q_k^*(r) \ \exp(-j1\theta) r \, \mathrm{d}r \, \mathrm{d}\theta$$

Generally, image is in digital form of size $M \times N$ of pixels $f(m, n)$ where m and n denote rows and columns, respectively. To compute the moment for digital image needs order approximation for the above equation.

$$\mathrm{xi} = \frac{2i + 1 - M}{\sqrt{M^2 + N^2}} \text{ and } \mathrm{yi} = \frac{2j + 1 - N}{\sqrt{M^2 + N^2}}$$

where $i = 0, 1 \ldots M - 1$; $k = 0, 1, \ldots N - 1$. The zeroth-order approximation is shown as follows:

$$M_{\mathrm{kl}} = \frac{1}{2\pi} \sum_{i=0}^{M-1} \sum_{k=0}^{N-1} f(\mathrm{xi}, \mathrm{yi}) q_k^*(rij) \ \exp(-jl\theta) \Delta x \Delta y$$

where

$$r = \sqrt{\mathrm{xi}^2 + \mathrm{yi}^2}, q_k(r) = \sqrt{\frac{1}{r}} \ \exp(j2k\pi r), \Delta x = \Delta y = \frac{2}{\sqrt{M^2 + N^2}}$$

$$\text{and } \theta = \tan^{-1} \frac{\mathrm{yk}}{\mathrm{xi}}$$

2.2 *Improved Exponential Fourier Moments*

Improved exponential Fourier moments [8] are thus given in this approach to solve problems about numerical instability involved in the process of calculating the exponential Fourier moments, and values of radial functions approach infinity at the origin.

$$M_{kl} = \frac{1}{2\pi z_k} \sum_{i=0}^{M-1} \sum_{k=0}^{N-1} f(\mathrm{xi}, \mathrm{yi}) q_k^*(rij) \ \exp(-jl\theta) \Delta x \Delta y$$

IEFM approach adds normalization constant at denominator $z_k = \exp\left(-4jk\pi^2\right)$, and radial function is $q_k(r) = \sqrt{\frac{1}{r}} \ \exp(-j2n\pi r)$ to maintain numerical stability.

2.3 Accurate and Fast Exponential Fourier Moments

In this approach, a new framework is proposed for calculating the A&F EFM by partitioning the radial and angular parts into equally spaced sectors. To calculate the coefficients, map the image functions into polar coordinates so that the transform becomes separable and the circular part can be evaluated fast employing fast Fourier transform. To keep the information about the finest patterns, the grid density should be high enough. Let M_r be radial and $M\varphi$ be angular size of the image according to the sampling theorem $M\varphi \geq 2l$ max, and the angular and radial components are partitioned into N equal parts to maintain the grid density as follows:

$$M_r = \frac{1}{N}\sum_{i=o}^{N-1} i \quad \text{and} \quad M_\theta = \frac{2\pi}{N}\sum_{j=o}^{N-1} j$$

Like this, we partitioned the whole image into the $N\hat{}2$ subregions. Now, the polar coordinates (M_r, M_θ) translate into rectangular coordinates as $x = (M_r \times M/2 \times \cos M_\theta)$ and $y = (M_r \times M/2 \times \sin M_\theta)$. Later that convert Cartesian coordinates to discrete domain given as $i = -\text{floor}(y) + M/2 + 1$, $j = \text{floor}(x) + M/2$. Finally, the polar image using the portioned data can be written as follows:

$$f\,p(M_r, M_\theta) = f(i, j)$$

Final expression to calculate accurate and fast exponential Fourier moments is

$$E_{\text{nm}} = \frac{1}{M^2}\sum_{i=0}^{M-1}\sum_{k=0}^{N-1} f(N_r, N_\theta) * \sqrt{N_r} \times \exp\left(-jm\frac{2\pi}{M}i\right)\exp\left(-jn\frac{2\pi}{M}k\right)$$

3 Results and Discussion

CBIR system is developed based on features, and it plays a significant role. In this CBIR system, shape base features are extracted by using exponential Fourier moments, improved exponential Fourier moments, and accurate and fast exponential Fourier moments, and also comparative analysis is performed based on average precision and retrieval time as performance metrics. Whole operation is performed on the Microsoft Windows 10 64-bit pc with 1.1 GHz and 4 GB internal memory, and algorithm is written with the help of MATLAB 2014a software with mathematical expression and loops.

Columbia Object Image Library (COIL)-100 and GT face databases are used to evaluate system performance. COIL-100 had 100 classes, and each class has 72 images of total 7200 images. GT face database has 50 different classes, and each class has 15 images of total 750 images in the GT face database.

In this literature comparison of three CBIR systems, performance is evaluated based on two performance metrics. Comparison of three CBIR systems is made by using two approaches.

(i) In the first approach, the average precision of each system is compared and decided which system is having the highest precision.
(ii) In the second approach, the retrieval time of each system is compared and decided which system is having the highest retrieval time.

3.1 *Comparison by Using Average Precision with Different Orders*

In this approach, three CBIR systems are compared by using average precision. In this CBIR system, average precision is calculated by using GT face database with exponential Fourier moments, improved exponential Fourier moments, and accurate and fast exponential Fourier moments for different distance measures. Exponential Fourier moment-based CBIR system's average precision values are given in Table 1.

Average precision of improved exponential Fourier moment-based CBIR system with different distance measures for different moment orders is given in Table 2.

Average precision of accurate and fast exponential Fourier moment-based CBIR system values is given in Table 3 with different distance measures.

From the above experimental results of three CBIR systems with different distances, the Braycurtis distance method gives the highest precision compared to remaining distance measures. Braycurtis distance measure has the highest precision with the same moment order of remaining distance measures. In three CBIR systems, Braycurtis distance measure average precision values are given in Table 4 (Fig. 3).

Table 1 Average precision of EFM-based CBIR system by using GT face database for different moment orders

S. no	Distance measures	Order					
		5	6	7	8	9	10
1	Euclidean	54.57	56.79	58.52	60.86	61.57	62.32
2	Manhattan	54.41	56.63	59.01	61.52	62.53	63.53
3	Canberra	54.35	56.66	59.03	61.55	63.06	64.04
4	Sqchord	53.95	56.46	58.86	61.15	61.6	62.92
5	Braycurtis	54.95	56.96	59.03	61.55	63.06	64.04
6	Sqchisq	53.91	56.42	58.87	61.16	61.69	62.43
7	Mahal	54.35	56.32	58.41	60.7	61.52	62.32

Table 2 Average precision of IEFM-based CBIR system by using GT face database for different moment orders

S. no	Distance measures	Order					
		5	6	7	8	9	10
1	Euclidean	54.90	57.0	58.86	60.06	62.32	63.52
2	Manhattan	54.81	57.03	59.29	61.79	62.75	63.83
3	Canberra	54.95	57.06	59.33	61.83	63.26	64.25
4	Sqchord	54.25	56.76	59.01	61.49	62.06	63.25
5	Braycurtis	54.95	57.06	59.33	61.83	63.26	64.25
6	Sqchisq	54.25	56.92	59.13	61.36	62.36	62.69
7	Mahal	54.85	56.62	59.03	61.99	62.01	62.19

Table 3 Average precision of A&F EFM-based CBIR system by using GT face database for different moment orders

S. no	Distance measures	Order					
		5	6	7	8	9	10
1	Euclidean	56.01	58.12	59.87	61.07	63.52	64.62
2	Manhattan	55.91	58.14	60.3	62.8	63.95	64.93
3	Canberra	56.05	58.17	60.34	62.84	64.46	65.35
4	Sqchord	55.35	57.87	60.02	62.5	63.26	64.35
5	Braycurtis	56.05	58.17	60.34	62.84	64.46	65.35
6	Sqchisq	55.35	58.03	60.14	62.37	63.56	63.79
7	Mahal	55.95	57.73	60.04	63	63.21	63.75

Table 4 Highest average precision of different EFM-based CBIR systems for different moment orders

Order	5	6	7	8	9	10
EFM-based CBIR	54.95	56.66	59.03	61.55	63.06	64.04
IEFM-based CBIR	54.95	57.06	59.33	61.83	63.26	64.25
A&F EFM-based CBIR	56.05	58.17	60.34	62.84	64.46	65.35

From the tabulated results, the average precision of the CBIR system increased with order. A&F EFM-based CBIR system gives highest average precision compared to the IEFM-based CBIR system, and also IEFM-based CBIR system gives more average precision compared to EFM-based CBIR system. From that experimental result, three moment-based CBIR system results of order 10 are compared. A&F EFM-based CBIR system average precision is 65.3, IEFM-based CBIR system average precision is 64.25, and EFM-based CBIR system average precision is 64.04; in these three systems, A&F EFM-based CBIR system has highest precision

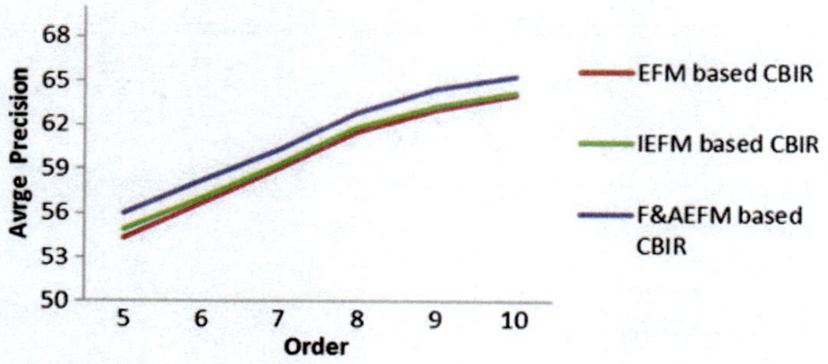

Fig. 3 Compression plot of three CBIR systems by using GT face database

of 65.30. In precision point of view, A&F EFM-based CBIR system is preferable compared to remaining two CBIR systems.

In this comparison process, the CIOL-100 database is also used to evaluate average precision performance with different distance measures of EFM-based CBIR system, IEFM-based CBIR, and A&F EFM-based CBIR system tabulated below. Average precision of exponential Fourier moment-based CBIR system with different distance measures for different moment orders is given in Table 5.

Average precision of improved exponential Fourier moment-based CBIR system with different distance measures for different moment orders is given in Table 6.

Average precision of accurate and fast exponential Fourier moment-based CBIR system is given in Table 7 with different distance measures.

From the above experimental results of three CBIR systems with different distances, the Braycurtis distance method gives the highest precision compared to remaining distance measures. Braycurtis distance measure has the highest precision with the same moment order of remaining distance measures. In three CBIR systems, Braycurtis distance measure average precision values are given in Table 8 (Fig. 4).

Table 5 Average precision of EFM-based CBIR system by using COIL-100 face database for different moment orders

S. no	Distance measures	Order					
		5	6	7	8	9	10
1	Euclidean	56.45	57.09	57.49	57.91	58.12	59.12
2	Manhattan	56.40	57.07	57.47	57.87	58.05	58.52
3	Canberra	57.01	57.99	58.33	58.67	59.12	59.99
4	Sqchord	56.10	56.97	57.47	57.89	58.07	59.02
5	Braycurtis	57.01	57.99	58.33	58.87	59.12	59.99
6	Sqchisq	56.13	56.99	57.29	57.76	58.16	58.67
7	Mahal	56.09	56.98	57.27	57.75	58.14	58.65

Table 6 Average precision of IEFM-based CBIR system by using COIL-100 face database for different moment orders

S. no	Distance measures	Order					
		5	6	7	8	9	10
1	Euclidean	5819	58.19	59.48	60.12	60.99	61.86
2	Manhattan	58.40	58.35	59.48	60.16	60.89	61.67
3	Canberra	59.02	59.17	59.57	60.23	61.12	61.99
4	Sqchord	58.21	58.27	58.39	60.02	60.96	61.23
5	Braycurtis	59.02	59.17	59.57	60.23	61.12	61.99
6	Sqchisq	58.13	58.17	58.47	60.96	60.56	61.12
7	Mahal	58.09	58.27	58.93	60.12	60.46	61.09

Table 7 Average precision of A&F EFM-based CBIR system

S. no	Distance measures	Order					
		5	6	7	8	9	10
1	Euclidean	59.63	60.19	61.6	62.24	63.29	64.07
2	Manhattan	59.92	60.55	61.6	62.28	63.19	63.88
3	Canberra	61.09	61.37	61.69	62.35	63.42	64.2
4	Sqchord	59.70	60.47	60.51	62.14	63.26	63.44
5	Braycurtis	61.09	61.37	61.69	62.35	63.42	64.2
6	Sqchisq	59.10	60.37	60.59	63.08	62.86	63.33
7	Mahal	59.21	60.47	61.05	62.24	62.76	63.3

Table 8 Highest average precision of different EFM-based CBIR systems for different moment orders

Order	5	6	7	8	9	10
EFM-based CBIR	57.01	57.99	58.33	58.87	59.12	59.99
IEFM-based CBIR	59.02	59.17	59.57	60.23	61.12	61.99
A&F EFM-based CBIR	61.09	61.37	61.69	62.35	63.42	64.2

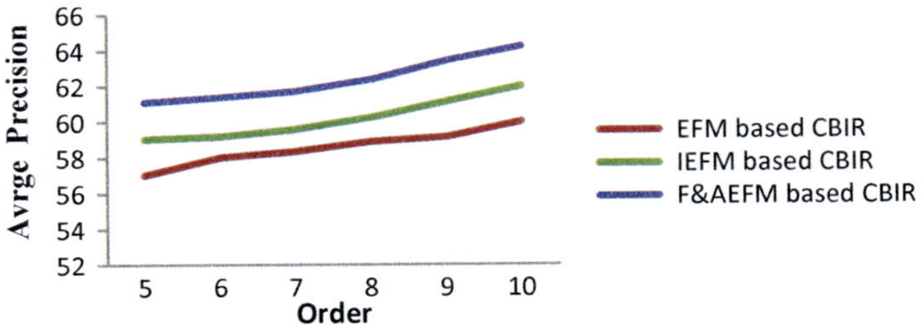

Fig. 4 Compression plot of three CBIR systems by using CIOL-100 database

Table 9 Retrieval time of different EFM-based CBIR systems by using GT face database

CBIR system	Image retrieval time for each image (ms)	Image retrieval efficiency for total database (s)
EFM-based CBIR system	1.7	1.49
IEFM-based CBIR system	1.7	1.49
A&F EFM-based CBIR system	1.02	0.765

From the tabulated results, the average precision of the system increased with the order. A&F EFM-based CBIR system gives highest average precision compared to the IEFM-based CBIR system, but IEFM-based CBIR system gives more average precision compared to the EFM-based CBIR system. From those experimental results, three moment-based CBIR system results of order 10 are compared. A&F EFM-based CBIR system average precision is 64.2, IEFM-based CBIR system average precision is 61.99, and EFM-based CBIR system average precision is 64.04. From these three systems, A&F EFM-based CBIR system has highest precision of 64.2; in the precision point of view, A&F EFM-based CBIR system is preferable compared to remaining two CBIR systems.

3.2 Comparison by Using Image Retrieval Time

Retrieval time is another performance metric. Retrieval times are how much time taken for the retrieval of images similar to quire image. Retrieval time of three CBIR systems for the GT face database is given in Table 9.

From the tabulated results, retrieval time of A&F EFM-based CBIR system is less compared to the IEFM-based CBIR system but IEFM-based CBIR system has the same retrieval time compared to EFM-based CBIR system. Experimental results of tenth-order three moment-based CBIR systems are compared. A&F EFM-based CBIR retrieval time of each image is 1.02 ms and for the total database is 0.765 s; IEFM-based CBIR system retrieval time of each image is 1.7 ms and for total database is 1.49 s; EFM-based CBIR system retrieval time of each image is 1.7 ms and for total database is 1.49 s. In these three systems, A&F EFM-based CBIR system has lowest retrieval time of 0.765 s for total database. A&F EFM-based CBIR system is preferable compared to EFM- and IEFM-based CBIR systems in the retrieval time prospective.

Retrieval time of three CBIR systems by using the COIL-100 database is given in Table 10.

From the tabulated results, retrieval time of A&F EFM-based CBIR system is less compared to the IEFM-based CBIR system, and also IEFM-based CBIR system has the same retrieval time compared to EFM-based CBIR system. From experimental results, the tenth-order three moment-based CBIR system results are compared. A&F

Table 10 Retrieval time of different EFM-based CBIR systems by using the COIL-100

CBIR system	Image retrieval time for each image (ms)	Image retrieval efficiency for total database (s)
EFM-based CBIR system	0.8	6
IEFM-based CBIR system	0.8	6
A&F EFM-based CBIR system	0.5	3.25

EFM-based CBIR system retrieval time of each image is 0.8 ms and for the total database is 6 s; IEFM-based CBIR system retrieval time of each image is 0.8 ms and for total database is 6 s; EFM-based CBIR system retrieval time of each image is 0.5 ms and for total database is 3.25 s. In these three systems, A&F EFM-based CBIR system has lowest retrieval time of 3.25 s; hence, A&F EFM-based CBIR system is preferable in the retrieval time point of view.

4 Conclusions

Three moment-based CBIR systems are compared by using the results; finally, it is concluded that accurate and fast exponential Fourier moment-based CBIR system gives better performance compared with improved exponential Fourier moment-based CBIR system and exponential Fourier moment-based CBIR system based on average precision and retrieval time.

References

1. Hu MK Visual pattern recognition by moment invariants. IRE Trans Inform Theory
2. Teh C-H, Chin RT (1988) On image analysis by the methods of moments. IEEE Trans Pattern Anal Mach Intell 10(4):496–513
3. Abu-Mostafa YS, Psaltis D (1985) Image normalization by complex moments. IEEE Trans Pattern Anal Mach Intell 1:46–55
4. Khotanzad A, Hong YH (1990) Invariant image recognition by Zernike moments. IEEE Trans Pattern Anal Mach Intell 12(5):489–497
5. Hu H-T et al (2014) Orthogonal moments based on exponent functions: exponent-Fourier moments. Pattern Recogn 47(8):2596–2606
6. Kim WY, Kim YS (2000) A region-based shape descriptor using Zernike moments. Signal Process: Image Commun 16:95–102
7. Xiao B, Li W-S, Wang G-Y (2015) Errata and comments on "Orthogonal moments based on exponent functions: exponent-Fouriermoments". Pattern Recogn 48(4):1571–1573
8. Hu H-T, Quan J, Shao C (2016) Errata and comments on "Errata and comments on orthogonal moments based on exponent functions: exponent-Fourier moments". Pattern Recogn 52:471–476
9. Satya PS, Urooj S (2017) Accurate and fast computation of exponent fourier moment, Springer (2017)

Quaternion Polar Complex Exponential Transform and Local Binary Pattern-Based Fusion Features for Content-Based Image Retrieval

D. Kishore and Ch. Srinivasa Rao

Abstract The presented work is a sincere effort to exhibit content-based image retrieval (CBIR) system utilized for extracting images from large databases. It makes use of image features like texture and color. Local binary pattern (LBP)-based operator brings about the information related to image texture by taking into consideration the surrounding pixel values. In spite of having its own advantages, this feature is not that superior at capturing the image's information related to color. The present work rectifies this disadvantage by incorporating an extra color feature by the name quaternion polar complex exponential transform (QPCET) in addition to the LBP-based feature in the image retrieval system. The integrated QPCET- and LBP-based CBIR system exhibits better average retrieval efficiency compared to other accessible techniques tested on different benchmark databases.

Keywords CBIR · Quaternion Polar Complex Exponential Transform · Local Binary Pattern

1 Introduction

In the recent past, ensemble of images has swiftly increased and seems to keep up that pace in future because of the extensive use of the Internet. Nowadays, most of the information exists in digital form which is collected from books and newspapers. With the help of the Internet, all the users are in a position to use the available information. In order to access and query the database of image that is indexed, one needs a competent technique. Location of appropriate information from such a

D. Kishore (✉)
Department of Electronics and Communications Engineering, Aditya College of Engineering and Technology, Surampalem, AP, India
e-mail: kishore.dannina@gmail.com

Ch. Srinivasa Rao
Department of Electronics and Communications Engineering, Jawaharlal Nehru Technological University Kakinada, UCEV, Kakinada, Andhra Pradesh, India
e-mail: chsrao.ece@jntukucev.ac.in

P. S. R. Chowdary et al. (eds.), *Microelectronics, Electromagnetics and Telecommunications*, Lecture Notes in Electrical Engineering 655,
https://doi.org/10.1007/978-981-15-3828-5_80

mammoth database is not at all easy. One of the appropriate techniques employed to retrieve images is CBIR [1]. CBIR facilitates image finding and extraction from a database depending on the features obtained from the image themselves. Corel-1k image database is utilized in this work, and features are extracted, and feature vectors are stored. Features are extracted using quaternion polar complex exponential transform and local binary pattern. In recent years, the color image processing uses quaternion most times, and the moment functions related to this quaternion are investigated [2]. A mathematician called Hamilton in 1843 introduced the generalization of complex numbers [3]. The quaternion theory has a major advantage that the image of color can be taken as the vector field and directly processed. Many researchers have reported that for the analysis of texture, LBP operator is a proven approach.

2 Proposed Method

The proposed structure of the architecture is shown in Fig. 1. The simulation-based experimentation and the workflow can be understood using the proposed architecture in Fig. 1.

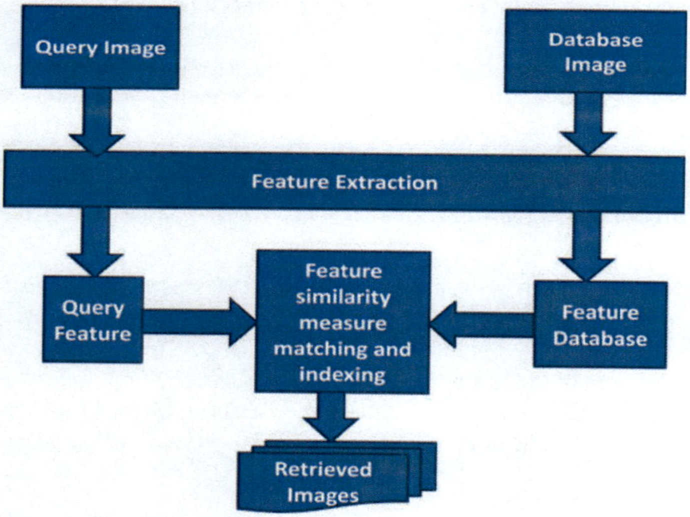

Fig. 1 CBIR architecture for proposed method

2.1 Quaternion Polar Complex Exponential Transform

In polar coordinates, taking $I(r, \theta)$ as color image in the RGB model, we define the QPCET's right side of the nth order with repetition l as [4, 5]

$$\begin{aligned} Q^R_{n,l} &= \frac{1}{\pi}\int_0^{2\pi}\int_0^1 I(r,\theta)\left[H_{n,l}(r,\theta)\right]^* r\mathrm{d}r\mathrm{d}\theta \\ &= \frac{1}{\pi}\int_0^{2\pi}\int_0^1 I(r,\theta)\exp\left(-\mu 2\pi n r^2\right)\exp(-\mu l\theta) r\mathrm{d}r\mathrm{d}\theta \end{aligned} \quad (1)$$

where $\mu = (p + q + r)/\sqrt{3}$.

As the basis functions of the polar complex exponential transform (PCET) are orthogonal, approximate reconstruction of $I(r, \theta)$ the color image is possible using restricted orders of QPCET coefficients ($n \le n_{\max}$, $l \le l_{\max}$). More the number of orders utilized, more the image precision of description.

$$\begin{aligned} I'(r,\theta) &= \sum_{n=-\infty}^{\infty}\sum_{l=-\infty}^{\infty} Q^R_{n,l} R_n(r)\exp(\mu l\theta) \\ &= \sum_{n=-\infty}^{\infty}\sum_{l=-\infty}^{\infty} Q^R_{n,l}\exp\left(\mu 2\pi n r^2\right)\exp(\mu l\theta) \\ &= \sum_{n=-n_{\max}}^{n_{\max}}\sum_{l=-l_{\max}}^{l_{\max}} Q^R_{n,l}\exp\left(\mu 2\pi n r^2\right)\exp(\mu l\theta) \end{aligned} \quad (2)$$

where $I'(r, \theta)$ is the color image that has been reconstructed. Over the interior of a circle with unit radius, the basis functions $R_n(r)\exp(\mu l\theta)$ of the QPCET are orthogonal and every order of the coefficients contributes independently to the reform of the color image.

Equation (2) is all about the product of quaternion being not commutative, so that the QPCET's left side of nth order can be defined with repletion l as

$$Q^L_{n,l} = \frac{1}{\pi}\int_0^{2\pi}\int_0^1 \exp(-\mu l\theta) I(r,\theta)\exp\left(-\mu 2\pi n r^2\right) r\mathrm{d}r\mathrm{d}\theta \quad (3)$$

In the proposed work, the QPCET's right side is denoted as $Q^R_{n,l}$.
Table 1 lists the details of the features employed in the study.

Table 1 Selected QPCET features list for various orders

Order	Modulus coefficients of QPCET	Number of moments	Accumulative no.
0	$\left\|M_{0,0}^R\right\|$	1	1
1	$\left\|M_{1,1}^R\right\|;\left\|M_{1,0}^R\right\|$	2	3
2	$\left\|M_{2,2}^R\right\|;\left\|M_{2,1}^R\right\|;\left\|M_{2,0}^R\right\|$	3	6
3	$\left\|M_{3,3}^R\right\|;\left\|M_{3,2}^R\right\|;\left\|M_{3,1}^R\right\|;\left\|M_{3,0}^R\right\|$	4	10
4	$\left\|M_{4,4}^R\right\|;\left\|M_{4,3}^R\right\|;\left\|M_{4,2}^R\right\|;\left\|M_{4,1}^R\right\|;\left\|M_{4,0}^R\right\|$	5	15
	…		
8	$\left\|M_{8,8}^R\right\|;\left\|M_{8,7}^R\right\|;\left\|M_{8,6}^R\right\|;\left\|M_{8,5}^R\right\|;\left\|M_{8,4}^R\right\|;\left\|M_{8,3}^R\right\|;\left\|M_{8,2}^R\right\|;\left\|M_{8,1}^R\right\|;\left\|M_{8,0}^R\right\|$	9	45
9	$\left\|M_{9,9}^R\right\|;\left\|M_{9,8}^R\right\|;\left\|M_{9,7}^R\right\|;\left\|M_{9,6}^R\right\|;\left\|M_{9,5}^R\right\|;\left\|M_{9,4}^R\right\|;\left\|M_{9,3}^R\right\|;\left\|M_{9,2}^R\right\|;\left\|M_{9,1}^R\right\|;\left\|M_{9,0}^R\right\|$	10	55
	…		

2.2 Local Binary Pattern-Based Feature

The LBP is extensively employed in image processing applications and computer vision [6, 7]. The textual information of an image is captured using LBP. The retrieval of images and their categorization uses the LBP code histogram as image descriptor feature. For a color image, the LBP computation is carried after converting it into a grayscale image. The LBP is carried out by taking into consideration the neighboring pixels' information and forming the LBP code by comparing the central pixel value with neighboring cells.

An image input sized $M \times N$ in RGB color space is transformed as inter-band average for the computation of LBP code as

$$g(x, y) = \frac{1}{3}[I_R(x \cdot y) + I_G(x \cdot y) + I_B(x \cdot y)] \tag{4}$$

The pixel position of the image is denoted by the symbol (x, y). R, G, B denote red, green and blue color space. The LBP code of image pixel (x, y) with center pixel g_c and neighboring pixel g_p is computed as:

$$\mathrm{LBP}_{P,R} = \sum_{p=0}^{P-1} s\left(g_p - g_c\right)2^p \tag{5}$$

where $s(x) = \begin{cases} 1, x \geq 0 \\ 0, x < 0 \end{cases}$.

The number of neighborhood pixels is denoted by P, radius of neighborhood is denoted by R, and index of neighboring pixel is denoted by p. The LBP histogram is derived after calculating the LBP code as

$$\mathrm{HIST}_{\mathrm{LBP}}(k) = \sum_{x=1}^{M}\sum_{y=1}^{N} h\left(\mathrm{LBP}_{P,R}(x, y), k\right); \quad k \in [o, \ldots K] \tag{6}$$

where $h(x, y) = \begin{cases} 1, \text{ if } \ x = y \\ 0 \ \text{ else} \end{cases}$.

The LBP code's upper value is denoted by K. The LBP histogram bin can be made compact further with the use of the constraint pertaining to pattern uniformity. The circular binary representation is constrained to a narrow alteration, i.e., $U(\mathrm{LBP}_{P,R} \leq 2)$. The computation time can be fastened by implementing the additional restriction by using lookup table. The uniformity constraint is defined as:

$$U\left(\mathrm{LBP}_{p,R}\right) = \left|s\left(g_{p-1} - g_c\right) - s(g_0 - g_c)\right| + \sum_{p=1}^{p-1}\left|s\left(g_p - g_c\right) - s\left(g_{p-1} - g_c\right)\right| \tag{7}$$

$P \times (P - 1) + 3$ are the dimensions related to the texture feature for uniformity constraint. For our executional convenience, the neighboring pixel number is restricted to 8 showing that the dimensions of feature are 59.

2.3 *Similarity Measure Based on Features of Fusion*

We made use of the modulus-based distance to estimate the similarity among query image and the database. The modulus based distance 'd'' between feature coefficients $m1$ and $m2$ is the absolute difference between them. The normalized modulus-based distance is

$$d = \frac{d'}{\max(m1, m2)} \tag{8}$$

'D' the overall modulus-based distance pertaining to the two feature vectors can be defined as

$$D = \frac{1}{N}\sqrt{\sum_{i=1}^{N} d^2(i)} \tag{9}$$

3 Results and Discussion

The performance efficiency of the technique put forward in this research paper is verified by using real outdoor scene images. To perform an experiment, as a test we make use of over 1000 images and 10 categories each comprising of 100 images of COREL database. Every image size in the database is 256×384 or 384×256 consisting of flowers, buses, beach, elephants, sunset, buildings, horses, dinosaurs, African people and dish.

The experimental results of the algorithm put forward are compared with the performance of the existing four image retrieval [8–11] methods. The performance criteria of the technique put forward here depend on precision and recall. Precision is the relevant image retrieved number divided by the total image number retrieved. Recall is the relevant image retrieved number divided by the total image number in the database. Further validation is tested by randomly selecting 100 query images from the chosen database. Every type of image extracts 10 images each, and the first 20 images are returned as retrieval results. For every image, the mean normal precision and the normal recall are calculated. The results obtained are considered as the performance standard for retrieval of the algorithm as depicted in Figs. 2 and 3.

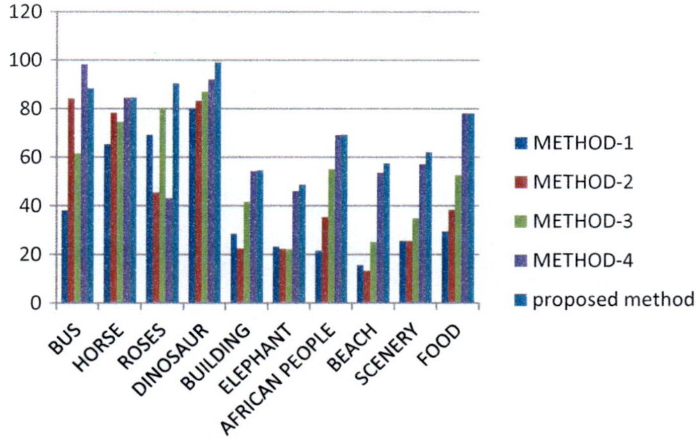

Fig. 2 Retrieval precision results based on each method

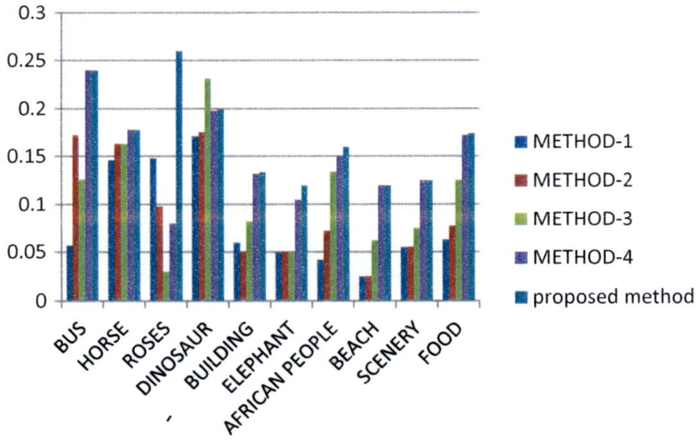

Fig. 3 Retrieval recall results based on each method

4 Conclusion

In this work, we put forward a novel technique of image retrieval by fusion of quaternion polar complex transform moments and features pertaining to texture which are extracted with the help of LBP. The obtained results showed the usefulness of the technique put forward with respect to some conventional methods existing in the literature. The extra required image feature, the strategies of fusion and various distances based on similarities can be explored in the future.

References

1. Rao CS (2012) Content based image retrieval fundamentals and AI. LAP LAMBART Publishing
2. Krishnamoorthi R, Sathiya Devi S (2013) A simple computational model for image retrieval with weighted multi-features based on orthogonal polynomials and genetic algorithms. Neurocomputing 116:165–181
3. Kantor IL, Solodovnikov AS (1989) Hypercomplex number: an elementary introduction to algebra. Springer, New York
4. Guo Liqiang, Dai Ming, Zhu Ming (2014) Quaternion moment and its invariants for color object classification. Inf Sci 273:132–143
5. Chen BJ, Shu HZ, Zhang H, Chen G, Toumoulin C, Dillenseger JI, Luo LM (2012) Quaternion Zernike moments and their invariants for color image analysis and object recognition. Signal Process 92:308–318
6. Gu J, Liu C (2013) Feature local binary patterns with applications to eye detection. Neurocomputing 113:138–152
7. Suruliandi A, Meena K, Rose RR (2012) Local binary pattern and its derivatives for face recognition. IET Comput Vis 6(5):480–488
8. Yap PT, Paramesran R (2006) CBIR using legendre chromaticity distribution moments. IEE Proc Vis Image Signal Process 153(1):17–24
9. Li X (2003) Image retrieval based on perceptive weighted color blocks. Pattern Recogn Lett 24(12):1935–1941
10. Wang XY, Zhang B-B, Yang H-Y (2014) CBIR by integrating color and texture features. Multimedia Tools Appl 68(3):545–569
11. Wang X, Li W, Yang H, Wang P, Li Y-W (2015) Quaternion polar complex exponential transform for invariant color image description. Appl Math Comput 256:951–967

FLM-Based Optimization Scheme for Ocular Artifacts Removal in EEG Signals

Shyam Prasad Devulapalli, Ch. Srinivasa Rao, and K. Satya Prasad

Abstract Neurophysiological analysis plays an important role in the diagnosis of various patients' conditions. This analysis depends on the bio-potentials developed over the brain of an individual patient. These developed bio-potentials are acquired through the electrodes placed at various points on the brain scalp. Sometimes these acquired bio-signals are influenced by other unwanted signals commonly known as artifacts. These artifacts are due to environmental condition, electromyogram (EMG), electrocardiogram (EKG) and electro-oculogram (EOG). The optimization of these artifacts plays vital role in medical diagnosis, which is also a challenging task. Henceforth, a hybrid scheme based on firefly–Levenberg–Marquardt (FLM) algorithm is introduced here for optimization of the above artifacts available in EEG signals. When an EEG signal is applied to an adaptive filter based on FLM, then the proposed scheme is able to reduce the aforesaid artifacts. Finally, the performance of proposed model is analyzed by considering signal-to-noise ratio (SNR), computational time and mean square error. From the experimental results, it has been observed that the proposed model is an effective scheme for optimization of ocular artifacts present in EEG signals.

Keywords Neurophysiological analysis · Electro-oculogram (EOG) · Environmental condition · Firefly scheme

S. P. Devulapalli (✉)
CVR College of Engineering, Hyderabad, India
e-mail: devula_shyam@yahoo.co.in

Ch. Srinivasa Rao
JNTUK UCEV, Vizianagaram, AP, India
e-mail: chsrao.ece@jntukucev.ac.in

K. Satya Prasad
VFSTR, Guntur-Tenali Rd, Vadlamudi, AP, India
e-mail: prasad_kodati@yahoo.co.in

P. S. R. Chowdary et al. (eds.), *Microelectronics, Electromagnetics and Telecommunications*, Lecture Notes in Electrical Engineering 655,
https://doi.org/10.1007/978-981-15-3828-5_81

1 Introduction

Numerous applications based on EEG are developed by different research scholars like smart wheel chair based on eye movement and voice operated [1–4], etc. Generally, by placing EEG electrodes on different parts of the brain scalp, the brain activity is observed for specific diagnosis [5]. The captured EEG signals from electrodes are impure as they are influenced by various unwanted signals known as artifacts. The main sources of these artifacts are improper positioning of electrode, noise in source power, muscle contraction, eye movement and environmental factors [5–8]. Therefore, the diagnosis of EEG signal is very complex due to the presence of these artifacts. Hence, a pure EEG signal is required for proper analysis by removing the unwanted noise or artifacts in the captured EEG [9] signal. In order to eliminate the artifacts, a suitable model which has low distortion of amplitude should be developed [10, 11]. Various schemes commonly used by researchers for eliminating the artifacts are independent component analysis (ICA) [4], wavelet transforms (WT) [12, 13], linear filtering [14, 15] and cascade adaptive filter [16]. Apart from these schemes, the other methods like principal component analysis are also used, but it has some limitations like orthogonal rotation limits.

The main objective of the proposed work is to plan and develop a model for artifacts removal based on FLM optimization algorithm along with neural-network-oriented adaptive filtering. In the primary stage, the captured EEG signal is fed to the proposed adaptive filtering for getting optimal weights through firefly (FF) and LM (Levenberg–Marquardt) algorithms. These hybridized two algorithms are then applied to neural network for obtaining the optimal weights for adaptive filtering. Finally, the proposed filter is used for artifacts removal from the captured EEG signal.

Original data from the electrodes are influenced by signal contaminations or various noise sources. As a result, this desired signal randomly fluctuates and is unsuitable for proper diagnosis. To meet the desired specifications, the noise or artifacts must be optimized. But, the main problem for EEG artifacts removal is threshold level selection, because of its uncertainty. All the bio-potentials originating from the living organs are time-varying in nature, so the artifacts associated with the bio-potentials like EEG are removed by the adaptive filters to some extent. But, there are some limitations in using the adaptive filters because of nonlinear issues. Hence, for smooth removal of artifacts from EEG signal, an optimal weight is highly necessary for the adaptive filtering which is a prime objective of the proposed work. Here, one input from EEG source signal $E(t)$ and other from artifact sources $\mathrm{Ar}(t)$ are applied to the adaptive filter for proper optimization. Before feeding the signal to the adaptive filter, artifact-included signal is applied to nonlinear dynamics for generating an interference signal $\mathrm{Int}(t)$. Then, a combination of both interference and EEG signal develops a primary input signal $\mathrm{Pr}(t)$ which is expressed as $\mathrm{Pr}(t) = E(t) + \mathrm{Pr}(t)$. Now, the $\mathrm{Ar}(t)$ is fed to the adaptive filter for getting filtered response $F(t)$ which is again subtracted from the resultant $\mathrm{Pr}(t)$, expressed as $O(t) = \mathrm{Pr}(t) - F(t)$. Here, $O(t)$ represents the output of adaptive structure for noise optimization.

As the acquired signal is nonlinear, a nonlinear autoregressive exogenous model (NAREX) should be followed which is the best scheme of time series analysis. This NAREX model comprises of multilayer feedforward network, recurrent loop and time delay as explained in the proposed work. Again this model contains three vector layers like input layer, hidden layer and the output layer. Here, input layer contains three vectors like exogenous input vector, delayed regressed output vector and delayed exogenous input vector. Finally, the output vector after completion of neural operation is developed as $L(n+1)$. An internal structure of NAREX model is presented in Fig. 1, and the mathematical expression for this model is presented as

$$L(n+1) = f(L(n), \ldots, L(n - D_L);\ V9n), \ldots, V(n - \mathrm{DV})$$

In Fig. 1, the detailed algorithm flowchart for the proposed model is presented, and it contains five stages where the various factors are associated for analysis. The intensity of firefly in stage-1 is calculated as

$$I_d = \frac{I_0}{1+\lambda d^2} \tag{1}$$

Fig. 1 Detailed algorithm flowchart for proposed hybrid model

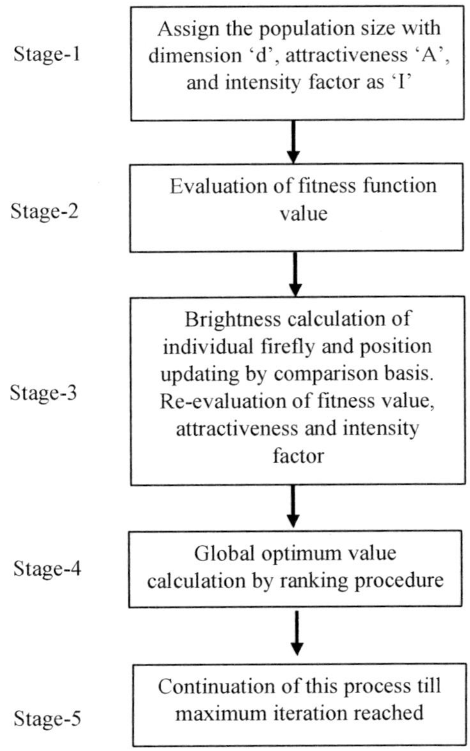

The term I_d is the intensity of firefly at a distance 'd'. Similarly, the attractiveness is calculated as

$$A_d = \frac{A_0}{1+\lambda d^2} \tag{2}$$

Weighted function used in firefly algorithm is presented as

$$W_{\text{ff}}^{t+1}=W^t + A_0 e^{-\lambda d^2}\left(W_{\text{cb}} - W^t\right) + \gamma\varepsilon \tag{3}$$

In Eq. (3), λ is the absorption coefficient of light, ε is a random number, and W_{cb} is the current best solution.

When firefly algorithm is applied, both the input vector and updated vector are combined as per Eq. (4).

$$O_{\text{ff}} = V(W, W_{\text{New}}) \tag{4}$$

Here in Eq. (4), O_{ff} is the updated output, W is input vector, and W_{New} is the new weighted vector developed from firefly algorithm.

2 Result and Discussion

The proposed model uses original signal which is obtained from the physio toolkit for carrying out the performance analysis. This proposed technique is executed with MATLAB by considering the original signals obtained from the aforesaid database. The interval for the signals to be analyzed is fixed at one minute and by adding external unwanted signal as artefacts for required analysis. Again the acquired bio-signals are sampled at 256 samples per second with a 16-bit resolution. During experimentation, several techniques like independent component analysis (ICA), wavelet transform analysis (WTA), neural network along with LM and the proposed NARX neural networks are considered for artefacts removal (Fig. 2 and Table 1).

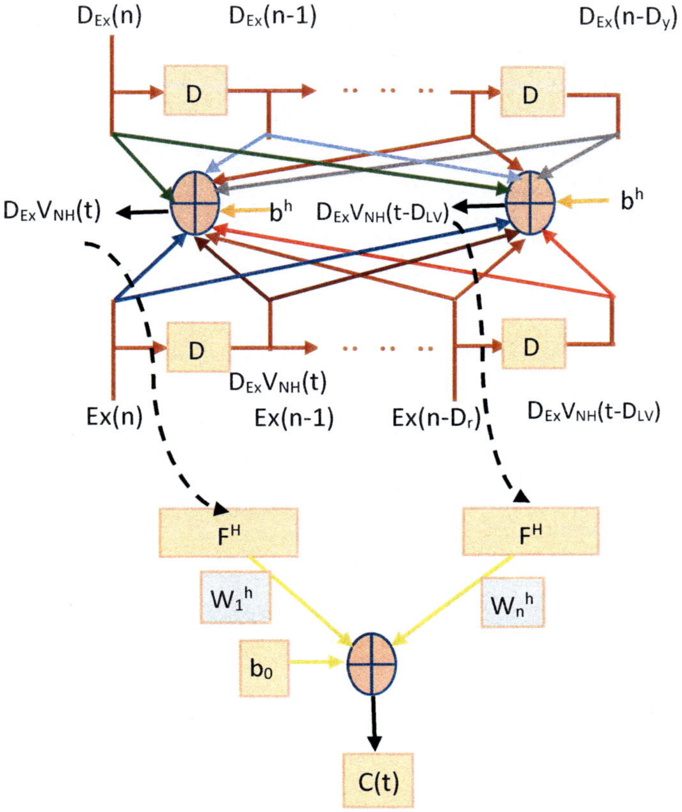

Fig. 2 Internal structure of nonlinear autoregressive exogenous model (NAREX)

Table 1 Performance analysis

Schemes	SNR	RMSE	MSE
Proposed FLM	43.025	0.12025	4421
Cascade filter	38.325	18.9852	50,214.4
WTA	12.236	63.2966	53,210.5
ICA	8.7528	67.2587	5407.6
NN-LM	43.012	0.210221	4236.4

References

1. Kirkove M, Francois C, Verly J (2014) Comparative evaluation of existing and new methods for correcting ocular artifacts in electroencephalographic recordings. Sig Process 98:102–120
2. Bagheri HM, Chitravas N, Kaiboriboon K, Lhatoo S, Loparo K (2014) Automated removal of EKG artifact from EEG data using independent component analysis and continuous wavelet transformation. IEEE Trans Biomed Eng 61:1634–1641

3. Looney D, Goverdovsky V, Kidmose P, Mandic DP (2014) Subspace denoising of EEG artefacts via multivariate EMD. In: Proceedings of 2014 IEEE international conference on acoustics, speech and signal processing, Florence, Italy, 4–9 May 2014
4. Daly I, Nicolaou N, Nasuto SJ, Warwick K (2013) Automated artifact removal from the electroencephalogram: a comparative study. Clin EEG Neuro-sci 44:291–306
5. Mateo J, Torres AM (2013) Eye interference reduction in electroencephalogram recordings using a radial basic function. Signal Process IET 7:565–576
6. Karthik GVS, Fathima SY, Rahman MZU, Ahamed SR, Lay-Ekuakille A (2013) Efficient signal conditioning techniques for brain activity in remote health monitoring network. IEEE Sens J 13:3276–3283
7. Damon C, Liutkus A, Gramfort A, Essid S (2013) Non-negative matrix factorization for single-channel EEG artifact rejection. In: Proceedings of 2013 IEEE international conference on acoustics, speech and signal processing (ICASSP), Vancouver, Canada, 26–31 May 2013, pp 1177–1181
8. Gorji HT, Taheri H, Koohpayezadeh A, Haddadnia J (2013) Ocular artifact detection and removing from EEG by wavelet families: a comparative study. J Inf Eng Appl 3(13):39–47
9. Mamun M, Al-Kadi M, Marufuzzaman M (2013) Effectiveness of wavelet denoising on lector-encephalogram signals. J Appl Res Technol 11(1):156–160
10. Daly I, Pichiorri F, Faller J, Kaiser V, Kreilinger A, Scherer R, Müller-Putz G (2012) What does clean EEG look like? In: Proceedings of the 34th annual international conference of the IEEE engineering in medicine and biology society, San Diego, CA, USA, 28 August–1 September 2012, pp 3963–3966
11. Gaidhane VH, Singh Vijander, Hote YV, Kumar M (2011) New approaches for image compression using neural network. J Intell Learn Syst Appl 3:220–229
12. Senthil Kumar P (2008) Removal of ocular artifacts in the EEG through wavelet transform without using an EOG reference channel. Int J Open Problems Compt Math 1(3), December 2008
13. Krishnaveni V, Jayaraman S, Aravind S, Hariharasudhan V, Ramadoss K Automatic identification and removal of ocular artifacts from EEG using wavelet transform
14. Suja Priyadharsini S, Edward Rajan S (2014) An efficient method for the removal of ECG artifact from measured EEG Signal using PSO algorithm. In: International journal of advances in soft computing and its applications. vol 6, No. 1, March 2014, ISSN 2074–8523
15. Liu T, Yao D (2006) Removal of the ocular artifacts from EEG data using a cascaded spatio-temporal processing. Comput Methods Programs Biomed 83(2):95–103 Epub 2006 Aug 1
16. Garcés Correa A, Laciar E, Patiño HD, Valentinuzzi ME Artifact removal from EEG signals using adaptive filters in cascade. Journal of Physics: Conference Series, Vol 90